HANDBOOK OF Media FOR Environmental Microbiology

RONALD M. ATLAS
University of Louisville

CRC Press
Boca Raton London New York Washington, D.C.

Library of Congress Cataloging-in-Publication Data

Atlas, Ronald M., 1946-
Handbook of media for environmental microbiology / by Ronald M. Atlas.
p. cm.
Includes index.
ISBN 0-8493-0603-5
1. Microbiology—Cultures and culture media—Handbooks, manuals, etc.
2. Sanitary microbiology—Handbooks, manuals, etc. 3. Water—Microbiology
Handbooks, manuals, etc. 4. Microbial ecology—Handbooks, manuals, etc. I. Title
QR66.3A848 1995
576—dc20 95-32261
CIP

Direct all inquiries to CRC Press LLC, 2000 N.W. Corporate Blvd., Boca Raton, Florida 33431.

Visit the CRC Press Web site at www.crcpress.com

International Standard Book Number 0-8493-0603-5
Library of Congress Card Number 95-32261
Printed in the United States of America 3 4 5 6 7 8 9 0
Printed on acid-free paper

Preface

The *Handbook of Media for Environmental Microbiology* began as a result of my frustration with not being able easily to locate the most basic information I needed to culture microorganisms. Too often my library didn't have information on commercial products. Frequently I couldn't find descriptions of many media even in most major research libraries because the information was never published or was contained only in technical manuals. There was a clear need for a comprehensive reference on the media used to cultivate microorganisms.

The *Handbook of Media for Environmental Microbiology* compiles in one place in a consistent style the formulations, methods of preparation, and uses for several thousand different media. The *Handbook of Media for Environmental Microbiology* includes descriptions of the media used for the testing of waters and wastewaters—including those recommended by the USEPA as standard methods. It also includes both classic and modern media used for the identification, cultivation, and maintenance of diverse bacteria. The compositions of various media can be compared so that alternate media can be used. Similar or identical media described by different names that actually have the same formulations can be readily identified. Using the *Handbook of Media for Environmental Microbiology* should save time and effort for anyone cultivating microorganisms.

The *Handbook of Media for Environmental Microbiology* is simple to use. The media are organized alphabetically. Synonyms for media are also listed and cross referenced within the manual. Each medium includes the composition, instructions for preparation, commercial sources, and uses. Using the *Handbook of Media for Environmental Microbiology*, the information needed to prepare media for the cultivation of microorganisms can easily be found.

About the Author:

Ronald M. Atlas is Professor of Biology at the University of Louisville. He received his B.S. from the State University of New York at Stony Brook in 1968 and his M.S. and Ph.D. from Rutgers in 1970 and 1972, respectively. He quickly advanced through the academic ranks at the University of Louisville after spending a year at the Jet Propulsion Laboratory in Pasadena, CA. Dr. Atlas has received a number of honors including: The University of Louisville Excellence in Research Award, Johnson and Johnson Fellowship for Biology, and he is listed in *American Men and Women of Science*. In 1991 he received the ASM Award in Applied and Environmental Microbiology.

He teaches a variety of courses in microbiology at the University of Louisville and has authored several textbooks in general microbiology and microbial ecology. He has written well over 100 research papers. He has conducted studies on the fate of oil in the sea. As part of these studies, he has extensively characterized marine bacterial populations and examined the diversity of microorganisms. He pioneered the field of bioremediation for marine oil spills. His recent studies have focused on the application of molecular techniques to environmental problems. His studies have included the development of "suicide vectors" for the containment of genetically engineered microorganisms and the use of gene probes and the polymerase chain reaction for environmental monitoring, including the detection of pathogens and indicator bacteria for water quality monitoring.

He has served on the NIH Recombinant Advisory Committee (RAC), as well as on various advisory boards for the Environmental Protection Agency (EPA). He is Chairperson of the Environmental Committee of the Public and Scientific Affairs Board of the American Society for Microbiology. He has been a national lecturer for Sigma Xi, an American Society for Microbiology Foundation lecturer, and an Australian Society for Microbiology national lecturer. He has served on the editorial boards of *Applied and Environmental Microbiology, Binary, Advances in Microbial Ecology, BioScience, Biotechniques* and *Journal of Industrial Microbiology*. He is editor of *CRC Critical Reviews in Microbiology*.

Table of Contents

Introduction

Organization

The media described in the *Handbook of Media for Environmental Microbiology* are organized alphabetically. Synonyms for media are listed and cross referenced. The description of each medium includes its name(s), composition, instructions for preparation, commercial sources, safety cautions where needed, and uses.

Names of Media

Media often have numerous names. In many cases media with identical compositions produced by different companies have different names. For example Trypticase Soy Agar produced by BBL Microbiology Systems, Tryptone Soy Agar produced by Oxoid Unipath, and Tryptic Soy Agar produced by Difco Laboratories have identical compositions. Many media also are known by acronyms. TSA, for example, is the common acronym for Trypticase Soy Agar. The *Handbook of Media for Environmental Microbiology* gives the various synonymous names and directs the reader to see the entry where the information about that medium is given. In cases where modifications to a medium yield a new medium, such media generally are listed with the original medium name, followed by the term modified—for example, TSA, Modified rather than Modified TSA. Media that do not have formal names are listed according to the organism grown on that medium—for example, *Bacillus stearothermophilus* Broth.

Trademarks

The names of some media, components of media, and other terms are registered trademarks. The trademarked items referred to in the *Handbook of Media for Environmental Microbiology* are listed below.

American Type Culture Collection® and ATCC® are trademarks of the American Type Culture Collection.

Bacto®, BiTek®, and Difco® are trademarks of Difco Laboratories.

Oxoid® and Lab–Lemco® are trademarks of Unipath Ltd.

Acidase®, BBL®, Biosate®, CTA Medium®, DTA Medium®, DCLS Agar®, Desoxycholate®, Desoxycholate Agar®, Desoxycholate Citrate Agar®, Enterococcosel®, Eugonagar®, Eugonbroth®, GC-Lect®, Gelysate®, IsoVitaleX®, Mycobactosel®, Mycophil®, Mycosel®, Myosate®, Phytone®, Polypeptone®, Selenite-F Enrichment®, Thiotone®, Trichosel®, Trypticase®, TSA II®, and TSI Agar® are trademarks of Becton Dickinson and Co.

Composition of Media

Media for the cultivation of microorganisms contain the substances necessary to support the growth of microorganisms. Due to the diversity of microorganisms and their diverse metabolic pathways, there are numerous media. Even slight differences in the composition of a medium can result in dramatically different growth characteristics of microorganisms.

The composition section of each medium describes the ingredients that make up the medium, their amounts, and the pH. It lists those ingredients in order of decreasing amount. Solids are listed first showing the weights to be added, followed by liquids showing the volumes to be included in the medium.

The composition uses generic terms where these are applicable. For example, pancreatic digest of casein is marketed by various manufacturers as trypticase, tryptone, and other commercial product names. While there may well be differences between these products, such differences are undefined. Variations also occur between batches of products produced as digests of animal tissues.

Media for the cultivation of microorganisms have a source of carbon for incorporation into biomass. For autotrophs the carbon source most often is carbon dioxide which may be supplied as bicarbonate within the medium. Carbohydrates, such as glucose, or other organic compounds, such as acetate, various lipids, proteins, hydrocarbons, and other organic compounds are included in media as sources of carbon for heterotrophs. These carbon sources may also serve as the supply of energy. Other compounds, such as ammonium ions, nitrite ions, elemental sulfur and reduced iron may be used as the sources of energy for the cultivation of autotrophs. Nitrogen also is required for microbial growth. It may be supplied as inorganic nitrogen compounds for the cultivation of some microorganisms but more commonly is supplied as proteins, peptones, or amino acids. Phosphates and metals—such as magnesium and iron, are also necessary components of microbiological media. Phosphates may also serve as buffers to maintain the pH of the medium within the growth tolerance limits of the microorganism being cultivated. Various additional growth factors may also be included in the media.

Agars

Agar is the most common solidifying agent used in microbiological media. Agar is a polysaccharide extract from marine algae. It melts at 84°C and solidifies at 38°C. Agar concentrations of 15.0g/L typically are used to form solid media. Lower concentrations of 7.5–10.0g/L are used to produce soft agars or semisolid media. Below are some agars used as solidifying agents in various media.

Agar Bacteriological (Agar No. 1)

An agar with low calcium and magnesium. Available from Oxoid Unipath.

Agar, Bacto

A purified agar with reduced pigmented compounds, salts, and extraneous matter. Available from Difco Laboratories.

Agar, BiTek™

Agar prepared as a special technical grade. Available from Difco Laboratories.

Agar, Flake

A technical grade agar. Available from Difco Laboratories.

Agar, Grade A

A select grade agar containing minerals. Available from BBL Microbiology Systems.

Agar, Granulated

A high grade granulated agar that has been filtered, decolorized, and purified. Available from BBL Microbiology Systems.

Agarose

A low sulfate neutral gelling fraction of agar that is a complex galactose polysaccharide of near neutral charge.

Agar, Purified

A very high grade agar that has been filtered, decolorized, and purified by washing and extraction of refined agars. It has reduced mineral content. Available from BBL Microbiology Systems.

Agar Technical (Agar No. 3)

A technical grade agar. Available from Difco Laboratories and Oxoid Unipath.

Ionagar

A purified agar. Available from Oxoid Unipath.

Noble Agar

An agar that has been extensively washed and is essentially free of impurities. Available from Difco Laboratories.

Purified Agar

An agar that has been extensively washed and extracted with water and organic solvent. Available from Difco Laboratories and Oxoid Unipath.

Peptones

Many complex media, that is, media in which not all the specific chemical components are known, contain peptones as the source of nitrogen. Peptones are hydrolyzed proteins formed by enzymatic or acidic digestion. Casein most often is used as the protein substrate for forming peptones. Below is a list of some of the peptones that are used as ingredients in various media.

Acidase™ Peptone

A hydrochloric acid hydrolysate of casein. It has a nitrogen content of 8% and is deficient in cystine and tryptophan. Available from BBL Microbiology Systems.

Bacto Casitone

A pancreatic digest of casein. Available from Difco Laboratories.

Bacto Peptamin

A peptic digest of animal tissues. Available from Difco Laboratories.

Bacto Peptone

An enzymatic digest of animal tissues. It has a high concentration of low molecular weight peptones and amino acids. Available from Difco Laboratories.

Bacto Proteose Peptone

An enzymatic digest of animal tissues. It has a high concentration of high molecular weight peptones. Available from Difco Laboratories.

Bacto Soytone

A enzymatic hydrolysate of soybean meal. Available from Difco Laboratories.

Bacto Tryptone

A pancreatic digest of casein. Available from Difco Laboratories.

Bacto Tryptose

An enzymatic hydrolysate containing numerous peptides including those of higher molecular weights. Available from Difco Laboratories.

Biosate™ Peptone

A hydrolysate of plant and animal proteins. Available from BBL Microbiology Systems.

Casein Hydrolysate

A hydrolysate of casein prepared with hydrochloric acid digestion under pressure and neutralized with sodium hydroxide. It contains total nitrogen of 7.6% and NaCl of 28.3%. Available from Oxoid Unipath.

Gelatone

A pancreatic digest of gelatin. Available from Difco Laboratories.

Gelysate™ Peptone
: A pancreatic digest of gelatin deficient in cystine and tryptophan and which has a low carbohydrate content. Available from Oxoid Unipath.

Lactoalbumin Hydrolysate
: A pancreatic digest of lactoalbumin, a milk whey protein. It has high levels of amino acids. It contains total nitrogen of 11.9% and NaCl of 1.4%. Available from Difco Laboratories and Oxoid Unipath.

Liver Digest Neutralized
: A papaic digest of liver that contains total nitrogen of 11.0% and NaCl of 1.6%. Available from Oxoid Unipath.

Mycological Peptone
: A peptone that contains total nitrogen of 9.5% and NaCl of 1.1%. Available from Oxoid Unipath.

Myosate™ Peptone
: A pancreatic digest of heart muscle. Available from BBL Microbiology Systems.

Neopeptone
: An enzymatic digest of protein. Available from Difco Laboratories.

Peptone Bacteriological Neutralized
: A mixed pancreatic and papaic digest of animal tissues. It contains total nitrogen of 14.0% and NaCl of 1.6%. Available from Difco Laboratories and Oxoid Unipath.

Peptone P
: A peptic digest of fresh meat that has a high sulfur content and contains total nitrogen of 11.12% and NaCl of 9.3%. Available from Difco Laboratories and Oxoid Unipath.

Peptonized Milk
: A pancreatic digest of high grade skim milk powder. It has a high carbohydrate and calcium concentration. It contains total nitrogen of 5.3% and NaCl of 1.6%. Available from Oxoid Unipath.

Phytone™ Peptone
: A papaic digest of soybean meal. It has a high vitamin and a high carbohydrate content. Available from BBL Microbiology Systems.

Polypeptone™ Peptone
: A mixture of peptones composed of equal parts of pancreatic digest of casein and peptic digest of animal tissue. Available from BBL Microbiology Systems.

Proteose Peptone
: A specialized peptone prepared from a mixture of peptones that contains a wide variety of high molecular weight peptides. It contains total nitrogen of 12.7% and NaCl of 8.0%. Available from Difco Laboratories and Oxoid Unipath.

Proteose Peptone No. 2
: An enzymatic digest of animal tissues with a high concentration of high molecular weight peptones. Available from Difco Laboratories.

Proteose Peptone No. 3
: An enzymatic digest of animal tissues. It has a high concentration of high molecular weight peptones. Available from Difco Laboratories.

Soya Peptone
: A papaic digest of soybean meal with a high carbohydrate concentration. It contains total nitrogen of 8.7% and NaCl of 0.4%. Available from Oxoid Unipath.

Soytone
: A papaic digest of soybean meal. Available from Difco Laboratories and Oxoid Unipath.

Special Peptone
: A mixture of peptones, including meat, plant and yeast digests. It contains a wide variety of peptides, nucleotides, and minerals. It contains total nitrogen of 11.7% and NaCl of 3.5%. Available from Oxoid Unipath.

Thiotone™ E Peptone
: An enzymatic digest of animal tissue. Available from BBL Microbiology Systems.

Trypticase™ Peptone
: A pancreatic digest of casein. It has a very low carbohydrate content and a relatively high tryptophan content. Available from BBL Microbiology Systems.

Tryptone
: A pancreatic digest of casein. It contains total nitrogen of 12.7% and NaCl of 0.4%. Available from Oxoid Unipath.

Tryptone T
: A pancreatic digest of casein with lower levels of calcium, magnesium, and iron than tryptone. It contains total nitrogen of 11.7% and NaCl of 4.9%. Available from Difco Laboratories and Oxoid Unipath.

Tryptose
: An enzymatic hydrolysate containing high molecular weight peptides. It contains total nitrogen of 12.2% and NaCl of 5.7%. Available from Difco Laboratories and Oxoid Unipath.

Meat and Plant Extracts

Meat and plant infusions are aqueous extracts that are commonly used as sources of nutrients for the cultivation of microorganisms. Such infusions contain amino acids and low molecular weight peptides, carbohydrates, vitamins, minerals, and trace metals. Extracts of animal tissues contain relatively high concentrations of water soluble protein components and glycogen. Extracts of plant tissues contain relatively high concentrations of carbohydrates.

With regard to infusions, many media list as an ingredient infusion from beef heart or another animal tissue. This ingredient is prepared by boiling a given amount of the animal tissue, for example 500.0g, and then using the liquid or, more commonly, drying the broth and using the solids from the infusion. The actual weight of the dry solids extracted from the hot water used to create the infusion varies and so the ingredient typically is simply listed as 500.0g beef heart infusion although the actual weight of solids recovered from the infusion and used in the medium is far less. Brain heart infusion is prepared from calf brains and beef heart.

Below is a list of some of the meat and plant extracts that are used as ingredients in various media.

Bacto Beef

A desiccated powder of lean beef. Available from Difco Laboratories.

Bacto Beef Extract

An extract of beef (paste). Available from Difco Laboratories.

Bacto Beef Extract Desiccated

An extract of desiccated beef. Available from Difco Laboratories.

Bacto Beef Heart for Infusion

A desiccated powder of beef heart. Available from Difco Laboratories.

Bacto Liver

A desiccated powder of beef liver. Available from Difco Laboratories.

Lab-Lemco

A meat extract powder. Available from Oxoid Unipath.

Liver Desiccated

Dehydrated ox livers. Available from Oxoid Unipath.

Malt Extract

A water soluble extract from germinated grain dried by low temperature evaporation. It has a high carbohydrate content. It contains total nitrogen of 1.1% and NaCl of 0.1%.

Growth Factors

Many microorganisms have specific growth factor requirements that must be included in media for their successful cultivation. Vitamins, amino acids, fatty acids, trace metals, and blood components often must be added to media. In some cases specific defined components are used to meet the growth factor requirements. Incorporation of growth factors are used to enrich, that is, to increase the numbers of particular species of microorganisms. Most often mixtures of growth factors are used in microbiological media. Acid hydrolysates of casein commonly are used as sources of amino acids. Extracts of yeast cells also are employed as sources of amino acids and vitamins for the cultivation of microorganisms. Many media, particularly those employed in the clinical laboratory, contain blood or blood components that serve as essential nutrients for fastidious microorganisms. X factor (heme) and V factor (nicotinamide adenine dinucleotide) often are supplied by adding hemoglobin (BBL and Difco Laboratories), IsoVitaleX (BBL Microbiology Systems), and/or Supplement VX (Difco Laboratories). Below is a list of some of the growth factors that are used as ingredients in various media.

Bacto Casamino Acids

A mixture of amino acids formed by acid hydrolysis of casein. Available from Difco Laboratories.

Bacto Vitamin Free Casamino Acids

A mixture of amino acids formed by acid hydrolysis of casein that is free of vitamins. Available from Difco Laboratories.

Campylobacter Growth Supplement

Sodium pyruvate, sodium metabisulfite, and $FeSO_4$.

Castenholtz Salts

Agar, $NaNO_3$, Na_2HPO_4, KNO_3, nitrilotriacetic acid, $MgSO_4{\cdot}7H_2O$, $CaSO_4{\cdot}2H_2O$, NaCl, $FeCl_3$, $MnSO_4$, H_3BO_3, $ZnSO_4$, $CoCl_2{\cdot}6H_2O$, Na_2MoO_4, $CuSO_4$, and H_2SO_4.

CVA Enrichment

Glucose, L-cysteine·HCl·H_2O, vitamin B_{12}, L-glutamine, L-cystine·2HCl, adenine, nicotinamide adenine dinucleotide, cocarboxylase, guanine·HCl, $Fe(NO_3)_3$, *p*-aminobenzoic acid, and thiamine·HCl.

Cysteine Sulfide Reducing Agent

L-Cysteine·HCl·H_2O and $Na_2S{\cdot}9H_2O$.

Dubos Medium Albumin

Albumin fraction V, glucose, and saline solution. Available from Difco Laboratories.

Dubos Oleic Albumin Complex
Alkalinized oleic acid, albumin fraction V, and saline solution. Available from Difco Laboratories.

Egg Yolk Emulsion
Chicken egg yolks and whole chicken egg. Available from Difco Laboratories and Oxoid Unipath.

Egg Yolk Emulsion, 50%
Chicken egg yolks, whole chicken egg, and saline solution. Available from Difco Laboratories.

EY Tellurite Enrichment
Egg yolk suspension with potassium tellurite. Available from Difco Laboratories and Oxoid Unipath.

Fresh Yeast Extract Solution
Live, pressed, starch-free, hydrolyzed Baker's yeast.

Fildes Enrichment
A peptic digest of sheep or horse blood that is a rich source of growth factors including hemin and nicotinamide adenine dinucleotide. Available from BBL Microbiology Systems, Difco Laboratories and Oxoid Unipath.

Hemin Solution
Hemin and NaOH.

Hoagland Trace Element Solution, Modified
H_3BO_3, $MnCl_2{\cdot}4H_2O$, $AlCl_3$, $CoCl_2$, $CuCl_2$, KI, $NiCl_2$, $ZnCl_2$, $BaCl_2$, Na_2MoO_4, $SeCl_4$, $SnCl_2{\cdot}2H_2O$, $NaVO_3{\cdot}H_2O$, KBr, and LiCl.

Hutner's Mineral Base
$MgSO_4{\cdot}7H_2O$, $CaCl_2{\cdot}2H_2O$, $FeSO_4{\cdot}7H_2O$, $(NH_4)_2MoO_4$, $FeSO_4{\cdot}7H_2O$, $ZnSO_4{\cdot}7H_2O$, EDTA, $MnSO_4{\cdot}7H_2O$, $Co(NO_3)_2{\cdot}6H_2O$, $CuSO_4{\cdot}5H_2O$, $Na_2B_4O_7{\cdot}10H_2O$, and nitrilotriacetic acid.

IsoVitaleX® Enrichment
Glucose, L-cysteine·HCl, L-glutamine, L-cystine, adenine, nicotinamide adenine dinucleotide, vitamin B_{12}, thiamine pyrophosphate, guanine·HCl, $Fe(NO_3)_3{\cdot}6H_2O$, *p*-aminobenzoic acid, and thiamine·HCl. Available from BBL Microbiology Systems.

Legionella Agar Enrichment
L-Cysteine and ferric pyrophosphate. Available from Difco Laboratories.

Legionella BCYE Growth Supplement
ACES buffer/KOH, ferric pyrophosphate, L-cysteine-HCl, and α-ketoglutarate. For theenrichment of *Legionella* species. Available from Oxoid Unipath.

Leptospira Enrichment
Lyophilized pooled rabbit serum containing hemoglobin that provides long chain fatty acids and B vitamins for growth of *Leptospira* species. Available from BBL Microbiology Systems and Difco Laboratories.

Metals "44"
$ZnSO_4{\cdot}7H_2O$, $FeSO_4{\cdot}7H_2O$, $MnSO_4{\cdot}7H_2O$, $CuSO_4{\cdot}5H_2O$, $Co(NO_3)_2{\cdot}6H_2O$, EDTA, and $Na_2B_4O_7{\cdot}10H_2O$.

Middlebrook ADC Enrichment
NaCl, bovine albumin fraction V, glucose and catalase. The albumin binds free fatty acids that may be toxic to mycobacteria. Available from BBL Microbiology Systems and Difco Laboratories.

Middlebrook OADC Enrichment
NaCl, bovine albumin, glucose, oleic acid, and catalase. The albumin binds free fatty acids that may be toxic to mycobacteria; the enrichment provides oleic acid used by *Mycobacterium tuberculosis* for growth. Available from BBL Microbiology Systems and Difco Laboratories.

Mycoplasma Enrichment without Penicillin
Horse serum, fresh autolysate of yeast—yeast extract, and thallium acetate. Provides cholesterol and nucleic acids for growth of *Mycoplasma* species. The thallium selectively inhibits other microorganisms. Available from BBL Microbiology Systems.

Mycoplasma Supplement
Yeast extract and horse serum. Available from Difco Laboratories.

Nitsch's Trace Elements
$MnSO_4$, H_3BO_3, $ZnSO_4$, Na_2MoO_4, $CuSO_4$, $CoCl_2{\cdot}6H_2O$, and H_2SO_4.

SLA Trace Elements
$FeCl_2{\cdot}4H_2O$, H_3BO_3, $CoCl_2{\cdot}6H_2O$, $ZnCl_2$, $MnCl_2{\cdot}4H_2O$, $NiCl_2{\cdot}6H_2O$, $CuCl_2{\cdot}2H_2O$, $Na_2MoO_4{\cdot}2H_2O$, and $Na_2SeO_3{\cdot}5H_2O$.

Soil Extract
African Violet soil and Na_2CO_3.

Trace Element Mixture
Ethylenediamine tetraacetic acid (EDTA), $ZnSO_4{\cdot}7H_2O$, $CaCl_2$, $MnCl_2{\cdot}4H_2O$, $FeSO_4{\cdot}7H_2O$, $CoCl_2{\cdot}6H_2O$, $CuSO_4{\cdot}5H_2O$, and $(NH_4)_6Mo_7O_{24}{\cdot}4H_2O$.

Trace Element Solution HO-LE
H_3BO_3, $MnCl_2{\cdot}4H_2O$, sodium tartrate, $FeSO_4{\cdot}7H_2O$, $CoCl_2{\cdot}6H_2O$, $CuCl_2{\cdot}2H_2O$, $Na_2MoO_4{\cdot}2H_2O$, and $ZnCl_2$.

Trace Elements Solution SL-6

H_3BO_3, $CoCl_2 \cdot 6H_2O$, $ZnSO_4 \cdot 7H_2O$, $MnCl_2 \cdot 4H_2O$, $Na_2MoO_4 \cdot H_2O$, $NiCl_2 \cdot 6H_2O$, and $CuCl_2 \cdot 2H_2O$.

Trace Elements Solution SL-7

$FeCl_2 \cdot 4H_2O$, $CoCl_2 \cdot 6H_2O$, $MnCl_2 \cdot 4H_2O$, $ZnCl_2$, H_3BO_3, $Na_2MoO_4 \cdot 2H_2O$, $NiCl_2 \cdot 6H_2O$, $CuCl_2 \cdot 2H_2O$, and HCl.

Trace Element Solution SL-8

Disodium EDTA, $FeCl_2 \cdot 4H_2O$, $CoCl_2 \cdot 6H_2O$, $MnCl_2 \cdot 4H_2O$, $NiCl_2 \cdot 6H_2O$, $ZnCl_2$, H_3BO_3, $NaMoO_4 \cdot 2H_2O$, and $CuCl_2 \cdot 2H_2O$.

Trace Elements Solution SL-10

$FeCl_2 \cdot 4H_2O$, $CoCl_2 \cdot 6H_2O$, $MnCl_2 \cdot 4H_2O$, $ZnCl_2$, $Na_2MoO_4 \cdot 2H_2O$, $NiCl_2 \cdot 6H_2O$, H_3BO_3, $CuCl_2 \cdot 2H_2O$ and HCl (25% solution).

Trace Metals A-5 Mix

H_3BO_3, $MnCl_2 \cdot 4H_2O$, $ZnSO_4 \cdot 7H_2O$, $CuSO_4 \cdot 5H_2O$, $Na_2MoO_4 \cdot 2H_2O$, and $Co(NO_3)_2 \cdot 6H_2O$.

VA Vitamin Solution

Nicotinamide, thiamine·HCl, *p*-aminobenzoic acid, biotin, calcium pantothenate, pyridoxine·2HCl, and cyanocobalamin.

Vitamin K_1 Solution

Vitamin K_1 and ethanol.

Vitox Supplement

Glucose, L-cysteine·HCl, L-glutamine, L-cystine, adenine sulfate, nicotinamide adenine dinucleotide, $Fe(NO_3)_3 \cdot 6H_2O$, cocarboxylase, guanine·HCl, *p*-aminobenzoic acid, vitamin B_{12}, and thiamine·HCl. Available from Oxoid Unipath.

Wolfe's Mineral Solution

$MgSO_4 \cdot 7H_2O$, nitriloacetic acid, NaCl, $MnSO_4 \cdot H_2O$, $FeSO_4 \cdot 7H_2O$, $CoCl_2 \cdot 6H_2O$, $CaCl_2$, $ZnSO_4 \cdot 7H_2O$, $CuSO_4 \cdot 5H_2O$, $AlK(SO_4)_2 \cdot 12H_2O$, H_3BO_3, and $Na_2MoO_4 \cdot 2H_2O$.

Wolfe's Vitamin Solution

Pyridoxine·HCl, thiamine·HCl, riboflavin, nicotinic acid, calcium pantothenate, *p*-aminobenzoic acid, thioctic acid, biotin, folic acid and cyanocobalamin.

Yeast Autolysate Growth Supplement

Yeast autolysate fractions, glucose, and $NaHCO_3$. Available from Oxoid Unipath.

Yeast Dialysate

Active, dried yeast.

Yeast Extract

A water soluble extract of autolyzed yeast cells. Available from BBL Microbiology Systems, Difco Laboratories, and Oxoid Unipath.

Yeastolate

A water soluble fraction of autolyzed yeast cells rich in vitamin B complex. Available from Difco Laboratories.

Selective Components

Many media contain selective components that inhibit the growth of nontarget microorganisms and favor the growth of specific organisms. Selective media are especially useful in the isolation of specific microorganisms from mixed populations. In many media for the study of microorganisms in nature, compounds are included in the media as sole sources of carbon or nitrogen so that only a few types of microorganisms can grow. Selective toxic compounds are also frequently used to select for the cultivation of particular microbial species. The isolation of a pathogen from a stool specimen, for example, where there is a high abundance of nonpathogenic normal microbiota, requires selective media. Often antimicrobics or other selectively toxic compounds are incorporated into media to suppress the growth of the background microbiota while permitting the cultivation of the target organism of interest. Bile salts, selenite, tetrathionate, tellurite, azide, phenylethanol, sodium lauryl sulfate, high sodium chloride concentrations, and various dyes—such as eosin, crystal violet, and methylene blue—are used as selective toxic chemicals. Antimicrobial agents used to suppress specific types of microorganisms include ampicillin, chloramphenicol, colistin, cycloheximide, gentamicin, kanamycin, nalidixic acid, sulfadiazine, and vancomycin. Various combinations of antimicrobics are effective in suppressing classes of microorganisms, such as enteric bacteria.

pH Buffers

Maintaining the pH of media usually is accomplished by the inclusion of suitable buffers. Since microorganisms grow optimally only within certain limits of a pH range, the pH generally is maintained within a few tenths of a pH unit. Phosphate buffers commonly are used. The pH is established by using varying volumes of equimolar concentrations of Na_2HPO_4 and NaH_2PO_4.

pH	Na_2HPO_4 (mL)	NaH_2PO_4 (mL)
5.4	3.0	97.0
5.6	5.0	95.0
5.8	7.8	92.2
6.0	12.0	88.0
6.2	18.5	81.5
6.4	26.5	73.5
6.6	37.5	62.5
6.8	50.0	50.0

pH	Na_2HPO_4 (mL)	NaH_2PO_4 (mL)
7.0	61.1	38.9
7.2	71.5	28.5
7.4	80.4	19.6
7.6	86.8	13.2
7.8	91.4	8.6
8.0	94.5	5.5

Differential Components

The differentiation of many microorganisms is based upon the production of acid from various carbohydrates and other carbon sources or the decarboxylation of amino acids. Some media include indicators, particularly of pH that permit the visual detection of changes in pH resulting from such metabolic reactions. Below is a list of some commonly used pH indicators and their color reactions.

pH Indicator	pH Range	Acid Color	Alkaline Color
m-Cresol Purple	0.5–2.5	Red	Yellow
Thymol Blue	1.2–2.8	Red	Yellow
Bromphenol Blue	3.0–4.6	Yellow	Blue
Bromcresol Green	3.8–5.4	Yellow	Blue
Chlorcresol Green	4.0–5.6	Yellow	Blue
Methyl Red	4.2–6.3	Red	Yellow
Chlorphenol Red	5.0–6.6	Yellow	Red
Bromcresol Purple	5.2–6.8	Yellow	Purple
Bromthymol Blue	6.0–7.6	Yellow	Blue
Phenol Red	6.8–8.4	Yellow	Red
Cresol Red	7.2–8.8	Yellow	Red
m-Cresol Purple	7.4–9.0	Yellow	Purple
Thymol Blue	8.0–9.6	Yellow	Blue
Cresolphthalein	8.2–9.8	Colorless	Red
Phenolphthalein	8.3–10.0	Colorless	Red

Preparation of Media

The ingredients in a medium are usually dissolved and the medium is then sterilized. When agar is used as a solidifying agent the medium must be heated gently, usually to boiling, to dissolve the agar. In some cases where interactions of components, such as metals, would cause precipitates, solutions must be prepared and occasionally sterilized separately before mixing the various solutions to prepare the complete medium. The pH often is adjusted prior to sterilization, but in some cases sterile acid or base is used to adjust the pH of the medium following sterilization. Many media are sterilized by exposure to elevated temperatures. The most common method is to autoclave the medium. Different sterilization procedures are employed when heat-labile compounds are included in the formulation of the medium.

Autoclaving

Autoclaving uses exposure to steam, generally under pressure, to kill microorganisms. Exposure for 15 min to steam at 115 psi—121°C is most commonly used. Such exposure kills vegetative bacterial cells and bacterial endospores. However, some substances do not tolerate such exposures and lower temperatures and different exposure times are sometimes employed. Media containing carbohydrates often are sterilized at 116–118°C in order to prevent the decomposition of the carbohydrate and the formation of toxic compounds that would inhibit microbial growth. Below is a list of pressure–temperature relationships.

Pressure—psi	Temperature—°C
0	100
1	101.9
2	103.6
3	105.3
4	106.9
5	108.4
6	109.8
7	111.3
8	112.6
9	113.9
10	115.2
11	116.4
12	117.6
13	118.8
14	119.9
15	121.0
16	122.0
17	123.0
18	124.0
19	125.0
20	126.0
21	126.9
22	127.8
23	128.7
24	129.6
25	130.4

Filtration

Filtration is commonly used to sterilize media containing heat-labile compounds. Liquid media are passed through sintered glass or membranes, typically made of cellulose acetate or nitrocellulose, with small pore sizes. A membrane with a pore size of 0.2mm will trap bacterial cells and, therefore, sometimes is called a bacteriological filter. By preventing the passage of microorganisms, filtration renders fluids free of bacteria and eukaryotic microorganisms, that is, free of living organisms, and hence sterile. Many carbohydrate solutions, antibiotic solutions, and vitamin solutions are filter sterilized and added to media that have been cooled to temperatures below 50°C.

Caution about Hazardous Components

Some media contain components that are toxic or carcinogenic. Appropriate safety precautions must be taken when using media with such components. Basic fuchsin and acid fuchsin are carcinogens and caution must be used in handling media with these compounds to avoid dangerous exposure that could lead to the development of malignancies. Thallium salts, sodium azide, sodium biselenite, and cyanide are among the toxic components found in some media. These compounds are poisonous and steps must be taken to avoid ingestion, inhalation, and skin contact. Azides also react with many metals, especially copper, to form explosive metal azides. The disposal of azides must avoid contact with copper or achieve sufficient dilution to avoid the formation of such hazardous explosive compounds. Media with sulfur-containing compounds may result in the formation of hydrogen sulfide which is a toxic gas. Care must be used to ensure proper ventilation. Media with human blood or human blood components must be handled with great caution to avoid exposure to human immunodeficiency virus and other pathogens that contaminate some blood supplies. Proper handling and disposal procedures must be followed with blood-containing as well as other media that are used to cultivate microorganisms.

References

Below is a list of references that can be consulted for further information about media used for the isolation, cultivation, and differentiation of microorganisms.

A Compilation of Culture Media for the Cultivation of Microorganisms. 1930. M. Levine and H. W. Schoenlein. Williams & Wilkins Co., Baltimore.

ATCC Catalogue of Bacteria and Bacteriophages. 1992. R. Gherna, P. Pienta, and R. Cote, eds. American Type Culture Collection, Rockville MD.

ATCC Catalogue of Filamentous Fungi. 1991. S. C. Jong and M. J. Edwards, eds. American Type Culture Collection, Rockville MD.

ATCC Catalogue of Recombinant DNA Materials. 1991. D. R. Maglott and W. C. Nierman, eds. American Type Culture Collection, Rockville MD.

Difco Manual: Dehydrated Culture Media and Reagents for Microbiology. Difco Laboratories, Detroit MI.

Manual of BBl Products and Laboratory Procedures. 1988. D. A. Power and P. J. McCuen, eds. Beckton Dickinson and Company, Cockeysville, MD.

The Oxoid Manual. 1990. E. Y. Bridson, ed. Unipath Ltd. Basingstoke, Hampshire England.

Standard Methods for the Examination of Water and Wastewater. 1992. A. E. Greenberg, L. S. Clesceri, and A. D. Eaton, eds. American Public Health Association, Washington DC.

A 1 Broth

Composition per liter:

Pancreatic digest of casein	20.0g
Lactose	5.0g
NaCl	5.0g
Salicin	0.5g
Triton™ X-100	1.0mL

pH 6.9 ± 0.1 at 25°C

Source: This medium is available as a premixed powder from Difco Laboratories and BBL Microbiology Systems.

Preparation of Medium: Add components to distilled/deionized water and bring volume to 1.0L. Mix thoroughly. Gently heat and bring to boiling. Distribute into test tubes containing an inverted Durham tube. Autoclave for 10 min at 15 psi pressure–121°C.

Use: For the detection of fecal coliforms in treated wastewater and seawater by a most-probable-number (MPN) method. Multiple dilutions of samples (3, 5, or 10 replicates per dilution) are added to tubes containing A 1 Broth. After incubation test tubes with gas accumulation in the Durham tubes are scored positive and those with no gas as negative. An MPN table is consulted to determine the most probable number of fecal coliforms.

A1 Minimal Medium

Composition per liter:

L-Asparagine	5.0g
$(NH_4)_2SO_4$	5.0g
Sodium pyruvate	5.0g
$MgSO_4 \cdot 7H_2O$	2.0g
Spermadine·3HCl	0.125g
L-Asparagine	0.10g
L-Isoleucine	0.10g
L-Methionine	0.10g
L-Phenylalanine	0.10g
L-Valine	0.10g
L-Leucine	0.05g
KH_2PO_4	0.013g
$FeCl_3 \cdot 6H_2O$	2.7mg
$CaCl_2$	1.1mg
Cyanocobalamin	1.0mg
Tris(hydroxymethyl)aminomethane buffer (0.01*M* solution, pH 7.6)	1.0L

pH 7.6 ± 0.2 at 25°C

Preparation of Medium: Add solid components to 1.0L of Tris buffer. Mix thoroughly. Filter sterilize. Aseptically distribute into tubes or flasks.

Use: For the cultivation of *Myxococcus xanthus*.

ACC Medium

Composition per liter:

Proteose peptone	20.0g
Agar	12.0g
Glycerol	1.5g
K_2SO_4	1.5g
$MgSO_4 \cdot 7H_2O$	1.5g
Antibiotic solution	10.0mL

pH 7.2 ± 0.2 at 25°C.

Antibiotic Solution:

Composition per 10.0mL:

Cycloheximide	0.075g
Ampicillin	0.050g
Chloramphenicol	0.0125g

Preparation of Antibiotic Solution: Add components to distilled/deionized water and bring volume to 10.0mL. Mix thoroughly. Filter sterilize.

Preparation of Medium: Add components, except antibiotic solution, to distilled/deionized water and bring volume to 990.0mL. Mix thoroughly. Gently heat and bring to boiling. Autoclave for 15 min at 15 psi pressure–121°C. Cool to 45–50°C. Aseptically add sterile antibiotic solution. Mix thoroughly. Pour into sterile Petri dishes or distribute into sterile tubes.

Use: For the selective isolation and cultivation of fluorescent *Pseudomonas* species.

Acetamide Agar

Composition per liter:

Agar	15.0g
Acetamide	10.0g
NaCl	5.0g
K_2HPO_4	1.0g
$NH_4H_2PO_4$	1.0g
$MgSO_4 \cdot 7H_2O$	0.2g
Bromthymol blue	0.08g

pH 6.9 ± 0.2 at 25°C

Preparation of Medium: Add components to distilled/deionized water and bring volume to 1.0L. Mix thoroughly. Gently heat and bring to boiling. Adjust pH. Distribute into tubes or flasks. Autoclave for 15 min at 15 psi pressure–121°C. Cool tubes in a slanted position to produce a long slant.

Use: For the differentiation of nonfermentative Gram-negative bacteria, especially *Pseudomonas aeruginosa*. Can be used as a confirmatory test for water analysis. Bacteria that deamidate acetamide turn the medium blue.

Acetamide Agar

Composition per liter:

Agar	15.0g
Acetamide	10.0g
NaCl	5.0g
K_2HPO_4	1.39g
KH_2PO_4	0.73g
$MgSO_4 \cdot 7H_2O$	0.5g
Phenol red	0.012g

pH 6.9 ± 0.2 at 25°C

Source: This medium is available as a premixed powder from BBL Microbiology Systems.

Preparation of Medium: Add components to distilled/deionized water and bring volume to 1.0L. Mix thoroughly. Gently heat and bring to boiling. Adjust pH. Distribute into tubes or flasks. Autoclave for 15 min at 15 psi pressure–121°C. Cool tubes in a slanted position to produce a long slant.

Use: For the differentiation of nonfermentative Gram-negative bacteria, especially *Pseudomonas aeruginosa*. Can be used as a confirmatory test for water analysis. Bacteria that deamidate acetamide turn the medium blue.

Acetamide Broth

Composition per liter:

Acetamide	10.0g
NaCl	5.0g
K_2HPO_4	1.39g
KH_2PO_4	0.73g
$MgSO_4 \cdot 7H_2O$	0.5g
Phenol red	0.012g

pH 6.9 ± 0.2 at 25°C

Preparation of Medium: Add components to distilled/deionized water and bring volume to 1.0L. Mix thoroughly. Adjust pH. Autoclave for 15 min at 15 psi pressure–121°C.

Use: For the differentiation of nonfermentative Gram-negative bacteria, especially *Pseudomonas aeruginosa*. Can be used as a confirmatory test for water analysis. Bacteria that deamidate acetamide turn the broth purplish red.

Acetamide Cetrimide Glycerol Mannitol Selective Medium

Composition per liter:

Agar	15.0g
K_2SO_4	10.0g
D-Mannitol	5.0g
$MgCl_2 \cdot 6H_2O$	1.4g
Cetrimide	0.3g
Peptone	0.2g
Acetamide solution	100.0mL
Glycerol	5.0mL

pH 7.0 ± 0.2 at 25°C

Acetamide Solution:
Composition per 100mL:

Acetamide	10.0g
Phenol red	0.012g

Preparation of Acetamide Solution: Add components to distilled/deionized water and bring volume to 100.0mL. Mix thoroughly. Filter sterilize.

Preparation of Medium: Add components, except acetamide solution, to distilled/deionized water and bring volume to 900.0mL. Mix thoroughly. Adjust pH to 7.0. Gently heat and bring to boiling. Autoclave for 20 min at 15 psi pressure–121°C. Cool to 45–50°C. Aseptically add sterile acetamide solution. Mix thoroughly. Pour into sterile Petri dishes.

Use: For the cultivation of *Pseudomonas aeruginosa, P. fluorescens, P. putida, P. alcaligenes, P. cepacia*, and *P. pseudoalcaligenes*.

Acetate Differential Agar (Sodium Acetate Agar) (Simmons' Citrate Agar, Modified)

Composition per liter:

Agar	20.0g
NaCl	5.0g
Sodium acetate	2.0g
$(NH_4)H_2PO_4$	1.0g
K_2HPO_4	1.0g
$MgSO_4 \cdot 7H_2O$	0.2g
Bromthymol blue	0.08g

pH 6.8 ± 0.2 at 25°C

Source: This medium is available as a premixed powder from Difco Laboratories and BBL Microbiology Systems.

Preparation of Medium: Add components to cold distilled/deionized water and bring volume to 1.0L. Mix thoroughly. Gently heat and bring to boiling. Distribute into tubes to produce a 1cm butt and 30cm slant. Autoclave for 15 min at 15 psi pressure–121°C. Cool tubes in a slanted position.

Use: For the differentiation of *Shigella* species from *Escherichia coli* and also for the differentiation of nonfermenting Gram-negative bacteria. Bacteria that can utilize acetate as the sole carbon source turn the medium blue.

Acetogen Medium

Composition per 421.8:

$NaHCO_3$	2.4g
NH_4Cl	0.2g
Yeast extract	0.2g
Stock salts solution #1	40.0mL
Potassium phosphate buffer	20.0mL
Clarified rumen fluid	20.0mL
Stock salts solution #2	4.0mL
Trace minerals	4.0mL
Vitamin solution	4.0mL
Reducing agent	4.0mL
Tungstate solution	0.4mL
Resazurin (0.1% solution)	0.4mL

Potassium Phosphate Buffer:

Composition per 830mL:

K_2HPO_4	15.68g
KH_2PO_4	4.72g

Preparation of Potassium Phosphate Buffer: Dissolve K_2HPO_4 in 600.0mL distilled/deionized water and KH_2PO_4 in 230.0mL distilled/deionized water. Mix the two solutions together and use.

Stock Salts Solution #1:

Composition per liter:

KCl	1.6g
NaCl	1.4g
$MgSO_4 \cdot 7H_2O$	0.2g

Preparation of Stock Salts Solution #1: Add components to distilled/deionized water and bring volume to 1.0L. Mix thoroughly.

Stock Salts Solution #2:

Composition per liter:

$CaCl_2 \cdot 2H_2O$	0.1g

Preparation of Stock Salts Solution #2: Add components to distilled/deionized water and bring volume to 1.0L. Mix thoroughly.

Trace Minerals:

Composition per liter:

Nitrilotriacetic acid	1.5g
$MgSO_4 \cdot 7H_2O$	3.0g
$MnSO_4 \cdot H_2O$	0.5g
NaCl	1.0g
$NiCl_2 \cdot 6H_2O$	0.1g
$FeSO_4 \cdot 7H_2O$	0.1g
$CoCl_2 \cdot 6H_2O$	0.1g
$CaCl_2$	0.1g
$ZnSO_4 \cdot 7H_2O$	0.1g
$Na_2SeO_3 \cdot 5H_2O$	0.01g
$CuSO_4 \cdot 5H_2O$	0.01g
$AlK(SO_4)_2 \cdot 12H_2O$	0.01g
H_3BO_3	0.01g
$Na_2MoO_4 \cdot 2H_2O$	0.01g

Preparation of Trace Minerals: Add nitrilotriacetic acid to 500.0mL of distilled/deionized water. Dissolve by adjusting pH to 6.5 with KOH. Bring volume to 1.0L with distilled/deionized water. Add remaining components. Mix thoroughly.

Vitamin Solution:

Composition per liter:

Pyridoxine·HCl	10.0mg
Ascorbic acid	5.0mg
Calcium pantothenate	5.0mg
Choline chloride	5.0mg
Lipoic acid	5.0mg
i-Inositol	5.0mg
Niacinamide	5.0mg
Nicotinic acid	5.0mg
p-Aminobenzoic acid	5.0mg
Pyridoxal·HCl	5.0mg
Riboflavin	5.0mg
Thiamine·HCl	5.0mg
Biotin	2.0mg
Folic acid	2.0mg
Vitamin B_{12}	0.1mg

Preparation of Vitamin Solution: Add components to distilled/deionized water and bring volume to 1.0L. Mix thoroughly. Store frozen.

Tungstate Solution:

Composition per liter:

$Na_2WO_4 \cdot 2H_2O$	99.0mg

Preparation of Tungstate Solution: Add components to distilled/deionized water and bring volume to 1.0L. Mix thoroughly.

Reducing Agent:

Composition per 110mL:

$Cysteine \cdot HCl \cdot H_2O$	2.5g
$Na_2S \cdot 9H_2O$	2.5g

Preparation of Reducing Agent: Add 110.0mL distilled/deionized water to a 250.0mL round bottom flask. Boil under N_2 gas for 1 min. Cool to room temperature. Add cysteine·HCl and dissolve. Adjust to pH 9 with 5*N* NaOH. Add washed $Na_2S \cdot 9H_2O$ and dissolve. Distribute in amounts needed into tubes or flasks. Autoclave for 10 min at 15 psi pressure–121°C.

Preparation of Medium: Add components, except $NaHCO_3$ and reducing agent, to distilled/deionized water and bring volume to 417.8mL. Mix thoroughly. Gently heat and bring to boiling under 80% N_2 + 20% CO_2. Cool to 45–50°C. Add $NaHCO_3$ and reducing agent. Distribute into tubes or flasks under 80% N_2 + 20% CO_2. Autoclave for 15 min at 15 psi pressure–121°C. After inoculation, exchange headspace with 80% H_2 + 20% CO_2.

Use: For the cultivation and maintenance of acetogenic anaerobes such as some *Clostridium* species.

Achromobacter Choline Medium

Composition per liter:

NaCl	30.0g
Agar	18.0g
Choline chloride	5.0g
K_2HPO_4	1.0g
$MgSO_4 \cdot 7H_2O$	0.5g
$FeSO_4 \cdot 7H_2O$	0.01g

Preparation of Medium: Add components to distilled/deionized water and bring volume to 1.0L. Mix well and warm gently until dissolved. Autoclave for 15 min at 15 psi pressure–121°C. Pour into sterile Petri dishes.

Use: For the cultivation and maintenance of *Achromobacter cholinophagum* and other bacteria that can utilize choline as a carbon source.

Achromobacter Medium (ATCC Medium 457)

Composition per liter:

K_2HPO_4	7.32g
Ammonium tartrate	4.6g
KH_2PO_4	1.09g
$MgSO_4 \cdot 7H_2O$	0.04g
$FeSO_4 \cdot 7H_2O$	0.04g
$CaCl_2 \cdot 2H_2O$	0.014g
$MgSO_4 \cdot 7H_2O$	0.002g

pH 7.5 ± 0.2 at 25°C

Preparation of Medium: Add components to distilled/deionized water and bring volume to 1.0L. Mix well and warm gently until dissolved. Distribute into test tubes or flasks. Autoclave for 15 min at 15 psi pressure–121°C.

Use: For the cultivation and maintenance of *Achromobacter* species and *Alcaligenes* species.

Achromobacter Medium (ATCC Medium 589)

Composition per liter:

Agar	20.0g
K_2HPO_4	7.0g
Methionine	5.0g
KH_2PO_4	2.0g
$(NH_4)_2SO_4$	1.0g
Sodium citrate	0.4g
$MgSO_4 \cdot 7H_2O$	0.1g

Preparation of Medium: Add components to distilled/deionized water and bring volume to 1.0L. Mix thoroughly. Gently heat and bring to boiling. Autoclave for 15 min at 15 psi pressure–121°C. Pour into sterile Petri dishes.

Use: For the cultivation and maintenance of *Achromobacter* species.

Achromobacter pestifer Medium

Composition per liter:

Agar	15.0g
Yeast extract	12.5g
Beef extract	10.0g
Peptone	10.0g
NaCl	5.0g

pH 7.2 ± 0.2 at 25°C

Preparation of Medium: Add components to distilled/deionized water and bring volume to1.0L. Mix thoroughly. Gently heat and bring to boiling. Distribute into tubes or flasks. Autoclave for 15 min at 15 psi pressure–121°C.

Use: For the cultivation and maintenance of *Achromobacter pestifer.*

Acidic Tomato Medium for *Leuconostoc*

Composition per liter:

Agar (if needed)	15.0g
Glucose	10.0g
Peptone	10.0g
Yeast extract	5.0g
$MgSO_4 \cdot 7H_2O$	0.20g
$MnSO_4 \cdot 4H_2O$	0.05g
Tomato juice	250.0mL

pH 4.8 ± 0.2 at 25°C

Preparation of Medium: Add solid components to 750.0mL distilled/deionized water. Add tomato juice. Mix well and warm gently until dissolved. Autoclave for 15 min at 15 psi pressure–121°C. Pour into sterile Petri dishes.

Use: For the cultivation and maintenance of *Leuconostoc oenos* and other *Leuconostoc* species.

Acidophilic *Bacillus stearothermophilus* Agar

Composition per liter:

Part A	400.0mL
Part B	600.0mL

pH 5.0 ± 0.2 at 25°C

Part A:

Composition per 400mL:

Soluble starch	10.0g
Pancreatic digest of casein	5.0g
Yeast extract	5.0g
KH_2PO_4	1.0g
$CaCl_2 \cdot 2H_2O$	0.5g
$MnCl_2 \cdot 4H_2O$	0.5g

Preparation of Part A: Add components to distilled/deionized water and bring volume to 400.0mL. Mix thoroughly. Gently heat and bring to boiling. Adjust pH to 4.7. Autoclave for 15 min at 15 psi pressure–121°C. Cool to 50°C.

Part B:

Composition per 600mL:

Agar	20.0g

Preparation of Part B: Add agar to distilled/deionized water and bring volume to 600.0mL. Autoclave for 15 min at 15 psi pressure–121°C. Cool to 50°C.

Preparation of Medium: Aseptically combine solution A and solution B. Mix thoroughly. Adjust pH to 5.0. Pour into sterile Petri dishes.

Use: For the cultivation and maintenance of *Bacillus stearothermophilus* and other acidophilic *Bacillus* species.

Acidophilic *Bacillus stearothermophilus* Broth

Composition per liter:

Soluble starch	10.0g
Pancreatic digest of casein	5.0g
Yeast extract	5.0g
KH_2PO_4	1.0g
$CaCl_2 \cdot 2H_2O$	0.5g
$MnCl_2 \cdot 4H_2O$	0.5g

pH 5.0 ± 0.2 at 25°C

Preparation of Medium: Dissolve all components except agar in 1.0L distilled/deionized water. Mix thoroughly. Gently heat and bring to boiling. Adjust to pH 5.0. Autoclave for 15 min at 15 psi pressure–121°C. Precipitate will dissolve after cooling and mixing.

Use: For the cultivation and maintenance of *Bacillus stearothermophilus* and other acidophilic *Bacillus* species.

Actidione® Agar (Cycloheximide Agar)

Composition per liter:

Glucose	50.0g
Agar	15.0g
Pancreatic digest of casein	5.0g
Yeast extract	4.0g
KH_2PO_4	0.55g
KCl	0.425g
$CaCl_2 \cdot 2H_2O$	0.125g
$MgSO_4 \cdot 7H_2O$	0.125g
Bromocresol green	22.0mg
Actidione® (cycloheximide)	10.0mg
$FeCl_3$	2.5mg

pH 5.5 ± 0.2 at 25°C

Source: Available as a prepared medium from Oxoid Unipath.

Preparation of Medium: Add components to distilled/deionized water and bring volume to 1.0L. Mix thoroughly. Gently heat and bring to boiling. Distribute into tubes or flasks. Autoclave for 15 min at 15 psi pressure–121°C. Pour into sterile Petri dishes or leave in tubes.

Use: For the enumeration and detection of bacteria in specimens containing large numbers of yeasts and molds.

Actinobacillus lignieresii Medium

Composition per 1010mL:

Agar	10.0g
Hartley's digest broth	900.0mL
Filde's enrichment	100.0mL
Antibiotic solution	10.0mL

pH 7.5 ± 0.2 at 25°C

Hartley's Digest Broth:

Composition per 10L:

Ox heart	3000.0g
Pancreatin	50.0g
Na_2CO_3, anhydrous (0.8% solution)	5.0L
HCl, concentrated	80.0mL

Preparation of Hartley's Digest Broth: Finely mince the ox heart. Add the meat to 5.0L of distilled/deionized water. Gently heat and bring to 80°C. Add Na_2CO_3 solution. Cool to 45°C. Add pancreatin and maintain at 45°C for 4 hr while stirring. Add the HCl and steam at 100°C for 30 min. Cool to room temperature. Adjust pH to 8.0 with 1*N* NaOH. Gently heat and bring to boiling. Continue boiling for 25 min. Filter while hot through Whatman #1 filter paper. Cool to room temperature. Adjust pH to 7.5.

Fildes Enrichment Solution:

Composition 206mL:

Pepsin	1.0g
NaCl (0.85% solution)	150.0mL
Sheep blood, defibrinated	50.0mL
HCl	6.0mL

Source: Fildes enrichment solution is available as a premixed powder from Difco Laboratories and Oxoid Unipath.

Preparation of Fildes Enrichment Solution: Combine components. Mix thoroughly. Incubate at 56°C for 4 hr. Bring pH to 7.0 with 20% NaOH. Adjust pH to 7.2 with HCl. Do not autoclave. Add 0.25 mL of chloroform and store at 4°C. Before use heat to 56°C to remove chloroform.

Antibiotic Solution:

Composition per 10mL:

Oleandomycin phosphate	0.02g
Neomycin sulfate	1.5mg

Preparation of Antibiotic Solution: Add components to distilled/deionized water and bring volume to 10.0mL. Mix thoroughly. Filter sterilize.

Preparation of Medium: Add agar to 900.0mL of Hartley's digest broth. Mix thoroughly. Gently heat and bring to boiling. Autoclave for 15 min at 15 psi pressure–121°C. Cool to 45–50°C. Aseptically add 100.0mL of Filde's enrichment and 10.0mL of antibiotic solution. Mix thoroughly. Pour into sterile Petri dishes or distribute into sterile tubes.

Use: For the isolation and cultivation of *Actinobacillus lignieresii.*

Actinomyces Agar

Composition per liter:

Agar	20.0g
K_2HPO_4	13.0g
Heart muscle, solids from infusion	10.0g
Peptic digest of animal tissue	10.0g
Glucose	5.0g
Yeast extract	5.0g
NaCl	5.0g
Pancreatic digest of casein	4.0g
KH_2PO_4	2.0g
$(NH_4)_2SO_4$	1.0g
L-Cysteine·HCl·H_2O	1.0g
Soluble starch	1.0g
$MgSO_4 \cdot 7H_2O$	0.2g
$CaCl_2 \cdot 2H_2O$	0.01g

pH 6.9 ± 0.2 at 25°C

Preparation of Medium: Add components to distilled/deionized water and bring volume to 1.0L. If a semisolid medium is desired, add 7.0g of agar instead of 20.0g. Mix thoroughly. Gently heat and bring to boiling. Distribute into tubes or flasks. Autoclave for 10 min at 15 psi pressure–121°C. Pour into sterile Petri dishes or leave in tubes.

Use: For the maintenance or cultivation of a variety of anaerobic bacteria including *Actinomyces* species, *Eubacterium* species, *Fusobacterium* species, *Propionibacterium* species and others.

Actinomyces Broth

Composition per liter:

K_2HPO_4	13.0g
Heart muscle, solids from infusion	10.0g
Peptic digest of animal tissue	10.0g
Glucose	5.0g
Yeast extract	5.0g
NaCl	5.0g
Pancreatic digest of casein	4.0g
KH_2PO_4	2.0g
$(NH_4)_2SO_4$	1.0g
L-Cysteine·HCl·H_2O	1.0g
Soluble starch	1.0g
$MgSO_4 \cdot 7H_2O$	0.2g
$CaCl_2 \cdot 2H_2O$	0.01g

pH 6.9 ± 0.2 at 25°C

Source: This medium is available as a premixed powder from BBL Microbiology Systems.

Preparation of Medium: Add components to distilled/deionized water and bring volume to 1.0L. Mix thoroughly. Distribute into tubes or flasks. Autoclave for 10 min at 15 psi pressure–121°C.

Use: For the maintenance or cultivation of a variety of anaerobic bacteria including *Actinomyces* species, *Eubacterium* species, *Fusobacterium* species, *Propionibacterium* species and others.

Actinomyces Broth

Composition per liter:

Beef heart, infusion from	500.0g
KH_2PO_4	15.0g
Peptic digest of animal tissue	10.0g
Glucose	5.0g
Yeast extract	5.0g
NaCl	5.0g
Pancreatic digest of casein	4.0g
KH_2PO_4	2.0g
$(NH_4)_2SO_4$	1.0g
L-Cysteine·HCl·H_2O	1.0g
Soluble starch	1.0g
$MgSO_4 \cdot 7H_2O$	0.2g
$CaCl_2 \cdot 2H_2O$	0.02g

pH 7.2 ± 0.2 at 25°C

Source: This medium is available as a premixed powder from Difco Laboratories.

Preparation of Medium: Add components to distilled/deionized water and bring volume to 1.0L. Mix thoroughly. Distribute into tubes or flasks. Autoclave for 10 min at 15 psi pressure–121°C.

Use: For the maintenance or cultivation of a variety of anaerobic bacteria including *Actinomyces* species, *Eubacterium* species, *Fusobacterium* species, *Propionibacterium* species and others.

Actinomyces Isolation Agar

Composition per liter:

Agar	15.0g
Glycerol	5.0g
Sodium propionate	4.0g
Sodium caseinate	2.0g
K_2HPO_4	0.5g
Asparagine	0.1g
$MgSO_4 \cdot 7H_2O$	0.1g
$FeSO_4 \cdot 7H_2O$	0.001g

Preparation of Medium: Add components to distilled/deionized water and bring volume to 1.0L. Mix thoroughly. Gently heat and bring to boiling. Distribute into tubes or flasks. Autoclave for 15 min at 15 psi pressure–121°C. Pour into sterile Petri dishes or leave in tubes.

Use: For the isolation and cultivation of *Actinomyces* species.

Actinomycete Growth Medium

Composition per liter:

Succinic acid	1.18g
L-Glutamine	0.29g
$CaCl_2 \cdot 2H_2O$	0.2g
KH_2PO_4	0.2g
$MgSO_4 \cdot 7H_2O$	0.2g
NaCl	0.1g
m-Inositol	0.090g
Ferric EDTA	0.037g
$MnSO_4 \cdot H_2O$	4.5mg
H_3BO_3	1.5mg
$ZnSO_4 \cdot 7H_2O$	1.5mg
Nicotonic acid	0.5mg
Pyridoxine-HCl	0.5mg
Thiamine-HCl	0.1mg
$CuSO_4 \cdot 5H_2O$	0.04mg
$Na_2MoO_4 \cdot 2H_2O$	0.025mg

pH 6.4 ± 0.2 at 25°C

Preparation of Medium: Add components to distilled/deionized water and bring volume to 1.0L. Mix thoroughly. Distribute into tubes or flasks. Autoclave for 15 min at 15 psi pressure–121°C.

Use: For the cultivation of actimomycetes.

Actinomycete Isolation Agar

Composition per liter:

Agar	15.0g
Sodium propionate	4.0g
Sodium caseinate	2.0g
K_2HPO_4	0.5g
Asparagine	0.1g
$MgSO_4 \cdot 7H_2O$	0.1g
$FeSO_4 \cdot 7H_2O$	1.0mg

pH 8.1± 0.2 at 25°C

Source: This medium is available as a premixed powder from Difco Laboratories.

Preparation of Medium: Add components to distilled/deionized water and bring volume to 1.0L. Mix thoroughly. Gently heat and bring to boiling. Add 5.0g of glycerol. Distribute into tubes or flasks. Autoclave for 15 min at 15 psi pressure–121°C.

Use: For the isolation and cultivation of aerobic *Actinomyces* from soil and water.

Actinoplanes Medium

Composition per liter:

Oatmeal, baby cereal	60.0g
Yeast	2.5g
K_2HPO_4	1.0g
KCl	0.5g
$MgSO_4 \cdot 7H_2O$	0.5g
$FeSO_4 \cdot 7H_2O$	0.01g

Preparation of Medium: Add components to distilled/deionized water and bring volume to 1.0L. Mix thoroughly. Distribute into tubes or flasks. Autoclave for 15 min at 15 psi pressure–121°C.

Use: For the cultivation and maintenance of *Actinoplanes* species.

Actinopolyspora Medium

Composition per liter:

Agar	20.0g
Maltose	10.0g
N-Z-Amine A	2.0g
Yeast extract	1.0g
Beef extract	1.0g

pH 7.3 ± 0.2 at 25°C

Preparation of Medium: Add components to distilled/deionized water and bring volume to 1.0L. Mix thoroughly. Gently heat and bring to boiling. Distribute into tubes or flasks. Autoclave for 15 min at 15 psi pressure–121°C. Pour into sterile Petri dishes or leave in tubes.

Use: For the cultivation and maintenance of *Actinopolyspora thermovinacea.*

Aeromonas Differential Agar (Dextrin Fuchsin Sulfite Agar)

Composition per liter:

Dextrin	15.0g
Agar	13.0g
Pancreatic digest of casein	10.0g
Na_2HPO_4	7.75g
NaCl	5.0g
Beef extract	3.0g
Na_2SO_3	1.6g
Acid fuchsin solution	50.0mL

pH 7.5 ± 0.2 at 25°C

Acid Fuchsin Solution:

Composition per 50mL:

Acid fuchsin	0.25g

Preparation of Acid Fuchsin Solution: Add Acid fuchsin to 50.0mL of 5% aqueous dioxan. Mix well to dissolve.

Caution: Acid fuchsin is a potential carcinogen and care must be taken to avoid inhalation of the powdered dye and contamination of the skin.

Preparation of Medium: Add components to distilled/deionized water and bring volume to 1.0L. Mix thoroughly. Gently heat while stirring and bring to boiling. Distribute into tubes or flasks. Autoclave for 15 min at 15 psi pressure–121°C. Pour into sterile Petri dishes or leave in tubes.

Use: For the isolation and differentiation of *Aeromonas* species from other Gram-negative rods such as *Pseudomonas* and Enterobacteriaceae. Specimens with low numbers of *Aeromonas* may first be enriched by growth in starch broth for 4–9 days. After 24 hours of growth on this agar, colonies are sprayed with Nadi reagent (1% solution of *N,N,N′,N′*-tetramethyl-*p*-phenylene-diammonium dichloride). A positive Nadi reaction (dextrin degradation) is indicated by a purple color at the periphery of the colony. Dextrin fermentation is also indicated by red colonies. *Aeromonas* species appear as large, convex dark red colonies with purple periphery.

Aeromonas hydrophila Medium

Composition per liter:

Inositol	10.0g
Pancreatic digest of casein	10.0g
L-Ornithine·HCl	5.0g
Proteose peptone	5.0g
Agar	3.0g
Yeast extract	3.0g
Mannitol	1.0g
Ferric ammonium citrate	0.5g
$Na_2S_2O_3 \cdot 5H_2O$	0.4g
Bromcresol purple	0.02g

pH 6.7 ± 0.2 at 25°C

Preparation of Medium: Add components to distilled/deionized water and bring volume to 1.0L. Mix thoroughly. Gently heat until dissolved. Adjust pH to 6.7. Distribute into tubes in 5.0mL volumes. Autoclave for 12 min at 15 psi pressure–121°C.

Use: For the isolation and cultivation of *Aeromonas hydrophila.*

Aeromonas Medium (Ryan's *Aeromonas* Medium)

Composition per liter:

Agar	12.5g
$Na_2S_2O_3$	10.67g
Proteose peptone	5.0g
NaCl	5.0g
Xylose	3.75g
L-Lysine·HCl	3.5g
Yeast extract	3.0g
Sorbitol	3.0g
Bile salts No, 3	3.0g
Inositol	2.5g
L-Arginine·HCl	2.0g
Lactose	1.5g
Ferric ammonium citrate	0.8g
Bromthymol blue	0.04g
Thymol blue	0.04g

pH 8.0 ± 0.1 at 25°C

Source: Available as a dehydrated powder from Oxoid Unipath.

Preparation of Medium: Add components to distilled/deionized water and bring volume to 1.0L. Mix thoroughly. Gently heat and bring to boiling. Do not autoclave. Cool to 50°C and aseptically add 5.0mg ampicillin. Pour into sterile Petri dishes.

Use: For the isolation and selective differentiation of *Aeromonas hydrophila* and other *Aeromonas* species. *Aeromonas* species appear as small (0.5–1.5mm), dark green colonies with darker centers.

AFPA (*Aspergillus flavus/parasiticus* Agar)

Composition per liter:

Yeast extract	20.0g
Agar	15.0g
Peptone	10.0g

Ferric ammonium citrate 0.5g
Dichloran (Botran®) 2.0mg
pH 6.3 ± 0.2 at 25°C

Source: Available as a dehydrated powder from Oxoid Unipath.

Preparation of Medium: Add components to distilled/deionized water and bring volume to 1.0L. Mix thoroughly. Gently heat while stirring and bring to boiling. Add 100.0mg chloramphenicol. Autoclave for 15 min at 15 psi pressure–121°C. Pour into sterile Petri dishes.

Use: For the selective isolation and enumeration of *Aspergillus flavus* and *A. parasiticus*. Colonies of these fungi appear with dark yellow-orange color on the reverse side.

Agrobacterium Mannitol Medium

Composition per liter:

Mannitol 10.0g
L-Glutamate 2.0g
KH_2PO_4 .... 0.5g
Yeast extract 0.3g
$MgSO_4 \cdot 7H_2O$ 0.2g
NaCl 0.2g
pH 7.0 ± 0.2 at 25°C

Preparation of Medium: Add components to distilled/deionized water and bring volume to 1.0L. Mix thoroughly. Adjust pH to 7.0. Autoclave for 15 min at 15 psi pressure–121°C.

Use: For the cultivation of *Agrobacterium rhizogenes*.

Agrobacterium Medium

Composition per liter:

Agar 18.0g
Erythritol 5.0g
$NaNO_3$ 2.5g
$CaCl_2$ 0.2g
$MgSO_4 \cdot 7H_2O$ 0.2g
NaCl 0.2g
KH_2PO_4 .... 0.1g
Ferric EDTA 1.3mg
Biotin 2μg
Supplement 10.0mL
pH 7.0 ± 0.2 at 25°C

Supplement:

Composition per liter:

Cycloheximide 0.25g
Bacitracin 0.1g
Na_2SeO_3 .... 0.1g
Tyrothricin 1.0mg

Preparation of Supplement: Add components to distilled/deionized water and bring volume to 10.0mL. Mix thoroughly. Filter sterilize.

Preparation of Medium: Add components, except supplement, to distilled/deionized water and bring volume to 990.0mL. Mix thoroughly. Adjust pH to 7.0 with 1*N* NaOH. Gently heat and bring to boiling. Autoclave for 15 min at 15 psi pressure–121°C. Cool to 45–50°C. Aseptically add sterile supplement. Mix thoroughly. Pour into sterile Petri dishes or distribute into sterile tubes.

Use: For the selective isolation and cultivation of *Agrobacterium* species biotype 2.

Agrobacterium Medium

Composition per liter:

Agar 20.0g
Mannitol 10.0g
$NaNO_3$ 4.0g
$MgCl_2$ 2.0g
Calcium propionate 1.2g
$Mg_3(PO_4)_2$ 0.2g
$MgSO_4$ 0.1g
$MgCO_3$ 0.075g
$NaHCO_3$ 0.075g
Supplement 100.0mL
pH 7.1 ± 0.2 at 25°C

Supplement:

Composition per 100mL:

Berberine 0.275g
Cycloheximide 0.2g
Bacitracin 0.1g
Na_2SeO_3 .... 0.1g
Penicillin G 0.06g
Streptomycin sulfate 0.03g
Tyrothricin 1.0mg

Preparation of Supplement: Add components to distilled/deionized water and bring volume to 100.0mL. Mix thoroughly. Filter sterilize.

Preparation of Medium: Add components, except supplement, to distilled/deionized water and bring volume to 900.0mL. Mix thoroughly. Gently heat and bring to boiling. Autoclave for 15 min at 15 psi pressure–121°C. Cool to 45–50°C. Aseptically add 100.0mL of sterile supplement. Mix thoroughly. Pour into sterile Petri dishes or distribute into sterile tubes.

Use: For the selective isolation and cultivation of *Agrobacterium* species.

Agrobacterium Medium

Composition per liter:

Agar	12.0g
Lactose	5.0g
Na_2HPO_4	1.8g
KNO_3	1.0g
$MgSO_4 \cdot 7H_2O$	0.1g
Supplement	100.0mL

pH 6.8 ± 0.2 at 25°C

Supplement:

Composition per 100mL:

$MnSO_4 \cdot 4H_2O$	3.35g
Ferric EDTA	2.5mg

Preparation of Supplement: Add components to distilled/deionized water and bring volume to 100.0mL. Mix thoroughly. Filter sterilize.

Preparation of Medium: Add components, except supplement, to distilled/deionized water and bring volume to 900.0mL. Mix thoroughly. Gently heat and bring to boiling. Autoclave for 1 min at 25 psi pressure–130°C. Cool to 45–50°C. Aseptically add sterile supplement. Mix thoroughly. Pour into sterile Petri dishes or distribute into sterile tubes.

Use: For the selective isolation and cultivation of *Agrobacterium* species.

Agrobacterium Medium D1

Composition per liter:

Agar	15.0g
Mannitol	15.0g
LiCl	6.0g
$NaNO_3$	5.0g
K_2HPO_4	2.0g
$MgSO_4 \cdot 7H_2O$	0.2g
Bromthymol blue	0.1g
$Ca(NO_3)_2 \cdot 4H_2O$	0.02g

pH 7.2 ± 0.2 at 25°C

Preparation of Medium: Add components to distilled/deionized water and bring volume to 1.0L. Mix thoroughly. Gently heat and bring to boiling. Distribute into tubes or flasks. Autoclave for 15 min at 15 psi pressure–121°C. Cool to 45–50°C. Adjust pH to 7.2. Pour into sterile Petri dishes or leave in tubes.

Use: For the selective isolation and cultivation of *Agrobacterium* species.

AKI Medium

Composition per liter:

Peptone	15.0g
NaCl	5.0g
Yeast extract	4.0g
Sodium bicarbonate solution	30.0mL

pH 7.2 ± 0.2 at 25°C

Sodium Bicarbonate Solution:

Composition per 100mL:

$NaHCO_3$	10.0g

Preparation of Sodium Bicarbonate Solution: Add sodium bicarbonate to distilled/deionized water and bring volume to 100.0mL. Mix thoroughly. Filter sterilize. Use freshly prepared solution.

Preparation of Medium: Add components, except sodium bicarbonate solution, to distilled/deionized water and bring volume to 970.0mL. Mix thoroughly. Autoclave for 15 min at 15 psi pressure–121°C. Cool to 45–50°C. Aseptically add sterile sodium bicarbonate solution. Mix thoroughly. Aseptically distribute into sterile tubes or flasks. Prepare medium freshly.

Use: For the cultivation of *Vibrio cholerae* and other *Vibrio* species.

Alcal Mannose Medium

Composition per liter:

K_2HPO_4	15.1g
KH_2PO_4	5.6g
Mannose	1.0g
Yeast extract	1.0g
Casamino acids	0.5g
$MgSO_4 \cdot 7H_2O$	0.4g
$CaCl_2 \cdot 2H_2O$	50.0mg
$FeSO_4 \cdot 7H_2O$	10.0mg

Preparation of Medium: Add components to distilled/deionized water and bring volume to 1.0L. Mix thoroughly. Distribute into tubes or flasks. Autoclave for 15 min at 15 psi pressure–121°C.

Use: For the cultivation of *Bacillus circulans.*

Alcaligenes Agar

Composition per liter:

Agar	10.0g
Peptone	5.0g
Ammonium lactate	3.0g
Meat extract	3.0g
Ferric citrate	0.2g

pH 7.0 ± 0.2 at 25°C

Preparation of Medium: Add ferric citrate to distilled/deionized water and bring volume to 100.0mL. In a separate flask, add remaining components to distilled/deionized water and bring volume to 900.0mL.

Mix thoroughly. Adjust pH to 7.0. Steam the two solutions for 20 min on three consecutive days. Aseptically combine the two solutions. Pour into sterile Petri dishes or distribute into sterile tubes.

Use: For the cultivation of *Alcaligenes* species.

Alcaligenes Medium

Composition per liter:

Peptone	5.0g
Beef extract	3.0g
Ferric citrate	0.2g
Ammonium lactate solution	3.0mL

pH 7.0 ± 0.2 at 25°C

Ammonium Lactate Solution:

Composition per 100mL:

Lactic acid	60.0g

Preparation of Ammonium Lactate Solution: Dissolve lactic acid in 100.0mL distilled/deionized water. Neutralize with NH_4OH to pH 7.0.

Preparation of Medium: Add peptone, beef extract, and ammonium lactate to distilled/deionized water and bring volume to 1.0L. Mix thoroughly. Gently heat and bring to boiling. Autoclave for 15 min at 15 psi pressure–121°C. Add ferric citrate aseptically. Mix thoroughly. Aseptically distribute into tubes or flasks.

Use: For the cultivation of *Alcaligenes tolerans.*

Alcaligenes N5 Medium

Composition per liter:

Sodium succinate·$2H_2O$	5.0g
KH_2PO_4	0.75g
NH_4Cl	0.67g
K_2HPO_4	0.61g
$MgSO_4 \cdot 7H_2O$	0.2g
$CaCl_2 \cdot 2H_2O$	0.03g
$MnCl_2 \cdot 4H_2O$	3.0mg
$FeCl_3$	2.4mg
$Na_2MoO_4 \cdot 2H_2O$	1.0mg

Preparation of Medium: Add components to distilled/deionized water and bring volume to 1.0L. Mix thoroughly. Gently heat while stirring and bring to boiling. Distribute into tubes or flasks. Autoclave for 15 min at 15 psi pressure–121°C.

Use: For the cultivation and maintenance of *Alcaligenes faecalis.*

Alcaligenes NA YE Medium (*Alcaligenes* Nutrient Agar Yeast Extract Medium)

Composition per liter:

Agar	15.0g
Pancreatic digest of gelatin	5.0g
Yeast extract	5.0g
Beef extract	3.0g

pH 7.0 ± 0.2 at 25°C

Preparation of Medium: Add components to distilled/deionized water and bring volume to 1.0L. Mix thoroughly. Gently heat while stirring and bring to boiling. Distribute into tubes or flasks. Autoclave for 15 min at 15 psi pressure–121°C. Pour into sterile Petri dishes or leave in tubes.

Use: For the cultivation and maintenance of *Alcaligenes* species.

Alcaligenes NB YE Agar (*Alcaligenes* Nutrient Broth Yeast Extract Agar)

Composition per liter:

Agar	15.0g
Pancreatic digest of gelatin	5.0g
Yeast extract	5.0g
Beef extract	3.0g

Preparation of Medium: Add components to distilled/deionized water and bring volume to 1.0L. Mix thoroughly. Gently heat while stirring and bring to boiling. Distribute into tubes or flasks. Autoclave for 15 min at 15 psi pressure–121°C. Pour into sterile Petri dishes or leave in tubes.

Use: For the cultivation and maintenance of *Alcaligenes faecalis.*

Alcaligenes NB YE Broth (Alcaligenes Nutrient Broth Yeast Extract Broth)

Composition per liter:

Pancreatic digest of gelatin	5.0g
Yeast extract	5.0g
Beef extract	3.0g

Preparation of Medium: Add components to distilled/deionized water and bring volume to 1.0L. Mix thoroughly. Gently heat while stirring and bring to boiling. Distribute into tubes or flasks. Autoclave for 15 min at 15 psi pressure–121°C.

Use: For the cultivation of *Alcaligenes faecalis.*

Alcaligenes NB YE Medium (*Alcaligenes* Nutrient Broth Yeast Extract Medium)

Composition per liter:

Pancreatic digest of gelatin	5.0g
Yeast extract	5.0g
Beef extract	3.0g

pH 7.0 ± 0.2 at 25°C

Preparation of Medium: Add components to distilled/deionized water and bring volume to 1.0L. Mix thoroughly. Distribute into tubes or flasks. Autoclave for 15 min at 15 psi pressure–121°C.

Use: For the cultivation and maintenance of *Alcaligenes* species.

Alginate Utilization Medium

Composition per liter:

Solution B	500.0mL
Solution A	400.0mL
Solution C	100.0mL

Solution A:

Composition per 400mL:

Marine salts	38.0g

Preparation of Solution A: Add marine salts to distilled/deionized water and bring volume to 400.0mL. Mix thoroughly. Autoclave for 15 min at 15 psi pressure–121°C.

Solution B:

Composition per 500mL:

Agar	20.0g
Sodium alginate	10.0g

Preparation of Solution B: Add components to distilled/deionized water and bring volume to 500.0mL. Mix thoroughly. Autoclave for 15 min at 15 psi pressure–121°C.

Solution C:

Composition per 100mL:

Tris·HCl buffer	0.067g
$NaNO_3$	0.047g
Ferric EDTA	66.5mg
Sodium glycerophosphate	6.67mg
Thiamine·HCl	67.0μg
Vitamin B_{12}	1.3μg
Biotin	0.67μg

Preparation of Solution C: Add components to distilled/deionized water and bring volume to 100.0mL. Mix thoroughly. Filter sterilize.

Preparation of Medium: Aseptically combine Solutions A, B and C. For liquid medium, omit agar from Solution B.

Use: For the cultivation of microorganisms that can utilize alginate as a carbon source. Growth on alginate (production of alginase) is a diagnostic test used in the differentiation of *Vibrio* species.

Alkaline *Bacillus* Medium

Composition per liter:

Agar	15.0g
Peptone	10.0g
Glucose	10.0g
Yeast extract	5.0g
K_2HPO_4	1.0g
Na_2CO_3 solution	100.0mL

pH 8.5–11.0 at 25°C

Na_2CO_3 Solution:

Composition per 100mL:

Na_2CO_3	10.0g

Preparation of Na_2CO_3 Solution: Add Na_2CO_3 to distilled/deionized water and bring volume to 100.0mL. Mix thoroughly. Filter sterilize.

Preparation of Medium: Add components, except Na_2CO_3 solution, to distilled/deionized water and bring volume to 900.0mL. Gently heat while stirring and bring to boiling. Autoclave for 15 min at 10 psi pressure–115°C. Cool to 45–50°C. Aseptically add sterile Na_2CO_3 solution. Mix thoroughly. Pour into sterile Petri dishes or distribute into sterile tubes.

Use: For the cultivation and maintenance of alkalophilic microorganisms such as *Bacillus alcalophilus*, *B. circulans* and other *Bacillus* species.

Alkaline Peptone Agar

Composition per liter:

NaCl	20.0g
Agar	15.0g
Peptone	10.0g

pH 8.5 ± 0.2 at 25°C

Preparation of Medium: Add components to distilled/deionized water and bring volume to 1.0L. Mix thoroughly. Gently heat and bring to boiling. Adjust pH to 8.5. Distribute into tubes. Autoclave for 15 min at 15 psi pressure–121°C. Allow tubes to cool in a slanted position.

Use: For the cultivation of *Vibrio cholerae* and other *Vibrio* species.

Alkaline Peptone Water

Composition per liter:

Peptone	10.0g
NaCl	5.0g

pH 9.0 ± 0.2 at 25°C

Preparation of Medium: Add components to distilled/deionized water and bring volume to 1.0L. Mix thoroughly. Adjust pH to 9.0. Distribute into tubes or flasks. Autoclave for 20 min at 15 psi pressure–121°C.

Use: For the cultivation of a variety of alkalophilic microorganisms, especially *Vibrio* species.

Alkaline Peptone Water

Composition per liter:

Peptone	10.0g
NaCl	5.0g

pH 8.4 ± 0.2 at 25°C

Preparation of Medium: Add components to distilled/deionized water and bring volume to 1.0L. Mix thoroughly. Adjust pH to 8.4. Distribute into tubes or flasks. Autoclave for 20 min at 15 psi pressure–121°C.

Use: For the cultivation of a variety of alkalophilic microorganisms.

Allen and Arnon Medium with Nitrate

Composition per 1000.25mL:

Noble agar	10.0g
KNO_3	0.253g
$NaNO_3$	0.212g
Solution A	25.0mL
Solution B	6.25mL

Solution A:

Composition per 2 liters:

$MgSO_4 \cdot 7H_2O$ (4% solution)	500.0mL
$CaCl_2 \cdot 2H_2O$ (1.2% solution)	500.0mL
NaCl (3.8% solution)	500.0mL
Microelements stock solution	500.0mL

Preparation of Solution A: Prepare individual solutions and combine.

Microelements Stock Solution:

Composition per 1090mL:

H_3BO_3	572.0mg
$MnCl_2 \cdot 4H_2O$	360.0mg
$ZnSO_4 \cdot 7H_2O$	44.0mg
MoO_3	36.0mg
$CuSO_4 \cdot 5H_2O$	15.8mg
$CoCl_2 \cdot 6H_2O$	8.0mg
NH_4VO_3	4.6mg
A & A FeEDTA solution	160.0mL

Preparation of Microelements Stock Solution: Add components to distilled/deionized water and bring volume to 1090.0mL. Mix well.

A & A FeEDTA Solution:

Composition per 550mL:

Disodium EDTA·$2H_2O$	8.0mg
$FeSO_4 \cdot 7H_2O$	4.6mg

Preparation of A & A FeEDTA Solution: Dissolve 5.2g KOH in 186.0mL distilled/deionized water. Add 20.4g disodium EDTA·$2H_2O$. Add 13.7g $FeSO_4 \cdot 7H_2O$ to 364.0mL distilled/deionized water. Combine the EDTA solution with the $FeSO_4$ solution. Sparge solution with filtered air until color changes. The pH of ferrous EDTA solution is about 3.5.

Solution B:

Composition per 500mL:

K_2HPO_4	28.0g

Preparation of Solution B: Add K_2HPO_4 to distilled/deionized water and bring volume to 500.0mL.

Preparation of Medium: Add agar, KNO_3, and $NaNO_3$ to distilled/deionized water and bring volume to 969.0mL. Mix thoroughly. Gently heat and bring to boiling. Add 25.0mL Solution A. Autoclave for 15 min at 15 psi pressure–121°C. Add 6.25mL Solution B aseptically after sterilization.

Use: For the cultivation and maintenance of *Anabaena* species and *Nostoc* species.

ALP Basal Medium (Aerobic Low Peptone Basal Medium)

Composition per liter:

Agar	15.0g
$(NH_4)_2SO_4$	1.0g
Pancreatic digest of casein	0.5g
Yeast extract	0.5g
$MgSO_4 \cdot 7H_2O$	0.2g
KCl	0.2g
Phenol red	0.02g
Substrate solution	50.0mL

pH 7.8 ± 0.2 at 25°C

Substrate Solution:

Composition per 50mL:

Substrate	0.1g

Preparation of Substrate Solution: Add substrate to distilled/deionized water and bring volume to 50.0mL. Use sugars, carbohydrates, *n*-butanol, other alcohols, or any acidogenic carbon source. Mix thoroughly. Filter sterilize.

Preparation of Medium: Add components, except substrate solution, to distilled/deionized water and bring volume to 950.0mL. Mix thoroughly. Gently heat and bring to boiling. Adjust pH to 7.8.

Distribute into screw-capped tubes in 3.0mL volumes. Autoclave for 15 min at 15 psi pressure–121°C. Cool to 45–50°C. Aseptically add 0.15mL of sterile substrate solution to each tube. Mix thoroughly. Allow tubes to cool in a slanted position.

Use: For the cultivation and differentiation of microorganisms based on their ability to utilize a variety of carbon sources such as carbohydrates, alcohols and other acidogenic substrates.

ALP Basal Medium (Aerobic Low Peptone Basal Medium)

Composition per liter:

Agar	15.0g
$(NH_4)_2SO_4$	1.0g
Pancreatic digest of casein	0.5g
Yeast extract	0.5g
Glucose	0.2g
$MgSO_4 \cdot 7H_2O$	0.2g
KCl	0.2g
Phenol red	0.02g
Substrate solution	50.0mL

pH 6.5 ± 0.2 at 25°C

Substrate Solution:

Composition per 50mL:

Substrate	0.1g

Preparation of Substrate Solution: Add substrate to distilled/deionized water and bring volume to 50.0mL. Use gelatin, aliphatic acids or any alkalogenic carbon source. Mix thoroughly. Filter sterilize.

Preparation of Medium: Add components, except substrate solution, to distilled/deionized water and bring volume to 950.0mL. Mix thoroughly. Gently heat and bring to boiling. Adjust pH to 6.5. Distribute into screw-capped tubes in 3.0mL volumes. Autoclave for 15 min at 15 psi pressure–121°C. Cool to 45–50°C. Aseptically add 0.15mL of sterile substrate solution to each tube. Mix thoroughly. Allow tubes to cool in a slanted position.

Use: For the cultivation and differentiation of microorganisms based on their ability to utilize a variety of carbon sources such as gelatin, aliphatic acids and other alkalophilic substrates.

AMB Agar

Composition per liter:

Agar	15.0g
Starch, soluble	5.0g
Pancreatic digest of casein	2.5g
$MgSO_4 \cdot 7H_2O$	0.5g
K_2HPO_4	0.25g

Preparation of Medium: Add components to distilled/deionized water and bring volume to 1.0L. Mix thoroughly. Gently heat and bring to boiling. Distribute into tubes or flasks. Autoclave for 15 min at 15 psi pressure–121°C. Pour into sterile Petri dishes or leave in tubes.

Use: For the cultivation of myxobacteria.

AMB Broth

Composition per liter:

Starch, soluble	5.0g
Pancreatic digest of casein	2.5g
$MgSO_4 \cdot 7H_2O$	0.5g
K_2HPO_4	0.25g

Preparation of Medium: Add components to distilled/deionized water and bring volume to 1.0L. Mix thoroughly. Distribute into tubes or flasks. Autoclave for 15 min at 15 psi pressure–121°C.

Use: For the cultivation of myxobacteria.

AMS Agar (Ammonium Mineral Salts Agar)

Composition per liter:

Agar	15.0g
$MgSO_4 \cdot 7H_2O$	1.0g
K_2HPO_4	0.7g
KH_2PO_4	0.54g
NH_4Cl	0.5g
$CaCl_2 \cdot 2H_2O$	0.2g
$FeSO_4 \cdot 7H_2O$	4.0mg
H_3BO_4	0.3mg
$CoCl_2 \cdot 6H_2O$	0.2mg
$ZnSO_4 \cdot 7H_2O$	0.1mg
$Na_2MoO_4 \cdot 2H_2O$	0.06mg
$MnCl_2 \cdot 4H_2O$	0.03mg
$NiCl_2 \cdot 6H_2O$	0.02mg
$CuCl_2 \cdot 2H_2O$	0.01mg

pH 6.8 ± 0.2 at 25°C

Preparation of Medium: Add components to distilled/deionized water and bring volume to 1.0L. Mix thoroughly. Gently heat and bring to boiling. Autoclave for 15 min at 15 psi–121°C. Add sterile methanol to a concentration of 0.5% aseptically to cooled basal medium.

Use: For the cultivation and maintenance of bacteria which can utilize methanol as a carbon source such as *Methylobacterium* species, *Methylomonas* species, and *Methylophilus* species.

AMS Agar without Methanol (Ammonium Mineral Salts Agar without Methanol)

Composition per liter:

Agar	15.0g
$MgSO_4 \cdot 7H_2O$	1.0g
K_2HPO_4	0.7g
KH_2PO_4	0.54g
NH_4Cl	0.5g
$CaCl_2 \cdot 2H_2O$	0.2g
$FeSO_4 \cdot 7H_2O$	4.0mg
H_3BO_4	0.3mg
$CoCl_2 \cdot 6H_2O$	0.2mg
$ZnSO_4 \cdot 7H_2O$	0.1mg
$Na_2MoO_4 \cdot 2H_2O$	0.06mg
$MnCl_2 \cdot 4H_2O$	0.03mg
$NiCl_2 \cdot 6H_2O$	0.02mg
$CuCl_2 \cdot 2H_2O$	0.01mg

pH 6.8 ± 0.2 at 25°C

Preparation of Medium: Add components to distilled/deionized water and bring volume to 1.0L. Mix thoroughly. Gently heat and bring to boiling. Autoclave for 15 min at 15 psi–121°C.

Use: For the cultivation and maintenance of *Methylosinus trichosporium* and other methane-oxidizing bacteria. Cultures are grown under an atmosphere of 50% methane.

AMS Medium

Composition per liter:

NaCl	26.0g
$MgSO_4 \cdot 7H_2O$	12.0g
Peptone	5.0g
Beef extract	3.0g
$CaCl_2 \cdot 2H_2O$	1.5g
KCl	0.7g

Preparation of Medium: Add components to distilled/deionized water and bring volume to 1.0L. Mix thoroughly. Gently heat and bring to boiling. Distribute into tubes or flasks. Autoclave for 15 min at 15 psi–121°C.

Use: For the cultivation of *Alteromonas espejiana.*

Amygdalin Medium

Composition per liter:

Peptone	10.0g
Beef extract	5.0g
NaCl	5.0g
Agar	3.0g
Amygdalin solution	200.0mL
Bromthymol blue (0.05% solution)	5.0mL

pH 7.0 ± 0.2 at 25°C

Amygdalin Solution:

Composition per 200mL:

Amygdalin	10.0g

Preparation of Amygdalin Solution: Add amygdalin to distilled/deionized water and bring volume to 200.0mL. Mix thoroughly. Filter sterilize.

Preparation of Medium: Add components, except amygdalin solution, to distilled/deionized water and bring volume to 800.0mL. Mix thoroughly. Gently heat and bring to boiling. Adjust pH to 7.0. Autoclave for 20 min at 15 psi pressure–121°C. Cool to 45–50°C. Aseptically add sterile amygdalin solution. Mix thoroughly. Aseptically distribute into sterile tubes with cotton plugs. Allow tubes to cool in a slanted position forming a short slant.

Use: For the cultivation and differentiation of *Serratia* species based on their ability to produce acid and HCN from amygdalin.

Anacker and Ordal Medium

Composition per liter:

Agar	10.0g
Pancreatic digest of casein	0.5g
Yeast extract	0.5g
Sodium acetate	0.2g
Beef extract	0.2g

pH 7.3 ± 0.1 at 25°C

Preparation of Medium: Add components to distilled/deionized water and bring volume to 1.0L. Mix thoroughly. Gently heat and bring to boiling. Distribute into tubes or flasks. Autoclave for 15 min at 15 psi–121°C. Pour into sterile Petri dishes or leave in tubes.

Use: For the cultivation and maintenance of *Flexibacter columnaris.*

Anacker and Ordal Medium, Enriched

Composition per liter:

Agar	10.0g
Pancreatic digest of casein	5.0g
Yeast extract	0.5g
Sodium acetate	0.2g
Beef extract	0.2g

pH 7.3 ± 0.1 at 25°C

Preparation of Medium: Add components to distilled/deionized water and bring volume to 1.0L. Mix thoroughly. Gently heat and bring to boiling. Distribute into tubes or flasks. Autoclave for 15 min at 15 psi–121°C. Pour into sterile Petri dishes or leave in tubes.

Use: For the cultivation and maintenance of *Flexibacter psychrophilus.*

Anaerobic Agar

Composition per liter:

Pancreatic digest of casein	20.0g
Agar	15.0g
NaCl	5.0g
Sodium thioglycollate	2.0g
Sodium formaldehyde sulfoxylate	1.0g

pH 7.2 ± 0.2 at 25°C

Preparation of Medium: Add components to distilled/deionized water and bring volume to 1.0L. Mix thoroughly. Gently heat and bring to boiling. Adjust pH to 7.2. Distribute into tubes until medium is 3 in. deep. Autoclave for 20 min at 15 psi pressure–121°C.

Use: For the anaerobic cultivation of *Bacillus* species and *Sporolactobacillus* species.

Anaerobic Agar

Composition per liter:

Pancreatic digest of casein	20.0g
Agar	15.0g
Yeast extract	15.0g
NaCl	5.0g
Sodium thioglycollate	2.0g
Sodium formaldehyde sulfoxylate	1.0g

pH 7.2 ± 0.2 at 25°C

Preparation of Medium: Add components to distilled/deionized water and bring volume to 1.0L. Mix thoroughly. Gently heat and bring to boiling. Adjust pH to 7.2. Distribute into tubes until medium is 3 in. deep. Autoclave for 20 min at 15 psi pressure–121°C.

Use: For the anaerobic cultivation of *Bacillus* species, especially *B. larvae, B. popilliae,* and *B. lentimorbus.*

Anaerobic Agar

Composition per liter:

Agar	20.0g
Pancreatic digest of casein	20.0g
Glucose	10.0g
NaCl	5.0g
Sodium thioglycollate	2.0g
Sodium formaldehyde sulfoxylate	1.0g
Methylene blue	2.0mg

pH 7.2 ± 0.2 at 25°C

Source: This medium is available as a premixed powder from Difco Laboratories.

Preparation of Medium: Add components to distilled/deionized water and bring volume to 1.0L. Mix thoroughly. Gently heat and bring to boiling. Adjust pH to 7.2. Distribute into tubes until medium is 3 in. deep. Autoclave for 15 min at 15 psi pressure–121°C.

Use: For the cultivation of a variety of anaerobic microorganisms, especially *Clostridium* species.

Anaerobic Agar

Composition per liter:

Pancreatic digest of casein	17.5g
Agar	15.0g
Glucose	10.0g
Papaic digest of soybean meal	2.5g
NaCl	2.5g
Sodium thioglycollate	2.0g
Sodium formaldehyde sulfoxylate	1.0g
L-Cystine	0.4g
Methylene Blue	2.0mg

pH 7.2 ± 0.2 at 25°C

Source: This medium is available as a premixed powder from BBL Microbiology Systems.

Preparation of Medium: Add components to distilled/deionized water and bring volume to 1.0L. Mix thoroughly. Gently heat and bring to boiling. Autoclave for 15 min at 15 psi–121°C. Use with Brewer anaerobic Petri dishes or in tubes or ordinary plates and incubate in anaerobic jars.

Use: For the cultivation of *Clostridium* species and for anaerobic microorganisms.

Anaerobic Broth

Composition per liter:

Pancreatic digest of casein	17.5g
Glucose	10.0g
NaCl	2.5g
Papaic digest of soybean meal	2.5g
Sodium thioglycollate	2.0g
Sodium formaldehyde sulfoxylate	1.0g
L-Cystine	0.4g
Methylene Blue	2.0mg

pH 7.2 ± 0.2 at 25°C

Preparation of Medium: Add components to distilled/deionized water and bring volume to 1.0L. Mix thoroughly. Gently heat and bring to boiling. Distribute into tubes or flasks. Autoclave for 15 min at 15 psi pressure–121°C.

Use: For the cultivation of a variety of anaerobic and microaerophilic microorganisms.

Anaerobic Cellulolytic Medium

Composition per liter:

NH_4Cl	1.0g
Cellobiose	1.0g
Yeast extract	1.0g
$MgSO_4$	0.5g
KCl	0.5g
L-Cysteine·HCl·H_2O	0.5g
K_2HPO_4	0.4g
Resazurin	1.0mg
Wolfe's mineral solution	20.0mL
Na_2CO_3 solution	10.0mL
$Na_2S{\cdot}9H_2O$ solution	10.0mL

pH 6.9 ± 0.1 at 25°C

Wolfe's Mineral Solution:

Composition per liter

$MgSO_4{\cdot}7H_2O$	3.0g
Nitrilotriacetic acid	1.5g
NaCl	1.0g
$MnSO_4{\cdot}H_2O$	0.5g
$FeSO_4{\cdot}7H_2O$	0.1g
$CoCl_2{\cdot}6H_2O$	0.1g
$CaCl_2$	0.1g
$ZnSO_4{\cdot}7H_2O$	0.1g
$CuSO_4{\cdot}5H_2O$	0.01g
$AlK(SO_4)_2{\cdot}12H_2O$	0.01g
H_3BO_3	0.01g
$Na_2MoO_4{\cdot}2H_2O$	0.01g

Preparation of Wolfe's Mineral Solution: Add nitrilotriacetic acid to approximately 500.0mL of distilled/deionized water and adjust pH to 6.5 with KOH to dissolve. Bring volume to 1.0L with distilled/deionized water. Add remaining compounds one at a time. Mix thoroughly.

Na_2CO_3 Solution:

Composition per 100mL:

Na_2CO_3	10.0g

Preparation of Na_2CO_3 Solution: Add Na_2CO_3 to distilled/deionized water and bring volume to 100.0mL. Mix thoroughly. Filter sterilize.

$Na_2S{\cdot}9H_2O$ Solution:

Composition per 100mL:

$Na_2S{\cdot}9H_2O$	15.0g

Preparation of $Na_2S{\cdot}9H_2O$ Solution: Add $Na_2S{\cdot}9H_2O$ to distilled/deionized water and bring volume to 100.0mL. Mix thoroughly. Filter sterilize.

Preparation of Medium: Add components, except Na_2CO_3 solution and $Na_2S{\cdot}9H_2O$ solution, to distilled/deionized water and bring volume to 980.0mL. Boil medium under a stream of 80% N_2 + 10% CO_2 + 10% H_2 until the resazurin indicator is colorless. Cool medium and distribute anaerobically into test tubes in 10.0mL volumes using 80% N_2 + 10% CO_2 + 10% H_2. Stopper the tubes anaerobically. Autoclave for 15 min at 15 psi pressure–121°C. Cool medium to room temperature. Aseptically add 0.1mL of sterile Na_2CO_3 solution and 0.1mL of sterile $Na_2S{\cdot}9H_2O$ solution to each tube. Mix thoroughly.

Use: For the cultivation and maintenance of microorganisms which can utilize cellobiose as sole carbon source such as *Clostridium cellulovorans.*

Anaerobic Egg Yolk Agar

Composition per 1080mL:

Agar	20.0g
Proteose peptone	20.0g
NaCl	5.0g
Pancreatic digest of casein	5.0g
Yeast extract	5.0g
Egg yolk emulsion, 50%	80.0mL

pH 7.0 ± 0.2 at 25°C

Egg Yolk Emulsion, 50%:

Composition per 100mL:

Chicken egg yolks	11
Whole chicken egg	1
NaCl (0.9% solution)	50.0mL

Preparation of Egg Yolk Emulsion, 50%: Soak eggs with 1:100 dilution of saturated mercuric chloride solution for 1 min. Crack 11 eggs and separate yolks from whites. Mix egg yolks with 1 chicken egg. Measure 50.0mL of egg yolk emulsion and add to 50.0mL of 0.9% NaCl solution. Mix thoroughly. Filter sterilize. Warm to 45–50°C.

Preparation of Medium: Add components, except egg yolk emulsion, 50%, to distilled/deionized water and bring volume to 1.0L. Mix thoroughly. Gently heat and bring to boiling. Autoclave for 15 min at 15 psi pressure–121°C. Cool to 45–50°C. Aseptically add 80.0mL of sterile egg yolk emulsion, 50%. Mix thoroughly. Pour into sterile Petri dishes or distribute into sterile tubes. Allow plates to dry at 35°C for 24 hr.

Use: For the cultivation of *Clostridium* species.

Anaerobic Egg Yolk Agar

Composition per liter:

Agar	20.0g
Proteose peptone	20.0g
Pancreatic digest of casein	5.0g
NaCl	5.0g
Yeast extract	5.0g
Egg yolk emulsion, 50%	20.0mL

pH 7.0 ± 0.2 at 25°C

Egg Yolk Emulsion, 50%:

Composition per 100mL:

Chicken egg yolks....................2
NaCl (0.9% solution)....................10.0mL

Preparation of Egg Yolk Emulsion, 50%: Soak eggs with 1:100 dilution of saturated mercuric chloride solution for 1 min. Crack eggs and separate yolks from whites. Measure 10.0mL of egg yolk emulsion and add to 10.0mL of 0.9% NaCl solution. Mix thoroughly. Filter sterilize. Warm to 45–50°C.

Preparation of Medium: Add components, except egg yolk emulsion, to distilled/deionized water and bring volume to 980.0mL. Mix thoroughly. Gently heat and bring to boiling. Autoclave for 15 min at 15 psi pressure–121°C. Cool to 45–50°C. Aseptically add sterile egg yolk emulsion. Mix thoroughly. Pour into sterile Petri dishes. Allow plates to dry at 35°C for 24 hr.

Use: For the cultivation of *Yersinia enterocolitica.*

Anaerobic D-Gluconate Medium

Composition per liter:

Agar....................15.0g
Pancreatic digest of casein....................10.0g
Yeast extract....................5.0g
D-Gluconate....................4.0g
$MgSO_4 \cdot 7H_2O$....................2.5g
$(NH_4)_2SO_4$....................1.4g
L-Cysteine·HCl·H_2O....................1.0g
$CaCl_2 \cdot 2H_2O$....................0.15g
$FeSO_4 \cdot 7H_2O$....................0.02g
Resazurin....................1.0mg
$NaHCO_3$ solution....................10.0mL

pH 7.1 ± 0.2 at 25°C

$NaHCO_3$ Solution:

Composition per 100mL:

$NaHCO_3$....................10.0g

Preparation of $NaHCO_3$ Solution: Add $NaHCO_3$ to distilled/deionized water and bring volume to 100.0mL. Mix thoroughly. Filter sterilize.

Preparation of Medium: Add components, except $NaHCO_3$ solution, to distilled/deionized water and bring volume to 990.0mL. Prepare anaerobically under 100% N_2. Autoclave for 15 min at 15 psi pressure–121°C. Aseptically add 10.0mL of the sterile $NaHCO_3$ solution. Mix thoroughly. Adjust pH to 7.1.

Use: For the cultivation and maintenance of microorganisms which can utilize D-gluconate as a carbon source such as *Bacteroides pectinophilus.*

Anaerobic Glucuronic Acid Medium

Composition per liter:

Agar....................15.0g
Pancreatic digest of casein....................10.0g
Yeast extract....................5.0g
Glucuronic acid....................4.0g
$MgSO_4 \cdot 7H_2O$....................2.5g
$(NH_4)_2SO_4$....................1.4g
L-Cysteine·HCl·H_2O....................1.0g
$CaCl_2 \cdot 2H_2O$....................0.15g
$FeSO_4 \cdot 7H_2O$....................0.02g
Resazurin....................1.0mg
$NaHCO_3$ solution....................10.0mL

pH 7.1 ± 0.2 at 25°C

$NaHCO_3$ Solution:

Composition per 100mL:

$NaHCO_3$....................10.0g

Preparation of $NaHCO_3$ Solution: Add $NaHCO_3$ to distilled/deionized water and bring volume to 100.0mL. Mix thoroughly. Filter sterilize.

Preparation of Medium: Add components, except $NaHCO_3$ solution, to distilled/deionized water and bring volume to 990.0mL. Prepare anaerobically under 100% N_2. Autoclave for 15 min at 15 psi pressure–121°C. Aseptically add 10.0mL of the sterile $NaHCO_3$ solution. Mix thoroughly. Adjust pH to 7.1.

Use: For the cultivation and maintenance of microorganisms which can utilize D-glucuronate as a carbon source such as *Bacteroides galacturonicus.*

Anaerobic TVLS Medium

Composition per liter:

Pancreatic digest of casein....................17.0g
Beef extract....................7.5g
Glucose....................6.0g
Enzymatic hydrolysate of soybean meal....................3.0g
Liver hydrolysate....................3.0g
NaCl....................2.5g
Na_2SO_3....................0.7g
Sodium thioglycollate....................0.5g
L-Cysteine·HCl·H_2O....................0.25g
Agar....................0.1g
Bovine serum....................100.0mL

pH 7.3 ± 0.2 at 25°C

Preparation of Medium: Add components, except bovine serum, to distilled/deionized water and bring volume to 900.0mL. Mix thoroughly. Gently heat and bring to boiling. Autoclave for 15 min at 15 psi–121°C. Cool to 45–50°C. Aseptically add 100.0mL bovine serum. Distribute into sterile tubes.

Use: For the isolation and cultivation of anaerobic microorganisms.

Anaerospirillum Medium

Composition per liter:

Polypeptone™	10.0g
Glucose	10.0g
Yeast extract	5.0g
Na_2CO_3	3.0g
NaCl	2.0g
K_2HPO_4	1.0g
$MgSO_4 \cdot 7H_2O$	0.2g

pH 6.5 ± 0.2 at 25°C

Preparation of Medium: Add components to distilled/deionized water and bring volume to 1.0L. Mix thoroughly. Gently heat and bring to boiling. Autoclave for 15 min at 15 psi–121°C. Distribute into tubes or flasks using anaerobic techniques and 100% CO_2 as gas phase.

Use: For the cultivation of *Anaerospirillum succiniciproducens.*

Andrade's Broth

Composition per liter:

Pancreatic digest of gelatin	10.0g
NaCl	5.0g
Beef extract	3.0g
Andrade's indicator	10.0mL
Carbohydrate solution	50.0mL

pH 7.4 ± 0.2 at 25°C

Source: Available as a prepared medium from BBL Microbiology Systems, in tubes containing adonitol, arabinose, cellobiose, glucose, dulcitol, fructose, galactose, inositol, lactose, maltose, mannitol, raffinose, rhamnose, salicin, sorbitol, sucrose, trehalose, or xylose.

Andrade's Indicator

Composition per 100mL:

NaOH (1*N* solution)	16.0mL
Acid fuchsin	0.1g

Preparation of Andrade's Indicator: Add Acid fuchsin to NaOH solution and bring volume to 100.0mL with distilled/deionized water.

Carbohydrate Solution:

Composition per 100mL:

Carbohydrate	10.0g

Preparation of Carbohydrate Solution: Add carbohydrate to distilled/deionized water and bring volume to 100.0mL. Adonitol, arabinose, cellobiose, glucose, dulcitol, fructose, galactose, inositol, lactose, maltose, mannitol, raffinose, rhamnose, salicin, sorbitol, sucrose, trehalose, xylose, or other carbohydrates may be used. Mix thoroughly. Filter sterilize.

Preparation of Medium: Add components, except carbohydrate solution, to distilled/deionized water and bring volume to 1.0L. Mix thoroughly. Gently heat and bring to boiling. Distribute in 10.0mL volumes into test tubes containing inverted Durham tubes. Autoclave for 15 min at 15 psi–121°C. Cool to 25°C. Add 0.5mL of sterile carbohydrate solution to each tube.

Caution: Acid fuchsin is a potential carcinogen and care must be taken to avoid inhalation of the powdered dye and contact with the skin.

Use: For the determination of carbohydrate fermentation reactions of microorganisms, particularly members of the Enterobacteriaceae. A Durham tube is used to collect gas produced during the fermentation reaction. Acid production is indicated by a pink color.

AO Agar

Composition per liter:

Agar	11.0g
Sodium acetate	0.5g
Pancreatic digest of casein	0.5g
Yeast extract	0.5g
Beef extract	0.2g

pH 7.2 ± 0.2 at 25°C

Preparation of Medium: Add components to distilled/deionized water and bring volume to 1.0L. Mix thoroughly. Gently heat and bring to boiling. Distribute into tubes or flasks. Autoclave for 15 min at 15 psi pressure–121°C. Pour into sterile Petri dishes or leave in tubes.

Use: For the isolation and cultivation of *Cytophaga* species, *Herpetosiphon* species, *Saprospira* species, and *Flexithrix* species.

AO Agar

Composition per liter:

Agar	4.0g
Sodium acetate	0.5g
Pancreatic digest of casein	0.5g
Yeast extract	0.5g
Beef extract	0.2g

pH 7.2 ± 0.2 at 25°C

Preparation of Medium: Add components to distilled/deionized water and bring volume to 1.0L. Mix thoroughly. Gently heat and bring to boiling. Distribute into tubes or flasks. Autoclave for 15 min at 15 psi pressure–121°C. Pour into sterile Petri dishes or leave in tubes.

Use: For the maintenance of *Cytophaga* species, *Herpetosiphon* species, *Saprospira* species, and *Flexithrix* species.

Aolpha Medium

Composition per 1041mL:

NaCl	100.0g
Agar	15.0g
$MgSO_4 \cdot 7H_2O$	9.5g
$MgCl_2 \cdot 6H_2O$	5.0g
KCl	5.0g
Peptone	5.0g
Yeast extract	1.0g
$CaCl_2 \cdot 2H_2O$	0.2g
$(NH_4)_2SO_4$	0.1g
KNO_3	0.1g
Metals solution	20.0mL
Phosphate solution	20.0mL
Vitamin solution	1.0mL

pH 7.0 ± 0.2 at 25°C

Metals Solution:

Composition per liter:

$MgSO_4 \cdot 7H_2O$	29.7g
Nitrilotriacetic acid	10.0g
$CaCl_2 \cdot 2H_2O$	3.3g
$FeSO_4 \cdot 7H_2O$	99.0mg
$Na_2MoO_4 \cdot 2H_2O$	12.7mg
Metals "44"	50.0mL

Preparation of Metals Solution: Solubilize nitrilotriacetic acid with KOH. Dissolve remaining ingredients. Adjust pH to 7.2 with KOH or H_2SO_4. Autoclave for 15 min at 15 psi–121°C. Add aseptically to sterile basal medium.

Metals "44":

Composition per 100mL:

$ZnSO_4 \cdot 7H_2O$	1.1g
$FeSO_4 \cdot 7H_2O$	0.5g
EDTA	0.25g
$MnSO_4 \cdot 7H_2O$	0.154g
$CuSO_4 \cdot 5H_2O$	0.04g
$Co(NO_3)_2 \cdot 6H_2O$	0.025g
$Na_2B_4O_7 \cdot 10H_2O$	0.018g

Preparation of Metals "44": Add components to distilled/deionized water and bring volume to 100.0mL. Mix thoroughly. Autoclave for 15 min at 15 psi–121°C. Add aseptically to sterile basal medium.

Phosphate Solution:

Composition per liter:

K_2HPO_4	2.5g
KH_2PO_4	2.5g

Preparation of Phosphate Solution: Add components to distilled/deionized water and bring volume to 1.0L. Mix thoroughly. Autoclave for 15 min at 15 psi–121°C. Add aseptically to sterile basal medium.

Vitamin Solution:

Composition per liter:

Pyridoxine·HCl	10.0mg
Calcium pantothenate	5.0mg
Nicotinamide	5.0mg
Riboflavin	5.0mg
Thiamine·HCl	5.0mg
Biotin	2.0mg
Folic acid	2.0mg
Cyanocobalamin	0.1mg

Preparation of Vitamin Solution: Add components to distilled/deionized water and bring volume to 1.0L. Mix thoroughly. Filter sterilize and add aseptically to sterile basal medium.

Preparation of Medium: Add components—except metals "44," phosphate solution and vitamin solution—to distilled/deionized water and bring volume to 1.0L. Mix thoroughly. Gently heat and bring to boiling. Adjust pH of basal medium to 7.0. Autoclave for 15 min at 15 psi–121°C. Cool to 50°C and aseptically add the metals "44," phosphate and vitamin solutions.

Use: For the cultivation and maintenance of *Halomonas meridiana* and other *Halomonas* species.

Aphanomyces Synthetic Medium

Composition per liter:

D-Glucose	5.0g
KH_2PO_4	2.0g
L-Asparagine	0.75g
$MgCl_2$	0.05g
$FeCl_3$	5.0mg
$MnCl_2$	5.0mg
$ZnCl_2$	5.0mg
L-Methionine	0.02mg

pH 5.5 ± 0.2 at 25°C

Preparation of Medium: Add components to distilled/deionized water and bring volume to 1.0L. Mix thoroughly. Autoclave for 15 min at 15 psi pressure–121°C. Adjust pH to 5.5. Aseptically distribute into sterile tubes or flasks.

Use: For the cultivation of *Aphanomyces* species.

Aplanobacterium Medium

Composition per liter:

Agar	20.0g
Glucose	10.0g

Peptone	5.0g
Yeast extract	5.0g

pH 7.2 ± 0.2 at 25°C

Preparation of Medium: Add components to distilled/deionized water and bring volume to 1.0L. Mix thoroughly. Gently heat and bring to boiling. Autoclave for 15 min at 15 psi–121°C. Pour into sterile Petri dishes or leave in tubes.

Use: For the cultivation and maintenance of *Xanthomonas* species.

Apple Juice Yeast Extract Medium (AJYE Medium)

Composition per 1200mL:

Agar	30.0g
Yeast extract	10.0g
Apple juice	1.0L

pH 4.8 ± 0.2 at 25°C

Preparation of Medium: Add yeast extract to 1.0L of apple juice. Mix thoroughly. Adjust pH to 4.8. Autoclave for 10 min at 9 psi pressure–114°C. Cool to 45–50°C. In a separate flask, add agar to 200.0mL of distilled/deionized water and bring volume to 1.0L. Mix thoroughly. Gently heat and bring to boiling. Autoclave for 15 min at 15 psi pressure–121°C. Cool to 45–50°C. Aseptically combine the sterile apple juice solution with the sterile agar solution. Mix thoroughly. Pour into sterile Petri dishes.

Use: For the cultivation of *Zymomonas* species.

Aquaspirillum Autotrophic Agar

Composition per liter:

Noble agar	15.0g
$Na_2HPO_4 \cdot 12H_2O$	9.0g
KH_2PO_4	1.5g
NH_4Cl	1.0g
$MgSO_4 \cdot 7H_2O$	0.2g
$CaCl_2 \cdot 2H_2O$	0.01g
Ferric ammonium citrate	5.0mg
$NaHCO_3$ solution	10.0mL
Trace elements solution	3.0mL

pH 7.1 ± 0.2 at 25°C

$NaHCO_3$ Solution:

Composition per 10mL:

$NaHCO_3$	0.5g

Preparation of $NaHCO_3$ Solution: Add the $NaHCO_3$ to distilled/deionized water and bring volume to 10.0mL. Mix thoroughly. Filter sterilize.

Trace Elements Solution:

Composition per liter:

H_3BO_3	30.0mg
$CoCl_2 \cdot 6H_2O$	20.0mg
$ZnSO_4 \cdot 7H_2O$	10.0mg
$MnCl_2 \cdot 4H_2O$	3.0mg
$Na_2MoO_4 \cdot 2H_2O$	3.0mg
$NiCl_2 \cdot 6H_2O$	2.0mg
$CuCl_2 \cdot 2H_2O$	1.0mg

Preparation of Trace Elements Solution: Add components to distilled/deionized water and bring volume to 1.0L. Mix thoroughly.

Preparation of Medium: Add components, except $NaHCO_3$ solution, to double distilled water and bring volume to 990.0mL. Mix thoroughly. Gently heat and bring to boiling. Autoclave for 15 min at 15 psi–121°C. Cool to 45–50°C. Aseptically add sterile $NaHCO_3$ solution. Mix thoroughly. Pour into sterile Petri dishes or distribute into sterile tubes. For autotrophic growth, incubate under 85% H_2 + 10% CO_2 + 5% O_2.

Use: For the autotrophic cultivation and maintenance of *Aquaspirillum autotrophicum*.

Aquaspirillum Autotrophic Broth

Composition per liter:

$Na_2HPO_4 \cdot 12H_2O$	9.0g
KH_2PO_4	1.5g
NH_4Cl	1.0g
$MgSO_4 \cdot 7H_2O$	0.2g
$CaCl_2 \cdot 2H_2O$	0.01g
Ferric ammonium citrate	5.0mg
$NaHCO_3$ solution	10.0mL
Trace elements solution	3.0mL

pH 7.1 ± 0.2 at 25°C

$NaHCO_3$ Solution:

Composition per 10mL:

$NaHCO_3$	0.5g

Preparation of $NaHCO_3$ Solution: Add the $NaHCO_3$ to distilled/deionized water and bring volume to 10.0mL. Mix thoroughly. Filter sterilize.

Trace Elements Solution:

Composition per liter:

H_3BO_3	30.0mg
$CoCl_2 \cdot 6H_2O$	20.0mg
$ZnSO_4 \cdot 7H_2O$	10.0mg
$MnCl_2 \cdot 4H_2O$	3.0mg
$Na_2MoO_4 \cdot 2H_2O$	3.0mg
$NiCl_2 \cdot 6H_2O$	2.0mg
$CuCl_2 \cdot 2H_2O$	1.0mg

Preparation of Trace Elements Solution: Add components to distilled/deionized water and bring volume to 1.0L. Mix thoroughly.

Preparation of Medium: Add components, except $NaHCO_3$ solution, to double distilled water and bring volume to 990.0mL. Mix thoroughly. Gently heat and bring to boiling. Autoclave for 15 min at 15 psi–121°C. Cool to 45–50°C. Aseptically add sterile $NaHCO_3$ solution. Mix thoroughly. Aseptically distribute into sterile tubes or flasks. To grow autotrophically, incubate under 85% H_2 + 10% CO_2 + 5% O_2.

Use: For the autotrophic cultivation of *Aquaspirillum autotrophicum.*

Aquaspirillum Heterotrophic Agar

Composition per liter:

Noble agar	15.0g
$Na_2HPO_4 \cdot 12H_2O$	9.0g
KH_2PO_4	1.5g
NH_4Cl	1.0g
Sodium succinate	1.0g
$MgSO_4 \cdot 7H_2O$	0.2g
$CaCl_2 \cdot 2H_2O$	0.01g
Ferric ammonium citrate	5.0mg
Trace elements solution	3.0mL

pH 7.1 ± 0.2 at 25°C

Trace Elements Solution:

Composition per liter:

H_3BO_3	30.0mg
$CoCl_2 \cdot 6H_2O$	20.0mg
$ZnSO_4 \cdot 7H_2O$	10.0mg
$MnCl_2 \cdot 4H_2O$	3.0mg
$Na_2MoO_4 \cdot 2H_2O$	3.0mg
$NiCl_2 \cdot 6H_2O$	2.0mg
$CuCl_2 \cdot 2H_2O$	1.0mg

Preparation of Trace Elements Solution: Add components to distilled/deionized water and bring volume to 1.0L. Mix thoroughly.

Preparation of Medium: Add components to double distilled water and bring volume to 1.0L. Mix thoroughly. Gently heat and bring to boiling. Autoclave for 15 min at 15 psi–121°C. Pour into sterile Petri dishes or distribute into sterile tubes.

Use: For the heterotrophic cultivation and maintenance of *Aquaspirillum autotrophicum.*

Aquaspirillum Heterotrophic Broth

Composition per liter:

$Na_2HPO_4 \cdot 12H_2O$	9.0g
KH_2PO_4	1.5g
NH_4Cl	1.0g
Sodium succinate	1.0g
$MgSO_4 \cdot 7H_2O$	0.2g
$CaCl_2 \cdot 2H_2O$	0.01g
Ferric ammonium citrate	5.0mg
Trace elements solution	3.0mL

pH 7.1 ± 0.2 at 25°C

Trace Elements Solution:

Composition per liter:

H_3BO_3	30.0mg
$CoCl_2 \cdot 6H_2O$	20.0mg
$ZnSO_4 \cdot 7H_2O$	10.0mg
$MnCl_2 \cdot 4H_2O$	3.0mg
$Na_2MoO_4 \cdot 2H_2O$	3.0mg
$NiCl_2 \cdot 6H_2O$	2.0mg
$CuCl_2 \cdot 2H_2O$	1.0mg

Preparation of Trace Elements Solution: Add components to distilled/deionized water and bring volume to 1.0L. Mix thoroughly.

Preparation of Medium: Add components to double distilled water and bring volume to 1.0L. Mix thoroughly. Gently heat and bring to boiling. Autoclave for 15 min at 15 psi–121°C. Aseptically distribute into sterile tubes or flasks.

Use: For the heterotrophic cultivation of *Aquaspirillum autotrophicum.*

Archaeoglobus Medium

Composition per liter:

NaCl	18.0g
$NaHCO_3$	5.0g
$MgCl_2 \cdot 6H_2O$	4.0g
$MgSO_4 \cdot 7H_2O$	3.45g
Sodium L-lactate	1.5g
Yeast extract	0.5g
KCl	0.34g
NH_4Cl	0.25g
$CaCl_2 \cdot 2H_2O$	0.14g
K_2HPO_4	0.14g
$Fe(NH_4)_2(SO_4)_2 \cdot 7H_2O$	2.0mg
Resazurin	1.0mg
$Na_2S \cdot 9H_2O$ solution	25.0mL
Trace elements solution	10.0mL

pH 6.9 ± 0.2 at 25°C

Trace Elements Solution:

Composition per liter:

$MgSO_4 \cdot 7H_2O$	3.0g
Nitrilotriacetic acid	1.5g
NaCl	1.0g
$MnSO_4 \cdot 2H_2O$	0.5g
$CoSO_4 \cdot 7H_2O$	0.18g
$ZnSO_4 \cdot 7H_2O$	0.18g
$FeSO_4 \cdot 7H_2O$	0.1g
$CaCl_2 \cdot 2H_2O$	0.1g
$NiCl_2 \cdot 6H_2O$	0.025g

$KAl(SO_4)_2 \cdot 12H_2O$	0.02g
$CuSO_4 \cdot 5H_2O$	0.01g
H_3BO_3	0.01g
$Na_2MoO_4 \cdot 2H_2O$	0.01g
$Na_2SeO_3 \cdot 5H_2O$	0.3mg

$Na_2S \cdot 9H_2O$ Solution:
Composition per 50mL:

$Na_2S \cdot 9H_2O$	1.0g

Preparation of $Na_2S \cdot 9H_2O$ Solution: Prepare and dispense solution anaerobically under with 80% N_2 + 20% CO_2. Add $Na_2S \cdot 9H_2O$ to distilled/deionized water and bring volume to 50.0mL. Mix thoroughly. Adjust pH to 7.0. Autoclave for 15 min at 15 psi–121°C.

Preparation of Trace Elements Solution: Add nitrilotriacetic acid to approximately 500.0mL distilled/deionized water. Dissolve by adding KOH and adjust pH to 6.5. Add remaining components. Bring volume to 1.0L with additional distilled/deionized water. Adjust pH to 7.0 with KOH.

Preparation of Medium: Add components, except $NaHCO_3$ and $Na_2S \cdot 9H_2O$ solution to distilled/deionized water and bring volume to 1.0L. Mix well and heat to boiling for a few minutes. Cool rapidly to room temperature while gassing with 80% N_2 + 20% CO_2. Add $NaHCO_3$ and adjust pH to 6.9. Distribute anaerobically under 80% N_2 + 20% CO_2 and pressurize sealed containers up to 2 bar pressure. Autoclave for 15 min at 15 psi–121°C. Prior to inoculation of cultures, add 0.25mL of sterile $Na_2S \cdot 9H_2O$ solution to each tube containing 9.75mL of sterile basal medium.

Use: For the cultivation and maintenance of *Archaeoglobus fulgidus*.

Archangium violaceum Medium

Composition per liter:

Monosodium glutamate	1.0g
L-Leucine	0.50g
L-Tyrosine	0.50g
L-Isoleucine	0.30g
L-Proline	0.25g
$MgSO_4 \cdot 7H_2O$	0.20g
L-Lysine	0.15g
L-Arginine	0.10g
L-Asparagine	0.10g
L-Serine	0.10g
L-Threonine	0.10g
L-Valine	0.10g
L-Alanine	0.05g
L-Glycine	0.05g
L-Histidine	0.05g
L-Methionine	0.05g
$Ca_3(PO_4)_2$	0.02g
KCl	0.02g
Tris(hydroxymethyl)aminomethane buffer (0.02M solution, pH 7.5)	1.0L

pH 7.5 ± 0.2 at 25°C

Preparation of Medium: Add solid components to 1.0L of Tris buffer. Mix thoroughly. Filter sterilize. Aseptically distribute into tubes or flasks.

Use: For the cultivation of *Archangium violaceum.*

Arginine Glucose Slants (AGS)

Composition per liter:

NaCl	20.0g
Agar	13.5g
Pancreatic digest of casein	10.0g
L-Arginine·HCl	5.0g
Peptone	5.0g
Yeast extract	3.0g
Glucose	1.0g
Ferric ammonium citrate	0.5g
$Na_2S_2O_3 \cdot 5H_2O$	0.3g
Bromcresol purple	0.02g

pH 6.8-7.0 at 25°C

Preparation of Medium: Add components to distilled/deionized water and bring volume to 1.0L. Mix thoroughly. Gently heat and bring to boiling. Distribute into tubes. Autoclave for 12 min at 15 psi pressure–121°C. Allow tubes to cool in a slanted position.

Use: For the cultivation and differentation of *Vibrio* species.

Armstrong *Fusarium* Medium

Composition per liter:

Glucose	20.0g
$Ca(NO_3)_2 \cdot 4H_2O$	8.4g
KH_2PO_4	1.09g
KCl	0.22g
$FeCl_3$	0.2μg
$MnSO_4$	0.2μg
$ZnSO_4$	0.2μg

Preparation of Medium: Add components to distilled/deionized water and bring volume to 1.0L. Mix thoroughly. Filter sterilize.

Use: For the cultivation of *Fusarium* species.

Arthrobacter Medium

Composition per liter:

Agar	10.0g
Casein	1.0g

Glucose 1.0g
K_2HPO_4 .... 1.0g
Yeast extract 0.7g
$MgSO_4·7H_2O$ 0.25g
$(NH_4)_2SO_4$ 0.25g

pH 6.9–7.0 at 25°C

Preparation of Medium: Add components to tap water and bring volume to 1.0L. Mix thoroughly. Gently heat and bring to boiling. Autoclave for 15 min at 15 psi pressure–121°C. Pour into sterile Petri dishes.

Use: For the isolation, cultivation and enumeration of *Arthrobacter* species from soil.

Arthrobacter Medium

Composition per liter:

Agar 15.0g
Peptone 10.0g
Yeast extract 10.0g
K_2HPO_4 .... 2.0g
Rhodotorulic acid (δ-*N*-acetyl-L-ornithine) or desferal 20.0μg

pH 7.4 ± 0.2 at 25°C

Preparation of Medium: Add components to distilled/deionized water and bring volume to 1.0L. Mix thoroughly. Gently heat and bring to boiling. Distribute into tubes or flasks. Autoclave for 15 min at 15 psi–121°C. Pour into sterile Petri dishes or leave in tubes.

Use: For the cultivation and maintenance of *Aureobacterium flavescens.*

Arthrobacter YCWD

Composition per liter:

Pancreatic digest of casein 10.0g
Yeast extract 1.0g

pH 7.2 ± 0.2 at 25°C

Preparation of Medium: Add components to distilled/deionized water and bring volume to 1.0L. Mix thoroughly. Gently heat and bring to boiling. Distribute into tubes or flasks. Autoclave for 15 min at 15 psi–121°C.

Use: For the cultivation and maintenance of *Arthrobacter* species.

Artificial Deep Lake Medium

Composition per liter:

NaCl 180.0g
$MgCl_2·6H_2O$ 75.0g
Noble agar 15.0g
Sodium succinate 10.0g
$MgSO_4·7H_2O$ 7.4g
KCl 7.4g
$CaCl_2·2H_2O$ 1.0g
Yeast extract 1.0g
Vitamin solution 10.0mL

pH 7.4 ± 0.2 at 25°C

Vitamin Solution:

Composition per liter:

Biotin 30.0mg
Cyanocobalamin 20.0mg
Thiamine·HCl 10.0mg

Preparation of Vitamin Solution: Add components to distilled/deionized water and bring volume to 1.0L. Mix thoroughly. Filter sterilize and add aseptically to sterile basal medium.

Preparation of Medium: Add components, except vitamin solution, to distilled/deionized water and bring volume to 990.0mL. Mix thoroughly. Gently heat and bring to boiling. Adjust medium to pH 7.4. Autoclave for 15 min at 15 psi–121°C. Cool to 50°C. Aseptically add 10.0mL of vitamin solution. Pour into sterile Petri dishes or leave in tubes.

Use: For the cultivation and maintenance *Halobacterium lacusprofundi*.

Ascospore Agar

Composition per liter :

Potassium acetate 30.0g
Yeast extract 2.5g
Glucose 1.0g

pH 6.4 ± 0.2 at 25°C

Preparation of Medium: Add components to distilled/deionized water and bring volume to 1.0L. Mix thoroughly. Gently heat and bring to boiling. Distribute into tubes or flasks. Autoclave for 15 min at 15 psi–121°C. Pour into sterile Petri dishes or leave in tubes.

Use: For the enrichment of ascosporogenous yeasts and their production of ascospores.

Ashbey's Nitrogen–Free Agar

Composition per liter:

Agar 15.0g
Mannitol 15.0g
$CaCl_2·2H_2O$ 0.2g
K_2HPO_4 .... 0.2g
$MgSO_4·7H_2O$ 0.2g
MoO_3 (10% solution) 0.1mL
$FeCl_3$ (10% solution 0.05mL

pH 7.2 ± 0.2 at 25°C

Preparation of Medium: Add components to distilled/deionized water and bring volume to 1.0L. Mix thoroughly. Gently heat and bring to boiling. Distribute into tubes or flasks. Autoclave for 15 min at 15 psi pressure–121°C. Pour into sterile Petri dishes or leave in tubes.

Use: For the isolation and cultivation of bacteria, such as *Azotobacter* species and cyanobacteria, that can utilize atmospheric N_2 as sole nitrogen source.

ASN-III Agar

Composition per liter:

NaCl	25.0g
$MgSO_4 \cdot 7H_2O$	3.5g
$MgCl_2 \cdot 6H_2O$	2.0g
$NaNO_3$	0.75g
$K_2HPO_4 \cdot 3H_2O$	0.75g
$CaCl_2 \cdot 2H_2O$	0.5g
KCl	0.5g
Na_2CO_3	0.02g
Citric acid	3.0mg
Ferric ammonium citrate	3.0mg
Magnesium EDTA	0.5mg
Vitamin B_{12}	10.0μg
Agar solution	100.0mL
A-5 Trace metals	1.0mL

pH 7.3 ± 0.2 at 25°C

Agar Solution:

Composition per 100mL:

Noble agar	10.0g

Preparation of Agar Solution: Add agar to glass distilled water and bring volume to 100.0mL. Mix thoroughly. Gently heat and bring to boiling. Autoclave for 15 min at 15 psi pressure–121°C. Cool to 45–50°C.

A-5 Trace Metals:

Composition per liter:

H_3BO_3	2.86g
$MnCl_2 \cdot 4H_2O$	1.81g
$ZnSO_4 \cdot 7H_2O$	0.222g
$CuSO_4 \cdot 5H_2O$	0.079g
$Co(NO_3)_2 \cdot 6H_2O$	0.049g
$Na_2MoO_4 \cdot 2H_2O$	0.039g

Preparation of A-5 Trace Metals: Add components to distilled/deionized water and bring volume to 1.0L. Mix thoroughly.

Preparation of Medium: Add components, except agar solution, to glass distilled water and bring volume to 900.0mL. Mix well and heat gently until dissolved. Filter sterilize. Warm to 45–50°C. Aseptically add agar solution. Mix thoroughly. Pour into sterile Petri dishes or distribute into sterile tubes.

Use: For the cultivation of *Xenococcus* species. Also, for the isolation of cyanobacteria from marine habitats.

ASN-III Broth

Composition per liter:

NaCl	25.0g
$MgSO_4 \cdot 7H_2O$	3.5g
$MgCl_2 \cdot 6H_2O$	2.0g
$NaNO_3$	0.75g
$K_2HPO_4 \cdot 3H_2O$	0.75g
$CaCl_2 \cdot 2H_2O$	0.5g
KCl	0.5g
Na_2CO_3	0.02g
Citric acid	3.0mg
Ferric ammonium citrate	3.0mg
Magnesium EDTA	0.5mg
Vitamin B_{12}	10.0μg
A-5 Trace metals	1.0mL

pH 7.3 ± 0.2 at 25°C

A-5 Trace Metals:

Composition per liter:

H_3BO_3	2.86g
$MnCl_2 \cdot 4H_2O$	1.81g
$ZnSO_4 \cdot 7H_2O$	0.222g
$CuSO_4 \cdot 5H_2O$	0.079g
$Co(NO_3)_2 \cdot 6H_2O$	0.049g
$Na_2MoO_4 \cdot 2H_2O$	0.039g

Preparation of A-5 Trace Metals: Add components to distilled/deionized water and bring volume to 1.0L. Mix thoroughly.

Preparation of Medium: Add components to glass distilled water and bring volume to 1.0L. Mix well and heat gently until dissolved. Filter sterilize.

Use: For the cultivation of *Xenococcus* species. Also, for the isolation of cyanobacteria from marine habitats.

Asparagine Broth

Composition per liter:

DL-Asparagine	30g
K_2HPO_4	1.0g
$MgSO_4 \cdot 7H_2O$	0.5g

pH 6.9–7.2 at 25°C

Preparation of Medium: Add components to distilled/deionized water and bring volume to 1.0L. Mix well until dissolved. Adjust pH to between 6.9 and 7.2. Distribute into tubes or flasks. Autoclave for 15 min at 15 psi pressure–121°C.

Use: For a presumptive test medium in the differentiation of nonfermentative Gram-negative bacteria, especially *Pseudomonas aeruginosa*. For use in the multiple tube technique in the microbiological analysis of recreational waters.

Aspergillus Differential Medium

Composition per liter:

Agar	15.0g
Pancreatic digest of casein	15.0g
Yeast extract	10.0g
Ferric citrate	0.5g

Preparation of Medium: Add components to distilled/deionized water and bring volume to 1.0L. Mix thoroughly. Gently heat and bring to boiling. Distribute into tubes in 7.0mL volumes. Autoclave for 15 min at 15 psi pressure–121°C. Allow tubes to cool in a slanted position.

Use: For the cultivation and differentiation of *Aspergillus flavus. A. flavus* appears as bright orange colonies.

Asticcacaulis Medium

Composition per liter:

Agar	15.0g
Pancreatic digest of casein	0.5g
Yeast extract	0.5g
Sodium acetate	0.2g

pH 6.8 ± 0.2 at 25°C

Preparation of Medium: Add components to distilled/deionized water and bring volume to 1.0L. Mix thoroughly. Gently heat and bring to boiling. Distribute into tubes or flasks. Autoclave for 15 min at 15 psi pressure–121°C. Pour into sterile Petri dishes or leave in tubes.

Use: For the isolation and cultivation of *Asticcacaulis* species.

ASTM Nutrient Salts Agar (American Society for Testing and Materials Nutrient Salts Agar)

Composition per liter:

Agar	15.0g
KH_2PO_4	0.7g
K_2HPO_4	0.7g
$MgSO_4·7H_2O$	0.7g
NH_4NO_3	1.0g
NaCl	5.0mg
$FeSO_4·7H_2O$	2.0mg
$ZnSO_4$	2.0mg
$MnSO_4·H_2O$	1.0mg

pH 6.5 ± 0.2 at 25°C

Preparation of Medium: Add components to tap water and bring volume to 1.0L. Mix thoroughly. Gently heat and bring to boiling. Distribute into tubes or flasks. Autoclave for 15 min at 15 psi pressure–121°C. Pour into sterile Petri dishes or leave in tubes.

Use: For determination of susceptibility of plastics to fungal degradation.

AT5N Medium

Composition per liter:

$CaCO_3$	10.0g
$(NH_4)_2SO_4$	1.5g
K_2HPO_4	0.5g
$MgSO_4$	50.0mg
$KHCO_3$	30.0mg
$CaCl_2·2H_2O$	20.0mg

Preparation of Medium: Add components to tap water and bring volume to 1.0L. Mix thoroughly. Gently heat and bring to boiling. Distribute into tubes or flasks. Autoclave for 15 min at 15 psi pressure–121°C.

Use: For the cultivation of bacteria that oxidize ammonia, especially those from wastewater.

Atlas Oil Agar

Composition per liter:

Bushnell-Haas agar	990.0mL
Oil	10.0mL

pH 7.0 ± 0.2 at 25°C

Bushnell-Haas Agar:
Composition per 990mL:

Agar	15.0g
KH_2PO_4	1.0g
K_2HPO_4	1.0g
NH_4NO_3	1.0g
$MgSO_4·7H_2O$	0.2g
$FeCl_3$	0.05g
$CaCl_2·2H_2O$	0.02g

Preparation of Bushnell-Haas Agar: Add components to distilled/deionized water and bring volume to 990.0mL. Mix thoroughly. Gently heat and bring to boiling. Autoclave for 15 min at 15 psi pressure–121°C. Cool to 60°C.

Preparation of Medium: Filter sterilize oil. Aseptically add 10.0mL of sterile oil to 990.0mL of cooled, sterile Bushnell-Haas agar. Put mixture into a sterile blender container. Blend on low speed to minimize the incorporation of air into the medium. Pour into sterile Petri dishes.

Use: For the cultivation and enumeration of hydrocarbon-utilizing bacteria by direct plating of water and sediment samples.

Aureomycin® Rose Bengal Glucose Peptone Agar

Composition per liter:

Agar	20.0g
Glucose	10.0g
Peptone	5.0g
KH_2PO_4	1.0g
$MgSO_4 \cdot 7H_2O$	0.5g
Rose Bengal	0.035g
Aureomycin solution	200.0mL

pH 5.4 ± 0.2 at 25°C

Aureomycin Solution:

Composition per 200mL:

Aureomycin·HCl	0.07g

Preparation of Aureomycin Solution: Add aureomycin·HCl to distilled/deionized water and bring volume to 200.0mL. Mix thoroughly. Filter sterilize.

Preparation of Medium: Add components, except aureomycin solution, to distilled/deionized water and bring volume to 800.0mL. Mix thoroughly. Gently heat and bring to boiling. Autoclave for 15 min at 15 psi pressure–121°C. Cool to 45–50°C. Aseptically add 200.0mL of sterile aureomycin solution. Mix thoroughly. Pour into sterile Petri dishes or distribute into sterile tubes.

Use: For the cultivation and enumeration of fungi isolated from sewage and polluted waters.

Azide Blood Agar

Composition per liter:

Agar	15.0g
Pancreatic digest of casein	5.0g
Peptic digest of animal tissue	5.0g
NaCl	5.0g
Beef extract	3.0g
NaN_3	0.2g
Sheep blood, defibrinated	50.0mL

pH 7.2 ± 0.2 at 25°C

Source: This medium is available as a premixed powder from Difco Laboratories, BBL Microbiology Systems and Oxoid Unipath.

Caution: Sodium azide is toxic. Azides also react with metals and disposal must be highly diluted.

Preparation of Medium: Add components, except sheep blood, to distilled/deionized water and bring volume to 950.0mL. Mix thoroughly. Gently heat and bring to boiling. Autoclave for 15 min at 15 psi pressure–121°C. Cool to 45–50°C. Aseptically add 50.0mL of sterile defibrinated sheep blood. Pour into sterile Petri dishes or distribute into sterile tubes. Allow tubes to cool in a slanted position.

Use: For the isolation and differentiation of streptococci and staphylococci from specimens containing mixed flora and from nonclinical specimens such as water and sewage.

Azide Blood Agar with Crystal Violet (Packer's Agar)

Composition per liter:

Agar	15.0g
Pancreatic digest of casein	5.0g
Peptic digest of animal tissue	5.0g
NaCl	5.0g
Beef extract	3.0g
NaN_3	0.9g
Crystal violet	2.0mg
Sheep blood, defibrinated	50.0mL

pH 7.2 ± 0.2 at 25°C

Caution: Sodium azide is toxic. Azides also react with metals and disposal must be highly diluted.

Preparation of Medium: Add components, except sheep blood, to distilled/deionized water and bring volume to 950.0mL. Mix thoroughly. Gently heat and bring to boiling. Autoclave for 15 min at 15 psi pressure–121°C. Cool to 45–50°C. Aseptically add 50.0mL of sterile defibrinated sheep blood. Pour into sterile Petri dishes or distribute into sterile tubes. Allow tubes to cool in a slanted position.

Use: For the isolation and enumeration of fecal streptococci from nonclinical specimens such as water and food. Also used for the isolation of *Streptococcus pneumoniae* and *Erysipelothrix rhusiopathiae*.

Azide Broth (Azide Glucose Broth) (Azide Dextrose Broth)

Composition per liter:

Pancreatic digest of casein	15.0g
Glucose	7.5g
Beef extract	4.5g
NaCl	7.5g
NaN_3	0.2g

pH 7.2 ± 0.2 at 25°C

Source: This medium is available as a premixed powder from Difco Laboratories and BBL Microbiology Systems.

Caution: Sodium azide is toxic. Azides also react with metals and disposal must be highly diluted.

Preparation of Medium: Add components to distilled/deionized water and bring volume to 1.0L. Mix thoroughly. Gently heat and bring to boiling. Distribute into tubes or flasks. Autoclave for 15 min at 15 psi pressure–121°C. Prepare double strength broth for samples larger than 1.0mL.

Use: For the detection and enrichment of fecal streptococci in water and sewage. Also used in the multiple-tube technique as a presumptive test for the presence of fecal streptococci.

Azide Broth, Rothe (Azide Glucose Broth, Rothe) (Azide Dextrose Broth, Rothe)

Composition per liter:

Peptone	20.0g
Glucose	5.0g
NaCl	5.0g
K_2HPO_4	2.7g
KH_2PO_4	2.7g
NaN_3	0.2g

pH 6.8 ± 0.2 at 25°C

Source: This medium is available as a premixed powder from Oxoid Unipath.

Caution: Sodium azide is toxic. Azides also react with metals and disposal must be highly diluted.

Preparation of Medium: Add components to distilled/deionized water and bring volume to 1.0L. Mix thoroughly. Gently heat and bring to boiling. Distribute into tubes or flasks. Autoclave for 15 min at 15 psi pressure–121°C. Prepare double strength broth for samples larger than 1.0mL.

Use: For the detection of enterococci in water and sewage.

Azide Citrate Broth

Composition per liter:

Pancreatic digest of casein	20.0g
Sodium citrate	10.0g
Yeast extract	5.0g
Glucose	5.0g
NaCl	5.0g
K_2HPO_4	4.0g
KH_2PO_4	1.5g
NaN_3	0.25g

pH 7.0 ± 0.2 at 25°C

Caution: Sodium azide is toxic. Azides also react with metals and disposal must be highly diluted.

Preparation of Medium: Add components to distilled/deionized water and bring volume to 1.0L. Mix thoroughly. Gently heat and bring to boiling. Distribute into tubes or flasks. Autoclave for 15 min at 15 psi pressure–118°C. Prepare double strength broth for samples larger than 1.0mL.

Use: For the detection and enrichment of fecal streptococci in water and sewage.

Azospirillum amazonense Medium (LGI Medium)

Composition per liter:

Sucrose	5.0g
Agar	1.75g
KH_2PO_4	0.6g
K_2HPO_4	0.2g
$MgSO_4 \cdot 7H_2O$	0.2g
$CaCl_2 \cdot 2H_2O$	0.02g
$FeCl_3$	0.01g
$Na_2MoO_4 \cdot 2H_2O$	2.0mg
Bromthymol blue (0.5% in 0.2*N* KOH)	5.0mL

pH 6.0 ± 0.2 at 25°C

Preparation of Medium: Add components to distilled/deionized water and bring volume to 1.0L. Mix thoroughly. Gently heat and bring to boiling. Distribute into tubes or flasks. Autoclave for 15 min at 15 psi pressure–121°C. Pour into sterile Petri dishes or leave in tubes.

Use: For the cultivation and maintenance of *Azospirillum amazonense.*

Azospirillum lipoferum Agar Medium

Composition per liter:

Glucose	20.0g
Agar	15.0g
K_2HPO_4	0.8g
$MgSO_4 \cdot 7H_2O$	0.5g
KH_2PO_4	0.2g
$FeCl_3 \cdot 6H_2O$	0.1g
Yeast extract	0.1g
$CaCl_2 \cdot 2H_2O$	0.02g
$Na_2MoO_4 \cdot 2H_2O$	0.02g

pH 6.9 ± 0.2 at 25°C

Preparation of Medium: Add components to distilled/deionized water and bring volume to 1.0L. Mix thoroughly. Gently heat and bring to boiling. Distribute into tubes or flasks. Autoclave for 15 min at 15 psi pressure–121°C. Pour into sterile Petri dishes or leave in tubes.

Use: For the cultivation of *Azospirillum lipoferum.*

Azospirillum lipoferum Agar Medium

Composition per liter:

Agar	15.0g
Calcium malate	10.0g
K_2HPO_4	0.8g
$MgSO_4 \cdot 7H_2O$	0.5g
KH_2PO_4	0.2g
$FeCl_3 \cdot 6H_2O$	0.1g
Yeast extract	0.1g
$CaCl_2 \cdot 2H_2O$	0.02g
$Na_2MoO_4 \cdot 2H_2O$	0.02g

pH 6.9 ± 0.2 at 25°C

Preparation of Medium: Add components to distilled/deionized water and bring volume to 1.0L. Mix thoroughly. Gently heat and bring to boiling. Distribute into tubes or flasks. Autoclave for 15 min at 15 psi pressure–121°C. Pour into sterile Petri dishes or leave in tubes.

Use: For the cultivation of *Azospirillum lipoferum.*

Azospirillum lipoferum Medium

Composition per liter:

Calcium malate	10.0g
K_2HPO_4	1.0g
$MgSO_4 \cdot 7H_2O$	0.5g
$CaCl_2 \cdot 2H_2O$	0.02g

pH 6.5 ± 0.2 at 25°C

Preparation of Medium: Add components to distilled/deionized water and bring volume to 1.0L. Mix thoroughly. Distribute into tubes or flasks. Autoclave for 15 min at 15 psi pressure–121°C.

Use: For the isolation and cultivation of *Azospirillum lipoferum.*

Azospirillum Medium

Composition per liter:

Sodium malate	5.0g
Agar	1.75g
KH_2PO_4	0.4g
$MgSO_4 \cdot 7H_2O$	0.2g
K_2HPO_4	0.1g
NaCl	0.1g
$FeCl_3$	0.01g
$CaCl_2 \cdot 2H_2O$	0.02g
$Na_2MoO_4 \cdot 2H_2O$	2.0mg
Bromthymol blue solution	5.0mL

pH 6.8 ± 0.2 at 25°C

Bromthymol Blue Solution:
Composition per 10mL:

Bromthymol blue	0.5g
Ethanol	10.0mL

Preparation of Bromthymol Blue Solution: Add bromthymol blue to 10.0mL of ethanol. Mix thoroughly.

Preparation of Medium: Add components to distilled/deionized water and bring volume to 1.0L. Mix thoroughly. Distribute into tubes or flasks. Autoclave for 15 min at 15 psi pressure–121°C.

Use: For the cultivation of *Azospirillum* species isolated from roots.

Azotobacter Agar, Modified I

Composition per liter:

Agar	15.0g
Sucrose	10.0g
Glucose	10.0g
$MgSO_4 \cdot 7H_2O$	0.2g
KH_2PO_4	0.15g
$CaSO_4 \cdot 2H_2O$	0.1g
K_2HPO_4	0.05g
$CaCl_2$	0.02g
Na_2MoO_4	2.0mg
$FeCl_3$	1.0mg
$Na_2MoO_4 \cdot 2H_2O$	1.0mg

pH 7.2 ± 0.2 at 25°C

Preparation of Medium: Add components to distilled/deionized water and bring volume to 1.0L. Mix thoroughly. Gently heat and bring to boiling. Adjust pH to 7.2. Distribute into tubes or flasks. Autoclave for 15 min at 15 psi pressure–121°C. Pour into sterile Petri dishes or leave in tubes.

Use: For the cultivation and maintenance of *Azotobacter* species.

Azotobacter Agar, Modified II

Composition per liter:

Sucrose	20.0g
Agar	15.0g
KH_2PO_4	0.15g
$MgSO_4 \cdot 7H_2O$	0.2g
K_2HPO_4	0.05g
$CaCl_2$	0.02g
Na_2MoO_4	2.0mg
$FeCl_3$	1.0mg
$Na_2MoO_4 \cdot 2H_2O$	1.0mg

pH 6.2 ± 0.2 at 25°C

Preparation of Medium: Add components to distilled/deionized water and bring volume to 1.0L. Mix

thoroughly. Gently heat and bring to boiling. Adjust pH to 6.2. Distribute into tubes or flasks. Autoclave for 15 min at 15 psi pressure–121°C. Pour into sterile Petri dishes or leave in tubes.

Use: For the cultivation and maintenance of *Azotobacter* species *and Beijerinckia derxii.*

Azotobacter Basal Agar

Composition per liter:

Agar	15.0g
K_2HPO_4	1.0g
$MgSO_4 \cdot 7H_2O$	0.2g
NaCl	0.2g
$FeSO_4 \cdot 7H_2O$	5.0mg
Soil extract	100.0mL

pH 7.2 ± 0.2 at 25°C

Soil Extract:

Composition per 200mL:

African violet soil	0.5g
Na_2CO_3	0.5g

Preparation of Soil Extract: Add components to tap water and bring volume to 200.0mL. Autoclave for 60 min at 15 psi pressure–121°C. Filter through Whatman filter paper.

Preparation of Medium: Add components including filtered soil extract to tap water and bring volume to 1.0L. Mix thoroughly. Gently heat and bring to boiling. Distribute into tubes or flasks. Autoclave for 15 min at 15 psi pressure–121°C. Pour into sterile Petri dishes or leave in tubes.

Use: For the cultivation of a variety of bacteria including *Azomonas* species, *Azotobacter* species, and others when a carbon source is added.

Azotobacter Basal Broth

Composition per liter:

K_2HPO_4	1.0g
$MgSO_4 \cdot 7H_2O$	0.2g
NaCl	0.2g
$FeSO_4 \cdot 7H_2O$	5.0mg
Soil extract	100.0mL

pH 7.2 ± 0.2 at 25°C

Soil Extract:

Composition per 200mL:

African violet soil	0.5g
Na_2CO_3	0.5g

Preparation of Soil Extract: Add components to tap water and bring volume to 200.0mL. Autoclave for 60 min at 15 psi pressure–121°C. Filter through Whatman filter paper.

Preparation of Medium: Add components including filtered soil extract to tap water and bring volume to 1.0L. Mix thoroughly. Distribute into tubes or flasks. Autoclave for 15 min at 15 psi pressure–121°C.

Use: For the cultivation of a variety of bacteria including *Azomonas* species, *Azotobacter* species, and others when a carbon source is added.

Azotobacter Broth, Modified I

Composition per liter:

Sucrose	10.0g
Glucose	10.0g
$MgSO_4 \cdot 7H_2O$	0.2g
KH_2PO_4	0.15g
$CaSO_4 \cdot 2H_2O$	0.1g
K_2HPO_4	0.05g
$CaCl_2$	0.02g
Na_2MoO_4	2.0mg
$FeCl_3$	1.0mg
$Na_2MoO_4 \cdot 2H_2O$	1.0mg

pH 7.2 ± 0.2 at 25°C

Preparation of Medium: Add components to distilled/deionized water and bring volume to 1.0L. Mix thoroughly. Gently heat and bring to boiling. Adjust pH to 7.2. Distribute into tubes or flasks. Autoclave for 15 min at 15 psi pressure–121°C.

Use: For the cultivation of *Azotobacter* species.

Azotobacter Broth, Modified II

Composition per liter:

Sucrose	20.0g
KH_2PO_4	0.15g
$MgSO_4 \cdot 7H_2O$	0.2g
K_2HPO_4	0.05g
$CaCl_2$	0.02g
Na_2MoO_4	2.0mg
$FeCl_3$	1.0mg
$Na_2MoO_4 \cdot 2H_2O$	1.0mg

pH 6.2 ± 0.2 at 25°C

Preparation of Medium: Add components to distilled/deionized water and bring volume to 1.0L. Mix thoroughly. Gently heat and bring to boiling. Adjust pH to 6.2. Distribute into tubes or flasks. Autoclave for 15 min at 15 psi pressure–121°C.

Use: For the cultivation of *Azotobacter* species and *Beijerinckia derxii.*

Azotobacter chroococcum Agar

Composition per liter:

Agar	20.0g
$CaCO_3$	20.0g
Glucose	20.0g
K_2HPO_4	0.8g
$MgSO_4 \cdot 7H_2O$	0.5g
KH_2PO_4	0.2g
$FeCl_3 \cdot 6H_2O$	0.1g
$Na_2MoO_4 \cdot 2H_2O$	0.05g

pH 7.4–7.6 at 25°C

Preparation of Medium: Add components to distilled/deionized water and bring volume to 1.0L. Mix thoroughly. Gently heat and bring to boiling. Distribute into tubes or flasks. Autoclave for 15 min at 15 psi pressure–121°C. Pour into sterile Petri dishes or leave in tubes.

Use: For the cultivation and maintenance of *Azotobacter chroococcum.*

Azotobacter chroococcum Agar

Composition per liter:

Agar	20.0g
Glucose	20.0g
K_2HPO_4	0.8g
$MgSO_4 \cdot 7H_2O$	0.5g
KH_2PO_4	0.2g
$FeCl_3 \cdot 6H_2O$	0.1g
$CaCl_2 \cdot 2H_2O$	0.05g
$Na_2MoO_4 \cdot 2H_2O$	0.05g

pH 7.4-7.6 ± 0.2 at 25°C

Preparation of Medium: Add components to distilled/deionized water and bring volume to 1.0L. Mix thoroughly. Gently heat and bring to boiling. Distribute into tubes or flasks. Autoclave for 15 min at 15 psi pressure–121°C. Pour into sterile Petri dishes or leave in tubes.

Use: For the cultivation and maintenance of *Azotobacter chroococcum.*

Azotobacter chroococcum Medium

Composition per liter:

$CaCO_3$	20.0g
Glucose	20.0g
K_2HPO_4	1.0g
$MgSO_4 \cdot 7H_2O$	0.5g

Preparation of Medium: Add components to distilled/deionized water and bring volume to 1.0L. Mix thoroughly. Distribute into tubes or flasks. Autoclave for 15 min at 15 psi pressure–121°C.

Use: For the cultivation of *Azotobacter chroococcum.*

Azotobacter Medium (ATCC Medium 14)

Composition per liter:

Sucrose	20.0g
Agar	15.0g
K_2HPO_4	0.8g
Yeast extract	0.5g
KH_2PO_4	0.2g
$MgSO_4 \cdot 7H_2O$	0.2g
$CaSO_4 \cdot 2H_2O$	0.1g
$FeCl_3$	1.0mg
$Na_2MoO_4 \cdot 2H_2O$	1.0mg

pH 7.2 ± 0.2 at 25°C

Preparation of Medium: Add components to distilled/deionized water and bring volume to 1.0L. Mix thoroughly. Gently heat and bring to boiling. Distribute into tubes or flasks. Autoclave for 15 min at 15 psi pressure–121°C. Pour into sterile Petri dishes or leave in tubes.

Use: For the cultivation of a variety of bacteria including *Azomonas* species, *Azotobacter* species, *Beijerinckia derxii*, *Pseudomonas azotocolligans*, and *Rhodococcus erythropolis*.

Azotobacter Medium (ATCC Medium 240)

Composition per liter:

Agar	15.0g
$MgSO_4 \cdot 7H_2O$	0.2g
KH_2PO_4	0.15g
K_2HPO_4	0.05g
$CaCl_2$	0.02g
$Na_2MoO_4 \cdot 2H_2O$	2.0mg
$FeCl_3$	1.0mg

pH 7.2 ± 0.2 at 25°C

Preparation of Medium: Add components to distilled/deionized water and bring volume to 1.0L. If required, add sucrose to a concentration of 1%. Mix thoroughly. Gently heat and bring to boiling. Distribute into tubes or flasks. Autoclave for 15 min at 15 psi pressure–121°C. Pour agar medium into sterile Petri dishes or leave in tubes.

Use: For the cultivation and maintenance of a variety of bacteria including *Azotobacter* species.

Azotobacter Medium (ATCC Medium 1771)

Composition per liter:

Agar	15.0g
Glucose	10.0g

KH_2PO_4	0.22g
$CaSO_4 \cdot 2H_2O$	0.1g
$MgSO_4 \cdot 7H_2O$	0.098g
NaCl	0.058g
K_2HPO_4	0.058g
$FeSO_4 \cdot 7H_2O$	5.0mg
$Na_2MoO_4 \cdot 2H_2O$	0.2mg

pH 7.2 ± 0.2 at 25°C

Preparation of Medium: Add components to distilled/deionized water and bring volume to 1.0L. Mix thoroughly. Gently heat and bring to boiling. Distribute into tubes or flasks. Autoclave for 15 min at 15 psi pressure–121°C. Pour into sterile Petri dishes or leave in tubes.

Use: For the cultivation and maintenance of a variety of bacteria including *Azotobacter* species.

Azotobacter paspali Medium

Composition per liter:

Agar	20.0g
Sucrose	20.0g
$CaCO_3$	1.0g
$MgSO_4 \cdot 7H_2O$	0.20g
KH_2PO_4	0.15g
K_2HPO_4	0.05g
$CaCl_2$	0.02g
$Na_2MoO_4 \cdot 2H_2O$	2.0mg
Bromthymol Blue solution	10.0mL
$FeCl_3$ (10% solution)	0.1mL

pH 7.0 ± 0.2 at 25°C

Bromthymol Blue Solution:

Composition per 10mL:

Bromthymol blue	0.5g
Ethanol	10.0mL

Preparation of Bromthymol Blue Solution: Add bromthymol blue to 10.0mL of ethanol. Mix thoroughly.

Preparation of Medium: Add components to distilled/deionized water and bring volume to 1.0L. Mix thoroughly. Gently heat and bring to boiling. Distribute into tubes or flasks. Autoclave for 15 min at 15 psi pressure–121°C. Pour into sterile Petri dishes or leave in tubes.

Use: For the cultivation and maintenance of *Azotobacter paspali*.

Azotobacter Supplement (ATCC Medium 11)

Composition per liter:

Agar	15.0g
K_2HPO_4	1.0g
$MgSO_4 \cdot 7H_2O$	0.2g
NaCl	0.2g
$FeSO_4 \cdot 7H_2O$	5.0mg
Soil extract	100.0mL
Glucose solution	100.0mL

pH 7.6 ± 0.2 at 25°C

Soil Extract:

Composition per 200mL:

African violet soil	0.5g
Na_2CO_3	0.5g

Preparation of Soil Extract: Add components to tap water and bring volume to 200.0mL. Autoclave for 60 min at 15 psi pressure–121°C. Filter through Whatman filter paper.

Glucose Solution:

Composition per 100mL:

Glucose	20.0g

Preparation of Glucose Solution: Add glucose to distilled/deionized water and bring volume to 100.0mL. Mix thoroughly. Filter sterilize.

Preparation of Medium: Add components, except glucose solution, to tap water and bring volume to 900.0mL. Mix thoroughly. Adjust pH to 7.6. Autoclave for 15 min at 15 psi pressure–121°C. Cool to 50–55°C. Aseptically add 100.0mL sterile glucose solution. Mix thoroughly. Pour into sterile Petri dishes or leave in tubes.

Use: For the cultivation of *Azomonas agilis* and *Azotobacter chroococcum*.

Azotobacter Supplement (ATCC Medium 12)

Composition per liter:

Agar	15.0g
K_2HPO_4	1.0g
$MgSO_4 \cdot 7H_2O$	0.2g
NaCl	0.2g
$FeSO_4 \cdot 7H_2O$	5.0mg
Soil extract	100.0mL
Mannitol solution	100.0mL

pH 7.6 ± 0.2 at 25°C

Soil Extract:

Composition per 200mL:

African violet soil	0.5g
Na_2CO_3	0.5g

Preparation of Soil Extract: Add components to distilled/deionized water and bring volume to 200.0mL. Autoclave for 60 min at 15 psi pressure–121°C. Filter through Whatman filter paper.

Mannitol Solution:

Composition per 100mL:

Mannitol 20.0g

Preparation of Mannitol Solution: Add mannitol to distilled/deionized water and bring volume to 100.0mL. Mix thoroughly. Filter sterilize.

Preparation of Medium: Add components, except mannitol solution, to tap water and bring volume to 900.0mL. Mix thoroughly. Adjust pH to7.6. Autoclave for 15 min at 15 psi pressure–121°C. Cool to 50–55°C. Aseptically add 100.0mL sterile mannitol solution. Mix thoroughly. Pour into sterile Petri dishes or leave in tubes.

Use: For the cultivation of *Azotobacter* species and *Azomonas* species.

Azotobacter Supplement (ATCC Medium 13)

Composition per liter:

Agar	15.0g
K_2HPO_4	1.0g
$MgSO_4 \cdot 7H_2O$	0.2g
NaCl	0.2g
$FeSO_4 \cdot 7H_2O$	5.0mg
Soil extract	100.0mL
Glucose solution	100.0mL

pH 6.0 ± 0.2 at 25°C

Soil Extract:

Composition per 200mL:

African violet soil	0.5g
Na_2CO_3	0.5g

Preparation of Soil Extract: Add components to tap water and bring volume to 200.0mL. Autoclave for 60 min at 15 psi pressure–121°C. Filter through Whatman filter paper.

Glucose Solution:

Composition per 100mL:

Glucose 20.0g

Preparation of Glucose Solution: Add glucose to distilled/deionized water and bring volume to 100.0mL. Mix thoroughly. Filter sterilize.

Preparation of Medium: Add components, except glucose solution, to tap water and bring volume to 900.0mL. Mix thoroughly. Adjust pH to 6.0. Autoclave for 15 min at 15 psi pressure–121°C. Cool to 50–55°C. Aseptically add 100.0mL sterile glucose solution. Mix thorougly. Pour into sterile Petri dishes or leave in tubes.

Use: For the cultivation of *Beijerinckia* species.

Azotobacter Supplement (ATCC Medium 15)

Composition per liter:

Agar	15.0g
K_2HPO_4	1.0g
$MgSO_4 \cdot 7H_2O$	0.2g
NaCl	0.2g
$FeSO_4 \cdot 7H_2O$	5.0mg
Soil extract	100.0mL
Mannitol solution	100.0mL

pH 6.0 ± 0.2 at 25°C

Soil Extract:

Composition per 200mL:

African violet soil	0.5g
Na_2CO_3	0.5g

Preparation of Soil Extract: Add components to distilled/deionized water and bring volume to 200.0mL. Autoclave for 60 min at 15 psi pressure–121°C. Filter through Whatman filter paper.

Mannitol Solution:

Composition per 100mL:

Mannitol 20.0g

Preparation of Mannitol Solution: Add mannitol to distilled/deionized water and bring volume to 100.0mL. Mix thoroughly. Filter sterilize.

Preparation of Medium: Add components, except mannitol solution, to tap water and bring volume to 900.0mL. Mix thoroughly. Adjust pH to 6.0. Autoclave for 15 min at 15 psi pressure–121°C. Cool to 50–55°C. Aseptically add 100.0mL sterile mannitol solution. Mix thoroughly. Pour into sterile Petri dishes or leave in tubes.

Use: For the cultivation of *Azomonas macrocytogenes*.

Azotobacter vinelandii Medium

Composition per liter:

Sodium benzoate	1.0g
K_2HPO_4	0.5g
Mannitol	0.5g

Preparation of Medium: Add components to distilled/deionized water and bring volume to 1.0L. Mix thoroughly. Distribute into tubes or flasks. Autoclave for 15 min at 15 psi pressure–121°C.

Use: For the cultivation of *Azotobacter vinelandii* from water samples.

Azotobacter vinelandii Medium

Composition per liter:

Sodium benzoate	1.0g
K_2HPO_4	0.5g
Ethanol	1.0mL

Preparation of Medium: Add components, except ethanol, to distilled/deionized water and bring volume to 999.0mL. Mix thoroughly. Autoclave for 15 min at 15 psi pressure–121°C. Cool to 45–50°C. Aseptically add 1.0mL of filter-sterilized ethanol. Mix thoroughly. Aseptically distribute into sterile tubes or flasks.

Use: For the cultivation of *Azotobacter vinelandii* from soil.

Baar's Medium for Sulfate Reducers

Composition per liter:

Sodium lactate	3.5g
$MgSO_4 \cdot 7H_2O$	2.0g
K_2HPO_4	1.0g
$CaSO_4$	1.0g
NH_4Cl	0.5g
Ferrous ammonium sulfate solution	10.0mL
Yeast extract solution	10.0mL

pH 7.5 ± 0.2 at 25°C

Ferrous Ammonium Sulfate Solution:

Composition per 10mL:

$Fe(NH_4)_2(SO_4)_2$	0.5g

Preparation of Ferrous Ammonium Sulfate Solution: Add $Fe(NH_4)_2(SO_4)_2$ to distilled/deionized water and bring volume to 10.0mL. Mix thoroughly. Autoclave for 15 min at 15 psi pressure–121°C.

Yeast Extract Solution:

Composition per 10mL:

Yeast extract	1.0g

Preparation of Yeast Extract Solution: Add yeast extract to distilled/deionized water and bring volume to 10.0mL. Mix thoroughly. Autoclave for 15 min at 15 psi pressure–121°C.

Preparation of Medium: Add components, except ferrous ammonium sulfate solution and yeast extract solution, to tap water and bring volume to 980.0mL. Mix thoroughly. Gently heat and bring to boiling. Autoclave for 15 min at 15 psi pressure–121°C. Cool to 45–50°C. Aseptically add 10.0mL of sterile ferrous ammonium sulfate solution and sterile yeast extract solution. Mix thoroughly. Aseptically distribute into tubes or flasks.

Use: For the cultivation and maintenance of *Desulfotomaculum nigrificans.*

Baar's Medium for Sulfate Reducers, Modified

Composition per 1020mL:

Component I	400.0mL
Component II	200.0mL
Component III	400.0mL
Ferrous ammonium sulfate solution	20.0mL

pH 7.5 ± 0.2 at 25°C

Component I:

Composition per 400mL:

Sodium citrate	5.0g
$MgSO_4$	2.0g
$CaSO_4$	1.0g
NH_4Cl	1.0g

Preparation of Component I: Add components to distilled/deionized water and bring volume to 400.0mL. Mix thoroughly. Adjust pH to 7.5. Autoclave for 15 min at 15 psi pressure–121°C.

Component II:

Composition per 200mL:

K_2HPO_4	0.5g

Preparation of Component II: Add K_2HPO_4 to distilled/deionized water and bring volume to 200.0mL. Mix thoroughly. Adjust pH to 7.5. Autoclave for 15 min at 15 psi pressure–121°C.

Component III:

Composition per 400mL:

Sodium lactate	3.5g
Yeast extract	1.0g

Preparation of Component III: Add components to distilled/deionized water and bring volume to 400.0mL. Mix thoroughly. Adjust pH to 7.5. Autoclave for 15 min at 15 psi pressure–121°C.

Ferrous Ammonium Sulfate Solution:

Composition per 20mL:

$Fe(NH_4)_2(SO_4)_2$	1.0g

Preparation of Ferrous Ammonium Sulfate Solution: Add $Fe(NH_4)_2(SO_4)_2$ to distilled/deionized water and bring volume to 20.0mL. Mix thoroughly. Filter sterilize.

Preparation of Medium: Aseptically combine component I, component II and component III. Mix thoroughly. Distribute 5.0mL volumes into tubes under 97% N_2 + 3% H_2. Add medium to tubes while still warm to exclude as much O_2 as possible. Aseptically add 0.1mL of sterile ferrous ammonium sulfate solution to 5.0mL of medium immediately prior to inoculation.

Use: For the cultivation and maintenance of *Desulfovibrio, Desulfobulbus, Desulfotomaculum,* and *Thermodesulfobacterium* species.

Baar's Medium for Sulfate Reducers, Modified with 2.5% NaCl

Composition per 1020mL:

Component I	400.0mL
Component II	200.0mL
Component III	400.0mL
Ferrous ammonium sulfate solution	20.0mL

pH 7.5 ± 0.2 at 25°C

Component I:
Composition per 400mL:

NaCl	25.0g
Sodium citrate	5.0g
$MgSO_4$	2.0g
$CaSO_4$	1.0g
NH_4Cl	1.0g

Preparation of Component I: Add components to distilled/deionized water and bring volume to 400.0mL. Mix thoroughly. Adjust pH to 7.5. Autoclave for 15 min at 15 psi pressure–121°C.

Component II:
Composition per 200mL:

K_2HPO_4	0.5g

Preparation of Component II: Add K_2HPO_4 to distilled/deionized water and bring volume to 200.0mL. Mix thoroughly. Adjust pH to 7.5. Autoclave for 15 min at 15 psi pressure–121°C.

Component III:
Composition per 400mL:

Sodium lactate	3.5g
Yeast extract	1.0g

Preparation of Component III: Add components to distilled/deionized water and bring volume to 400.0mL. Mix thoroughly. Adjust pH to 7.5. Autoclave for 15 min at 15 psi pressure–121°C.

Ferrous Ammonium Sulfate Solution:
Composition per 20mL:

$Fe(NH_4)_2(SO_4)_2$	1.0g

Preparation of Ferrous Ammonium Sulfate Solution: Add $Fe(NH_4)_2(SO_4)_2$ to distilled/deionized water and bring volume to 20.0mL. Mix thoroughly. Filter sterilize.

Preparation of Medium: Aseptically combine component I, component II and component III. Mix thoroughly. Distribute 5.0mL volumes into tubes under 97% N_2 + 3% H_2. Add medium to tubes while still warm to exclude as much O_2 as possible. Aseptically add 0.1mL of sterile ferrous ammonium sulfate solution to 5.0mL of medium immediately prior to inoculation.

Use: For the cultivation of *Desulfovibrio africanus* and other *Desulfovibrio* species which prefer 2.5% NaCl.

Bacillus Agar

Composition per liter:

Agar	20.0g
$(NH_4)_2SO_4$	1.3g
Glucose	1.0g
Yeast extract	1.0g
KH_2PO_4	0.37g
$MgSO_4 \cdot 7H_2O$	0.25g
$CaCl_2 \cdot 2H_2O$	0.07g
$FeCl_3$	0.02g

pH 4.0 ± 0.2 at 25°C

Preparation of Medium: Add components to distilled/deionized water and bring volume to 500.0mL. Mix thoroughly. Gently heat and bring to boiling. Adjust pH to 3.5. Prepare a separate agar solution by adding 20.0g/500.0mL of distilled/deionized water. Autoclave solutions separately for 15 min at 15 psi pressure–121°C. Cool to 50–55°C. Aseptically combine both solutions. This procedure avoids acid hydrolysis of the agar. Pour into sterile Petri dishes or leave in tubes.

Use: For the cultivation of acidophilic *Bacillus* species such as *B. acidocaldarius.*

Bacillus Agar, 1/4 Strength

Composition per liter:

Agar	18.0g
Yeast extract	2.5g
Pancreatic digest of casein	1.0g

pH 7.2 ± 0.2 at 25°C

Preparation of Medium: Add components to distilled/deionized water and bring volume to 1.0L. Mix thoroughly. Gently heat and bring to boiling. Distribute into tubes or flasks. Autoclave for 15 min at 15 psi pressure–121°C. Pour into sterile Petri dishes or leave in tubes.

Use: For the cultivation and maintenance of *Bacillus megaterium.*

Bacillus Broth

Composition per liter:

$(NH_4)_2SO_4$	1.3g
Glucose	1.0g

Yeast extract	1.0g
KH_2PO_4	0.37g
$MgSO_4 \cdot 7H_2O$	0.25g
$CaCl_2 \cdot 2H_2O$	0.07g
$FeCl_3$	0.02g

pH 4.0 ± 0.2 at 25°C

Preparation of Medium: Add components to distilled/deionized water and bring volume to 1.0L. Mix thoroughly. Gently heat and bring to boiling. Adjust pH to 4.0 with 10*N* H_2SO_4. Distribute into tubes or flasks. Autoclave for 15 min at 15 psi pressure–121°C.

Use: For the cultivation of acidophilic *Bacillus* species such as *B. acidocaldarius*.

Bacillus Broth, 1/4 Strength

Composition per liter:

Yeast extract	2.5g
Pancreatic digest of casein	1.0g

pH 7.2 ± 0.2 at 25°C

Preparation of Medium: Add components to distilled/deionized water and bring volume to 1.0L. Mix thoroughly. Distribute into tubes or flasks. Autoclave for 15 min at 15 psi pressure–121°C.

Use: For the cultivation of *Bacillus megaterium*.

Bacillus cereus Medium (BCM)

Composition per 110mL:

Agar	2.0g
D-Mannitol	1.0g
$(NH_4)_2PO_4$	0.1g
KCl	0.02g
$MgSO_4 \cdot 7H_2O$	0.02g
Yeast extract	0.02g
Bromcresol purple	4.0mg
Egg yolk emulsion, 20%	10.0mL

pH 7.0 ± 0.2 at 25°C

Egg Yolk Emulsion, 20%:

Composition per 100mL:

Chicken egg yolks	11
Whole chicken egg	1
NaCl (0.9% solution)	80.0mL

Preparation of Egg Yolk Emulsion, 20%: Soak eggs with 1:100 dilution of saturated mercuric chloride solution for 1 min. Crack eggs and separate yolks from whites. Mix egg yolks with 1 chicken egg. Measure 20.0mL of egg yolk emulsion and add to 80.0mL of 0.9% NaCl solution. Mix thoroughly. Filter sterilize. Warm to 45–50°C.

Preparation of Medium: Add components—except egg yolk emulsion, 20%—to distilled/deionized water and bring volume to 100.0mL. Mix thoroughly. Gently heat and bring to boiling. Autoclave for 15 min at 15 psi pressure–121°C. Cool to 45–50°C. Aseptically add 10.0mL of sterile egg yolk emulsion, 20%. Mix thoroughly. Pour into sterile Petri dishes or distribute into sterile tubes.

Use: For the cultivation of *Bacillus cereus*.

Bacillus Medium

Composition per liter:

Agar	25.0g
Peptone	6.0g
Pancreatic digest of casein	3.0g
Yeast extract	3.0g
Beef extract	1.5g
$MnSO_4 \cdot 4H_2O$	1.0μg

pH 7.0 ± 0.2 at 25°C

Preparation of Medium: Add components to distilled/deionized water and bring volume to 1.0L. Mix thoroughly. Gently heat and bring to boiling. Distribute into tubes or flasks. Autoclave for 15 min at 15 psi pressure–121°C. Pour into sterile Petri dishes or leave in tubes.

Use: For the cultivation of *Bacillus* species.

Bacillus Medium

Composition per liter:

$(NH_4)_2HPO_4$	1.0g
$MgSO_4 \cdot 7H_2O$	0.2g
KCl	0.2g
Yeast extract	0.2g
Glucose solution	50.0mL
Bromcresol purple solution	15.0mL

pH 7.0 ± 0.2 at 25°C

Glucose Solution:

Composition per 100mL:

Glucose	10.0g

Preparation of Glucose Solution: Add glucose to distilled/deionized water and bring volume to 100.0mL. Mix thoroughly. Filter sterilize.

Preparation of Medium: Add components, except glucose solution, to distilled/deionized water and bring volume to 1.0L. Mix thoroughly. Gently heat and bring to boiling. Distribute 9.5mL volumes into test tubes that contain an inverted Durham tube. Autoclave for 20 min at 15 psi pressure–121°C. Cool to 25°C. Aseptically add 0.5mL of sterile glucose to each tube. Mix thoroughly.

Use: For cultivation and differentiation of *Bacillus* species based on acid and gas production from glucose.

Bacillus Medium (ATCC Medium 21)

Composition per liter:

Glycerol	20.0g
L-Glutamic acid	4.0g
Citric acid	2.0g
K_2HPO_4	0.5g
Ferric ammonium citrate	0.5g
$MgSO_4$	0.5g

pH 7.4 ± 0.2 at 25°C

Preparation of Medium: Add components to tap water and bring volume to 1.0L. Mix thoroughly. Gently heat and bring to boiling. Distribute into tubes or flasks. Autoclave for 15 min at 15 psi pressure–121°C.

Use: For the cultivation of *Bacillus licheniformis.*

Bacillus Medium (ATCC Medium 455)

Composition per liter:

Soluble starch	30.0g
Agar	20.0g
Polypeptone™	5.0g
Yeast extract	5.0g

Preparation of Medium: Add components to distilled/deionized water and bring volume to 1.0L. Mix thoroughly. Gently heat and bring to boiling. Distribute into tubes or flasks. Autoclave for 15 min at 15 psi pressure–121°C. Swirl medium to resuspend starch. Pour into sterile Petri dishes or leave in tubes.

Use: For the cultivation and maintenance of *Bacillus subtilis.* Also used to detect amylase producing microorganisms.

Bacillus Medium (ATCC Medium 552)

Composition per liter:

Peptone	10.0g
Lactose	5.0g
NaCl	5.0g
Beef extract	3.0g
K_2HPO_4	2.0g

pH 7.2 ± 0.2 at 25°C

Preparation of Medium: Add components to distilled/deionized water and bring volume to 1.0L. Mix thoroughly. Gently heat and bring to boiling. Distribute into tubes or flasks. Autoclave for 15 min at 15 psi pressure–121°C.

Use: For the cultivation and maintenance of *Bacillus* species.

Bacillus pasteurii NH_4 YE Medium (Ammonium Yeast Extract Medium)

Composition per liter:

Yeast extract	20.0g
Agar	20.0g
$(NH_4)_2SO_4$	10.0g

pH 9.0 ± 0.2 at 25°C

Preparation of Medium: Add each component to a separate flask and bring volume of each to 333.0mL with 0.13M Tris buffer, pH 9.0. Autoclave ingredients separately for 15 min at 15 psi pressure–121°C. No growth occurs if components are sterilized together. Cool to 50–55°C and aseptically combine solutions. Pour into sterile Petri dishes.

Use: For the cultivation and maintenance of *Bacillus pasteurii.*

Bacillus popilliae Maintenance Medium

Composition per liter:

Agar	20.0g
Yeast extract	15.0g
Pancreatic digest of casein	5.0g
K_2HPO_4	3.0g
Glucose solution	10.0mL

pH 7.2 ± 0.2 at 25°C

Glucose Solution:

Composition per 10mL:

Glucose	2.0g

Preparation of Glucose Solution: Add glucose to distilled/deionized water and bring volume to 10.0mL. Mix thoroughly. Filter sterilize.

Preparation of Medium: Add components, except glucose solution, to distilled/deionized water and bring volume to 990.0mL. Mix thoroughly. Gently heat and bring to boiling. Autoclave for 15 min at 15 psi pressure–121°C. Cool to 45–50°C. Aseptically add sterile glucose solution. Mix thoroughly. Pour into sterile Petri dishes or distribute into sterile tubes.

Use: For the cultivation and maintenance of *Bacillus popilliae.*

Bacillus popilliae Medium

Composition per liter:

Yeast extract	10.0g
Acid hydrolysate of casein	7.95g
K_2HPO_4	3.0g
Beef extract	1.36g
Trehalose	1.0g
Starch	0.68g

pH 7.3 ± 0.1 at 25°C

Preparation of Medium: Add components to distilled/deionized water and bring to 1.0L. Mix thoroughly. Gently heat until dissolved. Do not overheat. Filter sterilize. Aseptically distribute into sterile tubes or flasks.

Use: For the cultivation of *Bacillus popilliae*.

Bacillus popilliae Medium

Composition per liter:

Yeast extract	15.0g
K_2HPO_4	3.0g

pH 7.2 ± 0.2 at 25°C

Preparation of Medium: Add components to distilled/deionized water and bring volume to 1.0L. Mix thoroughly. Distribute into tubes or flasks. Autoclave for 15 min at 15 psi pressure–121°C.

Use: For the cultivation of *Bacillus popilliae*.

Bacillus Pullulan Salts

Composition per liter:

Pullulan	2.50g
NaCl	1.00g
NH_4Cl	1.00g
KH_2PO_4	0.50g
$MgSO_4 \cdot 7H_2O$	0.50g
Yeast extract	0.10g
$CaCl_2 \cdot 2H_2O$	0.05g
Trace mineral solution	10.0mL
Vitamin solution	10.0mL

pH 6.0 ± 0.2 at 25°C

Trace Mineral Solution:

Composition per liter:

$CoCl_2 \cdot 6H_2O$	0.2g
$FeSO_4 \cdot 7H_2O$	0.13g
$ZnCl_2 \cdot 2H_2O$	0.1g
$MnCl_2 \cdot 4H_2O$	0.1g
$CaCl_2 \cdot 2H_2O$	20.0mg
Na_2SeO_3	20.0mg
$Na_2WO_4 \cdot 2H_2O$	20.0mg
$NaMoO_4 \cdot 2H_2O$	1.0mg
H_3BO_3	0.5mg
$CuSO_4 \cdot 5H_2O$	0.4mg
KI	0.1mg

Preparation of Trace Mineral Solution: Add components to distilled/deionized water and bring volume to 1.0L. Mix thoroughly.

Vitamin Solution:

Composition per liter:

Pyridoxine·HCl	10.0mg
Thiamine·HCl	5.0mg
Riboflavin	5.0mg
Nicotinic acid	5.0mg
Calcium pantothenate	5.0mg
p-Aminobenzoic acid	5.0mg
Thioctic acid	5.0mg
Biotin	2.0mg
Folic acid	2.0mg
Cyanocobalamin	0.1mg

Preparation of Vitamin Solution: Add components to distilled/deionized water and bring volume to 1.0L. Mix thoroughly. Filter sterilize.

Preparation of Medium: Add components, except vitamin solution, to distilled/deionized water and bring volume to 990.0L. Mix thoroughly. Gently heat and bring to boiling. Adjust pH to 6.0. Autoclave for 15 min at 15 psi pressure–121°C. Cool to 25°C. Aseptically add sterile vitamin solution. Mix thoroughly. Aseptically distribute into sterile tubes or flasks.

Use: For the cultivation and maintenance of *Bacillus* species which can degrade pullulan.

Bacillus schlegelii Agar

Composition per liter:

Noble agar	30.0g
$Na_2HPO_4 \cdot 2H_2O$	4.5g
KH_2PO_4	1.5g
Sodium pyruvate	1.5g
NH_4Cl	1.0g
$MgSO_4 \cdot 7H_2O$	0.2g
$MnSO_4 \cdot H_2O$	0.01g
$CaCl_2 \cdot 2H_2O$	0.01g
Ferric ammonium citrate	5.0mg
Agar solution	200.0mL
Sodium pyruvate	100.0mL
SL-6 trace elements	3.0mL

pH 7.1 ± 0.2 at 25°C

Agar Solution:

Composition per 200mL:

Noble agar	30.0g

Preparation of Agar Solution: Add agar to distilled/deionized water and bring volume to 200.0mL.

Mix thoroughly. Gently heat and bring to boiling. Autoclave for 15 min at 15 psi pressure–121°C. Cool to 45–50°C.

Pyruvate Solution:
Composition per 100mL:

Sodium pyruvate	1.5g

Preparation of Pyruvate Solution: Add sodium pyruvate to distilled/deionized water and bring volume to 100.0mL. Mix thoroughly. Filter sterilize. Warm to 45–50°C.

SL-6 Trace Elements Solution:
Composition per liter:

H_3BO_3	0.3g
$CoCl_2·6H_2O$	0.2g
$ZnSO_4·7H_2O$	0.10g
$MnCl_2·4H_2O$	0.03g
$Na_2MoO_4·H_2O$	0.03g
$NiCl_2·6H_2O$	0.02g
$CuCl_2·2H_2O$	0.01g

Preparation of SL-6 Trace Elements Solution: Add components to distilled/deionized water and bring volume to 1.0L. Mix thoroughly. Adjust pH to 3.4.

Preparation of Medium: Add components, except sodium pyruvate solution and agar solution, to distilled/deionized water and bring volume to 700.0mL. Mix thoroughly. Gently heat and bring to boiling. Adjust pH to 7.1. Autoclave for 15 min at 15 psi pressure–121°C. Cool to 50°C. Add sodium pyruvate solution and agar solution. Mix thoroughly. Pour into sterile Petri dishes or distribute into sterile tubes.

Use: For the cultivation and maintenance of *Bacillus schlegelii*.

Bacillus schlegelii Broth

Composition per liter:

$Na_2HPO_4·2H_2O$	4.5g
KH_2PO_4	1.5g
Sodium pyruvate	1.5g
NH_4Cl	1.0g
$MgSO_4·7H_2O$	0.2g
$MnSO_4·H_2O$	0.01g
$CaCl_2·2H_2O$	0.01g
Ferric ammonium citrate	5.0mg
Sodium pyruvate	100.0mL
SL-6 trace elements	3.0mL

pH 7.1 ± 0.2 at 25°C

Pyruvate Solution:
Composition per 100mL:

Sodium pyruvate	1.5g

Preparation of Pyruvate Solution: Add sodium pyruvate to distilled/deionized water and bring volume to 100.0mL. Mix thoroughly. Filter sterilize.

SL-6 Trace Elements Solution:
Composition per liter:

H_3BO_3	0.3g
$CoCl_2·6H_2O$	0.2g
$ZnSO_4·7H_2O$	0.10g
$MnCl_2·4H_2O$	0.03g
$Na_2MoO_4·H_2O$	0.03g
$NiCl_2·6H_2O$	0.02g
$CuCl_2·2H_2O$	0.01g

Preparation of SL-6 Trace Elements Solution: Add components to distilled/deionized water and bring volume to 1.0L. Mix thoroughly. Adjust pH to 3.4.

Preparation of Medium: Add components, except sodium pyruvate solution, to distilled/deionized water and bring volume to 900.0mL. Mix thoroughly. Gently heat and bring to boiling. Adjust pH to 7.1. Autoclave for 15 min at 15 psi pressure–121°C. Cool to 50°C. Aseptically add sodium pyruvate solution. Aseptically distribute into sterile tubes or flasks.

Use: For the cultivation and maintenance of *Bacillus schlegelii*.

Bacillus stearothermophilus Broth

Composition per liter:

Pancreatic digest of casein	10.0g
Yeast extract	5.0g
K_2HPO_4	2.0g

pH 7.2 ± 0.2 at 25°C

Preparation of Medium: Add components to distilled/deionized water and bring volume to 1.0L. Mix thoroughly. Distribute into tubes or flasks. Autoclave for 15 min at 15 psi pressure–121°C.

Use: For the cultivation of *Bacillus stearothermophilus*.

Bacillus stearothermophilus Defined Broth

Composition per 100mL:

Mineral salts solution	10.0mL
Potassium phosphate buffer	5.0mL
L-Glutamate·HCl (1% solution)	4.0mL
L-Leucine (1% solution)	1.64mL
L-Lysine·HCl (1% solution)	1.40mL
L-Serine (1% solution)	1.40mL
L-Aspartate (1% solution)	1.30mL

L-Valine (1% solution)	1.26mL
Biotin (0.01% solution)	1.0mL
Glucose (20% solution)	1.0mL
L-Isoleucine (1% solution)	1.0mL
L-Proline (1% solution)	1.0mL
Nicotinic acid (0.01% solution)	1.0mL
Thiamine·HCl (0.01% solution)	1.0mL
L-Phenylalanine (1% solution)	0.86mL
L-Alanine (1% solution)	0.84mL
L-Threonine (1% solution)	0.84mL
L-Arginine·HCl (1% solution)	0.64mL
L-Tyrosine (1% solution)	0.56mL
L-Methionine (1% solution)	0.52mL
Glycine (1% solution)	0.50mL
L-Asparagine·H_2O (1% solution)	0.50mL
L-Cystine (1% solution)	0.50mL
L-Glutamine (1% solution)	0.50mL
L-Histidine·HCl·H_2O (1% solution)	0.42mL
L-Tryptophan (1% solution)	0.30mL
$CaCl_2$ (5% solution)	0.01mL
$FeCl_3 \cdot 6H_2O$ (0.05% solution)	0.01mL
$MnCl_2$ (10m*M* solution)	0.01mL
$ZnSO_4 \cdot 7H_2O$ (5% solution)	0.01mL

pH 7.3 ± 0.2 at 25°C

Mineral Salts Solution:
Composition per liter:

NaCl	10.0g
NH_4Cl	10.0g
$MgSO_4$	4.0g

Preparation of Mineral Salts Solution: Add components to distilled/deionized water and bring volume to 1.0L. Mix thoroughly.

Potassium Phosphate Buffer:
Composition per 500mL:

K_2HPO_4	125.0g
KH_2PO_4	30.0g

Preparation of Potassium Phosphate Buffer: Add components to distilled/deionized water and bring volume to 500.0mL. Mix thoroughly.

Preparation of Medium: Add components to distilled/deionized water and bring volume to 100.0mL. Mix thoroughly. Filter sterilize.

Use: For the cultivation of *Bacillus stearothermophilus* in a chemically defined medium.

Bacillus stearothermophilus Sporulation Broth

Composition per liter:

Agar	20.0g
Pancreatic digest of gelatin	5.0g
Yeast extract	4.0g
Beef extract	3.0g
$MnCl_2 \cdot 4H_2O$	10.0μg

pH 7.2 ± 0.2 at 25°C

Preparation of Medium: Add components to distilled/deionized water and bring volume to 1.0L. Mix thoroughly. Gently heat and bring to boiling. Distribute into tubes or flasks. Autoclave for 15 min at 15 psi pressure–121°C. Pour into sterile Petri dishes or leave in tubes.

Use: For the cultivation and sporulation of *Bacillus stearothermophilus*.

Bacillus thuringiensis Medium

Composition per liter:

Glucose	3.0g
$(NH_4)_2SO_4$	2.0g
Yeast extract	2.0g
$K_2HPO_4 \cdot 3H_2O$	0.5g
$MgSO_4 \cdot 7H_2O$	0.2g
$CaCl_2 \cdot 2H_2O$	0.08g
$MnSO_4 \cdot 4H_2O$	0.05g

pH 7.3 ± 0.2 at 25°C

Preparation of Medium: Add components to distilled/deionized water and bring volume to 1.0L. Mix thoroughly. Adjust pH to 7.3. Distribute into tubes or flasks. Autoclave for 15 min at 15 psi pressure–121°C.

Use: For the cultivation of *Bacillus thuringiensis*.

Bacillus Xylose Salts

Composition per liter:

Yeast extract	5.0g
Xylose	5.0g
NaCl	1.0g
NH_4Cl	1.0g
KH_2PO_4	0.50g
$MgSO_4 \cdot 7H_2O$	0.50g
$CaCl_2 \cdot 2H_2O$	0.05g
Trace mineral solution	10.0mL
Vitamin solution	10.0mL

pH 4.0 ± 0.2 at 25°C

Trace Mineral Solution:
Composition per liter:

$CoCl_2 \cdot 6H_2O$	0.2g
$FeSO_4 \cdot 7H_2O$	0.13g
$ZnCl_2 \cdot 2H_2O$	0.1g
$MnCl_2 \cdot 4H_2O$	0.1g
$CaCl_2 \cdot 2H_2O$	20.0mg
Na_2SeO_3	20.0mg
$Na_2WO_4 \cdot 2H_2O$	20.0mg
$NaMoO_4 \cdot 2H_2O$	1.0mg
H_3BO_3	0.5mg

$CuSO_4 \cdot 5H_2O$	0.4mg
KI	0.1mg

Preparation of Trace Mineral Solution: Add components to distilled/deionized water and bring volume to 1.0L. Mix thoroughly.

Vitamin Solution:

Composition per liter:

Pyridoxine·HCl	10.0mg
Thiamine·HCl	5.0mg
Riboflavin	5.0mg
Nicotinic acid	5.0mg
Calcium pantothenate	5.0mg
p-Aminobenzoic acid	5.0mg
Thioctic acid	5.0mg
Biotin	2.0mg
Folic acid	2.0mg
Cyanocobalamin	0.1mg

Preparation of Vitamin Solution: Add components to distilled/deionized water and bring volume to 1.0L. Mix thoroughly.

Preparation of Medium: Add components to distilled/deionized water and bring volume to 1.0L. Mix thoroughly. Gently heat and bring to boiling. Adjust pH of medium to 4.0. Distribute into tubes or flasks. Autoclave for 15 min at 15 psi pressure–121°C.

Use: For the cultivation and maintenance of *Bacillus* species which can utilize xylose as a carbon source.

Bacterial Cell Agar (BCA)

Composition per liter:

Tryptose	17.36g
Agar	15.0g
NaCl	8.68g
Beef extract	5,2g
Yeast extract	1.7g

pH 7.3 ± 0.2 at 25°C

Preparation of Medium: Add components, except agar, to distilled/deionized water and bring volume to 1.0L. Mix thoroughly. Autoclave for 15 min at 15 psi pressure–121°C. Cool to 30°C. Inoculate with a culture of *Aeromonas hydrophila*. Incubate with shaking at 30°C for 72 hr. Centrifuge culture in 40.0mL volumes at 10,000 × *g* for 10 min. Wash the cells four times in sterile 0.85% saline. Resuspend the cell pellet in 25.0mL of distilled/deionized water. Autoclave for 15 min at 15 psi pressure–121°C. Cool to 45–50°C. In a separate flask, add 15.0g of agar to 1.0L of distilled/deionized water. Mix thoroughly. Gently heat and bring to boiling. Autoclave for 15 min at 15 psi pressure–121°C. Cool to 45–50°C. Aseptically combine 25.0mL of washed cells and 250.0mL of cooled, sterile agar solution. Mix thoroughly. Pour into sterile Petri dishes.

Use: For the cultivation of freshwater *Myxobacterium* species.

Bacterium Medium

Composition per liter:

Agar	20.0g
Peptone	6.0g
Yeast extract	3.0g
Beef extract	1.5g
Glucose	1.0g

Preparation of Medium: Add components to distilled/deionized water and bring volume to 1.0L. Mix thoroughly. Gently heat and bring to boiling. Distribute into tubes or flasks. Autoclave for 15 min at 15 psi pressure–121°C. Pour into sterile Petri dishes or leave in tubes.

Use: An archaic medium used for the cultivation and growth of bacteria originally classified in the genus *Bacterium* but now classified in the genera *Brevibacterium* and *Kurthia*.

Bacteroides Medium

Composition per liter:

Pancreatic digest of casein	27.0g
Yeast extract	3.0g
K_2HPO_4	2.5g
K_2CO_3	2.0g
NaCl	2.0g
Hemin solution	10.0mL
Vitamin K_1 solution	0.2mL

Hemin Solution:

Composition per 100mL:

Hemin	1.0g
NaOH (1*N* solution)	20.0mL

Preparation of Hemin Solution: Add hemin to 20.0mL of 1*N* NaOH solution. Mix thoroughly. Bring volume to 100.0mL with distilled/deionized water.

Vitamin K_1 Solution:

Composition per 100mL:

Vitamin K_1	1.0g
Ethanol	99.0mL

Preparation of Vitamin K_1 Solution: Add vitamin K_1 to 99.0mL of absolute ethanol. Mix thoroughly.

Preparation of Medium: Add components to distilled/deionized water and bring volume to 1.0L. Mix thoroughly. Distribute into tubes or flasks. Autoclave for 15 min at 15 psi pressure–121°C.

Use: For the cultivation of *Bacteroides asaccharolyticus* and *Bacteroides melaninogenicus*.

BAGG Broth (Buffered Azide Glucose Glycerol Broth)

Composition per liter:

Pancreatic digest of casein	10.0g
Peptic digest of animal tissue	10.0g
Glucose	5.0g
NaCl	5.0g
K_2HPO_4	4.0g
KH_2PO_4	1.5g
NaN_3	0.5g
Bromcresol purple	0.015g
Glycerol	5.0mL

pH 6.9 ± 0.2 at 25°C

Source: Available as a premixed powder from Difco Laboratories and BBL Microbiology Systems.

Caution: Sodium azide is toxic. Azides also react with metals and disposal must be highly diluted.

Preparation of Medium: Add 5.0mL glycerol to 900.0mL of distilled/deionized water. Add remaining components and bring volume to 1.0L. Mix thoroughly. Gently heat and bring to boiling. Distribute into tubes in 10.0mL volumes. Autoclave for 15 min at 10 psi pressure–116°C.

Use: For the cultivation of fecal streptococci. It is recommended for qualitative presumptive and confirmatory tests for fecal streptococci.

Baird–Parker Agar

Composition per liter:

Agar	17.0g
Glycine	12.0g
Sodium pyruvate	10.0g
Pancreatic digest of casein	10.0g
Beef extract	5.0g
LiCl	5.0g
Yeast extract	1.0g

pH 7.0 ± 0.2 at 25°C

Source: This medium is available as a premixed powder from Difco Laboratories, Oxoid Unipath, and BBL Microbiology Systems.

Preparation of Medium: Add components to distilled/deionized water and bring volume to 1.0L. Mix thoroughly. Gently heat and bring to boiling. Autoclave for 15 min at 15 psi pressure–121°C. Cool to 45–50°C. Pour into sterile Petri dishes.

Use: Used as a base for the preparation of Egg-Tellurite-Glycine-Pyruvate Agar for the selective isolation and enumeration of coagulase-positive staphylococci from soil, air and other materials.

Baird–Parker Agar

Composition per liter:

Agar	17.0g
Glycine	12.0g
Sodium pyruvate	10.0g
Pancreatic digest of casein	10.0g
Beef extract	5.0g
LiCl	5.0g
Yeast extract	1.0g
Sulfamethazine solution	10.0mL

pH 7.0 ± 0.2 at 25°C

Sulfamethazine Solution:

Composition per 10mL:

Sulfamethazine	0.05g

Preparation of Sulfamethazine Solution: Add sulfamethazine to distilled/deionized water and bring volume to 10.0mL. Mix thoroughly. Filter sterilize.

Preparation of Medium: Add components, except sulfamethazine solution, to distilled/deionized water and bring volume to 990.0mL. Mix thoroughly. Gently heat and bring to boiling. Autoclave for 15 min at 15 psi pressure–121°C. Cool to 45–50°C. Aseptically add sterile sulfamethazine solution. Mix thoroughly. Pour into sterile Petri dishes or distribute into sterile tubes.

Use: Used as a base for the preparation of Egg-Tellurite-Glycine-Pyruvate Agar for the selective isolation and enumeration of coagulase-positive staphylococci from soil, air and other materials.

Baird–Parker Agar, Supplemented

Composition per liter:

Agar	17.0g
Glycine	12.0g
Sodium pyruvate	10.0g
Pancreatic digest of casein	10.0g
Beef extract	5.0g
LiCl	5.0g
Yeast extract	1.0g
RPF supplement	100.0mL

pH 7.0 ± 0.2 at 25°C

RPF Supplement:

Composition per 100mL:

Bovine fibrinogen	3.75g
Trypsin inhibitor	25.0mg

K_2TeO_3 .... 25.0mg
Rabbit plasma 25.0mL

Caution: Potassium tellurite is toxic.

Preparation of RPF Supplement: Add components to distilled/deionized water and bring volume to 100.0mL. Mix thoroughly. Filter sterilize.

Preparation of Medium: Add components, except RPF supplement, to distilled/deionized water and bring volume to 900.0mL. Mix thoroughly. Gently heat and bring to boiling. Autoclave for 15 min at 15 psi pressure–121°C. Cool to 45–50°C. Aseptically add 100.0mL of filter-sterilized RPF supplement. Mix thoroughly but gently. Pour into sterile Petri dishes.

Use: For the selective isolation and enumeration of coagulase-positive staphylococci from soil, air and other materials. Also, for differentiation and identification of staphylococci on the basis of their ability to coagulate plasma. Colonies surrounded by an opaque zone of coagulated plasma are diagnostic for *Staphylococcus aureus*.

Balamuth Medium

Composition per 200mL:
Dehydrated egg yolk 36.0g
Dried liver concentrate 1.0g
Rice starch 0.2g
Potassium phosphate buffer, pH 7.5 .... 125.0mL
NaCl solution 125.0mL
pH 7.3 ± 0.2 at 25°C

NaCl Solution
Composition per 200mL:
NaCl 1.6g

Preparation of NaCl Solution: Add NaCl to distilled/deionized water and bring volume to 200.0mL. Mix thoroughly.

Potassium Phosphate Buffer, 0.067M
Composition per 200mL:
K_2HPO_4 (1*M* solution) 8.6mL
KH_2PO_4 (1*M* solution) 4.66mL

Preparation of Potassium Phosphate Buffer: Combine the K_2HPO_4 and KH_2PO_4 solutions. Bring volume to 200.0mL with distilled/deionized water. Adjust pH to 7.5.

Preparation of Medium: Add dehydrated egg yolk to 36.0mL of distilled/deionized water. Add 125.0mL of 0.8% NaCl. Mix thoroughly in a blender. Heat in a covered, double boiler until infusion reaches 80°C and maintain at this temperature for 20 min. Add 20.0mL of distilled/deionized H_2O. Filter through a layer of cheesecloth. To 90–100.0mL of filtrate add 0.8% NaCl solution to bring volume to 125.0mL. Autoclave for 20 min at 15 psi pressure–121°C. Cool to 4°C. Filter. To filtrate add an equal volume of 0.067*M* potassium phosphate buffer, pH 7.5. Add1.0g of dried liver concentrate. Mix thoroughly. Distribute into tubes or flasks in 10.0mL volumes. Autoclave for 20 min at 15 psi pressure–121°C. Prior to inoculation, add 0.01g of rice starch to each tube.

Use: For the cultivation and maintenance of *Entamoeba histolytica*.

BAM Agar

Composition per liter:
Agar 30.0g
Glucose 5.0g
KH_2PO_4 .... 3.0g
Yeast extract 1.0g
$MgSO_4 \cdot 7H_2O$ 0.5g
$CaCl_2 \cdot 2H_2O$ 0.25g
$(NH_4)_2SO_4$ 0.2g
Trace elements 1.0mL
pH 4.0 ± 0.2 at 25°C

Trace Elements:
Composition per liter:
$CaCl_2 \cdot 2H_2O$ 0.66g
$Na_2MoO_4 \cdot 2H_2O$ 0.30g
$ZnSO_4 \cdot 7H_2O$ 0.18g
$CoCl_2 \cdot 6H_2O$ 0.18g
$CuSO_4 \cdot 5H_2O$ 0.16g
$MnSO_4 \cdot 4H_2O$ 0.15g
H_3BO_3 .... 0.10g

Preparation of Trace Elements: Add components to 1.0L of distilled/deionized water. Mix thoroughly.

Preparation of Medium: Add components, except agar to distilled/deionized water and bring volume to 800.0mL. Mix thoroughly. Gently heat and bring to boiling. Adjust medium to pH 4.0 with H_2SO_4. Add agar to 200.0mL distilled/deionized water. Autoclave agar separately to avoid acid hydrolysis. Autoclave for 15 min at 15 psi pressure–121°C. Mix two solutions together. Pour into sterile Petri dishes or distribute into sterile tubes.

Use: For the cultivation and maintenance of *Bacillus acidoterrestris*.

BAM Broth

Composition per liter:
Glucose 5.0g
KH_2PO_4 .... 3.0g

Yeast extract	1.0g
$MgSO_4 \cdot 7H_2O$	0.5g
$CaCl_2 \cdot 2H_2O$	0.25g
$(NH_4)_2SO_4$	0.2g
Trace elements	1.0mL

pH 4.0 ± 0.2 at 25°C

Trace Elements:
Composition per liter:

$CaCl_2 \cdot 2H_2O$	0.66g
$Na_2MoO_4 \cdot 2H_2O$	0.30g
$ZnSO_4 \cdot 7H_2O$	0.18g
$CoCl_2 \cdot 6H_2O$	0.18g
$CuSO_4 \cdot 5H_2O$	0.16g
$MnSO_4 \cdot 4H_2O$	0.15g
H_3BO_3	0.10g

Preparation of Trace Elements: Add components to 1.0L of distilled/deionized water. Mix thoroughly.

Preparation of Medium: Add components to distilled/deionized water and bring volume to 1.0L. Mix thoroughly. Gently heat and bring to boiling. Adjust medium to pH 4.0 with H_2SO_4. Distribute into tubes or flasks. Autoclave for 15 min at 15 psi pressure–121°C.

Use: For the cultivation and maintenance of *Bacillus acidoterrestris.*

BAM SM Agar

Composition per liter:

Agar	20.0g
Glucose	5.0g
KH_2PO_4	3.0g
Yeast extract	6.0g
$MgSO_4 \cdot 7H_2O$	0.5g
$CaCl_2 \cdot 2H_2O$	0.25g
$(NH_4)_2SO_4$	0.2g
Trace elements	1.0mL

pH 4.0 ± 0.2 at 25°C

Trace Elements:
Composition per liter:

$CaCl_2 \cdot 2H_2O$	0.66g
$Na_2MoO_4 \cdot 2H_2O$	0.30g
$ZnSO_4 \cdot 7H_2O$	0.18g
$CoCl_2 \cdot 6H_2O$	0.18g
$CuSO_4 \cdot 5H_2O$	0.16g
$MnSO_4 \cdot 4H_2O$	0.15g
H_3BO_3	0.10g

Preparation of Trace Elements: Add components to 1.0L of distilled/deionized water. Mix thoroughly.

Preparation of Medium: Add components, except agar to distilled/deionized water and bring volume to 800.0mL. Mix thoroughly. Gently heat and bring to boiling. Adjust medium to pH 4.0 with H_2SO_4. Add agar to 200.0mL distilled/deionized water. Autoclave agar separately to avoid acid hydrolysis. Autoclave for 15 min at 15 psi pressure–121°C. Mix two solutions together. Pour into sterile Petri dishes or distribute into sterile tubes.

Use: For the cultivation and maintenance of *Bacillus cycloheptanicus.*

BAM SM Broth

Composition per liter:

Glucose	5.0g
KH_2PO_4	3.0g
Yeast extract	6.0g
$MgSO_4 \cdot 7H_2O$	0.5g
$CaCl_2 \cdot 2H_2O$	0.25g
$(NH_4)_2SO_4$	0.2g
Trace elements	1.0mL

pH 4.0 ± 0.2 at 25°C

Trace Elements:
Composition per liter:

$CaCl_2 \cdot 2H_2O$	0.66g
$Na_2MoO_4 \cdot 2H_2O$	0.30g
$ZnSO_4 \cdot 7H_2O$	0.18g
$CoCl_2 \cdot 6H_2O$	0.18g
$CuSO_4 \cdot 5H_2O$	0.16g
$MnSO_4 \cdot 4H_2O$	0.15g
H_3BO_3	0.10g

Preparation of Trace Elements: Add components to 1.0L of distilled/deionized water. Mix thoroughly.

Preparation of Medium: Add components to distilled/deionized water and bring volume to 1.0L. Mix thoroughly. Gently heat and bring to boiling. Adjust medium to pH 4.0 with H_2SO_4. Distribute into tubes or flasks. Autoclave for 15 min at 15 psi pressure–121°C.

Use: For the cultivation and maintenance of *Bacillus cycloheptanicus.*

Basal Mineral Medium

Composition per liter:

NH_4Cl	0.80g
K_2HPO_4	0.70g
$MgSO_4 \cdot 7H_2O$	0.01g
Disodium EDTA	9.2mg
$FeSO_4 \cdot 7H_2O$	7.0mg
$CaSO4 \cdot 2H_2O$	2.0mg
H_3BO_3	0.1mg
$ZnSO_4 \cdot 7H_2O$	0.1mg
$MnSO_4 \cdot 4H_2O$	0.02mg

$Co(NO_3)_2$ 0.01mg
$NaMoO_4 \cdot 2H_2O$ 0.01mg
$CuSO_4 \cdot 5H_2O$ 0.5μg

Preparation of Medium: Add components to distilled/deionized water and bring volume to 1.0L. Mix thoroughly. Filter sterilize.

Use: For the cultivation of *Beggiatoa* species.

Basal Synthetic Medium

Composition per liter:

L-Glutamic acid 20.0g
$(NH_4)_2SO_4$ 4.0g
K_2HPO_4 .... 1.88g
KH_2PO_4 .... 0.57g
$MgSO_4 \cdot 7H_2O$ 0.2g
Salt solution 10.0mL

Salt Solution:

Composition per liter:

$FeCl_3 \cdot 6H_2O$ 0.6g
$MnCl_2 \cdot 4H_2O$ 0.6g
$ZnCl_2$ 0.6g
$CuSO_4 \cdot 5H_2O$ 0.6g
$CaCl_2 \cdot 2H_2O$ 0.6g
NaCl 0.6g

Preparation of Salt Solution: Add components to 1.0L of distilled/deionized water. Mix thoroughly.

Preparation of Medium: Add components to distilled/deionized water and bring volume to 1.0L. Mix thoroughly. Gently heat and bring to boiling. Distribute into tubes or flasks. Autoclave for 15 min at 15 psi pressure–121°C.

Use: For the cultivation and maintenance of *Acinetobacter lwoffii.*

Basal Thermophile Medium

Composition per liter:

Solution 1 .... 850.0mL
Solution 2 .... 100.0mL
Solution 3 .... 50.0mL

Solution 1:

Composition per 850mL:

Pancreatic digest of casein 10.0g
K_2HPO_4 .... 1.5g
NH_4Cl 0.9g
KH_2PO_4 .... 0.75g
$MgCl_2 \cdot 6H_2O$ 0.2g
Trace element solution 9.0mL
Vitamin solution 5.0mL
Resazurin (0.2% solution) 1.0mL
$FeSO_4 \cdot 7H_2O$ (10% solution) 0.03mL

Preparation of Solution 1: Add components to distilled/deionized water and bring volume to 850.0mL. Mix thoroughly. Autoclave for 45 min at 15 psi pressure–121°C. Cool to 45–50°C.

Solution 2:

Composition per 100mL:

Yeast extract 3.0g

Preparation of Solution 2: Add yeast extract to distilled/deionized water and bring volume to 100.0mL. Mix thoroughly. Autoclave for 45 min at 15 psi pressure–121°C. Cool to 45–50°C.

Solution 3:

Composition per 50mL:

Glucose 5.0g

Preparation of Solution 3: Add glucose to distilled/deionized water and bring volume to 50.0mL. Mix thoroughly. Autoclave for 45 min at 15 psi pressure–121°C. Cool to 45–50°C.

Trace Element Solution:

Composition per liter:

Nitrilotriacetic acid 12.5g
NaCl 1.0g
$FeCl_3 \cdot 4H_2O$ 0.2g
$MnCl_2 \cdot 4H_2O$ 0.1g
$CaCl_2 \cdot 2H_2O$ 0.1g
$ZnCl_2$ 0.1g
$CuCl_2$ 0.02g
Na_2SeO_3 .... 0.02g
$CoCl_2 \cdot 6H_2O$ 0.017g
H_3BO_3 .... 0.01g
$Na_2MoO_4 \cdot 2H_2O$ 0.01g

Preparation of Trace Element Solution: Add nitrilotriacetic acid to 100.0mL of distilled/deionized water. Adjust pH to 6.5 with KOH. Add remaining components and bring volume to 1.0L. Mix thoroughly.

Wolfe's Vitamin Solution:

Composition per liter:

Pyridoxine·HCl 10.0mg
Thiamine·HCl 5.0mg
Riboflavin 5.0mg
Nicotinic acid 5.0mg
Calcium pantothenate 5.0mg
p-Aminobenzoic acid 5.0mg
Thioctic acid 5.0mg
Biotin 2.0mg
Folic acid 2.0mg
Cyanocobalamin 0.1mg

Preparation of Wolfe's Vitamin Solution: Add components to distilled/deionized water and bring volume to 1.0L. Mix thoroughly.

$Na_2S{\cdot}9H_2O$ Solution:
Composition per 100mL:

$Na_2S{\cdot}9H_2O$	10.0g

Preparation of $Na_2S{\cdot}9H_2O$ Solution: Add $Na_2S{\cdot}9H_2O$ to distilled/deionized water and bring volume to 100.0mL. Mix thoroughly. Autoclave for 15 min at 15 psi pressure–121°C.

Preparation of Medium: Aseptically combine solution 1, solution 2 and solution 3 under 100% N_2. Distribute into tubes in 10.0mL volumes under 100% N_2. Immediately prior to inoculation, aseptically add 0.1mL of sterile $Na_2S{\cdot}9H_2O$ solution to each tube.

Use: For the cultivation and maintenance of *Clostridium* species, *Fervidobacterium nodosum* and *Thermoanaerobium brockii.*

BC Medium (Medium for *Acetivibrio cellulolyticus*)

Composition per liter:

Cellulose powder	3.0g
$NaHCO_3$	2.0g
Mineral solution 1	75.0mL
Mineral solution 2	75.0mL
Cysteine-sulfide reducing solution	12.8mL
$FeSO_4{\cdot}7H_2O$ solution	10.0mL
Vitamin mixture	10.0mL
Wolfe's mineral solution	10.0mL
Resazurin (0.1% solution)	1.0mL

pH 7.6 ± 0.2 at 25°C

Caution: This medium contains sodium sulfide and may produce toxic H_2S gas. Prepare in a chemical fume hood.

Mineral Solution 1:
Composition per liter:

K_2HPO_4	3.9g

Preparation of Mineral Solution 1: Add K_2HPO_4 to distilled/deionized water and bring volume to 1.0L. Mix thoroughly.

Mineral Solution 2:
Composition per liter:

NH_4Cl	12.0g
Na_2SO_4	2.5g
KH_2PO_4	2.4g
$MgSO_4{\cdot}7H_2O$	1.2g
$CaCl_2{\cdot}2H_2O$	0.8g

Preparation of Mineral Solution 2: Add components to distilled/deionized water and bring volume to 1.0L. Mix thoroughly.

$FeSO_4{\cdot}7H_2O$ Solution:
Composition per 100mL:

$FeSO_4{\cdot}7H_2O$	0.2g

Preparation of $FeSO_4{\cdot}7H_2O$ Solution: Dissolve $FeSO_4{\cdot}7H_2O$ in 100.0mL distilled/deionized water. Add 3 drops concentrated HCl. Mix thoroughly.

Vitamin Mixture:
Composition per liter:

Pyridoxine·HCl	10.0mg
Thiamine·HCl	5.0mg
Cyanocobalamin	5.0mg
Lipoic acid (thioctic acid)	5.0mg
Biotin	2.0mg
p-Aminobenzoic acid	0.5mg

Preparation of Vitamin Mixture: Add components to distilled/deionized water and bring volume to 1.0L. Store below –20°C.

Wolfe's Mineral Solution:
Composition per liter

$MgSO_4{\cdot}7H_2O$	3.0g
Nitriloacetic acid	1.5g
$MnSO_4{\cdot}H_2O$	0.5g
NaCl	1.0g
$FeSO_4{\cdot}7H_2O$	0.1g
$CoCl_2{\cdot}6H_2O$	0.1g
$CaCl_2$	0.1g
$ZnSO_4{\cdot}7H_2O$	0.1g
$CuSO_4{\cdot}5H_2O$	0.01g
$AlK(SO_4)_2{\cdot}12H_2O$	0.01g
H_3BO_3	0.01g
$Na_2MoO_4{\cdot}2H_2O$	0.01g

Preparation of Wolfe's Mineral Solution: Add nitrilotriacetic acid to 500.0mL of distilled/deionized water and adjust to pH 6.5 with KOH to dissolve. Bring volume to 1.0L with distilled/deionized water. Add remaining components one at a time.

Cysteine-Sulfide Reducing Solution:
Composition per 200mL:

L-Cysteine·HCl·H_2O	2.5g
$Na_2S{\cdot}9H_2O$	2.5g

Preparation of Cysteine-Sulfide Reducing Solution: Add L-Cysteine·HCl·H_2O to 50.0mL distilled/deionized water. Quickly adjust pH to 10 with fresh 3*N* NaOH and flush under 100% N_2. Add $Na_2S{\cdot}9H_2O$. Bring volume to 200.0mL with distilled/deionized water. Boil under 100% N_2. Transfer anaerobically to tubes or flasks and stopper. Autoclave for 15 min at 15 psi pressure–121°C.

Preparation of Medium: Add cellulose and $NaHCO_3$ to distilled/deionized water and bring volume to 800.0mL. Add all other components except cysteine-sulfide reducing solution. Heat and boil under 90% N_2 + 10% CO_2. Cool and continue flushing under 90% N_2 + 10% CO_2. The pH should be 7.6 at room temperature; do not adjust. Add 8.0mL cysteine-sulfide reducing solution. Add 4.8mL more of cysteine-sulfide reducing solution. Distribute anaerobically into tubes in 7.0mL volumes and cap.

Use: For the cultivation and maintenance of *Acetivibrio cellulolyticus, A. cellulosolvens, Bacteroides cellulosolvens* and other cellulose degrading microorganisms.

BC Motility Medium (*Bacillus cereus* Motility Medium)

Composition per liter:

Pancreatic digest of casein	10.0g
Glucose	5.0g
Agar	3.0g
Na_2HPO_4	2.5g
Yeast extract	2.5g

pH 7.4 ± 0.2 at 25°C

Preparation of Medium: Add components to distilled/deionized water and bring volume to 1.0L. Mix thoroughly. Gently heat and bring to boiling. Distribute into tubes in 2.0mL volumes. Autoclave for 15 min at 15 psi pressure–121°C.

Use: For the cultivation and observation of motility of *Bacillus cereus.*

BCP Azide Broth (Bromcresol Purple Azide Broth)

Composition per liter:

Casein peptone	10.0g
Yeast extract	10.0g
D-Glucose	5.0g
NaCl	5.0g
K_2HPO_4	2.7g
KH_2PO_4	2.7g
NaN_3	0.5g
Bromcresol purple	0.032g

pH 6.9 ± 0.2 at 25°C

Caution: Sodium azide is toxic. Azides also react with metals and disposal must be highly diluted.

Preparation of Medium: Add components, except cysteine, to 900.0mL distilled/deionized water. Mix thoroughly. Gently heat to boiling. Distribute into tubes or flasks. Autoclave for 15 min at 15 psi pressure–121°C.

Use: For use in the confirmation test for the presence of fecal streptococci in water and wastewater.

BCP D Agar (Bromcresol Purple Deoxycholate Agar)

Composition per liter:

Agar	25.0g
Lactose	10.0g
Sucrose	10.0g
Pancreatic digest of casein	7.5g
Thiopeptone	7.5g
NaCl	5.0g
Yeast extract	2.0g
Sodium citrate	2.0g
Sodium deoxycholate	1.0g
Bromcresol purple	0.02g

pH 7.2 ± 0.2 at 25°C

Preparation of Medium: Add components to distilled/deionized water and bring volume to 1.0L. Mix thoroughly. Gently heat and bring to boiling. Pour into sterile Petri dishes without sterilization. Do not autoclave. Use the same day.

Use: For the isolation, cultivation, and differentiation of Gram-negative enteric bacilli. For the isolation and cultivation of *Salmonella, Shigella* and other nonlactose and nonsucrose fermenting microorganisms. Nonlactose/nonsucrose fermenting microorganisms appear as colorless or blue colonies. Lactose/sucrose fermenting microorganisms, such as coliform bacteria, appear as yellow-opaque white colonies surrounded by a zone of precipitated deoxycholate.

BCP DCLS Agar (Bromcresol Purple Deoxycholate Citrate Lactose Sucrose Agar)

Composition per liter:

Agar	14.0g
Sodium citrate	10.0g
Lactose	7.5g
Sucrose	7.5g
Pancreatic digest of casein	7.5g
Peptone	7.5g
NaCl	5.0g
$Na_2S_2O_3·5H_2O$	5.0g
Yeast extract	3.0g
Meat extract	3.0g
Sodium deoxycholate	2.5g
Bromcresol purple	0.02g

pH 7.2 ± 0.2 at 25°C

Preparation of Medium: Add components to distilled/deionized water and bring volume to 1.0L. Mix

thoroughly. Gently heat and bring to boiling. Pour into sterile Petri dishes without sterilization. Do not autoclave. Use the same day.

Use: For the differential isolation of Gram-negative enteric bacilli. For the isolation of *Salmonella, Shigella* and other nonlactose and nonsucrose fermenting microorganisms. Nonlactose/nonsucrose fermenting microorganisms appear as colorless or blue colonies. Lactose/sucrose fermenting microorganisms, such as coliform bacteria, appear as yellow-opaque white colonies surrounded by a zone of precipitated deoxycholate.

BCYE Agar (BCYE Alpha Base) (Buffered Charcoal Yeast Extract Agar)

Composition per liter:

Agar	15.0g
Yeast extract	10.0g
ACES buffer (2-[(2-Amino-2-oxoethyl)-amino]-ethane sulfonic acid)	10.0g
Charcoal, activated	2.0g
α-Ketogluatrate	1.0g
L-Cysteine·HCl·H_2O	0.4g
$Fe_4(P_2O_7)_3 \cdot 9H_2O$	0.25g

pH 6.9 ± 0.2 at 25°C

Source: This medium is available as a premixed powder from BBL Microbiology Systems.

Preparation of Medium: Add components, except cysteine, to distilled/deionized water and bring volume to 1.0L. Mix thoroughly. Adjust medium to pH 6.9 with 1*N* KOH. Heat gently and bring to boil for 1 minute. Autoclave for 15 min at 15 psi pressure–121°C. Cool to 50–55°C. Add 4.0mL of a 10% solution of L-cysteine·HCl·H_2O which has been filter-sterilized. Mix thoroughly. Pour into sterile Petri dishes with constant agitation to keep charcoal in suspension.

Use: For the isolation, cultivation, and maintenance of *Legionella pneumophila* and other *Legionella* species from environmental specimens.

BCYEα with Alb (Buffered Charcoal Yeast Extract Agar with Albumin)

Composition per liter:

Agar	15.0g
Yeast extract	10.0g
ACES buffer (2-[(2-Amino-2-oxoethyl)-amino]-ethane sulfonic acid)	10.0g
Charcoal, activated	2.0g
α-Ketogluatrate	1.0g
Bovine serum albumin solution	10.0mL
Cysteine·HCl·H_2O solution	10.0mL
$Fe_4(P_2O_7)_3 \cdot 9H_2O$ solution	10.0mL

pH 6.9 ± 0.2 at 25°C

Bovine Serum Albumin Solution:
Composition per 10mL:

Bovine serum albumin	0.1g

Preparation of Bovine Serum Albumin Solution: Add bovine serum albumin to distilled/deionized water and bring volume to 10.0mL. Mix thoroughly. Filter sterilize.

Cysteine·HCl·H_2O Solution:
Composition per 10mL:

L-Cysteine·HCl·H_2O	0.4g

Preparation of Cysteine·HCl·H_2O Solution: Add cysteine·HCl·H_2O to distilled/deionized water and bring volume to 10.0mL. Mix thoroughly. Filter sterilize.

$Fe_4(P_2O_7)_3 \cdot 9H_2O$ Solution:
Composition per 10mL:

$Fe_4(P_2O_7)_3 \cdot 9H_2O$	0.25g

Preparation of $Fe_4(P_2O_7)_3 \cdot 9H_2O$ Solution: Add $Fe_4(P_2O_7)_3 \cdot 9H_2O$ to distilled/deionized water and bring volume to 10.0mL. Mix thoroughly. Filter sterilize.

Preparation of Medium: Add components—except cysteine·HCl·H_2O solution, $Fe_4(P_2O_7)_3 \cdot 9H_2O$ solution, and bovine serum albumin solution—to distilled/deionized water and bring volume to 970.0mL. Mix thoroughly. Adjust medium to pH 6.9 with 1*N* KOH. Heat gently and bring to boil for 1 minute. Autoclave for 15 min at 15 psi pressure–121°C. Cool to 50–55°C. Aseptically add the cysteine·HCl·H_2O solution, $Fe_4(P_2O_7)_3 \cdot 9H_2O$ solution and 10.0mL of sterile bovine serum albumin solution. Mix thoroughly. Pour into sterile Petri dishes with constant agitation to keep charcoal in suspension.

Use: For the isolation, cultivation, and maintenance of *Legionella pneumophila* and other *Legionella* species from environmental specimens.

BCYEα without L-Cysteine (Buffered Charcoal Yeast Extract Agar without L-Cysteine)

Composition per liter:

Agar	15.0g
Yeast extract	10.0g

ACES buffer (2-[(2-Amino-2-oxoethyl)-
amino]-ethane sulfonic acid) 10.0g
Charcoal, activated 2.0g
α-Ketogluatrate 1.0g
$Fe_4(P_2O_7)_3 \cdot 9H_2O$ solution 10.0mL
pH 6.9 ± 0.2 at 25°C

$Fe_4(P_2O_7)_3 \cdot 9H_2O$ Solution:
Composition per 10mL:
$Fe_4(P_2O_7)_3 \cdot 9H_2O$ 0.25g

Preparation of $Fe_4(P_2O_7)_3 \cdot 9H_2O$ Solution: Add $Fe_4(P_2O_7)_3 \cdot 9H_2O$ to distilled/deionized water and bring volume to 10.0mL. Mix thoroughly. Filter sterilize.

Preparation of Medium: Add components, except $Fe_4(P_2O_7)_3 \cdot 9H_2O$ solution, to distilled/deionized water and bring volume to 990.0mL. Mix thoroughly. Adjust medium to pH 6.9 with 1*N* KOH. Heat gently and bring to boil for 1 minute. Autoclave for 15 min at 15 psi pressure–121°C. Cool to 50–55°C. Aseptically add 10.0mL of sterile $Fe_4(P_2O_7)_3 \cdot 9H_2O$ solution. Mix thoroughly. Pour into sterile Petri dishes with constant agitation to keep charcoal in suspension.

Use: For the isolation, cultivation, and maintenance of *Legionella pneumophila* and other *Legionella* species from environmental specimens.

BCYE Differential Agar (Buffered Charcoal Yeast Extract Differential Agar)

Composition per liter:
Agar 15.0g
Yeast extract 10.0g
ACES buffer (2-[(2-Amino-2-oxoethyl)-
amino]-ethane sulfonic acid) 10.0g
Charcoal, activated 2.0g
α-Ketogluatrate 1.0g
L-Cysteine·HCl·H_2O 0.4g
$Fe_4(P_2O_7)_3 \cdot 9H_2O$ 0.25g
Bromcresol purple 0.01g
Bromthymol blue 0.01g
pH 6.9 ± 0.2 at 25°C

Source: This medium is available as a premixed powder from BBL Microbiology Systems.

Preparation of Medium: Add components, except cysteine, to distilled/deionized water and bring volume to 1.0L. Mix thoroughly. Adjust medium to pH 6.9 with 1*N* KOH. Heat gently and bring to boil for 1 minute. Autoclave for 15 min at 15 psi pressure–121°C. Cool to 50–55°C. Add 4.0mL of a 10% solution of L-cysteine·HCl·H_2O which has been filter sterilized. Mix thoroughly. Pour into sterile Petri dishes with constant agitation to keep charcoal in suspension.

Use: For the isolation, cultivation, and maintenance of *Legionella pneumophila* and other *Legionella* species from environmental specimens. For the presumptive differential identification of *Legionella* species based on colony color and morphology. *L. pneumophila* appear as light blue/green colonies. *L. micdadei* appear as blue/gray or dark blue colonies.

BCYE Selective Agar with CCVC (Buffered Charcoal Yeast Extract Selective Agar with Cephalothin, Colistin, Vancomycin, and Cycloheximide)

Composition per 1014mL:
Agar 15.0g
Yeast extract 10.0g
ACES buffer (2-[(2-Amino-2-oxoethyl)-
amino]-ethane sulfonic acid) 10.0g
Charcoal, activated 2.0g
α-Ketogluatrate 1.0g
$Fe_4(P_2O_7)_3 \cdot 9H_2O$ 0.25g
Antibiotic solution 10.0mL
Cysteine·HCl·H_2O solution 4.0mL
pH 6.9 ± 0.2 at 25°C

Source: This medium is available as a premixed powder from BBL Microbiology Systems.

Cysteine·HCl·H_2O Solution:
Composition per 10mL:
L-Cysteine·HCl·H_2O 1.0g

Preparation of Cysteine·HCl·H_2O Solution: Add L-Cysteine·HCl·H_2O to distilled/deionized water and bring volume to 10.0mL. Mix thoroughly. Filter sterilize.

Antibiotic Solution:
Composition per 10mL:
Cycloheximide 80.0mg
Colistin 16.0mg
Cephalothin 4.0mg
Vancomycin 0.5mg

Preparation of Antibiotic Solution: Add components to distilled/deionized water and bring volume to 10.0mL. Mix thoroughly. Filter sterilize.

Preparation of Medium: Add components, except cysteine and antibiotic solution, to distilled/deionized water and bring volume to 1.0L. Mix thoroughly. Adjust medium to pH 6.9 with 1*N* KOH. Heat gently and bring to boil for 1 minute. Autoclave for 15 min at 15 psi pressure–121°C. Cool to

50–55°C. Add 4.0mL of L-cysteine·HCl·H_2O solution and 10.0mL of sterile antibiotic solution. Mix thoroughly. Pour into sterile Petri dishes with constant agitation to keep charcoal in suspension.

Use: For the isolation, cultivation, and maintenance of *Legionella pneumophila* and other *Legionella* species from environmental specimens. Used for the selective recovery of *L. pneumophila* while reducing contaminating microorganisms from environmental water samples.

BCYE Selective Agar with GPVA (Buffered Charcoal Yeast Extract Selective Agar with Glycine, Polymyxin B, Vancomycin, and Anisomycin)

Composition per 1014mL:

Agar	15.0g
Yeast extract	10.0g
ACES buffer (2-[(2-Amino-2-oxoethyl)-amino]-ethane sulfonic acid)	10.0g
Charcoal, activated	2.0g
α-Ketogluatrate	1.0g
$Fe_4(P_2O_7)_3 \cdot 9H_2O$	0.25g
Antibiotic solution	10.0mL
Cysteine·HCl·H_2O solution	4.0mL

pH 6.9 ± 0.2 at 25°C

Cysteine·HCl·H_2O Solution:
Composition per 10mL:

L-Cysteine·HCl·H_2O	1.0g

Preparation of Cysteine·HCl·H_2O Solution: Add L-Cysteine·HCl·H_2O to distilled/deionized water and bring volume to 10.0mL. Mix thoroughly. Filter sterilize.

Antibiotic Solution:
Composition per 10mL:

Glycine	3.0g
Anisomycin	0.08g
Vancomycin	5.0mg
Polymyxin B	100,000U

Preparation of Antibiotic Solution: Add components to distilled/deionized water and bring volume to 10.0mL. Mix thoroughly. Filter sterilize.

Preparation of Medium: Add components, except cysteine solution and antibiotic solution, to distilled/deionized water and bring volume to 1.0L. Mix thoroughly. Adjust medium to pH 6.9 with 1*N* KOH. Heat gently and bring to boil for 1 minute. Autoclave for 15 min at 15 psi pressure–121°C. Cool to 50–55°C. Add 4.0mL of L-cysteine·HCl·H_2O solution and 10.0mL of sterile antibiotic solution. Mix thoroughly. Pour into sterile Petri dishes with constant agitation to keep charcoal in suspension.

Use: For the isolation, cultivation, and maintenance of *Legionella pneumophila* and other *Legionella* species from environmental specimens. Used for the selective recovery of *L. pneumophila* while reducing contaminating microorganisms from potable water samples.

BCYE Selective Agar with GVPC (Buffered Charcoal Yeast Extract Selective Agar with Glycine, Vancomycin, Polymyxin B, and Cycloheximide)

Composition per 1014mL:

Agar	15.0g
Yeast extract	10.0g
ACES buffer (2-[(2-Amino-2-oxoethyl)-amino]-ethane sulfonic acid)	10.0g
Charcoal, activated	2.0g
α-Ketogluatrate	1.0g
$Fe_4(P_2O_7)_3 \cdot 9H_2O$	0.25g
Antibiotic solution	10.0mL
Cysteine·HCl·H_2O solution	4.0mL

pH 6.9 ± 0.2 at 25°C

Source: This medium is available as a premixed powder from Oxoid Unipath.

Cysteine·HCl·H_2O Solution:
Composition per 10mL:

L-Cysteine·HCl·H_2O	1.0g

Preparation of Cysteine·HCl·H_2O Solution: Add L-Cysteine·HCl·H_2O to distilled/deionized water and bring volume to 10.0mL. Mix thoroughly. Filter sterilize.

Antibiotic Solution:
Composition per 10mL:

Glycine	3.0g
Cycloheximide	0.08g
Vancomycin	1.0mg
Polymyxin B	79,200U

Preparation of Antibiotic Solution: Add components to distilled/deionized water and bring volume to 10.0mL. Mix thoroughly. Filter sterilize.

Preparation of Medium: Add components, except cysteine solution and antibiotic solution, to distilled/deionized water and bring volume to 1.0L. Mix thoroughly. Adjust medium to pH 6.9 with 1*N* KOH. Heat gently and bring to boil for 1 minute. Autoclave for 15 min at 15 psi pressure–121°C. Cool to

50–55°C. Add 4.0mL of L-cysteine·HCl·H_2O solution and 10.0mL of sterile antibiotic solution. Mix thoroughly. Pour into sterile Petri dishes with constant agitation to keep charcoal in suspension.

Use: For the isolation, cultivation, and maintenance of *Legionella pneumophila* and other *Legionella* species from environmental specimens. Used for the selective recovery of *L. pneumophila* while reducing contaminating microorganisms from potable water samples.

BCYE Selective Agar with PAC (Buffered Charcoal Yeast Extract Selective Agar with Polymyxin B, Anisomycin, and Cefamandole)

Composition per 1014mL:

Agar	15.0g
Yeast extract	10.0g
ACES buffer (2-[(2-Amino-2-oxoethyl)-amino]-ethane sulfonic acid)	10.0g
Charcoal, activated	2.0g
α-Ketogluatrate	1.0g
$Fe_4(P_2O_7)_3 \cdot 9H_2O$	0.25g
Antibiotic solution	10.0mL
Cysteine·HCl·H_2O solution	4.0mL

pH 6.9 ± 0.2 at 25°C

Source: This medium is available as a premixed powder from BBL Microbiology Systems.

Cysteine·HCl·H_2O Solution:
Composition per 10mL:

L-Cysteine·HCl·H_2O	1.0g

Preparation of Cysteine·HCl·H_2O Solution: Add L-Cysteine·HCl·H_2O to distilled/deionized water and bring volume to 10.0mL. Mix thoroughly. Filter sterilize.

Antibiotic Solution:
Composition per 10mL:

Polymyxin B	80,000U
Anisomycin	80.0mg
Cefamandole	2.0mg

Preparation of Antibiotic Solution: Add components to distilled/deionized water and bring volume to 10.0mL. Mix thoroughly. Filter sterilize.

Preparation of Medium: Add components, except cysteine and antibiotic solution, to distilled/deionized water and bring volume to 1.0L. Mix thoroughly. Adjust medium to pH 6.9 with 1*N* KOH. Heat gently and bring to boil for 1 minute. Autoclave for 15 min at 15 psi pressure–121°C. Cool to 50–55°C. Add 4.0mL of L-cysteine·HCl·H_2O solution and 10.0mL of sterile antibiotic solution. Mix thoroughly. Pour into sterile Petri dishes with constant agitation to keep charcoal in suspension.

Use: For the isolation, cultivation, and maintenance of *Legionella pneumophila* and other *Legionella* species from environmental specimens. Used for the selective recovery of *L. pneumophila* while reducing contaminating microorganisms from potable water samples.

BCYE Selective Agar with PAV (Buffered Charcoal Yeast Extract Selective Agar with Polymyxin B, Anisomycin, and Vancomycin) (Wadowsky–Yee Medium)

Composition per 1014mL:

Agar	15.0g
Yeast extract	10.0g
ACES buffer (2-[(2-Amino-2-oxoethyl)-amino]-ethane sulfonic acid)	10.0g
Charcoal, activated	2.0g
α-Ketogluatrate	1.0g
$Fe_4(P_2O_7)_3 \cdot 9H_2O$	0.25g
Antibiotic solution	10.0mL
Cysteine·HCl·H_2O solution	4.0mL

pH 6.9 ± 0.2 at 25°C

Source: This medium is available as a premixed powder from BBL Microbiology Systems.

Cysteine·HCl·H_2O Solution:
Composition per 10mL:

L-Cysteine·HCl·H_2O	1.0g

Preparation of Cysteine·HCl·H_2O Solution: Add L-Cysteine·HCl·H_2O to distilled/deionized water and bring volume to 10.0mL. Mix thoroughly. Filter sterilize.

Antibiotic Solution:
Composition per 10mL:

Polymyxin B	40,000U
Anisomycin	80.0mg
Vancomycin	0.5mg

Preparation of Antibiotic Solution: Add components to distilled/deionized water and bring volume to 10.0mL. Mix thoroughly. Filter sterilize.

Preparation of Medium: Add components, except cysteine and antibiotic solution, to distilled/deionized water and bring volume to 1.0L. Mix thoroughly. Adjust medium to pH 6.9 with 1*N* KOH. Heat gently and bring to boil for 1 minute. Autoclave for 15 min at 15 psi pressure–121°C. Cool to 50–55°C. Add 4.0mL of L-cysteine·HCl·H_2O solution and 10.0mL of sterile antibiotic solution. Mix thoroughly. Pour into sterile Petri dishes with constant agitation to keep charcoal in suspension.

Use: For the isolation, cultivation, and maintenance of *Legionella pneumophila* and other *Legionella* species from environmental specimens. Used for the selective recovery of *Legionella pneumophila* while reducing contaminating microorganisms from potable water samples.

Beef Extract Agar

Composition per liter:

Agar	15.0g
Peptone	5.0g
Beef extract	3.0g

pH 7.4 ± 0.2 at 25°C

Preparation of Medium: Add components to distilled/deionized water and bring volume to 1.0L. Mix thoroughly. Heat gently and bring to boiling. Distribute into tubes or flasks. Autoclave for 15 min at 15 psi pressure–121°C. Pour into Petri dishes or leave in tubes.

Use: For the cultivation and maintenance of a wide variety of microorganisms and recommended for culture of microorganisms from water.

Beef Extract Agar (ATCC Medium 225)

Composition per liter:

Agar	25.0g
Beef extract	10.0g
Peptone	10.0g
NaCl	5.0g

pH 7.2 ± 0.2 at 25°C

Preparation of Medium: Add components to distilled/deionized water and bring volume to 1.0L. Mix thoroughly. Heat gently and bring to boiling. Distribute into tubes or flasks. Autoclave for 15 min at 15 psi pressure–121°C. Pour into Petri dishes or leave in tubes.

Use: For the cultivation and maintenance of a wide variety of microorganisms including *Alcaligenes* species, *Pseudomonas aeruginosa* and *Bacillus sphaericus*.

Beef Extract Broth

Composition per liter:

Peptone	5.0g
Beef extract	3.0g

pH 7.4 ± 0.2 at 25°C

Preparation of Medium: Add components to distilled/deionized water and bring volume to 1.0L. Mix thoroughly. Heat gently and bring to boiling. Distribute into tubes or flasks. Autoclave for 15 min at 15 psi pressure–121°C.

Use: For the cultivation and maintenance of a wide variety of microorganisms and recommended for culture of microorganisms from water.

Beef Extract Broth (ATCC Medium 225)

Composition per liter:

Beef extract	10.0g
Peptone	10.0g
NaCl	5.0g

pH 7.2 ± 0.2 at 25°C

Preparation of Medium: Add components to distilled/deionized water and bring volume to 1.0L. Mix thoroughly. Heat gently and bring to boiling. Distribute into tubes or flasks. Autoclave for 15 min at 15 psi pressure–121°C.

Use: For the cultivation of a wide variety of microorganisms including *Alcaligenes* species, *Pseudomonas aeruginosa* and *Bacillus sphaericus*.

Beef Extract Peptone Serum Medium

Composition per liter:

Agar	25.0g
Beef extract	10.0g
Peptone	10.0g
NaCl	1.0g
Bovine serum	50.0mL

pH 8.5 ± 0.2 at 25°C

Preparation of Medium: Add components, except bovine serum, to distilled/deionized water and bring volume to 950.0mL. Mix thoroughly. Adjust pH to 8.5. Heat gently and bring to boiling. Autoclave for 15 min at 15 psi pressure–121°C. Cool to 50–55°C. Aseptically add 50.0mL of sterile bovine serum. Pour into sterile Petri dishes or leave in tubes.

Use: For the cultivation and maintenance of *Serratia marcescens*.

Beef Extract V

Composition per liter:

Beef extract	24 g

pH 9.0 at 25°C

Source: This medium is available as a premixed powder from BBL Microbiology Systems.

Preparation of Medium: Add component to distilled/deionized water and bring volume to 1.0L. Mix thoroughly. Adjust pH to 9.0 with NaOH. Autoclave for 15 min at 15 psi pressure–118–121°C.

Use: For use in the elution of viruses which have been adsorbed onto filters during filtration of water and wastewater samples.

Beef Extract with NaCl

Composition per liter:

Beef extract	10.0g
NaCl	5.0g

Preparation of Medium: Add components to distilled/deionized water and bring volume to 1.0L. Mix thoroughly. Distribute into tubes or flasks. Autoclave for 15 min at 15 psi pressure–121°C.

Use: For the cultivation of *Bacillus megaterium.*

Beef Infusion Agar

Composition per liter:

Ground defatted beef	453.6g
Agar	20.0g
Peptone	10.0g
NaCl	5.0g

pH 7.6 ± 0.2 at 25°C

Preparation of Medium: Add ground beef to 1.0L of distilled/deionized water. Let stand overnight at 4°C. Gently heat and bring to 80–90°C for 60 min. Let stand for 2 hr. Filter through muslin. To filtrate add peptone and salt. Mix thoroughly. Adjust pH to 7.6 with 4% NaOH. Filter through Whatman #1 filter paper. Bring volume of filtrate to 1.0L. Add agar. Gently heat and bring to boiling. Distribute into tubes or flasks. Autoclave for 15 min at 15 psi pressure–121°C. Pour into sterile Petri dishes or leave in tubes.

Use: For the cultivation of a variety of microorganisms.

Beef Infusion Broth

Composition per liter:

Ground defatted beef	453.6g
Peptone	10.0g
NaCl	5.0g

pH 7.6 ± 0.2 at 25°C

Preparation of Medium: Add ground beef to 1.0L of distilled/deionized water. Let stand overnight at 4°C. Gently heat and bring to 80–90°C for 60 min. Let stand for 2 hr. Filter through muslin. To filtrate add peptone and salt. Mix thoroughly. Adjust pH to 7.6 with 4% NaOH. Filter through Whatman #1 filter paper. Bring volume of filtrate to 1.0L. Add agar. Gently heat and bring to boiling. Distribute into tubes or flasks. Autoclave for 15 min at 15 psi pressure–121°C.

Use: For the cultivation of a variety of microorganisms.

Beef Liver Medium for Anaerobes

Composition per liter:

Beef liver, minced	500.0g
Peptone	10.0g
K_2HPO_4	1.0g

pH 8.0 ± 0.2 at 25°C

Preparation of Medium: Add beef liver to 1.0L of tap water. Soak for 12–24 hr at 4°C. Skim fat off top. Autoclave for 10 min at 15 psi pressure–121°C. Filter through cheesecloth. Save meat. To flitrate, add peptone and K_2HPO_4. Adjust pH to 8.0. Filter through paper. Add tap water and bring volume to 1.0L. Add a small amount of $CaCO_3$ to a flask or test tube. Add 0.5 in. of reserved liver. Cover meat with 2 in. of broth. Cap tubes and autoclave for 15 min at 15 psi pressure–121°C.

Use: For the cultivation and maintenance of a variety of *Clostridium* species.

Beggiatoa and *Thiothrix* Medium

Composition per liter:

$CaSO_4 \cdot 2H_2O$ (saturated solution)	20.0mL
NH_4Cl (4% solution)	5.0mL
Trace elements	5.0mL
K_2HPO_4 (1% solution)	1.0mL
$MgSO_4 \cdot 7H_2O$ (1% solution)	1.0mL

Trace Elements:

Composition per liter:

EDTA solution	20.0mL
$Co(NO_3)_2$ (0.01% solution)	10.0mL
$CuSO_4 \cdot 5H_2O$ (0.00005% solution)	10.0mL
H_3BO_3 (0.1% solution)	10.0mL
$MnSO_4 \cdot 4H_2O$ (0.02% solution)	10.0mL
$Na_2MoO_4 \cdot 2H_2O$ (0.01% solution)	10.0mL
$ZnSO_4 \cdot 7H_2O$ (0.1% solution)	10.0mL

Preparation of Trace Elements:: Add components to distilled/deionized water and bring volume to 1.0L. Mix thoroughly.

EDTA Solution:

Composition per 100mL:

$FeSO_4$	7.0g
EDTA	2.0g
HCl, concentrated	1.0mL

Preparation of EDTA Solution: Add EDTA and $FeSO_4$ to concentrated HCl. Mix thoroughly. Carefully add to distilled/deionized water and bring volume to 100.0mL.

Preparation of Medium: Add components to distilled/deionized water and bring volume to 1.0L. Mix thoroughly. Distribute into tubes or flasks. Autoclave for 15 min at 15 psi pressure–121°C.

Use: For the cultivation of *Beggiatoa* species and myxotrophic *Thiothrix* species.

Beggiatoa Medium (ATCC Medium 138)

Composition per liter:

Yeast extract	2.0g
Agar	2.0g
Sodium acetate	0.5g
$CaCl_2$	0.1g
Catalase	10,000U

pH 7.2 ± 0.2 at 25°C

Preparation of Medium: Add components, except catalase, to tap water and bring volume to 1.0L. Mix thoroughly. Autoclave for 15 min at 15 psi pressure–121°C. Cool to 45–50°C. Aspetically add 10,000 units of sterile catalase.

Use: For the cultivation and maintenance of *Beggiatoa alba* and *Vitreoscilla* species.

Beggiatoa Medium (ATCC Medium 1193)

Composition per liter:

Sodium sulfide	0.5g
Sodium acetate	0.01g
Yeast extract	0.01g
Nutrient broth	0.01g

pH 7.5 ± 0.2 at 25°C

Preparation of Medium: Add components to distilled/deionized water and bring volume to 1.0L. Mix thoroughly. Autoclave for 15 min at 15 psi pressure–121°C. Distribute into tubes or flasks.

Use: For the cultivation of *Beggiatoa alba.*

Beijerinckia Medium

Composition per liter:

Glucose	20.0g
KH_2PO_4	1.0g
$MgSO_4·7H_2O$	0.5g

pH 5.0 ± 0.2 at 25°C

Preparation of Medium: Add components to distilled/deionized water and bring volume to 1.0L. Mix thoroughly. Distribute into tubes or flasks. Autoclave for 15 min at 15 psi pressure–121°C.

Use: For the cultivation of *Beijerinckia* species.

Beijerinckia Medium

Composition per liter:

Glucose	20.0g
K_2HPO_4	0.8g
$MgSO_4·7H_2O$	0.5g
KH_2PO_4	0.2g
$CaCl_2$	0.05g
$FeCl_3·6H_2O$	0.025g
$Na_2MoO_4·2H_2O$	5.0mg

pH 6.9 ± 0.2 at 25°C

Preparation of Medium: Add components to distilled/deionized water and bring volume to 1.0L. Mix thoroughly. Distribute into tubes or flasks. Autoclave for 15 min at 15 psi pressure–121°C.

Use: For the isolation and cultivation of *Beijerinckia* species.

Beijerinck's *Thiobacillus* Medium

Composition per liter:

Noble agar	20.0g
Na_2HPO_4	0.2g
$MgCl_2$	0.1g
NH_4Cl	0.1g
$Na_2S_2O_3$ solution	100.0mL
$NaHCO_3$ solution	10.0mL

pH 7.0–7.2 at 25°C

$Na_2S_2O_3$ Solution:
Composition per 100mL:

$Na_2S_2O_3$	5.0g

Preparation of $Na_2S_2O_3$ Solution: Add $Na_2S_2O_3$ to distilled/deionized water and bring volume to 100.0mL. Mix thoroughly. Filter sterilize.

$NaHCO_3$ Solution:
Composition per 10mL:

$NaHCO_3$	1.0g

Preparation of $NaHCO_3$ Solution: Add $NaHCO_3$ to distilled/deionized water and bring volume to 10.0mL. Mix thoroughly. Filter sterilize.

Preparation of Medium: Add components, except $Na_2S_2O_3$ solution and $NaHCO_3$ solution, to distilled/deionized water and bring volume to 890.0mL. Mix thoroughly. Autoclave for 15 min at 15 psi pressure–121°C. Aseptically add 100.0mL of sterile

$Na_2S_2O_3$ solution and 10.0mL of sterile $NaHCO_3$ solution. Mix thoroughly. Pour into sterile Petri dishes or leave in tubes.

Use: For the cultivation and maintenance of *Thiobacillus thermophilica.*

Bennett's Medium

Composition per liter:

Agar	15.0g
Glucose	10.0g
Pancreatic digest of casein	2.0g
Yeast extract	1.0g
Beef extract	1.0g

pH 7.0 ± 0.2 at 25°C

Preparation of Medium: Add components to distilled/deionized water and bring volume to 1.0L. Mix thoroughly. Heat gently to boiling. Distribute into tubes or flasks. Autoclave for 15 min at 15 psi pressure–121°C. Pour into sterile Petri dishes or leave in tubes.

Use: For the cultivation and maintenance of a variety of soil microorganisms such as *Streptomyces* species, *Nocardia* species, *Flexibacter* species, *Micromonospora* species and others.

Bennett's Modified Agar Medium

Composition per liter:

Meer agar (washed agar)	20.0g
Dextrin	10.0g
Pancreatic digest of casein	2.0g
Yeast extract	1.0g
Beef extract	1.0g
$CoCl_2 \cdot 6H_2O$	0.01g

pH 7.0 ± 0.2 at 25°C

Preparation of Medium: Add components to distilled/deionized water and bring volume to 1.0L. Mix thoroughly. Heat gently to boiling. Distribute into tubes or flasks. Autoclave for 15 min at 15 psi pressure–121°C. Pour into sterile Petri dishes or leave in tubes.

Use: For the cultivation and maintenance of *Streptomyces* species.

Benzoate Medium

Composition per liter:

Noble agar	20.0g
NaCl	5.0g
$(NH_4)_2HPO_4$	3.0g
Sodium benzoate	3.0g
KH_2PO_4	1.2g
Yeast extract	0.5g
$MgSO_4 \cdot 7H_2O$	0.2g
Benzoate solution	25.0mL

Benzoate Solution:

Composition 25mL:

Sodium benzoate	3.0g

Preparation of Benzoate Solution: Add sodium benzoate to distilled/deionized water and bring volume to 25.0mL. Mix thoroughly. Filter sterilize.

Preparation of Medium: Add components except benzoate solution to distilled/deionized water and bring volume to 975.0mL. Mix thoroughly. Heat gently to boiling. Autoclave for 15 min at 15 psi pressure–121°C. Cool to 45–50°C. Aseptically add 25.0mL sterile benzoate solution. Mix thoroughly and pour into sterile Petri dishes or leave in tubes.

Use: For the cultivation of *Pseudomonas putida* and other microorganisms which can utilize benzoate as a carbon source.

Benzoate Medium II

Composition per 1.5 liters:

Noble agar	30.0g
$(NH_4)_2HPO_4$	3.0g
NaCl	1.67g
KH_2PO_4	1.2g
Yeast extract	0.5g
$MgSO_4 \cdot 7H_2O$	0.2g
$FeSO_4 \cdot 7H_2O$	0.1g

Benzoate Solution:

Composition 25mL:

Sodium benzoate	1.0g

Preparation of Benzoate Solution: Add sodium benzoate to distilled/deionized water and bring volume to 25.0mL. Mix thoroughly. Filter sterilize.

Preparation of Medium: Add components, except agar and sodium benzoate, to distilled/deionized water and bring volume to 600.0mL. Mix thoroughly. Autoclave for 15 min at 15 psi pressure–121°C. Cool to 45–50°C. In a separate flask add agar to distilled/deionized water and bring volume to 375.0mL. Mix thoroughly. Gently heat and bring to boiling. Autoclave for 15 min at 15 psi pressure–121°C. Cool to 45–50°C. Aseptically combine the two autoclave sterilized solutions. Mix thoroughly. Aseptically add the sterile benzoate solution. Mix thoroughly. Pour into sterile Petri dishes or leave in tubes.

Use: For the cultivation of *Pseudomonas putida* and other microorganisms which can utilize benzoate as a carbon source.

Benzoate Minimal Salts Medium

Composition per liter:

K_2HPO_4	10.0g
$NaNH_4HPO_4 \cdot 4H_2O$	3.5g
$MgSO_4 \cdot 7H_2O$	0.2g
Citric acid (anhydrous)	0.2g
Benzoate solution	25.0mL

pH 7.0 ± 0.2 at 25°C

Benzoate Solution:

Composition 25mL:

Sodium benzoate	2.5g

Preparation of Benzoate Solution: Add sodium benzoate to distilled/deionized water and bring volume to 25.0mL. Mix thoroughly. Filter sterilize.

Preparation of Medium: Add components to distilled/deionized water and bring volume to 950.0mL. Mix thoroughly. Adjust pH to 7.0. Autoclave for 15 min at 15 psi pressure–121°C. Cool to 45°C. Aseptically add 25.0mL sterile benzoate solution. Mix thoroughly. Aseptically distribute into sterile tubes or flasks.

Use: For the cultivation of microorganisms which can utilize benzoate as a carbon source.

BG 11 Agar (Medium BG 11 for Cyanobacteria)

Composition per liter:

Agar	10.0g
$NaNO_3$	1.5g
$MgSO_4 \cdot 7H_2O$	0.075g
K_2HPO_4	0.04g
$CaCl_2 \cdot 2H_2O$	0.036g
Na_2CO_3	0.02g
Citric acid	6.0mg
Ferric ammonium citrate	6.0mg
Disodium EDTA	1.0mg
Trace metal mix A5	1.0mL

pH 7.1 ± 0.2 at 25°C

Trace Metal Mix A5:

Composition per liter:

H_3BO_3	2.86g
$MnCl_2 \cdot 4H_2O$	1.81g
$Na_2MoO_4 \cdot 2H_2O$	0.39g
$ZnSO_4 \cdot 7H_2O$	0.222g
$CuSO_4 \cdot 5H_2O$	0.079g
$Co(NO_3)_2 \cdot 6H_2O$	0.049g

Preparation of Trace Metal Mix A5: Add components to distilled/deionized water and bring volume to 1.0L. Mix thoroughly.

Preparation of Medium: Add components to distilled/deionized water and bring volume to 1.0L. Mix thoroughly. Heat gently to boiling. Distribute into tubes or flasks. Autoclave for 15 min at 15 psi pressure–121°C. For solid medium, pour into sterile Petri dishes or leave in tubes.

Use: For the cultivation and maintenance of a variety of cyanobacteria including *Anabaena* species, *Calothrix* species, *Chaemisiphon* species, *Chorogloeopsis* species, *Chroococcidiopsis* species, *Cylindrospermum* species, *Dermocarpa* species, *Fischerella* species, *Gloebacter* species, *Gloeocapsa* species, *Gloeothece* species, *Nostoc* species, *Oscillatoria* species, *Phormidium* species, *Pleurocapsa* species *Pseudanabaena* species, *Scytonema* species, *Spirulina* species, *Synechococcus* species, *Synechocystis* species, and others.

BG 11 Marine Agar (Medium BG 11 for Marine Cyanobacteria)

Composition per liter:

Agar	10.0g
NaCl	10.0g
$NaNO_3$	1.5g
$MgSO_4 \cdot 7H_2O$	0.075g
K_2HPO_4	0.04g
$CaCl_2 \cdot 2H_2O$	0.036g
Na_2CO_3	0.02g
Citric acid	6.0mg
Ferric ammonium citrate	6.0mg
EDTA disodium salt	1.0mg
Vitamin B_{12} solution	100.0mL
Trace metal mix A5	1.0mL

pH 7.1 ± 0.2 at 25°C

Trace Metal Mix A5:

Composition per liter:

H_3BO_3	2.86g
$MnCl_2 \cdot 4H_2O$	1.81g
$Na_2MoO_4 \cdot 2H_2O$	0.39g
$ZnSO_4 \cdot 7H_2O$	0.222g
$CuSO_4 \cdot 5H_2O$	0.079g
$Co(NO_3)_2 \cdot 6H_2O$	0.049g

Preparation of Trace Metal Mix A5: Add components to distilled/deionized water and bring volume to 1.0L. Mix thoroughly.

Vitamin B_{12} Solution:

Composition per 100mL:

Vitamin B_{12}	1.0μg

Preparation of Vitamin B_{12} Solution: Add vitamin B_{12} to distilled/deionized water and bring volume to 100.0mL. Mix thoroughly. Filter sterilize.

Preparation of Medium: Add components, except vitamin B_{12} solution, to distilled/deionized water and bring volume to 900.0mL. Mix thoroughly. Heat gently to boiling. Autoclave for 15 min at 15 psi pressure–121°C. Aseptically add 100.0mL of sterile vitamin B_{12} solution. Mix thoroughly. Pour into sterile Petri dishes or leave in tubes.

Use: For the cultivation and maintenance of *Synechococcus* species. Also, used for the isolation of cyanobacteria from freshwater habitats.

BG 11 Marine Broth (Medium BG 11 for Marine Cyanobacteria)

Composition per liter:

NaCl	10.0g
$NaNO_3$	1.5g
$MgSO_4 \cdot 7H_2O$	0.075g
K_2HPO_4	0.04g
$CaCl_2 \cdot 2H_2O$	0.036g
Na_2CO_3	0.02g
Citric acid	6.0mg
Ferric ammonium citrate	6.0mg
EDTA disodium salt	1.0mg
Vitamin B_{12} solution	100.0mL
Trace metal mix A5	1.0mL

pH 7.1 ± 0.2 at 25°C

Trace Metal Mix A5:

Composition per liter:

H_3BO_3	2.86g
$MnCl_2 \cdot 4H_2O$	1.81g
$Na_2MoO_4 \cdot 2H_2O$	0.39g
$ZnSO_4 \cdot 7H_2O$	0.222g
$CuSO_4 \cdot 5H_2O$	0.079g
$Co(NO_3)_2 \cdot 6H_2O$	0.049g

Preparation of Trace metal mix A5: Add components to distilled/deionized water and bring volume to 1.0L. Mix thoroughly.

Vitamin B_{12} Solution:

Composition per 100mL:

Vitamin B_{12}	1.0μg

Preparation of Vitamin B_{12} Solution: Add vitamin B_{12} to distilled/deionized water and bring volume to 100.0mL. Mix thoroughly. Filter sterilize.

Preparation of Medium: Add components, except vitamin B_{12} solution, to distilled/deionized water and bring volume to 900.0mL. Mix thoroughly. Heat gently to boiling. Autoclave for 15 min at 15 psi pressure–121°C. Aseptically add 100.0mL of sterile vitamin B_{12} solution. Mix thoroughly. Distribute into sterile tubes or flasks.

Use: For the cultivation and maintenance of *Synechococcus* species. Also, used for the isolation of cyanobacteria from freshwater habitats.

BG 11 Medium (Medium BG 11 for Cyanobacteria)

Composition per liter:

Agar	10.0g
$NaNO_3$	1.5g
$MgSO_4 \cdot 7H_2O$	0.075g
K_2HPO_4	0.04g
$CaCl_2 \cdot 2H_2O$	0.036g
Na_2CO_3	0.02g
Citric acid	6.0mg
Ferric ammonium citrate	6.0mg
EDTA disodium salt	1.0mg
Trace metal mix A5	1.0mL

pH 7.1 ± 0.2 at 25°C

Trace Metal Mix A5:

Composition per liter:

H_3BO_3	2.86g
$MnCl_2 \cdot 4H_2O$	1.81g
$Na_2MoO_4 \cdot 2H_2O$	0.39g
$ZnSO_4 \cdot 7H_2O$	0.222g
$CuSO_4 \cdot 5H_2O$	0.079g
$Co(NO_3)_2 \cdot 6H_2O$	0.049g

Preparation of Trace Metal Mix A5: Add components to distilled/deionized water and bring volume to 1.0L. Mix thoroughly.

Preparation of Medium: Add components to distilled/deionized water and bring volume to 1.0L. Mix thoroughly. Gently heat and bring to boiling. Distribute into tubes or flasks. Autoclave for 15 min at 15 psi pressure–121°C. Pour into sterile Petri dishes or leave in tubes.

Use: For the cultivation and maintenance of *Anabaena* species, *Calothrix* species, *Chaemisiphon* species, *Chorogloeopsis* species, *Chroococcidiopsis* species, *Crinalium epipsammum*, *Cylindrospermum* species, *Dermocarpa* species, *Fischerella* species, *Gloebacter violaceus*, *Gloeocapsa* species, *Gloeothece* species, *Hapalosiphon fontinalis*, *Nostoc* species, *Oscillatoria* species, *Phormidium* species, *Pleurocapsa* species, *Pseudanabaena* species, *Scytonema* species, *Spirulina* species, *Synechococcus* species, *Synechocystis* species, and *Tolypothrix tenuis*.

BG 11 Uracil Agar

Composition per liter:

Agar 10.0g
Uracil....................2.8g
$NaNO_3$....................1.5g
$MgSO_4 \cdot 7H_2O$ 0.075g
K_2HPO_4....................0.04g
$CaCl_2 \cdot 2H_2O$....................0.036g
Na_2CO_3 ....0.02g
Citric acid....................6.0mg
Ferric ammonium citrate....................6.0mg
EDTA disodium salt....................1.0mg
Trace metal mix A5 ....1.0mL

pH 7.1 ± 0.2 at 25°C

Trace Metal Mix A5:
Composition per liter:

H_3BO_3 ....2.86g
$MnCl_2 \cdot 4H_2O$....................1.81g
$Na_2MoO_4 \cdot 2H_2O$ 0.39g
$ZnSO_4 \cdot 7H_2O$ 0.222g
$CuSO_4 \cdot 5H_2O$....................0.079g
$Co(NO_3)_2 \cdot 6H_2O$ 0.049g

Preparation of Trace Metal Mix A5: Add components to distilled/deionized water and bring volume to 1.0L. Mix thoroughly.

Preparation of Medium: Add components to distilled/deionized water and bring volume to 1.0L. Mix thoroughly. Heat gently to boiling. Distribute into tubes or flasks. Autoclave for 15 min at 15 psi pressure–121°C. Pour into sterile Petri dishes.

Use: For the cultivation and maintenance of *Anabaena variabilis*.

BG 11 Uracil Broth

Composition per liter:

Uracil....................2.8g
$NaNO_3$....................1.5g
$MgSO_4 \cdot 7H_2O$ 0.075g
K_2HPO_4....................0.04g
$CaCl_2 \cdot 2H_2O$....................0.036g
Na_2CO_3 ....0.02g
Citric acid....................6.0mg
Ferric ammonium citrate....................6.0mg
EDTA disodium salt....................1.0mg
Trace metal mix A5 ....1.0mL

pH 7.1 ± 0.2 at 25°C

Trace Metal Mix A5:
Composition per liter:

H_3BO_3 ....2.86g
$MnCl_2 \cdot 4H_2O$....................1.81g
$Na_2MoO_4 \cdot 2H_2O$ 0.39g
$ZnSO_4 \cdot 7H_2O$ 0.222g
$CuSO_4 \cdot 5H_2O$ 0.079g
$Co(NO_3)_2 \cdot 6H_2O$ 0.049g

Preparation of Trace Metal Mix A5: Add components to distilled/deionized water and bring volume to 1.0L. Mix thoroughly.

Preparation of Medium: Add components to distilled/deionized water and bring volume to 1.0L. Mix thoroughly. Heat gently to boiling. Distribute into tubes or flasks. Autoclave for 15 min at 15 psi pressure–121°C.

Use: For the cultivation and maintenance of *Anabena variabilis*.

Bile Esculin Agar

Composition per liter:

Oxgall....................20.0g
Agar....................15.0g
Pancreatic digest of gelatin....................5.0g
Beef extract....................3.0g
Esculin....................1.0g
Ferric citrate....................0.5g
Horse serum 50.0mL

pH 6.8 ± 0.2 at 25°

Source: This medium is available as a premixed powder from Oxoid Unipath and BBL Microbiology Systems.

Preparation of Medium: Add components, except horse serum, to distilled/deionized water and bring volume to 950.0L. Mix thoroughly and heat with frequent agitation until boiling. Autoclave for 15 min at 15 psi pressure–121°C. Cool to 45–50°C. Aseptically add 50.0mL of filter sterilized horse serum. Distribute into sterile Petri dishes or test tubes. Cool tubes in a slanted position.

Use: For differentiation between group D streptococci and non-group D streptococci. Also, to differentiate members of the Enterobacteriaceae, particularly *Klebsiella, Enterobacter* and *Serratia* from other enteric bacteria. Also, to differentiate *Listeria monocytogenes*. Bile tolerance and esculin hydrolysis (seen as a dark brown to black complex) are presumptive for enterococci (group D streptococci).

Birdseed Agar (*Guizotia abyssinica* Creatinine Agar) (Niger seed Agar) (Staib Agar)

Composition per liter:

Agar....................15.0g
Glucose 15.0g

Creatinine	5.0g
KH_2PO_4	3.0g
Biphenyl	1.0g
Chloramphenicol	0.5g
Guizotia abyssinica seed (niger seed) extract	1000.0mL

pH 6.7 ± 0.2 at 25°

Preparation of Medium: Prepare seed extract by grinding 50.0g of *Guizotia abyssinica* seed in 1.0L of distilled/deionized water. Boil for 30 min. Filter through cheesecloth and filter paper. Add remaining components to seed filtrate. Mix thoroughly and heat with frequent agitation until boiling. Distribute into flasks or tubes. Autoclave for 25 min at 15 psi pressure–110°C.

Use: For selective isolation and differentiation between *Cryptococcus neoformans* from other *Cryptococcus* species and other yeasts.

Bismuth Sulfite Agar

Composition per liter:

Agar	20.0g
$Bi_2(SO_3)_3$	8.0g
Pancreatic digest of casein	5.0g
Peptic digest of animal tissue	5.0g
Beef extract	5.0g
Glucose	5.0g
Na_2HPO_4	4.0g
$FeSO_4 \cdot 7H_2O$	0.3g

pH 7.5 ± 0.2 at 25°C

Source: This medium is available as a premixed powder from Difco Laboratories, Oxoid Unipath and BBL Microbiology Systems.

Preparation of Medium: Add components to distilled/deionized water and bring volume to 1.0L. Mix thoroughly and heat with frequent agitation until boiling. Boil for 1 min. Do not autoclave. Cool to 45–50°C. Pour into sterile Petri dishes while gently shaking flask to disperse precipitate. Use plates the same day as prepared.

Use: For the selective isolation and identification of *Salmonella typhi* and other enteric bacilli. *Salmonella typhi* appear as flat, black, "rabbit-eye" colonies surrounded by a zone of black with a metallic sheen.

Bismuth Sulfite Broth (m-Bismuth Sulfite Broth)

Composition per liter:

$Bi_2(SO_3)_3$	16.0g
Pancreatic digest of casein	10.0g
Peptic digest of animal tissue	10.0g
Beef extract	10.0g
Glucose	10.0g
Na_2HPO_4	8.0g
$FeSO_4 \cdot 7H_2O$	0.6g

pH 7.7 ± 0.2 at 25°C

Preparation of Medium: Add components to distilled/deionized water and bring volume to 1.0L. Mix thoroughly and heat with frequent agitation until boiling. Boil for 1 min. Do not autoclave. Cool to 45–50°C. Mix to disperse the precipitate and aseptically distribute into sterile tubes or flasks. Use 2.0–2.2mL of medium for each membrane filter.

Use: For the selective isolation of *Salmonella typhi* and other enteric bacilli and for the detection of *Salmonella* by the membrane filter method.

Blue–Green Agar

Composition per liter:

Agar (if needed)	10.0g
$NaNO_3$	1.5g
$MgSO_4 \cdot 7H_2O$	0.075g
K_2HPO_4	0.04g
$CaCl_2 \cdot 2H_2O$	0.036g
Na_2CO_3	0.02g
Citric acid	6.0mg
Ferric ammonium citrate	6.0mg
EDTA disodium salt	1.0mg
Vitamin B_{12} solution	50.0mL
Trace metal mix A5	1.0mL

pH 7.1 ± 0.2 at 25°C

Trace Metal Mix A5:

Composition per liter:

H_3BO_3	2.86g
$MnCl_2 \cdot 4H_2O$	1.81g
$Na_2MoO_4 \cdot 2H_2O$	0.39g
$ZnSO_4 \cdot 7H_2O$	0.222g
$CuSO_4 \cdot 5H_2O$	0.079g
$Co(NO_3)_2 \cdot 6H_2O$	0.049g

Preparation of Trace Metal Mix A5: Add components to distilled/deionized water and bring volume to 1.0L. Mix thoroughly.

Vitamin B_{12} Solution:

Composition per 50mL:

Vitamin B_{12}	0.01g

Preparation of Vitamin B_{12} Solution: Add vitamin B_{12} to distilled/deionized water and bring volume to 50.0mL. Mix thoroughly. Filter sterilize.

Preparation of Medium: Add components, except vitamin B_{12} solution, to glass distilled water and bring volume to 950.0mL. Mix thoroughly. Heat gently and bring to boiling. Autoclave for 15 min at

15 psi–121°C. Cool the basal medium to 45–50°C. Add vitamin B_{12} solution. Mix thoroughly. Pour into sterile Petri dishes or distribute into sterile tubes.

Use: For the cultivation and maintenance of *Synechococcus* species.

Blue–Green Broth

Composition per liter:

$NaNO_3$	1.5g
$MgSO_4 \cdot 7H_2O$	0.075g
K_2HPO_4	0.04g
$CaCl_2 \cdot 2H_2O$	0.036g
Na_2CO_3	0.02g
Citric acid	6.0mg
Ferric ammonium citrate	6.0mg
EDTA disodium salt	1.0mg
Vitamin B_{12} solution	50.0mL
Trace metal mix A5	1.0mL

pH 7.1 ± 0.2 at 25°C

Trace Metal Mix A5:
Composition per liter:

H_3BO_3	2.86g
$MnCl_2 \cdot 4H_2O$	1.81g
$Na_2MoO_4 \cdot 2H_2O$	0.39g
$ZnSO_4 \cdot 7H_2O$	0.222g
$CuSO_4 \cdot 5H_2O$	0.079g
$Co(NO_3)_2 \cdot 6H_2O$	0.049g

Preparation of Trace Metal Mix A5: Add components to distilled/deionized water and bring volume to 1.0L. Mix thoroughly.

Vitamin B_{12} Solution:
Composition per 50mL:

Vitamin B_{12}	0.01g

Preparation of Vitamin B_{12} Solution: Add vitamin B_{12} to distilled/deionized water and bring volume to 50.0mL. Mix thoroughly. Filter sterilize.

Preparation of Medium: Add components, except vitamin B_{12}, to glass distilled water and bring volume to 950.0mL. Mix thoroughly. Heat gently and bring to boiling. Autoclave for 15 min at 15 psi–121°C. Cool the basal medium to 45–50°C. Add vitamin B_{12} solution. Mix thoroughly. Distribute into sterile tubes or flasks.

Use: For the cultivation and maintenance of *Synechococcus* species.

Blue–Green Nitrogen-Fixing Agar

Composition per liter:

Noble agar	10.0g
$MgSO_4 \cdot 7H_2O$	0.075g
K_2HPO_4	0.04g
$CaCl_2 \cdot 2H_2O$	0.036g
Na_2CO_3	0.02g
Citric acid	6.0mg
Ferric ammonium citrate	6.0mg
EDTA disodium salt	1.0mg
Trace metal mix A5	1.0mL

pH 7.1 ± 0.2 at 25°C

Trace Metal Mix A5:
Composition per liter:

H_3BO_3	2.86g
$MnCl_2 \cdot 4H_2O$	1.81g
$Na_2MoO_4 \cdot 2H_2O$	0.39g
$ZnSO_4 \cdot 7H_2O$	0.222g
$CuSO_4 \cdot 5H_2O$	0.079g
$Co(NO_3)_2 \cdot 6H_2O$	0.049g

Preparation of Trace Metal Mix A5: Add components to distilled/deionized water and bring volume to 1.0L. Mix thoroughly.

Preparation of Medium: Add components to glass distilled water and bring volume to 1.0L. Mix thoroughly. Heat gently and bring to boiling. Autoclave for 15 min at 15 psi–121°C. Check pH after autoclaving and readjust if necessary. Pour into sterile Petri dishes or distribute into sterile tubes.

Use: For the cultivation and maintenance of *Calothrix, Fischerella* and *Nostoc* species.

Blue–Green Nitrogen-Fixing Broth

Composition per liter:

$MgSO_4 \cdot 7H_2O$	0.075g
K_2HPO_4	0.04g
$CaCl_2 \cdot 2H_2O$	0.036g
Na_2CO_3	0.02g
Citric acid	6.0mg
Ferric ammonium citrate	6.0mg
EDTA disodium salt	1.0mg
Trace metal mix A5	1.0mL

pH 7.1 ± 0.2 at 25°C

Trace Metal Mix A5:
Composition per liter:

H_3BO_3	2.86g
$MnCl_2 \cdot 4H_2O$	1.81g
$Na_2MoO_4 \cdot 2H_2O$	0.39g
$ZnSO_4 \cdot 7H_2O$	0.222g
$CuSO_4 \cdot 5H_2O$	0.079g
$Co(NO_3)_2 \cdot 6H_2O$	0.049g

Preparation of Trace Metal Mix A5: Add components to distilled/deionized water and bring volume to 1.0L. Mix thoroughly.

Preparation of Medium: Add components to glass distilled water and bring volume to 1.0L. Mix thoroughly. Heat gently and bring to boiling. Autoclave for 15 min at 15 psi–121°C. Check pH after autoclaving and readjust if necessary. Aseptically distribute into sterile tubes or flasks.

Use: For the cultivation and maintenance of *Calothrix*, *Fischerella* and *Nostoc* species.

BMPA–α Medium (Edelstein BMPA–α Medium)

Composition per liter:

Agar	13.0g
Yeast extract	10.0g
ACES buffer (2-[(2-Amino-2-oxoethyl)-amino]-ethane sulfonic acid)	2.0g
Charcoal, activated	2.0g
α-Ketogluatrate	0.2g
$Fe_4(P_2O_7)_3{\cdot}9H_2O$	0.05g
Antibiotic inhibitor	10.0mL
L-Cysteine·HCl·H_2O solution	10.0mL

pH 6.9 ± 0.2 at 25°C

Source: This medium is available as premixed vials from Oxoid Unipath.

Antibiotic Inhibitor:
Composition per 10mL:

Anisomycin	0.08g
Cefamandole	4.0mg
Polymyxin B	80,000U

Preparation of Antibiotic Inhibitor: Add components to distilled/deionized water and bring volume to 10.0mL. Mix thoroughly. Filter sterilize.

L-Cysteine·HCl·H_2O Solution:
Composition per 10mL:

L-Cysteine·HCl·H_2O	0.08g

Preparation of L-Cysteine·HCl·H_2O Solution: Add L-Cysteine·HCl·H_2O to distilled/deionized water and bring volume to 10.0mL. Mix thoroughly. Filter sterilize.

Preparation of Medium: Add components, except cysteine and antibiotic inhibitor, to distilled/deionized water and bring volume to 980.0mL. Mix thoroughly. Adjust medium to pH 6.9 with 1*N* KOH. Heat gently and bring to boiling for 1 min. Autoclave for 15 min at 15 psi pressure–121°C. Cool to 50–55°C. Add 10.0mL of the sterile L-cysteine·HCl·H_2O solution and 10.0mL of the sterile antibiotic solution. Mix thoroughly. Pour into sterile Petri dishes with constant agitation to keep charcoal in suspension.

Use: For the selective isolation and cultivation of *Legionella pneumophila* and other *Legionella* species.

BMPA–α Medium (Semiselective Medium for *Legionella pneumophila*)

Composition per liter:

Agar	15.0g
Yeast extract	10.0g
ACES buffer (2-[(2-Amino-2-oxoethyl)-amino]-ethane sulfonic acid)	10.0g
Charcoal, activated	2.0g
α-Ketogluatrate	1.0g
$Fe_4(P_2O_7)_3{\cdot}9H_2O$	0.25g
Antibiotic inhibitor	10.0mL
L-Cysteine·HCl·H_2O solution	10.0mL

pH 6.9 ± 0.2 at 25°C

Antibiotic Inhibitor:
Composition per 10mL:

Anisomycin	0.08g
Cefamandole	4.0mg
Polymyxin B	80,000U

Preparation of Antibiotic Inhibitor: Add components to distilled/deionized water and bring volume to 10.0mL. Mix thoroughly. Filter sterilize.

L-Cysteine·HCl·H_2O Solution:
Composition per 10mL:

L-Cysteine·HCl·H_2O	0.4g

Preparation of L-Cysteine·HCl·H_2O Solution: Add L-Cysteine·HCl·H_2O to distilled/deionized water and bring volume to 10.0mL. Mix thoroughly. Filter sterilize.

Preparation of Medium: Add components, except cysteine and antibiotic inhibitor, to distilled/deionized water and bring volume to 980.0mL. Mix thoroughly. Adjust medium to pH 6.9 with 1*N* KOH. Heat gently and bring to boiling for 1 min. Autoclave for 15 min at 15 psi pressure–121°C. Cool to 50–55°C. Add 10.0mL of the sterile L-cysteine·HCl·H_2O solution and 10.0mL of the sterile antibiotic solution. Mix thoroughly. Pour into sterile Petri dishes with constant agitation to keep charcoal in suspension.

Use: For the selective isolation and cultivation of *Legionella pneumophila* and other *Legionella* species.

Bonner–Addicott Medium

Composition per liter:

Agar	25.0g
Glucose	20.0g
$Ca(NO_3)_2{\cdot}4H_2O$	0.236g
KNO_3	0.081g
KCl	0.065g
$MgSO_4{\cdot}7H_2O$	0.036g

KH_2PO_4 .... 0.012g
Ferric tartrate 1.0mg

Preparation of Medium: Add components to distilled/deionized water and bring volume to 1.0L. Mix thoroughly. Gently heat and bring to boiling. Distribute into tubes or flasks. Autoclave for 15 min at 15 psi pressure–121°C. Pour into sterile Petri dishes or leave in tubes.

Use: For the cultivation of a variety of fungi.

Bovine Albumin Tween™ 80 Medium, Ellinghausen and McCullough, Modified (Albumin Fatty Acid Broth, *Leptospira* Medium)

Composition per liter:

Basal medium 900.0mL
Albumin fatty acid supplement 100.0mL

Basal Medium:

Composition per liter:

Na_2HPO_4, anhydrous 1.0g
NaCl 1.0g
KH_2PO_4, anhydrous 0.3g
NH_4Cl (25% solution) 1.0mL
Glycerol (10% solution) 1.0mL
Sodium pyruvate (10% solution) 1.0mL
Thiamine·HCl (0.5% solution) 1.0mL
pH 7.4 ± 0.2 at 25°C

Preparation of Basal Medium: Add components to distilled/deionized water and bring volume to 1.0L. Mix thoroughly. Adjust pH to 7.4. Gently heat and bring to boiling. Autoclave for 15 min at 15 psi pressure–121°C. Cool to 25°C.

Albumin Fatty Acid Supplement:

Composition per 200mL:

Bovine albumin fraction V 20.0g
Polysorbate (Tween™) 80 (10% solution) 25.0mL
$FeSO_4 \cdot 7H_2O$ (0.5% solution) 20.0mL
$CaCl_2 \cdot 2H_2O$ (1.5% solution) 2.0mL
$MgCl_2 \cdot 2H_2O$ (1.5% solution) 2.0mL
Vitamin B_{12} (0.2% solution) 2.0mL
$ZnSO_4 \cdot 7H_2O$ (0.4% solution) 2.0mL
$CuSO_4 \cdot 5H_2O$ (0.3% solution) 0.2mL

Preparation of Albumin Fatty Acid Supplement: Add bovine albumin to 100.0mL of distilled/deionized water. Mix thoroughly. Add remaining components while stirring. Adjust pH to 7.4. Bring volume to 200.0mL with distilled/deionized water. Filter sterilize. Store at −20°C.

Preparation of Medium: Aseptically combine 100.0mL of sterile albumin fatty acid supplement and 900.0mL of sterile basal medium. Mix thoroughly. Aseptically distribute into sterile tubes or flasks.

Use: For the cultivation of *Leptospira* species.

Bovine Albumin Tween™ 80 Semisolid Medium, Ellinghausen and McCullough, Modified (Albumin Fatty Acid Semisolid Medium, Modified)

Composition per liter:

Basal medium 900.0mL
Albumin fatty acid supplement 100.0mL

Basal Medium:

Composition per liter:

Agar 2.2g
Na_2HPO_4, anhydrous 1.0g
NaCl 1.0g
KH_2PO_4, anhydrous 0.3g
NH_4Cl (25% solution) 1.0mL
Glycerol (10% solution) 1.0mL
Sodium pyruvate (10% solution) 1.0mL
Thiamine·HCl (0.5% solution) 1.0mL
pH 7.4 ± 0.2 at 25°C

Preparation of Basal Medium: Add components to distilled/deionized water and bring volume to 1.0L. Mix thoroughly. Adjust pH to 7.4. Gently heat and bring to boiling. Autoclave for 15 min at 15 psi pressure–121°C. Cool to 25°C.

Albumin Fatty Acid Supplement:

Composition per 200mL:

Bovine albumin fraction V 20.0g
Polysorbate (Tween™) 80 (10% solution) 25.0mL
$FeSO_4 \cdot 7H_2O$ (0.5% solution) 20.0mL
$CaCl_2 \cdot 2H_2O$ (1.5% solution) 2.0mL
$MgCl_2 \cdot 2H_2O$ (1.5% solution) 2.0mL
Vitamin B_{12} (0.2% solution) 2.0mL
$ZnSO_4 \cdot 7H_2O$ (0.4% solution) 2.0mL
$CuSO_4 \cdot 5H_2O$ (0.3% solution) 0.2mL

Preparation of Albumin Fatty Acid Supplement: Add bovine albumin to 100.0mL of distilled/deionized water. Mix thoroughly. Add remaining components while stirring. Adjust pH to 7.4. Bring volume to 200.0mL with distilled/deionized water. Filter sterilize. Store at −20°C.

Preparation of Medium: Aseptically combine 100.0mL of sterile albumin fatty acid supplement and 900.0mL of sterile basal medium. Mix thoroughly. Aseptically distribute into sterile tubes or flasks.

Use: For the cultivation of *Leptospira* species.

Bovine Serum Albumin Tween™ 80 Agar (BSA Tween™ 80 Agar)

Composition per liter:

Basal medium	900.0mL
Albumin supplement	100.0mL

Basal Medium:

Composition per liter:

Agar	11.0g
Na_2HPO_4	1.0g
NaCl	1.0g
KH_2PO_4	0.3g
Glycerol (10% solution)	1.0mL
NH_4Cl (25% solution)	1.0mL
Sodium pyruvate (10% solution)	1.0mL
Thiamine (0.5% solution	1.0mL

Preparation of Basal Medium: Add components to distilled/deionized water and bring volume to 1.0L. Mix thoroughly. Adjust pH to 7.4. Autoclave for 15 min at 15 psi pressure–121°C. Cool to 25°C.

Albumin Supplement:

Composition per 100mL:

Bovine albumin	10.0g
Tween™ 80 (10% solution)	12.5mL
$FeSO_4$ (0.5% solution)	10.0mL
$MgCl_2$-$CaCl_2$ solution	1.0mL
Cyanocobalamin (0.02% solution)	1.0mL
$ZnSO_4$ (0.4% solution)	1.0mL

Preparation of Albumin Supplement: Add components to distilled/deionized water and bring volume to 100.0mL. Mix thoroughly. Adjust pH to 7.4. Filter sterilize.

$MgCl_2$–$CaCl_2$ Solution:

Composition per 100mL:

$CaCl_2 \cdot 2H_2O$	1.5g
$MgCl_2 \cdot 6H_2O$	1.5g

Preparation of $MgCl_2$–$CaCl_2$ Solution: Add components to distilled/deionized water and bring volume to 100.0mL. Mix thoroughly.

Preparation of Medium: To 900.0mL of cooled, sterile basal medium, aseptically add 100.0mL of sterile albumin supplement. Mix thoroughly. Aseptically distribute into sterile tubes or flasks.

Use: For the cultivation and maintenance of *Leptospira* species.

Bovine Serum Albumin Tween™ 80 Broth (BSA Tween™ 80 Broth)

Composition per liter:

Basal medium	900.0mL
Albumin supplement	100.0mL

pH 7.4 ± 0.2 at 25°C

Basal Medium:

Composition per liter:

Na_2HPO_4	1.0g
NaCl	1.0g
KH_2PO_4	0.3g
Glycerol (10% solution)	1.0mL
NH_4Cl (25% solution)	1.0mL
Sodium pyruvate (10% solution)	1.0mL
Thiamine (0.5% solution	1.0mL

Preparation of Basal Medium: Add components to distilled/deionized water and bring volume to 1.0L. Mix thoroughly. Adjust pH to 7.4. Autoclave for 15 min at 15 psi pressure–121°C. Cool to 25°C.

Albumin Supplement:

Composition per 100mL:

Bovine albumin	10.0g
Tween™ 80 (10% solution)	12.5mL
$FeSO_4$ (0.5% solution)	10.0mL
$MgCl_2$-$CaCl_2$ solution	1.0mL
Cyanocobalamin (0.02% solution)	1.0mL
$ZnSO_4$ (0.4% solution)	1.0mL

Preparation of Albumin Supplement: Add components to distilled/deionized water and bring volume to 100.0mL. Mix thoroughly. Adjust pH to 7.4. Filter sterilize.

$MgCl_2$–$CaCl_2$ Solution:

Composition per 100mL:

$CaCl_2 \cdot 2H_2O$	1.5g
$MgCl_2 \cdot 6H_2O$	1.5g

Preparation of $MgCl_2$–$CaCl_2$ Solution: Add components to distilled/deionized water and bring volume to 100.0mL. Mix thoroughly.

Preparation of Medium: To 900.0mL of cooled, sterile basal medium, aseptically add 100.0mL of sterile albumin supplement. Mix thoroughly. Aseptically distribute into sterile tubes or flasks.

Use: For the isolation and cultivation of *Leptospira* species.

Bovine Serum Albumin Tween™ 80 Soft Agar (BSA Tween™ 80 Soft Agar) (Semisolid BSA Tween™ 80 Medium)

Composition per liter:

Basal medium	900.0mL
Albumin supplement	100.0mL

Basal Medium:

Composition per liter:

Agar	2.0g
Na_2HPO_4	1.0g
NaCl	1.0g
KH_2PO_4	0.3g
Glycerol (10% solution)	1.0mL
NH_4Cl (25% solution)	1.0mL
Sodium pyruvate (10% solution)	1.0mL
Thiamine (0.5% solution	1.0mL

Preparation of Basal Medium: Add components to distilled/deionized water and bring volume to 1.0L. Mix thoroughly. Adjust pH to 7.4. Autoclave for 15 min at 15 psi pressure–121°C. Cool to 25°C.

Albumin Supplement:

Composition per 100mL:

Bovine albumin	10.0g
Tween™ 80 (10% solution)	12.5mL
$FeSO_4$ (0.5% solution)	10.0mL
$CaCl_2$-$MgCl_2$ solution	1.0mL
Cyanocobalamin (0.02% solution)	1.0mL
$ZnSO_4$ (0.4% solution)	1.0mL

Preparation of Albumin Supplement: Add components to distilled/deionized water and bring volume to 100.0mL. Mix thoroughly. Adjust pH to 7.4. Filter sterilize.

$MgCl_2$–$CaCl_2$ Solution:

Composition per 100mL:

$CaCl_2·2H_2O$	1.5g
$MgCl_2·6H_2O$	1.5g

Preparation of $MgCl_2$–$CaCl_2$ Solution: Add components to distilled/deionized water and bring volume to 100.0mL. Mix thoroughly.

Preparation of Medium: To 900.0mL of cooled, sterile basal medium aseptically add 100.0mL of sterile albumin supplement. Mix thoroughly. Aseptically distribute into sterile tubes or flasks.

Use: For the cultivation of *Leptospira* species.

Brackish Acetate

Composition per liter:

Sodium acetate	1.0g
KNO_3	1.0g
$NaH_2PO_4·2H_2O$	0.05g
Artifical seawater	250.0mL
Modified Hutner's basal salts	20.0mL
Vitamin solution	10.0mL

pH 7.2 ± 0.2 at 25°C

Artificial Seawater:

Composition per liter:

NaCl	23.5g
$MgCl_2$	5.0g
Na_2SO_4	3.9g
$CaCl_2$	1.1g
KCl	0.66g
$NaHCO_3$	0.19g
KBr	0.1g
H_3BO_3	0.026g
$SrCl_2$	0.024g
NaF	3.0mg

Preparation of Artificial Seawater: Add components to distilled/deionized water and bring volume to 100.0mL. Mix thoroughly.

Metals "44":

Composition per 100mL:

$ZnSO_4·7H_2O$	1.1g
$FeSO_4·7H_2O$	0.5g
EDTA	0.25g
$MnSO_4·7H_2O$	0.154g
$CuSO_4·5H_2O$	0.04g
$Co(NO_3)_2·6H_2O$	0.025g
$Na_2B_4O_7·10H_2O$	0.018g

Preparation of Metals "44": Add components to distilled/deionized water and bring volume to 100.0mL. Mix thoroughly. Autoclave for 15 min at 15 psi–121°C. Add aseptically to sterile basal medium.

Modified Hutner's Basal Salts:

Composition per liter:

$MgSO_4.7H_2O$	29.7g
Nitrilotriacetic acid	10.0g
$CaCl_2·2H_2O$	3.34g
$FeSO_4·7H_2O$	0.1g
$(NH_4)_2MoO_4$	9.25mg
Metals "44"	50.0mL

Preparation of Modified Hutner's Basal Salts: Dissolve the nitrilotriacetic acid first and neutralize the solution with KOH. Add other components and adjust the pH to 7.2 with KOH or H_2SO_4. There may be a slight precipitate. Store at 5°C.

Vitamin Solution:

Composition per liter:

Thiamine·HCl	5.0mg
D-Calcium pantothenate	5.0mg
Riboflavin	5.0mg
Biotin	2.0mg
Folic acid	2.0mg
Vitamin B_{12}	0.1mg

Preparation of Vitamin Solution: Add components to distilled/deionized water and bring volume to 1.0L. Mix thoroughly. Filter-sterilize and add aseptically to sterile basal medium.

Preparation of Medium: Add a few drops of H_2SO_4 to the distilled water to retard precipitation of the metal salts. Add components, except for Metals "44" and vitamin solutions, to 250.0mL of artificial seawater and 720.0mL of distilled/deionized water. Adjust pH to 7.2. Distribute into tubes or flasks. Sterilize by autoclaving for 15 min at 15 lb. pressure–121°C. Aseptically add Metals "44" and vitamin solutions. Mix thoroughly. Aseptically distribute into sterile tubes or flasks.

Use: For the cultivation of *Filomicrobium fusiforme*.

Brackish *Prosthecomicrobium* Medium

Composition per liter:

Agar	15.0g
Peptone	0.25g
Yeast extract	0.25g
Glucose	0.25g
Artificial seawater	250.0mL
Modified Hutner's basal salts	20.0mL
Vitamins	10.0mL

pH 7.2 ± 0.2 at 25°C

Artificial Seawater:

Composition per liter:

NaCl	23.477g
$MgCl_2$	4.981g
Na_2SO_4	3.917g
$CaCl_2$	1.102g
KCl	0.664g
$NaHCO_3$	0.192g
KBr	0.096g
H_3BO_3	0.026g
$SrCl_2$	0.024g
NaF	3.0mg

Preparation of Artificial Seawater: Add components to distilled/deionized water and bring volume to 100.0mL. Mix thoroughly.

Modified Hutner's Basal Salts:

Composition per liter:

$MgSO_4.7H_2O$	29.7g
Nitrilotriacetic acid	10.0g
$CaCl_2·2H_2O$	3.34g
$FeSO_4·7H_2O$	0.1g
$(NH_4)_2MoO_4$	9.25mg
Metals "44"	50.0mL

Preparation of Modified Hutner's Basal Salts: Dissolve the nitrilotriacetic acid first and neutralize the solution with KOH. Add other ingredients and readjust the pH with KOH and/or H_2SO_4 to 7.2. There may be a slight precipitate. Store at 5°C.

Metals "44":

Composition per 100mL:

$ZnSO_4·7H_2O$	1.1g
$FeSO_4·7H_2O$	0.5g
EDTA	0.25g
$MnSO_4·7H_2O$	0.154g
$CuSO_4·5H_2O$	0.04g
$Co(NO_3)_2·6H_2O$	0.025g
$Na_2B_4O_7·10H_2O$	0.018g

Preparation of Metals "44": Add components to distilled/deionized water and bring volume to 100.0mL. Mix thoroughly. Autoclave for 15 min at 15 psi–121°C. Add aseptically to sterile basal medium.

Vitamin Solution:

Composition per liter:

Thiamine·HCl	5.0mg
D-Calcium pantothenate	5.0mg
Riboflavin	5.0mg
Biotin	2.0mg
Folic acid	2.0mg
Vitamin B_{12}	0.1mg

Preparation of Vitamin Solution: Add components to distilled/deionized water and bring volume to 1.0L. Mix thoroughly. Filter-sterilize and add aseptically to sterile basal medium.

Preparation of Medium: Add a few drops of H_2SO_4 to the distilled water to retard precipitation of the metal salts. Add components, except for metals "44" and vitamin solutions, to 250.0mL of artificial seawater and 720.0mL of distilled/deionized water. Adjust pH to 7.2. Distribute into tubes or flasks. Sterilize by autoclaving for 15 min at 15 lb. pressure–121°C. Aseptically add metals "44" and vitamin solutions. Mix thoroughly. Aseptically distribute into sterile tubes or flasks.

Use: For the cultivation of *Prosthecomicrobium litoralum*.

Brain Heart Infusion Soil Extract Medium

Composition per liter:

Yeast extract	20.0g
Pancreatic digest of casein	16.0g
Brain/heart, solids from infusion	8.0g
Peptic digest of animal tissue	5.0g
NaCl	5.0g
Glucose	2.0g
Na_2HPO_4	2.5g
Soil extract	250.0mL
Vitamin B_{12} solution	1.0mL

pH 7.2 ± 0.2 at 25°C

Soil Extract:

Composition per 400mL:

African violet soil	1.0g
Na_2CO_3	1.0g

Preparation of Soil Extract: Autoclave for 60 min at 15 psi pressure–121°C. Filter through paper before using in medium.

Vitamin B_{12} Solution:

Composition per 1mL:

Vitamin B_{12}	2.0μg

Preparation of Vitamin B_{12} Solution: Add vitamin B_{12} to distilled/deionized water and bring volume to 1.0mL. Mix thoroughly. Filter sterilize.

Preparation of Medium: Add components, except glucose, yeast extract, and vitamin B_{12} solution, to tap water and bring volume to 799.0mL. Mix thoroughly. Autoclave for 15 min at 15 psi–121°C. Add yeast extract and glucose to 200.0mL of tap water. Filter sterilize and add aseptically to cooled sterile basal medium. Aseptically add 1.0mL of vitamin B_{12} solution. Mix thoroughly. Aseptically distribute into sterile tubes or flasks.

Use: For the cultivation of a wide variety of microorganisms, including bacteria, yeast and molds, especially fastidious species from soil. It is useful for the isolation of *Histoplasma capsulatum* and other pathogenic fungi including *Coccidioides immitis*.

Brain Heart Infusion with Thiamine

Composition per liter:

NaCl	30.0g
Pancreatic digest of gelatin	14.5g
Brain/heart, solids from infusion	6.0g
Peptic digest of animal tissue	6.0g
Glucose	3.0g
Na_2HPO_4	2.5g
Thiamine·HCl	1.0mg

pH 7.4 ± 0.2 at 25°C

Preparation of Medium: Add components to distilled/deionized water and bring volume to 1.0L. Mix thoroughly. Distribute into tubes or flasks. Autoclave for 15 min at 15 psi–121°C.

Use: For the cultivation *Bacillus larvae*.

Brain Heart Infusion with 3% Sodium Chloride

Composition per liter:

NaCl	30.0g
Pancreatic digest of gelatin	14.5g
Brain/heart, solids from infusion	6.0g
Peptic digest of animal tissue	6.0g
Glucose	3.0g
Na_2HPO_4	2.5g

pH 7.4 ± 0.2 at 25°C

Preparation of Medium: Add components to distilled/deionized water and bring volume to 1.0L. Mix thoroughly. Distribute into tubes or flasks. Autoclave for 15 min at 15 psi–121°C.

Use: For the cultivation of *Vibrio parahaemolyticus*.

Brevibacterium Medium (ATCC Medium 159)

Composition per liter:

Agar	25.0g
Glucose	20.0g
$CaCO_3$	20.0g
Yeast extract	10.0g

Preparation of Medium: Add components to distilled/deionized water and bring volume to 1.0L. Mix thoroughly. Gently heat to boiling. Distribute into tubes or flasks. Autoclave for 15 min at 15 psi–121°C. Pour into sterile Petri dishes or leave in tubes.

Use: For the cultivation and maintenance of *Brevibacterium* species.

Brevibacterium Medium (ATCC Medium 677)

Composition per liter:

Agar	30.0g
KH_2PO_4	2.0g
Na_2HPO_4	2.0g
$(NH_4)_2SO_4$	2.0g

Yeast extract 2.0g
Tween™ 60 .... 2.0g
$MgSO_4 \cdot 7H_2O$ 0.2g
$FeSO_4.7H_2O$ 0.1g
$MnSO_4$ 0.01g
n-Hexadecane 50.0mL

pH 7.0 ± 0.2 at 25°C

Preparation of Medium: Add components to distilled/deionized water and bring volume to 1.0L. Mix thoroughly. Blend for 30 minutes in a blender to disperse the *n*-hexadecane. Distribute into tubes or flasks. Autoclave for 20 min at 15 psi–121°C.

Use: For the cultivation and maintenance of *Brevibacterium alkanophilum* and other microorganisms which can utilize hexadecane as a carbon source.

Brevibacterium Medium (ATCC Medium 681)

Composition per liter:

Glucose 10.0g
Peptone 5.0g
Yeast extract 5.0g

pH 5.0–6.0 at 25°C

Preparation of Medium: Add components to distilled/deionized water and bring volume to 1.0L. Mix thoroughly. Distribute into tubes or flasks. Autoclave for 15 min at 15 psi–121°C.

Use: For the cultivation and maintenance of *Brevibacterium* spp. and *Enterobacter cloacae*.

Brewer Anaerobic Agar

Composition per liter:

Agar 20.0g
Proteose peptone No. 3 .... 10.0g
Glucose 10.0g
Pancreatic digest of casein 5.0g
Yeast extract 5.0g
NaCl 5.0g
Sodium thioglycollate 2.0g
Sodium formaldehyde sulfoxylate 1.0g
Resazurin 2.0mg

pH 7.2 ± 0.2 at 25°C

Source: This medium is available as a premixed powder from Difco Laboratories.

Preparation of Medium: Add components to distilled/deionized water and bring volume to 1.0L. Mix thoroughly. Distribute into tubes or flasks. Autoclave for 15 min at 15 psi–121°C.

Use: For the cultivation and maintenance of anaerobic and microaerophilic microorganisms.

Brilliant Green Agar, Modified

Composition per liter:

Agar 12.0g
Lactose 10.0g
Sucrose 10.0g
Beef extract 5.0g
Peptone 5.0g
NaCl 5.0g
NaCl 5.0g
Yeast extract 3.0g
Na_2HPO_4 .... 1.0g
NaH_2PO_4 .... 0.6g
Phenol red 0.09g
Brilliant green 4.7mg

pH 6.9 ± 0.2 at 25°C

Source: This medium is available as a premixed powder from Oxoid Unipath.

Preparation of Medium: Add components to distilled/deionized water and bring volume to1.0L. Mix thoroughly. Gently heat and bring to boiling. Do not autoclave. Cool to 45–50°C. Addition of 1.0g sodium sulfacetamide and 250.0mg sodium mandelate enhances inhibition of contaminating microorganisms. Pour into sterile Petri dishes.

Use: For the selective isolation of *Salmonella* other than *S. typhi* from feces and other specimens, and food and dairy products. *Salmonella* other than *S. typhi* appear as red/pink/white colonies surrounded by a zone of red in the agar indicating nonlactose/sucrose fermentation. *Proteus* or *Pseudomonas* species may appear as small red colonies. Lactose- or sucrose-fermenting bacteria appear as yellow-green colonies surrounded by a zone of yellow-green in the agar.

Brilliant Green Bile Agar

Composition per liter:

Noble agar 10.15g
Pancreatic digest of gelatin 8.25g
Lactose 1.9g
Na_2SO_3 .... 0.205g
$FeCl_3$ 0.0295g
Basic fuchsin 0.078g
Erioglaucine 0.065g
KH_2PO_4 .... 0.015g
Oxgall, dehydrated 2.95mg
Brilliant green 0.03mg

pH 6.9 ± 0.2 at 25°C

Source: This medium is available as a premixed powder from Difco Laboratories and BBL Microbiology Systems.

Caution: Basic fuchsin is a potential carcinogen and care must be taken to avoid inhalation of the powdered dye and contamination of the skin.

Preparation of Medium: Add components to distilled/deionized water and bring volume to1.0L. For plating 10.0mL samples, prepare the medium double strength. Mix thoroughly. Gently heat and bring to boiling. Distribute into tubes or flasks. Autoclave for 15 min at 15 psi pressure–121°C. Pour into sterile Petri dishes. Care should be taken to avoid exposure of the prepared medium to light.

Use: For the detection and enumeration of coliform bacteria in materials of sanitary importance such as water, sewage and foods. *Escherichia coli* appears as dark red colonies with a pink halo. *Enterobacter* species appear as pink colonies.

Brilliant Green Bile Broth (Brilliant Green Lactose Bile Broth)

Composition per liter:

Oxgall, dehydrated	20.0g
Lactose	10.0g
Pancreatic digest of gelatin	10.0g
Brilliant green	0.013g

pH 7.2 ± 0.2 at 25°C

Source: This medium is available as a premixed powder from Difco Laboratories, BBL Microbiology Systems and Oxoid Unipath.

Preparation of Medium: Add components to distilled/deionized water and bring volume to1.0L. Mix thoroughly. Distribute into tubes containing inverted Durham tubes, in 10.0mL amounts for testing 1.0mL or less of sample. Autoclave for 12 min (not longer than 15 min) at 15 psi pressure–121°C. After sterilization, cool the broth rapidly. Medium is sensitive to light.

Use: For the detection of coliform microorganisms in foods, dairy products, water and wastewater as well as in other materials of sanitary importance. Turbidity in the broth and gas in the Durham tube are positive indications of *Escherichia coli*.

Brilliant Green Bile Broth with MUG

Composition per liter:

Oxgall, dehydrated	20.0g
Lactose	10.0g
Pancreatic digest of gelatin	10.0g
MUG (4-methyl umbelliferyl-β-D-glucuronide)	0.05g
Brilliant green	0.013g

pH 7.2 ± 0.2 at 25°C

Source: This medium is available as a premixed powder from BBL Microbiology Systems.

Preparation of Medium: Add components to distilled/deionized water and bring volume to1.0L. Mix thoroughly. Distribute into tubes containing inverted Durham tubes, in 10.0mL amounts for testing 1.0mL or less of sample. Autoclave for 12 min (not longer than 15 min) at 15 psi pressure–121°C. After sterilization, cool the broth rapidly.

Use: For the detection of coliform microorganisms in foods, dairy products, water and wastewater as well as in other materials of sanitary importance. The presence of *E. coli* and other coliforms is determined by the presence of fluorescence in the tube.

Brilliant Green Broth (m–Brilliant Green Broth)

Composition per liter:

Proteose peptone No.3	20.0g
Lactose	20.0g
Sucrose	20.0g
NaCl	10.0g
Yeast extract	6.0g
Phenol red	0.16g
Brilliant green	0.025g

pH 6.9 ± 0.2 at 25°C

Source: This medium is available as a premixed powder from Difco Laboratories.

Preparation of Medium: Add components to distilled/deionized water and bring volume to 1.0L. Mix thoroughly. Gently heat with frequent mixing. Boil for 1 min. Cool to 25°C. Add 2.0mL to each sterile absorbent filter used.

Use: For the selective isolation and differentiation of *Salmonella* from polluted water by the membrane filter method.

Bromcresol Purple Broth

Composition per liter:

Peptone	10.0g
NaCl	5.0g
Beef extract	3.0g
Bromcresol purple	0.04g
Carbohydrate solution	10.0mL

pH 7.0 ± 0.2 at 25°C

Carbohydrate Solution:
Composition per 10mL:

Carbohydrate	5.0g

Preparation of Carbohydrate Solution: Add carbohydrate to distilled/deionized water and bring volume to 10.0mL. Mix thoroughly. Filter sterilize.

Preparation of Medium: Add components to distilled/deionized water and bring volume to 1.0L. Mix thoroughly. Gently heat and bring to boiling. Distribute into test tubes that contain an inverted Durham tube. Autoclave for 10 min at 15 psi pressure–121°C.

Use: For the differentiation of a variety of microorganisms based on their fermentation of specific carbohydrates. Bacteria that ferment the specific carbohydrate turn the medium yellow. When bacteria produce gas, the gas is trapped in the Durham tube.

Bromcresol Purple Dextrose Broth (BCP Broth)

Composition per liter:

Glucose	10.0g
Peptone	5.0g
Beef extract	3.0g
Bromcresol purple solution	2.0mL

pH 7.0 ± 0.2 at 25°C

Bromcresol Purple Solution:
Composition per 10mL:

Bromcresol purple	0.16g
Ethanol (95% solution)	10.0mL

Preparation of Bromcresol Purple Solution: Add bromcresol purple to 10.0mL of ethanol. Mix thoroughly.

Preparation of Medium: Add components to distilled/deionized water and bring volume to 1.0L. Mix thoroughly. Distribute into tubes in 12–15mL volumes. Autoclave for 15 min at 15 psi pressure–121°C.

Use: For the cultivation and differentiation of bacteria based on their ability to ferment glucose. Bacteria that ferment glucose turn the medium yellow.

Bromthymol Blue Agar

Composition per liter:

Agar	11.0g
Peptone	10.0g
NaCl	5.0g
Yeast extract	5.0g
Lactose (33% solution)	27.0mL
Bromthymol blue (1% solution)	10.0mL
Sodium thiosulfate (50% solution)	2.0mL
Glucose (33% solution)	1.2mL
Maranil solution (5% solution)	1.0mL

pH 7.7–7.8 at 25°C

Preparation of Medium: Add agar, peptone, NaCl, and yeast extract to distilled/deionized water and bring volume to 1.0L. Mix thoroughly. Adjust pH to 8.0. Autoclave for 20 min at 15 psi pressure–121°C. Cool to 45–50°C. Filter sterilize separately the lactose solution, bromthymol blue solution, sodium thiosulfate solution, glucose solution and maranil solution. To the cooled, sterile agar solution aseptically add 27.0mL of sterile lactose solution, 10.0mL of sterile bromthymol blue solution, 2.0mL of sterile sodium thiosulfate solution, 1.2mL of sterile glucose solution and1.0mL of sterile maranil solution. Mix thoroughly. Adjust pH to 7.7–7.8. Pour into sterile Petri dishes or distribute into sterile tubes.

Use: For the selective isolation and cultivation of members of the Enterobacteriaceae.

Bryant-Robinson Medium

Composition per 1010mL:

Glucose, cellobiose, or maltose	5.0g
L-Methionine	0.08g
Mineral solution	50.0mL
Na_2CO_3 solution	50.0mL
Hemin solution	10.0mL
Cysteine·HCl–Na_2S solution	10.0mL
Vitamin solution	5.0mL
VFA solution	4.5mL
Resazurin	1.0mL

pH 6.5 ± 0.2 at 25°C

Mineral Solution:
Composition per liter:

KH_2PO_4	18.0g
NaCl	18.0g
$(NH_4)_2SO_4$	8.0g
$CaCl_2·6H_2O$	0.53g
$MgCl_2·6H_2O$	0.4g
$CoCl_2·6H_2O$	0.2g
$MnCl_2·4H_2O$	0.2g
$FeSO_4·7H_2O$	0.08g

Preparation of Mineral Solution: Add components to distilled/deionized water and bring volume to 1.0L. Mix thoroughly.

Na_2CO_3 Solution:
Composition per 100mL:

Na_2CO_3	8.0g

Preparation of Na_2CO_3 Solution: Add Na_2CO_3 to O_2-free distilled/deionized water. Mix thoroughly. Gas with 100% CO_2 for 15 min. Autoclave for 15 min at 15 psi pressure–121°C.

Hemin Solution:
Composition per 100mL:

Hemin	0.01g
NaOH (0.002% solution)	100.0mL

Preparation of Hemin Solution: Add hemin to 100.0mL of NaOH solution. Mix thoroughly.

Cysteine·HCl–Na_2S Solution:

Composition per 100mL:

Cysteine·HCl	2.5g
$Na_2S·9H_2O$	2.5g

Preparation of Cysteine·HCl–Na_2S Solution: Add cysteine·HCl to distilled/deionized water and bring volume to 80.0mL. Mix thoroughly. Adjust pH to 11 with NaOH. Add $Na_2S·9H_2O$. Mix thoroughly. Bring volume to 100.0mL with distilled/deionized water. Gently heat and bring to boiling under 100% N_2. Cool to 25°C under 100% N_2. Autoclave for 15 min at 15 psi pressure–121°C.

Vitamin Solution:

Composition per 100mL:

Calcium pantothenate	0.02g
Nicotinamide	0.02g
Pyridoxine·HCl	0.02g
Riboflavin	0.02g
Thiamine·HCl	0.02g
p-Aminobenzoic acid	1.0mg
Biotin	0.25mg
Folic acid	0.25mg
Vitamin B_{12}	0.1mg

Preparation of Vitamin Solution: Add components to distilled/deionized water and bring volume to 100.0mL. Mix thoroughly. Filter sterilize.

Volatile Fatty Acid Solution:

Composition per liter:

Acetic acid	36.0mL
DL-α-Methylbutyric acid	2.0mL
Isovaleric acid	2.0mL
n-Valeric acid	2.0mL
Isobutyric acid	1.8mL

Preparation of Vitamin Solution: Add components to distilled/deionized water and bring volume to 1.0L. Mix thoroughly.

Preparation of Medium: Add components, except cysteine·HCl–Na_2S solution, to distilled/deionized water and bring volume to 1.0L. Mix thoroughly. Gently heat and bring to boiling. Continue boiling until resazurin turns colorless indicating reduction. Anaerobically distribute into tubes in 10.0mL volumes. Cap with butyl rubber stoppers. Place tubes in a press. Autoclave for 15 min at 15 psi pressure–121°C. Immediately prior to inoculation aseptically and anaerobically add 0.1mL of cysteine·HCl–Na_2S solution per tube.

Use: For the cultivation of *Bacteroides* species from rumens.

BSK Medium

Composition per 1260mL:

Bovine albumin Fraction V	50.0g
HEPES (*N*-[2-Hydroxyethyl]piperazine-*N*′-2-ethanesulfonic acid) buffer	6.0g
Neopeptone	5.0g
Glucose	5.0g
$NaHCO_3$	2.2g
Sodium pyruvate	0.8g
Sodium citrate	0.7g
N-Acetyl glucosamine	0.4g
Gelatin solution	200.0mL
CMRL 1066, without glutamine, without bicarbonate, 10X	100.0mL
Rabbit serum	72.0mL

pH 7.6–7.65 at 25°C

Gelatin Solution:

Composition per 200mL:

Gelatin	14.0g

Preparation of Gelatin Solution: Add gelatin to distilled/deionized water and bring volume to 200.0mL. Heat gently to boiling. Mix thoroughly. Filter sterilize.

CMRL 1066 Medium without Glutamine, without bicarbonate, 10X:

Composition per liter:

NaCl	6.8g
D-Glucose	1.0g
KCl	0.4g
L-Cysteine·HCl·H_2O	0.26g
$CaCl_2$, anhydrous	0.2g
$MgSO_4·7H_2O$	0.2g
$NaH_2PO_4·H_2O$	0.14g
Sodium acetate·$3H_2O$	0.083g
L-Glutamic acid	0.075g
L-Arginine·HCl	0.070g
L-Lysine·HCl	0.070g
L-Leucine	0.060g
Glycine	0.050g
Ascorbic acid	0.050g
L-Proline	0.040g
L-Tyrosine	0.040g
L-Aspartic acid	0.030g
L-Threonine	0.030g
L-Alanine	0.025g
L-Phenylalanine	0.025g
L-Serine	0.025g
L-Valine	0.025g
L-Cystine	0.020g
L-Histidine·HCl·H_2O	0.020g
L-Isoleucine	0.020g
Phenol red	0.020g
L-Methionine	0.015g

Deoxyadenosine	0.010g
Deoxycytidine	0.010g
Deoxyguanosine	0.010g
Glutathione, reduced	0.010g
Thymidine	0.010g
Hydroxy-L-proline	0.010g
L-Tryptophan	0.010g
Nicotinamide adenine dinucleotide	7.0mg
Tween™ 80	5.0mg
Sodium glucoronate·H_2O	4.2mg
Coenzyme A	2.5mg
Cocarboxylase	1.0mg
Flavin adenine dinucleotide	1.0mg
Nicotinamide adenine dinucleotide phosphate	1.0mg
Uridine triphosphate	1.0mg
Choline chloride	0.50mg
Cholesterol	0.20mg
5-Methyldeoxycytidine	0.10mg
Inositol	0.05mg
p-Aminobenzoic acid	0.05mg
Niacin	0.025mg
Niacinamide	0.025mg
Pyridoxine	0.025mg
Pyridoxal·HCl	0.025mg
Biotin	0.01mg
D-Calcium pantothenate	0.01mg
Folic acid	0.01mg
Riboflavin	0.01mg
Thiamine·HCl	0.01mg

pH 7.2 ± 0.2 at 25°C

Preparation of CMRL 1066 Medium without Glutamine, without bicarbonate, 10X: Add components to distilled/deionized water and bring volume to 1.0L. Mix thoroughly. Adjust pH to 7.2. Filter sterilize.

Preparation of Medium: Add components, except gelatin solution and rabbit serum, to 628.0mL of glass distilled water. Mix thoroughly. Adjust pH to 7.6–7.65. Add 200.0mL 7% aqueous gelatin solution. Filter-sterilize entire medium. Aseptically add 72.0mL of sterile rabbit serum.

Use: For the cultivation of a wide variety of microorganisms in a chemically defined medium. For the cultivation of *Borrelia* and *Spirochaeta* species.

BSR Medium

Composition per liter:

Beef heart, solids from infusion	500.0g
Sorbitol	70.0g
Sucrose	10.0g
Tryptose	10.0g
NaCl	5.0g
Fructose	1.0g
Glucose	1.0g
Phenol red	0.020g
Horse serum	100.0mL

pH 7.6 ± 0.2 at 25°C

Preparation of Medium: Add components, except horse serum, to distilled/deionized water and bring volume to 900.0mL. Mix thoroughly. Autoclave for 15 min at 15 psi pressure–121°C. Cool to 45–50°C. Aseptically add 100.0mL of horse serum. Mix thoroughly. Aseptically distribute into sterile tubes or flasks.

Use: For the isolation and cultivation of *Spiroplasma citri*.

BSTSY Agar

Composition per liter:

Pancreatic digest of casein	17.0g
Agar	15.0g
NaCl	5.0g
Yeast extract	4.0g
Papaic digest of soybean meal	3.0g
K_2HPO_4	2.5g
Glucose	2.5g
Bovine serum	100.0mL

pH 7.3 ± 0.2 at 25°C

Preparation of Medium: Add components, except bovine serum, to distilled/deionized water and bring volume to 900.0mL. Mix thoroughly. Gently heat and bring to boiling. Autoclave for 15 min at 15 psi pressure–121°C. Cool to 45–50°C. Aseptically add sterile bovine serum. Mix thoroughly. Pour into sterile Petri dishes or distribute into sterile tubes.

Use: For the isolation and cultivation of *Simonsiella* species and *Alysiella* species.

Buffered Charcoal Yeast Extract Differential Agar (DIFF/BCYE)

Composition per 1014mL:

Agar	17.0g
ACES (2-[(2-Amino-2-oxoethyl)-amino]-ethane sulfonic acid) buffer	10.0g
Yeast extract	10.0g
Charcoal, activated	1.5g
$Fe_4(P_2O_7)_3 \cdot 9H_2O$	0.25g
Bromcresol purple	0.01g
Bromthymol blue	0.01g
Antibiotic solution	10.0mL
Cysteine·HCl·H_2O solution	4.0mL

pH 6.9 ± 0.2 at 25°C

Antibiotic Solution:
Composition per 10mL:

Vancomycin	1.0mg
Polymyxin B	50,000U

Preparation of Antibiotic Solution: Add components to distilled/deionized water and bring volume to 10.0mL. Mix thoroughly. Filter sterilize.

Cysteine·HCl·H_2O Solution:
Composition per 10mL:

L-Cysteine·HCl·H_2O	1.0g

Preparation of Cysteine·HCl·H_2O Solution: Add L-Cysteine·HCl·H_2O to distilled/deionized water and bring volume to 10.0mL. Mix thoroughly. Filter sterilize.

Preparation of Medium: Add components, except cysteine solution and antibiotic solution, to distilled/deionized water and bring volume to 1.0L. Mix thoroughly. Adjust medium to pH 6.9 with 1*N* KOH. Heat gently and bring to boil for 1 minute. Autoclave for 15 min at 15 psi pressure–121°C. Cool to 50–55°C. Add 4.0mL of sterile L-cysteine·HCl·H_2O solution and 10.0mL of sterile antibiotic solution. Mix thoroughly. Pour into sterile Petri dishes with constant agitation to keep charcoal in suspension.

Use: For the isolation, cultivation, and maintenance of *Legionella pneumophila* and other *Legionella* species from environmental specimens. Used for the selective recovery of *L. pneumophila* while reducing contaminating microorganisms from environmental water samples.

Buffered Marine Yeast Medium

Composition per liter:

NaCl	24.0g
Agar	20.0g
Yeast extract	5.0g
1*M* Phosphate buffer, pH 6.8	20.0mL
Hutner's mineral base	20.0mL
KOH (1*N*)	7.0mL

pH 6.8 ± 0.2 at 25°C

1*M* Phosphate Buffer, pH 6.8:
Composition per liter:

$K_2H_2PO_4$	85.4g
$NaH_2PO_4 \cdot H_2O$	70.4g

Preparation of Phosphate Buffer: Add components to distilled/deionized water and bring volume to 1.0L. Mix thoroughly. Adjust pH to 6.8.

Hutner's Mineral Base:
Composition per liter:

$MgSO_4.7H_2O$	29.7g
Nitrilotriacetic acid	10.0g
$CaCl_2 \cdot 2H_2O$	3.34g
$FeSO_4 \cdot 7H_2O$	0.01g
$(NH_4)_2MoO_4$	9.25mg
Metals "44"	50.0mL

Preparation of Hutner's Mineral Base: Initially add a few drops of H_2SO_4 to the distilled water to retard precipitation. Dissolve the nitrilotriacetic acid first and neutralize the solution with KOH. Add other ingredients and adjust the pH to 7.2 with KOH and/or H_2SO_4. There may be a slight precipitate. Store at 5°C.

Metals "44":
Composition per 100mL:

$ZnSO_4 \cdot 7H_2O$	1.1g
$FeSO_4 \cdot 7H_2O$	0.5g
EDTA	0.25g
$MnSO_4 \cdot 7H_2O$	0.154g
$CuSO_4 \cdot 5H_2O$	0.04g
$Co(NO_3)_2 \cdot 6H_2O$	0.025g
$Na_2B_4O_7 \cdot 10H_2O$	0.018g

Preparation of Metals "44":

Preparation of Medium: Add components to distilled/deionized water and bring volume to 1.0L. Mix thoroughly. Distribute into tubes or flasks. Autoclave for 15 min at 15 psi pressure–121°C.

Use: For the cultivation and maintenance of *Pseudomonas* species.

Burke's Modified Nitrogen-Free Medium

Composition per liter:

$MgSO_4 \cdot 7H_2O$	0.2g
Na_2HPO_4	0.19g
$NaHCO_3$	0.05g
$CaSO_4 \cdot 2H_2O$	0.02g
KH_2PO_4	0.011g
$SrCl_2 \cdot 6H_2O$	0.01g
NaCl	0.01g
Adenine	0.01g
$FeSO_4.7H_2O$	6.0mg
Na_2MoO_3	0.5mg

pH 7.8 ± 0.2 at 25°C

Preparation of Medium: Add components to distilled/deionized water and bring volume to 1.0L. Mix thoroughly. Distribute into tubes or flasks. Autoclave for 15 min at 15 psi pressure–121°C.

Use: For the cultivation of *Azotobacter vinelandii*.

Burke's Modified Nitrogen-Free Medium

Composition per liter:

Noble agar	15.0g
Glucose	10.0g
Cellulose	10.0g
K_2HPO_4	1.0g
$CaCl_2 \cdot 2H_2O$	0.1g
$MgSO_4 \cdot 7H_2O$	0.02g
$FeSO_4 \cdot 7H_2O$	50.0mg
$Na_2MoO_4 \cdot 2H_2O$	25.0mg
Vitamin B_{12}	0.1mg
Vitamin solution	1.0mL

pH 7.2–7.3 ± 0.2 at 25°C

Vitamin Solution:

Composition per 50mL:

Thiamine·HCl	843.3mg
Pantothenic acid	595.8mg
Nicotinic acid	307.8mg
p-Aminobenzoic acid	68.6mg
Pyridoxamine·2HCl	60.3mg
Biotin	50.0mg
Folic acid	11.0mg
Vitamin B_{12}	3.5mg

Preparation of Vitamin Solution: Add components to distilled/deionized water and bring volume to 50.0mL. Mix thoroughly. Filter sterilize.

Preparation of Medium: Add components, except glucose, cellulose, K_2HPO_4, and vitamin solution to distilled/deionized water and bring volume to 850.0mL. Mix thoroughly. Adjust pH to 7.2–7.3. In three separate flasks add glucose, cellulose and K_2HPO_4 to 50.0mL of distilled/deionized water. Filter sterilize the vitamin solution. Autoclave the other solutions separately for 15 min at 15 psi pressure–121°C. Cool to 25°C. Aseptically combine all the solutions and mix thoroughly. Distribute into sterile tubes or flasks or pour into sterile Petri dishes.

Use: For the cultivation and maintenance of *Streptomyces* species.

Burke's Modified Nitrogen-Free Medium with Benzoate

Composition per liter:

Sodium benzoate	0.72g
$MgSO_4 \cdot 7H_2O$	0.2g
Na_2HPO_4	0.189g
$NaHCO_3$	0.05g
$CaSO_4 \cdot 2H_2O$	0.02g
KH_2PO_4	0.011g
$SrCl_2 \cdot 6H_2O$	0.01g
NaCl	0.01g
Adenine	0.01g
$FeSO_4 \cdot 7H_2O$	6.0mg
Na_2MoO_3	0.5mg

pH 7.8 ± 0.2 at 25°C

Preparation of Medium: Add components to distilled/deionized water and bring volume to 1.0L. Mix thoroughly. Distribute into tubes or flasks. Autoclave for 15 min at 15 psi pressure–121°C.

Use: For the cultivation of *Pseudomonas* species and other microorganisms which can utilize benzoate as the sole carbon source.

Bushnell-Haas Agar

Composition per liter:

Agar	15.0g
KH_2PO_4	1.0g
K_2HPO_4	1.0g
NH_4NO_3	1.0g
$MgSO_4 \cdot 7H_2O$	0.2g
$FeCl_3$	0.05g
$CaCl_2 \cdot 2H_2O$	0.02g

pH 7.0 ± 0.2 at 25°C

Preparation of Medium: Add components to distilled/deionized water and bring volume to 1.0L. Mix thoroughly. Gently heat and bring to boiling. Distribute into tubes or flasks. Autoclave for 15 min at 15 psi pressure–121°C. Pour into sterile Petri dishes or leave in tubes. Layer hydrocarbon on agar surface or add aseptically add sterile hydrocarbon to cooled agar prior to pouring plates.

Use: For examining fuels for microbial contamination and for studying the hydrocarbon utilization by microorganisms. Also for the cultivation of *Nocardia* species.

Bushnell-Haas Broth

Composition per liter:

KH_2PO_4	1.0g
K_2HPO_4	1.0g
NH_4NO_3	1.0g
$MgSO_4 \cdot 7H_2O$	0.2g
$FeCl_3$	0.05g
$CaCl_2 \cdot 2H_2O$	0.02g

pH 7.0 ± 0.2 at 25°C

Source: This medium is available as a premixed powder from Difco Laboratories.

Preparation of Medium: Add components to distilled/deionized water and bring volume to 1.0L. Mix thoroughly. Distribute into tubes or flasks. Autoclave for 15 min at 15 psi pressure–121°C. Layer hydrocarbon on broth surface or add directly to broth.

Use: For examining fuels for microbial contamination and for studying the hydrocarbon utilization by microorganisms. Also for the cultivation of *Nocardia* species.

Bushnell-Haas Medium

Composition per liter:

KH_2PO_4	1.0g
K_2HPO_4	1.0g
NH_4NO_3	1.0g
Cholesterol	0.3g
$MgSO_4 \cdot 7H_2O$	0.2g
$FeCl_3$	0.05g
$CaCl_2 \cdot 2H_2O$	0.02g

pH 7.0 ± 0.2 at 25°C

Source: This medium is available as a premixed powder from Difco Laboratories.

Preparation of Medium: Add components to distilled/deionized water and bring volume to 1.0L. Mix thoroughly. Distribute into tubes or flasks. Autoclave for 15 min at 15 psi pressure–121°C. Layer hydrocarbon on broth surface or add directly to broth.

Use: For the cultivation of *Nocardia* species.

C 3G *Spiroplasma* Medium

Composition per liter:

Sucrose	100.0g
Phenol red	10.0mg
PPLO broth without crystal violet	500.0mL
Horse serum	150.0mL
Fresh yeast extract solution	50.0mL
CMRL-1066 medium	5.0mL

pH 7.5 ± 0.2 at 25°C

PPLO Broth without Crystal Violet:

Composition per 500mL:

Beef heart, infusion from	11.52g
Peptone	2.32g
NaCl	1.15g

Source: PPLO broth without crystal violet is available as a premixed powder from Difco Laboratories.

Preparation of PPLO Broth without Crystal Violet: Add components to distilled/deionized water and bring volume to 500.0mL. Mix thoroughly.

Fresh Yeast Extract Solution:

Composition per 100mL:

Baker's yeast, live, pressed, starch-free	25.0g

Preparation of Fresh Yeast Extract Solution: Add the live Baker's yeast to 100.0mL of distilled/deionized water. Autoclave for 90 min at 15 psi pressure–121°C. Allow to stand. Remove supernatant solution. Adjust pH to 6.6–6.8. Filter sterilize.

CMRL-1066 Medium:

Composition per liter:

NaCl	6.8g
$NaHCO_3$	2.2g
D-Glucose	1.0g
KCl	0.4g
L-Cysteine·HCl·H_2O	0.26g
$CaCl_2$, anhydrous	0.2g
$MgSO_4 \cdot 7H_2O$	0.2g
$NaH_2PO_4 \cdot H_2O$	0.14g
L-Glutamine	0.1g
Sodium acetate·$3H_2O$	0.083g
L-Glutamic acid	0.075g
L-Arginine·HCl	0.070g
L-Lysine·HCl	0.070g
L-Leucine	0.060g
Glycine	0.050g
Ascorbic acid	0.050g
L-Proline	0.040g
L-Tyrosine	0.040g
L-Aspartic acid	0.030g
L-Threonine	0.030g
L-Alanine	0.025g
L-Phenylalanine	0.025g
L-Serine	0.025g
L-Valine	0.025g
L-Cystine	0.020g
L-Histidine·HCl·H_2O	0.020g
L-Isoleucine	0.020g
Phenol red	0.020g
L-Methionine	0.015g
Deoxyadenosine	0.010g
Deoxycytidine	0.010g
Deoxyguanosine	0.010g
Glutathione, reduced	0.010g
Thymidine	0.010g
Hydroxy-L-proline	0.010g
L-Tryptophan	0.010g
Nicotinamide adenine dinucleotide	7.0mg
Tween™ 80	5.0mg
Sodium glucoronate·H_2O	4.2mg
Coenzyme A	2.5mg
Cocarboxylase	1.0mg
Flavin adenine dinucleotide	1.0mg
Nicotinamide adenine dinucleotide phosphate	1.0mg
Uridine triphosphate	1.0mg
Choline chloride	0.50mg
Cholesterol	0.20mg
5-Methyldeoxycytidine	0.10mg
Inositol	0.05mg
p-Aminobenzoic acid	0.05mg
Niacin	0.025mg
Niacinamide	0.025mg
Pyridoxine	0.025mg

Pyridoxal·HCl	0.025mg
Biotin	0.01mg
D-Calcium pantothenate	0.01mg
Folic acid	0.01mg
Riboflavin	0.01mg
Thiamine·HCl	0.01mg

Source: CMRL-1066 medium is available as a premixed powder from GIBCO BRL.

Preparation of CMRL-1066 Medium: Add components to distilled/deionized water and bring volume to 1.0L. Mix thoroughly. Adjust pH to 7.2. Filter sterilize.

Preparation of Medium: Add components—except horse serum, fresh yeast extract, and CMRL medium—to distilled/deionized water and bring volume to 795.0mL. Adjust pH to 7.5. Autoclave for 15 min at 15 psi pressure–121°C. Aseptically add 150.0mL of sterile horse serum, 50.0mL of sterile fresh yeast extract solution and 5.0mL of sterile CMRL medium. Distribute into sterile tubes or flasks.

Use: For cultivation and maintenance of *Spiroplasma* species.

C 3N *Spiroplasma* Medium

Composition per 100mL:

Sucrose	12.0g
Phenol Red	10.0mg
PPLO Broth without crystal violet	50.0mL
Horse serum	20.0mL
Fresh yeast extract solution	5.0mL
CMRL-1066 medium	0.5mL

pH 7.5 ± 0.2 at 25°C

PPLO Broth without Crystal Violet:

Composition per 500mL:

Beef heart, infusion from	11.52g
Peptone	2.32g
NaCl	1.15g

Source: PPLO broth without Crystal Violet is available as a premixed powder from Difco Laboratories.

Preparation of PPLO Broth without Crystal Violet: Add components to distilled/deionized water and bring volume to 500.0mL. Mix thoroughly.

Fresh Yeast Extract Solution:

Composition per 100mL:

Baker's yeast, live, pressed, starch-free	25.0g

Preparation of Fresh Yeast Extract Solution: Add the live Baker's yeast to 100.0mL of distilled/deionized water. Autoclave for 90 min at 15 psi pressure–121°C. Allow to stand. Remove supernatant solution. Adjust pH to 6.6–6.8. Filter sterilize.

CMRL-1066 Medium:

Composition per liter:

NaCl	6.8g
$NaHCO_3$	2.2g
D-Glucose	1.0g
KCl	0.4g
L-Cysteine·HCl·H_2O	0.26g
$CaCl_2$, anhydrous	0.2g
$MgSO_4 \cdot 7H_2O$	0.2g
$NaH_2PO_4 \cdot H_2O$	0.14g
L-Glutamine	0.1g
Sodium acetate·$3H_2O$	0.083g
L-Glutamic acid	0.075g
L-Arginine·HCl	0.070g
L-Lysine·HCl	0.070g
L-Leucine	0.060g
Glycine	0.050g
Ascorbic acid	0.050g
L-Proline	0.040g
L-Tyrosine	0.040g
L-Aspartic acid	0.030g
L-Threonine	0.030g
L-Alanine	0.025g
L-Phenylalanine	0.025g
L-Serine	0.025g
L-Valine	0.025g
L-Cystine	0.020g
L-Histidine·HCl·H_2O	0.020g
L-Isoleucine	0.020g
Phenol Red	0.020g
L-Methionine	0.015g
Deoxyadenosine	0.010g
Deoxycytidine	0.010g
Deoxyguanosine	0.010g
Glutathione, reduced	0.010g
Thymidine	0.010g
Hydroxy-L-proline	0.010g
L-Tryptophan	0.010g
Nicotinamide adenine dinucleotide	7.0mg
Tween™ 80	5.0mg
Sodium glucoronate·H_2O	4.2mg
Coenzyme A	2.5mg
Cocarboxylase	1.0mg
Flavin adenine dinucleotide	1.0mg
Nicotinamide adenine dinucleotide phosphate	1.0mg
Uridine triphosphate	1.0mg
Choline chloride	0.50mg
Cholesterol	0.20mg
5-Methyldeoxycytidine	0.10mg
Inositol	0.05mg
p-Aminobenzoic acid	0.05mg
Niacin	0.025mg
Niacinamide	0.025mg
Pyridoxine	0.025mg

Pyridoxal·HCl 0.025mg
Biotin 0.01mg
D-Calcium pantothenate 0.01mg
Folic acid 0.01mg
Riboflavin 0.01mg
Thiamine·HCl 0.01mg

Source: CMRL-1066 medium is available as a premixed powder from GIBCO BRL.

Preparation of CMRL-1066 Medium: Add components to distilled/deionized water and bring volume to 1.0L. Mix thoroughly. Adjust pH to 7.2. Filter sterilize.

Preparation of Medium: Add components—except horse serum, fresh yeast extract, and CMRL medium—to distilled/deionized water and bring volume to 75.0mL. Adjust pH to 7.5. Autoclave for 15 min at 15 psi pressure–121°C. Aseptically add 20.0mL of sterile horse serum, 5.0mL of sterile yeast extract and 0.5mL of sterile CMRL medium. Distribute into sterile tubes or flasks.

Use: For cultivation and maintenance of *Spiroplasma kunkelii*.

C/10 Medium Reichenbach

Composition per liter:

Agar 15.0g
Pancreatic digest of casein 3.0g
$CaCl_2$ 1.0g

pH 7.2 ± 0.2 at 25°C

Preparation of Medium: Add components, except agar, to distilled/deionized water and bring volume to 1.0L. Adjust pH to 7.2. Add agar. Mix thoroughly. Gently heat to boiling. Distribute into tubes or flasks. Autoclave for 15 min at 15 psi pressure–121°C. Pour into sterile Petri dishes or leave in tubes.

Use: For cultivation and maintenance of *Flexibacter filiformis*.

CAL Agar (Cellobiose Arginine Lysine Agar) (*Yersinia* Isolation Agar)

Composition per liter:

Agar 20.0g
L-Arginine·HCl 6.5g
L-Lysine·HCl 6.5g
NaCl 5.0g
Cellobiose 3.5g
Yeast extract 3.0g
Sodium deoxycholate 1.5g
Neutral red 0.03g

Preparation of Medium: Add components to distilled/deionized water and bring volume to 1.0L. Mix thoroughly. Heat to boiling. Do not autoclave. Pour into sterile Petri dishes.

Use: For isolation and characterization of *Yersinia enterocolitica* from fecal specimens and enumeration of *Y. enterocolitica* from water and other liquid specimens.

CAL Broth (Cellobiose Arginine Lysine Broth)

Composition per liter:

L-Arginine·HCl 6.5g
L-Lysine·HCl 6.5g
NaCl 5.0g
Cellobiose 3.5g
Yeast extract 3.0g
Sodium deoxycholate 1.5g
Neutral red 0.03g

Preparation of Medium: Add components to distilled/deionized water and bring volume to 1.0L. Mix thoroughly. Heat to boiling. Do not autoclave. Distribute into sterile tubes in 6.0–8.0mL volumes.

Use: For isolation and characterization of *Yersinia enterocolitica* from fecal specimens and enumeration of *Y. enterocolitica* from water and other liquid specimens.

Camphor Minimal Medium

Composition per liter:

Agar 20.0g
K_2HPO_4 .... 4.4g
NH_4Cl 2.1g
KH_2PO_4 .... 1.7g
100× salt solution 10.0mL

100X Salt Solution:

Composition per liter:

$MgSO_4$ 19.5g
$FeSO_4 \cdot 7H_2O$ 5.0g
$MnSO_4 \cdot H_2O$ 5.0g
Ascorbic acid 1.0g
$CaCl_2 \cdot 2H_2O$ 0.3g

Preparation of 100X Salt Solution: Add components to distilled/deionized water and bring volume to 1.0L. Mix thoroughly.

Preparation of Medium: Add components to distilled/deionized water and bring volume to 1.0L. Gently heat and bring to boiling. Autoclave for 15 min at 15 psi pressure–121°C. Pour into sterile Petri dishes. Allow to cool to room temperature. Invert Petri dishes. Spread 0.2mL of 2M D-(+) camphor solution in methylene chloride (CH_2Cl_2) on the inside cover of each plate.

Use: For cultivation and maintenance of *Pseudomonas putida.*

Campylobacter Agar, Blaser's (Blaser's *Campylobacter* Agar)

Composition per liter:

Campylobacter agar base	990.0mL
Supplement B	10.0mL

pH 7.4 ± 0.2 at 25°C

***Campylobacter* Agar Base:**

Composition per liter:

Proteose peptone	15.0g
Agar	12.0g
NaCl	5.0g
Yeast extract	5.0g
Liver digest	2.5g

Source: *Campylobacter* agar base and *Campylobacter* antimicrobic supplement B are available as a premixed powder from Difco Laboratories.

Preparation of *Campylobacter* Agar Base: Add components to distilled/deionized water and bring volume to 990.0mL. Mix thoroughly. Gently heat and bring to boiling. Autoclave for 15 min at 15 psi pressure–121°C. Cool to 45–50°C.

Supplement B:

Composition per 10mL:

Cephalothin	15mg
Vancomycin	10mg
Trimethoprim	5mg
Amphotericin B	2mg
Polymyxin B	2,500U

Preparation of Supplement B: Add components to 10.0mL of distilled/deionized water. Filter sterilize.

Preparation of Medium: Prepare 990.0mL of *Campylobacter* Agar Base. Autoclave and cool to 45–50°C. Aseptically add 10.0mL of sterile Supplement B. Mix thoroughly. Pour into sterile Petri dishes.

Use: For the selective isolation of *Campylobacter jejuni* from fecal specimens, food and environmental specimens.

Campylobacter Agar, Skirrow's (Skirrow's *Campylobacter* Agar)

Composition per liter:

Campylobacter agar base	990.0mL
Supplement S	10.0mL

pH 7.4 ± 0.2 at 25°C

***Campylobacter* Agar Base:**

Composition per liter:

Proteose peptone	15.0g
Agar	12.0g
NaCl	5.0g
Yeast extract	5.0g
Liver digest	2.5g

Source: *Campylobacter* agar base and *Campylobacter* antimicrobic supplement S are available as a premixed powder from Difco Laboratories.

Preparation of *Campylobacter* Agar Base: Add components to distilled/deionized water and bring volume to 990.0mL. Mix thoroughly. Gently heat and bring to boiling. Autoclave for 15 min at 15 psi pressure–121°C. Cool to 45–50°C.

Supplement S:

Composition per 10mL:

Vancomycin	10mg
Trimethoprim	5mg
Polymyxin B	2,500U

Preparation of Supplement S: Add components to 10.0mL of distilled/deionized water. Filter sterilize.

Preparation of Medium: Prepare 990.0mL of *Campylobacter* Agar Base. Autoclave and cool to 45–50°C. Aseptically add 10.0mL of sterile Supplement S. Mix thoroughly. Pour into sterile Petri dishes.

Use: For the selective isolation of *Campylobacter jejuni* from fecal specimens, food and environmental specimens.

Campylobacter Blood-Free Selective Agar

Composition per liter:

Agar	12.0g
Beef extract	10.0g
Peptone	10.0g
Charcoal	4.0g
Casein hydrolysate	3.0g
Sodium deoxycholate	1.0g
$Fe_2SO_4 \cdot H_2O$	0.25g
Sodium pyruvate	0.25g
Cefoperazone solution	10.0mL

pH 7.4 ± 0.2 at 25°C

Cefoperazone Solution:

Composition per 10mL:

Sodium cefoperazone 0.032g

Preparation of Cefoperazone Solution: Add sodium cefoperazone to distilled/deionized water and bring volume to 10.0mL. Mix thoroughly. Filter sterilize.

Preparation of Medium: Add components, except cefoperazone solution, to distilled/deionized water and bring volume to 990.0mL. Mix throughly. Heat with frequent agitation and boil for 1 min to completely dissolve. Autoclave for 15 min at 15 psi–121°C. Cool to 50–55°C. Add 10.0mL of sterile cefoperazone solution. Addition of 10mg/L amphotericin B improves the selectivity of the medium. Mix thoroughly. Pour into sterile Petri dishes.

Use: For the selective isolation of *Campylobacter* species, especially *C. jejuni, C. coli* and *C. laridis.*

Campylobacter Charcoal Differential Agar (CCDA) (Preston Blood–Free Medium)

Composition per liter:

Agar 12.0g
Beef extract 10.0g
Peptone 10.0g
NaCl 5.0g
Charcoal 4.0g
Casein hydrolysate 3.0g
Sodium deoxycholate 1.0g
$FeSO_4$ 0.25g
Sodium pyruvate 0.25g
Cefoperazone solution 10.0mL

pH 7.5 ± 0.2 at 25°C

Cefoperazone Solution:

Composition per 10mL:

Sodium cefoperazone 0.032g

Preparation of Cefoperazone Solution: Add sodium cefoperazone to distilled/deionized water and bring volume to 10.0mL. Mix thoroughly. Filter sterilize.

Preparation of Medium: Add components, except cefoperazone solution, to distilled/deionized water and bring volume to 990.0mL. Mix thoroughly. Gently heat and bring to boiling. Autoclave for 15 min at 15 psi pressure–121°C. Cool to 45–50°C. Aseptically add 10.0mL of sterile cefoperazone solution. Mix thoroughly. Pour into sterile Petri dishes or distribute into sterile tubes.

Use: For the cultivation of *Campylobacter* species.

Campylobacter Enrichment Broth

Composition per liter:

Beef extract 10.0g
Peptone 10.0g
Yeast extract 6.0g
NaCl 5.0g
Horse blood, laked 50.0mL
FBP solution 4.0mL
Antibiotic solution 4.0mL

pH 7.5 ± 0.2 at 25·C

Horse Blood, Laked:

Composition per 50mL:

Horse blood, fresh 50.0mL

Preparation of Horse Blood, Laked: Add blood to a sterile polypropylene bottle. Freeze overnight at –20°C. Thaw at 8°C. Refreeze at –20°C. Thaw again at 8°C.

FBP Solution:

Composition per 100mL:

$FeSO_4$ 6.25g
$NaHSO_3$ 6.25g
Sodium pyruvate 6.25g

Preparation of FBP Solution: Add components to distilled/deionized water and bring volume to 100.0mL. Mix thoroughly. Filter sterilize.

Antibiotic Solution:

Composition per 10mL:

Cycloheximide 0.1g
Sodium cefoperazone 0.030g
Trimethoprim lactate 0.0125g
Rifampicin 0.010g

Preparation of Antibiotic Solution: Add components to distilled/deionized water and bring volume to 10.0mL. Mix thoroughly. Filter sterilize.

Preparation of Medium: Add components—except laked horse blood, FBP solution, and antibiotic solution—to distilled/deionized water and bring volume to 942.0mL. Mix thoroughly. Gently heat and bring to boiling. Autoclave for 15 min at 15 psi pressure–121°C. Cool to 45–50°C. Aseptically add 50.0mL of sterile laked horse blood, 4.0mL of FBP solution, and 4.0mL antibiotic solution. Mix thoroughly. Pour into sterile Petri dishes or distribute into sterile tubes.

Use: For the isolation and cultivation of *Campylobacter* species from dairy products.

Candida BCG Agar Base (*Candida* Bromcresol Green Agar Base)

Composition per liter:

Glucose	40.0g
Agar	15.0g
Peptone	10.0g
Yeast extract	1.0g
Bromcresol green	0.02g
Neomycin solution	10.0mL

pH 6.1 ± 0.1 at 25°C

Source: This medium is available as a premixed powder from Difco Laboratories.

Neomycin Solution:

Composition per 10mL:

Neomycin	0.5g

Preparation of Neomycin Solution: Add neomycin to distilled/deionized water and bring volume to 10.0mL. Mix thoroughly. Filter sterilize.

Preparation of Medium: Add components, except neomycin solution, to distilled/deionized water and bring volume to 1.0L. Mix thoroughly and heat gently until boiling. Autoclave for 15 min at 15 psi pressure–121°C. Cool to 50–55°C. Aseptically add 10.0mL sterile neomycin solution. Mix thoroughly. Pour into sterile Petri dishes or leave in tubes.

Use: For the selective isolation and identification of *Candida* species. It is a highly differential medium that is used for demonstrating morphological and biochemical reactions characterizing different *Candida* species. *C. albicans* appears as blunt conical colonies with smooth edges and yellow to blue-green color. *C. stellatoidea* appears as convex colonies with smooth edges and yellow to green color. *C. tropicalis* appears as convex colonies with wavy edges and yellow-green to green color with a dark blue-green base. *C. pseudotropicalis* appears as convex, shiny colonies with smooth edges and green color with a light green edge. *C. krusei* appears as low conical colonies with spreading edges and blue-green color. *C. stellatoidea* appears as convex colonies with smooth edges and yellow to green color.

Carbohydrate Fermentation Broth

Composition per liter:

Peptone	10.0g
NaCl	5.0g
Meat extract	3.0g
Carbohydrate solution	50.0mL
Andrade's indicator	10.0mL

pH 7.1 ± 0.2 at 25°C

Andrade's Indicator:

Composition per 100mL:

Acid fuchsin	0.1 g
NaOH (1*N* solution)	16.0mL

Preparation of Andrade's Indicator: Add components to distilled/deionized water and bring volume to 100.0mL. Mix thoroughly.

Carbohydrate Solution:

Composition per 100mL:

Carbohydrate	10.0g

Preparation of Carbohydrate Solution: Add carbohydrate to distilled/deionized water and bring volume to 100.0mL. Adonitol, arabinose, cellobiose, glucose, dulcitol, fructose, galactose, inositol, lactose, maltose, mannitol, raffinose, rhamnose, salicin, sorbitol, sucrose, trehalose, xylose, or other carbohydrates may be used. Mix thoroughly. Filter sterilize.

Caution: Acid Fuchsin is a potential carcinogen and care must be taken to avoid inhalation of the powdered dye and contact with the skin.

Preparation of Medium: Add components, except carbohydrate solution, to distilled/deionized water and bring volume to 1.0L. Mix thoroughly. Gently heat and bring to boiling. Distribute in 10.0mL volumes into test tubes containing inverted Durham tubes. Autoclave for 15 min at 15 psi–121°C. Cool to 25°C. Add 0.5mL of sterile carbohydrate solution to each tube.

Use: For the determination of carbohydrate fermentation reactions of microorganisms, particularly members of the Enterobacteriaceae. A Durham tube is used to collect gas produced during the fermentation reaction. Acid production is indicated by a pink reaction.

Carbon Assimilation Medium

Composition per liter:

Agar solution	500.0mL
Mineral base medium	500.0mL

pH 6.5 ± 0.1 at 25°C

Agar Solution:

Composition per liter:

Agar	32.0g

Preparation of Agar Solution: Add agar to distilled/deionized water and bring volume to 1.0L. Mix thoroughly. Gently heat and bring to boiling. Autoclave for 15 min at 15 psi pressure–121°C. Cool to 45–50°C.

Mineral Base Medium:

Composition per 500mL:

Carbohydrate	10.0g
NaCl	5.0g

NH_4HPO_4	1.0g
K_2HPO_4	1.0g
$MgSO_4 \cdot 7H_2O$, anhydrous	0.1g

Preparation of Mineral Base Medium: Add components to distilled/deionized water and bring volume to 500.0mL. Mix thoroughly. Gently heat until dissolved. Filter sterilize. Warm to 45–50°C.

Preparation of Medium: Combine 500.0mL of cooled, sterile agar solution and 500.0mL of sterile mineral base medium. Mix thoroughly. Aseptically distribute into sterile tubes. Allow tubes to cool in a slanted position.

Use: For the cultivation and differentiation of microorganisms based on their ability to utilize a particular carbon source.

Carbon Assimilation Medium, Auxanographic Method for Yeast Identification

Composition per liter:

Noble agar	20.00g
$(NH_4)_2SO_4$	0.5g
KH_2PO_4	0.1g
$MgSO_4 \cdot 7H_2O$	0.05g
NaCl	0.01g
$CaCl_2 \cdot 2H_2O$	0.01g
DL-Methionine	2.0mg
DL-Tryptophan	2.0mg
L-Histidine·HCl	1.0mg
Inositol	0.2mg
KI	0.01mg
H_3BO_3	0.05mg
$ZnSO_4 \cdot 7H_2O$	0.04mg
$MnSO_4 \cdot 4H_2O$	0.04mg
Thiamine·HCl	0.04mg
Pyroxidine·HCl	0.04mg
Niacin	0.04mg
Calcium pantothenate	0.04mg
p-Aminobenzoic acid	0.02mg
Riboflavin	0.02mg
$FeCl_3$	0.02mg
$Na_2MoO_4 \cdot 4H_2O$	0.02mg
$CuSO_4 \cdot 5H_2O$	4.0μg
Folic acid	0.2μg
Biotin	0.2μg

pH 4.5 ± 0.2 at 25°C

Preparation of Medium: Add components to distilled/deionized water and bring volume to 1.0L. Mix thoroughly. Gently heat and bring to boiling. Distribute into screw-capped tubes in 20.0mL volumes. Autoclave for 15 min at 15 psi pressure–121°C.

Use: For carbohydrate assimilation tests by the auxanographic method for identification of yeasts.

Carbon Utilization Test

Composition per liter:

Ionagar	10.0g
NH_4Cl	1.0g
$MgSO_4 \cdot 7H_2O$	0.5g
Ferric ammonium citrate	0.05g
$CaCl_2$	0.5mg
Sodium potassium phosphate buffer (0.33*M* solution, pH 6.8)	1.0L
Carbon source	10.0mL

pH 6.8 ± 0.2 at 25°C

Carbon Source:

Composition per 10mL:

Carbon source	1.0g

Preparation of Carbon Source: Add carbon source to distilled/deionized water and bring volume to 10.0mL. Mix thoroughly. Filter sterilize.

Preparation of Medium: Add components, except carbon source, to distilled/deionized water and bring volume to 990.0mL. Mix thoroughly. Gently heat and bring to boiling. Autoclave for 15 min at 15 psi pressure–121°C. Cool to 45–50°C. Aseptically add sterile carbon source. Mix thoroughly. Pour into sterile Petri dishes or distribute into sterile tubes.

Use: For the cultivation and differentiation of *Pseudomonas* species based on their ability to utilize a specific carbon source.

Caryophanon Medium

Composition per liter:

Agar	15.0g
Yeast extract	2.0g
Sodium acetate	1.0g
Pancreatic digest of casein	1.0g

pH 7.5 ± 0.2 at 25°C

Preparation of Medium: Add components to distilled/deionized water and bring volume to 1.0L. Mix thoroughly. Gently heat and bring to boiling. Distribute into tubes or flasks. Autoclave for 15 min at 15 psi pressure–121°C. Pour into sterile Petri dishes or leave in tubes.

Use: For cultivation and maintenance of *Caryophanon tenue* and other *Caryophanon* species.

CAS Medium

Composition per liter:

Pancreatic digest of casein	10.0g
$MgSO_4 \cdot 7H_2O$	1.0g
K_2HPO_4	0.25g

pH 6.8 ± 0.2 at 25°C

Preparation of Medium: Add components to distilled/deionized water and bring volume to 1.0L. Mix

thoroughly. Distribute into tubes or flasks. Autoclave for 15 min at 15 psi pressure–121°C.

Use: For the cultivation of myxobacteria.

Casamino Acids Glucose Medium (CAGV Medium)

Composition per liter:

Agar	20.0g
Glucose	1.0g
Vitamin-free casamino acids	1.0g
Mineral salts solution A	20.0mL
Vitamin solution	10.0mL

pH 7.2 ± 0.2 at 25°C

Mineral Salts Solution A:

Composition per liter:

$MgSO_4 \cdot 7H_2O$	29.7g
$NaMoO_4 \cdot 2H_2O$	12.67g
Nitrilotriacetic acid	10.0g
$CaCl_2 \cdot 2H_20$	3.34g
$FeSO_4 \cdot 7H_2O$	0.10g
Metallic salts solution B	50.0mL

Preparation of Mineral Salts Solution A: Add nitrilotriacetic acid to 500.0mL of distilled/deionized water. Dissolve by adjusting pH to 6.5 with KOH. Add remaining components. Readjust pH to 7.2 with H_2SO_4 or KOH. Add distilled/deionized water to 1.0L.

Metallic Salts Solution B:

Composition per 100mL:

$ZnSO_4 \cdot 7H_2O$	1.1g
$FeSO_4 \cdot 7H_20$	0.5g
Ethylenediaminetetraacetic acid	0.3g
$MnSO_4 \cdot H_2O$	0.3g
$CuSO_4 \cdot 5H_2O$	0.04g
$CoCL_2 \cdot 6H_20$	0.02g
$Na_2B_4O_7 \cdot 10H_20$	0.02g

Preparation of Metallic Salts Solution B: Add a few drops of H_2SO_4 to distilled/deionized water to inhibit precipitate formation. Add components to acidified distilled/deionized water and bring volume to 100.0mL. Mix thoroughly.

Vitamin Solution:

Pyridoxine·HCL	0.01g
Calcium pantothenate	5.0mg
Nicotinamide	5.0mg
Riboflavin	5.0mg
Thiamine·HCl	5.0mg
Biotin	2.0mg
Folic acid	2.0mg
Vitamin B_{12}	0.1mg

Preparation of Vitamin Solution: Add components to distilled/deionized water and bring volume to 1.0L. Mix thoroughly. Filter sterilize.

Preparation of Medium: Add components to distilled/deionized water and bring volume to 1.0L. Mix thoroughly. Gently heat and bring to boiling. Distribute into tubes or flasks. Autoclave for 15 min at 15 psi pressure–121°C. Pour into sterile Petri dishes or leave in tubes.

Use: For the cultivation and maintenance of *Microcyclus aquaticus.*

Casamino Acids Medium

Composition per liter:

Casamino acids	1.0g
Glucose	1.0g
Biotin	0.02mg
Modified Hutner's basal salts	20.0mL

Modified Hutner's Basal Salts:

Composition per liter:

$MgSO_4 \cdot 7H_2O$	29.7g
Nitrilotriacetic acid	10.0g
$CaCl_2 \cdot 2H_2O$	3.34g
$FeSO_4 \cdot 7H_2O$	0.1g
$(NH_4)_2MoO_4$	9.25mg
Metals "44"	50.0mL

Preparation of Modified Hutner's Basal Salts: Add nitrilotriacetic acid to 500.0mL of distilled/deionized water. Dissolve by adjusting pH to 6.5 with KOH. Add remaining components. Add distilled/deionized water to 1.0L.

Metals "44":

Composition per 100mL:

$ZnSO_4 \cdot 7H_2O$	1.1g
$FeSO_4 \cdot 7H_2O$	0.5g
EDTA	0.25g
$MnSO_4 \cdot 7H_2O$	0.154g
$CuSO_4 \cdot 5H_2O$	0.04g
$Co(NO_3)_2 \cdot 6H_2O$	0.025g
$Na_2B_4O_7 \cdot 10H_2O$	0.018g

Preparation of Metals "44": Acidify distilled/deionized water with a drop of H_2SO_4 to retard precipitation of salts. Add components to distilled/deionized water and bring volume to 100.0mL.

Preparation of Medium: Add components to distilled/deionized water and bring volume to 1.0L. Mix thoroughly. Distribute into tubes or flasks. Autoclave for 15 min at 15 psi pressure–121°C.

Use: For cultivation of *Ancylobacter aquaticus* and *Enhydrobacter aerosaccus.*

Casamino Acids Peptone Czapek's Agar

Composition per liter:

Sucrose	30.0g
Agar	15.0g
Peptone	2.0g
Casamino acids	1.0g
K_2HPO_4	1.0g
KCl	0.5g
$MgSO_4 \cdot 7H_2O$	0.5g
$FeSO_4 \cdot 7H_2O$	0.01g

Preparation of Medium: Add components to distilled/deionized water and bring volume to 1.0L. Mix thoroughly. Gently heat and bring to boiling. Distribute into tubes or flasks. Autoclave for 15 min at 15 psi pressure–121°C. Pour into sterile Petri dishes or leave in tubes.

Use: For the isolation and cultivation of *Actinomadura* species, *Actinopolyspora* species, *Excellospora* species and *Microspora* species.

Casamino Peptone Czapek Medium

Composition per liter:

Sucrose	30.0g
Agar	15.0g
Peptone	2.0g
Casamino acids	1.0g
K_2HPO_4	1.0g
KCl	0.5g
$MgSO_4 \cdot 7H_2O$	0.5g
$FeSO_4 \cdot 7H_2O$	0.01g

Preparation of Medium: Add components to distilled/deionized water and bring volume to 1.0L. Mix thoroughly. Gently heat to boiling. Distribute into tubes or flasks. Autoclave for 15 min at 15 psi pressure–121°C. Pour into sterile Petri dishes or leave in tubes.

Use: For cultivation and maintenance of *Actinoplanes* species, *Pseudonocardia compacta*, and *Streptomyces* species.

Casein Agar

Composition per liter:

Agar	10.0g
Skim milk	50.0mL

Preparation of Medium: Add components to distilled/deionized water and bring volume to 1.0L. Mix thoroughly. Gently heat and bring to boiling. Distribute into tubes or flasks. Autoclave for 15 min at 15 psi pressure–121°C. Pour into sterile Petri dishes or leave in tubes.

Use: For the cultivation and differentiation of aerobic actinomycetes based on casein utilization. Bacteria that utilize casein, such as *Streptomyces* and *Actinomadura* species, appear as colonies surrounded by a clear zone. *Nocardia asteroides, N. caviae,* and *Mycobacterium fortuitum* do not utilize casein.

Casein Medium

Composition per liter:

NaCl	250.0g
Agar	20.0g
$MgCl_2 \cdot 6H_2O$	20.0g
Casein hydrolysate	5.0g
Yeast extract	5.0g
KCl	2.0g
$CaCl_2 \cdot 2H_2O$	0.2g

pH 7.4 ± 0.2 at 25°C

Preparation of Medium: Add components to distilled/deionized water and bring volume to 950.0mL. Mix thoroughly. Gently heat to boiling. Adjust pH to 7.4. Bring volume to 1.0L with distilled/deionized water. Distribute into tubes or flasks. Autoclave for 15 min at 15 psi pressure–121°C. Pour into sterile Petri dishes or leave in tubes.

Use: For cultivation and maintenance of *Halobacterium* species and other halophilic bacteria.

Casitone Agar

Composition per liter:

Pancreatic digest of casein	20.0g
Agar	15.0g
$MgSO_4 \cdot 7H_2O$	1.0g
Potassium phosphate buffer, pH 7.2	1.0L

pH 7.2 ± 0.2 at 25°C

Preparation of Medium: Combine components. Mix thoroughly. Gently heat to boiling. Distribute into tubes or flasks. Autoclave for 15 min at 15 psi pressure–121°C. Pour into sterile Petri dishes or leave in tubes.

Use: For cultivation and maintenance of *Myxococcus* species.

Casitone Glycerol Yeast Autolysate Broth (CGY Autolysate Broth)

Composition per liter:

Pancreatic digest of casein	5.0g
Yeast autolysate	1.0g
Glycerol	10.0mL

Preparation of Medium: Add components to distilled/deionized water and bring volume to 1.0L. Mix thoroughly. Distribute into tubes or flasks. Autoclave for 15 min at 15 psi pressure–121°C.

Use: For the isolation, cultivation and enumeration, of iron and sulfur bacteria from the *Sphaerotilus* group.

Casitone Yeast Extract Agar

Composition per liter:

Agar	15.0g
Pancreatic digest of casein	5.0g
Yeast extract	3.0g
$MgSO_4 \cdot 7H_2O$	1.0g

Preparation of Medium: Add components to distilled/deionized water and bring volume to 1.0L. Mix thoroughly. Gently heat to boiling. Distribute into tubes or flasks. Autoclave for 15 min at 15 psi pressure–121°C. Pour into sterile Petri dishes or leave in tubes.

Use: For cultivation and maintenance of *Chitinophaga pinensis*.

Castenholtz D Medium (Medium D)

Composition per liter:

$NaNO_3$	0.7g
Na_2HPO_4	0.11g
KNO_3	0.10g
$MgSO_4 \cdot 7H_2O$	0.10g
Nitrilotriacetic acid	0.10g
$CaSO_4 \cdot 2H_2O$	0.06g
NaCl	8.0mg
$FeCl_3$ solution	1.0mL
Micronutrient solution	0.5mL

pH 7.5 ± 0.2 at 25°C

$FeCl_3$ Solution:

Composition per liter:

$FeCl_3 \cdot 6H_2O$	2.28g

Preparation of $FeCl_3$ Solution: Add $FeCl_3 \cdot 6H_2O$ to distilled/deionized water and bring volume to 1.0L. Mix thoroughly.

Micronutrient Solution:

Composition per liter:

$MnSO_4 \cdot H_2O$	2.28g
H_3BO_3	0.5g
$ZnSO_4 \cdot 7H_2O$	0.5g
$CoCl_2 \cdot 6H_2O$	0.025g
$CuSO_4 \cdot 5H_2O$	0.025g
$Na_2MoO_4 \cdot 2H_2O$	0.025g
H_2SO_4	0.5mL

Preparation of Micronutrient Solution: Add components to distilled/deionized water and bring volume to 1.0L. Mix thoroughly.

Preparation of Medium: Add nitrilotriacetic acid to 500.0mL of distilled/deionized water. Dissolve by adjusting pH to 6.5 with KOH. Add remaining components. Mix thoroughly. Readjust pH to 7.5. Bring volume to 1.0L with distilled/deionized water. Mix thoroughly. Distribute into tubes or flasks. Autoclave for 15 min at 15 psi pressure–121°C.

Use: For the isolation of cyanobacteria, including thermophilic species. For the cultivation of *Chloroflexus* species and *Fischerella* species.

Castenholtz D Medium, Modified (Medium D, Modified)

Composition per liter:

NaCl	160.0g
$NaNO_3$	0.69g
Na_2HPO_4	0.111g
KNO_3	0.103g
$MgSO_4 \cdot 7H_2O$	0.1g
Nitrilotriacetic acid	0.1g
$CaSO_4 \cdot 2H_2O$	0.06g
$FeCl_3$	0.3mg
Trace metal solution Castenholz	1.0mL

pH 7.5 ± 0.2 at 25°C

Trace Metal Solution Castenholz:

Composition per liter:

$MnSO_4 \cdot H_2O$	2.28g
H_3BO_3	0.5g
$ZnSO_4 \cdot 7H_2O$	0.5g
$Co(NO_3)_2 \cdot 6H_2O$	0.025g
$CuSO_4 \cdot 5H_2O$	0.025g
$Na_2MoO_4 \cdot 2H_2O$	0.025g
H_2SO_4	0.5mL

Preparation of Trace Metal Solution Castenholz: Add components to distilled/deionized water and bring volume to 1.0L. Mix thoroughly.

Preparation of Medium: Add nitrilotriacetic acid to 500.0mL of distilled/deionized water. Dissolve by adjusting pH to 6.5 with KOH. Add remaining components. Mix thoroughly. Readjust pH to 7.5. Bring volume to 1.0L with distilled/deionized water. Mix thoroughly. Distribute into screw-capped tubes or flasks. Autoclave for 15 min at 15 psi pressure–121°C.

Use: For the isolation of halophilic cyanobacteria.

Castenholtz DG Medium (Medium DG)

Composition per liter:

Glycyl-glycine buffer	0.8g
$NaNO_3$	0.7g
Na_2HPO_4	0.11g
KNO_3	0.10g
$MgSO_4 \cdot 7H_2O$	0.10g
Nitrilotriacetic acid	0.10g
$CaSO_4 \cdot 2H_2O$	0.06g
NaCl	8.0mg
$FeCl_3$ solution	1.0mL
Micronutrient solution	0.5mL

pH 7.5 ± 0.2 at 25°C

$FeCl_3$ Solution:

Composition per liter:

$FeCl_3 \cdot 6H_2O$	2.28g

Preparation of $FeCl_3$ Solution: Add $FeCl_3 \cdot 6H_2O$ to distilled/deionized water and bring volume to 1.0L. Mix thoroughly.

Micronutrient Solution:

Composition per liter:

$MnSO_4 \cdot H_2O$	2.28g
H_3BO_3	0.5g
$ZnSO_4 \cdot 7H_2O$	0.5g
$CoCl_2 \cdot 6H_2O$	0.025g
$CuSO_4 \cdot 5H_2O$	0.025g
$Na_2MoO_4 \cdot 2H_2O$	0.025g
H_2SO_4	0.5mL

Preparation of Micronutrient Solution: Add components to distilled/deionized water and bring volume to 1.0L. Mix thoroughly.

Preparation of Medium: Add nitrilotriacetic acid to 500.0mL of distilled/deionized water. Dissolve by adjusting pH to 6.5 with KOH. Add remaining components. Mix thoroughly. Readjust pH to 8.1. Bring volume to 1.0L with distilled/deionized water. Mix thoroughly. Distribute into tubes or flasks. Autoclave for 15 min at 15 psi pressure–121°C.

Use: For the isolation of cyanobacteria, including thermophilic species.

Castenholtz DGN Medium (Medium DGN)

Composition per liter:

Glycyl-glycine buffer	0.8g
$NaNO_3$	0.7g
NH_4Cl	0.2g
Na_2HPO_4	0.11g
KNO_3	0.10g
$MgSO_4 \cdot 7H_2O$	0.10g
Nitrilotriacetic acid	0.10g
$CaSO_4 \cdot 2H_2O$	0.06g
NaCl	8.0mg
$FeCl_3$ solution	1.0mL
Micronutrient solution	0.5mL

pH 7.5 ± 0.2 at 25°C

$FeCl_3$ Solution:

Composition per liter:

$FeCl_3 \cdot 6H_2O$	2.28g

Preparation of $FeCl_3$ Solution: Add $FeCl_3 \cdot 6H_2O$ to distilled/deionized water and bring volume to 1.0L. Mix thoroughly.

Micronutrient Solution:

Composition per liter:

$MnSO_4 \cdot H_2O$	2.28g
H_3BO_3	0.5g
$ZnSO_4 \cdot 7H_2O$	0.5g
$CoCl_2 \cdot 6H_2O$	0.025g
$CuSO_4 \cdot 5H_2O$	0.025g
$Na_2MoO_4 \cdot 2H_2O$	0.025g
H_2SO_4	0.5mL

Preparation of Micronutrient Solution: Add components to distilled/deionized water and bring volume to 1.0L. Mix thoroughly.

Preparation of Medium: Add nitrilotriacetic acid to 500.0mL of distilled/deionized water. Dissolve by adjusting pH to 6.5 with KOH. Add remaining components. Mix thoroughly. Readjust pH to 8.2. Bring volume to 1.0L with distilled/deionized water. Mix thoroughly. Distribute into tubes or flasks. Autoclave for 15 min at 15 psi pressure–121°C.

Use: For the isolation of cyanobacteria, including thermophilic species.

Castenholtz ND Medium (Medium ND)

Composition per liter:

Na_2HPO_4	0.11g
$MgSO_4 \cdot 7H_2O$	0.10g
Nitrilotriacetic acid	0.10g
$CaSO_4 \cdot 2H_2O$	0.06g
NaCl	8.0mg
$FeCl_3$ solution	1.0mL
Micronutrient solution	0.5mL

pH 7.5 ± 0.2 at 25°C

$FeCl_3$ Solution:

Composition per liter:

$FeCl_3 \cdot 6H_2O$	2.28g

Preparation of $FeCl_3$ Solution: Add $FeCl_3 \cdot 6H_2O$ to distilled/deionized water and bring volume to 1.0L. Mix thoroughly.

Micronutrient Solution:
Composition per liter:

$MnSO_4 \cdot H_2O$	2.28g
H_3BO_3	0.5g
$ZnSO_4 \cdot 7H_2O$	0.5g
$CoCl_2 \cdot 6H_2O$	0.025g
$CuSO_4 \cdot 5H_2O$	0.025g
$Na_2MoO_4 \cdot 2H_2O$	0.025g
H_2SO_4	0.5mL

Preparation of Micronutrient Solution: Add components to distilled/deionized water and bring volume to 1.0L. Mix thoroughly.

Preparation of Medium: Add nitrilotriacetic acid to 500.0mL of distilled/deionized water. Dissolve by adjusting pH to 6.5 with KOH. Add remaining components. Mix thoroughly. Readjust pH to 8.2. Bring volume to 1.0L with distilled/deionized water. Mix thoroughly. Distribute into tubes or flasks. Autoclave for 15 min at 15 psi pressure–121°C.

Use: For the isolation of cyanobacteria, including thermophilic species, that require reduced nitrogen concentrations.

Castenholz TYE Medium (Castenholz Trypticase™ Yeast Extract Medium)

Composition per liter:

Castenholz salts, 2X	500.0mL
1% TYE	100.0mL

pH 7.6 ± 0.2 at 25°C

Castenholz Salts, 2X:
Composition per liter:

Agar	30.0g
$NaNO_3$	1.4g
Na_2HPO_4	0.22g
KNO_3	0.21g
Nitrilotriacetic acid	0.2g
$MgSO_4 \cdot 7H_2O$	0.2g
$CaSO_4 \cdot 2H_2O$	0.12g
NaCl	0.016g
$FeCl_3$ (0.03% solution)	2.0mL
Nitsch's trace Elements	2.0mL

Preparation of Castenholz Salts, 2X: Add components to distilled/deionized water and bring volume to 1.0L. Mix thoroughly. Gently heat and bring to boiling. Adjust pH to 8.2. Autoclave for 15 min at 15 psi pressure–121°C.

Nitsch's Trace Elements:
Composition per liter:

$MnSO_4$	2.2g
H_3BO_3	0.5g
$ZnSO_4$	0.5g
$CoCl_2 \cdot 6H_2O$	0.046g
Na_2MoO_4	0.025g
$CuSO_4$	0.016g
H_2SO_4	0.5mL

Preparation of Nitsch's Trace Elements: Add components to distilled/deionized water and bring volume to 1.0L. Mix thoroughly.

1% TYE
Composition per liter:

Pancreatic digest of casein	10.0g
Yeast extract	10.0g

Preparation of 1% TYE: Add components to distilled/deionized water and bring volume to 1.0L. Mix thoroughly. Autoclave for 15 min at 15 psi pressure–121°C.

Preparation of Medium: Aseptically combine 500.0mL sterile Castenholz salts, 2X, 100.0mL sterile 1% TYE and 400.0mL sterile distilled/deionized water. Adjust pH to 7.6.

Use: For the cultivation and maintenance of *Thermonema lapsum* and *Thermus* species.

Castenholz TYE Medium with 2% Trypticase™ Yeast Extract

Composition per liter:

Castenholz salts, 2X	500.0mL
2% TYE	100.0mL

pH 7.6 ± 0.2 at 25°C

Castenholz Salts, 2X:
Composition per liter:

Agar	30.0g
$NaNO_3$	1.4g
Na_2HPO_4	0.22g
KNO_3	0.21g
$MgSO_4 \cdot 7H_2O$	0.2g
Nitrilotriacetic acid	0.2g
$CaSO_4 \cdot 2H_2O$	0.12g
NaCl	0.016g
$FeCl_3$ solution (0.03% solution)	2.0mL
Nitsch's trace Elements	2.0mL

Preparation of Castenholz Salts, 2X: Add components to distilled/deionized water and bring volume to 1.0L. Mix thoroughly. Gently heat and bring to boiling. Adjust pH to 8.2. Autoclave for 15 min at 15 psi pressure–121°C.

Nitsch's Trace Elements:
Composition per liter:

$MnSO_4$	2.2g
H_3BO_3	0.5g

$ZnSO_4$	0.5g
$CoCl_2 \cdot 6H_2O$	0.046g
Na_2MoO_4	0.025g
$CuSO_4$	0.016g
H_2SO_4	0.5mL

Preparation of Nitsch's Trace Elements: Add components to distilled/deionized water and bring volume to 1.0L. Mix thoroughly.

2% TYE

Composition per liter:

Pancreatic digest of casein	20.0g
Yeast extract	20.0g

Preparation of 2% TYE: Add components to distilled/deionized water and bring volume to 1.0L. Mix thoroughly. Autoclave for 15 min at 15 psi pressure–121°C.

Preparation of Medium: Aseptically combine 500.0mL sterile Castenholz salts, 2X, 100.0mL sterile 2% TYE and 400.0mL sterile distilled/deionized water. Adjust pH to 7.6.

Use: For the cultivation and maintenance of *Thermus* species.

Caulobacter Medium

Composition per liter:

Agar	10.0g
Peptone	2.0g
Yeast extract	1.0g
$MgSO_4 \cdot 7H_2O$	0.2g
Riboflavin	1.0mg

pH 7.0 ± 0.2 at 25°C

Preparation of Medium: Add components to tap water and bring volume to 1.0L. Mix thoroughly. Gently heat and bring to boiling. Distribute into tubes or flasks. Autoclave for 15 min at 15 psi pressure–121°C. Pour into sterile Petri dishes or leave in tubes.

Use: For the cultivation of *Caulobacter* species from fresh water.

Caulobacter Medium

Composition per liter:

Agar	10.0g
Peptone	0.5g
Seawater, filtered	1.0L

pH 7.0 ± 0.2 at 25°C

Preparation of Medium: Combine components. Mix thoroughly. Gently heat and bring to boiling. Distribute into tubes or flasks. Autoclave for 15 min at 15 psi pressure–121°C. Pour into sterile Petri dishes or leave in tubes.

Use: For the cultivation of *Caulobacter* species from marine isolates.

Caulobacter Medium

Composition per liter:

Glucose	1.0g
Peptone	1.0g
Yeast extract	1.0g
Salt solution	100.0mL

Salt Solution:

Composition per 100mL:

EDTA	0.1g
KNO_3	0.1g
K_2HPO_4	0.066g
$MgSO_4$	0.033g
$FeSO_4 \cdot 7H_2O$	9.3mg
$NaBO_3 \cdot 4H_2O$	2.63mg
$MgCl_2 \cdot 4H_2O$	1.81mg
$CaCl_2$	1.2mg
$(NH_4)_6Mo_7O_{24} \cdot 7H_2O$	1.0mg
$ZnSO_4 \cdot 7H_2O$	0.22mg
$CuSO_4 \cdot 5H_2O$	0.079mg
$Co(NO_3)_2 \cdot H_2O$	0.02mg

Preparation of Salt Solution: Add components to distilled/deionized water and bring volume to 100.0mL. Mix thoroughly.

Preparation of Medium: Add components to distilled/deionized water and bring volume to 1.0L. Mix thoroughly. Distribute into tubes or flasks. Autoclave for 15 min at 15 psi pressure–121°C.

Use: For the enrichment of *Stella* species from polluted waters.

Caulobacter Medium

Composition per liter:

Agar	10.0g
Peptone	2.0g
Yeast extract	1.0g
$MgSO_4 \cdot 7H_2O$	0.2g

Preparation of Medium: Add components to tap water and bring volume to 1.0L. Mix throughly. Gently heat to boiling. Distribute into tubes or flasks. Autoclave for 15 min at 15 psi–121°C. Pour into sterile Petri dishes or leave in tubes.

Use: For cultivation and maintenance of *Asticcacaulis excentricus, Caulobacter* species, *Labrys monachus, Pedomicrobium* species, *Pirellula staleyi, Pseudomonas carboxydohydrogena* and *Stella* species.

CC Medium

Composition per liter:

Agar	20.0g
KH_2PO_4	4.0g
Potato starch	0.5g
Solution 3	100.0mL
Solution 1	10.0mL

pH 7.3 ± 0.2 at 25°C

Solution 1:

Composition per liter:

$MgSO_4 \cdot 7H_2O$	20.0g
$CaCl_2 \cdot 2H_2O$	2.0g
$FeSO_4 \cdot 7H_2O$	0.4g
H_3BO_3	0.02g
$MnSO_4 \cdot 2H_2O$	0.015g
$NaMoO_4 \cdot 2H_2O$	0.015g
KJ	0.010g
$ZnSO_4$	4.0mg
$CoCl_2 \cdot 4H_2O$	0.4mg
$CuSO_4 \cdot 5H_2O$	0.4mg

Preparation of Solution 1: Add components to distilled/deionized water and bring volume to 1.0L. Mix thoroughly. Adjust pH with 10.0mL of 10% HCl solution.

Solution 3:

Composition per 100mL:

Pancreatic digest of casein	12.0g
Yeast extract	12.0g
L-Cysteine·HCl	0.5g
L-Asparagine	0.03g
DL-Tryptophan	0.02g
Solution 2	12.0mL

Preparation of Solution 3: Add components to distilled/deionized water and bring volume to 100.0mL. Mix thoroughly. Filter sterilize.

Solution 2:

Composition per 100mL:

p-Aminobenzoic acid	0.02g
Calcium pantothenate	0.02g
m-Inositol	0.02g
Pyridoxine·HCl	0.02g
Thiamine·HCl	0.02g
Nicotinamide	0.01g
Nicotinic acid	0.01g
Folic acid	5.0mg
Biotin	1.0mg
Vitamin B_{12}	1.0mL

Preparation of Solution 2: Add components to distilled/deionized water and bring volume to 100.0mL. Mix thoroughly.

Preparation of Medium: Add KH_2PO_4, to distilled/deionized water and bring volume to 250.0mL. Mix thoroughly. Adjust pH to 7.6 with NaOH. Add 10.0mL of solution 1. In a separate flask, add potato starch to 70.0mL of boiling distilled/deionized water. Add potato starch solution to other solution. Add agar. Bring volume to 900.0mL of distilled/deionized water. Autoclave for 15 min at 15 psi pressure–121°C. Cool to 45–50°C. Aseptically add 100.0mL of sterile solution 3. Mix thoroughly. Pour into sterile Petri dishes or distribute into sterile tubes.

Use: For the isolation and cultivation of *Actinomycetes* species.

CCPC Medium

Composition per liter:

Sucrose	30.0g
Peptone	2.0g
Casein hydrolysate	1.0g
$K_2HPO_4 \cdot 3H_2O$	1.0g
KCl	0.5g
$MgSO_4 \cdot 7H_2O$	0.5g
$FeSO_4 \cdot 7H_2O$	0.1g

pH 7.2 ± 0.2 at 25°C

Preparation of Medium: Add components to tap water and bring volume to 1.0L. Mix thoroughly. Distribute into tubes or flasks. Autoclave for 15 min at 15 psi pressure–121°C.

Use: For the cultivation of *Actinoplanes* species.

CCVC Medium (Cephalothin Cycloheximide Vancomycin Colistin Medium

Composition per liter:

BCYE-alpha base	990.0mL
Antibiotic supplement	10.0mL

pH 6.9 ± 0.2 at 25°C

Source: This medium is available as a premixed powder from BBL Microbiology Systems.

BCYE—Alpha Base:

Composition per liter:

Agar	15.0g
Yeast extract	10.0g
ACES buffer (2-[(2-Amino-2-oxoethyl)-amino]-ethane sulfonic acid)	10.0g
Charcoal, activated	2.0g
α-Ketogluatrate	1.0g
$Fe_4(P_2O_7)_3 \cdot 9H_2O$	0.25g
Cysteine·HCl·H_2O solution	10.0mL

Cysteine·HCl·H_2O Solution:

Composition per 10mL:

L-Cysteine·HCl·H_2O	0.4g

Preparation of Cysteine·HCl·H_2O Solution: Add cysteine·HCl·H_2O to distilled/deionized water and bring volume to 10.0mL. Mix thoroughly. Filter sterilize.

Preparation of BCYE—Alpha Base: Add components, except cysteine solution, to distilled/deionized water and bring volume to 990.0mL. Mix thoroughly. Adjust medium to pH 6.9 with 1*N* KOH. Heat gently and bring to boiling for 1 min. Autoclave for 15 min at 15 psi pressure–121°C. Cool to 50–55°C. Add 4.0mL of L-cysteine·HCl·H_2O solution. Mix thoroughly.

Antibiotic Supplement Solution:
Composition per 10.0mL:

Cycloheximide	80.0mg
Colistin	16.0mg
Cephalothin	4.0mg
Vancomycin	0.5mg

Preparation of Antibiotic Supplement Solution: Add components to 10.0mL of distilled/deionized water. Filter sterilize.

Preparation of Medium: To cooled BCYE-alpha base add 10.0mL sterile antibiotic supplement. Mix thoroughly. Adjust pH to 6.9 with sterile 1*N* KOH. Pour into sterile Petri dishes with constant agitation to keep charcoal in suspension.

Use: For the selective isolation and cultivation of *Legionella* species from environmental samples.

Cellulolytic Agar for Thermophiles

Composition per liter:

Agar	30.0g
K_2HPO_4	1.65g
NH_4SO_4	1.6g
Yeast extract	1.0g
NaCl	0.96g
Cysteine·HCl·H_2O	0.5g
$CaCl_2$	0.096g
$MgSO_4$	0.096g
Cellulose suspension	200.0mL
Resazurin (0.1% solution)	1.0mL

pH 7.2 ± 0.2 at 25°C

Cellulose Suspension:
Composition per 200mL:

Cellulose powder, Whatman CF11	8.0g

Preparation of Cellulose Suspension: Add cellulose powder to 200.0mL of distilled/deionized water and mix thoroughly.

Preparation of Medium: Prepare and dispense medium anaerobically in 100% N_2. Add components to distilled/deionized water and bring volume to 1.0L. Mix thoroughly. Adjust pH to 7.2 with 5*M* NaOH. Distribute into tubes or flasks. Autoclave for 15 min at 15 psi pressure–121°C.

Use: For cultivation of *Clostridium stercorarium* and other bacteria which can utilize cellulose as a carbon source.

Cellulolytic Agar with Sea Salts

Composition per liter:

Agar	20.0g
NH_4Cl	2.0g
K_2HPO_4	1.65g
Yeast extract	1.2g
Cysteine·HCl·H_2O	0.5g
Cellulose suspension	200.0mL
Filtered sea water	200.0mL
Mineral solution	150.0mL
Resazurin (0.1% solution)	1.0mL

pH 7.2 ± 0.2 at 25°C

Cellulose Suspension:
Composition per 200mL:

Cellulose powder, Whatman CF11	8.0g

Preparation of Cellulose Suspension: Add cellulose powder to 200.0mL of distilled/deionized water and mix thoroughly.

Mineral Solution:
Composition per liter:

NaCl	6.0g
$(NH_4)_2SO_4$	6.0g
$CaCl_2$	0.6g
$MgSO_4$	0.6g

Preparation of Mineral Solution: Add components to distilled/deionized water and bring volume to 1.0L. Mix thoroughly.

Preparation of Medium: Prepare and dispense medium anaerobically under 100% N_2. Add components to distilled/deionized water and bring volume to 1.0L. Mix thoroughly. Adjust pH to 7.2 with 5*M* NaOH. Distribute into tubes or flasks. Autoclave for 15 min at 15 psi pressure–121°C.

Use: For cultivation and maintenance of *Clostridium papyrosolvens* and other marine bacteria which can utilize cellulose as a carbon source.

Cellulolytic Broth for Thermophiles

Composition per liter:

K_2HPO_4	1.65g
NH_4SO_4	1.6g
Yeast extract	1.0g
NaCl	0.96g
Cysteine·HCl·H_2O	0.5g
$CaCl_2$	0.096g
$MgSO_4$	0.096g
Resazurin (0.1% solution)	1.0mL

pH 7.2 ± 0.2 at 25°C

Preparation of Medium: Prepare and dispense medium anaerobically in 100% N_2. Add components to distilled/deionized water and bring volume to 1.0L. Mix thoroughly. Adjust pH to 7.2 with 5*M* NaOH. Distribute into tubes or flasks which contain cellulose as a strip (4.5cm × 1.0cm) of Whatman No. 1 filter paper. Autoclave for 15 min at 15 psi pressure–121°C.

Use: For cultivation of *Clostridium stercorarium* and other bacteria which can utilize cellulose as a carbon source.

Cellulolytic Broth with Sea Salts

Composition per liter:

K_2HPO_4	1.65g
NH_4Cl	1.0g
Yeast extract	0.6g
Cysteine·HCl·H_2O	0.5g
Filtered sea water	200.0mL
Mineral solution	150.0mL
Resazurin (0.1% solution)	1.0mL

pH 7.2 ± 0.2 at 25°C

Mineral Solution:

Composition per liter:

NaCl	6.0g
$(NH_4)_2SO_4$	6.0g
$CaCl_2$	0.6g
$MgSO_4$	0.6g

Preparation of Mineral Solution: Add components to distilled/deionized water and bring volume to 1.0L. Mix thoroughly.

Preparation of Medium: Prepare and dispense medium anaerobically in 100% N_2 atmosphere. Add components to distilled/deionized water and bring volume to 1.0L. Adjust pH to 7.2 with 5*M* NaOH. Distribute into tubes or flasks which contain cellulose as a strip (4.5cm × 1.0cm) of Whatman No. 1 filter paper. Autoclave for 15 min at 15 psi pressure–121°C.

Use: For cultivation and maintenance of *Clostridium papyrosolvens* and other marine bacteria which can utilize cellulose as a carbon source.

Cellulolytic Clostridia Medium

Composition per liter:

Cellulose	20.0g
$CaCO_3$	2.0g
K_2HPO_4	1.0g
$(NH_4)_2SO_4$	1.0g
$MgSO_4·7H_2O$	0.5g
NaCl	0.5g
Resazurin	1.0mg

pH 7.1 ± 0.2 at 25°C

Preparation of Medium: Add components to distilled/deionized water and bring volume to 1.0L. Mix thoroughly. Distribute into tubes or flasks. Autoclave for 15 min at 15 psi pressure–121°C.

Use: For the isolation, cultivation and enrichment of cellulolytic *Clostridium* species.

Cellulolytic Medium with Rumen Fluid

Composition per liter:

Basal medium	975.0mL
Alkaline solution	25.0mL

pH 6.8 ± 0.2 at 25°C

Basal Medium:

Composition per 975mL:

Agar	15.0g
$NaHCO_3$	6.37g
Pancreatic digest of casein	5.0g
Cellobiose	5.0g
NaCl	0.90g
$(NH_4)_2SO_4$	0.90g
K_2HPO_4	0.45g
KH_2PO_4	0.45g
$MgSO_4·7H_2O$	0.18g
$CaCl_2$	0.09g
Resazurin	1.0mg
Rumen fluid, clarified	400.0mL

Preparation of Basal Medium: Add components to distilled/deionized water and bring volume to 975.0mL. Mix thoroughly. Gently heat and bring to boiling under a gas phase of 98% CO_2 + 2% H_2. Cool slightly.

Alkaline Solution:

Composition per 25mL:

Cysteine·HCl·H_2O	0.25g
$Na_2S·9H_2O$	0.25g

Preparation of Alkaline Solution: Add components to 25.0mL of distilled/deionized water. Mix thoroughly. Freshly prepare.

Preparation of Medium: Prepare 975.0mL of basal medium. Heat to boiling and cool as directed. Add 25.0mL of freshly prepared alkaline solution. Distribute into tubes using anaerobic techniques under a gas phase of 98% CO_2 + 2% H_2. Autoclave for 15 min at 15 psi pressure–121°C. Adjust pH to 6.8.

Use: For cultivation and maintenance of *Clostridium polysaccharolyticum*.

Cellulolytic Medium with Rumen Fluid and Soluble Starch

Composition per liter:

Basal medium	975.0mL
Alkaline solution	25.0mL

pH 6.8 ± 0.2 at 25°C

Basal Medium:

Composition per 975mL:

Agar	15.0g
$NaHCO_3$	6.37g
Pancreatic digest of casein	5.0g
Cellobiose	5.0g
Soluble starch	5.0g
NaCl	0.90g
$(NH_4)_2SO_4$	0.90g
K_2HPO_4	0.45g
KH_2PO_4	0.45g
$MgSO_4 \cdot 7H_2O$	0.18g
$CaCl_2$	0.09g
Resazurin	1.0mg

Preparation of Basal Medium: Add components to distilled/deionized water and bring volume to 975.0mL. Mix thoroughly. Gently heat and bring to boiling under a gas phase of 98% CO_2 + 2% H_2. Cool slightly.

Alkaline Solution:

Composition per 25mL:

Cysteine·HCl·H_2O	0.25g
$Na_2S \cdot 9H_2O$	0.25g

Preparation of Alkaline Solution: Add components to 25.0mL of distilled/deionized water. Mix thoroughly. Prepare freshly.

Preparation of Medium: Prepare 975.0mL of basal medium. Heat to boiling and cool as directed. Add 25.0mL of freshly prepared alkaline solution. Distribute into tubes using anaerobic techniques under a gas phase of 98% CO_2 + 2% H_2. Autoclave for 15 min at 15 psi pressure–121°C. Adjust pH to 6.8.

Use: For cultivation of *Selenomonas ruminantium* and *Succinimonas amylolytica*.

Cellulomonas PTYG Medium (*Cellulomonas* Peptone Tryptone Yeast Extract Glucose Medium)

Composition per liter:

Agar	15.0g
Glucose	5.0g
Peptone	5.0g
Pancreatic digest of casein	5.0g
Yeast extract	5.0g

Preparation of Medium: Add components to distilled/deionized water and bring volume to 1.0L. Mix thoroughly. Gently heat and bring to boiling. Distribute into tubes or flasks. Autoclave for 15 min at 15 psi pressure–121°C. Pour into sterile Petri dishes or leave in tubes.

Use: For cultivation and maintenance of *Cellulomonas* species.

Cellulose Broth

Composition per liter:

Cellulose, powdered	1.0g
K_2HPO_4	1.0g
$(NH_4)_2SO_4$	1.0g
$MgSO_4 \cdot 7H_2O$	0.2g
$CaCl_2 \cdot 2H_2O$	0.1g
$FeCl_3$	0.02g

pH 7.0–7.5 at 25°C

Preparation of Medium: Add cellulose to 100.0mL of distilled/deionized water. Mix thoroughly. In a separate flask, add remaining components to distilled/deionized water and bring volume to 900.0mL. Mix thoroughly. Autoclave both solutions separately for 15 min at 15 psi pressure–121°C. Cool to 45–50°C. Aseptically combine the two sterile solutions. Mix thoroughly. Aseptically distribute into sterile tubes or flasks.

Use: For the isolation and cultivation of *Cytophaga* species, *Herpetosiphon* species, *Saprospira* species, and *Flexithrix* species.

Cellulose Overlay Agar

Composition per plate:

Stan 5 agar	15.0mL
Cellulose overlay agar	5.0mL

Stan 5 Agar:

Composition per liter:

Solution B	650.0mL
Solution A	350.0mL

Solution A:

Composition per 350mL:

$CaCl_2 \cdot 2H_2O$	1.0g
$(NH_4)_2SO_4$	1.0g
$MgSO_4 \cdot 7H_2O$	1.0g
Trace element solution	1.0mL

Preparation of Solution A: Add components to distilled/deionized water and bring volume to 350.0mL. Mix thoroughly. Gently heat and bring to boiling. Autoclave for 15 min at 15 psi pressure–121°C. Cool to 45–50°C.

Trace Element Solution:

Composition per liter:

EDTA	8.0g
$MnCl_2 \cdot 4H_2O$	0.1g
$CoCl_2$	0.02g
KBr	0.02g
$ZnCl_2$	0.02g
$CuSO_4$	0.01g
H_3BO_3	0.01g
$NaMoO_4 \cdot 2H_2O$	0.01g
$BaCl_2$	5.0mg
LiCl	5.0mg
$SnCl_2 \cdot 2H_2O$	5.0mg

Preparation of Trace Element Solution: Add components to distilled/deionized water and bring volume to 1.0L. Mix thoroughly.

Solution B:

Composition per 650mL:

Agar	10.0g
K_2HPO_4	1.0g

Preparation of Solution B: Add components to distilled/deionized water and bring volume to 650.0mL. Mix thoroughly. Gently heat and bring to boiling. Autoclave for 15 min at 15 psi pressure–121°C. Cool to 45–50°C.

Preparation of Stan 5 Agar: Aseptically combine 350.0mL of cooled, sterile solution A and 650.0mL of cooled sterile solution B. Mix thoroughly.

Cellulose Overlay Agar:

Composition per liter:

Solution A	350.0mL
Solution B	650.0mL

Solution A:

Composition per 350mL:

$CaCl_2 \cdot 2H_2O$	1.0g
$(NH_4)_2SO_4$	1.0g
$MgSO_4 \cdot 7H_2O$	1.0g
Trace element solution	1.0mL

Preparation of Solution A: Add components to distilled/deionized water and bring volume to 350.0mL. Mix thoroughly. Gently heat and bring to boiling. Autoclave for 15 min at 15 psi pressure–121°C. Cool to 45–50°C.

Trace Element Solution:

Composition per liter:

EDTA	8.0g
$MnCl_2 \cdot 4H_2O$	0.1g
$CoCl_2$	0.02g
KBr	0.02g
$ZnCl_2$	0.02g
$CuSO_4$	0.01g
H_3BO_3	0.01g
$NaMoO_4 \cdot 2H_2O$	0.01g
$BaCl_2$	5.0mg
LiCl	5.0mg
$SnCl_2 \cdot 2H_2O$	5.0mg

Preparation of Trace Element Solution: Add components to distilled/deionized water and bring volume to 1.0L. Mix thoroughly.

Solution B:

Composition per 650mL:

Agar	10.0g
K_2HPO_4	1.0g

Preparation of Solution B: Add components to distilled/deionized water and bring volume to 650.0mL. Mix thoroughly. Gently heat and bring to boiling. Autoclave for 15 min at 15 psi pressure–121°C. Cool to 45–50°C.

Preparation of Cellulose Overlay Agar: Aseptically combine 350.0mL of cooled, sterile solution A and 650.0mL of cooled sterile solution B. Mix thoroughly.

Preparation of Medium: Pour cooled sterile Stan 5 agar into sterile Petri dishes in 15.0mL volumes. Allow agar to solidify. Overlay each plate with 5.0mL of cellulose overlay agar.

Use: For the cultivation of myxobacteria.

Centenum Medium

Composition per liter of tap water:

Agar	20.0g
Yeast extract	10.0g
Sodium pyruvate	2.2g
K_2HPO_4	1.0g

$MgSO_4$ 0.5g
Vitamin B_{12} 0.02mg

pH 7.0–7.2 at 25°C

Preparation of Medium: Add components to distilled/deionized water and bring volume to 1.0L. Mix thoroughly. Gently heat and bring to boiling. Distribute into tubes or flasks. Autoclave for 15 min at 15 psi pressure–121°C. Pour into sterile Petri dishes or leave in tubes.

Use: For cultivation and maintenance of *Rhodospirillum* species.

Cereal Agar

Composition per liter:

Cereal, precooked mixed 100.0g
Agar 15.0g

Preparation of Medium: Add components to distilled/deionized water and bring volume to 1.0L. Mix thoroughly. Gently heat and bring to boiling. Autoclave for 15 min at 15 psi pressure–121°C. Pour into sterile Petri dishes or distribute into sterile tubes. Allow tubes to cool in a slanted position.

Use: For the cultivation and sporulation of fungi.

Cetrimide Agar, Non-USP

Composition per liter:

Beef heart, solids from infusion 500.0g
Agar 15.0g
Tryptose 10.0g
NaCl 5.0g
Cetrimide 0.9g

pH 7.2 ± 0.2 at 25°C

Preparation of Medium: Add components to distilled/deionized water and bring volume to 1.0L. Mix thoroughly. Gently heat and bring to boiling. Distribute into tubes or flasks. Autoclave for 15 min at 13 psi pressure–118°C. Pour into sterile Petri dishes or leave in tubes.

Use: For the selective isolation, cultivation and identification of *Pseudomonas aeruginosa* and other Gram-negative nonfermentative bacteria.

Cetrimide Agar, USP (Pseudosel® Agar)

Composition per liter:

Pancreatic digest of gelatin 20.0g
Agar 13.6g
K_2SO_4 .... 10.0g
$MgCl_2$ 1.4g
Cetrimide 0.3g
Glycerol 10.0mL

pH 7.2 ± 0.2 at 25°C

Source: This medium is available as a premixed powder from Difco Laboratories and BBL Microbiology Systems.

Preparation of Medium: Add components to distilled/deionized water and bring volume to 1.0L. Mix thoroughly. Gently heat and bring to boiling. Distribute into tubes or flasks. Autoclave for 15 min at 13 psi pressure–118°C. Pour into sterile Petri dishes or leave in tubes.

Use: For the selective isolation, cultivation and identification of *Pseudomonas aeruginosa* and other Gram-negative nonfermentative bacteria.

CH 1 Medium

Composition per liter:

NaCl 250.0g
Tris 12.0g
Glycerol 10.0g
Hy-Case SF 5.0g
Yeast extract 5.0g
Solution 1 .... 50.0mL

Solution 1:

Composition per liter:

$MgCl_2 \cdot 6H_2O$ 40.0g
KCl 4.0g
$CaCl_2 \cdot 2H_2O$ 0.4g

pH 7.4 ± 0.2 at 25°C

Preparation of Solution 1: Add components to distilled/deionized water and bring volume to 1.0L. Mix thoroughly.

Preparation of Medium: Add components to distilled/deionized water and bring volume to 1.0L. Mix thoroughly. Adjust pH to 7.4. Distribute into tubes or flasks. Autoclave for 15 min at 15 psi pressure–121°C.

Use: For the cultivation of *Haloarcula vallismortis*.

Chapman Stone Agar

Composition per liter:

$(NH_4)_2SO_4$ 75.0g
NaCl 55.0g
Gelatin 30.0g
Agar 15.0g
D-Mannitol 10.0g
Pancreatic digest of casein 10.0g
K_2HPO_4 .... 5.0g
Yeast extract 2.0g

pH 7.0 ± 0.2 at 25°C

Source: This medium is available as a premixed powder from BBL Microbiology Systems and Difco Laboratories.

Preparation of Medium: Add components to distilled/deionized water and bring volume to 1.0L. Mix thoroughly. Autoclave for 10 min at 15 psi pressure–121°C. Pour into sterile Petri dishes while the medium is still hot. Add 25.0mL of medium per Petri dish.

Use: For the isolation of staphylococci from a variety of specimens.

CHCA Salts Medium (Cyclohexane Carboxylic Acid Salts Medium)

Composition per liter:

K_2HPO_4	3.5g
KH_2PO_4	1.5g
Cyclohexane carboxylic acid	1.0g
NH_4NO_3	1.0g
$MgSO_4 \cdot 7H_2O$	0.5g
$FeSO_4 \cdot 7H_2O$	0.1g
Yeast extract	0.1g
$CaCl_2 \cdot 2H_2O$	0.01g
$Na_2MoO_2 \cdot 2H_2O$	0.01g
$ZnSO_4 \cdot 7H_2O$	0.01g

pH 7.0 ± 0.2 at 25°C

Preparation of Medium: Add components to distilled/deionized water and bring volume to 1.0L. Mix thoroughly. Adjust pH to 7.0. Gently heat and bring to boiling. Distribute into tubes or flasks. Autoclave for 15 min at 15 psi pressure–121°C. Pour into sterile Petri dishes or leave in tubes.

Use: For the cultivation and maintenance of bacteria that can utilize cyclohexane carboxylic acid as a carbon source. For the cultivation and maintenance of *Arthrobacter globiformis*.

Chitin Agar

Composition per liter:

Agar	15.0g
Chitin, precipitated	3.0g
$(NH_4)_2SO_4$	2.0g
Na_2HPO_4	1.1g
KH_2PO_4	0.7g
$MgSO_4 \cdot 7H_2O$	0.2g
$FeSO_4$	1.0mg
$MnSO_4$	1.0mg

Chitin, Precipitated:

Composition:

Chitin	40.0g
HCl, concentrated	400.0mL

Preparation of Chitin, Precipitated: Add chitin to 400.0mL of cold concentrated HCl. Add this solution to 2.0L of distilled/deionized water at 5°C. Filter the solution through Whatman #1 filter paper. Dialyze the precipitated chitin against tap water for 12 hr. Adjust the pH to 7.0 with KOH.

Preparation of Medium: Add components to distilled/deionized water and bring volume to 1.0L. Mix thoroughly. Gently heat and bring to boiling. Distribute into tubes or flasks. Autoclave for 15 min at 15 psi pressure–121°C. Pour into sterile Petri dishes or leave in tubes.

Use: For the isolation and cultivation of *Cytophaga* species, *Herpetosiphon* species, *Saprospira* species, and *Flexithrix* species.

Chitin Agar

Composition per liter:

Agar	20.0g
Chitin	4.0g
K_2HPO_4	0.7g
$MgSO_4 \cdot 7H_2O$	0.5g
KH_2PO_4	0.3g
$FeSO_4 \cdot 7H_2O$	0.01g
$MnCl_2 \cdot 4H_2O$	0.001g
$ZnSO_4 \cdot 7H_2O$	0.001g

pH 8.0 ± 0.2 at 25°C

Preparation of Medium: Add components to distilled/deionized water and bring volume to 1.0L. Mix thoroughly. Gently heat and bring to boiling. Distribute into tubes or flasks. Autoclave for 15 min at 15 psi pressure–121°C. Pour into sterile Petri dishes or leave in tubes.

Use: For the selective isolation and cultivation of streptomycetes.

Chlorobiaceae Medium 1

Composition per 4990mL:

Solution 1	4.0L
O_2-free water	860.0mL
$NaHCO_3$ solution	100.0mL
$Na_2S \cdot 9H_2O$ solution	20.0mL
Trace element solution	5.0mL
Vitamin B_{12} solution	5.0mL

pH 6.8 ± 0.2 at 25°C

Solution 1:

Composition per 4L:

$MgSO_4 \cdot 7H_2O$	2.5g
KCl	1.7g
KH_2PO_4	1.7g
NH_4Cl	1.7g
$CaCl_2 \cdot 2H_2O$	1.25g

Preparation of Solution 1: Add components to distilled/deionized water and bring volume to 4.0L. Mix thoroughly. Autoclave for 45 min at 15 psi pressure–121°C. Cool to 25°C under 100% N_2. Saturate with CO_2 by stirring under 100% CO_2 for 30 min.

O_2-Free Water:

Composition per 860mL:

H_2O 860.0mL

Preparation of O_2-Free Water: Autoclave H_2O for 15 min at 15 psi pressure–121°C. Cool to 25°C under 100% N_2.

$NaHCO_3$ Solution:

Composition per 100mL:

$NaHCO_3$ 7.5g

Preparation of $NaHCO_3$ Solution: Add the $NaHCO_3$ to distilled/deionized water and bring volume to 100.0mL. Mix thoroughly. Gas with 100% CO_2 for 20 min. Filter sterilize with positive CO_2 pressure.

$Na_2S{\cdot}9H_2O$ Solution:

Composition per 100mL:

$Na_2S{\cdot}9H_2O$ 10.0g

Preparation of $Na_2S{\cdot}9H_2O$ Solution: Add $Na_2S{\cdot}9H_2O$ to distilled/deionized water. Mix thoroughly. Gas with 100% N_2 for 15 min in a screw-capped bottle. Tightly close cap. Autoclave for 15 min at 15 psi pressure–121°C. Cool to 25°C.

Trace Element Solution:

Composition per liter:

$FeCl_2{\cdot}4H_2O$	1.5g
$CoCl_2{\cdot}6H_2O$	0.19g
$MnCl_2{\cdot}4H_2O$	0.10g
$ZnCl_2$	0.07g
H_3BO_3	0.06g
$NaMoO_4{\cdot}2H_2O$	0.04g
$CuCl_2{\cdot}2H_2O$	0.02g
$NiCl_2{\cdot}6H_20$	0.02g
HCl (25% solution)	6.5mL

Preparation of Trace Element Solution: Add components to distilled/deionized water and bring volume to 1.0L. Mix thoroughly. Autoclave for 15 min at 15 psi pressure–121°C. Cool to 25°C.

Vitamin B_{12} Solution:

Composition per 100mL:

Vitamin B_{12} 2.0mg

Preparation of Vitamin B_{12} Solution: Add vitamin B_{12} to distilled/deionized water and bring volume to 100.0mL. Mix thoroughly. Filter sterilize.

Preparation of Medium: To 4.0L of sterile, CO_2-saturated solution 1, aseptically add the remaining components. Mix thoroughly. Adjust pH to 6.8. Aseptically distribute into sterile 100.0mL bottles using positive pressure of 95% N_2 + 5% CO_2. Completely fill bottles with medium except for a pea-sized air bubble.

Use: For the isolation and cultivation of members of the Chlorobiaceae.

Chlorobiaceae Medium 2

Composition per 1051mL:

Solution 1	950.0mL
$Na_2S{\cdot}9H_2O$ solution	60.0mL
$NaHCO_3$ solution	40.0mL
Vitamin B_{12} solution	1.0mL

pH 6.8 ± 0.2 at 25°C

Solution 1:

Composition per 950mL:

KH_2PO_4	1.0g
NH_4Cl	0.5g
$MgSO_4{\cdot}7H_2O$	0.4g
$CaCl_2{\cdot}2H_2O$	0.05g
Trace element solution SL-8	1.0mL

Preparation of Solution 1: Add components to distilled/deionized water and bring volume to 950.0mL. Mix thoroughly. Autoclave for 15 min at 15 psi pressure–121°C. Cool to 45–50°C.

Trace Element Solution SL-8:

Composition per liter:

Disodium EDTA	5.2g
$FeCl_2{\cdot}4H_2O$	1.5g
$CoCl_2{\cdot}6H_2O$	0.19g
$MnCl_2{\cdot}4H_2O$	0.10g
$ZnCl_2$	0.07g
H_3BO_3	0.06g
$NaMoO_4{\cdot}2H_2O$	0.04g
$CuCl_2{\cdot}2H_2O$	0.02g
$NiCl_2{\cdot}6H_20$	0.02g

Preparation of Trace Element Solution SL-8: Add components to distilled/deionized water and bring volume to 1.0L. Mix thoroughly.

$Na_2S{\cdot}9H_2O$ Solution:

Composition per 100mL:

$Na_2S{\cdot}9H_2O$ 5.0g

Preparation of $Na_2S{\cdot}9H_2O$ Solution: Add $Na_2S{\cdot}9H_2O$ to distilled/deionized water and bring volume to 100.0mL. Autoclave for 15 min at 15 psi pressure–121°C. Cool to 45–50°C.

$NaHCO_3$ Solution:

Composition per 100mL:

$NaHCO_3$ 5.0g

Preparation of $NaHCO_3$ Solution: Add $NaHCO_3$ to distilled/deionized water and bring volume to 100.0mL. Mix thoroughly. Filter sterilize.

Vitamin B_{12} Solution:

Composition per 100mL:

Vitamin B_{12}	2.0mg

Preparation of Vitamin B_{12} Solution: Add vitamin B_{12} to distilled/deionized water and bring volume to 100.0mL. Mix thoroughly. Filter sterilize.

Preparation of Medium: To 950.0mL of cooled, sterile solution 1, aseptically add 60.0mL of sterile $Na_2S{\cdot}9H_2O$ solution, 40.0mL of sterile $NaHCO_3$ solution, and 1.0mL of sterile vitamin B_{12} solution. Mix thoroughly. Adjust pH to 6.8 with sterile H_2SO_4 or Na_2CO_3. Aseptically distribute into sterile 50.0mL or 100.0mL bottles with metal screw caps and rubber seals. Completely fill bottles with medium except for a pea-sized air bubble.

Use: For the isolation and cultivation of freshwater and soil members of the Chlorobiaceae.

Chloroflexus Agar

Composition per liter:

Agar	15.0g
Glycyl-glycine	0.5g
Yeast extract	0.5g
Na_2S	0.5g
NH_4Cl	0.2g
$MgSO_4{\cdot}7H_2O$	0.1g
Nitrilotriacetic acid	0.1g
$NaNO_3$	0.689g
Na_2HPO_4	0.111g
KNO_3	0.103g
$CaSO_4{\cdot}2H_2O$	0.06g
NaCl	8.0mg
$FeCl_3$ solution	1.0mL
Micronutrient solution	1.0mL

pH 8.2–8.4 at 25°C

$FeCl_3$ Solution:

Composition per liter:

$FeCl_3$	0.29g

Preparation of $FeCl_3$ Solution: Add $FeCl_3$ to distilled/deionized water and bring volume to 1.0L. Mix thoroughly.

Micronutrient Solution:

Composition per liter:

$MnSO_4{\cdot}7H_2O$	2.28g
H_3BO_3	0.50g
$ZnSO_4{\cdot}7H_2O$	0.50g
$CoCl_2{\cdot}6H_2O$	0.045g
$CuSO_4{\cdot}2H_2O$	0.025g
$Na_2MoO_4{\cdot}2H_2O$	0.025g
H_2SO_4,concentrated	0.5mL

Preparation of Micronutrient Solution: Add components to distilled/deionized water and bring volume to 1.0L. Mix thoroughly.

Preparation of Medium: Add components, except Na_2S, to distilled/deionized water and bring volume to 1.0L. Mix thoroughly. Adjust pH to 8.2–8.4. Add Na_2S. Readjust pH to 8.2–8.4. Gently heat and bring to boiling. Distribute into tubes or flasks. Autoclave for 15 min at 15 psi pressure–121°C. Pour into sterile Petri dishes or leave in tubes.

Use: For the cultivation of *Chloroflexus aurantiacus*.

Chloroflexus Broth

Composition per liter:

Glycyl-glycine	0.5g
Yeast extract	0.5g
NH_4Cl	0.2g
$NaNO_3$	0.689g
Na_2S	0.5g
Na_2HPO_4	0.111g
KNO_3	0.103g
$MgSO_4{\cdot}7H_2O$	0.1g
Nitrilotriacetic acid	0.1g
$CaSO_4{\cdot}2H_2O$	0.06g
NaCl	8.0mg
$FeCl_3$ solution	1.0mL
Micronutrient solution	1.0mL

pH 8.2–8.4 at 25°C

$FeCl_3$ Solution:

Composition per liter:

$FeCl_3$	0.29g

Preparation of $FeCl_3$ Solution: Add $FeCl_3$ to distilled/deionized water and bring volume to 1.0L. Mix thoroughly.

Micronutrient Solution:

Composition per liter:

$MnSO_4{\cdot}7H_2O$	2.28g
H_3BO_3	0.50g
$ZnSO_4{\cdot}7H_2O$	0.50g
$CoCl_2{\cdot}6H_2O$	0.045g
$CuSO_4{\cdot}2H_2O$	0.025g
$Na_2MoO_4{\cdot}2H_2O$	0.025g
H_2SO_4 (concentrated)	0.5mL

Preparation of Micronutrient Solution: Add components to distilled/deionized water and bring volume to 1.0L. Mix thoroughly.

Preparation of Medium: Add components, except Na_2S, to distilled/deionized water and bring volume to 1.0L. Mix thoroughly. Adjust pH to 8.2–8.4. Add Na_2S. Readjust pH to 8.2–8.4. Filter sterilize. Distribute into sterile tubes or flasks.

Use: For the cultivation of *Chloroflexus aurantiacus*.

Chlorohydroxybenzoic Acid Medium

Composition per liter:

$K_2HPO_4 \cdot 3H_2O$	4.25g
NH_4Cl	2.0g
$NaH_2PO_4 \cdot H_2O$	1.0g
5-Chloro-2-hydroxybenzoic acid	0.5g
$MgSO_4 \cdot 7H_2O$	0.2g
Nitrilotriacetic acid	0.1g
$FeSO_4 \cdot 7H_2O$	0.012g
$MnSO_4 \cdot H_2O$	3.0mg
$ZnSO_4 \cdot 7H_2O$	3.0mg
$CoSO_4$	1.0mg

pH 7.0-7.4 at 25°C

Preparation of Medium: Add 5-chloro-2-hydroxybenzoic acid to 800.0mL of distilled/deionized water. Adjust pH to 7.0 with NaOH. Add remaining components and bring volume to 1.0L. Distribute into tubes or flasks. Autoclave for 15 min at 15 psi pressure–121°C.

Use: For the cultivation of bacteria that can utilize 5-chloro-hydroxybenzoic acid. For the cultivation of ATCC strain 35944.

CHO Medium Base (Carbohydrate Medium Base)

Composition per liter:

Pancreatic digest of casein	15.0g
Yeast extract	7.0g
NaCl	2.5g
Agar	0.75g
Sodium thioglycollate	0.5g
L-Cystine	0.25g
Ascorbic acid	0.1g
Bromthymol blue	0.01g

pH 7.0 ± 0.2 at 25°C

Preparation of Medium: Add components to distilled/deionized water and bring volume to 1.0L. Mix thoroughly. Gently heat and bring to boiling. Distribute into tubes or flasks. Autoclave for 15 min at 15 psi pressure–121°C. Cool to 45–50°C.

Use: Used as a basal medium to which carbohydrates are added for fermentation studies of anaerobic bacteria. Generally, 6.25mL of 10% filter-sterilized solution of carbohydrate is added to the sterile basal medium.

Chopped Meat Glucose Medium with NaCl

Composition per 1205mL:

NaCl	30.0g
Peptone	30.0g
K_2HPO_4	5.0g
Yeast extract	5.0g
Glucose	5.0g
L-Cysteine·HCl·H_2O	0.5g
Chopped meat extract filtrate	1.0L
Chopped meat extract solids	200.0mL
Resazurin (0.025% solution)	4.0mL

pH 7.0 ± 0.2 at 25°C

Chopped Meat Extract:

Composition per liter:

Beef or horse meat	500.0g
NaOH (1*N* solution)	25.0mL

Preparation of Chopped Meat Extract: Use lean beef or horse meat. Remove fat and connective tissue. Grind. Add meat and NaOH to distilled/deionized water and bring volume to 1.0L. Gently heat and bring to boiling while stirring. Cool to 25°C. Remove fat from surface. Filter. Reserve ground meat particles and filtrate. Add distilled/deionized water to filtrate and bring volume to 1.0L.

Preparation of Medium: To 1.0L of chopped meat extract filtrate, add the remaining components, except cysteine and chopped meat solids. Mix thoroughly. Gently heat to boiling. Cool to room temperature. Add the L-cysteine. Adjust pH to 7.0. Distribute 1 part chopped meat solids (by volume) and 5 parts of liquid (by volume) into tubes under O_2-free 97% N_2 + 3% H_2. Cap with rubber stoppers and place tubes in a press. Autoclave for 15 min at 15 psi pressure–121°C with fast exhaust.

Use: For the cultivation and maintenance of anaerobic halophilic bacteria.

Chopped Meat Medium, Modified

Composition per 1230mL:

Pancreatic digest of casein	30.0g
Peptone	30.0g
Agar	20.0g
K_2HPO_4	5.0g
Yeast extract	5.0g
L-Cysteine·HCl·H_2O	0.5g
Chopped meat extract filtrate	1.0L
Chopped meat extract solids	200.0mL
Hemin solution	10.0mL
Resazurin (0.025% solution)	4.0mL
Vitamin K_1 solution	0.2mL

pH 7.0 ± 0.2 at 25°C

Chopped Meat Extract:

Composition per liter:

Beef or horse meat	500.0g
NaOH (1*N* solution)	25.0mL

Preparation of Chopped Meat Extract: Use lean beef or horse meat. Remove fat and connective tissue. Grind. Add meat and NaOH to distilled/deionized water and bring volume to 1.0L. Gently heat and bring to boiling while stirring. Cool to 25°C. Remove fat from surface. Filter. Reserve ground meat particles and filtrate. Add distilled/deionized water to filtrate and bring volume to 1.0L.

Hemin Solution:
Composition per 100mL:

Hemin	0.05g
NaOH (1*N* solution)	1.0mL

Preparation of Hemin Solution: Add components to distilled/deionized water and bring volume to 100.0mL. Mix thoroughly.

Vitamin K_1 Solution:
Composition per 30mL:

Ethanol (95% solution)	30.0mL
Vitamin K_1	0.15mL

Preparation of Vitamin K_1 Solution: Mix components. Store solution protected from light at 5°C. Discard after one month.

Preparation of Medium: To 1.0L of chopped meat extract filtrate, add the remaining components, except cysteine, hemin solution, vitamin K_1 solution and chopped meat solids. Mix thoroughly. Gently heat to boiling. Cool to room temperature. Add the cysteine, hemin solution and vitamin K_1 solution. Adjust pH to 7.0. Distribute 1 part chopped meat solids (by volume) and 5 parts of liquid (by volume) into tubes under O_2-free 97% N_2 + 3% H_2. Cap with rubber stoppers and place tubes in a press. Autoclave for 15 min at 15 psi pressure–121°C with fast exhaust.

Use: For cultivation and maintenance of a variety of anaerobic bacteria including *Actinomyces* species, *Bacteroides* species, *Clostridium* species, *Eubacterium* species, *Fusobacterium* species, *Peptostreptococcus* species, *Porphyromonas* species, *Prevotella* species, *Propionibacterium* species, *Selenomonas* species and others.

Chopped Meat Medium with 10% Reduced Filtered Rumen Fluid

Composition per 1330mL:

Peptone	30.0g
K_2HPO_4	5.0g
Yeast extract	5.0g
L-Cysteine·HCl·H_2O	0.5g
Chopped meat extract filtrate	1.0L
Chopped meat extract solids	200.0mL
Rumen fluid, reduced and filtered	100.0mL
Resazurin (0.025% solution)	4.0mL

pH 7.0 ± 0.2 at 25°C

Chopped Meat Extract:
Composition per liter:

Beef or horse meat	500.0g
NaOH (1*N* solution)	25.0mL

Preparation of Chopped Meat Extract: Use lean beef or horse meat. Remove fat and connective tissue. Grind. Add meat and NaOH to distilled/deionized water and bring volume to 1.0L. Gently heat and bring to boiling while stirring. Cool to 25°C. Remove fat from surface. Filter. Reserve ground meat particles and filtrate. Add distilled/deionized water to filtrate and bring volume to 1.0L.

Preparation of Medium: To 1.0L of chopped meat extract filtrate, add the remaining components, except L-cysteine and chopped meat solids. Mix thoroughly. Gently heat to boiling. Cool to room temperature. Add the L-cysteine. Adjust pH to 7.0. Distribute 1 part chopped meat solids (by volume) and 5 parts of liquid (by volume) into tubes under O_2-free 97% N_2 + 3% H_2. Cap with rubber stoppers and place tubes in a press. Autoclave for 15 min at 15 psi pressure–121°C

Use: For cultivation and maintenance of *Eubacterium hallii*.

Chromatiaceae Medium 1

Composition per 4990mL:

Solution 1	4.0L
O_2-free water	860.0mL
$NaHCO_3$ solution	100.0mL
$Na_2S·9H_2O$ solution	20.0mL
Trace element solution	5.0mL
Vitamin B_{12} solution	5.0mL

pH 7.3 ± 0.2 at 25°C

Solution 1:
Composition per 4L:

$MgSO_4·7H_2O$	2.5g
KCl	1.7g
KH_2PO_4	1.7g
NH_4Cl	1.7g
$CaCl_2·2H_2O$	1.25g

Preparation of Solution 1: Add components to distilled/deionized water and bring volume to 4.0L. Mix thoroughly. Autoclave for 45 min at 15 psi pressure–121°C. Cool to 25°C under 100% N_2. Saturate with CO_2 by stirring under 100% CO_2 for 30 min.

O_2-Free Water:
Composition per 860mL:

H_2O	860.0mL

Preparation of O_2-Free Water: Autoclave H_2O-for 15 min at 15 psi pressure–121°C. Cool to 25°C under 100% N_2.

$NaHCO_3$ Solution:

Composition per 100mL:

$NaHCO_3$	7.5g

Preparation of $NaHCO_3$ Solution: Add the $NaHCO_3$ to distilled/deionized water and bring volume to 100.0mL. Mix thoroughly. Gas with 100% CO_2 for 20 min. Filter sterilize with positive CO_2 pressure.

$Na_2S·9H_2O$ Solution:

Composition per 100mL:

$Na_2S·9H_2O$	10.0g

Preparation of $Na_2S·9H_2O$ Solution: Add $Na_2S·9H_2O$ to distilled/deionized water. Mix thoroughly. Gas with 100% N_2 for 15 min in a screw-capped bottle. Tightly close cap. Autoclave for 15 min at 15 psi pressure–121°C. Cool to 25°C.

Trace Element Solution:

Composition per liter:

$FeCl_2·4H_2O$	1.5g
$CoCl_2·6H_2O$	0.19g
$MnCl_2·4H_2O$	0.10g
$ZnCl_2$	0.07g
H_3BO_3	0.06g
$NaMoO_4·2H_2O$	0.04g
$CuCl_2·2H_2O$	0.02g
$NiCl_2·6H_20$	0.02g
HCl (25% solution)	6.5mL

Preparation of Trace Element Solution: Add components to distilled/deionized water and bring volume to 1.0L. Mix thoroughly. Autoclave for 15 min at 15 psi pressure–121°C. Cool to 25°C.

Vitamin B_{12} Solution:

Composition per 100mL:

Vitamin B_{12}	2.0mg

Preparation of Vitamin B_{12} Solution: Add components to distilled/deionized water and bring volume to 100.0mL. Mix thoroughly. Filter sterilize.

Preparation of Medium: To 4.0L of sterile, CO_2-saturated solution 1, aseptically add the remaining components. Mix thoroughly. Adjust pH to 7.3. Aseptically distribute into sterile 100.0mL bottles using positive pressure of 95% N_2 + 5% CO_2. Completely fill bottles with medium except for a pea-sized air bubble.

Use: For the isolation and cultivation of members of the Chlorobiaceae.

Chromatiaceae Medium 2

Composition per 1051mL:

Solution 1	950.0mL
$Na_2S·9H_2O$ solution	60.0mL
$NaHCO_3$ solution	40.0mL
Vitamin B_{12} solution	1.0mL

pH 7.3 ± 0.2 at 25°C

Solution 1:

Composition per 950mL:

KH_2PO_4	1.0g
NH_4Cl	0.5g
$MgSO_4·7H_2O$	0.4g
$CaCl_2·2H_2O$	0.05g
Trace element solution SL-8	1.0mL

Preparation of Solution 1: Add components to distilled/deionized water and bring volume to 950.0mL. Mix thoroughly. Autoclave for 15 min at 15 psi pressure–121°C. Cool to 45–50°C.

Trace Element Solution SL-8:

Composition per liter:

Disodium EDTA	5.2g
$FeCl_2·4H_2O$	1.5g
$CoCl_2·6H_2O$	0.19g
$MnCl_2·4H_2O$	0.10g
$ZnCl_2$	0.07g
H_3BO_3	0.06g
$NaMoO_4·2H_2O$	0.04g
$CuCl_2·2H_2O$	0.02g
$NiCl_2·6H_20$	0.02g

Preparation of Trace Element Solution SL-8: Add components to distilled/deionized water and bring volume to 1.0L. Mix thoroughly.

$Na_2S·9H_2O$ Solution:

Composition per 100mL:

$Na_2S·9H_2O$	5.0g

Preparation of $Na_2S·9H_2O$ Solution: Add $Na_2S·9H_2O$ to distilled/deionized water and bring volume to 100.0mL. Autoclave for 15 min at 15 psi pressure–121°C. Cool to 45–50°C.

$NaHCO_3$ Solution:

Composition per 100mL:

$NaHCO_3$	5.0g

Preparation of $NaHCO_3$ Solution: Add $NaHCO_3$ to distilled/deionized water and bring volume to 100.0mL. Mix thoroughly. Filter sterilize.

Vitamin B_{12} Solution:

Composition per 100mL:

Vitamin B_{12}	2.0mg

Preparation of Vitamin B_{12} Solution: Add vitamin B_{12} to distilled/deionized water and bring volume to 100.0mL. Mix thoroughly. Filter sterilize.

Preparation of Medium: To 950.0mL of cooled, sterile solution 1, aseptically add 60.0mL of sterile $Na_2S{\cdot}9H_2O$ solution, 40.0mL of sterile $NaHCO_3$ solution, and 1.0mL of sterile vitamin B_{12} solution. Mix thoroughly. Adjust pH to 7.3 with sterile H_2SO_4 or Na_2CO_3. Aseptically distribute into sterile 50.0mL or 100.0mL bottles with metal screw caps and rubber seals. Completely fill bottles with medium except for a pea-sized air bubble.

Use: For the isolation and cultivation of freshwater and soil members of Chromatiaceae.

Chromatium Medium (ATCC Medium 37)

Composition per 127mL:

Solution 1	76.2mL
Solution 2 + Solution 3	44.8mL
Solution 4	6.0mL

Solution 1:

Composition per 2.5 liters:

$CaCl_2$	2.0g

Preparation of Solution 1: Add $CaCl_2$ to distilled/deionized water and bring volume to 2.5L. Distribute in 80.0mL volumes into 127.0mL screw-capped bottles. Autoclave for 15 min at 15 psi pressure–121°C.

Solution 2:

Composition per 100mL:

Sodium ascorbate	2.4g
KCl	1.0g
KH_2PO_4	1.0g
$MgCl_2{\cdot}6H_2O$	0.8g
NH_4Cl	0.8g
Heavy metal solution	50.0mL
Vitamin solution	15.0mL
Vitamin B_{12} solution	3.0mL

Preparation of Solution 2: Add components to distilled/deionized water and bring volume to 100.0mL. Mix thoroughly.

Solution 3:

Composition per 900mL:

$NaHCO_3$	4.5g

Preparation of Solution 3: Add $NaHCO_3$ to distilled/deionized water and bring volume to 900.0mL. Mix thoroughly. Bubble 100% CO_2 through the solution for 30 min. After CO_2 saturation of Solution 3, add Solution 2 and immediately filter the mixture through a Seitz filter (or a Millipore) using positive CO_2 pressure to push the liquid through.

Solution 4:

Composition per 200mL:

$Na_2S{\cdot}9H_2O$	3.0g

Preparation of Solution 4: Add $Na_2S{\cdot}9H_2O$ to distilled/deionized water and bring volume to 200.0mL. Add a magnetic stir bar to the flask. Autoclave for 15 min at 15 psi pressure–121°C. On a magnetic stirrer, slowly add 2.0mL of sterile 2M H_2SO_4. This partially neutralizes the solution. The solution should turn yellow. H_2S gas will be liberated—neutralization and distribution of the solution should be done as rapidly as possible under adequate ventilation.

Heavy Metal Solution:

Composition per liter:

Ethylenediamine tetraacetate (EDTA)	1.5g
$FeSO_4{\cdot}7H_2O$	0.2g
$ZnSO_4{\cdot}7H_2O$	0.1g
$MnCl_2{\cdot}4H_2O$	0.02g
Modified Hoagland trace element solution	6.0mL

Preparation of Heavy Metal Solution: Dissolve EDTA in approximately 800.0mL of distilled/deionized water. Add remaining components. Bring volume to 1.0L with distilled/deionized water. Mix thoroughly.

Vitamin B_{12} Solution:

Composition per 100mL:

Vitamin B_{12} (cyanocobalamin)	2.0mg

Preparation of Vitamin B_{12} Solution: Add vitamin B_{12} to distilled/deionized water and bring volume to 100.0mL. Mix thoroughly.

Vitamin Solution:

Composition per 100mL:

Pyridoxamine·2HCl	5.0mg
Nicotinic acid	2.0mg
Thiamine	1.0mg
Pantothenic acid	0.5mg
Biotin	0.2mg
p-Aminobenzoic acid	0.1mg

Preparation of Vitamin Solution: Add components to distilled/deionized water and bring volume to 100.0mL. Mix thoroughly.

Modified Hoagland Trace Element Solution:

Composition per 3.6L:

H_3BO_3	11.0g
$MnCl_2{\cdot}4H_2O$	7.0g
$AlCl_3$	1.0g
$CoCl_2$	1.0g
$CuCl_2$	1.0g
KI	1.0g
$NiCl_2$	1.0g
$ZnCl_2$	1.0g

$BaCl_2$	0.5g
KBr	0.5g
LiCl	0.5g
Na_2MoO_4	0.5g
$SeCl_4$	0.5g
$SnCl_2 \cdot 2H_2O$	0.5g
$NaVO_3 \cdot H_2O$	0.1g

Preparation of Modified Hoagland Trace Element Solution: Prepare each component as a separate solution. Dissolve each salt in approximately 100.0mL of distilled/deionized water. Adjust the pH of each solution to below 7.0. Combine all the salt solutions and bring the volume to 3.6L with distilled/deionized water. Adjust the pH to 3–4. A yellow precipitate may form after mixing. After a few days, it will turn into a fine white precipitate. Mix the solution thoroughly before using.

Preparation of Medium: To the 80.0mL of sterile solution 1 in screw-capped bottles, add combined solutions 2 and 3 immediately after filtration and fill bottles to capacity. Mix thoroughly. Aseptically remove 6.0mL of the medium from the bottles and replace it with 6.0mL of neutralized solution 4. Let stand for 24 hr. The medium should form a fine white precipitate before using. To inoculate, remove 6.0mL of the completed medium from the bottles and replace it with 6.0mL of inoculum.

Use: For the cultivation and maintenance of *Chromatium tepidum.*

Chromatium Medium (ATCC Medium 1449)

Composition per liter:

KH_2PO_4	0.5g
NH_4Cl	0.4g
$MgSO_4 \cdot 7H_2O$	0.2g
$CaCl_2 \cdot 2H_2O$	0.05g
Disodium EDTA	0.01g
Trace elements	1.0mL
$NaHCO_3$ solution	50.0mL
$Na_2S \cdot 9H_2O$ solution	50.0mL
Sodium pyruvate solution	50.0mL

pH 7.0 ± 0.2 at 25°C

Trace Elements:

Composition per liter:

Disodium EDTA	5.2g
$FeCl_2 \cdot 4H_2O$	1.5g
$CoCl_2 \cdot 6H_2O$	0.19g
$Na_2MoO_4 \cdot 2H_2O$	0.188g
$MnCl_2 \cdot 4H_2O$	0.1g
$ZnCl_2$	0.07g
$VOSO_4 \cdot 2H_2O$	0.03g
$NiCl_2 \cdot 6H_2O$	0.025g
H_3BO_3	6.0mg
$CuCl_2 \cdot 2H_2O$	2.0mg
Na_2SeO_3	2.0mg

Preparation of Trace Elements: Add components to distilled/deionized water and bring volume to 1.0L. Mix thoroughly.

$NaHCO_3$ Solution:

Composition per 50mL:

$NaHCO_3$	2.0g

Preparation of $NaHCO_3$ Solution: Add $NaHCO_3$ to distilled/deionized water and bring volume to 50.0mL. Filter sterilize. Use freshly prepared solution.

$Na_2S \cdot 9H_2O$ Solution:

Composition per 50mL:

$Na_2S \cdot 9H_2O$	1.0g

Preparation of $Na_2S \cdot 9H_2O$ Solution: Add $Na_2S \cdot 9H_2O$ to distilled/deionized water and bring volume to 50.0mL. Autoclave for 15 min at 15 psi pressure–121°C. Use freshly prepared solution.

Sodium Pyruvate Solution:

Composition per 50mL:

Sodium pyruvate	0.5g

Preparation of Sodium Pyruvate Solution: Add $NaHCO_3$ to distilled/deionized water and bring volume to 50.0mL. Filter sterilize. Use freshly prepared solution. Sodium acetate may be substituted for the sodium pyruvate.

Preparation of Medium: Add components, except $NaHCO_3$ solution, $Na_2S \cdot 9H_2O$ solution, and sodium pyruvate solution, to distilled deionized water and bring volume to 850.0mL. Autoclave for 15 min at 15 psi pressure–121°C. Cool to room temperature. Add the sterile $NaHCO_3$ solution, the sterile $Na_2S \cdot 9H_2O$ solution, and the sterile sodium pyruvate solution, in that order. Adjust the pH to 7.0. Distribute into screw-capped tubes or flasks. Fill to capacity.

Use: For cultivation and maintenance of *Chromatium* species.

Chromobacterium Medium

Composition per liter:

NaCl	30.0g
$MgCl_2$	10.8g
$MgSO_4$	5.4g
Peptone	5.0g
$CaCl_2$	1.0g
KCl	0.7g

pH 7.0 ± 0.2 at 25°C

Preparation of Medium: Add components to distilled/deionized water and bring volume to 1.0L. Mix thoroughly. Distribute into tubes or flasks. Autoclave for 15 min at 15 psi pressure–121°C.

Use: For the cultivation and maintenance of *Chromobacterium* species and *Alteromonas luteoviolacea*.

Chromogenic Substrate Broth

Composition per liter:

NaCl	10.0g
HEPES (*N*-[2-Hydroxyethyl] piperazine-*N*′-[2-ethanesulfonic acid]) buffer	6.9g
$(NH_4)_2SO_4$	5.0g
o-Nitrophenyl-*β*-D-galactopyranoside	0.50g
Solanium	0.50g
$MgSO_4$	0.10g
4-Methylumbelliferyl-*β*-D-glucuronide	0.075g
$CaCl_2$	0.05g
Na_2SO_3	0.04g
Amphotericin B	1.0mg
$MnSO_4$	0.5mg
$ZnSO_4$	0.5mg

Preparation of Medium: Add components to distilled/deionized water and bring volume to 1.0L. Mix thoroughly. Distribute into tubes or flasks. Autoclave for 15 min at 15 psi pressure–121°C.

Use: For the detection of coliform bacteria based on their hydrolysis of chromogenic substrates by production of *β*-D-galactopyranosidase. Bacteria that produce *β*-D-galactopyranosidase turn the medium yellow.

Chu's No. 10 Medium

Composition per liter:

Agar	15.0g
$Ca(NO_3)_2 \cdot 4H_2O$	0.232g
$Na_2SiO_3 \cdot 5H_2O$	0.044g
$MgSO_4 \cdot 7H_2O$	0.025g
Na_2CO_3	0.02g
K_2HPO_4	0.01g
Citric acid	3.5mg
Ferric citrate	3.5mg

Preparation of Medium: Add components to distilled/deionized water and bring volume to 1.0L. Mix thoroughly. Gently heat to boiling. Distribute into tubes or flasks. Autoclave for 15 min at 15 psi pressure–121°C. Pour into sterile Petri dishes or leave in tubes.

Use: For the cultivation and maintenance of *Anabaena* species and *Plectomena boryanum*.

Chu's No. 10 Medium, Modified

Composition per liter:

Agar	15.0g
$Ca(NO_3)_2 \cdot 4H_2O$	0.232g
$Na_2SiO_3 \cdot 5H_2O$	0.044g
$MgSO_4 \cdot 7H_2O$	0.025g
Na_2CO_3	0.02g
K_2HPO_4	0.01g
Citric acid	3.5mg
Ferric citrate	3.5mg
Metal solution	1.0mL

Metal Solution:

Composition per liter:

H_3BO_3	2.4g
$MnCl_2 \cdot 4H_2O$	1.4g
$ZnCl_2$	0.4g
$CoCl_2 \cdot 6H_2O$	0.02g
$CuCl_2 \cdot 2H_2O$	0.1mg

Preparation of Metal Solution: Add components to distilled/deionized water and bring volume to 1.0L. Mix thoroughly.

Preparation of Medium: Add components to distilled/deionized water and bring volume to 1.0L. Mix thoroughly. Gently heat to boiling. Distribute into tubes or flasks. Autoclave for 15 min at 15 psi pressure–121°C. Pour into sterile Petri dishes or leave in tubes.

Use: For the cultivation and maintenance of *Anabaena* species and *Plectomena boryanum*.

Chu's No. 11 Medium, Modified

Composition per liter:

$NaNO_3$	1.5g
$MgSO_4 \cdot 7H_2O$	0.08g
$Na_2SiO_3 \cdot 9H_2O$	0.06g
$CaCl_2 \cdot 2H_2O$	0.04g
$K_2HPO_4 \cdot 3H_2O$	0.04g
Na_2CO_3	0.02g
Citric acid	6.0mg
Ferric ammonium citrate	6.0mg
EDTA	1.0mg
Trace metal solution A5 with cobalt	1.0mL

pH 7.5 ± 0.2 at 25°C

Trace Metal Solution A5 With Cobalt:

Composition per liter:

H_3BO_3	2.86g
$MnCl_2 \cdot 4H_2O$	1.81g
$Na_2MoO_4 \cdot 2H_2O$	0.390g
$ZnSO_4 \cdot 7H_2O$	0.222g
$CuSO_4 \cdot H_2O$	0.079g
$Co(NO_3)_2 \cdot 6H_2O$	0.049g

Preparation of Trace Metal Solution A5 With Cobalt: Add components to distilled/deionized water and bring volume to 1.0L. Mix thoroughly.

Preparation of Medium: Add components to seawater and bring volume to 1.0L. Mix thoroughly. Gently heat and bring to boiling. Distribute into tubes or flasks. Autoclave for 15 min at 15 psi pressure–121°C.

Use: For the isolation and cultivation of cyanobacteria from marine habitats.

CIN Agar (*Yersinia* Selective Agar) (Cefsulodin Irgasan® Novobiocin Agar)

Composition per liter:

Mannitol	20.0g
Agar	12.0g
Pancreatic digest of gelatin	10.0g
Beef extract	5.0g
Peptic digest of animal tissue	5.0g
Sodium pyruvate	2.0g
Yeast extract	2.0g
NaCl	1.0g
Sodium deoxycholate	0.5g
Neutral red	0.03g
Cefsulodin	0.015g
Irgasan®(triclosan)	4.0mg
Novobiocin	2.5mg
Crystal violet	1.0mg

pH 7.4 ± 0.2 at 25°C

Source: This medium is available as a premixed powder from BBL Microbiology Systems.

Preparation of Medium: Add components, except cefsulodin and novobiocin, to distilled/deionized water and bring volume to 1.0L. Heat, mixing continuously, until boiling. Do not autoclave. Cool to 45–50°C. Aseptically add cefsulodin and novobiocin. Mix thoroughly. Pour into sterile Petri dishes or distribute into sterile tubes.

Use: For the selective isolation and differentiation of *Yersinia enterocolitica* based on mannitol fermentation. *Yersinia enterocolitica* appears as "bull's eye" colonies with deep red centers surrounded by a transparent periphery.

Cinnamate Medium

Composition per liter:

$NaHCO_3$	2.5g
$MgCl_2 \cdot 6H_2O$	2.03g
Cinnamic acid	1.48g
KH_2PO_4	1.36g
NH_4Cl	0.53g
$Na_2S \cdot 9H_2O$	0.24g
$CaCl_2 \cdot 2H_2O$	0.15g
Yeast extract	0.05g
Modified Wolfe's metals	10.0mL
Wolfe's vitamin solution	10.0mL

pH 7.5–7.7 at 25°C

Modified Wolfe's Metals:

Composition per liter:

Na_2SeO_3	10.0mg
$NaWO_4 \cdot 2H_2O$	10.0mg
$NiCl_2 \cdot 6H_2O$	10.0mg
Wolfe's metals solution	1.0L

Preparation of Modified Wolfe's Metals: Combine the components. Mix thoroughly.

Wolfe's Metals Solution:

Composition per liter:

$MgSO_4 \cdot 7H_2O$	3.0g
Nitriloacetic acid	1.5g
NaCl	1.0g
$MnSO_4 \cdot H_2O$	0.5g
$CaCl_2$	0.1g
$CoCl_2 \cdot 6H_2O$	0.1g
$FeSO_4 \cdot 7H_2O$	0.1g
$ZnSO_4 \cdot 7H_2O$	0.1g
$AlK(SO_4)_2 \cdot 12H_2O$	0.01g
$CuSO_4 \cdot 5H_2O$	0.01g
H_3BO_3	0.01g
$Na_2MoO_4 \cdot 2H_2O$	0.01g

Preparation of Wolfe's Metals Solution: Add nitrilotriacetic acid to 500.0mL of distilled/deionized water. Adjust pH to 6.5 with KOH. Add distilled/deionized water to 1.0L. Add remaining components. Mix thoroughly.

Wolfe's Vitamin Solution:

Composition per liter:

Pyridoxine·HCl	0.01g
p-Aminobenzoic acid	5.0mg
Calcium pantothenate	5.0mg
Nicotinic acid	5.0mg
Riboflavin	5.0mg
Thiamine·HCl	5.0mg
Thioctic acid	5.0mg
Biotin	2.0mg
Folic acid	2.0mg
Cyanocobalamin	100.0μg

Preparation of Wolfe's Vitamin Solution: Add components to distilled/deionized water and bring volume to 1.0L. Mix thoroughly.

Preparation of Medium: Add all components, except $NaHCO_3$ and $Na_2S \cdot 9H_2O$, to distilled/deionized water and bring volume to 1.0L. Gently heat and bring to boiling under 90% N_2 + 10% CO_2. Cool medium to room temperature while continuing to gas with 90%N_2 + 10%CO_2. Add $NaHCO_3$ and $Na_2S{\cdot}9H_2O$. Adjust pH to 7.5–7.7. Distribute into tubes under 90% N_2 + 10% CO_2 using anaerobic techniques. Autoclave for 15 min at 15 psi pressure–121°C.

Use: For the cultivation of anaerobic bacteria that can utilize cinnamic acid as a carbon source. As a basal medium for the cultivation of *Formivibrio citricus*.

Citrate Medium, Koser's Modified

Composition per liter:

NaCl	5.0g
Citric acid	2.0g
$(NH_4)H_2PO_4$	1.0g
K_2HPO_4	1.0g
$MgSO_4{\cdot}7H_2O$	0.2g

pH 6.8 ± 0.2 at 25°C

Preparation of Medium: Add components to distilled/deionized water and bring volume to 1.0L. Mix thoroughly. Adjust pH to 6.8. Distribute into tubes in 5.0mL volumes. Autoclave for 15 min at 15 psi pressure–121°C.

Use: For the cultivation and differentiation of bacteria based on their ability to utilize citrate as carbon source.

CK Agar

Composition per liter:

Agar	15.0g
Glucose	5.0g
KNO_3	2.0g
$CaCl_2$	1.5g
$MgSO_4{\cdot}7H_2O$	1.5g
K_2HPO_4	0.25g
Ferric citrate	0.02g

Preparation of Medium: Add components to distilled/deionized water and bring volume to 1.0L. Mix thoroughly. Gently heat and bring to boiling. Distribute into tubes or flasks. Autoclave for 15 min at 15 psi pressure–121°C. Pour into sterile Petri dishes or leave in tubes.

Use: For the cultivation of myxobacteria.

CK1 Medium

Composition per liter:

$MgSO_4{\cdot}7H_2O$	3.0g
KNO_3	2.0g
$CaCl_2$	1.4g
Ferric citrate	0.02g
Glucose solution	100.0mL
K_2HPO_4 solution	10.0mL

Glucose Solution:

Composition per 100mL:

D-Glucose	10.0g

Preparation of Glucose Solution: Add D-glucose to distilled/deionized water and bring volume to 100.0mL. Mix thoroughly. Autoclave for 15 min at 15 psi pressure–121°C. Cool to 25°C.

K_2HPO_4 Solution:

Composition per 10mL:

K_2HPO_4	2.5mg

Preparation of K_2HPO_4 Solution: Add K_2HPO_4 to distilled/deionized water and bring volume to 10.0mL. Mix thoroughly. Autoclave for 15 min at 15 psi pressure–121°C. Cool to 25°C.

Preparation of Medium: Add components, except glucose solution and K_2HPO_4 solution, to distilled/deionized water and bring volume to 890.0mL. Mix thoroughly. Autoclave for 15 min at 15 psi pressure–121°C. Cool to 25°C. Aseptically add sterile glucose solution and K_2HPO_4 solution. Mix thoroughly. Aseptically distribute into sterile tubes or flasks.

Use: For the cultivation of myxobacteria.

Clausen Medium (Dithionite Thioglycollate, HS T, Broth)

Composition per liter:

Pancreatic digest of casein	15.0g
Glucose	6.0g
Yeast extract	6.0g
Glycerol	5.0g
Papaic digest of soybean meal	3.0g
Tween™ 80	3.0g
NaCl	2.5g
K_2HPO_4	2.0g
L-Asparagine	1.25g
Sodium citrate	1.0g
Agar	0.75g
L-Cystine	0.5g
Sodium thioglycollate	0.5g
$MgSO_4{\cdot}7H_2O$	0.4g
Sodium dithionite	0.4g
Lecithin	0.3g
$CaCl_2{\cdot}2H_2O$	4.0mg
$MnCl_2{\cdot}4H_2O$	2.0mg
$CoSO_4{\cdot}7H_2O$	1.0mg

$CuSO_4 \cdot 5H_2O$	1.0mg
$FeSO_4 \cdot 7H_2O$	1.0mg
$ZnSO_4 \cdot 7H_2O$	1.0mg
Resazurin	1.0mg

pH 7.1 ± 0.2 at 25°C

Source: This medium is available as a premixed powder from Oxoid Unipath.

Preparation of Medium: Add Tween™ 80 and glycerol to distilled/deionized water and bring volume to 1.0L. Gently heat and bring to boiling. Distribute into tubes or flasks. Autoclave for 15 min at 15 psi pressure–121°C. The medium must not be resterilized.

Use: For sterility testing by the membrane filter method or the tube dilution method to determine the presence of microbial contamination in a variety of specimens.

CLED Agar with Andrade Indicator (Cystine Lactose Electrolyte Deficient Agar with Andrade Indicator)

Composition per liter:

Agar	15.0g
Pancreatic digest of casein	10.0g
Peptone	4.0g
Beef extract	3.0g
Cystine	0.128g
Bromthymol blue	0.02g
Andrade indicator	10.0mL

pH 7.5 ± 0.2

Source: This medium is available as a premixed powder from Oxoid Unipath.

Caution: Acid Fuchsin is a potential carcinogen and care must be taken to avoid inhalation of the powdered dye and contamination of the skin.

Andrade's Indicator:
Composition per 100mL:

NaOH (1*N* solution)	16.0mL
Acid fuchsin	0.1g

Preparation of Andrade's Indicator: Add Acid Fuchsin to NaOH solution and bring volume to 100.0mL with distilled/deionized water.

Preparation of Medium: Add components to distilled/deionized water and bring volume to 1.0L. Mix thoroughly. Gently heat while stirring and bring to boiling. Autoclave for 15 min at 15 psi pressure–121°C. Cool to 50–55°C. Pour into sterile Petri dishes or distribute into sterile tubes.

Use: For the differentiation of microorganisms based on colony characteristics.

Clostridia Medium

Composition per liter:

Sodium L-lactate	10.0g
Sodium acetate	8.0g
K_2HPO_4	0.5g
$(NH_4)_2 \cdot 7H_2O$	0.5g
Sodium thioglycollate	0.5g
Yeast extract	0.5g
$MgSO_4 \cdot 7H_2O$	0.1g
$FeSO_4 \cdot 7H_2O$	0.02g
p-Aminobenzoate	100.0μg
Biotin	0.1μg

pH 6.0–7.0 at 25°C

Preparation of Medium: Add components to distilled/deionized water and bring volume to 1.0L. Mix thoroughly. Adjust pH to 6.0–7.0. Distribute into tubes or flasks. Autoclave for 20 min at 15 psi pressure–121°C.

Use: For the isolation and cultivation of *Clostridium* species that ferment lactate and acetate.

Clostridium acidurici Medium

Composition per liter:

Uric acid	2.0g
Yeast extract	1.0g
K_2HPO_4	0.91g
KOH	0.67g
Sodium thioglycollate	0.5g
$MgSO_4 \cdot 7H_2O$	0.25g
$CaCl_2 \cdot 2H_2O$	0.015g
$FeSO_4 \cdot 7H_2O$	6.0mg
$NaHCO_3$ solution	25.0mL
Sodium thioglycollate solution	25.0mL

pH 7.0–7.5 at 25°C

$NaHCO_3$ Solution:
Composition per 25mL:

$NaHCO_3$	5.0g

Preparation of $NaHCO_3$ Solution: Add $NaHCO_3$ to distilled/deionized water and bring volume to 25.0mL. Mix thoroughly. Filter sterilize.

Sodium Thioglycollate Solution:
Composition per 25mL:

Sodium thioglycollate	0.5g

Preparation of Sodium Thioglycollate Solution: Add sodium thioglycollate to distilled/deionized water and bring volume to 25.0mL. Mix thoroughly. Autoclave solution separately for 15 min at 15 psi pressure–121°C.

Preparation of Medium: Add K_2HPO_4 and KOH to distilled/deionized water and bring volume to 1.0L. Add uric acid. Gently heat until boiling. Add the remaining components, except $NaHCO_3$ and sodium thioglycollate. Mix thoroughly. Adjust pH to 7.0–7.5. Distribute into tubes or flasks. Autoclave for 15 min at 15 psi pressure–121°C. Add 0.25mL of sterile $NaHCO_3$ solution and 0.25mL of sterile sodium thioglycollate solution for each 10.0mL of sterile basal medium.

Use: For the cultivation and maintenance of *Clostridium acidurici, C. purinolyticum* and other bacteria that can utilize uric acid as a carbon source.

Clostridium aerotolerans Medium

Composition per liter:

Agar	15.0g
Xylan	5.0g
Yeast extract	5.0g
Na_2CO_3	4.0g
NaCl	0.45g
$(NH_4)_2SO_4$	0.45g
K_2HPO_4	0.225g
KH_2PO_4	0.225g
Cysteine·HCl·H_2O	0.125g
$Na_2S·9H_2O$	0.125g
$MgSO_4·7H_2O$	0.09g
$CaCl_2$	0.045g

pH 7.0 ± 0.2 at 25°C

Preparation of Medium: Prepare medium anaerobically under 100% CO_2. Add components to distilled/deionized water and bring volume to 1.0L. Mix thoroughly. Gently heat while stirring and bring to boiling. Autoclave for 15 min at 15 psi pressure–121°C. Cool to 50–55°C. Adjust pH to 7.0. Pour into sterile Petri dishes or distribute into sterile tubes.

Use: For the cultivation and maintenance of *Clostridium aerotolerans*.

Clostridium Alginate Medium

Composition per liter of sea water:

Agar	15.0g
Sodium alginate	10.0g
K_2HPO_4	2.0g
Peptone	1.0g
Yeast extract	1.0g
Seawater	1.0L

pH 7.0–7.5 at 25°C

Preparation of Medium: Add K_2HPO_4 to 1.0L of seawater. Mix thoroughly. Gently heat while stirring to dissolve. Filter solution twice. Add remaining components. Mix thoroughly. Adjust pH to 7.0–7.5. Autoclave for 15 min at 15 psi pressure–121°C. Pour into sterile Petri dishes or distribute into sterile tubes.

Use: For the cultivation and maintenance of *Clostridium alginolyticum* and other bacteria which can utilize alginate as a carbon source.

Clostridium aminobutyricum Medium

Composition per liter:

K_2HPO_4	7.05g
Yeast extract	3.0g
KH_2PO_4	1.29g
$MgCl_2·6H_2O$	0.2g
$CaCl_2·2H_2O$	0.01g
$FeCl_3·6H_2O$	0.01g
Methylene blue	2.0mg
$MnSO_4·H_2O$	1.0mg
$Na_2MoO_4·2H_2O$	1.0mg
γ-Aminobutyrate solution	100.0mL
Na_2CO_3 solution	100.0mL
$Na_2S·9H_2O$ solution	50.0mL

pH 7.4–7.7 at 25°C

γ-Aminobutyrate Solution:

Composition per 100mL:

γ-Aminobutyrate	5.0g

Preparation of γ-Aminobutyrate Solution: Add γ-aminobutyrate to distilled/deionized water and bring volume to 100.0mL. Mix thoroughly. Filter sterilize.

Na_2CO_3 Solution:

Composition per 100mL:

Na_2CO_3	2.0g

Preparation of Na_2CO_3 Solution: Add Na_2CO_3 to distilled/deionized water and bring volume to 100.0mL. Mix thoroughly. Autoclave for 15 min at 15 psi pressure–121°C.

$Na_2S·9H_2O$ Solution:

Composition per 50mL:

$Na_2S·9H_2O$	0.3g

Preparation of $Na_2S·9H_2O$ Solution: Add $Na_2S·9H_2O$ to distilled/deionized water and bring volume to 50.0mL. Mix thoroughly. Autoclave for 15 min at 15 psi pressure–121°C.

Preparation of Medium: Add components, except γ-aminobutyrate solution, Na_2CO_3 solution, and $Na_2S·9H_2O$ solution, to distilled/deionized water and bring volume to 750.0mL. Autoclave for 15 min at 15 psi pressure–121°C. Cool under 80% N_2 + 10% CO_2 + 10% H_2. Aseptically add the sterile γ-aminobutyrate solution, Na_2CO_3 solution, and $Na_2S·9H_2O$ solution. Adjust pH to 7.4–7.7. Distribute using anaerobic technique into tubes or flasks.

Use: For the cultivation and maintenance of *Clostridium aminobutyricum* and other bacteria which can utilize aminobutyric acid as a carbon source.

Clostridium Cellulolytic Medium

Composition per liter:

Agar	20.0g
Cellulose	7.5g
$K_2HPO_4 \cdot 3H_2O$	2.9g
Yeast extract	2.0g
KH_2PO_4	1.5g
$(NH_4)_2SO_4$	1.3g
$FeSO_4$	1.25g
Cysteine·HCl·H_2O	1.0g
$MgCl_2 \cdot 6H_2O$	1.0g
$CaCl_2 \cdot 2H_2O$	0.15g
Resazurin	2.0mg

pH 7.5 ± 0.2 at 25°C

Preparation of Medium: Add components, except cysteine·HCl·H_2O, to distilled/deionized water and bring volume to 1.0L. Mix thoroughly. Heat to boiling. Adjust pH to 7.5. Prereduce under 100% N_2. Add cysteine·HCl·H_2O. Distribute into tubes under 100% N_2. Cap tubes with rubber stoppers. Autoclave for 15 min at 15 psi pressure–121°C.

Use: For the cultivation and maintenance of *Clostridium cellulolyticum* and other bacteria which can degrade cellulose.

Clostridium Cellulose Medium

Composition per liter:

Agar	20.0g
Filter paper (or 5.0g Avicel)	10.0g
$CaCO_3$	5.0g
Polypeptone™	5.0g
$Na_2CO_3 \cdot 10H_2O$	4.0g
K_2HPO_4	2.2g
Yeast extract	2.0g
KH_2PO_4	1.5g
$(NH_4)_2SO_4$	1.3g
$MgCl_2 \cdot 6H_2O$	1.0g
Cysteine·HCl·H_2O	0.5g
$CaCl_2$	0.15g
$FeSO_4 \cdot 7H_2O$	6.0mg

pH 7.0 ± 0.2 at 25°C

Preparation of Medium: Add components to distilled/deionized water and bring volume to 1.0L. Mix thoroughly. Heat to boiling. Autoclave for 15 min at 15 psi pressure–121°C. Pour into sterile Petri dishes or distribute into sterile tubes.

Use: For the cultivation and maintenance of *Clostridium cellulolyticum* and other bacteria which can degrade cellulose.

Clostridium kluyveri Medium

Composition per liter:

Potassium acetate	5.0g
Sodium thioglycollate	0.5g
K_2HPO_4	0.3g
NH_4Cl	0.25g
KH_2PO_4	0.2g
$MgSO_4 \cdot 7H_2O$	0.2g
$CaCl_2 \cdot 2H_2O$	0.01g
$FeSO_4 \cdot 7H_2O$	5.0mg
$MnSO_4 \cdot 4H_2O$	2.0mg
$Na_2MoO_4 \cdot 2H_2O$	2.0mg
p-Aminobenzoate	0.2mg
Biotin	0.01mg
Ethanol	20.0mL
Acetic acid, glacial	2.5mL

pH 7.0 ± 0.2 at 25°C

Preparation of Medium: Add components, except sodium thioglycollate to distilled/deionized water and bring volume to 1.0L. Gently heat and bring to boiling. Mix thoroughly. Add sodium thioglycollate immediately prior to sterilization. Mix thoroughly. Autoclave for 15 min at 15 psi pressure–121°C. Adjust pH to 7.0 with sterile 60% K_2CO_3 solution.

Use: For the isolation and cultivation of *Clostridium kluyveri.*

Clostridium kluyveri Medium

Composition per liter:

Part A	965.0mL
Part B	35.0mL

pH 7.0 ± 0.2 at 25°C

Part A:

Composition per 965mL:

Sodium acetate·$3H_2O$	7.5g
$(NH_4)_2SO_4$	2.65g
Agar	2.0g
Yeast extract	2.0g
Sodium thioglycollate	0.5g
p-Aminobenzoic acid	0.1mg
Biotin	5.0μg
Potassium phosphate buffer (2*M*, pH 7.0)	10.0mL
Salt solution	10.0mL

Preparation of Part A: Add components to distilled/deionized water and bring volume to 965.0mL. Mix thoroughly. Gently heat and bring to boiling. Autoclave for 15 min at 15 psi pressure–121°C. Cool to room temperature.

Salt Solution:

Composition per 100mL:

$MgSO_4 \cdot H_2O$	2.5g

$CaCl_2$	0.15g
$FeSO_4 \cdot 7H_2O$	0.15g
$MnSO_4 \cdot 2H_2O$	0.02g
$Na_2MoO_4 \cdot 2H_2O$	0.02g

Preparation of Salt Solution: Add components to distilled/deionized water and bring volume to 100.0mL. Mix thoroughly.

Part B:

Composition:

K_2CO_3 (1*M* solution)	20.0mL
Ethanol (95% solution)	15.0mL

Preparation of Part B: Prepare a 1*M* solution of K_2CO_3 and filter sterilize. Filter sterilize 25.0mL of 95% ethanol solution. Aseptically combine 20.0mL of sterile K_2CO_3 solution and 15.0mL of sterile ethanol.

Preparation of Medium: Add 35.0mL of sterile Part B to 965.0mL of sterile, cooled Part A. Adjust pH to 7.0 with HCl. Aseptically distribute into tubes under 97% N_2 + 3% H_2. Cap with rubber stoppers.

Use: For the cultivation and maintenance of *Clostridium kluyveri*.

Clostridium Medium

Composition per liter:

Sodium L-glutamate	10.0g
Sodium thioglycollate	0.5g
Yeast extract	0.5g
K_2HPO_4	0.2g
$MgSO_4 \cdot 7H_2O$	0.1g

pH 7.6 ± 0.2 at 25°C

Preparation of Medium: Add components to distilled/deionized water and bring volume to 1.0L. Mix thoroughly. Distribute into tubes or flasks. Autoclave for 15 min at 15 psi pressure–121°C.

Use: For the enrichment and isolation of glutamate-fermenting *Clostridium* species.

Clostridium Medium

Composition per liter:

Uric acid	2.0g
Yeast extract	1.2g
$MgSO_4 \cdot 7H_2O$	0.05g
$CaCl_2 \cdot 2H_2O$	5.0mg
$FeSO_4 \cdot 7H_2O$	2.0mg
Resazurin	1.0mg
KOH (10*N* solution)	3.0mL
$K_2HPO_4 \cdot 3H_2O$ (70% solution)	1.5mL
Mercaptoacetic acid	1.5mL

pH 7.2 ± 0.2 at 25°C

Preparation of Medium: Add KOH solution and $K_2HPO_4 \cdot 3H_2O$ solution to distilled/deionized water and bring volume to 500.0mL. Gently heat and bring to boiling. Mix thoroughly. Add uric acid slowly. Cool to 45–50°C. Add remaining components. Add mercaptoacetic acid immediately prior to sterilization. Bring volume to 1.0L with distilled/deionized water. Mix thoroughly. Autoclave for 15 min at 15 psi pressure–121°C. Adjust pH to 7.2 with sterile 60% K_2CO_3 solution.

Use: For the isolation and cultivation of purine-fermenting *Clostridium* species.

Clostridium Medium (ATCC Medium 39)

Composition per liter:

K_2HPO_4	7.0g
γ-Aminobutyric acid	5.0g
Yeast extract	3.0g
Agar	1.5g
KH_2PO_4	1.3g
$MgCl_2 \cdot 6H_2O$	0.2g
$CaCl_2 \cdot 2H_2O$	0.01g
$FeCl_3 \cdot 6H_2O$	0.01g
Methylene blue	2.0mg
$MnSO_4$	1.0mg
Na_2MoO_4	1.0mg
$Na_2S \cdot 9H_2O$ solution	10.0mL

$Na_2S \cdot 9H_2O$ solution:

Composition per 20mL:

$Na_2S \cdot 9H_2O$	0.6g

Preparation of $Na_2S \cdot 9H_2O$ Solution: Add $Na_2S \cdot 9H_2O$ to distilled/deionized water and bring volume to 20.0mL. Autoclave for 15 min at 15 psi pressure–121°C. Use freshly prepared solution.

Preparation of Medium: Add components, except $Na_2S \cdot 9H_2O$, to distilled/deionized water and bring volume to 1.0L. Mix thoroughly. Gently heat to boiling. Autoclave for 15 min at 15 psi pressure–121°C. Cool to 45–50°C. Distribute anaerobically into sterile tubes. Aseptically add 0.1mL of sterile 1.5% $Na_2S \cdot 9H_2O$ solution to each 5.0mL of the medium. Cap with rubber stoppers.

Use: For the cultivation and maintenance of a variety of *Clostridium* species.

Clostridium Medium (ATCC Medium 40)

Composition per liter:

K_2HPO_4	7.0g
δ-Aminovaleric acid·HCl (neutralized)	5.0g

Agar	1.5g
KH_2PO_4	1.3g
Yeast extract	1.0g
$MgCl_2 \cdot 6H_2O$	0.2g
$CaCl_2 \cdot 2H_2O$	0.01g
$FeCl_3 \cdot 6H_2O$	0.01g
Methylene blue	2.0mg
$MnSO_4$	1.0mg
Na_2MoO_4	1.0mg
$Na_2S \cdot 9H_2O$ solution	20.0mL

$Na_2S \cdot 9H_2O$ Solution:
Composition per 100mL:

$Na_2S \cdot 9H_2O$	1.5g

Preparation of $Na_2S \cdot 9H_2O$ Solution: Add $Na_2S \cdot 9H_2O$ to distilled/deionized water and bring volume to 100.0mL. Autoclave for 15 min at 15 psi pressure–121°C. Use freshly prepared solution.

Preparation of Medium: Add components, except $Na_2S \cdot 9H_2O$ solution, to distilled/deionized water and bring volume to 1.0L. Mix thoroughly. Gently heat to boiling. Autoclave for 15 min at 15 psi pressure–121°C. Cool to 45–50°C. Distribute anaerobically into sterile tubes. Aseptically add 0.1mL of sterile $Na_2S \cdot 9H_2O$ solution to each 5.0mL of the medium. Cap with rubber stoppers.

Use: For the cultivation and maintenance of a variety of *Clostridium* species.

Clostridium Medium (ATCC Medium 43)

Composition per liter:

Agar	15.0g
Yeast extract	5.0g
L-Arginine·HCl	2.0g
L-Lysine·HCl	2.0g
NH_4Cl	2.0g
Sodium formate	2.0g
K_2HPO_4	1.75g
$MgSO_4 \cdot 7H_2O$	0.2g
$CaCl_2 \cdot 2H_2O$	0.01g
$FeSO_4 \cdot 7H_2O$	0.01g
Methylene blue	2.0mg
$Na_2S \cdot 9H_2O$ solution	30.0mL

$Na_2S \cdot 9H_2O$ Solution:
Composition per 100mL:

$Na_2S \cdot 9H_2O$	1.0g

Preparation of $Na_2S \cdot 9H_2O$ Solution: Add $Na_2S \cdot 9H_2O$ to distilled/deionized water and bring volume to 100.0mL. Autoclave for 15 min at 15 psi pressure–121°C. Use freshly prepared solution.

Preparation of Medium: Add components, except $Na_2S \cdot 9H_2O$ solution, to tap water and bring volume to 1.0L. Mix thoroughly. Gently heat to boiling. Autoclave for 15 min at 15 psi pressure–121°C. Cool to 45–50°C. Distribute anaerobically into sterile tubes. Aseptically add 0.15mL of sterile $Na_2S \cdot 9H_2O$ solution to each 5.0mL of the medium. Cap with rubber stoppers.

Use: For the cultivation and maintenance of a variety of *Clostridium* species.

Clostridium Medium (ATCC Medium 163)

Composition per liter:

Agar	20.0g
Sodium glutamate	17.0g
Yeast extract	6.0g
Sodium thioglycollate	0.5g
Phosphate buffer (1.0 *M*, pH 7.4)	40.0mL
$MgSO_4$ (2.0*M* solution)	0.5mL
$FeSO_4$ (0.2*M* solution)	0.2mL
$CaCl_2$ (1.0*M* solution)	0.1mL
$CoCl_2$ (0.1*M* solution)	0.1mL
$MnCl_2$ (0.1*M* solution)	0.1mL
Na_2MoO_4 (0.1*M* solution)	0.1mL

Preparation of Medium: Add components to distilled/deionized water and bring volume to 1.0L. Mix thoroughly. Gently heat to boiling. Autoclave for 15 min at 15 psi pressure–121°C. Pour into sterile Petri dishes or distribute into sterile tubes.

Use: For the cultivation and maintenance of a variety of *Clostridium* species.

Clostridium Medium (ATCC Medium 511)

Composition per liter:

Yeast extract	4.0g
Alanine	3.0g
Peptone	3.0g
Cysteine	0.2g
$MgSO_4$	0.05g
$FeSO_4$	0.01g
Potassium phosphate buffer (1*M*, pH 7.1)	5.0mL
$CaSO_4$ (saturated solution)	2.5mL

Preparation of Medium: Add components to distilled/deionized water and bring volume to 1.0L. Mix thoroughly. Distribute into tubes or flasks. Autoclave for 15 min at 15 psi pressure–121°C.

Use: For the cultivation of a variety of *Clostridium* species.

Clostridium Medium (ATCC Medium 568)

Composition per liter:

Na_2CO_3	10.0g
Fructose	3.0g
K_2HPO_4	2.0g
Yeast extract	2.0g
$(NH_4)_2SO_4$	1.0g
$MgSO_4 \cdot 7H_2O$	0.5g
Sodium thioglycollate	0.05g
$CaSO_4$	0.015g
$FeSO_4 \cdot 7H_2O$	2.5mg
$MnSO_4 \cdot H_2O$	0.5mg
$Na_2MoO_4 \cdot 2H_2O$	0.5mg

pH 7.8 ± 0.2 at 25°C

Preparation of Medium: Add components to distilled/deionized water and bring volume to 1.0L. Mix thoroughly. Distribute into tubes or flasks. Autoclave for 15 min at 15 psi pressure–121°C.

Use: For the cultivation of a variety of *Clostridium* species.

Clostridium Medium (ATCC Medium 591)

Composition per liter:

Solution 1	600.0mL
Solution 2	400.0mL

pH 8.0 ± 0.2 at 25°C

Solution 1:

Composition per 600mL:

Peptone	5.0g

Preparation of Solution 1: Add component to distilled/deionized water and bring volume to 600.0mL. Mix thoroughly. Autoclave for 15 min at 15 psi pressure–121°C.

Solution 2:

Composition per 400mL:

$NaHCO_3$	20.0g
Fructose	10.0g
K_2HPO_4	10.0g
Sodium thioglycollate	0.75g
Vitamin solution	14.0mL
Trace elements solution	10.0mL

Preparation of Solution 2: Add components, except sodium thioglycollate, to distilled/deionized water and bring volume to 400.0mL. Mix thoroughly. Gas with 100% CO_2. Add sodium thioglycollate. Adjust pH to 8.0. Filter sterilize.

Vitamin Solution:

Composition per 100mL:

Thiamine	0.1g
Nicotinic acid	0.05g
Pyridoxine	0.05g
Pantothenic acid	0.025g
p-Aminobenzoic acid	5.0mg
Vitamin B_{12}	2.0mg
Biotin	1.0mg

Preparation of Vitamin Solution: Add components to distilled/deionized water and bring volume to 100.0mL. Mix thoroughly.

Trace Elements Solution:

Composition per liter:

EDTA	0.5g
$FeSO_4 \cdot 7H_2O$	0.2g
H_3BO_3	0.03g
$CoCl_2 \cdot 6H_2O$	0.02g
$ZnSO_4 \cdot 7H_2O$	0.01g
$MnCl_2 \cdot 4H_2O$	3.0mg
$Na_2MoO_4 \cdot 2H_2O$	3.0mg
$NiCl_2 \cdot 6H_2O$	2.0mg
$CuCl_2 \cdot 2H_2O$	1.0mg

Preparation of Trace Elements Solution: Add components to distilled/deionized water and bring volume to 1.0L. Mix thoroughly.

Preparation of Medium: Aseptically combine 600.0mL of sterile solution 1 and 400.0mL of sterile solution 2. Distribute into sterile tubes or flasks.

Use: For the cultivation and maintenance of a variety of *Clostridium* species.

Clostridium Selective Agar (Clostrisel Agar)

Composition per liter:

Pancreatic digest of casein	17.0g
Agar	14.0g
Glucose	6.0g
Papaic digest of soybean meal	3.0g
NaCl	2.5g
Sodium thioglycollate	1.8g
Sodium formaldehyde sulfoxylate	1.0g
L-Cystine	0.25g
NaN_3	0.15g
Neomycin sulfate	0.15g

pH 7.0 ± 0.2 at 25°C

Source: This medium is available as a premixed powder from BBL Microbiology Systems.

Preparation of Medium: Add components to distilled/deionized water and bring volume to 1.0L. Mix thoroughly. Gently heat while stirring and bring to boiling. Distribute into tubes or flasks. Autoclave for 15 min at 15 psi pressure–118°C. Pour into sterile Petri dishes or leave in tubes.

Caution: Sodium azide is toxic. Azides also react with metals and disposal must be highly diluted.

Use: For the selective isolation of pathogenic *Clostridium* species from specimens containing mixed flora, e.g., from soil, and other specimens.

Clostridium sphenoides Medium

Composition per liter:

Agar	15.0g
Trisodium citrate·$2H_2O$	14.7g
Yeast extract	4.0g
KH_2PO_4	3.4g
K_2HPO_4	2.0g
Peptone	2.0g
NaCl	0.6g
L-Cysteine·HCl	0.3g
$(NH_4)_2SO_4$	0.3g
$MgSO_4·7H_2O$	0.2g
$CaCl_2·2H_2O$	0.06g
Resazurin	1.0mg

pH 6.7–7.0 ± 0.2 at 25°C

Preparation of Medium: Add components, except L-cysteine·HCl, to distilled/deionized water and bring volume to 1.0L. Mix thoroughly. Gently heat and bring to boiling. Add L-cysteine·HCl. Distribute anaerobically into tubes in 5.0mL volumes. Autoclave for 20 min at 15 psi pressure–121°C. Cool to 45–50°C. Inoculate with serial dilution of mud specimens before agar solidifies.

Use: For the isolation of *Clostridium sphenoides* from mud.

Clostridium thermocellum Medium

Composition per liter:

Filter paper	18.75g
$Na_2HPO_4·12H_2O$	4.2g
Yeast extract	2.0g
KH_2PO_4	1.5g
NH_4Cl	0.5g
$MgCl_2·6H_2O$	0.18g
Reducing solution	40.0mL
Wolfe's modified mineral elixir	5.0mL
Resazurin (0.1% solution)	1.0mL
Vitamin solution	0.5mL

Caution: This medium contains Na_2S, and H_2S production will occur, especially upon prolonged boiling. H_2S is hazardous and preparation of this medium should be done in a chemical fume hood.

Reducing Solution:

Composition per 200mL:

Cysteine·HCl·H_2O	2.5g
$Na_2S·9H_2O$	2.5g
NaOH (0.2*N* solution)	200.0mL

Preparation of Reducing Solution: Gently heat the NaOH solution and bring to boiling. Gas with 95% N_2 + 5% H_2. Cool to room temperature. Add the cysteine·HCl·H_2O and $Na_2S·9H_2O$. Anaerobically distribute into tubes. Cap with rubber stoppers. Autoclave for 15 min at 15 psi pressure–121°C.

Vitamin Solution:

Composition per 500mL:

Pyridoxine HCl	0.1g
p-Aminobenzoic acid	0.05g
Calcium pantothenate	0.05g
Nicotinic acid	0.05g
Thioctic acid	0.05g
Biotin	0.02g
Folic acid	0.02g
Riboflavin	5.0mg
Thiamine·HCl	5.0mg
Vitamin B_{12}	1.0mg

Preparation of Vitamin Solution: Add components to distilled/deionized water and bring volume to 500.0mL. Mix thoroughly. Store solution in the dark at –10°C.

Wolfe's Modified Mineral Elixir:

Composition per liter:

$MgSO_4·7H_2O$	3.0g
Nitrilotriacetic acid	1.5g
NaCl	1.0g
$MnSO_4·H_2O$	0.5g
$CaCl_2$, anhydrous	0.1g
$Co(NO_3)_2·6H_2O$	0.1g
$FeSO_4·7H_2O$	0.1g
$ZnSO_4·7H_2O$	0.1g
$AlK(SO_4)_2$, anhydrous	0.01g
$CuSO_4·5H_2O$	0.01g
H_3BO_3	0.01g
$Na_2MoO_4·2H_2O$	0.01g
Na_2SeO_3, anhydrous	1.0mg

Preparation of Wolfe's Modified Mineral Elixir: Add nitrilotriacetic acid to 500.0mL of distilled/deionized water. Dissolve by adjusting pH to 6.5 with KOH. Add remaining components. Add distilled/deionized water to 1.0L.

Preparation of Medium: Add components, except reducing solution, to distilled/deionized water and bring volume to 1.0L. If medium is to be distributed into tubes, omit bulk filter paper and substitute one Whatman #1 filter paper strip (8mm × 70mm) per tube of broth. Gently heat and bring to boiling under 95% N_2 + 5% H_2. Continue boiling until color changes from blue to pink. Add the reducing solution. The pink color will disappear, indicating that the solution has been reduced. Distribute into tubes or flasks under 95% N_2 + 5% H_2 using anerobic techniques. If tubes are used, remember to add Whatman #1 filter paper strips, prior to the addition of broth. Cap tubes with rubber stoppers. Autoclave for 15 min at 15 psi pressure–121°C.

Use: For the cultivation and maintenance of *Clostridium thermocellum.*

Clostridium thermolacticum Medium

Composition per liter:

$KHCO_3$	4.5g
NaCl	2.25g
Sucrose	2.0g
Yeast extract	2.0g
$MgSO_4·7H_2O$	0.5g
NH_4Cl	0.5g
K_2HPO_4	0.348g
Cysteine·HCl·H_2O solution	0.3g
$Na_2S·9H_2O$ solution	0.3g
$CaCl_2·2H_2O$	0.25g
KH_2PO_4	0.227g
$FeSO_4·7H_2O$	2.0mg
Resazurin	1.0mg
Cysteine·HCl·H_2O solution	25.0mL
$Na_2S·9H_2O$ solution	25.0mL
Wolfe's vitamin solution	10.0mL
Trace element solution SL-6	3.0mL

pH 7.0–7.2 at 25°C

Wolfe's Vitamin Solution:
Composition per liter:

Pyridoxine·HCl	10.0mg
p-Aminobenzoic acid	5.0mg
Calcium pantothenate	5.0mg
Nicotinic acid	5.0mg
Riboflavin	5.0mg
Thiamine·HCl	5.0mg
Thioctic acid	5.0mg
Biotin	2.0mg
Folic acid	2.0mg
Cyanocobalamin	100.0μg

Preparation of Medium: Add components to distilled/deionized water and bring volume to 1.0L. Mix thoroughly.

Trace Elements Solution SL-6:
Composition per liter:

H_3BO_3	0.3g
$CoCl_2·6H_2O$	0.2g
$ZnSO_4·7H_2O$	0.10g
$MnCl_2·4H_2O$	0.03g
$Na_2MoO_4·H_2O$	0.03g
$NiCl_2·6H_2O$	0.02g
$CuCl_2·2H_2O$	0.01g

Preparation of Trace Elements Solution SL-6: Add components to distilled/deionized water and bring volume to 1.0L. Mix thoroughly. Adjust pH to 3.4.

Cysteine·HCl·H_2O Solution:
Composition per 10mL:

Cysteine·HCl·H_2O	0.12g

Preparation of Cysteine·HCl·H_2O Solution: Add cysteine·HCl·H_2O to distilled/deionized water and bring volume to 10.0mL. Gas tubes under 100% N_2 and tightly seal. Autoclave for 15 min at 15 psi pressure–121°C. Use freshly prepared solution.

$Na_2S·9H_2O$ Solution:
Composition per 10mL:

$Na_2S·9H_2O$	0.12g

Preparation of $Na_2S·9H_2O$ Solution: Add $Na_2S·9H_2O$ to distilled/deionized water and bring volume to 10.0mL. Gas tube under 100% N_2 and tightly seal. Autoclave for 15 min at 15 psi pressure–121°C. Use freshly prepared solution.

Preparation of Medium: Prepare anaerobically under 80% N_2 + 20% CO_2. Add components to distilled/deionized water and bring volume to 1.0L. Mix thoroughly. Distribute into tubes using anaerobic techniques. Autoclave for 15 min at 15 psi pressure–121°C. Prior to inoculation of cultures, inject 0.25mL of sterile cysteine·HCl·H_2O solution and 0.25mL of sterile $Na_2S·9H_2O$ solution per 10.0mL of medium.

Use: For the cultivation and maintenance of *Clostridium thermolacticum.*

CM Agar

Composition per liter:

Agar	20.0g
Polypeptone™	10.0g
Yeast extract	10.0g
NaCl	5.0g

pH 7.0 ± 0.2 at 25°C

Preparation of Medium: Add components to distilled/deionized water and bring volume to 1.0L. Mix thoroughly. Gently heat until boiling. Distribute into tubes or flasks. Autoclave for 15 min at 15 psi pressure–121°C. Pour into sterile Petri dishes or leave in tubes.

Use: For the cultivation and maintenance of *Bacillus subtilis.*

CM plus YE Medium

Composition per liter:

NaCl	200.0g
$MgSO_4·7H_2O$	20.0g
Yeast extract	10.0g
Casamino acids (vitamin free)	7.5g
Sodium citrate	3.0g
KCl	2.0g
$FeSO_4·7H_2O$ (4.98% solution)	1.0mL

pH 7.4 ± 0.2 at 25°C

Preparation of Medium: Add components to distilled/deionized water and bring volume to 1.0L. Mix thoroughly. Gently heat until boiling. Adjust pH to 7.4 with NaOH. Distribute into tubes or flasks. Autoclave for 10 min at 15 psi pressure–121°C.

Use: For the cultivation and maintenance of *Actinopolyspora halophila.*

CM4 Medium

Composition per liter:

Cellobiose	6.0g
Yeast extract	5.0g
K_2HPO_4	2.9g
KH_2PO_4	1.5g
$(NH_4)_2SO_4$	1.3g
NaCl	1.0g
$MgCl_2$	0.75g
Sodium thioglycollate	0.5g
$CaCl_2$	0.0132g
Resazurin (1.0% solution)	0.2mL
$FeSO_4$ (1.25% solution)	0.1mL

Preparation of Medium: Add components to distilled/deionized water and bring volume to 1.0L. Mix thoroughly. Gently heat until boiling. Boil until color changes from red to colorless, indicating reduced state. Cool. Distribute into tubes or flasks under 97% N_2 + 3% H_2. Cap with rubber stoppers. Autoclave for 15 min at 15 psi pressure–121°C. Pour into sterile Petri dishes or leave in tubes.

Use: For the cultivation and maintenance of *Clostridium* species and other bacteria that can utilize cellobiose as a carbon source.

CML Medium (Cooked Meat Liver Medium)

Composition per liter:

Cooked meat	57.0g
Glucose	10.0g
Tryptose	10.0g
Liver infusion broth	10.0mL

pH 6.9 ± 0.2 at 25°C

Liver Infusion Broth:

Composition per liter:

Beef liver, infusion from	500.0g
Proteose peptone	10.0g
NaCl	5.0g

Preparation of Liver Infusion Broth: Add components to distilled/deionized water and bring volume to 1.0L. Mix thoroughly.

Preparation of Medium: Add cooked meat to distilled/deionized water and bring volume to 1.0L. Chill to 4°C until liquid is clear. Filter through cheesecloth. Add remaining components to filtrate. Distribute into tubes or flasks. Autoclave for 10 min at 15 psi pressure–121°C.

Use: For the cultivation and maintenance of *Fusobacterium varium.*

CMRL 1066 Medium with Glutamine, 10X (Connaught Medical Research Laboratories Medium with Glutamine, 10X)

Composition per liter:

NaCl	6.8g
$NaHCO_3$	2.2g
D-Glucose	1.0g
KCl	0.4g
L-Cysteine·HCl·H_2O	0.26g
$CaCl_2$, anhydrous	0.2g
$MgSO_4·7H_2O$	0.2g
$NaH_2PO_4·H_2O$	0.14g
L-Glutamine	0.1g
Sodium acetate·$3H_2O$	0.083g
L-Glutamic acid	0.075g
L-Arginine·HCl	0.070g
L-Lysine·HCl	0.070g
L-Leucine	0.060g
Glycine	0.050g
Ascorbic acid	0.050g
L-Proline	0.040g
L-Tyrosine	0.040g
L-Aspartic acid	0.030g
L-Threonine	0.030g
L-Alanine	0.025g
L-Phenylalanine	0.025g
L-Serine	0.025g
L-Valine	0.025g
L-Cystine	0.020g
L-Histidine·HCl·H_2O	0.020g

L-Isoleucine	0.020g
Phenol Red	0.020g
L-Methionine	0.015g
Deoxyadenosine	0.010g
Deoxycytidine	0.010g
Deoxyguanosine	0.010g
Glutathione, reduced	0.010g
Thymidine	0.010g
Hydroxy-L-proline	0.010g
L-Tryptophan	0.010g
Nicotinamide adenine dinucleotide	7.0mg
Tween™ 80	5.0mg
Sodium glucoronate·H_2O	4.2mg
Coenzyme A	2.5mg
Cocarboxylase	1.0mg
Flavin adenine dinucleotide	1.0mg
Nicotinamide adenine dinucleotide phosphate	1.0mg
Uridine triphosphate	1.0mg
Choline chloride	0.50mg
Cholesterol	0.20mg
5-Methyldeoxycytidine	0.10mg
Inositol	0.05mg
p-Aminobenzoic acid	0.05mg
Niacin	0.025mg
Niacinamide	0.025mg
Pyridoxine	0.025mg
Pyridoxal·HCl	0.025mg
Biotin	0.01mg
D-Calcium pantothenate	0.01mg
Folic acid	0.01mg
Riboflavin	0.01mg
Thiamine·HCl	0.01mg

pH 7.2 ± 0.2 at 25°C

Source: This medium is available as a premixed powder from GIBCO BRL.

Preparation of Medium: Add components to distilled/deionized water and bring volume to 1.0L. Mix thoroughly. Adjust pH to 7.2. Filter sterilize.

Use: For the cultivation of a wide variety of microorganisms in a chemically defined basal medium.

Coagulase Agar Base

Composition per liter:

Agar	25.0g
Brain heart infusion	10.5g
Pancreatic digest of casein	10.5g
D-Mannitol	10.0g
Brain heart infusion	5.0g
NaCl	3.5g
Papaic digest of soybean meal	3.5g
Bromcresol Purple	0.02g
Rabbit plasma	100.0mL

pH 7.4 ± 0.2 at 25°C

Preparation of Medium: Add components, except rabbit plasma, to distilled/deionized water and bring volume to 1.0L. Mix thoroughly. Gently heat while stirring until boiling. Distribute into tubes or flasks. Autoclave for 15 min at 15 psi pressure–121°C. Cool to 45–50°C. Add rabbit plasma to a final concentration of 7–15%. Mix thoroughly. Pour into sterile Petri dishes in 18.0mL volume per plate.

Use: For the cultivation and differentiation of *Staphylococcus aureus* from other *Staphylococcus* species based on coagulase production.

Coagulase Mannitol Agar

Composition per liter:

Agar	14.5g
Pancreatic digest of casein	10.5g
D-Mannitol	10.0g
Brain heart infusion	5.0g
NaCl	3.5g
Papaic digest of soybean meal	3.5g
Bromcresol Purple	0.02g
Rabbit plasma with 0.15% EDTA	100.0mL

pH 7.3 ± 0.2 at 25°C

Source: This medium is available as a premixed powder from BBL Microbiology Systems.

Preparation of Medium: Add components, except rabbit plasma, to distilled/deionized water and bring volume to 1.0L. Mix thoroughly. Gently heat while stirring until boiling. Distribute into tubes or flasks. Autoclave for 15 min at 15 psi pressure–121°C. Cool to 45–50°C. Add rabbit plasma with 0.15% EDTA to a final concentration of 7–15%. Mix thoroughly. Pour into sterile Petri dishes in 18.0mL volume per plate.

Use: For the cultivation and differentiation of *Staphylococcus aureus* from other *Staphylococcus* species based on coagulase production and mannitol fermentation.

Colloidal Chitin Agar

Composition per liter:

Agar	20.0g
Chitin, colloidal	4.0g
K_2HPO_4	0.7g
$MgSO_4 \cdot 5H_2O$	0.5g
KH_2PO_4	0.3g
$FeSO_4 \cdot 7H_2O$	0.01g
$MnCl_2$	1.0mg
$ZnSO_4$	1.0mg

pH 7.0 ± 0.2 at 25°C

Preparation of Medium: Add components to distilled/deionized water and bring volume to 1.0L. Mix thoroughly. Gently heat and bring to boiling. Distribute into tubes or flasks. Autoclave for 15 min at 15 psi pressure–121°C. Pour into sterile Petri dishes or leave in tubes.

Use: For the isolation and cultivation of *Micromonospora* species from water, soil or sediment. For the germination of spores of *Micromonospora* species.

Cooke Rose Bengal Agar

Composition per liter:

Agar	20.0g
Glucose	10.0g
Enzymatic hydrolysate of soybean meal	5.0g
KH_2PO_4	1.0g
$MgSO_4 \cdot 7H_2O$	0.5g
Rose Bengal	35.0mg

pH 6.0 ± 0.2 at 25°C

Source: This medium is available as a premixed powder from Difco Laboratories.

Preparation of Medium: Add components to distilled/deionized water and bring volume to 1.0L. Mix thoroughly. Gently heat until boiling. Distribute into tubes or flasks. Autoclave for 15 min at 15 psi pressure–121°C. Pour into sterile Petri dishes or leave in tubes.

Use: For the isolation of fungi.

Cooked Meat Medium

Composition per liter:

Beef heart	454.0g
Proteose peptone	20.0g
NaCl	5.0g
Glucose	2.0g

pH 7.2 ± 0.2 at 25°C

Source: This medium is available as a premixed powder from Difco Laboratories.

Preparation of Medium: Finely chop beef heart. Add approximately 1.5g of heart particles to test tubes. Add remaining components to distilled/deionized water and bring volume to 1.0L. Mix thoroughly. Distribute into tubes in 10.0mL volumes. Autoclave for 15 min at 15 psi pressure–121°C. Slowly cool tubes to prevent expulsion of meat particles.

Use: For cultivation and maintenance of anaerobic microorganisms.

Cooked Meat Medium

Composition per liter:

Heart muscle	454.0g
Beef extract	10.0g
Peptone	10.0g
NaCl	5.0g
Glucose	2.0g

pH 7.2 ± 0.2 at 25°C

Source: This medium is available as a premixed powder from Oxoid Unipath.

Preparation of Medium: Finely chop beef heart. Add approximately 1.5g of heart particles to test tubes. Add remaining components to distilled/deionized water and bring volume to 1.0L. Mix thoroughly. Distribute into tubes in 10.0mL volumes. Autoclave for 15 min at 15 psi pressure–121°C. Slowly cool tubes to prevent expulsion of meat particles.

Use: For the cultivation and maintenance of aerobic and anaerobic microorganisms. Used for the cultivation of anaerobes, especially pathogenic *Clostridia.*

Cooked Meat Medium

Composition per liter:

Heart tissue granules	98.0g
Peptic digest of animal tissue	20.0g
NaCl	5.0g
Glucose	2.0g

pH 7.2 ± 0.2 at 25°C

Source: This medium is available as a premixed powder from BBL Microbiology Systems.

Preparation of Medium: Add approximately 1.0g of heart tissue granules to test tubes. Add remaining components to distilled/deionized water and bring volume to 1.0L. Mix thoroughly. Distribute into tubes in 10.0mL volumes. Autoclave for 15 min at 15 psi pressure–121°C. Slowly cool tubes to prevent expulsion of meat particles.

Use: For the cultivation of anaerobes, especially pathogenic *Clostridia.*

Cooked Meat Medium, Modified

Composition per liter:

Cooked meat medium	66.0g
Solution A	1.0L

pH 6.8 ± 0.2 at 25°C

Solution A:

Composition per liter:

Pancreatic digest of casein	10.0g
Glucose	2.0g

Soluble starch	1.0g
Sodium thioglycollate	1.0g
Neutral Red (1% aqueous)	5.0mL

Preparation of Solution A: Add components to distilled/deionized water and bring volume to 1.0L. Mix thoroughly. Gently heat until dissolved.

Preparation of Medium: Add 1.0g of cooked meat medium to each of 66 test tubes. Add 15.0mL of solution A to each test tube. Allow meat particles to rehydrate. Autoclave for 15 min at 15 psi pressure–121°C.

Use: For the cultivation of a variety of anaerobic microorganisms.

Cooked Meat Medium with Glucose, Hemin and Vitamin K

Composition per liter:

Heart tissue granules	98.0g
Peptic digest of animal tissue	20.0g
NaCl	5.0g
Glucose	5.0g
Yeast extract	5.0g
Hemin	5.0mg
Vitamin K	1.0mg

pH 7.2 ± 0.2 at 25°C

Source: This medium is available as a premixed powder from BBL Microbiology Systems.

Preparation of Medium: Add approximately 1.0g of heart tissue granules to test tubes. Add remaining components to distilled/deionized water and bring volume to 1.0L. Mix thoroughly. Distribute into tubes in 10.0mL volumes. Autoclave for 15 min at 15 psi pressure–121°C. Slowly cool tubes to prevent expulsion of meat particles.

Use: For the cultivation of anaerobes, especially pathogenic *Clostridia*.

Corn Meal Agar (Corn Meal Agar with Polysorbate 80)

Composition per liter:

Agar	15.0g
Corn meal, solids from infusion	2.0g
Tween™ 80	1.0g

pH 5.6–6.0 at 25°C

Source: This medium is available as a premixed powder from Difco Laboratories, BBL Microbiology Systems and Oxoid Unipath.

Preparation of Medium: Add components to distilled/deionized water and bring volume to 1.0L. Mix thoroughly. Gently heat until boiling. Distribute into tubes or flasks. Autoclave for 15 min at 15 psi pressure–121°C. Pour into sterile Petri dishes or leave in tubes.

Use: For the cultivation and maintenance of fungi. For the production of chlamydospores by *Candida albicans* and for the cultivation of phytopathological fungi.

Corn Meal Agar with Dextrose

Composition per liter:

Agar	15.0g
Corn meal, solids from infusion	2.0g
Glucose	2.0g
Tween™ 80	1.0g

pH 5.6–6.0 at 25°C

Source: This medium is available as a premixed powder from Difco Laboratories.

Preparation of Medium: Add components to distilled/deionized water and bring volume to 1.0L. Mix thoroughly. Gently heat until boiling. Distribute into tubes or flasks. Autoclave for 15 min at 15 psi pressure–121°C. Pour into sterile Petri dishes or leave in tubes.

Use: For the cultivation of phytopathological and other fungi.

Corn Meal *Phytophthora* Isolation Medium No. 1

Composition per liter:

Agar	15.0g
Corn meal, solids from infusion	2.0g
Vancomycin	0.2g
Pentachloronitrobenzene (PCNB)	0.1g
Pimaricin	0.01g

pH 5.6–6.0 at 25°C

Preparation of Medium: Add components, except pimaricin and vancomycin, to distilled/deionized water and bring volume to 1.0L. Mix thoroughly. Gently heat until boiling. Autoclave for 15 min at 15 psi pressure–121°C. Aseptically add pimaricin and vancomycin. Mix thoroughly. Pour into sterile Petri dishes.

Use: For the cultivation of *Phytophthora* species.

Corn Meal *Phytophthora* Isolation Medium No. 2

Composition per liter:

Agar	15.0g
Corn meal, solids from infusion	2.0g
Vancomycin	0.3g

Pentachloronitrobenzene (PCNB) 0.025g
Pimaricin 5.0mg

pH 5.6–6.0 at 25°C

Preparation of Medium: Add components, except pimaricin and vancomycin, to distilled/deionized water and bring volume to 1.0L. Mix thoroughly. Gently heat until boiling. Autoclave for 15 min at 15 psi pressure–121°C. Aseptically add pimaricin and vancomycin. Mix thoroughly. Pour into sterile Petri dishes.

Use: For the cultivation of *Phytophthora* species.

Corn Steep Liquor Medium

Composition per liter:

Glucose 60.0g
Corn steep liquor 40.0g
Urea 8.0g
KH_2PO_4 .... 5.0g
Fumaric acid 1.0g
$MgSO_4 \cdot 7H_2O$ 0.5g
Hutner's mineral base 20.0mL

pH 7.0 ± 0.2 at 25°C

Hutner's Mineral Base:

Composition per liter:

$MgSO_4 \cdot 7H_2O$ 29.7g
Nitrilotriacetic acid 10.0g
$CaCl_2 \cdot 2H_2O$ 3.34g
$FeSO_4 \cdot 7H_2O$ 99.0mg
$(NH_4)_2MoO_4$ 9.25mg
Metals "44" 50.0mL

Preparation of Hutner's Mineral Base: Add nitrilotriacetic acid to 500.0mL of distilled/deionized water. Dissolve by adjusting pH to 6.5 with KOH. Add remaining components. Add distilled/deionized water to 1.0L.

Metals "44":

Composition per 100mL:

$ZnSO_4 \cdot 7H_2O$ 1.1g
$FeSO_4 \cdot 7H_2O$ 0.5g
EDTA 0.25g
$MnSO_4 \cdot 7H_2O$ 0.154g
$CuSO_4 \cdot 5H_2O$ 0.04g
$Co(NO_3)_2 \cdot 6H_2O$ 0.025g
$Na_2B_4O_7 \cdot 10H_2O$ 0.018g

Preparation of Metals "44": Add components to distilled/deionized water and bring volume to 100.0mL. Mix thoroughly.

Preparation of Medium: Add components to distilled/deionized water and bring volume to 1.0L. Mix thoroughly. Distribute into tubes or flasks. Autoclave for 15 min at 15 psi pressure–121°C.

Use: For the cultivation of *Pseudomonas* species.

Cornstarch Soluble Medium (CSSM)

Composition per liter:

Cornstarch 42.0g
n-Butanol 18.0g
Yeast extract 10.0g
Asparagine·H_2O 2.0g
$(NH_4)_2SO_4$ 2.0g
NaCl 1.0g
KH_2PO_4 .... 0.75g
K_2HPO_4 .... 0.75g
Cysteine·HCl·H_2O 0.5g
$MgSO_4$ 0.02g
$FeSO_4 \cdot 7H_2O$ 0.01g
$MnSO_4 \cdot H_2O$ 0.01g

Preparation of Medium: Add components to distilled/deionized water and bring volume to 1.0L. Mix thoroughly. Gently heat until boiling. Boil and cool under 80% N_2 + 10% H_2 + 10% CO_2. Distribute anaerobically into tubes under the same gas mixture. Cap with butyl rubber stoppers. Autoclave for 15 min at 15 psi pressure–121°C.

Use: For the cultivation and maintenance of *Clostridium thermoamylolyticum*.

Cornstarch Soluble Medium (CSSM) (ATCC Medium 1500)

Composition per liter:

Cornstarch 42.0g
Yeast extract 10.0g
Asparagine·H_2O 2.0g
$(NH_4)_2SO_4$ 2.0g
NaCl 1.0g
KH_2PO_4 .... 0.75g
K_2HPO_4 .... 0.75g
Cysteine·HCl·H_2O 0.5g
$MgSO_4$ 0.02g
$FeSO_4 \cdot 7H_2O$ 0.01g
$MnSO_4 \cdot H_2O$ 0.01g

Preparation of Medium: Add components to distilled/deionized water and bring volume to 1.0L. Mix thoroughly. Gently heat until boiling. Boil and cool under 80% N_2 + 10% H_2 + 10% CO_2. Distribute anaerobically into tubes under the same gas mixture. Cap with butyl rubber stoppers. Autoclave for 15 min at 15 psi pressure–121°C.

Use: For the cultivation and maintenance of *Clostridium thermoamylolyticum*.

Creatinine Medium

Composition per liter:

Creatinine	5.0g
Agar	2.0g
Fumaric acid	2.0g
K_2HPO_4	2.0g
Yeast extract	1.0g
Salt solution	10.0mL

pH 6.8 ± 0.2 at 25°C

Salt Solution:

Composition per liter:

$MgSO_4$	12.2g
$FeSO_4 \cdot 7H_2O$	2.8g
$MnSO_4 \cdot H_2O$	1.7g
$CaCl_2 \cdot 2H_2O$	0.76g
NaCl	0.6g
$Na_2MoO_4 \cdot 2H_2O$	0.1g
$ZnSO_4 \cdot 7H_2O$	0.06g
HCl (0.1*N* solution)	1.0L

Preparation of Salt Solution: Dissolve salts in 1.0L of 0.1*N* HCl solution. Mix thoroughly.

Preparation of Medium: Add components to distilled/deionized water and bring volume to 1.0L. Mix thoroughly. Adjust pH to 6.8 with NaOH or KOH. Gently heat until boiling. Distribute into tubes or flasks. Autoclave for 15 min at 15 psi pressure–121°C.

Use: For the cultivation and maintenance of *Pseudomonas* species.

CT Agar (Caprylate Thallous Agar)

Composition per liter:

Solution A	500.0mL
Solution B	500.0mL

pH 7.2 ± 0.2 at 25°C

Solution A:

Composition per 500mL:

K_2HPO_4	2.61g
KH_2PO_4	0.68g
Thallous sulfate	0.25g
$MgSO_4 \cdot 7H_2O$	0.12g
$CaCl_2 \cdot 2H_2O$	0.016g
Trace element solution	10.0mL
Yeast extract	2.0mL
Caprylic acid	1.1mL

Preparation of Solution A: Add components to distilled/deionized water and bring volume to 500.0mL. Mix thoroughly. Adjust pH to 7.2 with NaOH. Autoclave for 20 min at 10 psi pressure–115°C.

Trace Element Solution:

Composition per liter:

H_3PO_4	1.96g
$FeSO_4 \cdot 7H_2O$	0.056g
$ZnSO_4 \cdot 4H_2O$	0.029g
$CuSO_4 \cdot 5H_2O$	0.025g
$MnSO_4 \cdot 4H_2O$	0.022g
H_3BO_3	6.2mg
$Co(NO_3)_2 \cdot 6H_2O$	3.0mg

Preparation of Trace Element Solution: Add components to distilled/deionized water and bring volume to 1.0L. Mix thoroughly. Store at 4°C.

Solution B:

Composition per liter:

Agar	15.0g
NaCl	7.0g
$(NH_4)_2SO_4$	1.0g

Preparation of Solution B: Add components to distilled/deionized water and bring volume to 500.0mL. Mix thoroughly. Gently heat and bring to boiling. Adjust pH to 7.2. Autoclave for 20 min at 10 psi pressure–115°C.

Preparation of Medium: Aseptically combine 500.0mL of sterile solution A and 500.0mL of sterile solution B. Mix thoroughly. Pour into sterile Petri dishes in 25.0–30.0mL volumes.

Use: For the isolation and cultivation of the *Serratia* species.

CT Agar

Composition per liter:

Agar	20.0g
Pancreatic digest of casein	20.0g
$MgSO_4 \cdot 7H_2O$	2.0g
Potassium phosphate buffer (0.02*M* solution, pH 7.6)	500.0mL

pH 7.6 ± 0.2 at 25°C

Preparation of Medium: Add agar, pancreatic digest of casein, and $MgSO_4 \cdot 7H_2O$ to distilled/deionized water and bring volume to 500.0mL. Mix thoroughly. Gently heat and bring to boiling. Autoclave agar–pancreatic digest of casein–$MgSO_4 \cdot 7H_2O$ solution and potassium phosphate buffer solution separately for 15 min at 15 psi pressure–121°C. Cool to 25°C. Aseptically combine the two solutions. Aseptically add sterile components. Mix thoroughly. Pour into sterile Petri dishes or distribute into sterile tubes.

Use: For the cultivation of myxobacteria.

CT Broth

Composition per liter:

Pancreatic digest of casein	20.0g
$MgSO_4 \cdot 7H_2O$	2.0g
Potassium phosphate buffer (0.02*M* solution, pH 7.6)	500.0mL

pH 7.6 ± 0.2 at 25°C

Preparation of Medium: Add pancreatic digest of casein and $MgSO_4 \cdot 7H_2O$ to distilled/deionized water and bring volume to 500.0mL. Mix thoroughly. Autoclave pancreatic digest of casein-$MgSO_4 \cdot 7H_2O$ solution and potassium phosphate buffer solution separately for 15 min at 15 psi pressure–121°C. Cool to 25°C. Aseptically combine the two solutions. Aseptically distribute into sterile tubes or flasks.

Use: For the cultivation of myxobacteria.

CT Medium

Composition per liter:

Agar	15.0g
Pancreatic digest of casein	10.0g
Yeast extract	3.5g
$MgSO_4$	0.96g

Preparation of Medium: Add components to distilled/deionized water and bring volume to 1.0L. Mix thoroughly. Gently heat until boiling. Distribute into tubes or flasks. Autoclave for 15 min at 15 psi pressure–121°C. Pour into sterile Petri dishes or leave in tubes.

Use: For the cultivation and maintenance of *Stigmatella aurantiaca.*

CTT Medium

Composition per liter:

Agar	15.0g
Pancreatic digest of casein	10.0g
Tris(hydroxymethyl)aminomethane buffer	1.21g
Potassium phosphate buffer (1m*M*, pH 7.6)	1.0L
Magnesium sulfate solution	10.0mL

pH 7.6 ± 0.2 at 25°C

Magnesium Sulfate Solution:

Composition per 10mL:

$MgSO_4 \cdot 7H_2O$	2.0g

Preparation of Magnesium Sulfate Solution: Add components to 10.0mL of distilled/deionized water. Mix thoroughly.

Preparation of Medium: Combine components. Mix thoroughly. Adjust pH to 7.6. Gently heat and bring to boiling. Distribute into tubes or flasks. Autoclave for 15 min at 15 psi pressure–121°C. Pour into sterile Petri dishes or leave in tubes.

Use: For the cultivation of myxobacteria.

CY Agar

Composition per liter:

Agar	15.0g
Pancreatic digest of casein	3.0g
$CaCl_2 \cdot 2H_2O$	1.0g
Yeast extract	1.0g
Cyanocobalamin	0.5mg

pH 7.2 ± 0.2 at 25°C

Preparation of Medium: Add components to distilled/deionized water and bring volume to 1.0L. Mix thoroughly. Gently heat and bring to boiling. Distribute into tubes or flasks. Autoclave for 15 min at 15 psi pressure–121°C. Pour into sterile Petri dishes or leave in tubes.

Use: For the cultivation of myxobacteria.

CYC Medium

Composition per liter:

Sucrose	30.0g
Casamino acids, vitamin-free	6.0g
$NaNO_3$	3.0g
Yeast extract	2.0g
K_2HPO_4	1.0g
$MgSO_4 \cdot 7H_2O$	0.5g
KCl	0.5g
$FeSO_4 \cdot 7H_2O$	0.01g
Antibiotic solution	10.0mL

pH 7.2 ± 0.2 at 25°C

Preparation of Medium: Add components to distilled/deionized water and bring volume to 1.0L. Mix thoroughly. Distribute into tubes or flasks. Autoclave for 15 min at 15 psi pressure–121°C.

Antibiotic Solution:

Composition per 10mL:

Cycloheximide	0.050g
Novobiocin	0.025g

Preparation of Antibiotic Solution: Add components to distilled/deionized water and bring volume to 10.0mL. Mix thoroughly. Filter sterilize.

Preparation of Medium: Add components, except antibiotic solution, to distilled/deionized water and bring volume to 990.0mL. Mix thoroughly. Gently heat and bring to boiling. Autoclave for 15 min at 15 psi pressure–121°C. Cool to 45–50°C. Aseptically add sterile antibiotic solution. Mix thoroughly. Aseptically distribute into sterile tubes.

Use: For the isolation and cultivation of *Thermoactinomyces* species.

Cyclohexanecarboxylic Acid Medium

Composition per liter:

K_2HPO_4	3.5g
Cyclohexanecarboxylic acid	2.0g
KH_2PO_4	1.5g
NH_4NO_3	1.0g
$MgSO_4 \cdot 7H_2O$	0.5g
Yeast extract	0.1g
$CaCl_2 \cdot 2H_2O$	0.01g
$FeCl_3 \cdot 6H_2O$	0.01g
$NaMoO_4 \cdot 7H_2O$	0.01g
$ZnSO_4 \cdot 7H_2O$	0.01g

pH 7.0 ± 0.2 at 25°C

Preparation of Medium: Add components to distilled/deionized water and bring volume to 1.0L. Mix thoroughly. Distribute into tubes or flasks. Autoclave for 15 min at 15 psi pressure–121°C.

Use: For the cultivation and maintenance of *Alcaligenes faecalis* and other bacteria that can utilize cyclohexanecarboxylic acid as a carbon source.

Cyclohexanone Medium

Composition per liter:

NH_4NO_3	3.0g
K_2HPO_4	0.25g
$MgSO_4 \cdot 7H_2O$	0.20g
$CaCl_2 \cdot 2H_2O$	0.01g
$FeCl_3 \cdot 6H_2O$	1.0mg
Cyclohexanone	1.0mL

Preparation of Medium: Add components, except cyclohexanone, to distilled/deionized water and bring volume to 999.0mL. Mix thoroughly. Distribute into tubes or flasks. Autoclave for 20 min at 15 psi pressure–121°C. Filter sterilize cyclohexanone. Aseptically add 1.0mL cyclohexanone. Mix thoroughly.

Use: For the cultivation and maintenance of *Nocardia* species and other bacteria that can utilize cyclohexanone as a carbon source.

CYE Agar (Charcoal Yeast Extract Agar)

Composition per liter:

Agar	17.0g
Yeast extract	10.0g
Charcoal, activated, acid-washed	2.0g
L-Cysteine·HCl·H_2O solution	10.0mL
$Fe_4(P_2O_7)_3$ solution	10.0mL

pH 6.9 ± .05 at 50°C

L-Cysteine·HCl·H_2O Solution:
Composition per 10mL:

L-Cysteine·HCl·H_2O	0.4g

Preparation of L-Cysteine·HCl·H_2O solution: Add L-Cysteine·HCl·H_2O to distilled/deionized water and bring volume to 10.0mL. Mix thoroughly. Filter sterilize.

$Fe_4(P_2O_7)_3$ Solution:
Composition per liter:

$Fe_4(P_2O_7)_3$	0.25g

Preparation of $Fe_4(P_2O_7)_3$ Solution: Add soluble $Fe_4(P_2O_7)_3$ to distilled/deionized water and bring volume to 10.0mL. Mix thoroughly. Filter sterilize. The soluble $Fe_4(P_2O_7)_3$ must be kept dry and in the dark. Do not use if brown or yellow. Prepare solutions freshly. Do not heat over 60°C to dissolve. The mixture dissolves readily in a 50°C water bath.

Preparation of Medium: Add components, except L-cysteine·HCl·H_2O solution and $Fe_4(P_2O_7)_3$ solution to distilled/deionized water and bring volume to 980.0mL. Mix thoroughly. Gently heat to boiling. Autoclave for 15 min at 15 psi pressure–121°C. Cool to 50° C. Add 10.0mL of sterile L-cysteine·HCl·H_2O solution. Add 10.0mL of sterile $Fe_4(P_2O_7)_3$ solution. Adjust pH to 6.9 at 50°C by adding 4.0–4.5mL of 1.0 *N* KOH. This is a critical step. Mix thoroughly. Pour in 20.0mL volumes into sterile Petri dishes. Swirl medium while pouring to keep charcoal in suspension.

Use: For the cultivation and maintenance of *Legionella* species and *Tatlockia micdadei.*

CYE Agar, Buffered (Charcoal Yeast Extract Agar, Buffered)

Composition per liter:

Agar	17.0g
ACES buffer (*N*-2-acetamido-2-aminoethane sulfonic acid)	10.0g
Yeast extract	10.0g
Charcoal, activated, acid-washed	2.0g
L-Cysteine·HCl·H_2O solution	10.0mL
$Fe_4(P_2O_7)_3$ solution	10.0mL

pH 6.9 ± .05 at 50°C

L-Cysteine·HCl·H_2O Solution:
Composition per 10mL:

L-Cysteine·HCl·H_2O	0.4g

Preparation of L-Cysteine·HCl·H_2O solution: Add L-Cysteine·HCl·H_2O to distilled/deionized water and bring volume to 10.0mL. Mix thoroughly. Filter sterilize.

$Fe_4(P_2O_7)_3$ solution:
Composition per liter:

$Fe_4(P_2O_7)_3$	0.25g

Preparation of $Fe_4(P_2O_7)_3$ solution: Add soluble $Fe_4(P_2O_7)_3$ to distilled/deionized water and bring volume to 10.0mL. Mix thoroughly. Filter sterilize. The soluble $Fe_4(P_2O_7)_3$ must be kept dry and in the dark. Do not use if brown or yellow. Prepare solutions freshly. Do not heat over 60°C to dissolve. The mixture dissolves readily in a 50°C water bath.

Preparation of Medium: Add components, except L-cysteine·HCl·H_2O solution and $Fe_4(P_2O_7)_3$ solution to distilled/deionized water and bring volume to 980.0mL. Mix thoroughly. Gently heat to boiling. Autoclave for 15 min at 15 psi pressure–121°C. Cool to 50°C. Add 10.0mL of sterile L-cysteine·HCl·H_2O solution. Add 10.0mL of sterile $Fe_4(P_2O_7)_3$ solution. Adjust pH to 6.9 at 50°C by adding 4.0–4.5mL of 1.0 *N* KOH. This is a critical step. Mix thoroughly. Pour in 20.0mL volumes into sterile Petri dishes. Swirl medium while pouring to keep charcoal in suspension.

Use: For the cultivation and maintenance of *Legionella* species and *Xylella fastidiosa.*

CYG Agar

Composition per liter:

Agar	15.0g
Pancreatic digest of casein	3.0g
$CaCl_2·2H_2O$	1.0g
Yeast extract	1.0g
Cyanocobalamin	0.5mg
Glucose solution	100.0mL

pH 7.2 ± 0.2 at 25°C

Glucose Solution:
Composition per 100mL:

D-Glucose	5.0g

Preparation of Glucose Solution: Add D-glucose to distilled/deionized water and bring volume to 100.0mL. Mix thoroughly. Autoclave for 15 min at 15 psi pressure–121°C. Cool to 25°C.

Preparation of Medium: Add components, except glucose solution, to distilled/deionized water and bring volume to 900.0mL. Mix thoroughly. Gently heat and bring to boiling. Autoclave for 15 min at 15 psi pressure–121°C. Cool to 45–50°C. Aseptically add sterile glucose solution. Mix thoroughly. Pour into sterile Petri dishes or distribute into sterile tubes.

Use: For the isolation and cultivation of *Cytophaga* species, *Herpetosiphon* species, *Saprospira* species, and *Flexithrix* species.

Cylindrocladium Isolation Medium

Composition per liter:

Agar	20.0g
Glucose	15.0g
KH_2PO_4	1.0g
KNO_3	0.5g
$MgSO_4·7H_2O$	0.5g
Yeast extract	0.5g
Chloramphenicol solution	10.0mL
Chlortetracycline solution	10.0mL
Thiabendazole solution	10.0mL
Tergitol NPX® (Union Carbide)	1.0mL

Chloramphenicol Solution:
Composition per 10mL:

Chloramphenicol	0.1g
Ethanol (95% solution)	10.0mL

Preparation of Chloramphenicol Solution: Add chloramphenicol to 10.0mL of ethanol. Mix thoroughly. Filter sterilize.

Chlortetracycline Solution:
Composition per 10mL:

Chlortetracycline	0.04g
Ethanol, absolute	5.0mL

Preparation of Chlortetracycline Solution: Add chlortetracycline to 5.0mL of ethanol. Mix thoroughly. Bring volume to 10.0mL with distilled/deionized water. Filter sterilize.

Thiabendazole Solution:
Composition per 10mL:

Thiabendazole	1.0mg

Preparation of Thiabendazole Solution: Add thiabendazole to distilled/deionized water and bring volume to 10.0mL. Mix thoroughly. Filter sterilize.

Preparation of Medium: Filter sterilize tergitol NPX. Add components—except tergitol NPX, thiabendazole solution, chloramphenicol solution and chlortetracycline solution—to distilled/deionized water and bring volume to 969.0mL. Mix thoroughly. Gently heat and bring to boiling. Autoclave for 15 min at 15 psi pressure–121°C. Cool to 45–50°C. Aseptically add sterile tergitol NPX, thiabendazole solution, chloramphenicol solution and chlortetracycline solution. Mix thoroughly. Pour into sterile Petri dishes or distribute into sterile tubes.

Use: For the isolation and cultivation of *Cylindrocladium* species.

CYT Agar

Composition per liter:

Agar	15.0g
Pancreatic digest of casein	1.0g
$CaCl_2 \cdot 2H_2O$	0.5g
$MgSO_4 \cdot 7H_2O$	0.5g
Yeast extract	0.5g

pH 7.2 ± 0.2 at 25°C

Preparation of Medium: Add components to distilled/deionized water and bring volume to 1.0L. Mix thoroughly. Gently heat and bring to boiling. Distribute into tubes or flasks. Autoclave for 15 min at 15 psi pressure–121°C. Pour into sterile Petri dishes or leave in tubes.

Use: For the isolation and cultivation of *Cytophaga* species, *Herpetosiphon* species, *Saprospira* species, and *Flexithrix* species.

Cytophaga Agarase Agar (ATCC Medium 793)

Composition per liter:

Agar	15.0g
KH_2PO_4	1.0g
$MgSO_4 \cdot 7H_2O$	0.5g
NH_4Cl	0.5g
$CaCl_2 \cdot H_2O$	0.02g
Vishniac and Santer trace element mixture	0.2mL

pH 7.2 ± 0.2 at 25°C

Vishniac and Santer Trace Element Mixture:
Composition per liter:

Ethylenediamine tetraacetic acid (EDTA)	50.0g
$ZnSO_4 \cdot 7H_2O$	22.0g
$CaCl_2$	5.54g
$MnCl_2 \cdot 4H_2O$	5.06g
$FeSO_4 \cdot 7H_2O$	4.99g
$CoCl_2 \cdot 6H_2O$	1.61g
$CuSO_4 \cdot 5H_2O$	1.57g
$(NH_4)_6Mo_7O_{24} \cdot 4H_2O$	1.10g

Preparation of Vishniac and Santer Trace Element Mixture: Add components to distilled/deionized water and bring volume to 1.0L. Adjust pH to 6.0 with KOH. Mix thoroughly.

Preparation of Medium: Add components to distilled/deionized water and bring volume to 1.0L. Adjust pH to 7.2. Mix thoroughly. Distribute into tubes or flasks. Autoclave for 15 min at 15 psi pressure–121°C. Pour into sterile Petri dishes or leave in tubes.

Use: For the cultivation and maintenance of *Cytophaga flevensis.*

Cytophaga Agarase Broth

Composition per liter:

Agar	1.0g
KH_2PO_4	1.0g
$MgSO_4 \cdot 7H_2O$	0.5g
NH_4Cl	0.5g
$CaCl_2$·H2O	0.02g
Vishniac and Santer trace element mixture	0.2mL

pH 7.2 ± 0.2 at 25°C

Vishniac and Santer Trace Element Mixture:
Composition per liter:

Ethylenediamine tetraacetic acid (EDTA)	50.0g
$ZnSO_4 \cdot 7H_2O$	22.0g
$CaCl_2$	5.54g
$MnCl_2 \cdot 4H_2O$	5.06g
$FeSO_4 \cdot 7H_2O$	4.99g
$CoCl_2 \cdot 6H_2O$	1.61g
$CuSO_4 \cdot 5H_2O$	1.57g
$(NH_4)_6Mo_7O_{24} \cdot 4H_2O$	1.10g

Preparation of Vishniac and Santer Trace Element Mixture: Add components to distilled/deionized water and bring volume to 1.0L. Adjust pH to 6.0 with KOH. Mix thoroughly.

Preparation of Medium: Add components to distilled/deionized water and bring volume to 1.0L. Adjust pH to 7.2. Mix thoroughly. Distribute into tubes or flasks. Autoclave for 15 min at 15 psi pressure–121°C.

Use: For the cultivation of *Cytophaga flevensis.*

Cytophaga fermentans Medium

Composition per liter:

NaCl	30.0g
Agar	5.0g
$NaHCO_3$	5.0g
KH_2PO_4	1.0g
NH_4Cl	1.0g
$MgCl_2 \cdot 6H_2O$	0.5g
Yeast extract	0.3g
$Na_2S \cdot 9H_2O$	0.1g
$CaCl_2$	0.04g
Ferric citrate (4m*M* solution)	5.0mL
Trace element solution	2.0mL

pH 7.0 ± 0.2 at 25°C

Trace Element Solution:
Composition per 100mL:

H_3BO_3	0.28g
$MnSO_4 \cdot 6H_2O$	0.21g
$Na_2MoO_4 \cdot 2H_2O$	0.075g
$Zn(NO_3)_2 \cdot 6H_2O$	0.025g
$CoCl_2 \cdot 6H_2O$	0.02g
$Cu(NO_3)_2 \cdot 3H_2O$	0.02g

Preparation of Trace Element Solution: Add components to distilled/deionized water and bring volume to 100.0mL. Mix thoroughly.

Preparation of Medium: Add components to distilled/deionized water and bring volume to 1.0L. Mix thoroughly. Gently heat and bring to boiling. Distribute into tubes or flasks. Autoclave for 15 min at 15 psi pressure–121°C.

Use: For the cultivation of agar-digesting *Cytophaga fermentans.*

Cytophaga Medium (ATCC Medium 420)

Composition per liter:

NaCl	20.0g
Yeast extract	10.0g
Agar	3.0g
$MgSO_4 \cdot 7H_2O$	1.0g
NH_4Cl	1.0g
K_2HPO_4	0.2g
$FeCl_3$	1.0μg

pH 7.5 ± 0.2 at 25°C

Preparation of Medium: Add components to tap water and bring volume to 1.0L. Adjust pH to 7.5. Mix thoroughly. Distribute into tubes or flasks. Autoclave for 15 min at 15 psi pressure–121°C.

Use: For the cultivation of *Cytophaga fermentens.*

Czapek Agar (ATCC Medium 312)

Composition per liter:

Sucrose	30.0g
Agar	15.0g
$NaNO_3$	3.0g
K_2HPO_4	1.0g
KCl	0.5g
$MgSO_4 \cdot 7H_20$	0.5g
$FeSO_4 \cdot 7H_2O$	0.01g

pH 7.3 ± 0.2 at 25°C

Preparation of Medium: Add components, except sucrose, to distilled/deionized water and bring volume to 900.0mL. Mix thoroughly. Distribute into tubes or flasks. In a separate flask, add sucrose to distilled/deionized water and bring volume to 100.0mL. Mix thoroughly. Autoclave both solutions separately for 15 min at 15 psi pressure–121°C. Cool to 50°C. Combine the sterile solutions. Mix thoroughly. Pour into sterile Petri dishes or distribute into sterile tubes.

Use: For the cultivation and maintenance of *Streptomyces* species. Also, for the cultivation of Actinoplanaceae.

Czapek Agar with Peptone (ATCC Medium 522)

Composition per liter:

Sucrose	30.0g
Agar	15.0g
Peptone	5.0g
$NaNO_3$	3.0g
K_2HPO_4	1.0g
KCl	0.5g
$MgSO_4 \cdot 7H_20$	0.5g
$FeSO_4 \cdot 7H_2O$	0.01g

pH 7.3 ± 0.2 at 25°C

Preparation of Medium: Add components, except sucrose, to distilled/deionized water and bring volume to 900.0mL. Mix thoroughly. Distribute into tubes or flasks. In a separate flask, add sucrose to distilled/deionized water and bring volume to 100.0mL. Mix thoroughly. Autoclave both solutions separately for 15 min at 15 psi pressure–121°C. Cool to 50°C. Combine the sterile solutions. Mix thoroughly. Pour into sterile Petri dishes or distribute into sterile tubes.

Use: For the cultivation and maintenance of *Streptomyces* species. Also, for the cultivation of Actinoplanaceae.

Czapek Dox Agar

Composition per liter:

Sucrose	30.0g
Agar	15.0g
$NaNO_3$	3.0g
K_2HPO_4	1.0g
$MgSO_4 \cdot 7H_2O$	0.5g
KCl	0.5g
$FeSO_4 \cdot 7H_2O$	0.01g

pH 7.3 ± 0.2 at 25°C

Preparation of Medium: Add components to distilled/deionized water and bring volume to 1.0L. Mix thoroughly. Distribute into tubes or flasks. Autoclave for 15 min at 15 psi pressure–121°C. Pour into sterile Petri dishes or leave in tubes.

Use: For the cultivation and maintenance of *Actinoplanes* species, *Amorphosporangium auranticolor, Ampullariella* species, *Spirillospora albida* and *Streptomyces armeniacus.*

Czapek Dox Agar with 3% Glucose

Composition per liter:

Glucose	30.0g
Agar	15.0g

$NaNO_3$	3.0g
K_2HPO_4	1.0g
KCl	0.5g
$MgSO_4 \cdot 7H_2O$	0.5g
$FeSO_4 \cdot 7H_2O$	0.01g

pH 7.3 ± 0.2 at 25°C

Preparation of Medium: Sterilize by autoclaving for 15 min at 15 lb pressure (121°C). Pour into sterile Petri dishes or leave in tubes.

Use: For the cultivation and maintenance of *Microbispora rosea* and *Streptomyces* species.

Czapek Dox Agar, Modified

Composition per liter:

Sucrose	30.0g
Agar	12.0g
$NaNO_3$	2.0g
Magnesium glycerophosphate	0.5g
KCl	0.5g
K_2SO_4	0.35g
$FeSO_4$	0.01g

pH 6.8 ± 0.2 at 25°C

Source: This medium is available as a premixed powder from Oxoid Unipath.

Preparation of Medium: Add components to distilled/deionized water and bring volume to 1.0L. Mix thoroughly. Distribute into tubes or flasks. Autoclave for 15 min at 15 psi pressure–121°C. Pour into sterile Petri dishes or leave in tubes.

Use: For the cultivation and maintenance of numerous fungal species. For chlamydospore production by *Candida albicans*.

Czapek Dox Broth

Composition per liter:

Sucrose	30.0g
$NaNO_3$	3.0g
K_2HPO_4	1.0g
$MgSO_4 \cdot 7H_2O$	0.5g
KCl	0.5g
$FeSO_4 \cdot 7H_2O$	0.01g

pH 7.3 ± 0.2 at 25°C

Source: This medium is available as a premixed powder from Difco Laboratories.

Preparation of Medium: Add components to distilled/deionized water and bring volume to 1.0L. Mix thoroughly. Distribute into tubes or flasks. Autoclave for 15 min at 15 psi pressure–121°C.

Use: For the cultivation and maintenance of a variety of fungal and bacterial species that can use nitrate as the sole nitrogen source.

Czapek Dox Liquid Medium, Modified

Composition per liter:

Sucrose	30.0g
$NaNO_3$	2.0g
Magnesium glycerophosphate	0.5g
KCl	0.5g
K_2SO_4	0.35g
$FeSO_4$	0.01g

pH 6.8 ± 0.2 at 25°C

Source: This medium is available as a premixed powder from Oxoid Unipath.

Preparation of Medium: Add components to distilled/deionized water and bring volume to 1.0L. Mix thoroughly. Distribute into tubes or flasks. Autoclave for 15 min at 15 psi pressure–121°C.

Use: For the cultivation of fungi and bacteria capable of utilizing sodium nitrate as the sole source of nitrogen.

Czapek Solution Agar

Composition per liter:

Sucrose	30.0g
Agar	15.0g
$NaNO_3$	2.0g
K_2HPO_4	1.0g
KCl	0.5g
$MgSO_4 \cdot 7H_20$	0.5g
$FeSO_4 \cdot 7H_2O$	0.01g

pH 7.3 ± 0.2 at 25°C

Source: This medium is available as a premixed powder from Difco Laboratories.

Preparation of Medium: Add components to distilled/deionized water and bring volume to 1.0L. Mix thoroughly. Distribute into tubes or flasks. Autoclave for 15 min at 15 psi pressure–121°C. Pour into sterile Petri dishes or leave in tubes.

Use: For the cultivation of *Aspergillus, Penicillium* and other fungi. For the cultivation and maintenance of microorganisms that can utilize nitrate as the sole nitrogen source.

DCLS Agar (Deoxycholate Citrate Lactose Sucrose Agar)

Composition per liter:

Agar	12.0g
Sodium citrate·$3H_2O$	10.5g
Lactose	5.0g
$Na_2S_2O_3$	5.0g
Sucrose	5.0g

Pancreatic digest of casein 3.5g
Peptic digest of animal tissue 3.5g
Beef extract 3.0g
Sodium deoxycholate 2.5g
Neutral red 0.03g

pH 7.2 ± 0.1 at 25°C

Source: This medium is available as a premixed powder from BBL Microbiology Systems, Difco Laboratories and Oxoid Unipath Unipath.

Preparation of Medium: Add components to distilled/deionized water and bring volume to 1.0L. Mix thoroughly. Gently heat while stirring and bring to boiling. Do not overheat. Do not autoclave. Pour into sterile Petri dishes in 20.0mL volumes.

Use: For the selective isolation of *Salmonella* species, *Shigella* species and *Vibrio* species.

Decarboxylase Base, Møller

Composition per liter:

Amino acid 10.0g
Beef extract 5.0g
Peptone 5.0g
Glucose 0.5g
Bromcresol purple 0.01g
Cresol red 5.0mg
Pyridoxal 5.0mg
Mineral oil 200.0mL

pH 6.0 ± 0.2 at 25°C

Source: This medium is available as a premixed powder Difco Laboratories.

Preparation of Medium: Add components, except mineral oil, to distilled/deionized water and bring volume to 1.0L. For amino acid, use L-arginine, L-lysine or L-ornithine. Mix thoroughly. Distribute into screw-capped tubes in 5.0mL volumes. Autoclave medium and mineral oil separately for 15 min at 15 psi pressure–121°C. After inoculation, overlay medium with 1.0mL of sterile mineral oil per tube.

Use: For the cultivation and differentiation of bacteria based on their ability to decarboxylate the amino acid. Bacteria that decarboxylate arginine, lysine or ornithine turn the medium turbid purple.

Decarboxylase Medium Base, Falkow

Composition per liter:

Amino acid 5.0g
Peptone 5.0g
Yeast extract 3.0g
Glucose 1.0g
Bromcresol purple 0.02g
Mineral oil 200.0mL

pH 6.8 ± 0.2 at 25°C

Source: This medium is available as a premixed powder from Difco Laboratories.

Preparation of Medium: Add components, except mineral oil, to distilled/deionized water and bring volume to 1.0L. For amino acid, use L-arginine, L-lysine or L-ornithine. Mix thoroughly. Distribute into screw-capped tubes in 5.0mL volumes. Autoclave medium and mineral oil separately for 15 min at 15 psi pressure–121°C. After inoculation, overlay medium with 1.0mL of sterile mineral oil per tube.

Use: For the cultivation and differentiation of bacteria based on their ability to decarboxylate the amino acid. Bacteria that decarboxylate arginine, lysine or ornithine turn the medium turbid purple.

Decarboxylase Medium, Ornithine Modified

Composition per liter:

L-Ornithine 10.0g
Meat peptone 5.0g
Yeast extract 3.0g
Bromcresol purple solution 5.0mL

pH 5.5 ± 0.2 at 25°C

Bromcresol Purple Solution:
Composition per 100mL:

Bromcresol purple 0.2g
Ethanol 50.0mL

Preparation of Bromcresol Purple Solution: Add bromcresol purple to ethanol. Mix thoroughly. Bring volume to 100.0mL with distilled/deionized water. Mix thoroughly. Filter sterilize.

Preparation of Medium: Add components to distilled/deionized water and bring volume to 1.0L. Mix thoroughly. Gently heat until dissolved. Adjust pH to 5.5 with HCl or NaOH. Distribute into screw-capped tubes. Autoclave for 15 min at 15 psi pressure–121°C.

Use: For the cultivation and differentiation of bacteria based on their ability to decarboxylate ornithine. Bacteria that decarboxylate ornithine turn the medium turbid purple.

Defined Glucose Medium EMSY-1

Composition per liter:

Na_2HPO_4 .... 1.79g
KH_2PO_4 .... 1.7g
Citric acid 0.5g

NH_4Cl	0.43g
$MgSO_4 \cdot 7H_2O$	0.41g
$CaCl_2 \cdot 2H_2O$	0.04g
NaCl	0.03g
$FeCl_3 \cdot 6H_2O$	4.84mg
Glucose solution	100.0mL
Yeast extract solution	10.0mL
TK6-3 solution	1.0mL

pH 7.2 ± 0.2 at 25°C

Glucose Solution:

Composition per 100mL:

Glucose	10.0g

Preparation of Glucose Solution: Add glucose to distilled/deionized water and bring volume to 100.0mL. Mix thoroughly. Filter sterilize.

Yeast Extract Solution:

Composition per 10mL:

Yeast extract	0.4g

Preparation of Yeast Extract Solution: Add yeast extract to distilled/deionized water and bring volume to 10.0mL. Mix thoroughly. Filter sterilize.

TK6-3 Solution:

Composition per liter:

$ZnSO_4 \cdot 7H_2O$	1.45g
$CuSO_4 \cdot 5H_2O$	0.76g
$MnSO_4 \cdot H_2O$	0.31g
H_3BO_3	0.19g
$Na_2MoO_4 \cdot 2H_2O$	0.17g
KI	0.04g
H_2SO_4 (1*N* solution)	1.0mL

Preparation of TK6-3 Solution: Add components to distilled/deionized water and bring volume to 1.0L. Mix thoroughly.

Preparation of Medium: Add components, except glucose solution and yeast extract solution, to distilled/deionized water and bring volume to 890.0mL. Mix thoroughly. Gently heat and bring to boiling. Autoclave for 15 min at 15 psi pressure–121°C. Cool rapidly to 25°C. Aseptically add 100.0mL of sterile glucose solution and 10.0mL of sterile yeast extract solution. Mix thoroughly. Aseptically distribute into sterile tubes or flasks.

Use: For the cultivation and maintenance of *Xanthomonas campestris.*

Defined Medium for *Rhodopseudomonas*

Composition per liter:

Malic acid	4.0g
$(NH_4)_2SO_4$	1.0g
K_2HPO_4	0.9g
KH_2PO_4	0.6g
$MgSO_4 \cdot 7H_2O$	0.2g
$CaCl_2 \cdot 2H_2O$	0.075g
EDTA	0.020g
$FeSO_4 \cdot 7H_2O$	0.012g
Thiamine	1.0mg
Biotin	0.015mg
Trace elements	1.0mL

pH 6.8 ± 0.2 at 25°C

Trace Elements:

Composition per 250mL:

H_3BO_3	0.7g
$MnSO_4 \cdot H_2O$	0.4g
$Na_2MoO_4 \cdot 2H_2O$	0.19g
$ZnSO_4 \cdot 7H_2O$	0.060g
$CoCl_2 \cdot 6H_2O$	0.050g
$Cu(NO_3)_2 \cdot 3H_2O$	0.010g

Preparation of Trace Elements: Add components to distilled/deionized water and bring volume to 250.0mL. Mix thoroughly.

Preparation of Medium: Add components to distilled/deionized water and bring volume to 1.0L. Mix thoroughly. Adjust pH to 6.8. Distribute into tubes or flasks. Autoclave for 15 min at 15 psi pressure–121°C.

Use: For the cultivation and maintenance of *Rhodobacter capsulatus.*

Defined Medium with Povidone Iodine

Composition per 1025mL:

Basal solution	1.0L
Solution B	10.0mL
Solution C	10.0mL
Solution A	5.0mL

Basal Solution:

Composition per liter:

Agar	20.0g
Na_2HPO_4	4.8g
KH_2PO_4	4.4g
NH_4Cl	1.0g
$MgSO_4 \cdot 7H_2O$	0.5g

Preparation of Basal Solution: Add components to distilled/deionized water and bring volume to 1.0L. Mix thoroughly. Gently heat and bring to boiling. Autoclave for 15 min at 15 psi pressure–121°C. Cool to 45–50°C.

Solution A:

Composition per 100mL:

Ferric ammonium citrate	1.0g
$CaCl_2 \cdot 2H_2O$	0.1g

Preparation of Solution A: Add components to distilled/deionized water and bring volume to 100.0mL. Mix thoroughly. Filter sterilize.

Solution B:

Composition per 100mL:

D-Glucose	10.0g

Preparation of Solution B: Add glucose to distilled/deionized water and bring volume to 100.0mL. Mix thoroughly. Filter sterilize.

Solution C:

Composition per 100mL:

Povidone-iodine	0.1g

Preparation of Solution C: Add povidone-iodine to distilled/deionized water and bring volume to 100.0mL. Mix thoroughly. Filter sterilize.

Preparation of Medium: To 1.0L of cooled, sterile basal solution aseptically add 5.0mL of sterile solution A, 10.0mL of sterile solution B, and 10.0mL of sterile solution C. Mix thoroughly. Pour into sterile Petri dishes or distribute into sterile tubes.

Use: For the cultivation and maintenance of *Pseudomonas aeruginosa* and *P. cepacia.*

Deoxycholate Agar

Composition per liter:

Agar	16.0g
Lactose	10.0g
NaCl	5.0g
Pancreatic digest of casein	5.0g
Peptic digest of animal tissue	5.0g
K_2HPO_4	2.0g
Ferric citrate	1.0g
Sodium citrate	1.0g
Sodium deoxycholate	1.0g
Neutral red	0.033g

pH 7.3 ± 0.2 at 25°C

Source: This medium is available as a premixed powder from BBL Microbiology Systems.

Preparation of Medium: Add components to distilled/deionized water and bring volume to 1.0L. Mix thoroughly. Gently heat and bring to boiling. Do not autoclave. Cool to 45–50°C. Pour into sterile Petri dishes.

Use: For the selective isolation, cultivation, enumeration and differentiation of Gram-negative enteric microorganisms. *Escherichia coli* appears as large, flat rose-red colonies. *Enterobacter* and *Klebsiella* species appear as large mucoid pale colonies with a pink center. *Proteus* and *Salmonella* species appear as large, colorless to tan colonies. *Shigella* species appear as colorless to pink colonies. *Pseudomonas* species appear as irregular colorless to brown colonies.

Deoxycholate Agar (Desoxycholate Agar)

Composition per liter:

Agar	15.0g
Lactose	10.0g
Peptone	10.0g
NaCl	5.0g
K_2HPO_4	2.0g
Ferric citrate	1.0g
Sodium citrate	1.0g
Sodium deoxycholate	1.0g
Neutral red	0.03g

pH 7.3 ± 0.2 at 25°C

Source: This medium is available as a premixed powder from Oxoid Unipath and Difco Laboratories.

Preparation of Medium: Add components to distilled/deionized water and bring volume to 1.0L. Mix thoroughly. Gently heat and bring to boiling. Do not autoclave. Cool to 50°C. Pour into sterile Petri dishes.

Use: For the selective isolation, cultivation, enumeration and differentiation of Gram-negative enteric microorganisms. *Escherichia coli* appears as large, flat rose-red colonies. *Enterobacter* and *Klebsiella* species appear as large mucoid pale colonies with a pink center. *Proteus* and *Salmonella* species appear as large, colorless to tan colonies. *Shigella* species appear as colorless to pink colonies. *Pseudomonas* species appear as irregular colorless to brown colonies.

Deoxycholate Lactose Agar

Composition per liter:

Agar	15.0g
Lactose	10.0g
NaCl	5.0g
Pancreatic digest of casein	5.0g
Peptic digest of animal tissue	5.0g
Sodium citrate	2.0g
Sodium deoxycholate	0.5g
Neutral red	0.033g

pH 7.1 ± 0.2 at 25°C

Source: This medium is available as a premixed powder from BBL Microbiology Systems and Difco Laboratories.

Preparation of Medium: Add components to distilled/deionized water and bring volume to 1.0L. Mix thoroughly. Gently heat and bring to boiling. Do not autoclave. Cool to 45–50°C. Pour into sterile Petri dishes. Dry the agar surface before use.

Use: For the selective isolation, cultivation, and differentiation of enteric pathogens, especially *Salmonella* and *Shigella* species. Lactose-fermenting bacteria appear as pink colonies that may or may not be surrounded by a zone of precipitated deoxycholate. Nonlactose-fermenting bacteria appear as colorless colonies that are surrounded by a clear orange-yellow zone. Also used for the enumeration of coliform bacteria from water.

Deoxycholate Lactose Sucrose Sorbitol Agar

Composition per liter:

Sodium citrate	20.0g
Agar	15.0g
D-Sorbitol	10.0g
Lactose	10.0g
Sucrose	5.0g
Pancreatic digest of casein	5.0g
Yeast extract	5.0g
Sodium deoxycholate	2.5g
Ferric citrate	1.0g
Neutral red	0.02g

pH 7.4 ± 0.2 at 25°C

Preparation of Medium: Add components to distilled/deionized water and bring volume to 1.0L. Mix thoroughly. Gently heat and bring to boiling. Do not overheat. Adjust pH to 7.4. Do not autoclave. Pour into sterile Petri dishes or distribute into sterile tubes.

Use: For the isolation and cultivation of the *Hafnia* species.

Derxia gummosa Medium

Composition per liter:

Agar	20.0g
Starch	20.0g
$MgSO_4 \cdot 7H_2O$	0.2g
KH_2PO_4	0.15g
$NaHCO_3$	0.1g
K_2HPO_4	0.05g
$CaCl_2$	0.02g
$Na_2MoO_4 \cdot 2H_2O$	2.0mg
Bromthymol blue solution	5.0mL
$FeCl_3 \cdot 6H_2O$ (10% solution)	0.1mL

pH 6.9 ± 0.2 at 25°C

Bromthymol Blue Solution:
Composition per 10mL:

Bromthymol blue	0.5g
Ethanol	10.0mL

Preparation of Bromthymol Blue Solution: Add bromthymol blue to 10.0mL of ethanol. Mix thoroughly.

Preparation of Medium: Add components to distilled/deionized water and bring volume to 1.0L. Mix thoroughly. Distribute into tubes or flasks. Autoclave for 15 min at 15 psi pressure–121°C.

Use: For the cultivation of *Derxia gummosa*.

Derxia Medium

Composition per liter:

Agar	20.0g
Glucose	20.0g
NH_4Cl	2.0g
K_2HPO_4	1.0g
$MgSO_4 \cdot 7H_2O$	0.2g
$CaSO_4$	5.0mg
$FeSO_4 \cdot 7H_2O$	5.0mg
$Na_2MoO_4 \cdot 2H_2O$	0.5mg

pH 6.7 ± 0.2 at 25°C

Preparation of Medium: Add components to distilled/deionized water and bring volume to 1.0L. Mix thoroughly. Gently heat and bring to boiling. Adjust pH to 6.7. Distribute into tubes or flasks. Autoclave for 15 min at 15 psi pressure–121°C. Pour into sterile Petri dishes or leave in tubes.

Use: For the cultivation and maintenance of *Derxia gummosa*.

Desulfobacterium indolicum Medium

Composition per 1002.4mL:

Solution A	900.0mL
Solution C	50.0mL
Solution D	30.0mL
Solution E—Wolfe's vitamin solution	10.0mL
Solution G	10.0mL
Solution B—Trace elements solution SL-10	1.0mL
Solution F	1.0mL
Solution H	0.4mL

pH 7.6 ± 0.2 at 25°C

Solution A:
Composition per 900mL:

NaCl	21.0g
$MgCl_2 \cdot 6H_2O$	3.0g
Na_2SO_4	3.0g
KCl	0.5g
NH_4Cl	0.3g
KH_2PO_4	0.2g
$CaCl_2 \cdot 2H_2O$	0.15g
Resazurin	1.0mg

Preparation of Solution A: Prepare and dispense solution anaerobically under 80% N_2 + 20% CO_2. Add components to distilled/deionized water and bring volume to 900.0mL. Mix thoroughly. Gently heat and bring to boiling. Continue boiling until resazurin turns colorless indicating reduction. Cap with rubber stoppers. Autoclave for 15 min at 15 psi pressure–121°C. Cool to 45–50°C.

Solution B—Trace Elements Solution SL-10:
Composition per liter:

$FeCl_2 \cdot 4H_2O$	1.5g
$CoCl_2 \cdot 6H_2O$	0.19g
$MnCl_2 \cdot 4H_2O$	0.10g
$ZnCl_2$	0.070g
$Na_2MoO_4 \cdot 2H_2O$	0.036g
$NiCl_2 \cdot 6H_2O$	0.024g
H_3BO_3	6.0mg
$CuCl_2 \cdot 2H_2O$	2.0mg
HCl (25% solution)	10.0mL

Preparation of Trace Elements Solution SL-10: Add the $FeCl_2 \cdot 4H_2O$ to 10.0mL of HCl solution. Mix thoroughly. Bring volume to approximately 900.0mL with distilled/deionized water. Mix thoroughly. Adjust pH to 6.0 with NaOH. Bring volume to 1.0L with distilled/deionized water. Filter sterilize. Aseptically gas under 100% N_2 for 20 min.

Solution C:
Composition per 50mL:

$NaHCO_3$	2.5g

Preparation of Solution C: Add $NaHCO_3$ to distilled/deionized water and bring volume to 50.0mL. Mix thoroughly. Filter sterilize. Aseptically gas under 80% N_2 + 20% CO_2 for 20 min.

Solution D:
Composition per 107.7mL:

Indole	0.3g
NaCl (30% solution)	7.0mL
$MgCl_2 \cdot 6H_2O$ (40% solution)	0.7mL

Preparation of Solution D: Prepare and dispense all solutions anaerobically under 100% N_2. Add indole to distilled/deionized water and bring volume to 100.0mL. Mix thoroughly. Gently heat while stirring until dissolved. Prepare the NaCl solution and the $MgCl_2 \cdot 6H_2O$ solution separately. Autoclave the three solutions separately for 15 min at 15 psi pressure–121°C. Cool to 45–50°C. To 100.0mL of sterile indole solution, aseptically and anaerobically add 7.0mL of sterile NaCl solution and 0.7mL of sterile $MgCl_2 \cdot 6H_2O$ solution. Mix thoroughly.

Solution E—Wolfe's Vitamin Solution:
Composition per liter:

Pyridoxine·HCl	0.01g
Thiamine·HCl	5.0mg
Riboflavin	5.0mg
Nicotinic acid	5.0mg
Calcium pantothenate	5.0mg
p-Aminobenzoic acid	5.0mg
Thioctic acid	5.0mg
Biotin	2.0mg
Folic acid	2.0mg
Cyanocobalamin	0.1mg

Preparation of Wolfe's Vitamin Solution: Add components to distilled/deionized water and bring volume to 1.0L. Mix thoroughly. Filter sterilize. Aseptically gas under 100% N_2 for 20 min.

Solution F:
Composition per liter:

$Na_2SeO_3 \cdot 5H_2O$	3.0mg
NaOH (0.01*M* solution)	1.0L

Preparation of Solution F: Add $Na_2SeO_3 \cdot 5H_2O$ to 1.0L of NaOH solution. Mix thoroughly. Filter sterilize. Aseptically gas under 100% N_2 for 20 min.

Solution G:
Composition per 10mL:

$Na_2S \cdot 9H_2O$	0.4g

Preparation of Solution G: Add $Na_2S \cdot 9H_2O$ to distilled/deionized water and bring volume to 10.0mL. Gas under 100% N_2 for 20 min. Cap with a rubber stopper. Autoclave for 15 min at 15 psi pressure–121°C. Cool to 25°C.

Solution H:
Composition per 10mL:

$Na_2S_2O_4$	0.5g

Preparation of Solution H: Add $Na_2S_2O_4$ to distilled/deionized water and bring volume to 10.0mL. Mix thoroughly. Filter sterilize. Aseptically gas under 100% N_2 for 20 min. Freshly prepare solution.

Preparation of Medium: To 900.0mL of cooled, sterile solution A, aseptically and anaerobically add in the following order: 1.0mL of sterile solution B, 50.0mL of sterile solution C, 10.0mL of sterile solution E, 1.0mL of sterile solution F, and 10.0mL of sterile solution G. Mix thoroughly. Immediately prior to inoculation aseptically and anaerobically add 30.0mL of sterile solution D and 0.4mL of sterile solution H. Mix thoroughly. Aseptically and anaerobically distribute into sterile tubes or flasks.

Use: For the cultivation and maintenance of *Desulfobacterium indolicum.*

Desulfobacterium Medium

Composition per 1002.4mL:

Solution A	930.0mL
Solution C	50.0mL
Solution D—Wolfe's vitamin solution	10.0mL
Solution F	10.0mL
Solution B—Trace elements solution SL-10	1.0mL
Solution E	1.0mL
Solution G	0.4mL

pH 7.0 ± 0.2 at 25°C

Solution A:

Composition per 930mL:

NaCl	21.0g
$MgCl_2 \cdot 6H_2O$	3.0g
Na_2SO_4	3.0g
KCl	0.5g
NH_4Cl	0.3g
KH_2PO_4	0.2g
$CaCl_2 \cdot 2H_2O$	0.15g
Resazurin	1.0mg

Preparation of Solution A: Prepare and dispense solution anaerobically under 80% N_2 + 20% CO_2. Add components to distilled/deionized water and bring volume to 930.0mL. Mix thoroughly. Gently heat and bring to boiling. Continue boiling until resazurin turns colorless indicating reduction. Cap with rubber stoppers. Autoclave for 15 min at 15 psi pressure–121°C. Cool to 45–50°C.

Solution B—Trace Elements Solution SL-10:

Composition per liter:

$FeCl_2 \cdot 4H_2O$	1.5g
$CoCl_2 \cdot 6H_2O$	0.19g
$MnCl_2 \cdot 4H_2O$	0.10g
$ZnCl_2$	0.070g
$Na_2MoO_4 \cdot 2H_2O$	0.036g
$NiCl_2 \cdot 6H_2O$	0.024g
H_3BO_3	6.0mg
$CuCl_2 \cdot 2H_2O$	2.0mg
HCl (25% solution)	10.0mL

Preparation of Trace Elements Solution SL-10: Add the $FeCl_2 \cdot 4H_2O$ to 10.0mL of HCl solution. Mix thoroughly. Bring volume to approximately 900.0mL with distilled/deionized water. Mix thoroughly. Adjust pH to 6.0 with NaOH. Bring volume to 1.0L with distilled/deionized water. Filter sterilize. Aseptically gas under 100% N_2 for 20 min.

Solution C:

Composition per 50mL:

$NaHCO_3$	2.5g

Preparation of Solution C: Add $NaHCO_3$ to distilled/deionized water and bring volume to 50.0mL. Mix thoroughly. Filter sterilize. Aseptically gas under 80% N_2 + 20% CO_2 for 20 min.

Solution D—Wolfe's Vitamin Solution:

Composition per liter:

Pyridoxine·HCl	0.01g
Thiamine·HCl	5.0mg
Riboflavin	5.0mg
Nicotinic acid	5.0mg
Calcium pantothenate	5.0mg
p-Aminobenzoic acid	5.0mg
Thioctic acid	5.0mg
Biotin	2.0mg
Folic acid	2.0mg
Cyanocobalamin	0.1mg

Preparation of Wolfe's Vitamin Solution: Add components to distilled/deionized water and bring volume to 1.0L. Mix thoroughly. Filter sterilize. Aseptically gas under 100% N_2 for 20 min.

Solution E:

Composition per liter:

$Na_2SeO_3 \cdot 5H_2O$	3.0mg
NaOH (0.01*M* solution)	1.0L

Preparation of Solution E: Add $Na_2SeO_3 \cdot 5H_2O$ to 1.0L of NaOH solution. Mix thoroughly. Filter sterilize. Aseptically gas under 100% N_2 for 20 min.

Solution F:

Composition per 10mL:

$Na_2S \cdot 9H_2O$	0.4g

Preparation of Solution F: Add $Na_2S \cdot 9H_2O$ to distilled/deionized water and bring volume to 10.0mL. Gas under 100% N_2 for 20 min. Cap with a rubber stopper. Autoclave for 15 min at 15 psi pressure–121°C. Cool to 25°C.

Solution G:

Composition per 10mL:

$Na_2S_2O_4$	0.5g

Preparation of Solution G: Add $Na_2S_2O_4$ to distilled/deionized water and bring volume to 10.0mL. Mix thoroughly. Filter sterilize. Aseptically gas under 100% N_2 for 20 min. Freshly prepare solution.

Preparation of Medium: To 900.0mL of cooled, sterile solution A, aseptically and anaerobically add in the following order: 1.0mL of sterile solution B, 50.0mL of sterile solution C, 10.0mL of sterile solution D, 1.0mL of sterile solution E, and 10.0mL of sterile solution F. Mix thoroughly. Immediately prior to inoculation aseptically and anaerobically add 0.4mL of sterile solution G. Mix thoroughly. Aseptically and anaerobically distribute into sterile tubes or flasks.

Use: For the cultivation and maintenance of *Desulfobacterium autotrophicum.*

Desulfobacterium Medium, Modified

Composition per 1002.4mL:

Solution A	920.0mL
Solution C	50.0mL
Solution D	10.0mL
Solution E—Wolfe's vitamin solution	10.0mL
Solution G	10.0mL
Solution B—Trace elements solution SL-10	1.0mL

Solution F 1.0mL
Solution H 0.4mL
pH 7.0 ± 0.2 at 25°C

Solution A:
Composition per 920mL:

NaCl	21.0g
$MgCl_2 \cdot 6H_2O$	3.0g
Na_2SO_4	3.0g
KCl	0.5g
NH_4Cl	0.3g
KH_2PO_4	0.2g
$CaCl_2 \cdot 2H_2O$	0.15g
Resazurin	1.0mg

Preparation of Solution A: Prepare and dispense solution anaerobically under 80% N_2 + 20% CO_2. Add components to distilled/deionized water and bring volume to 920.0mL. Mix thoroughly. Gently heat and bring to boiling. Continue boiling until resazurin turns colorless indicating reduction. Cap with rubber stoppers. Autoclave for 15 min at 15 psi pressure–121°C. Cool to 45–50°C.

Solution B—Trace Elements Solution SL-10:
Composition per liter:

$FeCl_2 \cdot 4H_2O$	1.5g
$CoCl_2 \cdot 6H_2O$	0.19g
$MnCl_2 \cdot 4H_2O$	0.10g
$ZnCl_2$	0.070g
$Na_2MoO_4 \cdot 2H_2O$	0.036g
$NiCl_2 \cdot 6H_2O$	0.024g
H_3BO_3	6.0mg
$CuCl_2 \cdot 2H_2O$	2.0mg
HCl (25% solution)	10.0mL

Preparation of Trace Elements Solution SL-10: Add the $FeCl_2 \cdot 4H_2O$ to 10.0mL of HCl solution. Mix thoroughly. Bring volume to approximately 900.0mL with distilled/deionized water. Mix thoroughly. Adjust pH to 6.0 with NaOH. Bring volume to 1.0L with distilled/deionized water. Filter sterilize. Aseptically gas under 100% N_2 for 20 min.

Solution C:
Composition per 50mL:

$NaHCO_3$	2.5g

Preparation of Solution C: Add $NaHCO_3$ to distilled/deionized water and bring volume to 50.0mL. Mix thoroughly. Filter sterilize. Aseptically gas under 80% N_2 + 20% CO_2 for 20 min.

Solution D:
Composition per 10mL:

Sodium acetate·$3H_2O$	2.5g

Preparation of Solution D: Prepare and dispense solution anaerobically under 80% N_2 + 20% CO_2. Add sodium acetate to distilled/deionized water and bring volume to 10.0mL. Mix thoroughly. Cap with rubber stopper. Autoclave for 15 min at 15 psi pressure–121°C. Cool to 45–50°C.

Solution E—Wolfe's Vitamin Solution:
Composition per liter:

Pyridoxine·HCl	0.01g
Thiamine·HCl	5.0mg
Riboflavin	5.0mg
Nicotinic acid	5.0mg
Calcium pantothenate	5.0mg
p-Aminobenzoic acid	5.0mg
Thioctic acid	5.0mg
Biotin	2.0mg
Folic acid	2.0mg
Cyanocobalamin	0.1mg

Preparation of Wolfe's Vitamin Solution: Add components to distilled/deionized water and bring volume to 1.0L. Mix thoroughly. Filter sterilize. Aseptically gas under 100% N_2 for 20 min.

Solution F:
Composition per liter:

$Na_2SeO_3 \cdot 5H_2O$	3.0mg
NaOH (0.01*M* solution)	1.0L

Preparation of Solution F: Add $Na_2SeO_3 \cdot 5H_2O$ to 1.0L of NaOH solution. Mix thoroughly. Filter sterilize. Aseptically gas under 100% N_2 for 20 min.

Solution G:
Composition per 10mL:

$Na_2S \cdot 9H_2O$	0.4g

Preparation of Solution G: Add $Na_2S \cdot 9H_2O$ to distilled/deionized water and bring volume to 10.0mL. Gas under 100% N_2 for 20 min. Cap with a rubber stopper. Autoclave for 15 min at 15 psi pressure–121°C. Cool to 25°C.

Solution H:
Composition per 10mL:

$Na_2S_2O_4$	0.5g

Preparation of Solution H: Add $Na_2S_2O_4$ to distilled/deionized water and bring volume to 10.0mL. Mix thoroughly. Filter sterilize. Aseptically gas under 100% N_2 for 20 min. Freshly prepare solution.

Preparation of Medium: To 920.0mL of cooled, sterile solution A, aseptically and anaerobically add in the following order: 1.0mL of sterile solution B, 50.0mL of sterile solution C, 10.0mL of sterile solution D, 10.0mL of sterile solution E, 1.0mL of sterile solution F, and 10.0mL of sterile solution G. Mix thoroughly. Immediately prior to inoculation aseptically and anaerobically add 0.4mL of sterile solution H. Mix thoroughly. Aseptically and anaerobically distribute into sterile tubes or flasks.

Use: For the cultivation and maintenance of *Desulfobacter curvatus* and *Desulfobacter latus*.

Desulfobacterium phenolicum Medium

Composition per 1002.4mL:

Solution A	930.0mL
Solution C	50.0mL
Solution E—Wolfe's vitamin solution	10.0mL
Solution G	10.0mL
Solution D	4.0mL
Solution B—Trace elements solution SL-10	1.0mL
Solution F	1.0mL
Solution H	0.4mL

pH 7.0 ± 0.2 at 25°C

Solution A:

Composition per 920mL:

NaCl	21.0g
$MgCl_2 \cdot 6H_2O$	3.0g
Na_2SO_4	3.0g
KCl	0.5g
NH_4Cl	0.3g
KH_2PO_4	0.2g
$CaCl_2 \cdot 2H_2O$	0.15g
Resazurin	1.0mg

Preparation of Solution A: Prepare and dispense solution anaerobically under 80% N_2 + 20% CO_2. Add components to distilled/deionized water and bring volume to 920.0mL. Mix thoroughly. Gently heat and bring to boiling. Continue boiling until resazurin turns colorless indicating reduction. Cap with rubber stoppers. Autoclave for 15 min at 15 psi pressure–121°C. Cool to 45–50°C.

Solution B—Trace Elements Solution SL-10:

Composition per liter:

$FeCl_2 \cdot 4H_2O$	1.5g
$CoCl_2 \cdot 6H_2O$	0.19g
$MnCl_2 \cdot 4H_2O$	0.10g
$ZnCl_2$	0.070g
$Na_2MoO_4 \cdot 2H_2O$	0.036g
$NiCl_2 \cdot 6H_2O$	0.024g
H_3BO_3	6.0mg
$CuCl_2 \cdot 2H_2O$	2.0mg
HCl (25% solution)	10.0mL

Preparation of Trace Elements Solution SL-10: Add $FeCl_2 \cdot 4H_2O$ to 10.0mL of HCl solution. Mix thoroughly. Bring volume to 900.0mL with distilled/deionized water. Mix thoroughly. Adjust pH to 6.0 with NaOH. Bring volume to 1.0L with distilled/deionized water. Filter sterilize. Aseptically gas under 100% N_2 for 20 min.

Solution C:

Composition per 50mL:

$NaHCO_3$	2.5g

Preparation of Solution C: Add $NaHCO_3$ to distilled/deionized water and bring volume to 50.0mL. Mix thoroughly. Filter sterilize. Aseptically gas under 80% N_2 + 20% CO_2 for 20 min.

Solution D:

Composition per 10mL:

Sodium benzoate	1.0g
Phenol	0.1g

Preparation of Solution D: Add components to distilled/deionized water and bring volume to 10.0mL. Mix thoroughly. Filter sterilize. Aseptically gas under 100% N_2 for 20 min.

Solution E—Wolfe's Vitamin Solution:

Composition per liter:

Pyridoxine·HCl	0.01g
Thiamine·HCl	5.0mg
Riboflavin	5.0mg
Nicotinic acid	5.0mg
Calcium pantothenate	5.0mg
p-Aminobenzoic acid	5.0mg
Thioctic acid	5.0mg
Biotin	2.0mg
Folic acid	2.0mg
Cyanocobalamin	0.1mg

Preparation of Wolfe's Vitamin Solution: Add components to distilled/deionized water and bring volume to 1.0L. Mix thoroughly. Filter sterilize. Aseptically gas under 100% N_2 for 20 min.

Solution F:

Composition per liter:

$Na_2SeO_3 \cdot 5H_2O$	3.0mg
NaOH (0.01*M* solution)	1.0L

Preparation of Solution F: Add $Na_2SeO_3 \cdot 5H_2O$ to 1.0L of NaOH solution. Mix thoroughly. Filter sterilize. Aseptically gas under 100% N_2 for 20 min.

Solution G:

Composition per 10mL:

$Na_2S \cdot 9H_2O$	0.4g

Preparation of Solution G: Add $Na_2S \cdot 9H_2O$ to distilled/deionized water and bring volume to 10.0mL. Gas under 100% N_2 for 20 min. Cap with a rubber stopper. Autoclave for 15 min at 15 psi pressure–121°C. Cool to 25°C.

Solution H:

Composition per 10mL:

$Na_2S_2O_4$	0.5g

Preparation of Solution H: Add $Na_2S_2O_4$ to distilled/deionized water and bring volume to 10.0mL. Mix thoroughly. Filter sterilize. Aseptically gas under 100% N_2 for 20 min. Freshly prepare solution.

Preparation of Medium: To 920.0mL of cooled, sterile solution A, aseptically and anaerobically add in the following order: 1.0mL of sterile solution B, 50.0mL of sterile solution C, 10.0mL of sterile solution D, 10.0mL of sterile solution E, 1.0mL of sterile solution F, and 10.0mL of sterile solution G. Mix thoroughly. Immediately prior to inoculation aseptically and anaerobically add 0.4mL of sterile solution H. Mix thoroughly. Aseptically and anaerobically distribute into sterile tubes or flasks.

Use: For the cultivation and maintenance of *Desulfobacterium phenolicum.*

Desulfobulbus Medium

Composition per liter:

Sodium propionate	1.85g
Na_2SO_4	1.5g
$(NH_4)_2SO_4$	1.24g
NaCl	0.6g
Cysteine·HCl·H_2O	0.5g
KH_2PO_4	0.3g
$MgSO_4 \cdot 7H_2O$	0.12g
$CaCl_2 \cdot 2H_2O$	0.08g
Trace minerals	10.0mL
Vitamin solution	10.0mL

pH 7.0 ± 0.2 at 25°C

Trace Minerals:

Composition per liter:

Nitrilotriacetic acid	12.8g
$FeSO_4 \cdot 7H_2O$	0.3g
$CoCl_2 \cdot 2H_2O$	0.1g
$MnCl_2 \cdot 4H_2O$	0.1g
$ZnCl_2$	0.1g
$CuCl_2$	0.02g
H_3BO_3	0.01g
Na_2MoO_4	0.01g
$NiSO_4 \cdot 6H_2O$	2.6mg
Na_2SeO_3	1.7mg

Preparation of Trace Minerals: Add nitrilotriacetic acid to 500.0mL of distilled/deionized water. Dissolve by adjusting pH to 6.5 with KOH. Add remaining components. Add distilled/deionized water to 1.0L.

Wolfe's Vitamin Solution:

Composition per liter:

Pyridoxine·HCl	0.01g
Thiamine·HCl	5.0mg
Riboflavin	5.0mg
Nicotinic acid	5.0mg
Calcium pantothenate	5.0mg
p-Aminobenzoic acid	5.0mg
Thioctic acid	5.0mg
Biotin	2.0mg
Folic acid	2.0mg
Cyanocobalamin	0.1mg

Preparation of Wolfe's Vitamin Solution: Add components to distilled/deionized water and bring volume to 1.0L. Mix thoroughly.

Preparation of Medium: Add components, except cysteine·HCl·H_2O, to distilled/deionized water and bring volume to 1.0L. Mix thoroughly. Gently heat and bring to boiling. Cool to 25°C under 85% N_2 + 15% CO_2. Add cysteine·HCl·H_2O. Mix thoroughly. Adjust pH to 7.2 with $KHCO_3$. Anaerobically distribute into tubes or flasks under 85% N_2 + 15% CO_2. Autoclave for 15 min at 15 psi pressure–121°C. Adjust pH to 7.0, if necessary.

Use: For the cultivation and maintenance of *Desulfobulbus elongatus.*

Desulfomonile tiedjei Medium

Composition per liter:

$NaHCO_3$	3.0g
PIPES (Piperazine-*N,N′*-bis-2-ethanesulfonic acid) buffer	1.5g
Na_2SO_4	1.42g
Yeast extract	1.0g
Mineral solution	20.0mL
Trace metal solution	10.0mL
$Na_2S_2O_4$ solution	10.0mL
Vitamin solution	10.0mL
Sodium pyruvate solution	10.0mL
Resazurin (0.1% solution)	1.0mL

pH 7.3 ± 0.2 at 25°C

Mineral Solution:

Composition per liter:

NH_4Cl	50.0g
NaCl	40.0g
$MgCl_2 \cdot 6H_2O$	8.3g
KCl	5.0g
KH_2PO_4	5.0g
$CaCl_2 \cdot 2H_2O$	1.0g

Preparation of Mineral Solution: Add components to distilled/deionized water and bring volume to 1.0L. Mix thoroughly.

Trace Metal Solution:

Composition per liter:

Nitrilotriacetic acid	2.0g
$MnSO_4.H_2O$	1.0g
$Fe(NH_4)_2(SO_4)_2 \cdot 6H_2O$	0.8g
$CoCl_2 \cdot 6H_2O$	0.2g
$ZnSO_4 \cdot 7H_2O$	0.2g

$CuCl_2 \cdot 2H_2O$	0.02g
$Na_2MoO_4 \cdot H_2O$	0.02g
Na_2SeO_4	0.02g
Na_2WO_4	0.02g
$NiCl_2 \cdot 6H_2O$	0.02g

Preparation of Trace Metal Solution: Add nitrilotriacetic acid to 500.0mL of distilled/deionized water. Dissolve by adjusting pH to 6.5 with KOH. Add remaining components. Add distilled/deionized water to 1.0L.

Vitamin Solution:

Composition per liter:

Nicotinamide	0.050g
1,4-Naphthoquinone	0.020g
p-Aminobenzoic acid	5.0mg
Biotin	5.0mg
Calcium pantothenate	5.0mg
Cyanocobalamin	5.0mg
Folic acid	5.0mg
Hemin	5.0mg
Pyridoxine·HCl	5.0mg
Riboflavin	5.0mg
Thioctic acid	5.0mg
NaOH (0.1*N* solution)	5.0mL

Preparation of Vitamin Solution: Add thioctic acid, 1,4-naphthoquinone, and hemin to 5.0mL of 0.1*N* NaOH solution. Mix thoroughly. Bring volume to 1.0L with distilled/deionized water. Add remaining components. Mix thoroughly.

$Na_2S_2O_4$ Solution:

Composition per 10mL:

$Na_2S_2O_4$	0.087g

Preparation of $Na_2S_2O_4$ Solution: Add $Na_2S_2O_4$ to distilled/deionized water and bring volume to 10.0mL. Filter sterilize. Prepare freshly.

Sodium Pyruvate Solution:

Composition per 10mL:

Sodium pyruvate	4.4g

Preparation of Sodium Pyruvate Solution: Add sodium pyruvate to distilled/deionized water and bring volume to 10.0mL. Filter sterilize.

Preparation of Medium: Add PIPES buffer, Na_2SO_4, yeast extract, mineral solution and trace metal solution to distilled/deionized water and bring volume to 970.0mL. Mix thoroughly. Adjust pH to 7.3 with HCl. Add $NaHCO_3$ and resazurin. Gently heat and bring to boiling under 80% N_2 + 20% CO_2. Replace headspace with 2 atm pressure of the same gas phase. Autoclave for 15 min at 15 psi pressure–121°C. Cool to 25°C. Anaerobically and aseptically add sterile vitamin solution, sodium pyruvate solution and $Na_2S_2O_4$ solution. Mix thoroughly.

Use: For the cultivation and maintenance of *Desulfomonile tiedjei*.

Desulfonema limicola Medium

Composition per 1009mL:

Solution A	850.0mL
Solution C	100.0mL
Solution H	20.0mL
Solution D	10.0mL
Solution F—Wolfe's vitamin solution	10.0mL
Solution I	10.0mL
Solution G	6.6mL
Solution B—Trace elements solution SL-10	1.0mL
Solution E	1.0mL
Solution J	0.4mL

pH 7.6 ± 0.2 at 25°C

Solution A:

Composition per 920mL:

NaCl	13.0g
$MgCl_2 \cdot 6H_2O$	2.2g
Na_2SO_4	3.0g
KCl	0.5g
NH_4Cl	0.3g
KH_2PO_4	0.2g
$CaCl_2 \cdot 2H_2O$	0.15g
Resazurin	0.5mg

Preparation of Solution A: Prepare and dispense solution anaerobically under 80% N_2 + 20% CO_2. Add components to distilled/deionized water and bring volume to 920.0mL. Mix thoroughly. Gently heat and bring to boiling. Continue boiling until resazurin turns colorless indicating reduction and a pH of 6.0 is reached. Cap with rubber stoppers. Autoclave for 15 min at 15 psi pressure–121°C. Cool to 25°C.

Solution B—Trace Elements Solution SL-10:

Composition per liter:

$FeCl_2 \cdot 4H_2O$	1.5g
$CoCl_2 \cdot 6H_2O$	0.19g
$MnCl_2 \cdot 4H_2O$	0.10g
$ZnCl_2$	0.070g
$Na_2MoO_4 \cdot 2H_2O$	0.036g
$NiCl_2 \cdot 6H_2O$	0.024g
H_3BO_3	6.0mg
$CuCl_2 \cdot 2H_2O$	2.0mg
HCl (25% solution)	10.0mL

Preparation of Trace Elements Solution SL-10: Add the $FeCl_2 \cdot 4H_2O$ to 10.0mL of HCl solution. Mix thoroughly. Bring volume to approximately 900.0mL with distilled/deionized water. Mix thoroughly. Adjust pH to 6.0 with NaOH. Bring volume to 1.0L with distilled/deionized water. Filter sterilize. Aseptically gas under 100% N_2 for 20 min.

Solution C:

Composition per 100mL:

$NaHCO_3$ 5.0g

Preparation of Solution C: Add $NaHCO_3$ to distilled/deionized water and bring volume to 100.0mL. Mix thoroughly. Filter sterilize. Aseptically gas under 80% N_2 + 20% CO_2 for 20 min.

Solution D:

Composition per 10mL:

Sodium acetate·$3H_2O$ 2.5g

Preparation of Solution D: Prepare and dispense solution anaerobically under 80% N_2 + 20% CO_2. Add sodium acetate to distilled/deionized water and bring volume to 10.0mL. Mix thoroughly. Cap with rubber stopper. Autoclave for 15 min at 15 psi pressure–121°C. Cool to 25°C.

Solution E:

Composition per 1mL:

Disodium succinate 0.1g

Preparation of Solution E: Add disodium succinate to distilled/deionized water and bring volume to 1.0mL. Mix thoroughly. Gas with 80% N_2 + 20% CO_2. Cap with rubber stopper. Autoclave for 15 min at 15 psi pressure–121°C. Cool to 25°C.

Solution F—Wolfe's Vitamin Solution:

Composition per liter:

Pyridoxine·HCl 0.01g
Thiamine·HCl 5.0mg
Riboflavin 5.0mg
Nicotinic acid 5.0mg
Calcium pantothenate 5.0mg
p-Aminobenzoic acid 5.0mg
Thioctic acid 5.0mg
Biotin 2.0mg
Folic acid 2.0mg
Cyanocobalamin 0.1mg

Preparation of Wolfe's Vitamin Solution: Add components to distilled/deionized water and bring volume to 1.0L. Mix thoroughly. Filter sterilize. Aseptically gas under 100% N_2 for 20 min.

Solution G:

Composition per 6.6mL:

$AlCl_3 \cdot 6H_2O$ (4.9% solution) 5.0mL
Na_2CO_3 (10.6% solution) 1.6mL

Preparation of Solution G: Combine both solutions. Mix thoroughly. Gas with 100% N_2. Cap with a rubber stopper. Autoclave for 15 min at 15 psi pressure–121°C. Cool to 25°C.

Solution H:

Composition per 10mL:

Rumen fluid, clarified 20.0mL

Preparation of Solution H: Gas rumen fluid under 100% N_2 for 20 min. Cap with a rubber stopper. Autoclave for 15 min at 15 psi pressure–121°C. Cool to 25°C.

Solution I:

Composition per 10mL:

$Na_2S \cdot 9H_2O$ 0.4g

Preparation of Solution I: Add $Na_2S \cdot 9H_2O$ to distilled/deionized water and bring volume to 10.0mL. Gas under 100% N_2 for 20 min. Cap with a rubber stopper. Autoclave for 15 min at 15 psi pressure–121°C. Cool to 25°C.

Solution J:

Composition per 10mL:

$Na_2S_2O_4$ 0.5g

Preparation of Solution J: Add $Na_2S_2O_4$ to distilled/deionized water and bring volume to 10.0mL. Mix thoroughly. Filter sterilize. Aseptically gas under 100% N_2 for 20 min. Freshly prepare solution.

Preparation of Medium: To 850.0mL of cooled, sterile solution A, aseptically and anaerobically add in the following order: 1.0mL of sterile solution B, 1000.0mL of sterile solution C, 10.0mL of sterile solution D, 1.0mL of sterile solution E, 10.0mL of sterile solution F, and 6.6mL of sterile solution G, 20.0mL of sterile solution H and 10.0mL of sterile solution I. Mix thoroughly. Immediately prior to inoculation aseptically and anaerobically add 0.4mL of sterile solution J. Mix thoroughly. Aseptically and anaerobically distribute into sterile tubes or flasks.

Use: For the cultivation and maintenance of *Desulfonema limicola.*

Desulfonema magnum Medium

Composition per 1001mL:

Solution A 890.0mL
Solution C 50.0mL
Solution J 20.0mL
Solution D 10.0mL
Solution G—Wolfe's vitamin solution 10.0mL
Solution K 10.0mL
Solution I 6.6mL
Solution B—
  Trace elements solution SL-10 .... 1.0mL
Solution E 1.0mL
Solution F 1.0mL
Solution H 1.0mL
Solution L 0.4mL

pH 6.9 ± 0.2 at 25°C

Solution A:

Composition per 890mL:

NaCl 21.0g
$MgCl_2 \cdot 6H_2O$ 5.5g

Na_2SO_4	3.0g
$CaCl_2 \cdot 2H_2O$	1.35g
KCl	0.5g
NH_4Cl	0.3g
KH_2PO_4	0.2g
Resazurin	0.5mg

Preparation of Solution A: Prepare and dispense solution anaerobically under 80% N_2 + 20% CO_2. Add components to distilled/deionized water and bring volume to 890.0mL. Mix thoroughly. Gently heat and bring to boiling. Continue boiling until resazurin turns colorless indicating reduction and a pH of 6.0 is reached. Cap with rubber stoppers. Autoclave for 15 min at 15 psi pressure–121°C. Cool to 25°C.

Solution B—Trace Elements Solution SL-10:
Composition per liter:

$FeCl_2 \cdot 4H_2O$	1.5g
$CoCl_2 \cdot 6H_2O$	0.19g
$MnCl_2 \cdot 4H_2O$	0.10g
$ZnCl_2$	0.070g
$Na_2MoO_4 \cdot 2H_2O$	0.036g
$NiCl_2 \cdot 6H_2O$	0.024g
H_3BO_3	6.0mg
$CuCl_2 \cdot 2H_2O$	2.0mg
HCl (25% solution)	10.0mL

Preparation of Trace Elements Solution SL-10: Add the $FeCl_2 \cdot 4H_2O$ to 10.0mL of HCl solution. Mix thoroughly. Bring volume to approximately 900.0mL with distilled/deionized water. Mix thoroughly. Adjust pH to 6.0 with NaOH. Bring volume to 1.0L with distilled/deionized water. Filter sterilize. Aseptically gas under 100% N_2 for 20 min.

Solution C:
Composition per 50mL:

$NaHCO_3$	2.5g

Preparation of Solution C: Add $NaHCO_3$ to distilled/deionized water and bring volume to 50.0mL. Mix thoroughly. Filter sterilize. Aseptically gas under 80% N_2 + 20% CO_2 for 20 min.

Solution D:
Composition per 10mL:

Sodium benzoate	0.6g

Preparation of Solution D: Add sodium benzoate to distilled/deionized water and bring volume to 10.0mL. Mix thoroughly. Gas with 100% N_2 for 10 min. Cap with a rubber stopper. Autoclave for 15 min at 15 psi pressure–121°C. Cool to 25°C.

Solution E:
Composition per 1mL:

$Na_2SeO_3 \cdot 5H_2O$	3.5μg

Preparation of Solution E: Add $Na_2SeO_3 \cdot 5H_2O$ to distilled/deionized water and bring volume to 1.0mL. Mix thoroughly. Gas with 100% N_2 for 10 min. Cap with a rubber stopper. Autoclave for 15 min at 15 psi pressure–121°C. Cool to 25°C.

Solution F:
Composition per 1mL:

Disodium succinate	0.1g

Preparation of Solution F: Add disodium succinate to distilled/deionized water and bring volume to 1.0mL. Mix thoroughly. Gas with 100% N_2. Cap with a rubber stopper. Autoclave for 15 min at 15 psi pressure–121°C. Cool to 25°C.

Solution G—Wolfe's Vitamin Solution:
Composition per liter:

Pyridoxine·HCl	0.01g
Thiamine·HCl	5.0mg
Riboflavin	5.0mg
Nicotinic acid	5.0mg
Calcium pantothenate	5.0mg
p-Aminobenzoic acid	5.0mg
Thioctic acid	5.0mg
Biotin	2.0mg
Folic acid	2.0mg
Cyanocobalamin	0.1mg

Preparation of Wolfe's Vitamin Solution: Add components to distilled/deionized water and bring volume to 1.0L. Mix thoroughly. Filter sterilize. Aseptically gas under 100% N_2 for 20 min.

Solution H:
Composition per 1mL:

Vitamin B_{12}	0.05mg

Preparation of Solution H: Add disodium succinate to distilled/deionized water and bring volume to 1.0mL. Mix thoroughly. Gas with 100% N_2. Cap with a rubber stopper. Autoclave for 15 min at 15 psi pressure–121°C. Cool to 25°C.

Solution I:
Composition per 6.6mL:

$AlCl_3 \cdot 6H_2O$ (4.9% solution)	5.0mL
Na_2CO_3 (10.6% solution)	1.6mL

Preparation of Solution I: Combine both solutions. Mix thoroughly. Gas with 100% N_2. Cap with a rubber stopper. Autoclave for 15 min at 15 psi pressure–121°C. Cool to 25°C.

Solution J:
Composition per 10mL:

Rumen fluid, clarified	20.0mL

Preparation of Solution J: Gas rumen fluid under 100% N_2 for 20 min. Cap with a rubber stopper. Autoclave for 15 min at 15 psi pressure–121°C. Cool to 25°C.

Solution K:
Composition per 10mL:
$Na_2S·9H_2O$ 0.4g

Preparation of Solution K: Add $Na_2S·9H_2O$ to distilled/deionized water and bring volume to 10.0mL. Gas under 100% N_2 for 20 min. Cap with a rubber stopper. Autoclave for 15 min at 15 psi pressure–121°C. Cool to 25°C.

Solution L:
Composition per 10mL:
$Na_2S_2O_4$ 0.5g

Preparation of Solution L: Add $Na_2S_2O_4$ to distilled/deionized water and bring volume to 10.0mL. Mix thoroughly. Filter sterilize. Aseptically gas under 100% N_2 for 20 min. Use freshly prepared solution.

Preparation of Medium: To 890.0mL of cooled, sterile solution A, aseptically and anaerobically add in the following order: 1.0mL of sterile solution B, 50.0mL of sterile solution C, 10.0mL of sterile solution D, 1.0mL of sterile solution E, 1.0mL of sterile solution F, and 10.0mL of sterile solution G, 1.0mL of sterile solution H and 6.6mL of sterile solution I, 20.0mL of sterile solution J, and 10.0mL of sterile solution K. Mix thoroughly. Immediately prior to inoculation aseptically and anaerobically add 0.4mL of sterile solution L. Mix thoroughly. Aseptically and anaerobically distribute into sterile tubes or flasks.

Use: For the cultivation and maintenance of *Desulfonema magnum.*

Desulfovibrio Medium

Composition per 1056.5mL:
$(NH_4)_2SO_4$ 5.3g
Sodium acetate 2.0g
NaCl 1.0g
KH_2PO_4 .... 0.5g
$MgSO_4·7H_2O$ 0.2g
$CaCl_2·2H_2O$ 0.1g
Na_2CO_3 solution 50.0mL
Solution 1 .... 10.0mL
Solution 2 .... 1.0mL
pH 7.2 ± 0.2 at 25°C

Solution 1:
Composition per liter:
Nitrilotriacetic acid 12.8g
$FeCl_2·4H_2O$ 0.3g
$CoCl_2·6H_2O$ 0.17g
$MnCl_2·4H_2O$ 0.1g
$ZnCl_2$ 0.1g
$CuCl_2$ 0.02g
H_3BO_3 .... 0.01g
$Na_2MoO_4·2H_2O$ 0.01g

Preparation of Solution 1: Add nitrilotriacetic acid to 500.0mL of distilled/deionized water. Dissolve by adjusting pH to 6.5 with NaOH. Add remaining components. Readjust pH to 7.2 with H_2SO_4 or NaOH. Add distilled/deionized water to 1.0L.

Solution 2:
Composition per 100mL:
Resazurin 0.2g

Preparation of Solution 2: Add resazurin to distilled/deionized water and bring volume to 100.0mL. Mix thoroughly.

Na_2CO_3 Solution:
Composition per 100mL:
Na_2CO_3 .... 8.0g

Preparation of Na_2CO_3 Solution Solution: Add Na_2CO_3 to distilled/deionized water and bring volume to 100.0mL. Mix thoroughly. Filter sterilize. Gas with 100% N_2 for 20 min.

HCl Solution:
Composition per 100mL:
HCl 25.0mL

Preparation of HCl Solution: Add HCl to distilled/deionized water and bring volume to 100.0mL. Mix thoroughly. Autoclave for 15 min at 15 psi pressure–121°C. Cool to 25°C. Gas with 100% N_2 for 20 min.

$Na_2S_2O_4$ Solution:
Composition per 100mL:
$Na_2S_2O_4$ 8.7g

Preparation of $Na_2S_2O_4$ Solution: Add $Na_2S_2O_4$ to distilled/deionized water and bring volume to 100.0mL. Mix thoroughly. Autoclave for 15 min at 15 psi pressure–121°C. Cool to 25°C. Gas with 100% N_2 for 20 min.

Preparation of Medium: Add components—except Na_2CO_3 solution, HCl solution, and $Na_2S_2O_4$ solution—to distilled/deionized water and bring volume to 1.0L. Mix thoroughly. Gently heat and bring to boiling. Autoclave for 15 min at 15 psi pressure–121°C. Cool to 45–50°C. Anaerobically and aseptically add 50.0mL of sterile Na_2CO_3 solution, 5.5mL of sterile HCl solution, and 1.0mL of sterile $Na_2S_2O_4$ solution. Mix thoroughly. Anaerobically and aseptically distribute into sterile tubes or flasks.

Use; For the isolation, cultivation and enrichment of *Desulfovibrio* species.

Desulfovibrio Medium

Composition per liter of tap water:
Agar 15.0g
Glucose 5.0g

Peptone 5.0g
Beef extract 3.0g
$MgSO_4$ 1.5g
Na_2SO_4 .... 1.5g
Yeast extract 0.2g
$Fe(NH_4)_2(SO_4)_2$ 0.1g

pH 7.0 ± 0.2 at 25°C

Preparation of Medium: Sterilize by autoclaving for 15 min at 15 lb pressure (121°C).

Use: For the cultivation and maintenance of *Desulfomaculum nigrificans, Desulfovibiro desulfuricans,* and *Desulfovibrio gigas.*

Desulfovibrio Medium with Lactate

Composition per liter:

Agar 15.0g
Lactate 10.0g
Glucose 5.0g
Peptone 5.0g
Beef extract 3.0g
$MgSO_4$ 1.5g
Na_2SO_4 .... 1.5g
Yeast extract 0.2g
$Fe(NH_4)_2(SO_4)_2$ 0.1g

pH 7.0 ± 0.2 at 25°C

Preparation of Medium: Add components to tap water and bring volume to 1.0L. Mix thoroughly. Gently heat and bring to boiling. Distribute into tubes or flasks. Autoclave for 15 min at 15 psi pressure–121°C. Pour into sterile Petri dishes or leave in tubes.

Use: For the cultivation and maintenance of *Desulfovibrio desulfuricans.*

Desulfovibrio Medium with NaCl

Composition per liter:

NaCl 30.0g
Agar 15.0g
Glucose 5.0g
Peptone 5.0g
Beef extract 3.0g
$MgSO_4$ 1.5g
Na_2SO_4 .... 1.5g
Yeast extract 0.2g
$Fe(NH_4)_2(SO_4)_2$ 0.1g

pH 7.0 ± 0.2 at 25°C

Preparation of Medium: Add components to tap water and bring volume to 1.0L. Mix thoroughly. Gently heat and bring to boiling. Distribute into tubes or flasks. Autoclave for 15 min at 15 psi pressure–121°C. Pour into sterile Petri dishes or leave in tubes.

Use: For the cultivation and maintenance of *Desulfovibrio desulfuricans* and *D. salexigens.*

Desulfovibrio sulfodismutans Medium

Composition per 1002mL:

Solution A 920.0mL
Solution C 50.0mL
Solution D 10.0mL
Solution F 10.0mL
Solution G 10.0mL
Solution B—
  Trace elements solution SL-10 .... 1.0mL
Solution E 1.0mL

pH 7.1–7.4 at 25°C

Solution A:

Composition per 920mL:

NaCl 1.0g
KCl 0.5g
$MgCl_2·6H_2O$ 0.4g
NH_4Cl 0.3g
KH_2PO_4 .... 0.2g
$CaCl_2·2H_2O$ 0.15g

Preparation of Solution A: Prepare and dispense solution anaerobically under 80% N_2 + 20% CO_2. Add components to distilled/deionized water and bring volume to 920.0mL. Mix thoroughly. Gently heat and bring to boiling. Continue boiling until resazurin turns colorless indicating reduction. Cap with rubber stoppers. Autoclave for 15 min at 15 psi pressure–121°C. Cool to 45–50°C.

Solution B—Trace Elements Solution SL-10:

Composition per liter:

$FeCl_2·4H_2O$ 1.5g
$CoCl_2·6H_2O$ 0.19g
$MnCl_2·4H_2O$ 0.10g
$ZnCl_2$ 0.070g
$Na_2MoO_4·2H_2O$ 0.036g
$NiCl_2·6H_2O$ 0.024g
H_3BO_3 .... 6.0mg
$CuCl_2·2H_2O$ 2.0mg
HCl (25% solution) 10.0mL

Preparation of Trace Elements Solution SL-10: Add the $FeCl_2·4H_2O$ to 10.0mL of HCl solution. Mix thoroughly. Bring volume to approximately 900.0mL with distilled/deionized water. Mix thoroughly. Adjust pH to 6.0 with NaOH. Bring volume to 1.0L with distilled/deionized water. Filter sterilize. Aseptically gas under 100% N_2 for 20 min.

Solution C:

Composition per 50mL:

$NaKCO_3$ 2.5g

Preparation of Solution C: Add $NaHCO_3$ to distilled/deionized water and bring volume to 50.0mL. Mix thoroughly. Filter sterilize. Aseptically gas under 80% N_2 + 20% CO_2 for 20 min.

Solution D:

Composition per 10mL:

Sodium acetate·$3H_2O$	0.3g

Preparation of Solution D: Prepare and dispense solution anaerobically under 80% N_2 + 20% CO_2. Add sodium acetate to distilled/deionized water and bring volume to 10.0mL. Mix thoroughly. Cap with rubber stopper. Autoclave for 15 min at 15 psi pressure–121°C. Cool to 45–50°C.

Solution E:

Composition per liter:

Calcium pantothenate	0.050mg
Biotin	0.010mg

Preparation of Solution E: Add components to distilled/deionized water and bring volume to 1.0L. Mix thoroughly. Filter sterilize. Aseptically gas under 100% N_2 for 20 min.

Solution F:

Composition per 10mL:

$Na_2S·9H_2O$	0.4g

Preparation of Solution F: Add $Na_2S·9H_2O$ to distilled/deionized water and bring volume to 10.0mL. Gas under 100% N_2 for 20 min. Cap with a rubber stopper. Autoclave for 15 min at 15 psi pressure–121°C. Cool to 25°C.

Solution G:

Composition per 10mL:

$Na_2S_2O_5$	1.05g

Preparation of Solution G: Add $Na_2S_2O_5$ to distilled/deionized water and bring volume to 10.0mL. Mix thoroughly. Filter sterilize. Aseptically gas under 100% N_2 for 20 min. Freshly prepare solution.

Preparation of Medium: To 920.0mL of cooled, sterile solution A, aseptically and anaerobically add in the following order: 1.0mL of sterile solution B, 50.0mL of sterile solution C, 10.0mL of sterile solution D, 1.0mL of sterile solution E, 10.0mL of sterile solution F, and 10.0mL of sterile solution G. Mix thoroughly. Aseptically and anaerobically distribute into sterile tubes or flasks.

Use: For the cultivation and maintenance of *Desulfovibrio sulfodismutans.*

Desulfurococcus Medium

Composition per 1300mL:

Solution C	500.0mL
Sloution B	450.0mL
Solution A	300.0mL
Solution D	50.0mL

Solution A:

Composition per 300mL:

$(NH_4)_2SO_4$	1.3g
KH_2PO_4	0.28g
$MgSO_4·7H_2O$	0.25g
$CaCl_2·2H_2O$	0.07g
$FeSO_4·7H_2O$	0.028g
$Na_2B_4O_7·10H_2O$	4.5mg
$MnCl_2·4H_2O$	1.8mg
$ZnSO_4·7H_2O$	0.220mg
$CuCl_2·2H_2O$	0.050mg
$Na_2MoO_4·2H_2O$	0.030mg
$VOSO_4·2H_2O$	0.030mg
$CoSO_4·7H_2O$	0.010mg

Preparation of Solution A: Add components to distilled/deionized water and bring volume to 300.0mL. Mix thoroughly. Gently heat and bring to boiling. Autoclave for 15 min at 15 psi pressure–121°C. Cool to 25°C. Gas under 100% N_2 for 20 min.

Solution B:

Composition per 450mL:

Sulfur	5.0g

Preparation of Solution B: Add sulfur to distilled/deionized water and bring volume to 450.0mL. Autoclave for 30 min at 0 psi pressure–100°C on three consecutive days. Gas under 100% N_2 for 20 min.

Solution C:

Composition per 500mL:

Pancreatic digest of casein	2.0g
Yeast extract	2.0g
Resazurin	1.0mg

Preparation of Solution C: Add components to distilled/deionized water and bring volume to 500.0mL. Mix thoroughly. Gently heat and bring to boiling. Autoclave for 15 min at 15 psi pressure–121°C. Cool to 25°C. Gas under 100% N_2 for 20 min.

Solution D:

$Na_2S·9H_2O$	0.5g

Preparation of Solution D: Add components to distilled/deionized water and bring volume to 50.0mL. Mix thoroughly. Autoclave for 15 min at 15 psi pressure–121°C. Cool to 25°C. Gas under 100% N_2 for 20 min.

Preparation of Medium: Aseptically combine solutions A–D under nitrogen gas. Seal containers with butyl rubber stoppers.

Use: For the cultivation and maintenance of *Desulfurococcus mobilis* and *D. mucosus.*

Desulfuromonas Medium

Composition per 1051mL:

Elemental sulfur slurry	10.0g
Solution 1	1.0L
Solution 3	40.0mL
Solution 4	6.0mL
Solution 5	5.0mL
Solution 2	1.0mL

pH 7.2 ± 0.2 at 25°C

Solution 1:

Composition per liter:

NaCl	20.0g
$MgCl_2 \cdot 6H_2O$	3.0g
KH_2PO_4	1.0g
NH_4Cl	0.3g
$CaCl_2 \cdot 2H_2O$	0.1g
HCl (2*N* solution)	4.0mL

Preparation of Solution 1: Add components to distilled/deionized water and bring volume to 1.0L. Mix thoroughly. Autoclave for 15 min at 15 psi pressure–121°C. Cool to 25°C.

Solution 2:

Composition per liter:

Disodium EDTA	5.2g
$CoCl_2 \cdot 6H_2O$	1.9g
$FeCl_2 \cdot 4H_2O$	1.5g
$MnCl_2 \cdot 4H_2O$	1.0g
$ZnCl_2$	0.7g
H_3BO_3	0.62g
$Na_2MoO_4 \cdot 2H_2O$	0.36g
$NiCl_2 \cdot 6H_2O$	0.24g
$CuCl_2 \cdot 2H_2O$	0.17g

pH 6.5 ± 0.2 at 25°C

Preparation of Solution 2: Add components to distilled/deionized water and bring volume to 1.0L. Mix thoroughly. Adjust pH to 6.5. Autoclave for 15 min at 15 psi pressure–121°C. Cool to 25°C.

Solution 3:

Composition per 100mL:

$NaHCO_3$	10.0g

Preparation of Solution 3: Add $NaHCO_3$ to distilled/deionized water and bring volume to 100.0mL. Mix thoroughly. Autoclave for 15 min at 15 psi pressure–121°C. Cool to 25°C.

Solution 4:

Composition per 100mL:

$Na_2S \cdot 9H_2O$	5.0g

Preparation of Solution 4: Add $Na_2S \cdot 9H_2O$ to distilled/deionized water and bring volume to 100.0mL. Mix thoroughly. Autoclave for 15 min at 15 psi pressure–121°C. Cool to 25°C.

Solution 5:

Composition per 200mL:

Pyridoxamine·HCl	0.01g
Nicotinic acid	4.0mg
p-Aminobenzoic acid	2.0mg
Thiamine	2.0mg
Cyanocobalamin	1.0mg
Pantothenic acid	1.0mg
Biotin	0.5mg

Preparation of Solution 5: Add components to distilled/deionized water and bring volume to 200.0mL. Mix thoroughly. Filter sterilize.

Elemental Sulfur Slurry:

Composition per 10g:

Sulfur flowers	10.0g

Preparation of Elemental Sulfur Slurry: Add highly purified sulfur flowers to a mortar and grind to a fine powder. Add sufficient distilled/deionized water to produce a slurry. Distribute into 100 mL screw-capped bottles in 20.0mL volumes. Autoclave for 30 min at 10 psi pressure–115°C. Decant supernatant solution. Reserve sulfur slurry.

Preparation of Medium: To 1.0L of cooled, sterile solution 1, aseptically add 1.0mL of sterile solution 2, 40.0mL of sterile solution 3, 6.0mL of sterile solution 4 and 5.0mL of sterile solution 5. Mix thoroughly. Adjust pH to 7.2. Aseptically distribute into sterile 50.0mL screw-capped bottles. Fill bottles completely with medium except for a pea-sized air bubble. Aseptically add a pea-sized piece of sulfur slurry to each 50.0mL of medium.

Use: For the isolation and cultivation of marine *Desulfuromonas* species.

Desulfuromonas Medium

Composition per 1031mL:

Elemental sulfur slurry	10.0g
Solution 1	1.0L
Solution 3	20.0mL
Solution 4	6.0mL
Solution 5	5.0mL
Solution 2	1.0mL

pH 7.2 ± 0.2 at 25°C

Solution 1:

Composition per liter:

KH_2PO_4	1.0g
$MgCl_2 \cdot 6H_2O$	0.4g
NH_4Cl	0.3g
$CaCl_2 \cdot 2H_2O$	0.1g
HCl (2*N* solution)	4.0mL

Preparation of Solution 1: Add components to distilled/deionized water and bring volume to 1.0L.

Mix thoroughly. Autoclave for 15 min at 15 psi pressure–121°C. Cool to 25°C.

Solution 2:

Composition per liter:

Disodium EDTA	5.2g
$CoCl_2 \cdot 6H_2O$	1.9g
$FeCl_2 \cdot 4H_2O$	1.5g
$MnCl_2 \cdot 4H_2O$	1.0g
$ZnCl_2$	0.7g
H_3BO_3	0.62g
$Na_2MoO_4 \cdot 2H_2O$	0.36g
$NiCl_2 \cdot 6H_2O$	0.24g
$CuCl_2 \cdot 2H_2O$	0.17g

Preparation of Solution 2: Add components to distilled/deionized water and bring volume to 1.0L. Mix thoroughly. Adjust pH to 6.5. Autoclave for 15 min at 15 psi pressure–121°C. Cool to 25°C.

Solution 3:

Composition per 100mL:

$NaHCO_3$	10.0g

Preparation of Solution 3: Add $NaHCO_3$ to distilled/deionized water and bring volume to 100.0mL. Mix thoroughly. Autoclave for 15 min at 15 psi pressure–121°C. Cool to 25°C.

Solution 4:

Composition per 100mL:

$Na_2S \cdot 9H_2O$	5.0g

Preparation of Solution 4: Add $Na_2S \cdot 9H_2O$ to distilled/deionized water and bring volume to 100.0mL. Mix thoroughly. Autoclave for 15 min at 15 psi pressure–121°C. Cool to 25°C.

Solution 5:

Composition per 200mL:

Pyridoxamine·HCl	0.01g
Nicotinic acid	4.0mg
p-Aminobenzoic acid	2.0mg
Thiamine	2.0mg
Cyanocobalamin	1.0mg
Pantothenic acid	1.0mg
Biotin	0.5mg

Preparation of Solution 5: Add components to distilled/deionized water and bring volume to 200.0mL. Mix thoroughly. Filter sterilize.

Elemental Sulfur Slurry:

Composition per 10g:

Sulfur flowers	10.0g

Preparation of Elemental Sulfur Slurry: Add highly purified sulfur flowers to a mortar and grind to a fine powder. Add sufficient distilled/deionized water to produce a slurry. Distribute into 100.0mL screw-capped bottles in 20.0mL volumes. Autoclave for 30 min at 10 psi pressure–115°C. Decant supernatant solution. Reserve sulfur slurry.

Preparation of Medium: To 1.0L of cooled, sterile solution 1, aseptically add 1.0mL of sterile solution 2, 40.0mL of sterile solution 3, 6.0mL of sterile solution 4 and 5.0mL of sterile solution 5. Mix thoroughly. Adjust pH to 7.2. Aseptically distribute into sterile 50.0mL screw-capped bottles. Fill bottles completely with medium except for a pea-sized air bubble. Aseptically add a pea-sized piece of sulfur slurry to each 50.0mL of medium.

Use: For the isolation and cultivation of freshwater *Desulfuromonas* species.

Diaminopimelic Acid Medium

Composition per liter:

Pancreatic digest of gelatin	5.0g
Beef extract	3.0g
Diaminopimelic acid	0.050g

pH 6.9 ± 0.2 at 25°C

Preparation of Medium: Add components to distilled/deionized water and bring volume to 1.0L. Mix thoroughly. Distribute into tubes or flasks. Autoclave for 15 min at 15 psi pressure–121°C.

Use: For the cultivation and maintenance of *Bacillus megaterium*.

Dibenzothiophene Mineral Medium

Composition per liter:

Beef extract	10.0g
Na_2HPO_4	3.0g
KH_2PO_4	2.0g
NH_4Cl	2.0g
Dibenzothiophene	0.5g
$MgCl_2 \cdot 6H_2O$	0.2g
$FeCl_3 \cdot 6H_2O$	0.028g

Preparation of Medium: Add components to distilled/deionized water and bring volume to 1.0L. Mix thoroughly. Distribute into tubes or flasks. Autoclave for 15 min at 15 psi pressure–121°C.

Use: For the cultivation of bacteria that can metabolize dibenzothiophene.

Dichloromethane Medium for *Hyphomicrobium*

Composition per liter:

$K_2HPO_4 \cdot 3H_2O$	4.1g
KH_2PO_4	1.4g

$MgSO_4 \cdot 7H_2O$ 0.2g
$(NH_4)_2SO_4$ 0.2g
Dichloromethane (methylene chloride) 1.0mL
Trace elements solution 1.0mL

pH 7.2 ± 0.2 at 25°C

Trace Elements Solution:

Composition per liter:

$Ca(NO_3)_2$ 25.0g
$FeSO_4 \cdot 7H_2O$ 1.0g
H_3BO_3 .... 1.0g
$MnSO_4 \cdot H_2O$ 1.0g
$Co(NO_3)_2 \cdot 6H_2O$ 0.25g
$CuCl_2 \cdot 2H_2O$ 0.25g
$(NH_4)_6Mo_7O_{24} \cdot 4H_2O$ 0.25g
$ZnCl_2$ 0.25g
NH_4VO_3 .... 0.1g

Preparation of Medium: Filter sterilize dichloromethane. Add components, except dichloromethane, to distilled/deionized water and bring volume to 999.0mL. Mix thoroughly. Gently heat and bring to boiling. Adjust pH to 7.2. Autoclave for 15 min at 15 psi pressure–121°C. Cool to 45–50°C. Aseptically add sterile dichloromethane. Mix thoroughly. Aseptically distribute into sterile tubes or flasks.

Use: For the cultivation and maintenance of *Hyphomicrobium* species.

Dichotomicrobium Medium

Composition per liter:

Agar 18.0g
Sodium DL-malate 1.0g
Yeast extract 1.0g
Articial seawater, 3X solution 980.0mL
Hutner's mineral base 20.0mL

pH 7.1 ± 0.1 at 25°C

Artificial Seawater, 3X Solution:

Composition per 5 liters:

NaCl 352.14g
$MgCl_2 \cdot 6H_2O$ 159.30g
Na_2SO_4 .... 58.76g
$CaCl_2 \cdot 2H_2O$ 21.75g
KCl 9.96g
$NaHCO_3$ 2.88g
KBr 1.44g
H_3BO_3 .... 0.39g

Preparation of Artificial Seawater, 3X Solution: Add components to distilled/deionized water and bring volume to 5.0L. Mix thoroughly.

Hutner's Mineral Base:

Composition per liter:

$MgSO_4 \cdot 7H_2O$ 29.7g
Nitrilotriacetic acid 10.0g
$CaCl_2 \cdot 2H_2O$ 3.34g
$FeSO_4 \cdot 7H_2O$ 0.099g
$(NH_4)_2MoO_4$ 9.25mg
Metals "44" 50.0mL

Preparation of Hutner's Mineral Base: Add nitrilotriacetic acid to 500.0mL of distilled/deionized water. Dissolve by adjusting pH to 6.5 with KOH. Add remaining components. Readjust pH to 7.2 with H_2SO_4 or KOH. Add distilled/deionized water to 1.0L. Store at 5°C.

Metals "44":

Composition per 100mL:

$ZnSO_4 \cdot 7H_2O$ 1.1g
$FeSO_4 \cdot 7H_2O$ 0.5g
EDTA 0.25g
$MnSO_4 \cdot 7H_2O$ 0.154g
$CuSO_4 \cdot 5H_2O$ 0.04g
$Co(NO_3)_2 \cdot 6H_2O$ 0.025g
$Na_2B_4O_7 \cdot 10H_2O$ 0.018g

Preparation of Metals "44": Add a few drops of H_2SO_4 to distilled/deionized water to inhibit precipitate formation. Add components to acidified distilled/deionized water and bring volume to 100.0mL. Mix thoroughly.

Preparation of Medium: Combine components. Mix thoroughly. Gently heat and bring to boiling. Adjust pH to 7.1. Distribute into tubes or flasks. Autoclave for 15 min at 15 psi pressure–121°C. Pour into sterile Petri dishes or leave in tubes.

Use: For the cultivation and maintenance of *Dichotomicrobium thermohalophilium.*

Dictyoglomus Medium

Composition per liter:

Soluble starch 5.0g
$Na_2HPO_4 \cdot 12H_2O$ 4.2g
Polypeptone™ 2.0g
Yeast extract 2.0g
KH_2PO_4 .... 1.5g
$Cysteine \cdot HCl \cdot H_2O$ 1.0g
Na_2CO_3 .... 1.0g
NH_4Cl 0.5g
$MgCl_2 \cdot 6H_2O$ 0.38g
$CaCl_2$ 0.05g
$Fe(NH_4)_2(SO_4)_2 \cdot 6H_2O$ 0.039g
Resazurin 2.0mg
Trace metals 10.0mL
Wolfe's vitamin solution 10.0mL

pH 7.2 ± 0.2 at 25°C

Trace Metals:

Composition per liter:

$CoCl_2 \cdot 6H_2O$	0.29g
$ZnSO_4 \cdot 7H_2O$	0.28g
$Na_2MoO_4 \cdot 2H_2O$	0.24g
$MnCl_2 \cdot 4H_2O$	0.20g
Na_2SeO_3	0.017g

Preparation of Trace Metals: Add components to distilled/deionized water and bring volume to 1.0L. Adjust pH to 6.0 with KOH. Mix thoroughly.

Wolfe's Vitamin Solution:

Composition per liter:

Pyridoxine·HCl	0.01g
Thiamine·HCl	5.0mg
Riboflavin	5.0mg
Nicotinic acid	5.0mg
Calcium pantothenate	5.0mg
p-Aminobenzoic acid	5.0mg
Thioctic acid	5.0mg
Biotin	2.0mg
Folic acid	2.0mg
Cyanocobalamin	0.1mg

Preparation of Wolfe's Vitamin Solution: Add components to distilled/deionized water and bring volume to 1.0L. Mix thoroughly. Filter sterilize.

Preparation of Medium: Prepare and dispense medium under 100% N_2. Add components, except Wolfe's vitamin solution, to distilled/deionized water and bring volume to 990.0mL. Mix thoroughly. Gently heat and bring to boiling. Continue boiling until resazurin turns colorless indicating reduction. Autoclave for 15 min at 15 psi pressure–121°C. Cool to 25°C under 100% N_2. Aseptically add sterile Wolfe's vitamin solution. Mix thoroughly. Adjust pH to 7.2 if necessary. Aseptically and anaerobically distribute into sterile tubes or flasks.

Use: For the cultivation and maintenance of *Dictyoglomus thermophilum.*

Dilute Potato Medium

Composition per 1090mL:

Glucose	1.0g
Na_2HPO_4	0.12g
$Ca(NO_3)_2 \cdot 4H_2O)$	0.05g
Peptone	0.05g
Potato decoction	100.0mL

pH 6.8 ± 0.2 at 25°C

Potato Decoction:

Composition per liter:

Potato	20.0g

Preparation of Potato Decoction: Peel and dice potato. Add to 1.0L of distilled/deionized water. Gently heat and bring to boiling. Continue boiling for 30 min. Filter through Whatman #1 filter paper. Bring volume of filtrate to 1.0L with distilled/deionized water.

Preparation of Medium: Add components to distilled/deionized water and bring volume to 1090.0mL. Mix thoroughly. Adjust pH to 6.8. Distribute into tubes or flasks. Autoclave for 15 min at 15 psi pressure–121°C.

Use: For the cultivation and maintenance of *Rhizobacter daucus.*

Diphasic Medium for Amoeba (Charcoal Agar Slants)

Composition per liter:

Agar slants	1.0L
Buffered saline overlay	1.0L

pH 7.4 ± 0.2 at 25°C

Agar Slants:

Composition per liter:

Agar	10.0g
Charcoal, activated	10.0g
Pancreatic digest of casein	5.0g
KH_2PO_4	4.0g
Na_2HPO_4	3.0g
Asparagine	2.0g
Sodium citrate	1.0g
Ferric ammonium citrate	0.1g
$MgSO_4 \cdot 7H_2O$	0.1g
Cholesterol solution	25.0mL
Glycerol	10.0mL

Cholesterol Solution:

Composition per 25mL:

Cholesterol	0.25g
Acetone	25.0mL

Preparation of Cholesterol Solution: Add cholesterol to 25.0mL of acetone. Mix thoroughly.

Preparation of Agar Slants: Add components, except agar, charcoal and cholesterol solution, to distilled/deionized water and bring volume to 1.0L. Mix thoroughly. Gently heat to dissolve. Do not boil. Add agar, charcoal and cholesterol solution. Mix thoroughly. Gently heat and bring to boiling. Distribute into tubes in 3.0mL volumes. Autoclave for 15 min at 15 psi pressure–121°C. Resuspend charcoal. Allow tubes to cool in a slanted position with short butts or no butts.

Buffered Saline Overlay:

Composition per liter:

NaCl	5.0g
Solution B	810.0mL
Solution A	190.0mL

Solution A:

Composition per liter:

KH_2PO_4, anhydrous	9.07g

Preparation of Medium: Add KH_2PO_4 to distilled/deionized water and bring volume to 1.0L. Mix thoroughly.

Solution B:

Composition per liter:

Na_2HPO_4, anhydrous	9.46g

Preparation of Solution B: Add Na_2HPO_4 to distilled/deionized water and bring volume to 1.0L. Mix thoroughly.

Preparation of Buffered Saline Overlay: Combine 810.0mL of solution A and 190.0mL of solution B. Add the NaCl. Mix thoroughly. Autoclave for 15 min at 15 psi pressure–121°C. Cool to 25°C. Store at 4°C.

Preparation of Medium: To each agar slant, aseptically add 3.0mL of sterile buffered saline overlay.

Use: For the cultivation and maintenance of *Amoebae* species.

DM Medium

Composition per liter:

Starch, soluble	5.0g
$MgSO_4 \cdot 7H_2O$	0.50g
K_2HPO_4	0.25g

Preparation of Medium: Add components to distilled/deionized water and bring volume to 1.0L. Mix thoroughly. Distribute into tubes or flasks. Autoclave for 15 min at 15 psi pressure–121°C.

Use: For the cultivation of myxobacteria.

DNase Agar

Composition per liter:

Tryptose	20.0g
Agar	12.0g
NaCl	5.0g
Deoxyribonucleic acid	2.0g

pH 7.3 ± 0.2 at 25°C

Source: This medium is available as a premixed powder from Oxoid Unipath.

Preparation of Medium: Add components to distilled/deionized water and bring volume to 1.0L. Mix thoroughly. Gently heat and bring to boiling. Distribute into tubes or flasks. Autoclave for 15 min at 15 psi pressure–121°C. Pour into sterile Petri dishes or leave in tubes.

Use: For the differentiation of microorganisms, especially *Staphylococcus* species and *Serratia marcescens*, based on their production of deoxyribonuclease.

DNase Medium

Composition per liter:

Agar	15.0g
Pancreatic digest of casein	10.0g
Peptic digest of animal tissue	10.0g
L-Arabinose	10.0g
NaCl	5.0g
Deoxyribonucleic acid	2.0g
Methyl green	0.09g
Phenol red	0.05g
Antibiotic solution	10.0mL

pH 7.3 ± 0.2 at 25°C

Antibiotic Solution:

Composition per 10mL:

Cephalothin	0.01g
Ampicillin	5.0mg
Colistimethate	5.0mg
Amphotericin B	2.5mg

Preparation of Antibiotic Solution: Add components to distilled/deionized water and bring volume to 10.0mL. Mix thoroughly. Filter sterilize.

Preparation of Medium: Add components, except antibiotic solution, to distilled/deionized water and bring volume to 990.0mL. Mix thoroughly. Gently heat and bring to boiling. Autoclave for 15 min at 15 psi pressure–121°C. Cool to 45–50°C. Aseptically add sterile components. Mix thoroughly. Pour into sterile Petri dishes or distribute into sterile tubes.

Use: For the isolation and cultivation of *Serratia marcescens*.

DNase Test Agar

Composition per liter:

Agar	15.0g
Pancreatic digest of casein	15.0g
NaCl	5.0g
Papaic digest of soybean meal	5.0g
Deoxyribonucleic acid	2.0g

pH 7.3 ± 0.2 at 25°C

Source: This medium is available as a premixed powder from BBL Microbiology Systems and Difco Laboratories.

Preparation of Medium: Add components to distilled/deionized water and bring volume to 1.0L. Mix thoroughly. Gently heat while stirring and bring to boiling. Distribute into tubes or flasks. Autoclave for 15 min at 13 psi pressure–118°C. Pour into sterile Petri dishes or leave in tubes.

Use: For the differentiation of microorganisms, especially *Staphylococcus* species and *Serratia marcescens*, based on their production of deoxyribonuclease.

DNase Test Agar with Methyl Green

Composition per liter:

Agar	15.0g
Pancreatic digest of casein	10.0g
Peptic digest of animal tissue	10.0g
NaCl	5.0g
Deoxyribonucleic acid	2.0g
Methyl green	0.05g

pH 7.3 ± 0.2 at 25°C

Source: This medium is available as a premixed powder from Difco Laboratories.

Preparation of Medium: Add components to distilled/deionized water and bring volume to 1.0L. Mix thoroughly. Gently heat while stirring and bring to boiling. Distribute into tubes or flasks. Autoclave for 15 min at 13 psi pressure–118°C. Pour into sterile Petri dishes or leave in tubes.

Use: For the differentiation of microorganisms, especially *Staphylococcus* species and *Serratia marcescens*, based on their production of deoxyribonuclease.

DNase Test Agar with Toluidine Blue

Composition per liter:

Agar	15.0g
Pancreatic digest of casein	10.0g
Peptic digest of animal tissue	10.0g
NaCl	5.0g
Deoxyribonucleic acid	2.0g
Toluidine blue	0.1g

pH 7.3 ± 0.2

Preparation of Medium: Add components to distilled/deionized water and bring volume to 1.0L. Mix thoroughly. Gently heat while stirring and bring to boiling. Distribute into tubes or flasks. Autoclave for 15 min at 13 psi pressure–118°C. Pour into sterile Petri dishes or leave in tubes.

Use: For the differentiation of microorganisms, especially *Staphylococcus* species and *Serratia marcescens*, based on their production of deoxyribonuclease.

DNB Medium

Composition per liter:

Nutrient broth	2.4g
Yeast extract	1.5g

Preparation of Medium: Add components to distilled/deionized water and bring volume to 1.0L. Mix thoroughly. Distribute into tubes or flasks. Autoclave for 15 min at 15 psi pressure–121°C.

Use: For the cultivation of *Bdellovibrio bacteriovorus* and ATCC strain 43826.

Dorset Egg Medium

Composition per liter:

Homogenized whole egg	950.0mL
Glycerol	50.0mL

pH 6.8–7.4 at 25°C

Source: This medium is available as a prepared medium from Difco Laboratories.

Homogenized Whole Egg:
Composition per liter:

Whole eggs	18–24

Preparation of Homogenized Whole Egg: Use fresh eggs, less than 1 week old. Scrub the shells with soap. Let stand in a soap solution for 30 min. Rinse in running water. Soak eggs in 70% ethanol for 15 min. Break the eggs into a sterile container. Homogenize by shaking. Filter through four layers of sterile cheesecloth into a sterile graduated cylinder. Measure out 1.0L.

Preparation of Medium: Filter sterilize glycerol. Combine glycerol and homogenized whole egg. Mix thoroughly. Distribute into sterile screw-capped tubes. Place tubes in a slanted position. Inspissate at 85°C (moist heat) for 45 min.

Use: For the maintenance of *Mycobacterium* species.

DTC Agar

Composition per liter:

Agar	20.0g
Pancreatic digest of casein	15.0g
NaCl	5.0g
Papaic digest of soybean meal	5.0g
Deoxyribonucleic acid	2.0g
Toluidine blue O	0.1g
Cephalothin solution	10.0mL

Cephalothin Solution:

Composition per 10mL:

Cephalothin	1.0g

Preparation of Cephalothin Solution: Add cephalothin to distilled/deionized water and bring volume to 10.0mL. Mix thoroughly. Filter sterilize.

Preparation of Medium: Add components, except cephalothin solution, to distilled/deionized water and bring volume to 990.0mL. Mix thoroughly. Gently heat and bring to boiling. Autoclave for 15 min at 15 psi pressure–121°C. Cool to 45–50°C. Aseptically add sterile cephalothin solution. Mix thoroughly. Pour into sterile Petri dishes.

Use: For the isolation and cultivation of *Serratia* species. *Serratia* appear as colonies with red halos.

Dubos Agar with Filter Paper

Composition per liter:

Agar	15.0g
K_2HPO_4	1.0g
KCl	0.5g
$MgSO_4 \cdot 7H_2O$	0.5g
$NaNO_3$	0.5g
$FeSO_4 \cdot 7H_2O$	0.01g

pH 7.2 ± 0.2 at 25°C

Preparation of Medium: Add components to distilled/deionized water and bring volume to 1.0L. Mix thoroughly. Gently heat and bring to boiling. Adjust pH to 7.2. Autoclave for 15 min at 15 psi pressure–121°C. Pour into sterile Petri dishes. Lay sterile strips of Whatman #1 filter paper on the surface of the agar.

Use: For the cultivation and maintenance of *Cytophaga hutchinsonii.*

E Agar (m–E Agar)

Composition per liter:

Yeast extract	30.0g
Agar	15.0g
NaCl	15.0g
Pancreatic digest of gelatin	10.0g
Esculin	1.0g
Nalidixic acid	0.25g
NaN_3	0.15g
Cycloheximide	0.05g
TTC solution	15.0mL

pH 7.1 ± 0.2 at 25°C

Source: This medium is available as a premixed powder from BBL Microbiology Systems and Difco Laboratories.

TTC Solution:

Composition per 15mL:

2,3,5-Triphenyltetrazolium chloride	0.15g

Preparation of TTC Solution: Add triphenyltetrazolium chloride to distilled/deionized water and bring volume to 15.0mL. Mix thoroughly. Filter sterilize.

Preparation of Medium: Add components, except TTC solution, to distilled/deionized water and bring volume to 1.0L. Mix thoroughly. Gently heat and bring to boiling. Autoclave for 15 min at 15 psi pressure–121°C. Cool to 45–50°C. Aseptically add sterile TTC solution. Mix thoroughly. Pour into sterile Petri dishes or distribute into sterile tubes.

Caution: Sodium azide is toxic. Azides also react with metals and disposal must be highly diluted.

Use: For the isolation, cultivation and enumeration of enterococci in water by the membrane filter method. It is used in conjunction with Esculin Iron Agar.

E Medium for Anaerobes

Composition per 100mL:

Glucose	0.05g
L-Cysteine·HCl·H_2O	0.05g
Maltose	0.05g
$(NH_4)_2SO_4$	0.05g
Peptone	0.05g
Soluble starch	0.05g
Yeast extract	0.05g
Salts solution	50.0mL
Rumen fluid	30.0mL
Resazurin solution	0.4mL

pH 7.0 ± 0.2 at 25°C

Salts Solution:

Composition per liter:

$NaHCO_3$	10.0g
NaCl	2.0g
K_2HPO_4	1.0g
KH_2PO_4	1.0g
$CaCl_2$, anhydrous	0.2g
$MgSO_4$	0.2g

Preparation of Salts Solution: Add $CaCl_2$ and $MgSO_4$ to approximately 300.0mL of distilled/deionized water. Mix thoroughly. Bring volume to 800.0mL with distilled/deionized water. Add remaining components. Mix thoroughly. Bring volume to 1.0L with distilled/deionized water. Mix thoroughly. Store at 4°C.

Rumen Fluid:

Composition per 100mL:

Rumen fluid	100.0mL

Preparation of Rumen Fluid: Obtain the rumen contents from a cow that has been fed an alfalfa-hay ration. Filter rumen contents through two layers of cheesecloth. Store under 100% CO_2 in the refrigerator. The particulate material will settle out. Use only the supernatant liquid.

Resazurin Solution:

Composition per 44mL:

Resazurin....................0.011g

Preparation of Resazurin Solution: Add resazurin to distilled/deionized water and bring volume to 44.0mL. Mix thoroughly.

Preparation of Medium: Add components, except cysteine·HCl·H_2O, to distilled/deionized water and bring volume to 100.0mL. Mix thoroughly. Gently heat and bring to boiling. Continue boiling until resazurin turns colorless indicating reduction. Cool in an ice-water bath under 100% CO_2. Add the cysteine·HCl·H_2O. Adjust pH to 7.0 with 8*N* NaOH or 5*N* HCl. Anaerobically distribute into tubes under O_2-free 100% N_2. Cap tubes with butyl rubber stoppers. Place tubes in a press. Autoclave for 12 min at 15 psi pressure–121°C with fast exhaust.

Use: For the cultivation and maintenance of *Bacteroides ruminicola, Bacteroides succinogenes, Butyrivibrio fibrisolvens, Clostridium methylpentosum, Eubacterium ruminantium, Lachnospira multipara, Micromonospora ruminantium, Prevotella ruminicola, Propionibacterium acidipropionici, Selenomonas ruminantium, Selenomonas suis, Succinivibrio dextrinosolvens, Treponema bryantii,* and *Treponema succinifaciens.*

E Medium for Anaerobes with 0.1% Cellobiose

Composition per 100mL:

Cellobiose....................0.1g
Glucose....................0.05g
L-Cysteine·HCl·H_2O....................0.05g
Maltose....................0.05g
$(NH_4)_2SO_4$....................0.05g
Peptone....................0.05g
Soluble starch....................0.05g
Yeast extract....................0.05g
Salts solution....................50.0mL
Rumen fluid....................30.0mL
Resazurin solution....................0.4mL

pH 7.0 ± 0.2 at 25°C

Salts Solution:

Composition per liter:

$NaHCO_3$....................10.0g
NaCl....................2.0g
K_2HPO_4....................1.0g
KH_2PO_4....................1.0g
$CaCl_2$, anhydrous....................0.2g
$MgSO_4$....................0.2g

Preparation of Salts Solution: Add $CaCl_2$ and $MgSO_4$ to approximately 300.0mL of distilled/deionized water. Mix thoroughly. Bring volume to 800.0mL with distilled/deionized water. Add remaining components. Mix thoroughly. Bring volume to 1.0L with distilled/deionized water. Mix thoroughly. Store at 4°C.

Rumen Fluid:

Composition per 100mL:

Rumen fluid....................100.0mL

Preparation of Rumen Fluid: Obtain the rumen contents from a cow that has been fed an alfalfa-hay ration. Filter rumen contents through two layers of cheesecloth. Store under 100% CO_2 in the refrigerator. The particulate material will settle out. Use only the supernatant liquid.

Resazurin Solution:

Composition per 44mL:

Resazurin....................0.011g

Preparation of Resazurin Solution: Add resazurin to distilled/deionized water and bring volume to 44.0mL. Mix thoroughly.

Preparation of Medium: Add components, except cysteine·HCl·H_2O, to distilled/deionized water and bring volume to 100.0mL. Mix thoroughly. Gently heat and bring to boiling. Continue boiling until resazurin turns colorless indicating reduction. Cool in an ice-water bath under 100% CO_2. Add the cysteine·HCl·H_2O. Adjust pH to 7.0 with 8*N* NaOH or 5*N* HCl. Anaerobically distribute into tubes under O_2-free 100% N_2. Cap tubes with butyl rubber stoppers. Place tubes in a press. Autoclave for 12 min at 15 psi pressure–121°C with fast exhaust.

Use: For the cultivation and maintenance of *Eubacterium cellulosolvens* and *Fibrobacter inyrdyinslid.*

E Medium for Anaerobes with Filtered Rumen Fluid and 0.1% Cellobiose

Composition per 100mL:

Cellobiose....................0.1g
Glucose....................0.05g
L-Cysteine·HCl·H_2O....................0.05g
Maltose....................0.05g
$(NH_4)_2SO_4$....................0.05g
Peptone....................0.05g

Soluble starch..0.05g
Yeast extract..0.05g
Salts solution 50.0mL
Rumen fluid, filtered 30.0mL
Resazurin solution 0.4mL

pH 7.0 ± 0.2 at 25°C

Salts Solution:

Composition per liter:

$NaHCO_3$..10.0g
NaCl..2.0g
K_2HPO_4..1.0g
KH_2PO_4..1.0g
$CaCl_2$ (anhydrous)...0.2g
$MgSO_4$...0.2g

Preparation of Salts Solution: Add $CaCl_2$ and $MgSO_4$ to approximately 300.0mL of distilled/deionized water. Mix thoroughly. Bring volume to 800.0mL with distilled/deionized water. Add remaining components. Mix thoroughly. Bring volume to 1.0L with distilled/deionized water. Mix thoroughly. Store at 4°C.

Rumen Fluid:

Composition per 100mL:

Rumen fluid..100.0mL

Preparation of Rumen Fluid: Obtain the rumen contents from a cow that has been fed an alfalfa-hay ration. Filter rumen contents through two layers of cheesecloth. Store under 100% CO_2 in the refrigerator. The particulate material will settle out. Use only the supernatant liquid. Filter through a 0.20μm filter.

Resazurin Solution:

Composition per 44mL:

Resazurin..0.011g

Preparation of Resazurin Solution: Add resazurin to distilled/deionized water and bring volume to 44.0mL. Mix thoroughly.

Preparation of Medium: Add components, except cysteine·HCl·H_2O, to distilled/deionized water and bring volume to 100.0mL. Mix thoroughly. Gently heat and bring to boiling. Continue boiling until resazurin turns colorless indicating reduction. Cool in an ice-water bath under 100% CO_2. Add the cysteine·HCl·H_2O. Adjust pH to 7.0 with 8*N* NaOH or 5*N* HCl. Anaerobically distribute into tubes under O_2-free 100% N_2. Cap tubes with butyl rubber stoppers. Place tubes in a press. Autoclave for 12 min at 15 psi pressure–121°C with fast exhaust.

Use: For the cultivation and maintenance of the *Fibrobacter* species.

E Medium for Anaerobes with 0.3% Phloroglucinol

Composition per 110.4mL:

$(NH_4)_2SO_4$...0.5g
L-Cysteine·HCl·H_2O...0.05g
Soluble starch..0.05g
Salts solution 50.0mL
Rumen fluid 30.0mL
Phloroglucinol solution.....................................30.0mL
Resazurin solution 0.4mL

pH 6.6 ± 0.2 at 25°C

Salts Solution:

Composition per liter:

$NaHCO_3$..10.0g
NaCl..2.0g
K_2HPO_4..1.0g
KH_2PO_4..1.0g
$CaCl_2$, anhydrous 0.2g
$MgSO_4$...0.2g

Preparation of Salts Solution: Add $CaCl_2$ and $MgSO_4$ to approximately 300.0mL of distilled/deionized water. Mix thoroughly. Bring volume to 800.0mL with distilled/deionized water. Add remaining components. Mix thoroughly. Bring volume to 1.0L with distilled/deionized water. Mix thoroughly. Store at 4°C.

Rumen Fluid:

Composition per 100mL:

Rumen fluid..100.0mL

Preparation of Rumen Fluid: Obtain the rumen contents from a cow that has been fed an alfalfa-hay ration. Filter rumen contents through two layers of cheesecloth. Store under 100% CO_2 in the refrigerator. The particulate material will settle out. Use only the supernatant liquid.

Phloroglucinol Solution:

Composition per 100mL:

Phloroglucinol..1.0g

Preparation of Phloroglucinol Solution: Add phloroglucinol to distilled/deionized water and bring volume to 100.0mL. Mix thoroughly. Filter sterilize. Keep away from light.

Resazurin Solution:

Composition per 44mL:

Resazurin..0.011g

Preparation of Resazurin Solution: Add resazurin to distilled/deionized water and bring volume to 44.0mL. Mix thoroughly.

Preparation of Medium: Add components, except cysteine·HCl·H_2O, to distilled/deionized water and bring volume to 100.0mL. Mix thoroughly.

Gently heat and bring to boiling. Continue boiling until resazurin turns colorless indicating reduction. Cool in an ice-water bath under 100% CO_2. Add the cysteine·HCl·H_2O. Adjust pH to 6.6 with 8*N* NaOH or 5*N* HCl. Anaerobically distribute into tubes under O_2-free 100% N_2. Cap tubes with butyl rubber stoppers. Place tubes in a press. Autoclave for 12 min at 15 psi pressure–121°C with fast exhaust.

Use: For the cultivation and maintenance of *Coprococcus* species.

E Medium for Anaerobes with 0.2% Rutin

Composition per 110.4mL:

$(NH_4)_2SO_4$	0.5g
L-Cysteine·HCl·H_2O	0.05g
Soluble starch	0.05g
Salts solution	50.0mL
Rumen fluid	30.0mL
Rutin solution	30.0mL
Resazurin solution	0.4mL

pH 6.6 ± 0.2 at 25°C

Salts Solution:
Composition per liter:

$NaHCO_3$	10.0g
NaCl	2.0g
K_2HPO_4	1.0g
KH_2PO_4	1.0g
$CaCl_2$, anhydrous	0.2g
$MgSO_4$	0.2g

Preparation of Salts Solution: Add $CaCl_2$ and $MgSO_4$ to approximately 300.0mL of distilled/deionized water. Mix thoroughly. Bring volume to 800.0mL with distilled/deionized water. Add remaining components. Mix thoroughly. Bring volume to 1.0L with distilled/deionized water. Mix thoroughly. Store at 4°C.

Rumen Fluid:
Composition per 100mL:

Rumen fluid	100.0mL

Preparation of Rumen Fluid: Obtain the rumen contents from a cow that has been fed an alfalfa-hay ration. Filter rumen contents through two layers of cheesecloth. Store under 100% CO_2 at 4°C. The particulate material will settle out. Use the liquid.

Rutin Solution:
Composition per 100mL:

Rutin	0.2g

Preparation of Rutin Solution: Add rutin to distilled/deionized water and bring volume to 100.0mL. Mix thoroughly. Filter sterilize.

Resazurin Solution:
Composition per 44mL:

Resazurin	0.011g

Preparation of Resazurin Solution: Add resazurin to distilled/deionized water and bring volume to 44.0mL. Mix thoroughly.

Preparation of Medium: Add components, except cysteine·HCl·H_2O, to distilled/deionized water and bring volume to 100.0mL. Mix thoroughly. Gently heat and bring to boiling. Continue boiling until resazurin turns colorless indicating reduction. Cool in an ice-water bath under 100% CO_2. Add the cysteine·HCl·H_2O. Adjust pH to 6.6 with 8*N* NaOH or 5*N* HCl. Anaerobically distribute into tubes under O_2-free 100% N_2. Cap tubes with butyl rubber stoppers. Place tubes in a press. Autoclave for 12 min at 15 psi pressure–121°C with fast exhaust.

Use: For the cultivation of *Butyrivibrio* species.

E Medium for Anaerobes, Modified

Composition per 103.6mL:

L-Cysteine·HCl·H_2O	0.05g
$(NH_4)_2SO_4$	0.05g
Peptone	0.05g
Yeast extract	0.05g
Salts solution	50.0mL
Rumen fluid	30.0mL
Potassium phosphate buffer (1*M*, pH 6.5)	2.8mL
Hemin solution	1.0mL
Glucose-maltose solution	1.4mL
Starch solution	1.4mL
Resazurin (0.025% solution)	0.4mL
Vitamin K_3 solution	0.2mL

pH 6.5 ± 0.2 at 25°C

Salts Solution:
Composition per liter:

$NaHCO_3$	10.0g
NaCl	2.0g
K_2HPO_4	1.0g
KH_2PO_4	1.0g
$CaCl_2$, anhydrous	0.2g
$MgSO_4$	0.2g

Preparation of Salts Solution: Add $CaCl_2$ and $MgSO_4$ to approximately 300.0mL of distilled/deionized water. Mix thoroughly. Bring volume to 800.0mL with distilled/deionized water. Add remaining components. Mix thoroughly. Bring volume to 1.0L with distilled/deionized water. Mix thoroughly. Store at 4°C.

Rumen Fluid:
Composition per 100mL:

Rumen fluid	100.0mL

Preparation of Rumen Fluid: Obtain the rumen contents from a cow that has been fed an alfalfa-hay ration. Filter rumen contents through two layers of cheesecloth. Store under 100% CO_2 in the refrigerator. The particulate material will settle out. Use only the supernatant liquid.

Hemin Solution:
Composition per 100mL:

Hemin	0.050g
NaOH (1*N* solution)	1.0mL

Preparation of Hemin Solution: Add hemin to 1.0mL of 1*N* NaOH solution. Mix thoroughly. Bring volume to 100.0mL with distilled/deionized water. Autoclave for 15 min at 15 psi pressure–121°C. Cool to 45–50°C.

Glucose-Maltose Solution:
Composition per 10mL:

Glucose	0.5g
Maltose	0.5g

Preparation of Glucose-Maltose Solution: Add components to distilled/deionized water and bring volume to 10.0mL. Mix thoroughly. Filter sterilize.

Starch Solution:
Composition per 10mL:

Starch, soluble	0.5g

Preparation of Starch Solution: Add starch to distilled/deionized water and bring volume to 10.0mL. Mix thoroughly. Autoclave for 15 min at 15 psi pressure–121°C.

Resazurin Solution:
Composition per 44mL:

Resazurin	0.011g

Preparation of Resazurin Solution: Add resazurin to distilled/deionized water and bring volume to 44.0mL. Mix thoroughly.

Vitamin K_3 Solution:
Composition per 25mL:

Vitamin K_3 (menadione)	0.0125g
Ethanol, absolute	25.0mL

Preparation of Vitamin K_1 Solution: Add vitamin K_3 to 99.0mL of ethanol. Mix thoroughly.

Preparation of Medium: Filter sterilize potassium phosphate buffer. Add components—except cysteine·HCl·H_2O, vitamin K_3 solution, potassium phosphate buffer, glucose-maltose solution, and starch solution—to distilled/deionized water and bring volume to 98.0mL. Mix thoroughly. Gently heat and bring to boiling. Continue boiling until resazurin turns colorless indicating reduction. Cool in an ice-water bath under O_2-free 97% N_2 + 3% H_2. Add the cysteine·HCl·H_2O and vitamin K_3 solution. Mix thoroughly. Adjust pH to 6.5 with 8*N* NaOH or 5*N* HCl. Anaerobically distribute into tubes under O_2-free 97% N_2 + 3% H_2 in 7.0mL volumes. Cap tubes with butyl rubber stoppers. Place tubes in a press. Autoclave for 12 min at 15 psi pressure–121°C with fast exhaust. Immediately prior to inoculation, aseptically add 0.2mL of filter-sterilized potassium phosphate buffer, 0.1mL of sterile glucose-maltose solution, and 0.1mL of sterile starch solution to each tube. Mix thoroughly.

Use: For the cultivation and maintenance of *Bacteroides* species, *Butyrivibrio fibrisolvens, Clostridium methylpentosum, Eubacterium ruminantium, Lachnospira multipara, Micromonospora ruminantium, Prevotella ruminicola, Propionibacterium acidipropionici, Selenomonas* species, *Succinivibrio dextrinosolvens*, and *Treponema* species.

EC Broth (*Escherichia coli* Broth) (EC Medium)

Composition per liter:

Pancreatic digest of casein	20.0g
Lactose	5.0g
NaCl	5.0g
K_2HPO_4	4.0g
Bile salts mixture	1.5g
KH_2PO_4	1.5g

pH 6.9 ± 0.2 at 25°C

Source: This medium is available as a premixed powder from BBL Microbiology Systems and Difco Laboratories.

Preparation of Medium: Add components to distilled/deionized water and bring volume to 1.0L. Mix thoroughly. Distribute into test tubes that contain an inverted Durham tube. Autoclave for 12 min at 15 psi pressure–121°C. Cool broth as quickly as possible.

Use: For the cultivation and differentiation of coliform bacteria at 37°C and of *Escherichia coli* at 45.5°C.

EC Broth with MUG

Composition per liter:

Pancreatic digest of casein	20.0g
Lactose	5.0g
NaCl	5.0g
K_2HPO_4	4.0g
Bile salts mixture	1.5g
KH_2PO_4	1.5g
4-Methylumbeliferyl-β-D-glucuronide (MUG)	0.05g

pH 6.9 ± 0.2 at 25°C

Source: This medium is available as a premixed powder from BBL Microbiology Systems.

Preparation of Medium: Add components to distilled/deionized water and bring volume to 1.0L. Mix thoroughly. Distribute into test tubes that contain an inverted Durham tube in 10.0mL volumes. Autoclave for 15 min at 15 psi pressure–121°C.

Use: For the detection of *Escherichia coli* in water and food samples by a fluorogenic procedure.

ECM Agar

Composition per liter:

Agar	15.0g
NaCl	6.0g
Escherichia coli cells, washed	1.0g
$MgSO_4·7H_2O$	0.50g

Preparation of Medium: Add components to distilled/deionized water and bring volume to 1.0L. Mix thoroughly. Gently heat and bring to boiling. Distribute into tubes or flasks. Autoclave for 15 min at 15 psi pressure–121°C. Pour into sterile Petri dishes or leave in tubes.

Use: For the cultivation of myxobacteria.

Ectothiorhodospira halochloris Medium

Composition per liter:

NaCl	180.0g
Na_2SO_4	20.0g
$NaHCO_3$	14.0g
Na_2CO_3	6.0g
$Na_2S·9H_2O$	1.0g
Sodium succinate	1.0g
NH_4Cl	0.8g
KH_2PO_4	0.5g
Yeast extract	0.5g
$MgCl_2·6H_2O$	0.10g
$CaCl_2·2H_2O$	0.05g
Vitamin solution VA	1.0mL
Trace element solution SLA	1.0mL

pH 8.5–8.7 at 25°C

Vitamin Solution VA:
Composition per liter:

Nicotinamide	0.04g
Thiamine dichloride	0.03g
p-Aminobenzoic acid	0.02g
Biotin	0.01g
Calcium pantothenate	0.01g
Pyridoxal chloride	0.01g
Vitamin B_{12}	5.0mg

Preparation of Vitamin Solution VA: Add components to distilled/deionized water and bring volume to 1.0L. Mix thoroughly.

Trace Element Solution SLA:
Composition per liter:

$FeCl_2·4H_2O$	1.8g
H_3BO_3	0.5g
$CoCl_2·6H_2O$	0.25g
$ZnCl_2$	0.10g
$MnCl_2·4H_2O$	0.07g
$NaMoO_4·2H_2O$	0.03g
$CuCl_2·2H_2O$	0.01g
Na_2SeO_3	0.01g
$NiCl_2·6H_2O$	0.01g

Preparation of Trace Element Solution SLA: Add components to distilled/deionized water and bring volume to 1.0L. Mix thoroughly. Adjust pH to 3.0 with 2*N* HCl.

Preparation of Medium: Add components, except trace element solution SLA, to distilled/deionized water and bring volume to 999.0mL. Mix thoroughly. Filter sterilize. Aseptically add 1.0mL of sterile trace element solution SLA. Mix thoroughly. Aseptically distribute into flasks or bottles. Completely fill bottles with medium except for a pea-sized air bubble.

Use: For the enrichment and isolation of *Ectothiorhodospira halochloris*.

Ectothiorhodospira halophila Medium

Composition per liter:

NaCl	200.0g
NH_4Cl	0.4g
$(NH_4)_2SO_4$	0.1g
Na_2CO_3 solution	100.0mL
Tris buffer (1*M* solution, pH 7.5)	30.0mL
Solution C	5.0mL
Potassium phosphate buffer (1*M* solution, pH 7.5)	3.0mL
Additional solution	2.5mL

pH 7.4–8.0 at 25°C

Na_2CO_3 Solution:
Composition per 100mL:

Na_2CO_3	10.0g

Preparation of Na_2CO_3 Solution: Add Na_2CO_3 to distilled/deionized water and bring volume to 100.0mL. Mix thoroughly. Autoclave for 15 min at 15 psi pressure–121°C. Cool to 25°C.

Solution C:

Composition per liter:

$MgCl_2·6H_2O$	24.0g
$CaCl_2·2H_2O$	3.3g
$FeCl_3·4H_2O$	1.1g
$(NH_4)_6Mo_7O_{24}·4H_2O$	0.1g
Nitrilotriacetic acid	10mg
Trace elements solution	50.0mL

Preparation of Solution C: Add nitrilotriacetic acid to 500.0mL of distilled/deionized water. Dissolve by adjusting pH to 6.5 with KOH. Add remaining components. Readjust pH to 7.2 with H_2SO_4 or KOH. Add distilled/deionized water to 1.0L.

Trace Elements Solution:

Composition per 100mL:

$ZnCl_2$	0.52g
EDTA	0.25g
$MnCl_2·4H_2O$	0.08g
$FeCl_3·4H_2O$	0.03g
$Co(NO_3)_2·6H_2O$	0.02g
$CuCl_2·2H_2O$	0.02g
H_3BO_3	0.01g

Preparation of Trace Elements Solution: Add components to distilled/deionized water and bring volume to 1.0L. Mix thoroughly. Adjust pH to 3.0 with 2*N* HCl.

Additional Solution:

Composition per 50mL:

$NaS_2O_3·6H_2O$	6.0g
Sodium succinate	5.0g
Sodium ascorbate	1.0g

Preparation of Additional Solution: Add components to distilled/deionized water and bring volume to 50.0mL. Mix thoroughly. Filter sterilize.

Preparation of Medium: Add components, except Na_2CO_3 solution and additional solution, to distilled/deionized water and bring volume to 900.0mL. Autoclave for 15 min at 15 psi pressure–121°C. Cool to 45–50°C. Aseptically adjust pH to 7.4–7.8 with filter-sterilized HCl. Aseptically distribute into 50.0mL screw-capped bottles. Fill each bottle almost to the top, leaving a space of 2.8mL in the neck. Aseptically add 2.5mL of sterile additional solution to each bottle. Mix thoroughly.

Use: For the isolation and cultivation of *Ectothiorhodospira halophila*.

Ectothiorhodospira halophila Medium

Composition per liter:

NaCl	220.0g
Potassium succinate	1.0g
$Na_2S·9H_2O$	0.1g
K_2HPO_4 solution	20.0mL
$NaHCO_3$ solution	20.0mL
Solution C	20.0mL
$(NH_4)_2SO_4$ solution	5.0mL
Vitamin solution	0.5mL

pH 7.4–8.0 at 25°C

K_2HPO_4 Solution:

Composition per liter:

K_2HPO_4	125.0g

Preparation of K_2HPO_4 Solution: Add K_2HPO_4 to distilled/deionized water and bring volume to 1.0L. Mix thoroughly.

$NaHCO_3$ Solution:

Composition per liter:

$NaHCO_3$	100.0g

Preparation of $NaHCO_3$ Solution: Add $NaHCO_3$ to distilled/deionized water and bring volume to 1.0L. Mix thoroughly.

Solution C:

Composition per liter:

$MgCl_2·6H_2O$	24.0g
$CaCl_2·2H_2O$	3.3g
$FeCl_3·4H_2O$	1.1g
$(NH_4)_6Mo_7O_{24}·4H_2O$	0.1g
Nitrilotriacetic acid	10.0mg
Trace elements solution	50.0mL

Preparation of Solution C: Add components to distilled/deionized water and bring volume to 1.0L. Mix thoroughly.

Trace Elements Solution:

Composition per 100mL:

$ZnCl_2$	0.52g
EDTA	0.25g
$MnCl_2·4H_2O$	0.08g
$FeCl_3·4H_2O$	0.03g
$Co(NO_3)_2·6H_2O$	0.02g
$CuCl_2·2H_2O$	0.02g
H_3BO_3	0.01g

Preparation of Trace Elements Solution: Add components to distilled/deionized water and bring volume to 1.0L. Mix thoroughly. Adjust pH to 3.0 with 2*N* HCl.

$(NH_4)_2SO_4$ Solution:

Composition per liter:

$(NH_4)_2SO_4$	100.0g

Preparation of $(NH_4)_2SO_4$ Solution: Add components to distilled/deionized water and bring volume to 1.0L. Mix thoroughly.

Vitamin Solution:

Composition per liter:

Nicotinic acid	2.0mg
Thiamine	1.0mg
p-Aminobenzoic acid	0.2mg
Biotin	0.02mg
Vitamin B_{12}	1.0μg

Preparation of Vitamin Solution: Add components to distilled/deionized water and bring volume to 1.0L. Mix thoroughly.

Preparation of Medium: Add components to distilled/deionized water and bring volume to 1.0L. Mix thoroughly. Adjust pH to 7.4–8.0. Filter sterilize. Aseptically distribute into flasks or bottles. Completely fill bottles with medium except for a pea-sized air bubble.

Use: For the isolation and cultivation of *Ectothiorhodospira halophila.*

Ectothiorhodospira Medium

Composition per liter:

NaCl	180.0g
Na_2SO_4	20.0g
$NaHCO_3$	14.0g
Na_2CO_3	6.0g
Sodium succinate	1.0g
NH_4Cl	0.8g
KH_2PO_4	0.5g
$MgCl_2{\cdot}6H_2O$	0.10g
$CaCl_2{\cdot}2H_2O$	0.05g
Feeding solution	10.0mL
Trace element solution SLA	1.0mL
Vitamin solution VA	1.0mL

pH 8.5–8.7 at 25°C

Feeding Solution:

Composition per 100mL:

NaCl	10.0g
$NaHCO_3$	10.0g
$Na_2S{\cdot}9H_2O$	5.0g

Preparation of Feeding Solution: Add components to distilled/deionized water and bring volume to 100.0mL. Mix thoroughly. Filter sterilize.

Trace Element Solution SLA:

Composition per liter:

$FeCl_2{\cdot}4H_2O$	1.8g
H_3BO_3	0.5g
$CoCl_2{\cdot}6H_2O$	0.25g
$ZnCl_2$	0.1g
$MnCl_2{\cdot}4H_2O$	0.07g
$NaMoO_4{\cdot}2H_2O$	0.03g
$CuCl_2{\cdot}2H_2O$	0.01g
Na_2SeO_3	0.01g
$NiCl_2{\cdot}6H_2O$	0.01g

Preparation of Trace Element Solution SLA: Add components to distilled/deionized water and bring volume to 1.0L. Mix thoroughly. Adjust pH to 3.0 with 2*N* HCl.

Vitamin Solution VA:

Composition per liter:

Nicotinamide	0.04g
Thiamine dichloride	0.03g
p-Aminobenzoic acid	0.02g
Biotin	0.01g
Calcium pantothenate	0.01g
Pyridoxal chloride	0.01g
Vitamin B_{12}	5.0mg

Preparation of Vitamin Solution VA: Add components to distilled/deionized water and bring volume to 1.0L. Mix thoroughly.

Preparation of Medium: Add components, except trace element solution SLA and feeding solution, to distilled/deionized water and bring volume to 999.0mL. Mix thoroughly. Filter sterilize. Aseptically add 1.0mL of sterile trace element solution SLA. Mix thoroughly. Aseptically distribute into flasks or bottles. Completely fill bottles with medium except for a pea-sized air bubble. Prior to inoculation aseptically remove sufficient amount of medium to permit the addition of feeding medium. Add 1.0mL of feeding solution per each 100.0mL of medium.

Use: For the isolation and cultivation of *Ectothiorhodospira halochloris* and *E. halophila.*

Ectothiorhodospira Medium

Composition per liter:

NaCl	130.0g
Na_2SO_4	10.0g
Sodium acetate	2.0g
KH_2PO_4	0.8g
Sodium carbonate buffer, (1*M*, pH 9.0)	200.0mL
$MgCl_2{\cdot}6H_2O$ solution	10.0mL
$Na_2S{\cdot}9H_2O$ solution	6.0mL
$CaCl_2{\cdot}2H_2O$ solution	5.0mL
NH_4Cl solution	4.0mL
SLA trace elements	1.0mL
VA vitamin solution	1.0mL

pH 9.0 ± 0.2 at 25°C

$MgCl_2{\cdot}6H_2O$ Solution:

Composition per 10mL:

$MgCl_2{\cdot}6H_2O$	0.1g

Preparation of $MgCl_2{\cdot}6H_2O$ Solution: Add $MgCl_2{\cdot}6H_2O$ to distilled/deionized water and bring volume to 10.0mL. Mix thoroughly. Filter sterilize.

$Na_2S{\cdot}9H_2O$ Solution:

Composition per 10mL:

$Na_2S{\cdot}9H_2O$ 0.5g

Preparation of $Na_2S{\cdot}9H_2O$ Solution: Add $Na_2S{\cdot}9H_2O$ to distilled/deionized water and bring volume to 10.0mL. Mix thoroughly. Filter sterilize. Use freshly prepared solution.

$CaCl_2{\cdot}2H_2O$ Solution:

Composition per 10mL:

$CaCl_2{\cdot}2H_2O$ 0.1g

Preparation of $CaCl_2{\cdot}2H_2O$ Solution: Add $CaCl_2{\cdot}2H_2O$ to distilled/deionized water and bring volume to 10.0mL. Mix thoroughly. Filter sterilize.

NH_4Cl Solution:

Composition per 10mL:

NH_4Cl 2.0g

Preparation of NH_4Cl Solution: Add NH_4Cl to distilled/deionized water and bring volume to 10.0mL. Mix thoroughly. Filter sterilize.

SLA Trace Elements:

Composition per liter:

$FeCl_2{\cdot}4H_2O$	1.8g
H_3BO_3	0.5g
$CoCl_2{\cdot}6H_2O$	0.25g
$ZnCl_2$	0.1g
$MnCl_2{\cdot}4H_2O$	0.07g
$Na_2MoO_4{\cdot}2H_2O$	0.03g
$CuCl_2{\cdot}2H_2O$	0.01g
$Na_2SeO_3{\cdot}5H_2O$	0.01g
$NiCl_2{\cdot}6H_2O$	0.01g

Preparation of SLA Trace Elements: Add components to distilled/deionized water and bring volume to 1.0L. Mix thoroughly. Adjust pH to 2–3.

VA Vitamin Solution:

Composition per 500mL:

Nicotinamide	0.175g
Thiamine·HCl	0.15g
p-Aminobenzoic acid	0.10g
Biotin	0.05g
Calcium pantothenate	0.05g
Pyridoxine·2HCl	0.05g
Cyanocobalamin	0.025g

Preparation of VA Vitamin Solution: Add components to distilled/deionized water and bring volume to 500.0mL. Mix thoroughly.

Preparation of Medium: Add components—except $MgCl_2{\cdot}6H_2O$ solution, $Na_2S{\cdot}9H_2O$ solution, $CaCl_2{\cdot}2H_2O$ solution, and NH_4Cl solution—to distilled/deionized water and bring volume to 975.0mL. Mix thoroughly. Gently heat and bring to boiling. Autoclave for 15 min at 15 psi pressure–121°C. Cool to 25°C. Aseptically add 10.0mL of sterile $MgCl_2{\cdot}6H_2O$ solution, 6.0mL of sterile $Na_2S{\cdot}9H_2O$ solution, 5.0mL of sterile $CaCl_2{\cdot}2H_2O$ solution, and 4.0mL of sterile NH_4Cl solution. Mix thoroughly. Aseptically distribute into culture bottles. Incubate for 2 days before inoculation.

Use: For the cultivation and maintenance of *Ectothiorhodospira abdelmalekii* and *E. halochloris.*

Ectothiorhodospira Medium, Modified

Composition per liter:

NaCl	30.0g
Na_2SO_4	10.0g
Sodium acetate	2.0g
KH_2PO_4	0.8g
Sodium carbonate buffer (1*M*, pH 9.0)	200.0mL
$MgCl_2{\cdot}6H_2O$ solution	10.0mL
$Na_2S{\cdot}9H_2O$ solution	6.0mL
$CaCl_2{\cdot}2H_2O$ solution	5.0mL
NH_4Cl solution	4.0mL
SLA trace elements	1.0mL
VA vitamin solution	1.0mL

pH 9.0 ± 0.2 at 25°C

$MgCl_2{\cdot}6H_2O$ Solution:

Composition per 10mL:

$MgCl_2{\cdot}6H_2O$ 0.1g

Preparation of $MgCl_2{\cdot}6H_2O$ Solution: Add $MgCl_2{\cdot}6H_2O$ to distilled/deionized water and bring volume to 10.0mL. Mix thoroughly. Filter sterilize.

$Na_2S{\cdot}9H_2O$ Solution:

Composition per 10mL:

$Na_2S{\cdot}9H_2O$ 0.5g

Preparation of $Na_2S{\cdot}9H_2O$ Solution: Add $Na_2S{\cdot}9H_2O$ to distilled/deionized water and bring volume to 10.0mL. Mix thoroughly. Filter sterilize. Use freshly prepared solution.

$CaCl_2{\cdot}2H_2O$ Solution:

Composition per 10mL:

$CaCl_2{\cdot}2H_2O$ 0.1g

Preparation of $CaCl_2{\cdot}2H_2O$ Solution: Add $CaCl_2{\cdot}2H_2O$ to distilled/deionized water and bring volume to 10.0mL. Mix thoroughly. Filter sterilize.

NH_4Cl Solution:

Composition per 10mL:

NH_4Cl 2.0g

Preparation of NH_4Cl Solution: Add NH_4Cl to distilled/deionized water and bring volume to 10.0mL. Mix thoroughly. Filter sterilize.

SLA Trace Elements:

Composition per liter:

$FeCl_2 \cdot 4H_2O$	1.8g
H_3BO_3	0.5g
$CoCl_2 \cdot 6H_2O$	0.25g
$ZnCl_2$	0.1g
$MnCl_2 \cdot 4H_2O$	0.07g
$Na_2MoO_4 \cdot 2H_2O$	0.03g
$CuCl_2 \cdot 2H_2O$	0.01g
$Na_2SeO_3 \cdot 5H_2O$	0.01g
$NiCl_2 \cdot 6H_2O$	0.01g

Preparation of SLA Trace Elements: Add components to distilled/deionized water and bring volume to 1.0L. Mix thoroughly. Adjust pH to 2–3.

VA Vitamin Solution:

Composition per 500mL:

Nicotinamide	0.175g
Thiamine·HCl	0.15g
p-Aminobenzoic acid	0.1g
Biotin	0.05g
Calcium pantothenate	0.05g
Pyridoxine·2HCl	0.05g
Cyanocobalamin	0.025g

Preparation of VA Vitamin Solution: Add components to distilled/deionized water and bring volume to 500.0mL. Mix thoroughly.

Preparation of Medium: Add components—except $MgCl_2 \cdot 6H_2O$ solution, $Na_2S \cdot 9H_2O$ solution, $CaCl_2 \cdot 2H_2O$ solution, and NH_4Cl solution—to distilled/deionized water and bring volume to 975.0mL. Mix thoroughly. Gently heat and bring to boiling. Autoclave for 15 min at 15 psi pressure–121°C. Cool to 25°C. Aseptically add 10.0mL of sterile $MgCl_2 \cdot 6H_2O$ solution, 6.0mL of sterile $Na_2S \cdot 9H_2O$ solution, 5.0mL of sterile $CaCl_2 \cdot 2H_2O$ solution, and 4.0mL of sterile NH_4Cl solution. Mix thoroughly. Aseptically distribute into culture bottles. Incubate for 2 days before inoculation.

Use: For the cultivation and maintenance of *Ectothiorhodospira vacuotalta.*

Egg Yolk Agar

Composition per liter:

Proteose peptone No. 2	40.0g
Agar	25.0g
Na_2HPO_4	5.0g
Glucose	2.0g
NaCl	2.0g
KH_2PO_4	1.0g
$MgSO_4 \cdot 7H_2O$	0.1g
Egg yolk emulsion	100.0mL
Hemin solution	1.0mL

pH 7.6 ± 0.2 at 25°C

Hemin Solution:

Composition per 100mL:

Hemin	0.5g
NaOH (1*N* solution)	20.0mL

Preparation of Hemin Solution: Add hemin to 20.0mL of 1*N* NaOH solution. Mix thoroughly. Bring volume to 100.0mL with distilled/deionized water.

Egg Yolk Emulsion:

Composition:

Chicken egg yolks	11
Whole chicken egg	1

Preparation of Egg Yolk Emulsion: Soak eggs with 1:100 dilution of saturated mercuric chloride solution for 1 min. Crack eggs and separate yolks from whites. Mix egg yolks with 1 chicken egg.

Preparation of Medium: Add components, except egg yolk emulsion, to distilled/deionized water and bring volume to 900.0mL. Mix thoroughly. Gently heat and bring to boiling. Autoclave for 15 min at 15 psi pressure–121°C. Cool to 45–50°C. Aseptically add sterile egg yolk emulsion. Mix thoroughly. Pour into sterile Petri dishes.

Use: For the isolation, cultivation and differentiation of *Clostridium* species and some other anaerobic bacteria.

Egg Yolk Agar, Modified

Composition per liter:

Agar	20.0g
Pancreatic digest of casein	15.0g
Vitamin K_1	10.0g
NaCl	5.0g
Papaic digest of soybean meal	5.0g
Yeast extract	5.0g
L-Cystine	0.4g
Hemin	5.0mg
Egg yolk emulsion	100.0mL

Source: Available as a prepared medium from BBL Microbiology Systems.

Egg Yolk Emulsion:

Composition:

Chicken egg yolks	11
Whole chicken egg	1

Preparation of Egg Yolk Emulsion: Soak eggs with 1:100 dilution of saturated mercuric chloride solution for 1 min. Crack eggs and separate yolks from whites. Mix egg yolks with 1 chicken egg.

Preparation of Medium: Add components, except egg yolk emulsion, to distilled/deionized water and bring volume to 900.0mL. Mix thoroughly.

Gently heat and bring to boiling. Autoclave for 15 min at 15 psi pressure–121°C. Cool to 45–50°C. Aseptically add sterile egg yolk emulsion. Mix thoroughly. Pour into sterile Petri dishes.

Use: For the isolation, cultivation and differentiation of *Clostridium* species and some other anaerobic bacteria.

EIA Substrate

Composition per liter:

Agar	15.0g
Esculin	1.0g
Ferric citrate	0.5g

pH 7.1 ± 0.1 at 25°C

Source: This medium is available as a premixed powder from Difco Laboratories.

Preparation of Medium: Add components to distilled/deionized water and bring volume to 1.0L. Mix thoroughly. Gently heat and bring to boiling. Adjust pH to 7.1. Distribute into tubes or flasks. Autoclave for 15 min at 15 psi pressure–121°C. Pour into sterile Petri dishes.

Use: For the cultivation and enumeration of marine enterococci by the membrane filter method.

Eijkman Lactose Medium

Composition per liter:

Pancreatic digest of casein	15.0g
K_2HPO_4	10.0g
KH_2PO_4	4.0g
Lactose	3.0g
NaCl	2.5g

pH 6.8 ± 0.1 at 25°C

Preparation of Medium: Add components to distilled/deionized water and bring volume to 1.0L. Mix thoroughly. Distribute into test tubes that contain an inverted Durham tube. Autoclave for 15 min at 15 psi pressure–121°C.

Use: For the cultivation and differentiation of *Escherichia coli* from other coliform organisms based on their ability to ferment lactose and produce gas.

Eijkman Lactose Medium

Composition per liter:

Tryptose	15.0g
NaCl	5.0g
K_2HPO_4	4.0g
Lactose	3.0g
KH_2PO_4	1.5g

pH 6.8 ± 0.1 at 25°C

Preparation of Medium: Add components to distilled/deionized water and bring volume to 1.0L. Mix thoroughly. Distribute into test tubes that contain an inverted Durham tube. Autoclave for 15 min at 15 psi pressure–121°C.

Use: For the cultivation and differentiation of *Escherichia coli* from other coliform organisms based on their ability to ferment lactose and produce gas.

EMB Agar (Eosin Methylene Blue Agar)

Composition per liter:

Agar	13.5g
Pancreatic digest of casein	10.0g
Lactose	5.0g
Sucrose	5.0g
K_2HPO_4	2.0g
Eosin Y	0.4g
Methylene blue	0.065g

pH 7.2 ± 0.2 at 25°C

Source: This medium is available as a premixed powder from BBL Microbiology Systems and Difco Laboratories.

Preparation of Medium: Add components to distilled/deionized water and bring volume to 1.0L. Mix thoroughly. Gently heat and bring to boiling. Distribute into tubes or flasks. Autoclave for 15 min at 15 psi pressure–121°C. Pour into sterile Petri dishes.

Use: For the isolation, cultivation and differentiation of Gram-negative enteric bacteria based on lactose fermentation. Bacteria that ferment lactose, especially the coliform bacterium *Escherichia coli*, appear as colonies with a green metallic sheen or blue-black to brown color. Bacteria that do not ferment lactose appear as colorless or transparent light purple colonies.

EMB Agar Base

Composition per liter:

Agar	15.0g
Peptone	10.0g
K_2HPO_4	2.0g
Eosin Y	0.4g
Methylene blue	0.065g

pH 7.3 ± 0.2 at 25°C

Preparation of Medium: Add components to distilled/deionized water and bring volume to 1.0L. Mix thoroughly. Gently heat and bring to boiling. Distribute into tubes or flasks. Autoclave for 15 min at 15 psi pressure–121°C. Pour into sterile Petri dishes.

Use: For the isolation, cultivation and differentiation of Gram-negative enteric bacteria based on lactose fermentation. Bacteria that ferment lactose, especially the coliform bacterium *Escherichia coli*, appear as colonies with a green metallic sheen or blue-black to brown color. Bacteria that do not ferment lactose appear as colorless or transparent light purple colonies.

EMB Agar, Modified (Eosin Methylene Blue Agar, Modified)

Composition per liter:

Agar	15.0g
Lactose	10.0g
Pancreatic digest of gelatin	10.0g
K_2HPO_4	2.0g
Eosin Y	0.4g
Methylene blue	0.065g

pH 6.8 ± 0.2 at 25°C

Source: This medium is available as a premixed powder from Oxoid.

Preparation of Medium: Add components to distilled/deionized water and bring volume to 1.0L. Mix thoroughly. Gently heat and bring to boiling. Distribute into tubes or flasks. Autoclave for 15 min at 15 psi pressure–121°C. Cool to 60°C. Shake medium to oxidize methylene blue. Pour into sterile Petri dishes. Swirl flask while pouring plates to distribute precipitate.

Use: For the isolation, cultivation and differentiation of Gram-negative enteric bacteria based on lactose fermentation. Bacteria that ferment lactose, especially the coliform bacterium *Escherichia coli*, appear as colonies with a green metallic sheen or blue-black to brown color. Bacteria that do not ferment lactose appear as colorless or transparent light purple colonies.

Emerson Agar

Composition per liter:

Agar	20.0g
Glucose	10.0g
Beef extract	4.0g
Pancreatic digest of gelatin	4.0g
NaCl	2.5g
Yeast extract	1.0g

pH 7.0 ± 0.2 at 25°C

Source: This medium is available as a premixed powder from BBL Microbiology Systems.

Preparation of Medium: Add components to distilled/deionized water and bring volume to 1.0L. Mix thoroughly. Gently heat and bring to boiling. Distribute into tubes or flasks. Autoclave for 15 min at 13 psi pressure–118°C. Pour into sterile Petri dishes or leave in tubes.

Use: For the isolation, cultivation, and maintenance of members of the Actinomycetaceae, Streptomycetaceae and molds. For the cultivation and maintenance of *Arthrobacter* species, *Microbispora rosea, Micromonospora coerulea, Mycobacterium* species, *Nocardia asteroides, Nocardiopsis dassonvillei, Pseudonocardia thermophila, Staphylococcus epidermidis, Streptomyces flaveus, Streptomyces olivaceus, Streptomyces thermoviolaceus, Streptomyces thermovulgaris*, and *Streptomyces vendargensis*.

Emerson Yp Ss Agar

Composition per liter:

Agar	20.0g
Soluble starch	15.0g
Yeast extract	4.0g
K_2HPO_4	1.0g
$MgSO_4 \cdot 7H_2O$	0.5g

pH 7.0 ± 0.2 at 25°C

Source: This medium is available as a premixed powder from Difco Laboratories.

Preparation of Medium: Add components to distilled/deionized water and bring volume to 1.0L. Mix thoroughly. Gently heat and bring to boiling. Distribute into tubes or flasks. Autoclave for 15 min at 13 psi pressure–118°C. Pour into sterile Petri dishes or leave in tubes.

Use: For the cultivation and maintenance of *Allomyces* and other fungi.

Endo Agar

Composition per liter:

Agar	15.0g
Lactose	10.0g
Peptic digest of animal tissue	10.0g
K_2HPO_4	3.5g
Na_2SO_3	2.5g
Basic fuchsin	0.5g

pH 7.4 ± 0.2 at 25°C

Source: This medium is available as a premixed powder from BBL Microbiology Systems and Difco Laboratories.

Caution: Basic fuchsin is a potential carcinogen and care must be taken to avoid inhalation of the powdered dye and contact with the skin.

Preparation of Medium: Add components to distilled/deionized water and bring volume to 1.0L. Mix thoroughly. Gently heat and bring to boiling. Autoclave for 15 min at 15 psi pressure–121°C. Cool to 45–50°C. Pour into sterile Petri dishes. Swirl flask while pouring plates to keep precipitate in suspension. Protect from the light.

Use: For the selective isolation, cultivation and differentiation of coliform and other enteric microorganisms based on their ability to ferment lactose. Lactose-fermenting bacteria appear as dark red colonies with a gold metallic sheen. Lactose-nonfermenting bacteria appear as colorless or translucent colonies.

Endo Agar

Composition per liter:

Agar	10.0g
Lactose	10.0g
Peptic digest of animal tissue	10.0g
K_2HPO_4	3.5g
Na_2SO_3	2.5g
Basic fuchsin solution	4.0mL

pH 7.5 ± 0.2 at 25°C

Source: This medium is available as a premixed powder from Oxoid.

Caution: Basic Fuchsin is a potential carcinogen and care must be taken to avoid inhalation of the powdered dye and contact with the skin.

Basic Fuchsin Solution:

Composition per 10mL:

Basic fuchsin	1.0g
Ethanol (95% solution)	10.0mL

Preparation of Basic Fuchsin Solution: Add basic fuchsin to 10.0mL of ethanol. Mix thoroughly.

Preparation of Medium: Add components to distilled/deionized water and bring volume to 1.0L. Mix thoroughly. Gently heat and bring to boiling. Autoclave for 15 min at 15 psi pressure–121°C. Cool to 45–50°C. Pour into sterile Petri dishes. Swirl flask while pouring plates to keep precipitate in suspension. Protect from the light.

Use: For the selective isolation, cultivation and differentiation of coliform and other enteric microorganisms based on their ability to ferment lactose. Lactose-fermenting bacteria appear as dark red colonies with a gold metallic sheen. Lactose-nonfermenting bacteria appear as colorless or translucent colonies.

Endo Agar, LES (Endo Agar, Laurance Experimental Station) (m-Endo Agar, LES) (m–LES, Endo Agar)

Composition per liter:

Agar	14.0g
Lactose	9.4g
Peptones (pancreatic digest of casein 65% and yeast extract 35%)	7.5g
NaCl	3.7g
Pancreatic digest of casein	3.7g
Peptic digest of animal tissue	3.7g
K_2HPO_4	3.3g
Na_2SO_3	1.6g
Yeast extract	1.2g
KH_2PO_4	1.0g
Basic fuchsin	0.8g
Sodium lauryl sulfate	0.05g
Ethanol	20.0mL

pH 7.2 ± 0.2 at 25°C

Source: This medium is available as a premixed powder from BBL Microbiology Systems and Difco Laboratories.

Caution: Basic fuchsin is a potential carcinogen and care must be taken to avoid inhalation of the powdered dye and contact with the skin.

Preparation of Medium: Add ethanol to approximately 900.0mL of distilled/deionized water. Add remaining components. Bring volume to 1.0L with distilled/deionized water. Mix thoroughly. Gently heat and bring to boiling. Autoclave for 15 min at 15 psi pressure–121°C. Pour into sterile 60mm Petri dishes in 4.0mL volumes. Protect from the light.

Use: For the cultivation and enumeration of coliform bacteria by the membrane filter method.

Endo Agar, LES (m-Endo Agar, LES)

Composition per liter:

Agar	10.0g
Lactose	9.4g
Tryptose	7.5g
NaCl	3.7g
Peptone	3.7g
Pancreatic digest of casein	3.7g
K_2HPO_4	3.3g
Na_2SO_3	1.6g
Yeast extract	1.2g
KH_2PO_4	1.0g
Sodium deoxycholate	0.1g
Sodium lauryl sulfate	0.05g
Basic fuchsin solution	8.0mL

pH 7.2 ± 0.2 at 25°C

Caution: Basic fuchsin is a potential carcinogen and care must be taken to avoid inhalation of the powdered dye and contact with the skin.

Basic Fuchsin Solution:

Composition per 10mL:

Basic fuchsin	1.0g
Ethanol (95% solution)	10.0mL

Preparation of Basic Fuchsin Solution: Add Basic fuchsin to 10.0mL of ethanol. Mix thoroughly.

Preparation of Medium: Add components to distilled/deionized water and bring volume to 1.0L. Mix thoroughly. Gently heat and bring to boiling. Autoclave for 15 min at 15 psi pressure–121°C. Cool to 45–50°C. Pour into sterile Petri dishes. Swirl flask while pouring plates to keep precipitate in suspension. Protect from the light.

Use: For the cultivation and enumeration of coliform bacteria from water by the membrane filter method.

Endo Broth (m-Endo Broth)

Composition per liter:

Lactose	12.5g
Peptone	10.0g
NaCl	5.0g
Pancreatic digest of casein	5.0g
Peptic digest of animal tissue	5.0g
K_2HPO_4	4.375g
Na_2SO_3	2.1g
Yeast extract	1.5g
KH_2PO_4	1.375g
Basic fuchsin	1.05g
Sodium deoxycholate	0.1g
Ethanol (95% solution)	20.0mL

pH 7.2 ± 0.1 at 25°C

Source: This medium is available as a premixed powder from BBL Microbiology Systems and Difco Laboratories.

Caution: Basic fuchsin is a potential carcinogen and care must be taken to avoid inhalation of the powdered dye and contact with the skin.

Preparation of Medium: Add ethanol to approximately 900.0mL of distilled/deionized water. Add remaining components. Bring volume to 1.0L with distilled/deionized water. Mix thoroughly. Gently heat and bring to boiling. Rapidly cool broth below 45°C. Do not autoclave. Use 1.8–2.0mL for each filter pad. Protect from the light. Prepare broth freshly.

Use: For the cultivation and enumeration of coliform bacteria from water by the membrane filter method.

Enriched Nutrient Broth

Composition per liter:

Beef heart, infusion from	250.0g
Tryptose	5.0g
Pancreatic digest of gelatin	3.38g
NaCl	2.5g
Yeast extract	2.5g
Beef extract	2.02g

pH 7.2 ± 0.1 at 25°C

Preparation of Medium: Add components to distilled/deionized water and bring volume to 1.0L. Mix thoroughly. Gently heat and bring to boiling. Distribute into tubes or flasks. Autoclave for 15 min at 15 psi pressure–121°C. Pour into sterile Petri dishes or leave in tubes.

Use: For the cultivation and maintenance of *Bacillus licheniformis, Bacillus polymyxa, Bacillus subtilis, Escherichia coli, Listonella anguillarum, Micrococcus luteus, Pseudomonas aeruginosa, Salmonella choleraesuis, Staphylococcus aureus, Staphylococcus epidermidis, Streptococcus* species, and *Vibrio cholerae.*

Enrichment Broth for *Aeromonas hydrophila*

Composition per liter:

NaCl	5.0g
Maltose	3.5g
Yeast extract	3.0g
Bile salts No. 3	1.0g
L-Cysteine·HCl·H_2O	0.3g
Bromthymol blue	0.03g
Novobiocin	5.0mg

pH 7.0 ± 0.2 at 25°C

Preparation of Medium: Add components to distilled/deionized water and bring volume to 1.0L. Mix thoroughly. Distribute into tubes or flasks. Autoclave for 15 min at 15 psi pressure–121°C.

Use: For the cultivation and enrichment of *Aeromonas hydrophila.*

Entamoeba Medium (Endamoeba Medium)

Composition per liter:

Liver infusion	272.0g
Rice powder	14.2g
Agar	11.0g
Proteose peptone	5.5g
Sodium glycerophosphate	3.0g
NaCl	2.7g
Horse serum	50.0mL

pH 7.0 ± 0.2 at 25°C

Source: This medium is available as a premixed powder from Difco Laboratories.

Rice Powder:

Composition per 15g:

Rice powder	15.0g

Preparation of Rice Powder: Sterilize rice powder at 160°C for 60 min. Do not overheat or rice powder will scorch.

Preparation of Medium: Add components, except horse serum and rice powder, to distilled/deionized water and bring volume to 994.0mL. Mix thoroughly. Gently heat and bring to boiling. Distribute into tubes in 7.0mL volumes. Autoclave for 15 min at 15 psi pressure–121°C. Allow tubes to cool in a slanted position. Aseptically add enough sterile horse serumm to each tube to cover about half the slant. Aseptically add 0.1g of sterile rice powder to each tube.

Use: For the cultivation of *Entamoeba histolytica.*

Enteric Fermentation Base (Fermentation Base for *Campylobacter*)

Composition per liter:

Peptic digest of animal tissue	10.0g
NaCl	5.0g
Beef extract	3.0g
Carbohydrate solution	100.0mL
Andrade's indicator	10.0mL

pH 7.2 ± 0.1 at 25°C

Source: This medium is available as a premixed powder from Difco Laboratories.

Carbohydrate Solution:

Composition per 100mL:

Carbohydrate	10.0g

Preparation of Carbohydrate Solution: Add carbohydrate to distilled/deionized water and bring volume to 100.0mL. Mix thoroughly. Filter sterilize. Glucose, lactose, mannitol, sucrose, adonitol, arabinose, cellobiose, dulcitol, glycerol, inositol, salicin, xylose or other carbohydrates may be used. For the preparation of expensive carbohydrate solutions (adonitol, arabinose, cellobiose, dulcitol, glycerol, inositol, salicin or xylose), 5.0g of carbohydrate per 100.0mL of distilled/deionized water may be used.

Andrade's Indicator:

Composition per 100mL:

NaOH (1*N* solution)	16.0mL
Acid fuchsin	0.1 g

Caution: Acid fuchsin is a potential carcinogen and care must be taken to avoid inhalation of the powdered dye and contact with the skin.

Preparation of Andrade's Indicator: Add components to distilled/deionized water and bring volume to 100.0mL. Mix thoroughly.

Preparation of Medium: Add components, except carbohydrate solution, to distilled/deionized water and bring volume to 900.0mL. Mix thoroughly. Gently heat and bring to boiling. Distribute into tubes that contain an inverted Durham tube in 9.0mL volumes. Autoclave for 15 min at 15 psi pressure–121°C. Cool to 25°C. Aseptically add 1.0mL of sterile carbohydrate solution per tube. Mix thoroughly.

Use: For the cultivation and differentiation of a variety of bacteria based on their ability to ferment different carbohydrates. Bacteria that produce acid from carbohydrate fermentation turn the medium dark pink to red. Bacteria that produce gas have a bubble trapped in the Durham tube.

Enterobacter Medium

Composition per 800mL:

Casein hydrolysate	2.0g
K_2HPO_4	1.4g
K_2SO_4	1.0g
Yeast extract	1.0g
KH_2PO_4	0.6g
$MgSO_4$	0.5g
Glycerol	20.0mL

Preparation of Medium: Add components to distilled/deionized water and bring volume to 800.0mL. Mix thoroughly. Distribute into tubes or flasks. Autoclave for 15 min at 15 psi pressure–121°C.

Use: For the cultivation and maintenance of *Enterobacter* species and *Klebsiella pneumoniae*.

Enterococci Confirmatory Agar

Composition per liter:

Agar	15.0g
Glucose	5.0g
Pancreatic digest of casein	5.0g
Yeast extract	5.0g
NaN_3	0.4g
Methylene blue	10.0mg

pH 8.0 ± 0.2 at 25°C

Source: This medium is available as a premixed powder from Difco Laboratories.

Caution: Sodium azide is toxic. Azides also react with metals and disposal must be highly diluted.

Preparation of Medium: Add components to distilled/deionized water and bring volume to 1.0L. Mix thoroughly. Gently heat and bring to boiling. Distribute into tubes. Autoclave for 15 min at 15 psi pressure–121°C. Allow tubes to cool in a slanted position. Add sufficient amount of Enterococci confirmatory broth to cover half the slant.

Use: For the identification of enterococci from water by the confirmatory test.

Enterococci Confirmatory Broth

Composition per liter:

NaCl	65.0g
Glucose	5.0g
Pancreatic digest of casein	5.0g
Yeast extract	5.0g
NaN_3	0.4g
Methylene blue	10.0mg
Penicillin	650U

pH 8.0 ± 0.2 at 25°C

Source: This medium is available as a premixed powder from Difco Laboratories.

Caution: Sodium azide is toxic. Azides also react with metals and disposal must be highly diluted.

Preparation of Medium: Add components, except penicillin, to distilled/deionized water and bring volume to 1.0L. Mix thoroughly. Gently heat and bring to boiling. Autoclave for 15 min at 15 psi pressure–121°C. Cool to 25°C. Aseptically add penicillin. Mix thoroughly. Add sufficient amount of Enterococci confirmatory broth to cover half the slant of Enterococci confirmatory agar.

Use: For the identification of enterococci from water by the confirmatory test.

Enterococci Presumptive Broth

Composition per liter:

Glucose	5.0g
Pancreatic digest of casein	5.0g
Yeast extract	5.0g
NaN_3	0.4g
Bromthymol blue	32.0mg

pH 8.4 ± 0.2 at 25°C

Source: This medium is available as a premixed powder from Difco Laboratories.

Caution: Sodium azide is toxic. Azides also react with metals and disposal must be highly diluted.

Preparation of Medium: Add components to distilled/deionized water and bring volume to 1.0L. Mix thoroughly. Distribute into tubes or flasks. Autoclave for 15 min at 15 psi pressure–121°C.

Use: For the isolation and identification of enterococci from water by the presumptive test. Bacteria that produce acid and turn the medium yellow and turbid after incubation at 45°C are presumptive enterococci.

Enterococcosel™ Agar

Composition per liter:

Pancreatic digest of casein	17.0g
Agar	13.5g
Oxgall	10.0g
NaCl	5.0g
Yeast extract	5.0g
Peptic digest of animal tissue	3.0g
Esculin	1.0g
Sodium citrate	1.0g
Ferric ammonium citrate	0.5g
NaN_3	0.25g

pH 7.1 ± 0.2 at 25°C

Source: This medium is available as a premixed powder from BBL Microbiology Systems.

Caution: Sodium azide is toxic. Azides also react with metals and disposal must be highly diluted.

Preparation of Medium: Add components to distilled/deionized water and bring volume to 1.0L. Mix thoroughly. Gently heat while stirring and bring to boiling. Distribute into tubes or flasks. Autoclave for 15 min at 15 psi pressure–121°C. Pour into sterile Petri dishes or leave in tubes.

Use: For the rapid, selective isolation, cultivation and enumeration of fecal group D streptococci (enterococci). For the cultivation of staphylococci and *Listeria monocytogenes*.

Enterococcosel™ Broth

Composition per liter:

Pancreatic digest of casein	17.0g
Oxgall	10.0g
NaCl	5.0g
Yeast extract	5.0g
Peptic digest of animal tissue	3.0g
Esculin	1.0g
Sodium citrate	1.0g
Ferric ammonium citrate	0.5g
NaN_3	0.25g

pH 7.1 ± 0.2 at 25°C

Source: This medium is available as a premixed powder from BBL Microbiology Systems.

Caution: Sodium azide is toxic. Azides also react with metals and disposal must be highly diluted.

Preparation of Medium: Add components to distilled/deionized water and bring volume to 1.0L. Mix thoroughly. Gently heat while stirring until dissolved. Distribute into tubes or flasks. Autoclave for 15 min at 15 psi pressure–121°C.

Use: For the cultivation and differentiation of group D streptococci (enterococci).

Enterococcus Agar (m-*Enterococcus* Agar) (Azide Agar)

Composition per liter:

Pancreatic digest of casein	15.0g
Agar	10.0g
Papaic digest of soybean meal	5.0g
Yeast extract	5.0g
KH_2PO_4	4.0g
Glucose	2.0g
NaN_3	0.4g
Triphenyltetrazolium chloride	0.1g

pH 7.2 ± 0.2 at 25°C

Source: This medium is available as a premixed powder from BBL Microbiology Systems and Difco Laboratories.

Caution: Sodium azide is toxic. Azides also react with metals and disposal must be highly diluted.

Preparation of Medium: Add components to distilled/deionized water and bring volume to 1.0L. Mix thoroughly. Gently heat and bring to boiling. Cool to 45–50°C. Do not autoclave. Pour into sterile Petri dishes.

Use: For the isolation, cultivation and enumeration of entercocci in water, sewage and feces by the membrane filter method. Also used for the direct plating of specimens for detection and enumeration of fecal streptococci.

Erwinia amylovora Selective Medium

Composition per liter:

Agar	20.0g
Mannitol	10.0g
L-Asparagine	3.0g
Sodium taurocholate	2.5g
K_2HPO_4	2.0g
Nicotinic acid	0.5g
$MgSO_4·7H_2O$	0.2g
Nitrilotriacetic acid	10.0mL
Actidione (cycloheximide) solution	10.0mL
Bromthymol blue	9.0mL
Neutral red	2.5mL
$TlNO_3$ solution	1.75mL
Tergitol™ 7	0.1mL

pH 7.2–7.3 at 25°C

Cycloheximide Solution:

Composition per 10mL:

Cycloheximide	0.05g

Preparation of Cycloheximide Solution: Add cycloheximide to distilled/deionized water and bring volume to 10.0mL. Mix thoroughly. Filter sterilize.

$TlNO_3$ Solution:

Composition per 10mL:

$TlNO_3$	0.1g

Preparation of $TlNO_3$ Solution: Add $TlNO_3$ to distilled/deionized water and bring volume to 10.0mL. Mix thoroughly. Filter sterilize.

Preparation of Medium: Add components, except $TlNO_3$ solution and cycloheximide solution, to distilled/deionized water and bring volume to 988.25mL. Mix thoroughly. Adjust pH to 7.2–7.3. Gently heat and bring to boiling. Autoclave for 15 min at 15 psi pressure–121°C. Cool to 45–50°C. Aseptically add 1.75mL of sterile $TlNO_3$ solution and 10.0mL of sterile cycloheximide solution. Mix thoroughly. Pour into sterile Petri dishes or distribute into sterile tubes.

Use: For the isolation and cultivation of *Erwinia amylovora.*

Erwinia Medium D3

Composition per liter:

Agar	15.0g
Arabinose	10.0g
Sucrose	10.0g
LiCl	7.0g
Casein hydrolysate	5.0g
NaCl	5.0g
$MgSO_4·7H_2O$	0.3g
Acid fuchsin	0.1g
Bromthymol blue	0.06g
Sodium dodecyl sulfate	0.05g

pH 7.0 ± 0.2 at 25°C

Caution: Acid Fuchsin is a potential carcinogen and care must be taken to avoid inhalation of the powdered dye and contact with the skin.

Preparation of Medium: Add components to distilled/deionized water and bring volume to 1.0L. Mix thoroughly. Gently heat and bring to boiling. Adjust pH to 8.2. Autoclave for 15 min at 15 psi pressure–121°C. Cool to 45–50°C. The pH after autoclaving should be 7.0. Pour into sterile Petri dishes or distribute into sterile tubes.

Use: For the isolation and cultivation of *Erwinia* species.

Erwinia Selective Medium

Composition per liter:

Agar	15.0g
$(NH_4)_2SO_4$	5.0g
K_2HPO_4	2.0g
Eosin Y	0.4g
Methylene blue	0.065g
Glycerol	10.0mL
Antibiotic solution	10.0mL

Antibiotic Solution:

Composition per 10mL:

Cycloheximide	0.25g
Novobiocin	0.04g
Neomycin sulfate	0.04g

Preparation of Antibiotic Solution: Add components to distilled/deionized water and bring volume to 10.0mL. Mix thoroughly. Filter sterilize.

Preparation of Medium: Add components, except antibiotic solution, to distilled/deionized water and bring volume to 990.0mL. Mix thoroughly. Gently heat and bring to boiling. Autoclave for 15 min at 15 psi pressure–121°C. Cool to 45–50°C. Aseptically add sterile antibiotic solution. Mix thoroughly. Pour into sterile Petri dishes or distribute into sterile tubes.

Use: For the selective isolation and cultivation of *Erwinia* species.

Esculin Agar

Composition per liter:

Agar	15.0g
Pancreatic digest of casein	13.0g
NaCl	5.0g
Yeast extract	5.0g
Heart muscle, solids from infusion	2.0g
Esculin	1.0g
Ferric citrate	0.5g

pH 7.3 ± 0.2 at 25°C

Preparation of Medium: Add components to distilled/deionized water and bring volume to 1.0L. Mix thoroughly. Gently heat and bring to boiling. Distribute into screw-capped tubes in 3.0mL volumes. Autoclave for 15 min at 15 psi pressure–121°C. Allow tubes to cool in a slanted position.

Use: For the cultivation and differentiation of bacteria based on their ability to hydrolyze esculin and produce H_2S. Bacteria that hydrolyze esculin appear as colonies surrounded by a reddish-brown to dark brown zone. Bacteria that produce H_2S appear as black colonies.

Esculin Broth

Composition per liter:

Beef heart, solids from infusion	500.0g
Tryptose	10.0g
NaCl	5.0g
Agar	1.0g
Esculin	1.0g

pH 7.0 ± 0.2 at 25°C

Preparation of Medium: Add components to distilled/deionized water and bring volume to 1.0L. Mix thoroughly. Gently heat and bring to boiling. Distribute into screw-capped tubes in 7.0mL volumes. Autoclave for 15 min at 15 psi pressure–121°C.

Use: For the cultivation and differentiation of bacteria based on their ability to hydrolyze esculin. Bacteria that hydrolyze esculin turn the medium brown-black to black.

Esculin Iron Agar

Composition per liter:

Agar	15.0g
Esculin	1.0g
Ferric ammonium citrate	0.5g

pH 7.1 ± 0.2 at 25°C

Source: This medium is available as a premixed powder from BBL Microbiology Systems.

Preparation of Medium: Add components to distilled/deionized water and bring volume to 1.0L. Mix thoroughly. Gently heat and bring to boiling. Distribute into tubes or flasks. Autoclave for 15 min at 15 psi pressure–121°C. Pour into sterile Petri dishes.

Use: For the cultivation and identification of enterococci based on their ability to hydrolyze esculin. Used in conjunction with E agar and the membrane filter method.

Esculin Mannitol Agar

Composition per liter:

Agar	13.5g
Polypeptone™	10.0g
D-Mannitol	10.0g
Pancreatic digest of casein	5.0g
Yeast extract	5.0g
NaCl	5.0g
Heart peptone	3.0g
Cornstarch	1.0g
Esculin	1.0g
Ferric ammonium citrate	0.5g
Phenol red	0.025g
Nalidixic acid solution	10.0mL
Colistin solution	10.0mL

pH 7.3 ± 0.2 at 25°C

Nalidixic Acid Solution:

Composition per 10mL:

Nalidixic acid 0.015g

Preparation of Nalidixic Acid Solution: Add nalidixic acid to distilled/deionized water and bring volume to 10.0mL. Mix thoroughly. Filter sterilize.

Colistin Solution:

Composition per 10mL:

Colistin 0.01g

Preparation of Colistin Solution: Add colistin to distilled/deionized water and bring volume to 10.0mL. Mix thoroughly. Filter sterilize.

Preparation of Medium: Add components, except nalidixic acid solution and colistin solution, to distilled/deionized water and bring volume to 980.0mL. Mix thoroughly. Gently heat and bring to boiling. Autoclave for 15 min at 15 psi pressure–121°C. Cool to 45–50°C. Aseptically add sterile nalidixic acid solution and colistin solution. Mix thoroughly. Pour into sterile Petri dishes or distribute into sterile tubes.

Use: For the selective isolation, cultivation and differentiation of *Staphylococcus aureus* and group D streptococci based on mannitol fermentation and hydrolysis of esculin. Bacteria that ferment mannitol appear as yellow colonies surrounded by a yellow zone. Bacteria that hydrolyze esculin appear as dark brown to black colonies surrounded by a dark brown to black zone.

ETGPA (Egg Tellurite Glycine Pyruvate Agar)

Composition per liter:

Agar 17.0g
Glycine 12.0g
Sodium pyruvate 10.0g
Pancreatic digest of casein 10.0g
Beef extract 5.0g
LiCl 5.0g
Yeast extract 1.0g
Egg yolk emulsion 50.0mL
K_2TeO_3 solution 10.0mL

pH 7.0 ± 0.2 at 25°C

Source: This medium is available as a premixed powder from BBL Microbiology Systems.

Egg Yolk Emulsion:

Composition:

Chicken egg yolks 11
Whole chicken egg 1

Preparation of Egg Yolk Emulsion: Soak egg with 1:100 dilution of saturated mercuric chloride solution for 1 min. Crack eggs and separate yolks from whites. Mix egg yolks with 1 chicken egg.

K_2TeO_3 Solution:

Composition per 100mL:

K_2TeO_3 .... 1.0g

Preparation of K_2TeO_3 Solution: Add K_2TeO_3 to distilled/deionized water and bring volume to 100.0mL. Mix thoroughly. Filter sterilize.

Caution: Potassium tellurite is toxic.

Preparation of Medium: Add components to distilled/deionized water and bring volume to 940.0mL. Mix thoroughly. Gently heat and bring to boiling. Autoclave for 15 min at 15 psi pressure–121°C. Cool to 45–50°C. Add 10.0mL of sterile 1% tellurite solution and 50.0mL of sterile egg yolk emulsion. If desired, add sulfamethazine to a final concentration of 50mg/mL. Mix thoroughly but gently and pour into sterile Petri dishes.

Use: For the selective isolation and enumeration of coagulase-positive staphylococci from soil, air and other materials. For the differentiation and identification of staphylococci on the basis of their ability to clear egg yolk. Addition of sulfamethazine inhibits the growth of *Proteus*. Gray-black colonies surrounded by a clear zone are diagnostic for *Staphylococcus aureus*.

Ethyl Violet Azide Broth (EVA Broth)

Composition per liter:

Pancreatic digest of casein 13.5g
Yeast extract 6.5g
Glucose 5.0g
NaCl 5.0g
K_2HPO_4 .... 2.7g
KH_2PO_4 .... 2.7g
NaN_3 .... 0.4g
Ethyl violet 0.83mg

pH 7.0 ± 0.2 at 25°C

Source: This medium is available as a premixed powder from BBL Microbiology Systems.

Preparation of Medium: Add components to distilled/deionized water and bring volume to 1.0L. Mix thoroughly. Gently heat and bring to boiling. Distribute into tubes in 10.0mL volumes. Autoclave for 15 min at 15 psi pressure–121°C.

Caution: Sodium azide is toxic. Azides also react with metals and disposal must be highly diluted.

Use: For the isolation, cultivation and enumeration of enterococci from water and other specimens. Fecal enterococci turn the medium turbid with a purple sediment on the bottom of the tube.

Ethyl Violet Azide Broth (EVA Broth)

Composition per liter:

Tryptose	20.0g
Glucose	5.0g
NaCl	5.0g
K_2HPO_4	2.7g
KH_2PO_4	2.7g
NaN_3	0.4g
Ethyl Violet	0.83mg

pH 7.0 ± 0.2 at 25°C

Source: This medium is available as a premixed powder from Difco Laboratories.

Preparation of Medium: Add components to distilled/deionized water and bring volume to 1.0L. Mix thoroughly. Gently heat and bring to boiling. Distribute into tubes in 10.0mL volumes. Autoclave for 15 min at 15 psi pressure–121°C.

Caution: Sodium azide is toxic. Azides also react with metals and disposal must be highly diluted.

Use: For the isolation, cultivation and enumeration of enterococci from water and other specimens. Fecal enterococci turn the medium turbid with a purple sediment on the bottom of the tube.

Ethyl Violet Azide Broth (EVA Broth)

Composition per liter:

Tryptose	20.0g
Glucose	5.0g
NaCl	5.0g
K_2HPO_4	2.7g
KH_2PO_4	2.7g
NaN_3	0.3g
Ethyl violet	0.5mg

pH 6.8 ± 0.2 at 25°C

Source: This medium is available as a premixed powder from Oxoid.

Preparation of Medium: Add components to distilled/deionized water and bring volume to 1.0L. Mix thoroughly. Gently heat and bring to boiling. Distribute into tubes in 10.0mL volumes. Autoclave for 15 min at 15 psi pressure–121°C.

Caution: Sodium azide is toxic. Azides also react with metals and disposal must be highly diluted.

Use: For the isolation, cultivation and enumeration of enterococci from water and other specimens. Fecal enterococci turn the medium turbid with a purple sediment on the bottom of the tube.

Eubacterium acidaminophilum Medium

Composition per liter:

KH_2PO_4	1.0g
NH_4Cl	0.5g
$MgCl_2 \cdot 6H_2O$	0.4g
$CaCl_2 \cdot 2H_2O$	0.1g
$NaHCO_3$ solution	20.0mL
Sodium selenite solution	10.0mL
$Na_2S \cdot 9H_2O$ solution	3.0mL
Vitamin solution	1.0mL
Trace element solution	1.0mL

pH 7.2–7.3 at 25°C

Sodium Selenite Solution:

Composition per liter:

NaOH	0.5g
$Na_2SeO_3 \cdot 5H_2O$	0.03g

Preparation of Sodium Selenite Solution: Add components to distilled/deionized water and bring volume to 1.0L. Mix thoroughly. Filter sterilize. Gas with 100% N_2 for 20 min.

$Na_2S \cdot 9H_2O$ Solution:

Composition per 10mL:

$Na_2S \cdot 9H_2O$	1.0g

Preparation of $Na_2S \cdot 9H_2O$ Solution: Add $Na_2S \cdot 9H_2O$ to distilled/deionized water and bring volume to 10.0mL. Mix thoroughly. Gas with 100% N_2 for 20 min. Autoclave for 15 min at 15 psi pressure–121°C. Cool to 45–50°C.

Vitamin Solution:

Composition per 100mL:

p-Aminobenzoic acid	4.0mg
Biotin	1.0mg

Preparation of Vitamin Solution: Add components to distilled/deionized water and bring volume to 100.0mL. Mix thoroughly. Filter sterilize.

Trace Elements Solution SL-7:

Composition per liter:

$FeCl_2 \cdot 4H_2O$	1.5g
$CoCl_2 \cdot 6H_2O$	0.19g
$MnCl_2 \cdot 4H_2O$	0.1g
$ZnCl_2$	0.07g
$Na_2MoO_4 \cdot 2H_2O$	0.036g
$NiCl_2 \cdot 6H_2O$	0.024g
H_3BO_3	6.0mg
$CuCl_2 \cdot 2H_2O$	2.0mg
HCl (25% solution)	10.0mL

Preparation of Trace Elements Solution SL-7: Add the $FeCl_2 \cdot 4H_2O$ to the HCl. Add distilled/deionized water and bring volume to 1.0L. Add remaining components. Mix thoroughly. Autoclave for 15 min at 15 psi pressure–121°C under 100% N_2. Cool to 25°C.

Preparation of Medium: Add components—except $NaHCO_3$ solution, $Na_2S \cdot 9H_2O$ solution, vitamin solution, and sodium selenite solution—to distilled/deionized water and bring volume to 966.0mL. Mix thoroughly. Gently heat and bring to boiling. Autoclave for 15 min at 15 psi pressure–121°C. Cool to 45–50°C under 80% N_2 and 20% CO_2. Aseptically add 20.0mL of sterile $NaHCO_3$ solution, 3.0mL of sterile $Na_2S \cdot 9H_2O$ solution, 1.0mL of sterile vitamin solution, and 10.0mL of sterile sodium selenite solution. Mix thoroughly. Adjust pH to 7.2–7.3 with dilute sterile HCl or Na_2CO_3, if necessary. Aseptically and anaerobically distribute into sterile tubes under 80% N_2 and 20% CO_2. Cap tubes with rubber stoppers.

Use: For the cultivation and maintenance of *Eubacterium acidaminophilum.*

Eubacterium angustum Medium

Composition per liter:

$NaHCO_3$	5.0g
Tris(hydroxymethyl)aminomethane buffer	3.0g
Uric acid	3.0g
Yeast extract	1.0g
Cysteine·HCl·H_2O	0.5g
NH_4Cl	0.5g
K_2HPO_4	0.1g
$MgSO_4 \cdot 7H_2O$	0.05g
$Na_2S \cdot 9H_2O$	0.05g
Resazurin	1.0mg
Wolfe's mineral solution	10.0mL
Selenium solution	1.0mL

pH 7.9 ± 0.2 at 25°C

Selenium Solution:

Composition per liter:

$Na_2SeO_3 \cdot 5H_2O$	3.0mg
NaOH (0.01*M* solution)	1.0L

Preparation of Selenium Solution: Add $Na_2SeO_3 \cdot 5H_2O$ to 1.0L of NaOH solution. Mix thoroughly. Filter sterilize. Aseptically gas under 100% N_2 for 20 min.

Wolfe's Mineral Solution:

Composition per liter

$MgSO_4 \cdot 7H_2O$	3.0g
Nitriloacetic acid	1.5g
NaCl	1.0g
$MnSO_4 \cdot H_2O$	0.5g
$FeSO_4 \cdot 7H_2O$	0.1g
$CoCl_2 \cdot 6H_2O$	0.1g
$CaCl_2$	0.1g
$ZnSO_4 \cdot 7H_2O$	0.1g
$CuSO_4 \cdot 5H_2O$	0.01g
$AlK(SO_4)_2 \cdot 12H_2O$	0.01g
H_3BO_3	0.01g
$Na_2MoO_4 \cdot 2H_2O$	0.01g

Preparation of Wolfe's Mineral Solution: Add nitrilotriacetic acid to 500.0mL of distilled/deionized water. Dissolve by adjusting pH to 6.5 with KOH. Add remaining components. Add distilled/deionized water to 1.0L.

Preparation of Medium: Add the uric acid and the tris(hydroxymethyl)aminomethane buffer to 900.0mL of distilled/deionized water. Mix thoroughly. Gently heat while stirring until dissolved. Add remaining components, except cysteine·HCl·H_2O, $NaHCO_3$, and $Na_2S \cdot 9H_2O$. Gently heat and bring to boiling. Cool to 25°C under 80% N_2 + 10% CO_2 + 10% H_2. Add cysteine·HCl·H_2O, $NaHCO_3$, and $Na_2S \cdot 9H_2O$. Mix thoroughly. Anaerobically distribute into tubes under 80% N_2 + 10% CO_2 + 10% H_2. Cap tubes with rubber stoppers. Place tubes in a press. Autoclave for 15 min at 15 psi pressure–121°C with fast exhaust.

Use: the cultivation and maintenance of *Eubacterium angustum.*

Eubacterium Medium

Composition per liter:

Pancreatic digest of casein	20.0g
Agar	15.0g
Meat extract	15.0g
Glucose	5.0g
$Na_2HPO_4 \cdot 12H_2O$	4.0g
Cysteine·HCl	0.5g

pH 7.4 ± 0.2 at 25°C

Preparation of Medium: Add components to distilled/deionized water and bring volume to 1.0L. Mix thoroughly. Gently heat and bring to boiling. Distribute into tubes or flasks. Autoclave for 15 min at 15 psi pressure–121°C. Pour into sterile Petri dishes or leave in tubes. Use freshly prepared medium.

Use: For the cultivation of *Eubacterium* species.

Eubacterium Medium

Composition per liter:

Beef brain powder	33.33g
Pancreatic digest of casein	15.0g
Yeast extract	10.0g
Glucose	5.5g
Yeast extract	5.0g
NaCl	2.5g

Sodium thioglycollate	1.8g
L-Cystine	0.5g

pH 7.0 ± 0.2 at 25°C

Preparation of Medium: Add components, except beef brain powder, to distilled/deionized water and bring volume to 1.0L. Mix thoroughly. Gently heat and bring to boiling under 97% N_2 + 3% H_2. Continue boiling for 15–20 min. Adjust pH to 7.0. Cool to 25°C under 97% N_2 + 3% H_2. Anaerobically distribute into tubes in 9.0mL volumes. Add 0.3g of beef brain powder to each tube. Cap tubes with rubber stoppers. Place tubes in a press. Autoclave for 15 min at 15 psi pressure–121°C with fast exhaust.

Use: For the cultivation of *Eubacterium* species.

Extracted Hay Medium

Composition:

Hay or grass	50.0g

Preparation of Medium: Add hay or grass to 1.0L of distilled/deionized water. Gently heat and bring to boiling. Continue boiling for 30 min. Rinse with cold water twice. Add 1.0L of of distilled/deionized water, boil 30 min and rinse. Repeat this process at least five times. Dry the extracted hay or grass. Add 10–30 blades of extracted hay or grass to a large test tube. Autoclave for 15 min at 15 psi pressure–121°C.

Use: For the isolation and cultivation of *Beggiatoa* species and myxotrophic *Thiothrix* species.

Fay and Barry Medium

Composition per liter:

Amino acid	10.0g
Peptone	5.0g
Yeast extract	3.0g
Bromcresol purple solution	5.0mL

pH 5.5 ± 0.2 at 25°C

Bromcresol Purple Solution:
Composition per 100mL:

Bromcresol purple	0.2g
Ethanol	50.0mL

Preparation of Bromcresol Purple Solution: Add bromcresol purple to 50.0mL of absolute ethanol. Add distilled/deionized water and bring volume to 100.0mL. Mix thoroughly.

Preparation of Medium: Add components to distilled/deionized water and bring volume to 1.0L. The amino acid may be L-arginine, L-ornithine, or L-lysine, depending on which amino acid decarboxylase activity is being measured. Mix thoroughly. Distribute into tubes or flasks. Autoclave for 15 min at 15 psi pressure–121°C.

Use: For determination of decarboxylase activities of *Aeromonas* species.

FC Agar (Fecal Coliform Agar) (m-FC Agar) (m–Fecal Coliform Agar)

Composition per liter:

Agar	15.0g
Lactose	12.5g
NaCl	5.0g
Proteose peptone No. 3	5.0g
Yeast extract	3.0g
Bile salts	1.5g
Aniline blue	0.1g
Rosolic acid solution	10.0mL

pH 7.4 ± 0.2 at 25°C

Source: This medium is available as a premixed powder from BBL Microbiology Systems and Difco Laboratories.

Rosolic Acid Solution:
Composition per 100mL:

Rosolic acid	1.0g

Preparation of Rosolic Acid Solution: Add rosolic acid to 0.2*N* NaOH and bring volume to 100.0L. Mix thoroughly.

Preparation of Medium: Add 10.0mL rosolic acid solution to 950.0mL distilled/deionized water. Mix thoroughly. Add other components and bring volume to 1.0L with distilled/deionized water. Mix thoroughly. Gently heat and bring to boiling with frequent mixing. Do not autoclave. Pour into sterile Petri dishes or leave in tubes.

Use: For the cultivation of fecal coliform bacteria from waters and the enumeration of coliform bacteria using the membrane filtration method.

FC Agar (Fecal Coliform Agar) (m-FC Agar) (m–Fecal Coliform Agar)

Composition per liter:

Agar	15.0g
Lactose	12.5g
Tryptose	10.0g
NaCl	5.0g
Proteose peptone No. 3	5.0g
Yeast extract	3.0g
Bile salts	1.5g

Aniline blue..0.1g
Rosolic acid solution....................................10.0mL

pH 7.4 ± 0.2 at 25°C

Rosolic Acid Solution:
Composition per 100mL:
Rosolic acid...1.0g

Preparation of Rosolic Acid Solution: Add rosolic acid to 0.2*N* NaOH and bring volume to 100.0L. Mix thoroughly.

Preparation of Medium: Add 10.0mL rosolic acid solution to 950.0mL distilled/deionized water. Mix thoroughly. Add other components and bring volume to 1.0L with distilled/deionized water. Mix thoroughly. Gently heat and bring to boiling with frequent mixing. Do not autoclave. Pour into sterile Petri dishes or leave in tubes.

Use: For the cultivation of fecal coliform bacteria from waters and the enumeration of coliform bacteria using the membrane filtration method.

FC Broth (Fecal Coliform Broth) (m-FC Broth) (m–Fecal Coliform Broth)

Composition per liter:
Lactose..12.5g
Tryptose 10.0g
NaCl...5.0g
Proteose peptone No. 3 ....5.0g
Yeast extract..3.0g
Bile salts..1.5g
Aniline blue..0.1g
Rosolic acid solution.....................................10.0mL

pH 7.4 ± 0.2 at 25°C

Rosolic Acid Solution:
Composition per 100mL:
Rosolic acid...1.0g

Preparation of Rosolic Acid Solution: Add rosolic acid to 0.2*N* NaOH and bring volume to 100.0L. Mix thoroughly.

Preparation of Medium: Add 10.0mL rosolic acid solution to 950.0mL distilled/deionized water. Mix thoroughly. Add other components and bring volume to 1.0L with distilled/deionized water. Mix thoroughly. Gently heat and bring to boiling with frequent mixing. Do not autoclave. Pour into sterile Petri dishes or leave in tubes.

Use: For the cultivation of fecal coliform bacteria from waters and the enumeration of coliform bacteria using the membrane filtration method.

FC Broth (Fecal Coliform Broth) (m-FC Broth) (m–Fecal Coliform Broth)

Composition per liter:
Lactose..12.5g
NaCl...5.0g
Proteose peptone No. 3 ....5.0g
Yeast extract..3.0g
Bile salts..1.5g
Aniline blue..0.1g
Rosolic acid solution.....................................10.0mL

pH 7.4 ± 0.2 at 25°C

Source: This medium is available as a premixed powder from BBL Microbiology Systems and Difco Laboratories.

Rosolic Acid Solution:
Composition per 100mL:
Rosolic acid...1.0g

Preparation of Rosolic Acid Solution: Add rosolic acid to 0.2*N* NaOH and bring volume to 100.0L. Mix thoroughly.

Preparation of Medium: Add 10.0mL rosolic acid solution to 950.0mL distilled/deionized water. Mix thoroughly. Add other components and bring volume to 1.0L with distilled/deionized water. Mix thoroughly. Gently heat and bring to boiling with frequent mixing. Do not autoclave. Pour into sterile Petri dishes or leave in tubes.

Use: For the cultivation of fecal coliform bacteria from waters and the enumeration of coliform bacteria using the membrane filtration method.

Fecal Coliform Agar, Modified (m–Fecal Coliform Agar, Modified) (FCIC)

Composition per liter:
Agar...15.0g
Inositol 10.0g
Tryptose 10.0g
Proteose peptone No. 3 ....5.0g
NaCl...5.0g
Yeast extract..3.0g
Bile salts No. 3 ....1.5g
Aniline blue..0.1g

pH 7.4 ± 0.2 at 25°C

Preparation of Medium: Add components and bring volume to 1.0L. Mix thoroughly. Gently heat and bring to boiling. Do not autoclave. Cool to 50°C.

Adjust pH to 7.4. Pour into sterile Petri dishes in 20.0mL volumes. Allow surface of plates to dry before using.

Use: For the isolation, cultivation and enumeration of *Klebsiella* species using the membrane filter method.

Fecal Coliform Agar, Modified

Composition per liter:

Agar	15.0g
Lactose	12.5g
Tryptose	10.0g
Proteose peptone No. 3	5.0g
NaCl	5.0g
Yeast extract	3.0g
Bile salts No. 3	1.5g
Aniline blue	0.1g

pH 7.4 ± 0.2 at 25°C

Preparation of Medium: Add components and bring volume to 1.0L. Mix thoroughly. Gently heat and bring to boiling. Do not autoclave. Cool to 50°C. Adjust pH to 7.4. Pour into sterile Petri dishes in 20.0mL volumes. Allow surface of plates to dry before using.

Use: For the isolation, cultivation and identification of stressed fecal coliform microorganisms based on their ability to ferment lactose. Lactose-fermenting bacteria turn the medium blue.

Feodorov Medium

Composition per liter:

Mannitol or glucose	20.0g
Marine salts mixture	18.0g
$CaCO_3$	0.5g
K_2HPO_4	0.3g
$MgSO_4$	0.3g
$CaHPO_4$	0.2g
K_2SO_4	0.2g
$FeCl_3$	0.1g
Trace elements solution	1.0mL

Trace Elements Solution:
Composition per 100mL:

H_3BO_3	0.5g
$(NH_4)_6Mo_7O_{24}\cdot 4H_2O$	0.5g
KI	0.05g
NaBr	0.05g
$Al_2(SO_4)_3\cdot 18H_2O$	0.03g
$ZnSO_4$	0.02g

Preparation of Trace Elements Solution: Add components to distilled/deionized water and bring volume to 100.0mL. Mix thoroughly.

Preparation of Medium: Add components to distilled/deionized water and bring volume to 1.0L. Mix thoroughly. Distribute into tubes or flasks. Autoclave for 15 min at 15 psi pressure–121°C.

Use: For the cultivation and maintenance of *Azotobacter vinelandii*.

Fermentation Basal Medium

Composition per liter:

Agar	15.0g
$(NH_4)_2HPO_4$	1.0g
$MgSO_4\cdot 7H_2O$	0.2g
KCl	0.02g
Carbohydrate solution	100.0mL
Bromcresol purple solution	20.0mL

pH 7.0 ± 0.2 at 25°C

Carbohydrate Solution:
Composition per 100mL:

Carbohydrate	10.0g

Preparation of Carbohydrate Solution: Add carbohydrate to distilled/deionized water and bring volume to 100.0mL. Mix thoroughly. Filter sterilize.

Bromcresol Purple Solution:
Composition per 100mL:

Bromcresol purple	0.04g
Ethanol	50.0mL

Preparation of Bromcresol Purple Solution: Add bromcresol purple to 50.0mL of absolute ethanol. Add distilled/deionized water and bring volume to 100.0mL. Mix thoroughly.

Preparation of Medium: Add components, except carbohydrate solution, to distilled/deionized water and bring volume to 900.0mL. Mix thoroughly. Gently heat and bring to boiling. Autoclave for 15 min at 15 psi pressure–121°C. Cool to 45°–50°C. Aseptically add 100.0mL sterile carbohydrate solution. Various carbohydrates are used for different fermentation tests. Mix thoroughly. Pour into sterile Petri dishes or distribute into sterile tubes.

Use: For differentiation of aerobic actinomycetes based upon carbohydrate fermentation. Actinomycetes that produce acid from carbohydrates turn the medium yellow.

Fermentation Broth (CHO Medium)

Composition per liter:

Pancreatic digest of casein	15.0g
Yeast extract	7.0g
NaCl	2.5g
Agar	0.75g

Sodium thioglycollate	0.5g
L-Cystine	0.25g
Ascorbic acid	0.1g
Bromthymol Blue	0.01g
Carbohydrate or starch solution	100.0mL

pH 7.0 ± 0.1 at 25°C

Source: This medium is available as a premixed powder from Difco Laboratories.

Carbohydrate Solution:
Composition per 100mL:

Carbohydrate	6.0g

Preparation of Carbohydrate Solution: Add carbohydrate to distilled/deionized water and bring volume to 10.0mL. Mix thoroughly. Filter sterilize.

Starch Solution:
Composition per 100mL:

Starch	2.5g

Preparation of Starch Solution: Add starch to distilled/deionized water and bring volume to 100.0mL. Mix thoroughly. Filter sterilize.

Preparation of Medium: Add components, except carbohydrate solution, to distilled/deionized water and bring volume to 900.0mL. Mix thoroughly. Distribute into tubes or flasks. Autoclave for 15 min at 15 psi pressure–121°C. Cool to 45–50°C. Aseptically add 100.0mL sterile carbohydrate solution. Mix thoroughly. Aseptically distribute into sterile tubes or flasks. Loosen caps on tubes. Place in an anaerobic chamber under an atmosphere of 85% N_2, 10% H_2, and 5% CO_2. Fasten the caps securely or maintain in an anaerobic chamber.

Use: For differentiation of anaerobic bacteria based upon carbohydrate fermentation. Bacteria that ferment carbohydrates turn the medium yellow.

Ferrous Sulfide Agar

Composition per 1200mL:

Agar layer	1.0L
Liquid overlay	200.0mL

Agar Layer:
Composition per liter:

Agar	30.0g
FeS washed precipitate supension	500.0mL

Preparation of Agar Layer: Add agar to distilled/deionized water and bring volume to 500.0mL. Mix thoroughly. Gently heat and bring to boiling. Autoclave for 15 min at 15 psi pressure–121°C. Cool to 45°–50°C. Heat FeS washed precipitate suspension to 45°–50°C. Mix thoroughly. Aseptically add 500.0mL sterile FeS washed precipitate supension to 500.0mL sterile agar at 45°–50°C. Mix thoroughly.

FeS Washed Precipitate Suspension:
Composition per 500mL:

$Fe(NH_4)_2(SO_4)_2 \cdot 6H_2O$	78.4g
$Na_2S \cdot 9H_2O$	15.6g

Preparation of FeS Washed Precipitate Suspension: Add $Na_2S \cdot 9H_2O$ and $Fe(NH_4)_2(SO_4)_2$ to 500.0mL boiling distilled/deionized water. Let precipitate settle from the hot solution in a completely filled and stoppered bottle. Wash precipitate four times by decanting supernatant and replacing each time with 500.0mL boiling distilled/deionizedwater. Store FeS washed precipitate suspension in a completely filled 500.0mL glass stoppered bottle.

Liquid Overlay:
Composition per liter:

$(NH_4)_2Cl$	1.0g
K_2HPO_4	0.5g
$MgSO_4 \cdot 7H_2O$	0.2g
$CaCl_2$	0.1g

Preparation of Liquid Overlay: Add components to distilled/deionized water and bring volume to 1.0L. Mix thoroughly. Autoclave for 15 min at 15 psi pressure–121°C. Cool to 25°C. Aseptically bubble 100% CO_2 for 15 sec.

Preparation of Medium: Aseptically distribute agar layer into sterile tubes in 10.0mL volumes. Allow tubes to cool in a slanted poistion. Aseptically add 2.0mL of sterile liquid overlay to each tube.

Use: For enumeration, enrichment and isolation of iron and sulfur bacteria, including *Gallionella ferruginea.*

Fervidobacterium Medium

Composition per liter:

Pancreatic digest of casein	10.0g
Glucose	5.0g
Yeast extract	3.0g
K_2HPO_4	1.5g
NH_4Cl	0.9g
KH_2PO_4	0.75g
$MgCl_2 \cdot 6H_2O$	0.2g
$Na_2S \cdot 9H_2O$ solution	10.0mL
Trace elements solution	9.0mL
Wolfe's vitamin solution	5.0mL
Resazurin (0.2% solution)	1.0mL
$FeSO_4 \cdot 7H_2O$ (10% solution)	0.03mL

pH 7.0 ± 0.1 at 25°C

Trace Elements Solution:
Composition per liter:

Nitrilotriacetic acid	12.5g
NaCl	1.0g
$FeCl_3 \cdot 4H_2O$	0.2g

$CaCl_2 \cdot 2H_2O$	0.1g
$MnCl_2 \cdot 4H_2O$	0.1g
$ZnCl_2$	0.1g
$CuCl_2$	0.02g
Na_2SeO_3	0.02g
$CoCl_2 \cdot 6H_2O$	0.017g
H_3BO_3	0.01g
$Na_2MoO_4 \cdot 2H_2O$	0.01g

Preparation of Trace Elements Solution: Add nitrilotriacetic acid to 500.0mL of distilled/deionized water. Dissolve by adjusting pH to 6.5 with KOH. Add remaining components. Add distilled/deionized water to 1.0L. Filter sterilize. Maintain under an atmosphere of 100% N_2.

Wolfe's Vitamin Solution:
Composition per liter:

Pyridoxine·HCl	0.01g
Thiamine·HCl	5.0mg
Riboflavin	5.0mg
Nicotinic acid	5.0mg
Calcium pantothenate	5.0mg
p-Aminobenzoic acid	5.0mg
Thioctic acid	5.0mg
Biotin	2.0mg
Folic acid	2.0mg
Cyanocobalamin	0.1mg

Preparation of Wolfe's Vitamin Solution: Add components to distilled/deionized water and bring volume to 1.0L. Mix thoroughly. Filter sterilize. Maintain under an atmosphere of 100% N_2.

$Na_2S \cdot 9H_2O$ Solution:
Composition per 10mL:

$Na_2S \cdot 9H_2O$	0.5g

Preparation of $Na_2S \cdot 9H_2O$ Solution: Add $Na_2S \cdot 9H_2O$ to distilled/deionized water and bring volume to 10.0mL. Mix thoroughly. Filter sterilize. Maintain under an atmosphere of 100% N_2.

Preparation of Medium: Add components, except sodium sulfide solution, trace elements solution, and Wolfe's vitamin solution, to distilled/deionized water and bring volume to 976.0mL. Mix thoroughly. Autoclave for 15 min at 15 psi pressure–121°C. Cool under an atmosphere of 100% N_2. Aseptically add 9.0mL of trace elements solution and 5.0mL of Wolfe's vitamin solution under an atmosphere of 100% N_2. Mix thoroughly. Aseptically distribute into sterile tubes or flasks under an atmosphere of 100% N_2. Add $Na_2S \cdot 9H_2O$ solution just prior to use to a concentration 0.1%.

Use: For cultivation and maintenance of *Clostridium* species, *Fervidobacterium nodosum, F. islandicum* and *Thermoanaerobium brockii.*

F–G Agar
(Feeley–Gorman Agar)

Composition per liter:

Casein, acid hydrolyzed	17.5g
Agar	17.0g
Beef extract	3.0g
Starch	1.5g
Cysteine solution	10.0mL
$Fe_4(P_2O_7)_3$ solution	10.0mL

pH 6.9 ± 0.05 at 25°C

Cysteine Solution:
Composition per 10mL:

L-Cysteine $HCl \cdot H_2O$	0.4g

Preparation of Cysteine Solution: Add L-Cysteine $HCl \cdot H_2O$ to distilled/deionized water and bring volume to 10.0mL. Mix thoroughly. Filter sterilize.

$Fe_4(P_2O_7)_3$ Solution:
Composition per 10mL:

$Fe_4(P_2O_7)_3$	0.25g

Preparation of $Fe_4(P_2O_7)_3$ Solution: Add $Fe_4(P_2O_7)_3$ to distilled/deionized water and bring volume to 10.0mL. Mix thoroughly. Filter sterilize.

Preparation of Medium: Add components, except cysteine solution and $Fe_4(P_2O_7)_3$ solution, to distilled/deionized water and bring volume to 980.0mL. Mix thoroughly. Gently heat and bring to boiling. Autoclave for 15 min at 15 psi pressure–121°C. Cool to 45°–50°C. Aseptically add 10.0mL cysteine solution. Mix thoroughly. Aseptically add 10.0mL $Fe_4(P_2O_7)_3$ solution. Mix thoroughly. Adjust pH to 6.9. Pour into sterile Petri dishes or distribute into sterile tubes.

Use: For the isolation and cultivation of *Legionella pneumophila.*

F–G Agar with Selenium
(Feeley–Gorman Agar with Selenium)

Composition per liter:

Casein, acid hydrolyzed	17.5g
Agar	17.0g
Beef extract	3.0g
Starch	1.5g
Cysteine solution	10.0mL
$Fe_4(P_2O_7)_3$ solution	10.0mL
$Na_2SeO_3 \cdot 5H_2O$ solution	10.0mL

pH 6.9 ± 0.05 at 25°C

Cysteine Solution:
Composition per 10mL:

L-Cysteine $HCl \cdot H_2O$	0.4g

Preparation of Cysteine Solution: Add L-Cysteine HCl·H_2O to distilled/deionized water and bring volume to 10.0mL. Mix thoroughly. Filter sterilize.

$Fe_4(P_2O_7)_3$ Solution:

Composition per 10mL:

$Fe_4(P_2O_7)_3$	0.25g

Preparation of $Fe_4(P_2O_7)_3$ Solution: Add $Fe_4(P_2O_7)_3$ to distilled/deionized water and bring volume to 10.0mL. Mix thoroughly. Filter sterilize.

Na_2SeO_3·$5H_2O$ Solution:

Composition per 10mL:

Na_2SeO_3·$5H_2O$	0.010g

Preparation of Na_2SeO_3·$5H_2O$ Solution: Add Na_2SeO_3·$5H_2O$ to distilled/deionized water and bring volume to 10.0mL. Mix thoroughly. Filter sterilize.

Preparation of Medium: Add components—except cysteine solution, $Fe_4(P_2O_7)_3$ solution, and Na_2SeO_3·$5H_2O$ solution—to distilled/deionized water and bring volume to 970.0mL. Mix thoroughly. Gently heat and bring to boiling. Autoclave for 15 min at 15 psi pressure–121°C. Cool to 45°–50°C. Aseptically add 10.0mL of sterile cysteine solution. Mix thoroughly. Aseptically add 10.0mL of sterile $Fe_4(P_2O_7)_3$ solution and 10.0mL of sterile Na_2SeO_3·$5H_2O$ solution. Mix thoroughly. Adjust pH to 6.9. Pour into sterile Petri dishes or distribute into sterile tubes.

Use: For the isolation and cultivation of *Legionella pneumophila*.

F–G Broth (Feeley–Gorman Broth)

Composition per liter:

Casein, acid hydrolyzed	17.5g
Beef extract	3.0g
Starch	1.5g
Cysteine solution	10.0mL
$Fe_4(P_2O_7)_3$ solution	10.0mL

pH 6.9 ± 0.05 at 25°C

Cysteine Solution:

Composition per 10mL:

L-Cysteine HCl·H_2O	0.4g

Preparation of Cysteine Solution: Add L-Cysteine HCl·H_2O to distilled/deionized water and bring volume to 10.0mL. Mix thoroughly. Filter sterilize.

$Fe_4(P_2O_7)_3$ Solution:

Composition per 10mL:

$Fe_4(P_2O_7)_3$	0.25g

Preparation of $Fe_4(P_2O_7)_3$ Solution: Add $Fe_4(P_2O_7)_3$ to distilled/deionized water and bring volume to 10.0mL. Mix thoroughly. Filter sterilize.

Preparation of Medium: Add components, except cysteine solution and $Fe_4(P_2O_7)_3$ solution, to distilled/deionized water and bring volume to 980.0mL. Mix thoroughly. Gently heat and bring to boiling. Autoclave for 15 min at 15 psi pressure–121°C. Cool to 45°–50°C. Aseptically add 10.0mL cysteine solution. Mix thoroughly. Aseptically add 10.0mL $Fe_4(P_2O_7)_3$ solution. Mix thoroughly. Adjust pH to 6.9. Aseptically distribute into sterile tubes or flasks.

Use: For the cultivation of *Legionella pneumophila*.

fGTC Agar

Composition per liter:

Pancreatic digest of casein	15.0g
Agar	15.0g
Papaic digest of soybean meal	5.0g
NaCl	5.0g
KH_2PO_4	5.0g
Amylose azure	3.0g
Galactose	1.0g
Thallous acetate	0.5g
MUG (4-Methylumbelliferyl-α-D-galactoside	0.1g
$NaHCO_3$ solution	20.0mL
Gentamicin solution	2.5mL
Tween™ 80	0.75mL

pH 7.3 ± 0.2 at 25°C

Gentamicin Solution:

Composition per 10mL:

Gentamicin	0.01g

Preparation of Gentamicin Solution: Add gentamicin to distilled/deionized water and bring volume to 10.0mL. Mix thoroughly.

$NaHCO_3$ Solution:

Composition per 20mL:

$NaHCO_3$	2.0g

Preparation of $NaHCO_3$ Solution: Add the $NaHCO_3$ to distilled/deionized water and bring volume to 20.0mL. Mix thoroughly. Filter sterilize. Use freshly prepared solution.

Preparation of Medium: Add components, except $NaHCO_3$ solution, to distilled/deionized water and bring volume to 980.0mL. Mix thoroughly. Gently heat and bring to boiling. Autoclave for 15 min at 15 psi pressure–121°C. Cool to 50°C. Aseptically add sterile $NaHCO_3$ solution. Mix thoroughly. Pour into sterile Petri dishes.

Use: For the cultivation, differentiation and enumeration of *Enterococcus* species based on starch hydrolysis and production of fluorescence. Bacteria that hydrolyze starch, such as *Streptococcus bovis*, appear as colonies surrounded by a clear zone. Bacteria that produce fluorescence, such as *Streptococcus bovis* and *Enterococcus faecium*, appear as colonies surrounded by a zone of of bright bluish fluorescence when viewed under a long-wave UV lamp. Other bacteria, such as *Enterococcus faecalis*, *Enterococcus avium*, or *Streptococcus equinus*, do not hydrolyze starch or produce fluorescence.

Fish Peptone Agar

Composition per liter:

Agar	5.0g
Maltose	5.0g
NaCl	5.0g
Peptone	5.0g
Pancreatic digest of casein	5.0g
Yeast extract	5.0g
Trout tissue extract solution	50.0mL

pH 7.0 ± 0.2 at 25°C

Trout Tissue Extract Solution:

Composition per liter:

Fish (brook trout)	500.0g
Pepsin	1.0g
HCl, concentrated	15.0mL

Preparation of Trout Tissue Extract: Add 1.0L distilled/deionized water to brook trout and blend for 20–30 min. Add 1.0g pepsin and 15.0mL concentrated HCl to digest the trout proteins. Incubate for 12 hr at 45°C. Adjust pH to 7.0. Allow solids to settle. Filter sterilize. Do not autoclave. Store at 5°C.

Preparation of Medium: Add components, except trout tissue extract, to distilled/deionized water and bring volume to 950.0L. Mix thoroughly. Gently heat and bring to boiling. Autoclave for 15 min at 13 psi pressure–118°C. Cool to 45°–50°C. Aseptically add 50.0mL sterile trout tissue extract. Mix thoroughly. Pour into sterile Petri dishes or distribute into sterile tubes.

Use: For the cultivation and maintenance of *Aeromonas salmonicida*.

Fish Peptone Broth

Composition per liter:

Maltose	5.0g
NaCl	5.0g
Peptone	5.0g
Pancreatic digest of casein	5.0g
Yeast extract	5.0g
Trout tissue extract solution	50.0mL

pH 7.0 ± 0.2 at 25°C

Trout Tissue Extract Solution:

Composition per liter:

Fish (brook trout)	500.0g
Pepsin	1.0g
HCl, concentrated	15.0mL

Preparation of Trout Tissue Extract: Add 1.0L distilled/deionized water to brook trout and blend for 20–30 min. Add 1.0 g pepsin and 15.0mL concentrated HCl to digest the trout proteins. Incubate for 12 hr at 45°C. Adjust pH to 7.0. Allow solids to settle. Filter sterilize. Do not autoclave. Store at 5°C.

Preparation of Medium: Add components, except trout tissue extract, to distilled/deionized water and bring volume to 950.0L. Mix thoroughly. Gently heat and bring to boiling. Autoclave for 15 min at 10 psi pressure–118°C. Cool to 45°–50°C. Aseptically add 50.0mL sterile trout tissue extract. Mix thoroughly. Aseptically distribute into sterile tubes or flasks.

Use: For the cultivation of *Aeromonas salmonicida*.

Flagella Broth

Composition per liter:

Tryptose or biosate	10.0g
NaCl	2.5g
K_2HPO_4	1.0g

pH 7.0 ± 0.1 at 25°C

Preparation of Medium: Add components to distilled/deionized water and bring volume to 1.0L. Mix thoroughly. Gently heat and bring to boiling. Distribute into tubes or flasks. Autoclave for 15 min at 15 psi pressure–121°C. Pour into sterile Petri dishes or leave in tubes.

Use: For the cultivation of flagella-producing bacteria.

Flavobacterium M1 Agar

Composition per liter:

Agar	15.0g
Proteose peptone	5.0g
NaCl	3.0g
Beef extract	2.0g
Yeast extract	1.0g

pH 7.0–7.2 at 25°C

Preparation of Medium: Add components to distilled/deionized water and bring volume to 1.0L. Mix thoroughly. Gently heat and bring to boiling. Distribute into tubes or flasks. Autoclave for 15 min at 15 psi pressure–121°C. Pour into sterile Petri dishes or leave in tubes.

Use: For the cultivation and maintenance of *Flavobacterium indolthelicum*.

Flavobacterium Medium (ATCC Medium 65)

Composition per liter:

Agar	12.0g
Sodium caseinate	2.0g

Peptone......1.0g
K_2HPO_4......0.5g
Yeast extract......0.5g

pH 7.4 ± 0.2 at 25°C

Preparation of Medium: Add components to distilled/deionized water and bring volume to 1.0L. Mix thoroughly. Gently heat and bring to boiling. Distribute into tubes or flasks. Autoclave for 15 min at 15 psi pressure–121°C. Pour into sterile Petri dishes or leave in tubes.

Use: For the cultivation and maintenance of *Flavobacterium aquatile.*

Flavobacterium Medium (ATCC Medium 1687)

Composition per liter:

Sodium glutamate......4.0g
K_2HPO_4......0.65g
$NaNO_3$......0.5g
KH_2PO_4......0.19g
$MgSO_4·7H_2O$......0.1g
$FeSO_4$ solution......2.0mL

pH 7.4 ± 0.2 at 25°C

$FeSO_4$ Solution:

Composition per 10mL:

$FeSO_4·7H_2O$......0.03g

Preparation of $FeSO_4$ Solution: Add $FeSO_4$ to distilled/deionized water and bring volume to 10.0mL. Mix thoroughly. Filter sterilize.

Preparation of Medium: Add components, except $FeSO_4$ solution, to distilled/deionized water and bring volume to 998.0mL. Mix thoroughly. Autoclave for 15 min at 15 psi pressure–121°C. Cool to 25°C. Aseptically add 2.0mL sterile $FeSO_4$ solution. Mix thoroughly. Adjust pH to 7.4. Aseptically distribute into sterile tubes or flasks.

Use: For the cultivation of *Flavobacterium* species.

Flavobacterium Medium M1

Composition per liter:

Agar......12.0g
Proteose peptone......5.0g
NaCl......3.0g
Beef extract......2.0g
Yeast extract......0.2g

pH 7.2–7.4 at 25°C

Preparation of Medium: Add components to distilled/deionized water and bring volume to 1.0L. Mix thoroughly. Gently heat and bring to boiling. Distribute into tubes or flasks. Autoclave for 15 min at 15 psi pressure–121°C. Pour into sterile Petri dishes or leave in tubes.

Use: For the isolation and cultivation of *Flavobacterium* species.

Fletcher Medium

Composition per liter:

Agar......1.5g
NaCl......0.5g
Peptone......0.3g
Beef extract......0.2g
Rabbit serum......50.0mL

pH 7.9 ± 0.1 at 25°C

Source: This medium is available as a premixed powder from Difco Laboratories.

Preparation of Medium: Add components, except rabbit serum, to distilled/deionized water and bring volume to 950.0mL. Mix thoroughly. Gently heat and bring to boiling. Autoclave for 15 min at 15 psi pressure–121°C. Cool to 50°–55°C. Aseptically add 50.0mL sterile rabbit serum. Mix thoroughly. Aseptically distribute into sterile tubes or flasks.

Use: For the isolation, cultivation and maintenance of cultures of *Leptospira* species.

Fletcher Medium with Fluorouracil (Flurouracil *Leptospira* Medium)

Composition per liter:

Agar......1.5g
NaCl......0.5g
Peptone......0.3g
Beef extract......0.2g
Rabbit serum......50.0mL
Fluorouracil solution......20.0mL

pH 7.9 ± 0.1 at 25°C

Fluorouracil Solution:

Composition per 100mL:

Fluorouracil......10.0g

Preparation of Fluorouracil Solution: Add fluorouracil to 50.0mL distilled/deionized water. Add 1.0mL 2*N* NaOH and bring volume to 100.0mL. Gently heat to 56°C for 2 hr. Adjust pH to 7.4–7.6 with NaOH. Mix thoroughly. Filter sterilize.

Preparation of Medium: Add components, except rabbit serum and fluorouracil solution, to distilled/deionized water and bring volume to 930.0mL. Mix thoroughly. Gently heat and bring to boiling. Autoclave for 15 min at 15 psi pressure–121°C. Cool to 50°–55°C. Aseptically add 80.0mL sterile rabbit serum. Mix thoroughly. Aseptically distribute into sterile tubes or flasks. Immediately prior to use add 0.1mL fluorouracil solution per 5.0mL medium.

Use: For the isolation, cultivation and maintenance of cultures of *Leptospira* species.

Fletcher's Semisolid Medium

Composition per 2120mL:

Agar	1.5g
NaCl	0.5g
Peptone	0.3g
Beef extract	0.2g
Rabbit serum	240.0mL

pH 7.9 ± 0.1 at 25°C

Preparation of Medium: Add components, except rabbit serum, to distilled/deionized water and bring volume to 1880.0mL. Mix thoroughly. Gently heat and bring to boiling. Autoclave for 15 min at 15 psi pressure–121°C. Cool to 50°–55°C. Aseptically add 240.0mL sterile rabbit serum. Mix thoroughly. Aseptically distribute into sterile tubes or flasks.

Use: For the isolation, cultivation and maintenance of cultures of *Leptospira* species.

Flexibacter Agar

Composition per liter:

Agar	15.0g
Monosodium glutamate	5.0g
Pancreatic digest of casein	1.0g
Vitamin-free casamino acids	1.0g
Sodium glycerophosphate	0.1g
Vitamin B_{12}	1.0μg
Trace element solution HO-LE	1.0mL

Trace Element Solution HO-LE:
Composition per liter:

H_3BO_3	2.85g
$MnCl_2·4H_2O$	1.8g
Sodium tartrate	1.77g
$FeSO_4·7H_2O$	1.36g
$CoCl_2·6H_2O$	0.04g
$CuCl_2.2H_2O$	0.027g
$Na_2MoO_4·2H_2O$	0.025g
$ZnCl_2$	0.020g

Preparation of Trace Element Solution HO-LE: Add components to distilled/deionized water and bring volume to 1.0L. Mix thoroughly. Filter sterilize.

Preparation of Medium: Add components to filtered seawater and bring volume to 999.0mL. Mix thoroughly. Gently heat and bring to boiling. Autoclave for 15 min at 15 psi pressure–121°C. Cool to 45°–50°C. Aseptically add 1.0mL sterile trace element solution HO-LE. Mix thoroughly. Pour into sterile Petri dishes or distribute into sterile tubes.

Use: For the cultivation and maintenance of *Flexibacter polymorphus.*

Flexibacter Broth

Composition per liter:

Monosodium glutamate	5.0g
Pancreatic digest of casein	1.0g
Vitamin-free casamino acids	1.0g
Sodium glycerophosphate	0.1g
Vitamin B_{12}	1.0μg
Trace element solution HO-LE	1.0mL

Trace Element Solution HO-LE:
Composition per liter:

H_3BO_3	2.85g
$MnCl_2·4H_2O$	1.8g
Sodium tartrate	1.77g
$FeSO_4·7H_2O$	1.36g
$CoCl_2·6H_2O$	0.04g
$CuCl_2.2H_2O$	0.027g
$Na_2MoO_4·2H_2O$	0.025g
$ZnCl_2$	0.020g

Preparation of Trace Element Solution HO-LE: Add components to distilled/deionized water and bring volume to 1.0L. Mix thoroughly. Filter sterilize.

Preparation of Medium: Add components to filtered seawater and bring volume to 999.0mL. Mix thoroughly. Autoclave for 15 min at 15 psi pressure–121°C. Cool to 45°–50°C. Aseptically add 1.0mL sterile trace element solution HO-LE. Mix thoroughly. Aseptically distribute into sterile tubes or flasks.

Use: For the cultivation of *Flexibacter polymorphus.*

Flexibacter Medium (ATCC Medium 1559)

Composition per liter:

Solution B	700.0mL
Solution A	300.0mL

Solution A:
Composition per 300 mL:

Pancreatic digest of casein	0.5g
Yeast extract	0.5g
Beef extract	0.2g
Sodium acetate	0.2g

Preparation of Solution A: Add components to distilled/deionized water and bring volume to 1.0L. Mix thoroughly. Autoclave for 15 min at 15 psi pressure–121°C. Cool to 45°–50°C.

Solution B:

Aged seawater .... 700.0 mL

Preparation of Solution B: Allow seawater to sit for 7 days. Autoclave for 15 min at 15 psi pressure–121°C. Cool to 45°–50°C.

Preparation of Medium: Aseptically add 300.0mL of sterile solution A to 700.0mL of sterile solution B at 45°–50°C. Mix thoroughly. Aseptically distribute into sterile tubes or flasks.

Use: For the cultivation of *Flexibacter maritimus.*

Flexibacterium Medium

Composition per 1060mL:

Yeast extract 1.0g
$Ca(NO_3)_2 \cdot 4H_2O$ 0.1g
K_2HPO_4 .... 0.02g
Seawater, filtered 1.0L
Glucose solution 50.0mL
Trace elements 10.0mL
pH 7.0 ± 0.2 at 25°C

Glucose Solution:

Composition per 50mL:

Glucose 1.0g

Preparation of Glucose Solution: Add glucose to distilled/deionized water and bring volume to 50.0mL. Mix thoroughly. Autoclave for 15 min at 15 psi pressure–121°C. Cool to 25°C.

Trace Elements:

Composition per liter:

$FeSO_4 \cdot 7H_2O$ 0.5mg
$ZnSO_4 \cdot 7H_2O$ 0.3mg
H_3BO_3 .... 0.1mg
$CoCl_2 \cdot 6H_2O$ 0.1mg
$CuSO_4 \cdot 5H_2O$ 0.1mg
$MnSO_4 \cdot 4H_2O$ 0.1mg
$Na_2MoO_4 \cdot 2H_2O$ 0.1mg

Preparation of Trace Elements: Add components to distilled/deionized water and bring volume to 1.0L. Mix thoroughly.

Preparation of Medium: Combine components, except glucose solution. Mix thoroughly. Adjust pH to 7.2. Gently heat and bring to boiling. Autoclave for 15 min at 15 psi pressure–121°C. Cool to 45°–50°C. Aseptically add sterile glucose solution. Mix thoroughly. Aseptically distribute into sterile tubes or bottles.

Use: For the cultivation of *Flexibacter litorale* and *F. marinum.*

Flexibacterium Medium

Composition per 1050mL:

Tris(hydroxymethyl)aminomethane buffer 1.0g
Yeast extract 1.0g
$CaCl_2 \cdot 2H_2O$ 0.1g
KCl 0.1g
$MgSO_4 \cdot 7H_2O$ 0.1g
Sodium glycerophosphate 0.1g
$NaNO_3$ 0.1g
Cobalamin 1.0μg
Glucose solution 50.0mL
Trace elements 10.0mL
pH 7.5 ± 0.2 at 25°C

Glucose Solution:

Composition per 50mL:

Glucose 1.0g

Preparation of Glucose Solution: Add glucose to distilled/deionized water and bring volume to 50.0mL. Mix thoroughly. Autoclave for 15 min at 15 psi pressure–121°C. Cool to 25°C.

Trace Elements:

Composition per liter:

$FeSO_4 \cdot 7H_2O$ 0.5mg
$ZnSO_4 \cdot 7H_2O$ 0.3mg
H_3BO_3 .... 0.1mg
$CoCl_2 \cdot 6H_2O$ 0.1mg
$CuSO_4 \cdot 5H_2O$ 0.1mg
$MnSO_4 \cdot 4H_2O$ 0.1mg
$Na_2MoO_4 \cdot 2H_2O$ 0.1mg

Preparation of Trace Elements: Add components to distilled/deionized water and bring volume to 1.0L. Mix thoroughly.

Preparation of Medium: Add components, except glucose solution, to distilled/deionized water and bring volume to 1.0L. Mix thoroughly. Adjust pH to 7.5. Autoclave for 15 min at 15 psi pressure–121°C. Cool to 45°–50°C. Aseptically add sterile glucose solution. Mix thoroughly. Aseptically distribute into sterile tubes or flasks.

Use: For the cultivation of *Saprospira thermalis, Flexibacter elegans* and *F. rubrum.*

Flexibacterium Medium

Composition per 1050mL:

Tris(hydroxymethyl)aminomethane buffer 1.0g
Yeast extract 1.0g
Glycerol 1.0g
$CaCl_2 \cdot 2H_2O$ 0.1g
KCl 0.1g
$MgSO_4 \cdot 7H_2O$ 0.1g
Sodium glycerophosphate 0.1g
$NaNO_3$ 0.1g

Cobalamin	1.0μg
Trace elements	10.0mL

pH 7.5 ± 0.2 at 25°C

Trace Elements:
Composition per liter:

$FeSO_4·7H_2O$	0.5mg
$ZnSO_4·7H_2O$	0.3mg
H_3BO_3	0.1mg
$CoCl_2·6H_2O$	0.1mg
$CuSO_4·5H_2O$	0.1mg
$MnSO_4·4H_2O$	0.1mg
$Na_2MoO_4·2H_2O$	0.1mg

Preparation of Trace Elements: Add components to distilled/deionized water and bring volume to 1.0L. Mix thoroughly.

Preparation of Medium: Add components to distilled/deionized water and bring volume to 1.0L. Mix thoroughly. Adjust pH to 7.5. Distribute into sterile tubes or flasks. Autoclave for 15 min at 15 psi pressure–121°C.

Use: For the cultivation of *Saprospira albida*.

Flexiligladius Medium

Composition per liter:

Beef heart, solids from infusion	50.0g
Agar	15.0g
Tryptose	1.0g
NaCl	0.5g

pH 7.4 ± 0.2 at 25°C

Preparation of Medium: Add components to distilled/deionized water and bring volume to 1.0L. Mix thoroughly. Gently heat and bring to boiling. Distribute into tubes or flasks. Autoclave for 15 min at 15 psi pressure–121°C. Pour into sterile Petri dishes or leave in tubes.

Use: For the cultivation and maintenance of *Ensifer adhaerens*.

FLN Medium (Fluorescence Lactose Nitrate Medium)

Composition per liter:

Lactose	20.0g
Agar	15.0g
Proteose peptone No. 3	10.0g
KNO_3	2.0g
K_2HPO_4	1.5g
$MgSO_4·7H_2O$	1.5g
$NaNO_2$	0.5g
Phenol red.	0.02g

pH 7.2 ±0.2 at 25°C

Preparation of Medium: Add components to distilled/deionized water and bring volume to 1.0L. Mix thoroughly. Gently heat and bring to boiling. Distribute into tubes or flasks. Autoclave for 15 min at 15 psi pressure–121°C. Pour into sterile Petri dishes or leave in tubes.

Use: For the differentiation of pseudomonads from other nonfermentative bacilli. Lactose fermentation is indicated by the medium turning yellow. *Pseudomonas cepacia* often produces acid from lactose. Denitrification from nitrate or nitrite is indicated by the formation of gas bubbles in the solid medium. *P. aeruginosa*, *P. mendocina*, and *P. denitrificans* are positive for denitrification. Fluorescein production is indicated by fluorescence under UV light. *P. aeruginosa* is positive for fluorescein production; *P. denitrificans* does not produce fluorescein.

Flo Agar

Composition per liter:

Agar	14.0g
Pancreatic digest of casein	10.0g
Peptic digest of animal tissue	10.0g
K_2HPO_4	1.5g
$MgSO_4·7H_2O$	1.5g

pH 7.2 ± 0.2 at 25°C

Source: This medium is available as a premixed powder from BBL Microbiology Systems.

Preparation of Medium: Add components to distilled/deionized water and bring volume to 1.0L. Mix thoroughly. Gently heat and bring to boiling. Distribute into tubes or flasks. Autoclave for 15 min at 15 psi pressure–121°C. Pour into sterile Petri dishes or leave in tubes.

Use: For cultivation of fluorescent *Pseudomonas* species.

Fluid Thioglycollate Medium

Composition per liter:

Pancreatic digest of casein	15.0g
Glucose	5.5g
Yeast extract	5.0g
NaCl	2.5g
Agar	0.75g
L-Cystine	0.5g
Sodium thioglycollate	0.5g
Resazurin	1.0mg

pH 7.1 ± 0.2 at 25°C

Source: This medium is available as a premixed powder from Difco Laboratories.

Preparation of Medium: Add components to distilled/deionized water and bring volume to 1.0L. Mix thoroughly. Gently heat and bring to boiling. Distribute into tubes or flasks. Autoclave for 15 min at 15 psi pressure–121°C. If medium becomes oxidized before use (resazurin turns red) heat in a boiling water bath to expel absorbed O_2. Cool to 25°C.

Use: For the cultivation of anaerobic, microaerophilic and aerobic microorganisms.

Fluorescent Pectolytic Agar (FPA Medium)

Composition per liter:

Proteose peptone No. 3	20.0g
Agar	15.0g
Pectin	5.0g
K_2HPO_4	1.5g
$MgSO_4 \cdot 7H_2O$	0.73g
Antibiotic solution	10.0mL

Antibiotic Solution:

Composition per 10mL:

Cycloheximide	0.075g
Novobiocin	0.045g
Penicillin G	75,000U
Ethanol	1.0mL

Preparation of Antibiotic Solution: Add components to 1.0mL of ethanol. Mix thoroughly. Let stand for 30 min. Bring volume to 10.0mL with distilled/deionized water. Mix thoroughly. Filter sterilize.

Preparation of Medium: Add components, except antibiotic solution, to distilled/deionized water and bring volume to 990.0mL. Mix thoroughly. Gently heat and bring to boiling. Autoclave for 15 min at 15 psi pressure–121°C. Cool to 45°–50°C. Aseptically add sterile antibiotic solution. Mix thoroughly. Pour into sterile Petri dishes.

Use: For the cultivation of fluorescent *Pseudomonas* species that are pectinolytic.

FN Medium (Fluorescence Denitrification Medium)

Composition per liter:

Agar	15.0g
Proteose peptone No. 3	10.0g
KNO_3	2.0g
K_2HPO_4	1.5g
$MgSO_4 \cdot 7H_2O$	1.5g
$NaNO_2$	0.5g

pH 7.2 ±0.2 at 25°C

Preparation of Medium: Add components to distilled/deionized water and bring volume to 1.0L. Mix thoroughly. Gently heat and bring to boiling. Distribute into tubes or flasks. Autoclave for 15 min at 15 psi pressure–121°C. Pour into sterile Petri dishes or leave in tubes.

Use: For the differentiation of pseudomonads from other nonfermentative bacilli. Denitrification from nitrate or nitrite is indicated by the formation of gas bubbles in the solid medium. *P. aeruginosa, P. mendocina,* and *P. denitrificans* are positive for denitrification. Fluorescein production is indicated by fluorescence under UV light. *P. aeruginosa* is positive for fluorescein production; *P. denitrificans* does not produce fluorescein.

Folic Acid Agar

Composition per liter:

Noble agar	15.0g
$K_2HPO_4 \cdot 3H_2O$	1.2g
Folic acid	1.0g
KH_2PO_4	0.5g
Salts A	5.0mL
Salts B	1.5mL

Salts A:

Composition per 100mL:

$MgSO_4 \cdot 7H_2O$	1.0g
$CaCl_2 \cdot 2H_2O$	0.1g
$FeSO_4 \cdot 7H_2O$	0.1g

Preparation of Salts A Solution: Add components to distilled/deionized water and bring volume to 100.0mL. Mix thoroughly. Maintain for 3 days at 25°C to dissolve. Filter sterilize the supernatant.

Salts B:

Composition per 100mL:

$MnSO_4$	0.1g
Na_2MoO_4	0.1g

Preparation of Salts B Solution: Add components to distilled/deionized water and bring volume to 100.0mL. Mix thoroughly. Filter sterilize.

Preparation of Medium: Add components, except salts A solution and salts B solution, to distilled/deionized water and bring volume to 994.5mL. Mix thoroughly. Gently heat and bring to boiling. Autoclave for 15 min at 15 psi pressure–121°C. Cool to 45°–50°C. Aseptically add 5.0mL sterile salts A solution and 1.5mL sterile salts B solution. Mix thoroughly. Pour into sterile Petri dishes or distribute into sterile tubes.

Use: For the cultivation and maintenance of *Pseudomonas* species.

Fomes annosus Isolation Medium No. 1

Composition per liter:

Agar	20.0g
Peptone	5.0g
KH_2PO_4	0.5g
$MgSO_4$	0.25g
Pentachloronitrobenzene (PCNB)	0.19mg
Streptomycin	0.1mg
Ethanol (95% solution)	20.0mL
Lactic acid (50% solution)	2.0mL

Preparation of Medium: Filter sterilize ethanol and lactic acid. Add components, except ethanol and lactic acid, to distilled/deionized water and bring volume to 978.0mL. Mix thoroughly. Gently heat and bring to boiling. Autoclave for 15 min at 15 psi pressure–121°C. Cool to 45°–50°C. Aseptically add sterile ethanol and lactic acid. Mix thoroughly. Pour into sterile Petri dishes or distribute into sterile tubes

Use: For the cultivation of *Fomes annosus*.

Fomes annosus Isolation Medium No. 2

Composition per liter:

Malt extract	20.0g
Agar	17.0g
Streptomycin sulfate solution	10.0mL
Phenylphenol solution	2.5mL
Lactic acid (50% solution)	1.0mL

Streptomycin Sulfate Solution:
Composition per 10mL:

Streptomycin sulfate	0.1g

Preparation of Streptomycin Sulfate Solution: Add streptomycin sulfate to distilled/deionized water and bring volume to 10.0mL. Mix thoroughly. Filter sterilize.

Phenylphenol Solution:
Composition per 20mL:

o-Phenylphenol	0.48g
Ethanol (95% solution)	20.0mL

Preparation of Phenylphenol Solution: Add *o*-phenylphenol to 20.0mL of ethanol. Mix thoroughly. Filter sterilize.

Preparation of Medium: Filter sterilize lactic acid. Add components—except lactic acid, phenylphenol solution, and streptomycin sulfate solution—to distilled/deionized water and bring volume to 986.5mL. Mix thoroughly. Gently heat and bring to boiling. Autoclave for 15 min at 15 psi pressure–121°C. Cool to 45°–50°C. Aseptically add sterile lactic acid, phenylphenol solution, and streptomycin sulfate solution. Mix thoroughly. Pour into sterile Petri dishes or distribute into sterile tubes.

Use: For the cultivation of *Fomes annosus*.

Formivibrio citricus Medium

Composition per liter:

Trisodium citrate	2.94g
$NaHCO_3$	2.5g
$MgCl_2 \cdot 6H_2O$	2.03g
KH_2PO_4	1.36g
NH_4Cl	0.53g
$Na_2S \cdot 9H_2O$	0.24g
$CaCl_2 \cdot 2H_2O$	0.15g
Modified Wolfe's metals solution	10.0mL
Wolfe's vitamins solution	10.0mL

pH 7.5–7.7 at 25°C

Modified Wolfe's Metals Solution:
Composition per liter

$NaWO_4 \cdot 2H_2O$	10.01g
Na_2SeO_3	0.01g
$NiCl_2 \cdot 6H_2O$	0.01g
Wolfe's mineral solution	1.0L

Preparation of Modified Wolfe's Metals Solution: Add components to Wolfe's mineral solution and bring volume to 10.0mL. Mix thoroughly.

Wolfe's Mineral Solution:
Composition per liter

$MgSO_4 \cdot 7H_2O$	3.0g
Nitriloacetic acid	1.5g
NaCl	1.0g
$MnSO_4 \cdot H_2O$	0.5g
$FeSO_4 \cdot 7H_2O$	0.1g
$CoCl_2 \cdot 6H_2O$	0.1g
$CaCl_2$	0.1g
$ZnSO_4 \cdot 7H_2O$	0.1g
$CuSO_4 \cdot 5H_2O$	0.01g
$AlK(SO_4)_2 \cdot 12H_2O$	0.01g
H_3BO_3	0.01g
$Na_2MoO_4 \cdot 2H_2O$	0.01g

Preparation of Wolfe's Mineral Solution: Add nitrilotriacetic acid to 500.0mL of distilled/deionized water. Dissolve by adjusting pH to 6.5 with KOH. Add remaining components. Add distilled/deionized water to 1.0L.

Wolfe's Vitamin Solution:
Composition per liter:

Pyridoxine·HCl	0.01g
Thiamine·HCl	5.0mg
Riboflavin	5.0mg
Nicotinic acid	5.0mg
Calcium pantothenate	5.0mg
p-Aminobenzoic acid	5.0mg

Thioctic acid	5.0mg
Biotin	2.0mg
Folic acid	2.0mg
Cyanocobalamin	0.1mg

Preparation of Wolfe's Vitamin Solution: Add components to distilled/deionized water and bring volume to 1.0L. Mix thoroughly.

Preparation of Medium: Add components, except $NaHCO_3$ and sodium sulfide, to distilled/deionized water and bring volume to 1.0L. Mix thoroughly. Gently heat and bring to boiling. Cool under an atmosphere of 90% N_2 + 10% CO_2. Add bicarbonate and sodium sulfide. Mix thoroughly. Adjust pH to 7.5–7.7. Anaerobically distribute into tubes under 90% N_2 + 10% CO_2. Cap tubes with rubber stoppers. Place tubes in a press. Autoclave at 121°C for 15 minutes with fast exhaust.

Use: For the cultivation and maintenance of *Formivibrio citricus.*

Fowells Acetate Agar

Composition per liter:

Agar	20.0g
Sodium acetate trihydrate	5.0g

pH 6.5–7.0 at 25°C

Preparation of Medium: Add sodium acetate trihydrate to distilled/deionized water and bring volume to 1.0L. Adjust pH to 6.5–7.0. Add agar. Mix thoroughly. Gently heat and bring to boiling. Distribute into tubes or flasks. Autoclave for 15 min at 15 psi pressure–121°C. Pour into sterile Petri dishes or leave in tubes.

Use: For the cultivation of fungi with the production of ascospores.

FPA Medium

Composition per liter:

Proteose peptone No. 3	20.0g
Agar	15.0g
Pectin, citrus	5.0g
K_2HPO_4	1.5g
$MgSO_4 \cdot 7H_2O$	1.5g
Antibiotic solution	10.0mL

pH 7.0 ± 0.2 at 25°C

Antibiotic Solution:

Composition per 10.0mL:

Cycloheximide	0.075g
Novobiocin	0.045g
Penicillin G	75,000U

Preparation of Antibiotic Solution: Add components to distilled/deionized water and bring volume to 10.0mL. Mix thoroughly. Filter sterilize.

Preparation of Medium: Add components, except agar and antibiotic solution, to distilled/deionized water and bring volume to 990.0mL. Mix thoroughly. Adjust pH to 7.0. Add agar. Mix thoroughly. Gently heat and bring to boiling. Autoclave for 15 min at 15 psi pressure–121°C. Cool to 45°–50°C. Aseptically add sterile antibiotic solution. Mix thoroughly. Pour into sterile Petri dishes or distribute into sterile tubes.

Use: For the isolation and cultivation of *Pseudomonas* species that cause soft-rot.

Frankia Isolation Medium

Composition per liter:

Sucrose	40.0g
$Ca(NO_3)_2 \cdot 4H_2O$	0.242g
KNO_3	0.085g
KCl	0.061g
$MgSO_4 \cdot 7H_2O$	0.042g
KH_2PO_4	0.020g
$MnSO_4 \cdot H_2O$	4.5mg
$FeCl_3 \cdot 6H_2O$	2.5mg
H_3BO_3	1.5mg
$ZnSO_4 \cdot 7H_2O$	1.5mg
Nicotinic acid	0.5mg
Pyridoxine·HCl	0.5mg
$Na_2MoO_4 \cdot 2H_2O$	0.25mg
Thiamine·HCl	0.1mg
$CuSO_4 \cdot 5H_2O$	0.04mg
Mannitol solution	10.0mL
Supplement solution	10.0mL

pH 5.5 ± 0.2 at 25°C

Mannitol Solution:

Composition per 100mL:

Mannitol	11.84g

Preparation of Mannitol Solution: Add mannitol to distilled/deionized water and bring volume to 100.0mL. Mix thoroughly. Filter sterilize.

Supplement Solution:

Composition per 10mL:

L-Glutamic acid	0.185g
L-Arginine	0.174g
L-Glutamine	0.146g
L-Aspartic acid	0.133g
L-Asparagine	0.132g
Glycine	0.075g
Urea	0.060g
Naphthaloneacetic acid	2.0mg
Zeatin	1.0μg

Preparation of Supplement Solution: Add components to distilled/deionized water and bring volume to 10.0mL. Mix thoroughly. Filter sterilize.

Preparation of Medium: Add components, except mannitol solution and supplement solution, to distilled/deionized water and bring volume to 980.0mL. Mix thoroughly. Adjust pH to 5.5. Gently heat and bring to boiling. Autoclave for 15 min at 15 psi pressure–121°C. Cool to 45°–50°C. Aseptically add 10.0mL of sterile mannitol solution and 10.0mL of sterile supplement solution. Mix thoroughly. Aseptically distribute into sterile tubes or flasks.

Use: For the isolation and cultivation of *Frankia* species from root nodules.

Frankia Medium

Composition per liter:

Sucrose	20.0g
Agar	10.0g
Edamin	1.0g
Mannitol	1.0g
$CaCO_3$	0.5g
K_2HPO_4	0.5g
Yeast extract	0.5g
$MgCl_2 \cdot 7H_2O$	0.2g
NaCl	0.1g
H_3BO_3	0.1g
$MnSO_4 \cdot H_2O$	0.025g
$ZnSO_4 \cdot 7H_2O$	0.010g
Nicotinic acid	0.5mg
$Na_2MoO_4 \cdot 2H_2O$	0.25mg
Pyridoxine·HCl	0.1mg
Thiamine·HCl	0.1mg
$CuSO_4 \cdot 5H_2O$	0.025mg

pH 7.0 ± 0.2 at 25°C

Preparation of Medium: Add components to distilled/deionized water and bring volume to 1.0L. Mix thoroughly. Gently heat and bring to boiling. Adjust pH to 7.0. Distribute into tubes or flasks. Autoclave for 15 min at 15 psi pressure–121°C. Pour into sterile Petri dishes or leave in tubes.

Use: For the cultivation of *Frankia* species from root nodules.

Frankia Medium

Composition per 1200mL:

Na_2HPO_4	1.0g
NaCl	1.0g
KH_2PO_4	0.3g
Glycerol	1.0mL
NH_4Cl	1.0mL
Thiamine	1.0mL
Albumin fatty acid supplement	200.0mL

pH 7.4 ± 0.2 at 25°C

Albumin Fatty Acid Supplement:

Composition per 200mL:

Bovine albumin fraction V	20.0g
Tween™ 80	25.0mL
$FeSO_4 \cdot 7H_2O$	20.0mL
$CaCl_2 \cdot 2H_2O$	2.0mL
$MgCl_2 \cdot 6H_2O$	2.0mL
Vitamin B_{12}	2.0mL
$ZnSO_4 \cdot 7H_2O$	2.0mL
$CuSO_4 \cdot 5H_2O$	0.2mL

Preparation of Albumin Fatty Acid Supplement: Add components to distilled/deionized water and bring volume to 200.0mL. Mix thoroughly. Filter sterilize.

Preparation of Medium: Add components, except albumin fatty acid supplement, to distilled/deionized water and bring volume to 800.0mL. Mix thoroughly. Gently heat and bring to boiling. Autoclave for 15 min at 15 psi pressure–121°C. Cool to 45°–50°C. Aseptically add 200.0mL of sterile albumin fatty acid supplement. Mix thoroughly. Aseptically distribute into sterile tubes or flasks.

Use: For the cultivation and maintenance of *Frankia* species.

Frankia Medium

Composition per liter:

Basal medium	900.0mL
Albumin fatty acid supplement	100.0mL

pH 7.4 ± 0.2 at 25°C

Basal Medium:

Composition per liter:

Na_2HPO_4	1.0g
NaCl	1.0g
KH_2PO_4	0.3g
Glycerol (10% solution)	1.0mL
NH_4Cl (25% solution)	1.0mL
Thiamine (0.5% solution)	1.0mL

Preparation of Medium: Add components to distilled/deionized water and bring volume to 1.0L. Mix thoroughly. Adjust pH to 7.4. Distribute into tubes or flasks. Autoclave for 15 min at 15 psi pressure–121°C.

Albumin Fatty Acid Supplement:

Composition per 200mL:

Bovine albumin, fraction V	20.0g
Tween™ 80 (10% solution)	25.0mL
$FeSO_4 \cdot 7H_2O$ (0.5% solution)	20.0mL
$CaCl_2 \cdot 2H_2O$ (1% solution)	2.0mL
$MgCl_2 \cdot 6H_2O$ (1% solution)	2.0mL
Vitamin B_{12} (0.2% solution)	2.0mL
$ZnSO_4 \cdot 7H_2O$ (0.4% solution)	2.0mL
$CuSO_4 \cdot 5H_2O$ (0.3% solution)	0.2mL

Preparation of Albumin Fatty Acid Supplement: Add albumin to 100.0mL distilled/deionized water. Mix thoroughly. Slowly add Tween™ 80 and then other components. Adjust pH to 7.4. Add distilled/deionized water and bring volume to 200.0mL. Mix thoroughly. Autoclave for 15 min at 15 psi pressure–121°C.

Preparation of Medium: Aseptically combine 900.0mL sterile basal medium and 100.0mL sterile albumin fatty acid supplement. Mix thoroughly. Aseptically distribute into sterile tubes or flasks.

Use: For the cultivation and maintenance of *Frankia* species.

Fraser Broth

Composition per liter:

NaCl	20.0g
Na_2HPO_4	12.0g
Beef Extract	5.0g
Proteose peptone	5.0g
Pancreatic digest of casein	5.0g
Yeast extract	5.0g
LiCl	3.0g
KH_2PO_4	1.35g
Aesculin	1.0g
Fraser supplement solution	10.0mL

pH 7.2 ± 0.2 at 25°C

Source: This medium is available as a premixed powder from Difco Laboratories and Oxoid Unipath.

Fraser Supplement Solution:
Composition per 10mL:

Ferric ammonium citrate	0.5g
Acriflavine·HCl	0.25g
Nalidixic acid	0.1g
Ethanol	5.0mL

Preparation of Fraser Supplement Solution Solution: Add components to distilled/deionized water and bring volume to 10.0mL. Mix thoroughly. Filter sterilize.

Preparation of Medium: Add components, except Fraser supplement solution, to distilled/deionized water and bring volume to 990.0mL. Mix thoroughly. Gently heat and bring to boiling. Autoclave for 15 min at 15 psi pressure–121°C. Cool to 45°–50°C. Aseptically add sterile Fraser supplement solution. Mix thoroughly. Aseptically distribute into sterile tubes or flasks.

Use: For the isolation of *Listeria* species from environmental species.

Fraser Secondary Enrichment Broth

Composition per liter:

NaCl	20.0g
Na_2HPO_4	12.0g
Beef extract	5.0g
Proteose peptone	5.0g
Pancreatic digest of casein	5.0g
Yeast extract	5.0g
LiCl	3.0g
KH_2PO_4	1.35g
Esculin	1.0g
Acriflavin solution	10.0mL
Ferric ammonium citrate solution	10.0mL
Nalidixic acid solution	1.0mL

Ferric Ammonium Citrate Solution:
Composition per 10mL:

Ferric ammonium citrate	0.5g

Preparation of Ferric Ammonium Citrate Solution: Add ferric ammonium citrate to distilled/deionized water and bring volume to 10.0mL. Mix thoroughly. Filter sterilize.

Acriflavin Solution:
Composition per 10mL:

Acriflavin	0.025g

Preparation of Acriflavin Solution: Add acriflavin to distilled/deionized water and bring volume to 10.0mL. Mix thoroughly. Filter sterilize.

Nalidixic Acid Solution:
Composition per 10mL:

Nalidixic acid	0.04g
NaOH (0.1*N* solution)	10.0mL

Preparation of Nalidixic Acid Solution: Add nalidixic acid to 10.0mL of NaOH solution. Mix thoroughly. Filter sterilize.

Preparation of Medium: Add components, except acriflavin solution and ferric ammonium citrate solution, to distilled/deionized water and bring volume to 980.0mL. Mix thoroughly. Gently heat and bring to boiling. Distribute into tubes in 10.0mL volumes. Autoclave for 12 min at 15 psi pressure–121°C. Cool rapidly to 25°C. Immediately prior to inoculation, aseptically add 0.1mL of sterile acriflavin solution and 0.1mL of ferric ammonium citrate solution to each tube. Mix thoroughly.

Use: For the isolation, cultivation and enrichment of *Listeria monocytogenes* from environmental specimens based on esculin hydrolysis. Bacteria that hydrolyze esculin appear as black colonies.

FSM Selective Medium

Composition per liter:

Agar	18.0g
Peptone	10.0g
Glucose	4.0g
$(NH_4)_2SO_4$	1.32g
K_2HPO_4	1.18g
Casamino acids	1.0g
Yeast extract	1.0g
KH_2PO_4	0.44g
$MgSO_4 \cdot 7H_2O$	0.2g
$FeC_6H_5O_7 \cdot 5H_2O$	3.0mg
Citric acid	1.9mg
$ZnSO_4 \cdot 7H_2O$	1.6mg
$MnSO_4 \cdot H_2O$	1.5mg
2,3,5-Triphenyltetrazolium·HCl solution	10.0mL
Benomyl solution	10.0mL
Polymyxin B solution	10.0mL
Chloroneb solution	10.0mL
Dichloran solution	10.0mL
Bacitracin solution	10.0mL
Cycloheximide solution	10.0mL
Pentachloronitrobenzene solution	10.0mL
Pimaricin solution	10.0mL
Tyrothricin solution	10.0mL
Vancomycin solution	10.0mL
Chloromycetin solution	10.0mL
Penicillin G solution	10.0mL

2,3,5-Triphenyltetrazolium·HCl Solution:

Composition per 10mL:

2,3,5-Triphenyltetrazolium·HCl	0.5mg

Preparation of 2,3,5-Triphenyltetrazolium-HCl Solution: Add 2,3,5-triphenyltetrazolium·HCl to distilled/deionized water and bring volume to 10.0mL. Mix thoroughly. Autoclave for 7 min at 15 psi pressure–121°C.

Benomyl Solution:

Composition per 10mL:

Benomyl	0.5mg

Preparation of Benomyl Solution: Add benomyl to distilled/deionized water and bring volume to 10.0mL. Mix thoroughly. Filter sterilize.

Polymyxin B Solution:

Composition per 10mL:

Polymyxin B	0.1mg

Preparation of Polymyxin B Solution: Add polymyxin B to distilled/deionized water and bring volume to 10.0mL. Mix thoroughly. Filter sterilize.

Chloroneb Solution:

Composition per 10mL:

Chloroneb	0.1mg

Preparation of Chloroneb Solution: Add chloroneb to distilled/deionized water and bring volume to 10.0mL. Mix thoroughly. Filter sterilize.

Dichloran Solution:

Composition per 10mL:

Dichloran	0.1mg

Preparation of Dichloran Solution: Add dichloran to distilled/deionized water and bring volume to 10.0mL. Mix thoroughly. Filter sterilize.

Bacitracin Solution:

Composition per 10mL:

Bacitracin	0.05mg

Preparation of Bacitracin Solution: Add bacitracin to distilled/deionized water and bring volume to 10.0mL. Mix thoroughly. Filter sterilize.

Cycloheximide Solution:

Composition per 10mL:

Cycloheximide	0.05mg

Preparation of Cycloheximide Solution: Add cycloheximide to distilled/deionized water and bring volume to 10.0mL. Mix thoroughly. Filter sterilize.

Pentachloronitrobenzene Solution:

Composition per 10mL:

Pentachloronitrobenzene	0.03mg

Preparation of Pentachloronitrobenzene Solution: Add pentachloronitrobenzene to distilled/deionized water and bring volume to 10.0mL. Mix thoroughly. Filter sterilize.

Pimaricin Solution:

Composition per 10mL:

Pimaricin	0.02mg

Preparation of Pimaricin Solution: Add pimaricin to distilled/deionized water and bring volume to 10.0mL. Mix thoroughly. Filter sterilize.

Tyrothricin Solution:

Composition per 10mL:

Tyrothricin	0.02mg

Preparation of Tyrothricin Solution: Add tyrothricin to distilled/deionized water and bring volume to 10.0mL. Mix thoroughly. Filter sterilize.

Vancomycin Solution:

Composition per 10mL:

Vancomycin	0.01mg

Preparation of Vancomycin Solution: Add vancomycin to distilled/deionized water and bring volume to 10.0mL. Mix thoroughly. Filter sterilize.

Chloromycetin Solution:

Composition per 10mL:

Chloromycetin	5.0μg

Preparation of Chloromycetin Solution: Add chloromycetin to distilled/deionized water and bring volume to 10.0mL. Mix thoroughly. Filter sterilize.

Penicillin G Solution:

Composition per 10mL:

Penicillin G	1.0μg

Preparation of Penicillin G Solution: Add penicillin G to distilled/deionized water and bring volume to 10.0mL. Mix thoroughly. Filter sterilize.

Preparation of Medium: Add components—except 2,3,5-triphenyltetrazolium chloride solution, benomyl solution, polymyxin B solution, chloroneb solution, dichloran solution, bacitracin solution, cycloheximide solution, pentachloronitrobenzene solution, pimaricin solution, tyrothricin solution, vancomycin solution, chloromycetin solution, and penicillin G solution—to distilled/deionized water and bring volume to 870.0mL. Mix thoroughly. Gently heat and bring to boiling. Autoclave for 15 min at 15 psi pressure–121°C. Cool to 45°–50°C. Aseptically add 10.0mL each of sterile 2,3,5-triphenyltetrazolium chloride solution, benomyl solution, polymyxin B solution, chloroneb solution, dichloran solution, bacitracin solution, cycloheximide solution, pentachloronitrobenzene solution, pimaricin solution, tyrothricin solution, vancomycin solution, chloromycetin solution, and penicillin G solution. Mix thoroughly. Pour into sterile Petri dishes. Dry plates for 24 hr at 30°C.

Use: For the isolation and cultivation of *Pseudomonas solanacearum* from soil.

FTX Broth

Composition per 1001mL:

Sodium glutamate	10.0g
Glucose	2.0g
Tris	2.0g
Sodium glycerophosphate	0.1g
Artificial seawater	1.0L
Trace element solution	1.0mL

pH 8.0 ± 0.2 at 25°C

Artificial Seawater:

Composition per liter:

NaCl	24.7g
$MgSO_4 \cdot 7H_2O$	6.3g
$MgCl_2 \cdot 6H_2O$	4.6g
$CaCl_2$	1.0g
KCl	0.7g
$NaHCO_3$	0.2g

Preparation of Artificial Seawater: Add components to distilled/deionized water and bring volume to 1.0L. Mix thoroughly.

Trace Element Solution:

Composition per liter:

Disodium EDTA	8.0g
$MnCl_2 \cdot 4H_2O$	0.1g
$CoCl_2 \cdot 6H_2O$	0.02g
KBr	0.02g
KI	0.02g
$ZnCl_2$	0.02g
$CuSO_4$	0.01g
H_3BO_3	0.01g
$Na_2MoO_4 \cdot 2H_2O$	0.01g
LiCl	5.0mg
$SnCl_2 \cdot 2H_2O$	5.0mg

Preparation of Trace Element Solution: Add components to distilled/deionized water and bring volume to 1.0L. Mix thoroughly.

Preparation of Medium: Add components to 1.0L of artificial seawater. Mix thoroughly. Gently heat and bring to boiling. Distribute into tubes or flasks. Autoclave for 15 min at 15 psi pressure–121°C.

Use: For the isolation and cultivation of *Cytophaga* species, *Herpetosiphon* species, *Saprospira* species, and *Flexithrix* species.

Furoate Agar

Composition per liter:

Agar	20.0g
2-Furoic acid	2.0g
K_2HPO_4	1.0g
NH_4Cl	1.0g
$MgSO_4 \cdot 7H_2O$	0.1g

pH 7.0 ± 0.2 at 25°C

Preparation of Medium: Add components to distilled/deionized water and bring volume to 1.0L. Mix thoroughly. Gently heat and bring to boiling. Distribute into tubes or flasks. Autoclave for 15 min at 15 psi pressure–121°C. Pour into sterile Petri dishes or leave in tubes.

Use: For the cultivation and maintenance of *Bacillus megaterium* and *Pseudomonas* species.

Fusobacterium Medium

Composition per liter:

Agar	15.0g
Pancreatic digest of casein	15.0g
Glucose	5.0g
NaCl	5.0g
Yeast extract	5.0g
L-Cysteine	0.75g
Crystal violet	0.01g

Bovine serum	50.0mL
Streptomycin solution	10.0mL

pH 7.2 ± 0.2 at 25°C

Streptomycin Solution:
Composition per 10mL:

Streptomycin	0.01g

Preparation of Streptomycin Solution: Add streptomycin to distilled/deionized water and bring volume to 10.0mL. Mix thoroughly. Filter sterilize.

Preparation of Medium: Add components, except bovine serum and streptomycin solution, to distilled/deionized water and bring volume to 940.0mL. Mix thoroughly. Gently heat and bring to boiling. Autoclave for 15 min at 15 psi pressure–121°C. Cool to 45–50°C. Aseptically add 50.0mL of sterile bovine serum and 10.0mL of sterile streptomycin solution. Mix thoroughly. Pour into sterile Petri dishes or distribute into sterile tubes.

Use: For the cultivation of *Fusobacterium* species.

G Medium

Composition per liter:

$(NH_4)_2SO_4$	2.0g
Yeast extract	2.0g
Glucose	1.0g
K_2HPO_4	0.6g
KH_2PO_4	0.4g
$MgSO_4 \cdot 7H_2O$	0.2g
$CaCl_2$	0.08g
$MnSO_4 \cdot H_2O$	0.05g
$CuSO_4 \cdot 5H_2O$	5.0mg
$ZnSO_4 \cdot 7H_2O$	5.0mg
$FeSO_4 \cdot 7H_2O$	0.5mg

pH 7.8 ± 0.2 at 25°C

Preparation of Medium: Add components to distilled/deionized water and bring volume to 1.0L. Mix thoroughly. Distribute into tubes or flasks. Autoclave for 15 min at 15 psi pressure–121°C.

Use: For the cultivation and maintenance of *Bacillus cereus.*

Gelatin Agar

Composition per liter:

Gelatin	30.0g
Agar	15.0g
Pancreatic digest of casein	10.0g
NaCl	10.0g

pH 7.2 ± 0.2 at 25°C

Preparation of Medium: Add components to distilled/deionized water and bring volume to 1.0L. Mix thoroughly. Gently heat and bring to boiling. Distribute into tubes or flasks. Autoclave for 15 min at 15 psi pressure–121°C.

Use: For the cultivation of bacteria isolated from foods and their differentiation based on proteolytic activity.

Gelatin Agar

Composition per liter:

Agar	15.0g
Gelatin	15.0g
Peptone	4.0g
Yeast extract	1.0g

pH 7.2 ± 0.2 at 25°C

Preparation of Medium: Add components to distilled/deionized water and bring volume to 1.0L. Mix thoroughly. Gently heat and bring to boiling. Distribute into tubes or flasks. Autoclave for 15 min at 15 psi pressure–121°C. Pour into sterile Petri dishes or leave in tubes.

Use: For the cultivation of a variety of heterotrophic bacteria based upon their utilization of gelatin.

Gelatin Infusion Broth

Composition per liter:

Beef heart, solids from infusion	500.0g
Gelatin	40.0g
Tryptose	10.0g
NaCl	5.0g

pH 7.4 ± 0.2 at 25°C

Preparation of Medium: Add components to distilled/deionized water and bring volume to 1.0L. Mix thoroughly. Gently heat and bring to boiling. Distribute into tubes or flasks. Autoclave for 15 min at 15 psi pressure–121°C. Pour into sterile Petri dishes or leave in tubes.

Use: For the cultivation and differentiation of a variety of heterotrophic bacteria based upon their production of gelatinase. The gelatinase liquefies the medium.

Gelatin Medium

Composition per liter:

Gelatin	4.0g

pH 7.0 ± 0.2 at 25°C

Preparation of Medium: Add gelatin to distilled/deionized water and bring volume to 1.0L. Mix thoroughly. Gently heat and bring to boiling. Distribute into tubes. Autoclave for 15 min at 15 psi pressure–121°C.

Use: For the cultivation and differentiation of *Nocardia* and *Streptomyces* species based on utilization of gelatin. *Nocardia asteroides* usually exhibits no growth. *Nocardia brasiliensis* shows good growth and round compact colonies. *Streptomyces* species show varying degrees of growth.

Gelatin Medium

Composition per liter:

Gelatin	120.0g
Pancreatic digest of casein	13.0g
Sodium chloride	5.0g
Yeast extract	5.0g
Heart muscle, solids from infusion	2.0g
Sodium thioglycollate	0.5g

pH 7.0 ± 0.2 at 25°C

Preparation of Medium: Add components to distilled/deionized water and bring volume to 1.0L. Mix thoroughly. Gently heat and bring to boiling. Distribute into tubes. Autoclave for 15 min at 15 psi pressure–121°C. Pour into sterile Petri dishes or leave in tubes.

Use: For the cultivation of gelatin-utilizing *Clostridium* species.

Gelatinase Test Medium

Composition per liter:

Gelatin	3.0g
ACES buffer	1.0g
Yeast extract	1.0g
Charcoal, activated	0.15g
α-Ketoglutarate monopotassium salt	0.1g
L-Cysteine·HCl·H_2O (4% solution)	1.0mL
KOH (85% solution)	1.0mL
$Fe_4(P_2O_7)_3$ solution	1.0mL

pH 6.9 ± 0.2 at 25°C

L-Cysteine·HCl·H_2O Solution:
Composition per 10mL:

L-Cysteine·HCl·H_2O	0.4g

Preparation of L-Cysteine·HCl·H_2O Solution: Add L-cysteine·HCl·H_2O to distilled/deionized water and bring volume to 10.0mL. Mix thoroughly. Filter sterilize.

$Fe_4(P_2O_7)_3$ Solution:
Composition per 10mL:

$Fe_4(P_2O_7)_3$	0.15g

Preparation of $Fe_4(P_2O_7)_3$ Solution: Add $Fe_4(P_2O_7)_3$ to distilled/deionized water and bring volume to 10.0mL. Mix thoroughly. Filter sterilize.

Preparation of Medium: Add ACES buffer to distilled/deionized water and bring volume to 899.0mL. Mix thoroughly. Gently heat to 50°C. Add 1.0mL KOH solution. Mix thoroughly. Add charcoal, yeast extract and α-ketoglutarate. Add 80.0mL distilled/deionized water to wash sides of flask. Mix thoroughly. Autoclave for 15 min at 15 psi pressure–121°C. Cool to 50°C. Aseptically add 10.0mL sterile cysteine solution and 10.0mL sterile $Fe_4(P_2O_7)_3$ solution. Mix thoroughly. Adjust pH to 6.9. Aseptically distribute into sterile screw-cap tubes.

Use: For the cultivation and differentiation of gelatinase producing bacteria.

General Salts Medium for Estuarine Methanogens

Composition per 410.8mL:

Agar	8.0g
NaCl	3.6g
$NaHCO_3$	2.0g
Complete salts solution	200.0mL
Cysteine-sulfide reducing agent	16.0mL
Wolfe's mineral solution	4.0mL
Vitamin solution	4.0mL
Yeast extract–trypticase solution	4.0mL
Sodium acetate (25% solution)	2.0mL
$Fe(NH_4)_2SO_4$ (0.2% solution)	0.4mL
Resazurin (0.1% solution)	0.4mL

pH 7.0 ± 0.2 at 25°C

Complete Salts Solution:
Composition per liter:

$MgSO_4·7H_2O$	6.9g
$MgCl_2·6H_2O$	5.5g
KCl	0.67g
NH_4Cl	0.5g
$CaCl_2·2H_2O$	0.28g
K_2HPO_4	0.28g

Preparation of Complete Salts Solution: Add components to distilled/deionized water and bring volume to 1.0L. Mix thoroughly.

Cysteine-Sulfide Reducing Agent:
Composition per 400mL:

L-Cysteine·HCl·H_2O	5.0g
Na_2S (12.5% solution)	40.0mL
NaOH (1*N* solution)	30.0mL

Preparation of Cysteine-Sulfide Reducing Agent: Add distilled/deionized water to a 500.0mL round bottom flask. Add freshly prepared NaOH solution. Gently heat and bring to boiling under 100% N_2. Remove gassing probe. Add cysteine·HCl·H_2O. Add freshly prepared Na_2S solution. Renew gassing for several minutes. Cap with rubber stoppers. Distribute into 8.0mL/18mm Hungate tubes.

Yeast Extract Trypticase Solution:
Composition per 100mL:

Yeast extract	20.0g
Pancreatic digest of casein	20.0g

Preparation of Yeast Extract Trypticase Solution: Add components to distilled/deionized water and bring volume to 100.0mL. Mix thoroughly.

Wolfe's Mineral Solution:
Composition per liter:

$MgSO_4 \cdot 7H_2O$	3.0g
Nitriloacetic acid	1.5g
NaCl	1.0g
$MnSO_4 \cdot H_2O$	0.5g
$FeSO_4 \cdot 7H_2O$	0.1g
$CoCl_2 \cdot 6H_2O$	0.1g
$CaCl_2$	0.1g
$ZnSO_4 \cdot 7H_2O$	0.1g
$CuSO_4 \cdot 5H_2O$	0.01g
$AlK(SO_4)_2 \cdot 12H_2O$	0.01g
H_3BO_3	0.01g
$Na_2MoO_4 \cdot 2H_2O$	0.01g

Preparation of Wolfe's Mineral Solution: Add nitrilotriacetic acid to 500.0mL of distilled/deionized water. Dissolve by adjusting pH to 6.5 with KOH. Add remaining components. Add distilled/deionized water to 1.0L.

Preparation of Medium: Add components, except cysteine-sulfide reducing agent, to distilled/deionized water and bring volume to 410.8mL. Mix thoroughly. Adjust pH to 7.0. Gently heat and bring to boiling under 80% N_2 + 20% CO_2. Add cysteine-sulfide reducing agent. Continue boiling until resazurin turns colorless indicating reduction. Distribute anaerobically into culture tubes with aluminum crimp seals. Autoclave for 15 min at 15 psi pressure–121°C.

Use: For the cultivation and maintenance of *Methanococcus deltae, Methanococcus vannielii, Methanococcus voltae, Methanogenium cariaci, Methanogenium marisnigri,* and *Methanogenium olentangyi.*

Geodermatophilus obscurus Medium

Composition per liter:

Agar	20.0g
Malt extract, purified solids	15.0g
Starch, soluble	10.0g
Sucrose	10.0g
Yeast extract	5.0g
$CaCO_3$	2.0g

Preparation of Medium: Add components to distilled/deionized water and bring volume to 1.0L. Mix thoroughly. Gently heat and bring to boiling. Distribute into tubes or flasks. Autoclave for 15 min at 15 psi pressure–121°C. Pour into sterile Petri dishes or leave in tubes.

Use: For the isolation and cultivation of *Geodermatophilus obscurus.*

George's Medium, Modified

Composition per liter:

Agar	15.0g
Peptone	1.0g
KNO_3	0.2g
K_2HPO_4	0.02g
$MgSO_4 \cdot 7H_2O$	0.02g
Ferric citrate	0.035mg

Preparation of Medium: Add components to tap water and bring volume to 1.0L. Mix thoroughly. Gently heat and bring to boiling. Distribute into tubes. Autoclave for 15 min at 15 psi pressure–121°C. Pour into sterile Petri dishes or leave in tubes.

Use: For the cultivation of a variety of algae.

Gluconate Peptone Broth

Composition per liter:

Potassium gluconate	40.0g
Casein peptone	1.5g
K_2HPO_4	1.0g
Yeast extract	1.0g

pH 7.0 ± 0.2 at 25°C

Preparation of Medium: Add components to distilled/deionized water and bring volume to 1.0L. Mix thoroughly. Distribute into tubes or flasks. Autoclave for 15 min at 15 psi pressure–121°C.

Use: For the cultivation and differentiation of Gram-negative bacteria based on their ability to oxidize gluconate to 2-ketogluconate. For the differentiation of fluorescent *Pseudomonas* species. After inoculation with bacteria and 48 hr of growth in this medium, Benedict's reagent is added. Bacteria that produce the reducing sugar, 2-ketogluconate turn the reagent yellow-orange to orange-red.

Glucose Agar, 9K

Composition per liter:

$(NH_4)_2SO_4$	3.0g
KH_2PO_4	0.5g
$MgSO_4 \cdot 7H_2O$	0.5g
KCl	0.1g
$Ca(NO_3)_2$	0.0125g
$FeSO_4 \cdot 7H_2O$	0.01mg

Agar solution....................500.0mL
Glucose solution....................100.0mL

pH 4.5 ± 0.2 at 25°C

Agar Solution:

Composition per 500mL:

Agar....................15.0g

Preparation of Agar Solution: Add agar to distilled/deionized water and bring volume to 500.0mL. Mix thoroughly. Autoclave for 15 min at 15 psi pressure–121°C. Cool to 55°C.

Glucose Solution:

Composition per 100mL:

Glucose....................10.0g

Preparation of Glucose Solution: Add glucose to distilled/deionized water and bring volume to 100.0mL. Mix thoroughly. Filter sterilize.

Preparation of Medium: Add components, except agar solution and glucose solution, to distilled/deionized water and bring volume to 400.0mL. Mix thoroughly. Adjust pH to 4.5 with H_2SO_4. Autoclave for 15 min at 15 psi pressure–121°C. Cool to 55°C. Aseptically add 500.0mL sterile agar solution and 100.0mL sterile glucose solution. Mix thoroughly. Pour into sterile Petri dishes or distribute into sterile tubes.

Use: For the cultivation of *Thiobacillus acidophilus.*

Glucose Broth

Composition per 800 mL:

Agar....................10.0g
Beef extract....................10.0g
Peptone....................10.0g
NaCl....................5.0g
Glucose solution....................200.0mL

pH 7.0 ± 0.2 at 25°C

Glucose Solution:

Composition per 200mL:

Glucose....................20.0g

Preparation of Glucose Solution: Add glucose to distilled/deionized water and bring volume to 200.0mL. Mix thoroughly. Filter sterilize.

Preparation of Medium: Add components, except glucose solution, to distilled/deionized water and bring volume to 800.0mL. Mix thoroughly. Gently heat and bring to boiling. Adjust pH to 7.0. Autoclave for 15 min at 15 psi pressure–121°C. Cool to 45°C. Aseptically add 200.0mL sterile glucose solution. Mix thoroughly. Aseptically distribute into sterile tubes or flasks.

Use: For the cultivation of *Pseudomonas* species.

Glucose Broth, 9K

Composition per liter:

Glucose....................10.0g
$(NH_4)_2SO_4$....................3.0g
KH_2PO_4....................0.5g
$MgSO_4 \cdot 7H_2O$....................0.5g
KCl....................0.1g
$Ca(NO_3)_2$....................0.0125g
$FeSO_4 \cdot 7H_2O$....................0.01mg
Glucose solution....................100.0mL

pH 3.5 ± 0.2 at 25°C

Glucose Solution:

Composition per 100mL:

Glucose....................10.0g

Preparation of Glucose Solution: Add glucose to distilled/deionized water and bring volume to 100.0mL. Mix thoroughly. Filter sterilize.

Preparation of Medium: Add components, except glucose solution, to distilled/deionized water and bring volume to 900.0mL. Mix thoroughly. Adjust pH to 3.5 with H_2SO_4. Autoclave for 15 min at 15 psi pressure–121°C. Cool to 25°C. Aseptically add 100.0mL sterile glucose solution. Mix thoroughly. Aseptically distribute into sterile tubes or flasks.

Use: For the cultivation of *Thiobacillus acidophilus.*

Glucose Nitrogen–Free Salt Agar

Composition per liter:

Agar....................15.0g
$CaCO_3$....................1.0g
K_2HPO_4....................1.0g
$MgSO_4 \cdot 7H_2O$....................0.2g
NaCl....................0.2g
$FeSO_4 \cdot 7H_2O$....................0.1g
$Na_2MoO_4 \cdot 2H_2O$....................5.0mg
Glucose solution....................100.0mL

pH 7.0 ± 0.2 at 25°C

Glucose Solution:

Composition per 100mL:

Glucose....................10.0g

Preparation of Glucose Solution: Add glucose to distilled/deionized water and bring volume to 100.0mL. Mix thoroughly. Filter sterilize.

Preparation of Medium: Add components, except glucose solution, to distilled/deionized water and bring volume to 900.0mL. Mix thoroughly. Gently heat and bring to boiling. Autoclave for 15 min at 15 psi pressure–121°C. Cool to 45–50°C. Aseptically add 100.0mL of sterile glucose solution. Mix thoroughly. Pour into sterile Petri dishes or distribute into sterile tubes.

Use: For the cultivation of *Azotobacter* species.

Glucose Nitrogen–Free Salt Solution

Composition per liter:

$CaCO_3$	1.0g
K_2HPO_4	1.0g
$MgSO_4 \cdot 7H_2O$	0.2g
NaCl	0.2g
$FeSO_4 \cdot 7H_2O$	0.1g
$Na_2MoO_4 \cdot 2H_2O$	5.0mg
Glucose solution	100.0mL

pH 7.0 ± 0.2 at 25°C

Glucose Solution:
Composition per 100mL:

Glucose	10.0g

Preparation of Glucose Solution: Add glucose to distilled/deionized water and bring volume to 100.0mL. Mix thoroughly. Filter sterilize.

Preparation of Medium: Add components, except glucose solution, to distilled/deionized water and bring volume to 900.0mL. Mix thoroughly. Gently heat and bring to boiling. Autoclave for 15 min at 15 psi pressure–121°C. Cool to 45–50°C. Aseptically add 100.0mL of sterile glucose solution. Mix thoroughly. Aseptically distribute into sterile tubes or flasks.

Use: For the cultivation of *Azotobacter* species.

Glucose Peptone Agar

Composition per liter:

Peptone	20.0g
Agar	15.0g
Glucose	10.0g
NaCl	5.0g

pH 7.2 ± 0.2 at 25°C

Preparation of Medium: Add components to distilled/deionized water and bring volume to 1.0L. Mix thoroughly. Gently heat and bring to boiling. Distribute into tubes or flasks. Autoclave for 15 min at 15 psi pressure–121°C. Pour into sterile Petri dishes or leave in tubes.

Use: For the cultivation of *Agrobacterium* species.

Glucose Phosphate Broth

Composition per liter:

Peptone	10.0g
K_2HPO_4	5.0g
Glucose	5.0g

pH 7.5 ± 0.2 at 25°C

Preparation of Medium: Add components, except glucose, to distilled/deionized water and bring volume to 1.0L. Mix thoroughly. Gently heat and bring to boiling. Filter while hot through Whatman filter paper. Cool to 25°C. Adjust pH to 7.5. Add 5.0g glucose. Mix thoroughly. Distribute into sterile tubes or flasks. Autoclave for 10 min at 10 psi pressure–115°C.

Use: For the cultivation of a variety of nonfastidious heterotrophic microorganisms.

Glucose Yeast Chalk Agar

Composition per liter:

Chalk	40.0g
Agar	15.0g
Glucose	5.0g
Yeast extract	5.0g

Preparation of Medium: Add components to distilled/deionized water and bring volume to 1.0L. Mix thoroughly. Gently heat and bring to boiling. Distribute into tubes or flasks. Autoclave for 15 min at 15 psi pressure–121°C. Pour into sterile Petri dishes or leave in tubes.

Use: For the cultivation and maintenance of *Xanthomonas* species.

Glucose Yeast Extract Medium (ATCC Medium 985)

Composition per liter:

Agar	15.0g
Yeast extract	3.0g
Glucose	1.0g

pH 7.0 ± 0.2 at 25°C

Preparation of Medium: Add components to distilled/deionized water and bring volume to 1.0L. Mix thoroughly. Gently heat and bring to boiling. Distribute into tubes or flasks. Autoclave for 15 min at 15 psi pressure–121°C. Pour into sterile Petri dishes or leave in tubes.

Use: For the cultivation and maintenance of *Acinetobacter tartarogenes, Agrobacterium viscosum,* and *Pseudomonas* species.

Glucose Yeast Extract Medium (ATCC Medium 1742)

Composition per liter:

Agar	15.0g
Glucose	5.0g
Yeast extract	3.5g

Preparation of Medium: Add components to distilled/deionized water and bring volume to 1.0L. Mix thoroughly. Gently heat and bring to boiling. Distribute into tubes or flasks. Autoclave for 15 min at 15 psi pressure–121°C. Pour into sterile Petri dishes or leave in tubes.

Use: For the cultivation and maintenance of *Xanthobacter* species.

Glucose Yeast Extract Peptone Medium (GYP Medium)

Composition per liter:

Glucose	20.0g
Agar	10.0g
Peptone	5.0g
Yeast extract	5.0g
$CaCO_3$	0.1g

Preparation of Medium: Add components to distilled/deionized water and bring volume to 1.0L. Mix thoroughly. Gently heat and bring to boiling. Distribute into tubes or flasks. Autoclave for 15 min at 15 psi pressure–121°C. Pour into sterile Petri dishes or leave in tubes.

Use: For the isolation and cultivation of *Sporolactobacillus* species.

Glucose Yeast Medium with Calcium Carbonate

Composition per liter:

Agar	15.0g
$CaCO_3$	7.5g
Peptone	5.0g
Yeast extract	5.0g
Glucose	3.0g

pH 7.0 ± 0.2 at 25°C

Preparation of Medium: Add components to distilled/deionized water and bring volume to 1.0L. Mix thoroughly. Gently heat and bring to boiling. Distribute into tubes or flasks. Adjust pH to 6.3. Autoclave for 15 min at 15 psi pressure–121°C. Pour into sterile Petri dishes or leave in tubes.

Use: For the cultivation and maintenance of *Erwinia herbicola*.

Glucose Yeast Peptone Medium

Composition per liter:

Agar	20.0g
Glucose	5.0g
Peptone	5.0g
Yeast extract	3.0g

pH 7.0 ± 0.2 at 25°C

Preparation of Medium: Add components to distilled/deionized water and bring volume to 1.0L. Mix thoroughly. Gently heat and bring to boiling. Distribute into tubes or flasks. Autoclave for 15 min at 15 psi pressure–121°C. Pour into sterile Petri dishes or leave in tubes.

Use: For the cultivation and maintenance of a variety of heterotrophic microorganisms.

Glutamate Medium

Composition per liter:

Solution A	500.0mL
Solution B	250.0mL
Solution C	250.0mL

Solution A:

Composition per 500mL:

Mannitol	10.0g
K_2HPO_4	0.22g

Preparation of Solution A: Add components to distilled/deionized water and bring volume to 500.0mL. Mix thoroughly. Autoclave for 15 min at 15 psi pressure–121°C. Cool to 25°C.

Solution B:

Composition per 250mL:

$MgSO_4 \cdot 7H_2O$	0.1g
$CaCl_2 \cdot 6H_2O$	0.08g
$FeCl_3 \cdot 6H_2O$	0.05g

Preparation of Solution B: Add components to distilled/deionized water and bring volume to 250.0mL. Mix thoroughly. Autoclave for 15 min at 15 psi pressure–121°C. Cool to 25°C.

Solution C:

Composition per 250mL:

Sodium glutamate	1.1g
Calcium pantothenate	0.5mg
Thiamine·HCl	0.1mg
Biotin	0.5μg

Preparation of Solution C: Add components to distilled/deionized water and bring volume to 250.0mL. Mix thoroughly. Filter sterilize.

Preparation of Medium: Aseptically combine 500.0mL of cooled, sterile solution A, 250.0mL of cooled, sterile solution B, and 250.0mL of sterile solution C. Mix thoroughly. Aseptically distribute into sterile tubes or flasks.

Use: For the isolation of *Rhizobium* species.

Glutamate Medium (ATCC Medium 820)

Composition per liter:

Agar	15.0g
Sodium glutamate	5.0g
KH_2PO_4	1.0g
$MgSO_4 \cdot 7H_2O$	0.2g
KCl	0.1g
Glucose solution	100.0mL

pH 6.5 ± 0.2 at 25°C

Glucose Solution:

Composition per 100mL:

Glucose	10.0g

Preparation of Glucose Solution: Add glucose to distilled/deionized water and bring volume to 100.0mL. Mix thoroughly. Filter sterilize.

Preparation of Medium: Add components, except glucose solution, to distilled/deionized water and bring volume to 900.0mL. Mix thoroughly. Gently heat and bring to boiling. Autoclave for 15 min at 15 psi pressure–121°C. Cool to 45–50°C. Aseptically add 100.0mL sterile glucose solution. Mix thoroughly. Pour into sterile Petri dishes or distribute into sterile tubes.

Use: For the cultivation and maintenance of *Pseudomonas* species.

Glycerol Agar

Composition per 1070mL:

Agar	15.0g
Peptone	5.0g
Beef extract	3.0g
Soil extract	1.0L
Glycerol	70.0mL

pH 7.0 ± 0.2 at 25°C

Soil Extract:

Composition per liter:

Soil, air-dried	1.0Kg

Preparation of Soil Extract: Sift soil through a No. 9 mesh screen. Add to 2.4L of tap water. Mix thoroughly. Autoclave for 60 min at 15 psi pressure–121°C. Cool to 25°C. Filter through Whatman #1 filter paper. Bring volume to 1.0L with tap water.

Preparation of Medium: Combine components. Mix thoroughly. Gently heat and bring to boiling. Autoclave for 15 min at 15 psi pressure–121°C. Pour into sterile Petri dishes or distribute into sterile tubes.

Use: For the selective isolation and cultivation of *Nocardia* species and *Rhodococcus* species.

Glycerol Agar

Composition per liter:

Beef heart, solids from infusion	250.0g
Glycerol	60.0g
Agar	15.0g
Pancreatic digest of gelatin	5.0g
Beef extract	3.0g
Tryptose	5.0g
NaCl	2.5g

pH 7.3 ± 0.2 at 25°C

Preparation of Medium: Add components to distilled/deionized water and bring volume to 1.0L. Mix thoroughly. Gently heat and bring to boiling. Distribute into tubes or flasks. Autoclave for 15 min at 15 psi pressure–121°C. Pour into sterile Petri dishes or leave in tubes.

Use: For the cultivation and maintenance of *Bacillus subtilis, Enterococcus faecalis, Erwinia chrysanthemi, Gordona rubropertinctus, Mycobacterium* species, *Nocardia brevicatena, Rhodococcus equi,* and *R. rhodochrous.*

Glycerol Arginine Agar

Composition per liter:

Agar	15.0g
Glycerol	12.5g
Arginine	1.0g
K_2HPO_4	1.0g
NaCl	1.0g
$MgSO_4 \cdot 7H_2O$	0.5g
$Fe_2(SO_4)_3 \cdot 6H_2O$	0.01g
$CuSO_4 \cdot 5H_2O$	1.0mg
$MnSO_4 \cdot H_2O$	1.0mg
$ZnSO_4 \cdot 7H_2O$	1.0mg

Preparation of Medium: Add components to distilled/deionized water and bring volume to 1.0L. Mix thoroughly. Gently heat and bring to boiling. Distribute into tubes or flasks. Autoclave for 15 min at 15 psi pressure–121°C. Pour into sterile Petri dishes or leave in tubes.

Use: For the selective isolation and cultivation of streptomycetes.

Glycerol Beef Extract Medium

Composition per liter:

Agar	15.0g
Peptone	10.0g
NaCl	5.0g
Beef extract	3.0g
Glycerol	40.0mL

Preparation of Medium: Add components to distilled/deionized water and bring volume to 1.0L. Mix thoroughly. Gently heat and bring to boiling. Distribute into tubes or flasks. Autoclave for 15 min at 15 psi pressure–121°C. Pour into sterile Petri dishes or leave in tubes.

Use: For the cultivation and maintenance of *Corynebacterium alkanolyticum, Pseudomonas mallei, P. pseudomallei,* and *Rhodococcus* species.

Glycerol Chalk Agar

Composition per liter:

NaCl	30.0g
Agar	15.0g
Glycerol	10.0g
$CaCO_3$	5.0g
Peptone	5.0g
Yeast extract	3.0g

Preparation of Medium: Add components to distilled/deionized water and bring volume to 1.0L. Mix thoroughly. Gently heat and bring to boiling. Autoclave for 15 min at 15 psi pressure–121°C. Pour into sterile Petri dishes. Swirl flask while dispensing medium to keep $CaCO_3$ in suspension.

Use: For the cultivation of *Photobacterium* species and *Lucibacterium* species.

Glycerol Enriched Medium

Composition per liter:

Glycerol	30.0g
Peptone	20.0g
Yeast extract	10.0g

Preparation of Medium: Add components to distilled/deionized water and bring volume to 1.0L. Mix thoroughly. Gently heat and bring to boiling. Distribute into tubes or flasks. Autoclave for 15 min at 15 psi pressure–121°C. Pour into sterile Petri dishes or leave in tubes.

Use: For the cultivation of a variety of heterotrophic microorganisms.

Glycerol Glycine Agar

Composition per liter:

Agar	15.0g
Glycine	2.5g
K_2HPO_4	1.0g
NaCl	1.0g
$CaCO_3$	0.1g
$FeSO_4$	0.1g
$MgSO_4$	0.1g
Glycerol	20.0mL

pH 7.0 ± 0.2 at 25°C

Preparation of Medium: Add components, except glycerol, to distilled/deionized water and bring volume to 1.0L. Mix thoroughly. Gently heat and bring to boiling. Add glycerol. Mix thoroughly. Distribute into tubes or flasks. Autoclave for 15 min at 15 psi pressure–121°C. Pour into sterile Petri dishes or leave in tubes.

Use: For the cultivation of *Streptomyces* species.

Glycocholate Mineral Medium

Composition per liter:

Agar	15.0g
K_2HPO_4	3.5g
$(NH_4)_2SO_4$	2.0g
Sodium glycocholate	2.0g
KH_2PO_4	1.5g
$MgSO_4 \cdot 7H_2O$	0.1g
Yeast extract	0.1g
$CaCl_2 \cdot 2H_2O$	0.01g
$FeSO_4 \cdot 7H_2O$	0.5mg

pH 7.0 ± 0.2 at 25°C

Preparation of Medium: Add components to distilled/deionized water and bring volume to 1.0L. Mix thoroughly. Gently heat and bring to boiling. Distribute into tubes or flasks. Autoclave for 15 min at 15 psi pressure–121°C. Pour into sterile Petri dishes or leave in tubes.

Use: For the cultivation and maintenance of *Pseudomonas putida* and *Pseudomonas* species.

GN Broth, Hajna

Composition per liter:

Pancreatic digest of casein	10.0g
Peptic digest of animal tissue	10.0g
NaCl	5.0g
Sodium citrate	5.0g
K_2HPO_4	4.0g
D-Mannitol	2.0g
KH_2PO_4	1.5g
Glucose	1.0g
Sodium deoxycholate	0.5g

pH 7.0 ± 0.2 at 25°C

Source: This medium is available as a premixed powder from BBL Microbiology Systems and Difco Laboratories.

Preparation of Medium: Add components to distilled/deionized water and bring volume to 1.0L. Mix thoroughly. Gently heat and bring to boiling. Distribute into tubes or flasks. Autoclave for 15 min at 13 psi pressure–118°C. Pour into sterile Petri dishes or leave in tubes.

Use: For the selective cultivation of *Salmonella* and *Shigella* species.

Gorham's Medium for Algae

Composition per liter:

$NaNO_3$	4.96g
$MgSO_4 \cdot 7H_2O$	0.075g
$Na_2SiO_3 \cdot 9H_2O$	0.058g
K_2HPO_4	0.039g
$CaCl_2 \cdot 2H_2O$	0.036g
Na_2CO_3	0.020g
Citric acid	6.0mg
EDTA	6.0mg
Ferric citrate	6.0mg

pH 7.5 ± 0.5 at 25°C

Preparation of Medium: Add components to distilled/deionized water and bring volume to 1.0L. Mix thoroughly. Distribute into tubes or flasks. Autoclave for 15 min at 15 psi pressure–121°C.

Use: For the cultivation and maintenance of *Anabaena flos-aquae* and *Microcystis aeruginosa.*

GPVA Medium

Composition per liter:

Agar	15.0g
Yeast extract	10.0g
ACES buffer (2-[(2-Amino-2-oxoethyl)-amino]-ethane sulfonic acid)	10.0g
Glycine	3.0g
Charcoal, activated	2.0g
α-Ketogluatrate	1.0g
$Fe_4(P_2O_7)_3 \cdot 9H_2O$	0.25g
Antibiotic inhibitor solution	10.0mL

pH 6.9 ± 0.2 at 25°C

Antibiotic Inhibitor Solution:

Composition per 10mL:

Anisomycin	0.08g
Vancomycin	5.0mg
Polymyxin B	100,000U

Preparation of Antibiotic Inhibitor Solution: Add components to distilled/deionized water and bring volume to 10.0mL. Mix thoroughly. Filter sterilize.

Preparation of Medium: Add components, except antibiotic inhibitor solution, to distilled/deionized water and bring volume to 990.0mL. Mix thoroughly. Gently heat and bring to boiling. Autoclave for 15 min at 15 psi pressure–121°C. Cool to 45–50°C. Adjust pH to 6.9. Aseptically add 10.0mL sterile antibiotic inhibitor solution. Mix thoroughly. Pour into sterile Petri dishes or distribute into sterile tubes.

Use: For the isolation and cultivation of *Legionella* species from environmental waters.

Green Top Agar

Composition per liter:

Agar	15.0g
Yeast extract	2.0g
Sodium acetate	1.0g
Pancreatic digest of casein	1.0g
Soil extract	50.0mL

pH 7.4 ± 0.2 at 25°C

Soil Extract:

Composition per 200mL:

African violet soil	0.5g
Na_2CO_3	0.5g

Preparation of Soil Extract: Add components to tap water and bring volume to 200.0mL. Autoclave for 60 min at 15 psi pressure–121°C. Filter through Whatman filter paper.

Preparation of Medium: Add components to distilled/deionized water and bring volume to 1.0L. Mix thoroughly. Gently heat and bring to boiling. Distribute into tubes or flasks. Autoclave for 15 min at 15 psi pressure–121°C. Pour into sterile Petri dishes or leave in tubes.

Use: For the cultivation and maintenance of *Bacillus macroides, Caryophanon latum, Lampropedia lyalina,* and *Vittreoscilla stercoraria.*

GSP Agar

Composition per liter:

Starch, soluble	20.0g
Agar	15.0g
Sodium glutamate	10.0g
K_2HPO_4	2.0g
$MgSO_4 \cdot 7H_2O$	0.5g
Phenol red	0.36g

pH 7.2–7.4 at 25°C

Preparation of Medium: Add components to distilled/deionized water and bring volume to 1.0L. Mix thoroughly. Distribute into tubes or flasks. Autoclave for 15 min at 15 psi pressure–121°C.

Use: For the selective isolation and cultivation of *Pseudomonas* species.

Gum Tragacanth Gum Arabic Medium

Composition per 100mL:

Gum tragacanth	2.0g
Gum arabic	1.0g

Preparation of Medium: Add components to distilled/deionized water and bring volume to 100.0mL. Mix thoroughly. Gently heat and bring to boiling. Autoclave for 15 min at 15 psi pressure–121°C. Pour into sterile Petri dishes.

Use: For the cultivation of aciduric flat sour sporeformers from foods. For the isolation and cultivation of *Desulfotomaculum nigrificans* from foods.

GYPT Medium (Glucose Yeast Extract Peptone Thioglycollate Medium)

Composition per liter:

Agar	8.0g
Yeast extract	6.0g
Glucose	5.0g
Peptone	2.0g
Sodium thioglycollate	0.5g

pH 7.4 ± 0.2 at 25°C

Preparation of Medium: Add components to distilled/deionized water and bring volume to 1.0L. Mix thoroughly. Adjust pH to 7.4. Gently heat and bring to boiling. Anaerobically distribute into tubes under 97% N_2 + 3% H_2. Cap tubes with rubber stoppers. Place tubes in a press. Autoclave for 15 min at 15 psi pressure–121°C with fast exhaust.

Use: the cultivation and maintenance of *Spirochaeta stenostrepta*.

H Medium

Composition per liter:

Pancreatic digest of casein	10.0g
NaCl	8.0g

Preparation of Medium: Add components to distilled/deionized water and bring volume to 1.0L. Mix thoroughly. Distribute into tubes or flasks. Autoclave for 15 min at 15 psi pressure–121°C.

Use: For the cultivation of *Escherichia coli* and a variety of other bacteria.

H Top Agar

Composition per liter:

Pancreatic digest of casein	10.0g
NaCl	8.0g
Agar	7.0g

Preparation of Medium: Add components to distilled/deionized water and bring volume to 1.0L. Mix thoroughly. Gently heat and bring to boiling. Autoclave for 15 min at 15 psi pressure–121°C. Pour into sterile Petri dishes that contain H agar.

Use: For the cultivation of *Escherichia coli* and a variety of other bacteria.

Hagedorn and Holt Selective Medium

Composition per liter:

NaCl	20.0g
Agar	15.0g
Yeast extract	2.0g
Pancreatic digest of casein	1.7g
Agar	1.5g
NaCl	0.5g
Papaic digest of soybean meal	0.3g
K_2HPO_4	0.25g
Glucose	0.25g
Cycloheximide	0.1g
Methyl red	0.15mg

pH 7.3 ± 0.2 at 25°C

Preparation of Medium: Add components to distilled/deionized water and bring volume to 1.0L. Mix thoroughly. Gently heat and bring to boiling. Distribute into tubes or flasks. Autoclave for 15 min at 15 psi pressure–121°C. Pour into sterile Petri dishes or leave in tubes.

Use: For the selective isolation of *Arthrobacter* species in soil.

Haliscomenobacter Medium

Composition per liter:

Agar	10.0g
$(NH_4)_2SO_4$	0.5g
Glucose	0.15g
$CaCO_3$	0.1g
KCl	0.05g
K_2HPO_4	0.05g
$MgSO_4 \cdot 7H_2O$	0.05g
$Ca(NO_3)_2$	0.01g
Vitamin solution	10.0mL

Vitamin Solution:

Composition per 10mL:

Thiamine	0.4mg
Vitamin B_{12}	0.05mg

Preparation of Vitamin Solution: Add components to distilled/deionized water and bring volume to 10.0mL. Mix thoroughly. Filter sterilize.

Preparation of Medium: Add components, except vitamin solution, to distilled/deionized water and bring volume to 990.0mL. Mix thoroughly.

Gently heat and bring to boiling. Autoclave for 15 min at 15 psi pressure–121°C. Cool to 45–50°C. Aseptically add sterile vitamin solution. Mix thoroughly. Pour into sterile Petri dishes or distribute into sterile tubes.

Use: For the isolation of *Haliscomenobacter* species from activated sludge.

Haliscomenobacter Medium (DSM 134)

Glutamic acid	1.31g
$MgSO_4 \cdot 7H_2O$	0.075g
$CaCl_2 \cdot 2H_2O$	0.050g
K_2HPO_4	0.040g
$Na_2HPO_4 \cdot 2H_2O$	0.040g
KH_2PO_4	0.027g
$FeCl_3 \cdot 6H_2O$	5.0mg
$MnSO_4 \cdot H_2O$	3.0mg
Pancreatic digest of casein	1.7mg
NaCl	0.5mg
Papaic digest of soybean meal	0.3mg
K_2HPO_4	0.25mg
Vitamin solution	10.0mL
Glucose solution	5.0mL
Trace element solution SL-6	1.0mL

pH 7.5 ± 0.2 at 25°C

Vitamin Solution:
Composition per 10mL:

Thiamine	0.4mg
Vitamin B_{12}	0.01mg

Preparation of Vitamin Solution: Add components to distilled/deionized water and bring volume to 10.0mL. Mix thoroughly. Filter sterilize.

Glucose Solution:
Composition per 5mL:

D-Glucose	2.0g

Preparation of Glucose Solution: Add glucose to distilled/deionized water and bring volume to 5.0mL. Mix thoroughly. Autoclave for 15 min at 15 psi pressure–121°C.

Trace Elements Solution SL-6:
Composition per liter:

H_3BO_3	0.3g
$CoCl_2 \cdot 6H_2O$	0.2g
$ZnSO_4 \cdot 7H_2O$	0.1g
$MnCl_2 \cdot 4H_2O$	0.03g
$Na_2MoO_4 \cdot H_2O$	0.03g
$NiCl_2 \cdot 6H_2O$	0.02g
$CuCl_2.2H_2O$	0.01g

Preparation of Trace Elements Solution SL-6: Add components to distilled/deionized water and bring volume to 1.0L. Mix thoroughly. Adjust pH to 3.4.

Preparation of Medium: Add components, except vitamin solution and glucose solution, to distilled/deionized water and bring volume to 985.0mL. Mix thoroughly. Adjust pH to 7.5. Gently heat and bring to boiling. Autoclave for 15 min at 15 psi pressure–121°C. Cool to 25°C. Aseptically add sterile vitamin solution and glucose solution. Mix thoroughly. Aseptically distribute into sterile tubes or flasks.

Use: For the cultivation and maintenance of *Haliscomenobacter hydrossis.*

Haloanaerobium Medium

Composition per 1066mL:

NaCl	130.0g
Pancreatic digest of casein	10.0g
Yeast extract	10.0g
$MgSO_4 \cdot H_2O$	5.0g
KCl	1.0g
Thioglycollate-ascorbate reducing agent	30.9mL
Glucose solution	25.75mL
NaOH solution	10.3mL
Wolfe's vitamin solution	10.0mL
Wolfe's mineral solution	10.0mL

pH 7.0 ± 0.2 at 25°C

Glucose Solution:
Composition per 30mL:

D-Glucose	3.0g

Preparation of Glucose Solution: Add D-glucose to distilled/deionized water and bring volume to 30.0mL. Mix thoroughly. Filter sterilize.

NaOH Solution:
Composition per 20mL:

NaOH	1.6g

Preparation of NaOH Solution: Add NaOH to distilled/deionized water and bring volume to 20.0mL. Mix thoroughly. Autoclave for 15 min at 15 psi pressure–121°C.

Wolfe's Vitamin Solution:
Composition per liter:

Pyridoxine·HCl	0.01g
Thiamine·HCl	5.0mg
Riboflavin	5.0mg
Nicotinic acid	5.0mg
Calcium pantothenate	5.0mg
p-Aminobenzoic acid	5.0mg
Thioctic acid	5.0mg
Biotin	2.0mg
Folic acid	2.0mg
Cyanocobalamin	0.1mg

Preparation of Wolfe's Vitamin Solution: Add components to distilled/deionized water and bring volume to 1.0L. Mix thoroughly.

Wolfe's Mineral Solution:
Composition per liter

$MgSO_4 \cdot 7H_2O$	3.0g
Nitriloacetic acid	1.5g
NaCl	1.0g
$MnSO_4 \cdot H_2O$	0.5g
$FeSO_4 \cdot 7H_2O$	0.1g
$CoCl_2 \cdot 6H_2O$	0.1g
$CaCl_2$	0.1g
$ZnSO_4 \cdot 7H_2O$	0.1g
$CuSO_4 \cdot 5H_2O$	0.01g
$AlK(SO_4)_2 \cdot 12H_2O$	0.01g
H_3BO_3	0.01g
$Na_2MoO_4 \cdot 2H_2O$	0.01g

Preparation of Wolfe's Mineral Solution: Add nitrilotriacetic acid to 500.0mL of distilled/deionized water. Dissolve by adjusting pH to 6.5 with KOH. Add remaining components. Add distilled/deionized water to 1.0L.

Thioglycollate-Ascorbate Reducing Agent:
Composition per 100mL:

Ascorbic acid	1.0g
Sodium thioglycollate	1.0g

Preparation of Thioglycollate-Ascorbate Reducing Agent: Add components to distilled/deionized water and bring volume to 100.0mL. Mix thoroughly. Adjust pH to 7.0. Filter sterilize.

Preparation of Medium: Add components, except thioglycollate-ascorbate reducing agent, glucose and NaOH solutions, to distilled/deionized water and bring volume to 990.0mL. Mix thoroughly. Gently heat and bring to boiling. Anaerobically distribute into tubes under 97% N_2 + 3% H_2 in 9.7mL volumes. Cap tubes with rubber stoppers. Autoclave for 15 min at 15 psi pressure–121°C. Cool to 25°C. Immediately prior to inoculation, aseptically add 0.3mL of sterile thioglycollate-ascorbate reducing agent, 0.25mL of sterile glucose solution and 0.1mL of sterile NaOH solution to each tube.

Use: For the cultivation and maintenance of *Haloanaerobium praevalens.*

Haloanaerobium praevalens Medium

Composition per liter:

NaCl	130.0g
Agar	20.0g
Yeast extract	2.00g
Pancreatic digest of casein	2.00g
NH_4Cl	0.50g
$MgSO_4 \cdot 7H_2O$	0.50g
K_2HPO_4	0.35g
$CaCl_2 \cdot 2H_2O$	0.25g
KH_2PO_4	0.23g
$FeSO_4 \cdot 7H_2O$	2.0mg
$NaHCO_3$ solution	20.0mL
Cysteine-sulfide reducing agent	20.0mL
Wolfe's vitamin solution	10.0mL
Methanol	10.0mL
Resazurin (0.025% solution)	4.0mL
Trace elements SL-6	3.0mL

pH 6.8 ± 0.2 at 25°C

$NaHCO_3$ Solution:
Composition per 20mL:

$NaHCO_3$	850.0mg

Preparation of $NaHCO_3$ Solution: Add $NaHCO_3$ to distilled/deionized water and bring volume to 20.0mL. Mix thoroughly. Filter sterilize. Gas with 100% CO_2 for 20 min.

Cysteine-Sulfide Reducing Agent:
Composition per 20mL:

L-Cysteine·HCl·H_2O	0.3g
$Na_2S \cdot 9H_2O$	0.3g

Preparation of Cysteine-Sulfide Reducing Agent: Add L-Cysteine·HCl·H_2O to 10.0mL of distilled/deionized water. Mix thoroughly. In a separate tube, add $Na_2S \cdot 9H_2O$ to 10.0mL of distilled/deionized water. Mix thoroughly. Gas both solutions with 100% N_2 and cap tubes. Autoclave both solutions for 15 min at 15 psi pressure–121°C using fast exhaust. Cool to 50°C. Aseptically combine the two solutions under 100% N_2.

Wolfe's Vitamin Solution:
Composition per liter:

Pyridoxine·HCl	10.0mg
Thiamine·HCl	5.0mg
Riboflavin	5.0mg
Nicotinic acid	5.0mg
Calcium pantothenate	5.0mg
p-Aminobenzoic acid	5.0mg
Thioctic acid	5.0mg
Biotin	2.0mg
Folic acid	2.0mg
Cyanocobalamin	100.0μg

Preparation of Wolfe's Vitamin Solution: Add components to distilled/deionized water and bring volume to 1.0L. Mix thoroughly. Filter sterilize.

Trace Elements Solution SL-6:
Composition per liter:

H_3BO_3	0.3g
$CoCl_2 \cdot 6H_2O$	0.2g
$ZnSO_4 \cdot 7H_2O$	0.1g
$MnCl_2 \cdot 4H_2O$	0.03g
$Na_2MoO_4 \cdot H_2O$	0.03g
$NiCl_2 \cdot 6H_2O$	0.02g
$CuCl_2 .2H_2O$	0.01g

Preparation of Trace Elements Solution SL-6: Add components to distilled/deionized water and bring volume to 1.0L. Mix thoroughly. Adjust pH to 3.4.

Preparation of Medium: Add components, except $NaHCO_3$ solution, cysteine-sulfide reducing agent, Wolfe's vitamin solution and methanol, to distilled/deionized water and bring volume to 940.0mL. Mix thoroughly. Autoclave for 15 min at 15 psi pressure–121°C. Cool under 80% N_2 + 20% CO_2. Aseptically and anaerobically add the sterile $NaHCO_3$ solution, the sterile cysteine-sulfide reducing agent, the sterile Wolfe's vitamin solution and filter-sterilized methanol. Mix thoroughly. Adjust pH to 6.8. Aseptically and anaerobically distribute into sterile tubes or flasks.

Use: For the cultivation and maintenance of *Haloanaerobium praevalens.*

Haloarcula Medium

Composition per 1001mL:

NaCl	250.0g
$MgSO_4 \cdot 7H_2O$	20.0g
Agar	15.0g
Sodium citrate	3.0g
KCl	2.0g
$CaCl_2$	0.2g
Peptone solution	100.0mL
Trace elements solution	1.0mL

pH 7.4 ± 0.1 at 25°C

Peptone Solution:
Composition per 100mL:

Bacteriological peptone, Oxoid Unipath	10.0g

Preparation of Peptone Solution: Add peptone to distilled/deionized water and bring volume to 100.0mL. Mix thoroughly. Filter sterilize.

Trace Elements Solution:
Composition per 100mL:

$FeCl_2 \cdot 4H_2O$	0.36g
$MnCl_2 \cdot 4H_2O$	0.022g

Preparation of Trace Elements Solution: Add components to distilled/deionized water and bring volume to 100.0mL. Mix thoroughly. Filter sterilize.

Preparation of Medium: Add components, except peptone solution and trace elements solution, to distilled/deionized water and bring volume to 900.0mL. Mix thoroughly. Gently heat and bring to boiling. Adjust pH to 7.4 with NaOH. Autoclave for 15 min at 15 psi pressure–121°C. Cool to 45–50°C. Aseptically add 100.0mL of sterile peptone solution and 1.0mL of sterile trace elements solution. Mix thoroughly. Pour into sterile Petri dishes or distribute into sterile tubes.

Use: For the cultivation and maintenance of *Haloanaerobium praevalens.*

Halobacteria Medium

Composition per liter:

NaCl	220.0g
Agar	10.0g
$MgSO_4 \cdot 7H_2O$	10.0g
Casein hydrolysate	5.0g
KCl	5.0g
Disodium citrate	3.0g
KNO_3	1.0g
Yeast extract	1.0g
$CaCl_2 \cdot 6H_2O$	0.2g

pH 7.2–7.4 at 25°C

Preparation of Medium: Add components to distilled/deionized water and bring volume to 1.0L. Mix thoroughly. Gently heat until dissolved. Adjust pH to 7.2–7.4. Distribute into tubes or flasks. Autoclave for 15 min at 15 psi pressure–121°C.

Use: For the cultivation and enumeration of halobacteria.

Halobacteriaceae Medium 1

Composition per liter:

Salt, crude solar	250.0g
$MgSO_4 \cdot 7H_2O$	20.0g
KCl	5.0g
Pancreatic digest of casein	5.0g
Yeast extract	5.0g
$CaCl_2 \cdot 6H_2O$	0.2g

pH 7.0 ± 0.2 at 25°C

Preparation of Medium: Add components to distilled/deionized water and bring volume to 1.0L. Mix thoroughly. Gently heat until dissolved. Adjust pH to 7.0. Distribute into tubes or flasks. Autoclave for 15 min at 15 psi pressure–121°C.

Use: For axenic cultivation of members of the Halobacteriacea.

Halobacteriaceae Medium 2

Composition per liter:

NaCl	250.0g
$MgSO_4 \cdot 7H_2O$	20.0g
Yeast extract	10.0g
Casamino acids	7.5g
Trisodium citrate	3.0g
KCl	2.0g
$FeCl_2$	2.3mg

pH 7.5–7.8 at 25°C

Preparation of Medium: Add components to distilled/deionized water and bring volume to 1.0L. Mix thoroughly. Gently heat until dissolved. Adjust pH to 7.5–7.8. Distribute into tubes or flasks. Autoclave for 15 min at 15 psi pressure–121°C.

Use: For the axenic cultivation of halobacteria and halococci.

Halobacteriaceae Medium 3

Composition per liter:

NaCl	240.0g
L-Glutamine	15.0g
KCl	5.0g
K_2SO_4	5.0g
$MgCl_2 \cdot 6H_2O$	5.0g
$MgSO_4$, anhydrous	5.0g
NH_4Cl	5.0g
Pancreatic digest of casein	5.0g
Yeast extract	5.0g
K_2HPO_4	0.5g
L-Arginine	0.5g
L-Isoleucine	0.25g
L-Leucine	0.25g
L-Lysine	0.25g
L-Proline	0.25g
L-Valine	0.25g
Cytidylic acid	0.2g
$CaCl_2 \cdot 2H_2O$	0.1g
L-Methionine	0.1g
L-Tyrosine	0.1g
L-Phenylalanine	0.05g
$FeCl_2 \cdot 6H_2O$	5.0mg

pH 6.8 ± 0.2 at 25°C

Preparation of Medium: Add components to distilled/deionized water and bring volume to 1.0L. Mix thoroughly. Gently heat until dissolved. Adjust pH to 6.8. Distribute into tubes or flasks. Autoclave for 15 min at 15 psi pressure–121°C.

Use: For the cultivation of some halobacteria and halococci.

Halobacteriaceae Medium 4

Composition per liter:

NaCl	250.0g
$MgSO_4 \cdot 7H_2O$	20.0g
NH_4Cl	5.0g
L-Glutamic acid	1.3g
DL-Valine	1.0g
Glycerol	1.0g
L-Lysine	0.85g
L-Leucine	0.80g
DL-Serine	0.61g
DL-Threonine	0.50g
DL-Isoleucine	0.44g
DL-Alanine	0.43g
L-Arginine	0.40g
DL-Methionine	0.37g
DL-Phenylalanine	0.26g
L-Tyrosine	0.20g
Adenylic acid	0.1g
KNO_3	0.1g
Uridylic acid	0.1g
Glycine	0.06g
KH_2PO_4	0.05g
K_2HPO_4	0.05g
L-Cysteine	0.05g
L-Proline	0.05g
Sodium citrate	0.05g
$FeCl_2$	2.3mg
$CaCl_2 \cdot 2H_2O$	0.7mg
$ZnSO_4 \cdot 7H_2O$	0.44mg
$MnSO_4 \cdot H_2O$	0.3mg
$CuSO_4 \cdot 5H_2O$	0.05mg

pH 6.2 ± 0.2 at 25°C

Preparation of Medium: Add components to distilled/deionized water and bring volume to 1.0L. Mix thoroughly. Gently heat until dissolved. Adjust pH to 6.2. Distribute into tubes or flasks. Autoclave for 15 min at 15 psi pressure–121°C.

Use: For the cultivation of members of the Halobacteriaceae.

Halobacterium denitrificans Medium

Composition per liter:

NaCl	176.0g
Agar	20.0g
$MgCl_2 \cdot 6H_2O$	20.0g
HEPES (*N*-2-Hydroxyethylpiperazine-*N′*-2-ethanesulfonic acid) buffer	11.9g
Yeast extract	5.0g
Hy-Case SF (Humko-Sheffield)	2.0g
KCl	2.0g
$CaCl_2 \cdot 2H_2O$	0.1g

pH 6.7 ± 0.2 at 25°C

Preparation of Medium: Add components to distilled/deionized water and bring volume to 1.0L. Mix thoroughly. Gently heat and bring to boiling. Adjust pH to 6.7. Distribute into tubes or flasks. Autoclave for 15 min at 15 psi pressure–121°C. Pour into sterile Petri dishes or leave in tubes.

Use: For the aerobic cultivation and maintenance of *Haloferax (Halobacterium) denitrificans.*

Halobacterium denitrificans Medium

Composition per liter:

NaCl	176.0g
Agar	20.0g
$MgCl_2 \cdot 6H_2O$	20.0g
HEPES (N-2-Hydroxyethylpiperazine-N′-2-ethanesulfonic acid) buffer	11.9g
KNO_3	5.0g
Yeast extract	5.0g
Hy-Case SF (Humko-Sheffield)	2.0g
KCl	2.0g
$CaCl_2 \cdot 2H_2O$	0.1g

pH 6.7 ± 0.2 at 25°C

Preparation of Medium: Add components to distilled/deionized water and bring volume to 1.0L. Mix thoroughly. Gently heat and bring to boiling. Adjust pH to 6.7. Distribute into tubes or flasks. Autoclave for 15 min at 15 psi pressure–121°C. Pour into sterile Petri dishes or leave in tubes.

Use: For the anerobic cultivation and maintenance of *Haloferax (Halobacterium) denitrificans.*

Halobacterium Medium (ATCC Medium 974)

Composition per 100mL:

Solution 1	75.0mL
Solution 2	25.0mL

pH 6.8 ± 0.2 at 25°C

Solution 1:

Composition per 75mL:

NaCl	12.5g
$MgCl_2 \cdot 6H_2O$	5.0g
K_2SO_4	0.5g
$CaCl_2 \cdot 6H_2O$	0.02g

Preparation of Solution 1: Add components to distilled/deionized water and bring volume to 75.0mL. Mix thoroughly. Adjust pH to 6.8. Autoclave for 15 min at 15 psi pressure–121°C. Cool to 45–50°C.

Solution 2:

Composition per 25mL:

Agar	2.0g
Pancreatic digest of casein	0.5g
Yeast extract	0.5g

Preparation of Solution 2: Add components to distilled/deionized water and bring volume to 25.0mL. Mix thoroughly. Adjust pH to 6.8. Autoclave for 15 min at 15 psi pressure–121°C. Cool to 45–50°C.

Preparation of Medium: Aseptically combine sterile solution 1 and sterile solution 2. Mix thoroughly. Pour into sterile Petri dishes or distribute into sterile tubes.

Use: For the cultivation and maintenance of *Haloferax volcanii.*

Halobacterium Medium (ATCC Medium 1270)

Composition per liter:

NaCl	194.0g
$MgSO_4$	24.0g
$MgCl_2$	16.0g
KCl	5.0g
Yeast extract	5.0g
$CaCl_2$	1.0g
NaBr	0.5g
$NaHCO_3$	0.2g

pH 7.3 ± 0.2 at 25°C

Preparation ofPreparation of Medium: Add components to distilled/deionized water and bring volume to 1.0L. Mix thoroughly. Distribute into tubes or flasks. Autoclave for 15 min at 15 psi pressure–121°C.

Use: For the cultivation of *Haloarcula hispanica* and *Haloferax gibbonsii.*

Halobacterium Medium (ATCC Medium 213)

Composition per liter:

Solution 1	500.0mL
Solution 2	500.0mL

pH 7.0 ± 0.2 at 25°C

Solution 1:

Composition per 500mL:

Yeast extract	10.0g
Pancreatic digest of casein	2.5g

Preparation of Solution 1: Add components to distilled/deionized water and bring volume to 500.0mL. Mix thoroughly. Gently heat and bring to boiling. Adjust pH to 7.0. Autoclave for 15 min at 15 psi pressure–121°C. Cool to 45–50°C.

Solution 2:

Composition per 500mL:

NaCl	250.0g
Agar	20.0g
$MgSO_4 \cdot 7H_2O$	10.0g
KCl	5.0g
$CaCl_2 \cdot 6H_2O$	0.2g

Preparation of Solution 2: Add components to distilled/deionized water and bring volume to 500.0mL. Mix thoroughly. Gently heat and bring to boiling. Autoclave for 15 min at 15 psi pressure–121°C. Cool to 45–50°C.

Preparation of Medium: Aseptically combine sterile solution 1 and sterile solution 2. Mix thoroughly. Aseptically distribute into sterile tubes or flasks.

Use: For the cultivation of *Halobacterium salinarium*.

Halobacterium Medium

Composition per liter:

Solution 1	500.0mL
Solution 2	500.0mL

pH 7.0 ± 0.2 at 25°C

Solution 1:

Composition per 500mL:

Yeast extract	10.0g
Pancreatic digest of casein	2.5g

Preparation of Solution 1: Add components to distilled/deionized water and bring volume to 500.0mL. Mix thoroughly. Gently heat and bring to boiling. Adjust pH to 7.0. Autoclave for 15 min at 15 psi pressure–121°C. Cool to 45–50°C.

Solution 2:

Composition per 500mL:

NaCl	250.0g
$MgSO_4 \cdot 7H_2O$	10.0g
KCl	5.0g
$CaCl_2 \cdot 6H_2O$	0.2g

Preparation of Solution 2: Add components to distilled/deionized water and bring volume to 500.0mL. Mix thoroughly. Gently heat and bring to boiling. Autoclave for 15 min at 15 psi pressure–121°C. Cool to 45–50°C.

Preparation of Medium: Aseptically combine sterile solution 1 and sterile solution 2. Mix thoroughly. Aseptically distribute into sterile tubes or flasks.

Use: For the cultivation of *Halobacterium salinarium*.

Halobacterium Medium (ATCC Medium 1176)

Composition per liter:

NaCl	156.0g
$MgSO_4 \cdot 7H_2O$	20.0g
$MgCl_2 \cdot 6H_2O$	13.0g
Yeast extract	5.0g
KCl	4.0g
$CaCl_2 \cdot 6H_2O$	1.0g
Glucose	1.0g
NaBr	0.5g
$NaHCO_3$	0.2g

pH 7.0 ± 0.2 at 25°C

Preparation of Medium: Add components to distilled/deionized water and bring volume to 1.0L. Mix thoroughly. Distribute into tubes or flasks. Autoclave for 15 min at 15 psi pressure–121°C.

Use: For the cultivation of *Haloferax mediterranei*.

Halobacterium pharaonis Medium

Composition per liter:

NaCl	250.0g
Agar	20.0g
Casamino acids	15.0g
Trisodium citrate·$2H_2O$	3.0g
Glutamic acid	2.5g
$MgSO_4 \cdot 7H_2O$	2.5g
KCl	2.0g

pH 8.5 ± 0.2 at 25°C

Preparation of Medium: Add components to distilled/deionized water and bring volume to 1.0L. Mix thoroughly. Gently heat and bring to boiling. Adjust pH to 6.0. Autoclave for 15 min at 15 psi pressure–121°C. Cool to 50°C. Readjust pH to 8.5. Pour into sterile Petri dishes or distribute into sterile tubes.

Use: For the cultivation and maintenance of *Natronobacterium (Halobacterium) pharaonis*.

Halobacterium Starch Medium

Composition per liter:

$MgCl_2 \cdot 6H_2O$	160.0g
NaCl	125.0g
K_2SO_4	5.0g
Soluble starch	2.0g
Peptone	1.0g
Yeast extract	1.0g
$CaCl_2 \cdot 2H_2O$	0.13g

pH 7.0 ± 0.2 at 25°C

Preparation of Medium: Add components to distilled/deionized water and bring volume to 1.0L. Mix thoroughly. Distribute into tubes or flasks. Autoclave for 15 min at 15 psi pressure–121°C.

Use: For the cultivation and maintenance of *Haloarcula marismortui* and *Halobacterium sodomense*.

Halobacterium volcanii Medium

Composition per 100mL:

NaCl	25.0g
$MgSO_4 \cdot 7H_2O$	1.0g
KCl	0.5g
Glycine	0.2g
$CaCl_2 \cdot 6H_2O$	0.02g
Yeast autolysate	1.0mL

pH 7.0 ± 0.2 at 25°C

Preparation of Medium: Add components to distilled/deionized water and bring volume to 100.0mL. Mix thoroughly. Adjust pH to 7.0. Distribute into tubes or flasks. Autoclave for 15 min at 15 psi pressure–121°C.

Use: For the specific enrichment of *Halobacterium volcanii.*

Halobacterium volcanii Medium

Composition per liter:

NaCl	125.0g
$MgCl_2 \cdot 6H_2O$	50.0g
K_2SO_4	5.0g
Pancreatic digest of casein	5.0g
Yeast extract	5.0g
$CaCl_2 \cdot 6H_2O$	0.2g

pH 6.8 ± 0.2 at 25°C

Preparation of Medium: Add components to distilled/deionized water and bring volume to 1.0L. Mix thoroughly. Gently heat until dissolved. Adjust pH to 6.8. Distribute into tubes or flasks. Autoclave for 15 min at 15 psi pressure–121°C.

Use: For the cultivation of *Halobacterium volcanii.*

Halobacteroides Medium

Composition per 990mL:

NaCl	88.0g
$MgCl_2 \cdot 6H_2O$	20.0g
$CaCl_2 \cdot 2H_2O$	7.4g
Yeast extract	5.0g
KCl	3.7g
L-Cysteine·HCl·H_2O	0.5g
Resazurin	1.0mg
Glucose (10% solution)	50.0mL
Sodium PIPES (Piperazine-*N,N*′-bis-2-ethane sulfonate buffer, 1*M*, pH 6.8–7.0)	40.0mL

pH 6.8–7.0 at 25°C

Preparation of Medium: Filter sterilize glucose solution and PIPES buffer solution separately. Add remaining components—except glucose solution, PIPES buffer solution, and cysteine·HCl·H_2O—to distilled/deionized water and bring volume to 900.0mL. Mix thoroughly. Gently heat and bring to boiling under 100% N_2. Add cysteine·HCl·H_2O. Mix thoroughly. Autoclave for 15 min at 15 psi pressure–121°C. Cool to 45–50°C. Aseptically add 50.0mL of sterile glucose solution and 40.0mL of sterile PIPES buffer solution. Mix thoroughly. Aseptically and anaerobically distribute into sterile tubes or flasks.

Use: For the cultivation and maintenance of *Halobacteroides halobius* and *Sporohalobacter marismortui.*

Halobius Medium

Composition per liter:

NaCl	116.0g
Agar	20.0g
$MgSO_4 \cdot 7H_2O$	20.0g
Yeast extract	10.0g
Vitamin-free casamino acids	7.5g
Sodium citrate	3.0g
KCl	2.0g
$FeCl_2$	0.023g

pH 6.2 ± 0.2 at 25°C

Preparation of Medium: Add components to distilled/deionized water and bring volume to 1.0L. Mix thoroughly. Gently heat and bring to boiling. Distribute into tubes or flasks. Autoclave for 15 min at 15 psi pressure–121°C. Pour into sterile Petri dishes or leave in tubes.

Use: For the cultivation and maintenance of *Micrococcus halobius.*

Halodurans Medium

Composition per liter:

NaCl	150.0g
Agar	20.0g
$MgSO_4 \cdot 7H_2O$	20.0g
Yeast extract	10.0g
Casamino acids	7.5g
Sodium citrate	3.0g
KCl	2.0g
$FeCl_2$	0.023g

pH 7.0 ± 0.2 at 25°C

Preparation of Medium: Add components to distilled/deionized water and bring volume to 1.0L. Mix thoroughly. Gently heat and bring to boiling. Distribute into tubes or flasks. Autoclave for 15 min at 15 psi pressure–121°C. Pour into sterile Petri dishes or leave in tubes.

Use: For the cultivation and maintenance of *Micrococcus varians.*

Halomethanococcus Medium (*Methanohalophilus* Medium)

Composition per 1030mL:

Trimethylamine·HCl	2.5g
Na_2CO_3	2.0g
$NaHCO_3$	2.0g
Casamino acids	0.5g

L-Cysteine·HCl·H_2O 0.5g
$Na_2S·9H_2O$ solution 0.5g
NH_4Cl 0.5g
Pancreatic digest of casein 0.5g
Yeast extract 0.5g
K_2HPO_4 .... 0.2g
Ammonium-2-mercaptoethanesulfonate 1.0mg
Artificial brine 1.0L
Wolfe's vitamin solution 10.0mL
Wolfe's mineral solution 10.0mL
Volatile acids solution 10.0mL
pH 7.1 ± 0.2 at 25°C

$Na_2S·9H_2O$ Solution:
Composition per 10mL:
$Na_2S·9H_2O$ 2.0g

Preparation of $Na_2S·9H_2O$ Solution: Add $Na_2S·9H_2O$ to distilled/deionized water and bring volume to 10.0mL. Mix thoroughly. Filter sterilize.

Artificial Brine:
Composition per liter:
NaCl 80.7g
$MgCl_2·6H_2O$ 35.1g
Na_2SO_4 .... 12.9g
KCl 5.7g
$CaCl_2$ 0.55g
$LiCl_2$ 0.13g
H_3BO_3 .... 0.12g

Preparation of Artificial Brine: Add components to distilled/deionized water and bring volume to 1.0L. Mix thoroughly.

Wolfe's Vitamin Solution:
Composition per liter:
Pyridoxine·HCl 0.01g
Thiamine·HCl 5.0mg
Riboflavin 5.0mg
Nicotinic acid 5.0mg
Calcium pantothenate 5.0mg
p-Aminobenzoic acid 5.0mg
Thioctic acid 5.0mg
Biotin 2.0mg
Folic acid 2.0mg
Cyanocobalamin 0.1mg

Preparation of Wolfe's Vitamin Solution: Add components to distilled/deionized water and bring volume to 1.0L. Mix thoroughly. Filter sterilize. Store at 4°C.

Wolfe's Mineral Solution:
Composition per liter:
$MgSO_4·7H_2O$ 3.0g
Nitriloacetic acid 1.5g
NaCl 1.0g
$MnSO_4·H_2O$ 0.5g
$FeSO_4·7H_2O$ 0.1g
$CoCl_2·6H_2O$ 0.1g
$CaCl_2$ 0.1g
$ZnSO_4·7H_2O$ 0.1g
$CuSO_4·5H_2O$ 0.01g
$AlK(SO_4)_2·12H_2O$ 0.01g
H_3BO_3 .... 0.01g
$Na_2MoO_4·2H_2O$ 0.01g

Preparation of Wolfe's Mineral Solution: Add nitrilotriacetic acid to 500.0mL of distilled/deionized water. Dissolve by adjusting pH to 6.5 with KOH. Add remaining components. Add distilled/deionized water to 1.0L.

Volatile Acids Solution:
Composition per liter:
α-Methylbutyric acid 0.5mL
Isobutyric acid 0.5mL
Isovaleric acid 0.5mL
Valeric acid 0.5mL

Preparation of Volatile Acids Solution: Add components to distilled/deionized water and bring volume to 1.0L. Mix thoroughly.

Preparation of Medium: Combine components, except the cysteine·HCl·H_2O, trimethylamine·HCl, and $Na_2S·9H_2O$ solution. Mix thoroughly. Gently heat and bring to boiling. Add cysteine·HCl·H_2O. Mix thoroughly. Cool in an ice water bath under 80% N_2 + 20% CO_2. Add trimethylamine·HCl. Mix thoroughly. Adjust pH to 7.1. Aseptically and anaerobically distribute into tubes in 10.0mL volumes under 80% N_2 + 20% CO_2. Autoclave for 15 min at 15 psi pressure–121°C. Immediately prior to inoculation, aseptically add 0.25mL of sterile $Na_2S·9H_2O$ solution to each tube. Mix thoroughly.

Use: For the cultivation and maintenance of *Methanohalophilus mahii.*

Halomonas Medium

Composition per liter:
NaCl 80.0g
$MgSO_4·7H_2O$ 20.0g
Casamino acids with vitamins 7.5g
Proteose peptone No. 3 .... 5.0g
Sodium citrate 3.0g
Yeast extract 1.0g
K_2HPO_4 .... 0.5g
$Fe(NH_4)_2(SO_4)_2·6H_2O$ 0.05g
pH 7.0 ± 0.2 at 25°C

Preparation of Medium: Add components to distilled/deionized water and bring volume to 1.0L. Mix thoroughly. Adjust pH to 7.0 with KOH. Distribute into tubes or flasks. Autoclave for 15 min at 15 psi pressure–121°C.

Use: For the cultivation and maintenance of *Halomonas elongata.*

Halophile Medium

Composition per liter:

NaCl	100.0g
KCl	5.0g
$MgCl_2 \cdot 6H_2O$	5.0g
$MgSO_4 \cdot 7H_2O$	5.0g
NH_4Cl	5.0g
Peptone solution (15% solution)	30.0mL
Yeast extract solution (15% solution)	30.0mL
Ferric citrate solution (1% solution)	10.0mL
Trace element solution	5.0mL

Trace Element Solution:
Composition per liter:

$ZnSO_4 \cdot 7H_2O$	0.22g
$MgCl_2 \cdot 4H_2O$	0.18g
$CoCl_2 \cdot 6H_2O$	0.01g
$Na_2MoO_4 \cdot H_2O$	6.3mg
$CuSO_4 \cdot 5H_2O$	1.0mg

Preparation of Trace Element Solution: Add components to distilled/deionized water and bring volume to 1.0L. Mix thoroughly.

Preparation of Medium: Add components to distilled/deionized water and bring volume to 1.0L. Mix thoroughly. Gently heat and bring to boiling. Distribute into tubes or flasks. Autoclave for 15 min at 15 psi pressure–121°C.

Use: For the cultivation of *Rhodospirillum salinarum.*

Halophilic Agar (HA)

Composition per liter:

NaCl	250.0g
$MgSO_4 \cdot 7H_2O$	25.0g
Agar	20.0g
Casamino acids	10.0g
Yeast extract	10.0g
Proteose peptone	5.0g
Trisodium citrate	3.0g
KCl	2.0g

pH 7.2 ± 0.2 at 25°C

Preparation of Medium: Combine the ingredients with distilled water and heat to boiling to dissolve completely. Autoclave at 121°C for 15 min.

Use: For the isolation and cultivation of halophilic microorganisms from foods, such as *Pseudomonas* species and *Flavobacterium* species from fish and salted foods.

Halophilic Broth (HB)

Composition per liter:

NaCl	250.0g
$MgSO_4 \cdot 7H_2O$	25.0g
Casamino acids	10.0g
Yeast extract	10.0g
Proteose peptone	5.0g
Trisodium citrate	3.0g
KCl	2.0g

pH 7.2 ± 0.2 at 25°C

Preparation of Medium: Add components to distilled/deionized water and bring volume to 1.0L. Mix thoroughly. Gently heat and bring to boiling. Distribute into tubes or flasks. Autoclave for 15 min at 15 psi pressure–121°C.

Use: For the isolation and cultivation of halophilic microorganisms from foods, such as *Pseudomonas* species and *Flavobacterium* species from fish and salted foods.

Halophilic *Clostridium* Agar

Composition per liter:

L-Cysteine·HCl·H_2O	0.5g
Solution 1	1.0L
Solution 2	100.0mL

pH 6.2–7.0 at 25°C

Solution 1:
Composition per liter:

NaCl	105.0g
Agar	20.0g
KCl	7.5g
$CaCO_3$	5.0g
L-Glutamic acid	4.0g
Soluble starch	2.0g
Casamino acids	2.0g
Nutrient broth	2.0g
Yeast extract	2.0g
$FeSO_4 \cdot 7H_2O$	2.0mg
Resazurin	1.0mg
NaOH (2.5 *N* solution)	12.5mL
Wolfe's vitamin solution	10.0mL
Wolfe's mineral solution	10.0mL

Preparation of Solution 1: Add components, except $CaCO_3$, to distilled/deionized water and bring volume to 1.0L. Mix thoroughly. Gently heat and bring to boiling. When all components have dissolved, add the $CaCO_3$. Mix thoroughly.

Wolfe's Vitamin Solution:
Composition per liter:

Pyridoxine·HCl	0.01g
Thiamine·HCl	5.0mg

Riboflavin	5.0mg
Nicotinic acid	5.0mg
Calcium pantothenate	5.0mg
p-Aminobenzoic acid	5.0mg
Thioctic acid	5.0mg
Biotin	2.0mg
Folic acid	2.0mg
Cyanocobalamin	0.1mg

Preparation of Wolfe's Vitamin Solution: Add components to distilled/deionized water and bring volume to 1.0L. Mix thoroughly.

Wolfe's Mineral Solution:
Composition per liter

$MgSO_4 \cdot 7H_2O$	3.0g
Nitriloacetic acid	1.5g
NaCl	1.0g
$MnSO_4 \cdot H_2O$	0.5g
$FeSO_4 \cdot 7H_2O$	0.1g
$CoCl_2 \cdot 6H_2O$	0.1g
$CaCl_2$	0.1g
$ZnSO_4 \cdot 7H_2O$	0.1g
$CuSO_4 \cdot 5H_2O$	0.01g
$AlK(SO_4)_2 \cdot 12H_2O$	0.01g
H_3BO_3	0.01g
$Na_2MoO_4 \cdot 2H_2O$	0.01g

Preparation of Wolfe's Mineral Solution: Add nitrilotriacetic acid to 500.0mL of distilled/deionized water. Dissolve by adjusting pH to 6.5 with KOH. Add remaining components. Add distilled/deionized water to 1.0L.

Solution 2:
Composition per 100mL:

$MgCl_2 \cdot 6H_2O$	20.3g
$CaCl_2 \cdot 2H_2O$	7.35g

Preparation of Solution 2: Add components to distilled/deionized water and bring volume to 100.0mL. Mix thoroughly. Gas with 100% N_2 for 20 min. Autoclave for 15 min at 15 psi pressure–121°C. Cool to 45–50°C.

Preparation of Medium: Gently heat 1.0L of solution 1 and bring to boiling under 100% N_2. Add the cysteine·HCl·H_2O. Continue boiling until resazurin turns colorless indicating reduction. Volume of solution 1 should be about 900.0mL. Anaerobically distribute into tubes in 9.0mL volumes under 100% N_2. Cap tubes with rubber stoppers. Place tubes in a press. Autoclave for 15 min at 15 psi pressure–121°C with fast exhaust. Cool to 50°C. Aseptically add 1.0mL of sterile solution 2 to each tube. In the presence of $CaCO_3$, the pH may be higher than 7.0. Do not adjust pH.

Use: For the cultivation and maintenance of *Sporohalobacter lortetii.*

Halophilic *Clostridium* Broth

Composition per liter:

L-Cysteine·HCl·H_2O	0.5g
Solution 1	1.0L
Solution 2	100.0mL

pH 6.2–7.0 at 25°C

Solution 1:
Composition per liter:

NaCl	105.0g
KCl	7.5g
L-Glutamic acid	4.0g
Casamino acids	2.0g
Nutrient broth	2.0g
Yeast extract	2.0g
$FeSO_4 \cdot 7H_2O$	2.0mg
Resazurin	1.0mg
NaOH (2.5*N* solution)	12.5mL
Wolfe's vitamin solution	10.0mL
Wolfe's elements solution	10.0mL

Preparation of Solution 1: Add components to distilled/deionized water and bring volume to 1.0L. Mix thoroughly.

Wolfe's Vitamin Solution:
Composition per liter:

Pyridoxine·HCl	0.01g
Thiamine·HCl	5.0mg
Riboflavin	5.0mg
Nicotinic acid	5.0mg
Calcium pantothenate	5.0mg
p-Aminobenzoic acid	5.0mg
Thioctic acid	5.0mg
Biotin	2.0mg
Folic acid	2.0mg
Cyanocobalamin	0.1mg

Preparation of Wolfe's Vitamin Solution: Add components to distilled/deionized water and bring volume to 1.0L. Mix thoroughly.

Wolfe's Mineral Solution:
Composition per liter

$MgSO_4 \cdot 7H_2O$	3.0g
Nitriloacetic acid	1.5g
NaCl	1.0g
$MnSO_4 \cdot H_2O$	0.5g
$FeSO_4 \cdot 7H_2O$	0.1g
$CoCl_2 \cdot 6H_2O$	0.1g
$CaCl_2$	0.1g
$ZnSO_4 \cdot 7H_2O$	0.1g
$CuSO_4 \cdot 5H_2O$	0.01g
$AlK(SO_4)_2 \cdot 12H_2O$	0.01g
H_3BO_3	0.01g
$Na_2MoO_4 \cdot 2H_2O$	0.01g

Preparation of Wolfe's Mineral Solution: Add nitrilotriacetic acid to 500.0mL of distilled/deionized water. Dissolve by adjusting pH to 6.5 with KOH. Add remaining components. Add distilled/deionized water to 1.0L.

Solution 2:

Composition per 100mL:

$MgCl_2 \cdot 6H_2O$	20.3g
$CaCl_2 \cdot 2H_2O$	7.35g

Preparation of Solution 2: Add components to distilled/deionized water and bring volume to 100.0mL. Mix thoroughly. Gas with 100% N_2 for 20 min. Autoclave for 15 min at 15 psi pressure–121°C.

Preparation of Medium: Gently heat 1.0L of solution 1 and bring to boiling under 100% N_2. Add the cysteine·HCl·H_2O. Continue boiling until resazurin turns colorless indicating reduction. Volume of solution 1 should be about 900.0mL. Anaerobically distribute into tubes in 9.0mL volumes under 100% N_2. Cap tubes with rubber stoppers. Place tubes in a press. Autoclave for 15 min at 15 psi pressure–121°C with fast exhaust. Cool to 25°C. Aseptically add 1.0mL of sterile solution 2 to each tube. Adjust pH to 6.2–7.0 if necessary with sterile O_2-free NaOH or HCl.

Use: For the cultivation and maintenance of *Sporohalobacter lortetii.*

Halophilic *Halobacterium* Medium

Composition per liter:

NaCl	200.0g
$MgSO_4 \cdot 7H_2O$	37.0g
$CaCl_2 \cdot 2H_2O$	0.7g
KCl	0.5g
$MnCl_2 \cdot 4H_2O$	0.05g
Yeast extract	100.0mL

pH 7.0 ± 0.2 at 25°C

Preparation of Medium: Add components to distilled/deionized water and bring volume to 1.0L. Mix thoroughly. Gently heat until dissolved. Adjust pH to 7.0. Distribute into tubes or flasks. Autoclave for 15 min at 15 psi pressure–121°C.

Use: For the cultivation of extremely halophilic *Halobacterium* species.

Halophilic Synthetic Medium

Composition per liter:

Glucose	0.1g
KNO_3	0.05g
$FePO_4$	0.01g
Artificial seawater	100.0mL

Artificial Seawater:

Composition per 100mL:

NaCl	2.4g
$MgCl_2 \cdot 6H_2O$	1.1g
Na_2SO_4	0.4g
$CaCl_2 \cdot 6H_2O$	0.2g
KCl	0.07g
$NaHCO_3$	0.02g
KBr	0.01g
$SrCl_2 \cdot 6H_2O$	4.0mg
H_3BO_3	3.0mg
$Na_2SiO_3 \cdot 9H_2O$	0.5mg
NaF	0.3mg

Preparation of Artificial Seawater: Add components to distilled/deionized water and bring volume to 100.0mL. Mix thoroughly.

Preparation of Medium: Add components to distilled/deionized water and bring volume to 1.0L. Mix thoroughly. Distribute into tubes or flasks. Autoclave for 15 min at 15 psi pressure–121°C.

Use: For the cultivation of halophilic bacteria.

Hayflick Medium

Composition per 107.5mL:

Mycoplasma broth base	70.0mL
Horse serum	20.0mL
Fresh yeast extract solution	10.0mL
Penicillin solution	5.0mL
Thallous acetate solution	2.5mL

pH 7.8 ± 0.2 at 25°C

Mycoplasma Broth Base:

Pancreatic digest of casein	7.0g
NaCl	5.0g
Beef extract	3.0g
Yeast extract	3.0g
Beef heart, solids from infusion	2.0g

pH 7.8 ± 0.2 at 25°C

Preparation of Mycoplasma Broth Base: Add components to distilled/deionized water and bring volume to 1.0L. Gently heat and bring to boiling. Mix thoroughly. Distribute into tubes or flasks. Autoclave for 15 min at 15 psi pressure–121°C.

Fresh Yeast Extract Solution:

Composition per 100mL:

Baker's yeast, live, pressed, starch-free	25.0g

Preparation of Fresh Yeast Extract Solution: Add live Baker's yeast to 100.0mL of distilled/deionized water. Autoclave for 90 min at 15 psi pressure–121°C. Allow to stand. Remove supernatant solution. Adjust pH to 6.6–6.8. Filter sterilize.

Penicillin Solution:
Composition per 5mL:

Penicillin	20,000U

Preparation of Penicillin Solution: Add penicillin to distilled/deionized water and bring volume to 5.0mL. Mix thoroughly. Filter sterilize.

Thallous Acetate Solution:
Composition per 10mL:

Thallous acetate	0.1g

Preparation of Thallous Acetate Solution: Add thallous acetate to distilled/deionized water and bring volume to 10.0mL. Mix thoroughly. Filter sterilize.

Use: For the cultivation of *Mycoplasma* species.

Hayflick Medium, Modified

Composition per 1212mL:

Beef heart, infusion from	500.0g
Tryptose	10.0g
Noble agar	9.6g
NaCl	5.0g
Horse serum, normal	200.0mL
Fresh yeast extract solution	100.0mL
Calf thymus DNA solution	12.0mL

pH 7.8 ± 0.2 at 25°C

Fresh Yeast Extract Solution:
Composition per 100mL:

Baker's yeast live, pressed, starch-free	25.0g

Preparation of Fresh Yeast Extract Solution: Add the live Baker's yeast to 100.0mL of distilled/deionized water. Autoclave for 90 min at 15 psi pressure–121°C. Allow to stand. Remove supernatant solution. Adjust pH to 6.6–6.8. Filter sterilize.

Calf Thymus DNA Solution:
Composition per 20mL:

Calf thymus DNA	0.04g

Preparation of Calf Thymus DNA Solution: Add calf thymus DNA to distilled/deionized water and bring volume to 20.0mL. Mix thoroughly. Filter sterilize.

Preparation of Medium: Add components, except horse serum, fresh yeast extract solution, and calf thymus DNA solution, to distilled/deionized water and bring volume to 900.0mL. Mix thoroughly. Gently heat and bring to boiling. Autoclave for 15 min at 15 psi pressure–121°C. Cool to 45–50°C. Aseptically add 200.0mL of sterile horse serum, 100.0mL of sterile fresh yeast extract solution, and 12.0mL of sterile calf thymus DNA solution. Mix thoroughly. Aseptically distribute into sterile tubes.

Use: For the cultivation and maintenance of *Mycoplasma mustelae*.

HE Medium (Hay Extract Medium)

Composition per liter:

Agar	10.0g
Peptone	1.0g
Yeast extract	1.0g
Hay extract solution	500.0mL

pH 6.5 ± 0.2 at 25°C

Hay Extract Solution:
Composition per liter:

Hay, dried	50.0g

Preparation of Hay Extract Solution: Add dried barn hay to distilled/deionized water and bring volume to 1.0L. Mix thoroughly. Gently heat and bring to boiling. Filter through Whatman #40 filter paper.

Preparation of Medium: Add components to distilled/deionized water and bring volume to 1.0L. Mix thoroughly. Gently heat and bring to boiling. Distribute into tubes or flasks. Autoclave for 15 min at 15 psi pressure–121°C. Pour into sterile Petri dishes or leave in tubes.

Use: For the isolation and cultivation of *Spirochaeta aurantia*.

Heart Infusion Agar

Composition per liter:

Beef heart, infusion from	500.0g
Agar	15.0g
Tryptose	10.0g
NaCl	5.0g

pH 7.4 ± 0.2 at 25°C

Source: This medium is available as a premixed powder from Difco Laboratories.

Preparation of Medium: Add components to distilled/deionized water and bring volume to 1.0L. Mix thoroughly. Gently heat and bring to boiling. Distribute into tubes or flasks. Autoclave for 15 min at 15 psi pressure–121°C. Pour into sterile Petri dishes or leave in tubes.

Use: For the isolation and cultivation of a wide variety of fastidious microorganisms. It can also be used as a base for the preparation of blood agar in determining hemolytic reactions. For the cultivation and maintenance of *Bacillus anthracis, B. cereus, B. mycoides, Serratia rubidaea, Staphylococcus aureus, Tsatumella ptyseos,* and *Vibrio vulnificus*.

Heart Infusion Broth with Glucose

Composition per liter:

Beef heart, infusion from	500.0g
Tryptose	10.0g
NaCl	5.0g
Glucose	1.0g

pH 7.4 ± 0.2 at 25°C

Preparation of Medium: Add components to distilled/deionized water and bring volume to 1.0L. Mix thoroughly. Distribute into tubes or flasks. Autoclave for 15 min at 15 psi pressure–121°C.

Use: For the cultivation and maintenance of *Arthrobacter* species, *Bacillus* species, and *Pseudomonas* species.

Hektoen Enteric Agar

Composition per liter:

Agar	13.5g
Lactose	12.0g
Peptic digest of animal tissue	12.0g
Sucrose	12.0g
Bile salts	9.0g
NaCl	5.0g
$Na_2S_2O_3$	5.0g
Yeast extract	3.0g
Salicin	2.0g
Ferric ammonium citrate	1.5g
Acid fuchsin	0.1g
Bromthymol blue	0.064g

pH 7.6 ± 0.2 at 25°C

Source: This medium is available as a premixed powder from BBL Microbiology Systems, Difco Laboratories and Oxoid Unipath.

Caution: Acid fuchsin is a potential carcinogen and care must be taken to avoid inhalation of the powdered dye and contact with the skin.

Preparation of Medium: Add components to distilled/deionized water and bring volume to 1.0L. Mix thoroughly. Gently heat while stirring until components are dissolved. Do not autoclave. Pour into sterile Petri dishes. Allow agar to solidify with the Petri dish covers partially off.

Use: For the isolation and cultivation of Gram-negative enteric microorganisms from a variety of clinical and nonclinical specimens based on lactose or sucrose fermentation and H_2S production. It is used for the isolation and differentiation of *Salmonella* and *Shigella.* Bacteria that ferment lactose or sucrose appear as yellow to orange colonies. Bacteria that produce H_2S appear as colonies with black centers.

Hershey's Tris–Buffered Salts Medium

Composition per liter:

Tris(hydroxymethyl)aminomethane buffer (0.1*M* solution)	12.1g
NaCl	5.4g
KCl	3.0g
NH_4Cl	1.1g
$MgCl_2$	0.095g
KH_2PO_4	0.087g
Na_2SO_4	0.023g
$CaCl_2$	0.011g
$FeCl_3$	0.16mg
Glucose solution	100.0mL

pH 7.4 ± 0.2 at 25°C

Glucose Solution:

Composition per 100mL:

Glucose	2.0g

Preparation of Glucose Solution: Add glucose to distilled/deionized water and bring volume to 100.0mL. Mix thoroughly. Autoclave for 15 min at 15 psi pressure–121°C. Cool to 25°C.

Preparation of Medium: Add components, except glucose solution, to distilled/deionized water and bring volume to 900.0mL. Mix thoroughly. Gently heat and bring to boiling. Autoclave for 15 min at 15 psi pressure–121°C. Cool to 25°C. Aseptically add sterile glucose solution. Mix thoroughly. Aseptically distribute into sterile tubes or flasks.

Use: For the cultivation of a variety of heterotrophic microorganisms.

Heterotrophic Medium for *Hydrogenomonas*

Composition per liter:

Agar	15.0g
Tryptose	5.0g
Cornstarch	2.0g
Sodium succinate·$6H_2O$	2.0g
Sodium glutamate	1.0g
Yeast extract	1.0g
Sodium citrate·$2H_2O$	0.5g
Sodium acetate·$3H_2O$	0.3g
KH_2PO_4	0.2g
$MgSO_4$	0.1g

pH 6.8–7.2 at 25°C

Preparation of Medium: Add components to distilled/deionized water and bring volume to 1.0L. Mix thoroughly. Gently heat and bring to boiling. Distribute into tubes or flasks. Autoclave for 15 min at 15 psi pressure–121°C. Pour into sterile Petri dishes or leave in tubes.

Use: For the heterotrophic cultivation of *Hydrogenomonas* species.

Heterotrophic Medium for Hydrogen–Oxidizing Bacteria

Composition per 1010mL:

Solution A	900.0mL
Solution C	100.0mL
Solution B	10.0mL

Solution A:

Composition per 900mL:

Noble agar	17.0g
$Na_2HPO_4 \cdot 12H_2O$	9.0g
KH_2PO_4	1.5g
NH_4Cl	1.0g
$MgSO_4 \cdot 7H_2O$	0.2g
Trace elements solution SL-6	1.0mL

Preparation of Solution A: Add components to distilled/deionized water and bring volume to 900.0mL. Mix thoroughly. Autoclave for 15 min at 15 psi pressure–121°C. Cool to 45–50°C.

Trace Elements Solution SL-6:

Composition per liter:

H_3BO_3	0.3g
$CoCl_2 \cdot 6H_2O$	0.2g
$ZnSO_4 \cdot 7H_2O$	0.1g
$MnCl_2 \cdot 4H_2O$	0.03g
$Na_2MoO_4 \cdot H_2O$	0.03g
$NiCl_2 \cdot 6H_2O$	0.02g
$CuCl_2.2H_2O$	0.01g

Preparation of Trace Elements Solution SL-6: Add components to distilled/deionized water and bring volume to 1.0L. Mix thoroughly. Adjust pH to 3.4.

Solution B:

Composition per 10mL:

$CaCl_2 \cdot 2H_2O$	0.01g
Ferric ammonium citrate	5.0mg

Preparation of Solution B: Add components to distilled/deionized water and bring volume to 10.0mL. Mix thoroughly. Autoclave for 15 min at 15 psi pressure–121°C. Cool to 45–50°C.

Solution C:

Composition per 100mL:

Sodium 3-hydroxybutyrate	2.0g

Preparation of Solution C: Add sodium 3-hydroxybutyrate to distilled/deionized water and bring volume to 100.0mL. Mix thoroughly. Filter sterilize. Warm to 45–50°C.

Preparation of Medium: Aseptically combine 900.0mL of sterile solution A, 10.0mL of sterile solution B and 100.0mL of sterile solution C. Mix thoroughly. Pour into sterile Petri dishes or distribute into sterile tubes.

Use: For the heterotrophic cultivation and maintenance of *Xanthobacter agilis*.

Hickey–Tresner Agar

Composition per liter:

Agar	15.0g
Dextrin	10.0g
Pancreatic digest of casein	2.0g
Meat extract	1.0g
Yeast extract	1.0g
$CaCl_2$	2.0mg

pH 7.2 ± 0.2 at 25°C

Preparation of Medium: Add components to distilled/deionized water and bring volume to 1.0L. Mix thoroughly. Gently heat and bring to boiling. Distribute into tubes or flasks. Autoclave for 15 min at 15 psi pressure–121°C. Pour into sterile Petri dishes or leave in tubes.

Use: For the cultivation and maintenance of *Thermomonospora curvata*.

Histoplasma capsulatum Agar

Composition per liter:

Agar	12.5g
Glucose	10.0g
Citric acid	10.0g
Potato starch	2.0g
α-Ketoglutaric acid	1.0g
L-Cystine·HCl·H_2O	1.0g
Glutathione, reduced	0.5g
L-Asparagine	0.1g
L-Tryptophan	0.02g
Solution 1	250.0mL
Solution 3	40.0mL
Solution 2	10.0mL
Solution 4	10.0mL
Solution 8	10.0mL
Solution 5	1.0mL
Solution 6	0.1mL
Solution 7	0.1mL

pH 6.5 ± 0.2 at 25°C

Solution 1:

Composition per liter:

KH_2PO_4	8.0g
$(NH_4)_2SO_4$	8.0g
$MgSO_4 \cdot 7H_2O$	0.86g
$CaCl_2$, anhydrous	0.08g
$ZnSO_4 \cdot 7H_2O$	0.05g

Preparation of Solution 1: Add components to distilled/deionized water and bring volume to

500.0mL. Mix thoroughly. Bring volume to 1.0L with distilled/deionized water. Store at 5°C.

Solution 2:

Composition per liter:

$FeSO_4 \cdot 7H_2O$	5.70g
$MnCl_2 \cdot 6H_2O$	0.80g
$NaMoO_4 \cdot 2H_2O$	0.15g
HCl, concentrated	1.0mL

Preparation for Solution 2: Add 1.0mL of concentrated HCl to 100.0mL of distilled water in a 1.0L volumetric flask. Dissolve each component completely in the sequence given. Bring volume to 1.0L with distilled/deionized water. Store at 5°C. Discard if red color or red precipitate appears.

Solution 3:

Composition per 100mL:

Casein, acid-hydrolyzed, vitamin-free	10.0g

Preparation for Solution 3: Add casein to distilled/deionized water and bring volume to 100.0mL. Do not use enzymatically digested casein.

Solution 4:

Composition per liter:

Calcium pantothenate	0.2g
Inositol	0.2g
Riboflavin	0.2g
Thiamine·HCl	0.2g
Nicotinamide	0.1g
Biotin	0.01g

Preparation for Solution 4: Add components to distilled/deionized water and bring volume to 1.0L. Mix thoroughly. Store at –20°C.

Solution 5:

Composition per 100mL:

Hemin	0.2g
NH_4OH, concentrated	0.3mL

Preparation for Solution 5: Add hemin to approximately 30.0mL of distilled/deionized water. Add NH_4OH. Mix thoroughly until dissolved. Bring volume to 100.0mL with distilled/deionized water. Store at 5° C.

Solution 6:

Composition per 10mL:

DL-Thioctic acid	0.01g
Ethanol (95% solution)	10.0mL

Preparation for Solution 6: Add DL-thioctic acid to 10.0mL of ethanol. Mix thoroughly. Store at –20°C.

Solution 7:

Composition per 10mL:

Coenzyme A	0.01g
$Na_2S \cdot 5H_2O$ (0.05% solution)	0.2mL

Preparation for Solution 7: Prepare $Na_2S \cdot 5H_2O$ solution in freshly boiled distilled/deionized water. Add coenzyme A to 9.8mL of distilled/deionized water. Mix thoroughly. Add freshly prepared $Na_2S \cdot 5H_2O$ solution. Mix thoroughly. Store the solution at –20°C.

Solution 8:

Composition per 100mL:

Oleic acid	0.10g

Preparation for Solution 8: Add oleic acid to 50.0mL of distilled/deionized water. Adjust pH to 9.0 with NaOH. Gently heat until dissolved. Bring volume to 100.0mL with distilled/deionized water. Store at 5°C.

Preparation of Medium: Add components—except agar, potato starch, and solution 8—to distilled/deionized water and bring volume to 400.0mL. Mix thoroughly. Adjust pH to 6.5 with 20% KOH solution. Filter sterilize. In a separate flask add potato starch to 50.0mL of distilled/deionized water. Add the starch solution to 450.0mL of boiling distilled/deionized water. Add 10.0mL of solution 8 and the agar. Mix thoroughly. Autoclave for 15 min at 15 psi pressure–121°C. Cool to 70°C. Aseptically combine the two sterile solutions. Pour into sterile Petri dishes or distribute into sterile tubes.

Use: For cultivation and maintenance of *Histoplasma capsulatum* in the yeast phase. Also for the cultivation of *H. duboisii, Blastomyces dermatitidis* and *Sprotrichum schenckii*.

Histoplasma capsulatum Agar

Composition per liter:

Agar	15.0g
Glucose	10.0g
Potato starch	2.0g
α-Ketoglutaric acid	1.0g
L-Cystine·HCl·H_2O	1.0g
Glutathione, reduced	0.5g
L-Asparagine	0.1g
L-Tryptophan	0.02g
Solution 1	250.0mL
Solution 3	40.0mL
Solution 2	10.0mL
Solution 4	10.0mL
Solution 8	10.0mL
Solution 5	1.0mL
Solution 6	0.1mL
Solution 7	0.1mL

pH 6.5 ± 0.2 at 25°C

Solution 1:

Composition per liter:

KH_2PO_4	8.0g
$(NH_4)_2SO_4$	8.0g

$MgSO_4 \cdot 7H_2O$ 0.86g
$CaCl_2$, anhydrous 0.08g

Preparation of Solution 1: Add components to distilled/deionized water and bring volume to 500.0mL. Mix thoroughly. Bring volume to 1.0L with distilled/deionized water. Store at 5°C.

Solution 2:

Composition per liter:

$FeSO_4 \cdot 7H_2O$ 5.70g
$MnCl_2 \cdot 6H_2O$ 0.80g
$NaMoO_4 \cdot 2H_2O$ 0.15g
HCl, concentrated 1.0mL

Preparation for Solution 2: Add the 1.0mL of concentrated HCl to 100.0mL of distilled water in a 1.0L volumetric flask. Dissolve each component completely in the sequence given. Bring volume to 1.0L with distilled/deionized water. Store at 5°C. Discard if red color or red precipitate appears.

Solution 3:

Composition per 100mL:

Casein, acid-hydrolyzed, vitamin-free 10.0g

Preparation for Solution 3: Add casein to distilled/deionized water and bring volume to 100.0mL. Do not use enzymatically digested casein.

Solution 4:

Composition per liter:

Calcium pantothenate 0.2g
Inositol 0.2g
Riboflavin 0.2g
Thiamine·HCl 0.2g
Nicotinamide 0.1g
Biotin 0.01g

Preparation for Solution 4: Add components to distilled/deionized water and bring volume to 1.0L. Mix thoroughly. Store at –20°C.

Solution 5:

Composition per 100mL:

Hemin 0.2g
NH_4OH, concentrated 0.3mL

Preparation for Solution 5: Add hemin to approximately 30.0mL of distilled/deionized water. Add NH_4OH. Mix thoroughly until dissolved. Bring volume to 100.0mL with distilled/deionized water. Store at 5°C.

Solution 6:

Composition per 10mL:

DL-Thioctic acid 0.01g
Ethanol (95% solution) 10.0mL

Preparation for Solution 6: Add DL-thioctic acid to 10.0mL of ethanol. Mix thoroughly. Store at –20°C.

Solution 7:

Composition per 10mL:

Coenzyme A 0.01g
$Na_2S \cdot 5H_2O$ (0.05% solution) 0.2mL

Preparation for Solution 7: Prepare $Na_2S \cdot 5H_2O$ solution in freshly boiled distilled/deionized water. Add coenzyme A to 9.8mL of distilled/deionized water. Mix thoroughly. Add freshly prepared $Na_2S \cdot 5H_2O$ solution. Mix thoroughly. Store the solution at –20°C.

Solution 8:

Composition per 100mL:

Oleic acid 0.10g

Preparation for Solution 8: Add oleic acid to 50.0mL of distilled/deionized water. Adjust pH to 9.0 with NaOH. Gently heat until dissolved. Bring volume to 100.0mL with distilled/deionized water. Store at 5°C.

Preparation of Medium: Add components—except agar, potato starch, and solution 8—to distilled/deionized water and bring volume to 400.0mL. Mix thoroughly. Adjust pH to 6.5 with 20% KOH solution. Filter sterilize. In a separate flask add potato starch to 50.0mL of distilled/deionized water. Add the starch solution to 450.0mL of boiling distilled/deionized water. Add 10.0mL of solution 8 and the agar. Mix thoroughly. Autoclave for 15 min at 15 psi pressure–121°C. Cool to 70°C. Aseptically combine the two sterile solutions. Pour into sterile Petri dishes or distribute into sterile tubes.

Use: For cultivation and maintenance of *Histoplasma capsulatum* in the mycelial phase.

Histoplasma capsulatum Broth

Composition per liter:

Glucose 10.0g
Citric acid 10.0g
α-Ketoglutaric acid 1.0g
L-Cystine·HCl·H_2O 1.0g
Potato starch 0.5g
Glutathione, reduced 0.5g
L-Asparagine 0.1g
L-Tryptophan 0.02g
Solution 1 .... 250.0mL
Solution 3 .... 40.0mL
Solution 2 .... 10.0mL
Solution 4 .... 10.0mL
Solution 5 .... 1.0mL
Solution 8 .... 1.0mL
Solution 6 .... 0.1mL
Solution 7 .... 0.1mL

pH 6.5 ± 0.2 at 25°C

Solution 1:

Composition per liter:

KH_2PO_4	8.0g
$(NH_4)_2SO_4$	8.0g
$MgSO_4 \cdot 7H_2O$	0.86g
$CaCl_2$, anhydrous	0.08g
$ZnSO_4 \cdot 7H_2O$	0.05g

Preparation of Solution 1: Add components to distilled/deionized water and bring volume to 500.0mL. Mix thoroughly. Bring volume to 1.0L with distilled/deionized water. Store at 5°C.

Solution 2:

Composition per liter:

$FeSO_4 \cdot 7H_2O$	5.70g
$MnCl_2 \cdot 6H_2O$	0.80g
$NaMoO_4 \cdot 2H_2O$	0.15g
HCl, concentrated	1.0mL

Preparation for Solution 2: Add 1.0mL of concentrated HCl to 100mL of distilled water in a 1.0L volumetric flask. Dissolve each component completely in the sequence given. Bring volume to 1.0L with distilled/deionized water. Store at 5°C. Discard if red color or red precipitate appears.

Solution 3:

Composition per 100mL:

Casein, acid-hydrolyzed, vitamin-free	10.0g

Preparation for Solution 3: Add casein to distilled/deionized water and bring volume to 100.0mL. Do not use enzymatically-digested casein.

Solution 4:

Composition per liter:

Calcium pantothenate	0.2g
Inositol	0.2g
Riboflavin	0.2g
Thiamine·HCl	0.2g
Nicotinamide	0.1g
Biotin	0.01g

Preparation for Solution 4: Add components to distilled/deionized water and bring volume to 1.0L. Mix thoroughly. Store at –20°C.

Solution 5:

Composition per 100mL:

Hemin	0.2g
NH_4OH, concentrated	0.3mL

Preparation for Solution 5: Add hemin to approximately 30.0mL of distilled/deionized water. Add NH_4OH. Mix thoroughly until dissolved. Bring volume to 100.0mL with distilled/deionized water. Store at 5°C.

Solution 6:

Composition per 10mL:

DL-Thioctic acid	0.01g
Ethanol (95% solution)	10.0mL

Preparation for Solution 6: Add DL-thioctic acid to 10.0mL of ethanol. Mix thoroughly. Store at –20°C.

Solution 7:

Composition per 10mL:

Coenzyme A	0.01g
$Na_2S \cdot 5H_2O$ (0.05% solution)	0.2mL

Preparation for Solution 7: Prepare $Na_2S \cdot 5H_2O$ solution in freshly boiled distilled/deionized water. Add coenzyme A to 9.8mL of distilled/deionized water. Mix thoroughly. Add freshly prepared $Na_2S \cdot 5H_2O$ solution. Mix thoroughly. Store the solution at –20°C.

Solution 8:

Composition per 100mL:

Oleic acid	0.10g

Preparation for Solution 8: Add oleic acid to 50.0mL of distilled/deionized water. Adjust pH to 9.0 with NaOH. Gently heat until dissolved. Bring volume to 100.0mL with distilled/deionized water. Store at 5°C.

Preparation of Medium: Add components—except potato starch, and solution 8—to distilled/deionized water and bring volume to 400.0mL. Mix thoroughly. Adjust pH to 6.5 with 20% KOH solution. Filter sterilize. In a separate flask add potato starch to 50.0mL of distilled/deionized water. Add the starch solution to 450.0mL of boiling distilled/deionized water. Add 1.0mL of solution 8. Mix thoroughly. Autoclave for 15 min at 15 psi pressure–121°C. Cool to 70°C. Aseptically combine the two sterile solutions. Pour into sterile Petri dishes or distribute into sterile tubes.

Use: For cultivation of *Histoplasma capsulatum* in the yeast phase. Also for the cultivation of *H. duboisii*, *Blastomyces dermatitidis* and *Sprotrichum schenckii*.

HO-LE Trace Element Solution

Composition per liter:

H_3BO_3	2.85g
$MnCl_2 \cdot 4H_2O$	1.8g
Sodium tartrate	1.77g
$FeSO_4$	1.36g
$CoCl_2 \cdot 6H_2O$	0.04g
$CuCl_2 \cdot 2H_2O$	0.026g
$Na_2MoO_4 \cdot 2H_2O$	0.025g
$ZnCl_2$	0.021g

Preparation of Medium: Add components to distilled/deionized water and bring volume to 1.0L. Mix thoroughly. Distribute into tubes or flasks. Autoclave for 15 min at 15 psi pressure–121°C.

Use: For use as an enrichment to other media that require trace minerals.

HP 101 Halophile Medium

Composition per liter:

NaCl	100.0g
Agar	20.0g
Peptone	10.0g
$MgSO_4 \cdot 7H_2O$	4.3g
$NaNO_3$	2.0g
Yeast extract	1.0g

pH 7.2 ± 0.2 at 25°C

Preparation of Medium: Add components to distilled/deionized water and bring volume to 1.0L. Mix thoroughly. Gently heat and bring to boiling. Distribute into tubes or flasks. Autoclave for 15 min at 15 psi pressure–121°C. Pour into sterile Petri dishes or leave in tubes.

Use: For the cultivation and maintenance of *Pseudomonas* species.

HPC Agar (Heterotrophic Plate Count Agar) (m–HPC Agar)

Composition per liter:

Gelatin	25.0g
Pancreatic digest of gelatin	20.0g
Agar	15.0g
Glycerol	10.0mL

pH 7.1 ± 0.2 at 25°C

Source: Available from Difco Laboratories.

Preparation of Medium: Add components, except glycerol, to distilled/deionized water and bring volume to 990.0mL. Mix thoroughly. Gently heat and bring to boiling. Add glycerol. Mix thoroughly. Autoclave for 15 min at 15 psi pressure–121°C. Cool to 45–50°C. Pour into sterile Petri dishes.

Use: For the the cultivation and enumeration of microorganisms from potable water sources, swimming pools and other water specimens, by the membrane filter method and heterotrophic plate count technique.

Hugh–Leifson's Glucose Broth

Composition per liter:

NaCl	30.0g
Glucose	10.0g
Agar	3.0g
Peptone	2.0g
Yeast extract	0.5g
Bromcresol purple	0.015g

pH 7.4 ± 0.2 at 25°C

Preparation of Medium: Add components to distilled/deionized water and bring volume to 1.0L. Mix thoroughly. Gently heat while stirring and bring to boiling. Adjust pH to 7.4. Distibute into tubes or flasks. Autoclave for 15 min at 15 psi pressure–121°C.

Use: For the cultivation and differentiation of bacteria based on their ability to ferment glucose. Bacteria that ferment glucose turn the medium yellow.

Hungate's Habitat–Simulating Medium

Composition per 1140.2mL:

Rumen fluid	333.0mL
Mineral solution A	167.0mL
Mineral solution B	167.0mL
$NaHCO_3$ solution	53.0mL
Cysteine·HCl solution	10.6mL
Substrate solution	10.6mL
Resazurin solution	1.0mL

Mineral Solution A:

Composition per liter:

NaCl	6.0g
KH_2PO_4	3.0g
$(NH_4)_2SO_4$	3.0g
$CaCl_2$	0.6g
$MgSO_4$	0.6g

Preparation of Solution A: Add components to distilled/deionized water and bring volume to 1.0L. Mix thoroughly.

Mineral Solution B:

K_2HPO_4	3.0

Preparation of Solution B: Add K_2HPO_4 to distilled/deionized water and bring volume to 1.0L. Mix thoroughly.

Resazurin Solution:

Composition per 100mL:

Resazurin	0.1g

Preparation of Resazurin Solution: Add resazurin to distilled/deionized water and bring volume to 100.0L. Mix thoroughly.

Cysteine·HCl Solution:

Composition per 100mL:

Cysteine·HCl	3.0g

Preparation of Cysteine·HCl Solution: Add Cysteine·HCl to O_2-free distilled/deionized water and bring volume to 100.0L. Mix thoroughly. Gently heat and bring to boiling. Continue boiling for 2 min. Cool to 25°C under 100% N_2. Seal tube with a stopper that is wired in place. Autoclave for 15 min at 15 psi pressure–121°C. Cool to 25°.

$NaHCO_3$ Solution:
Composition per 10mL:

$NaHCO_3$	1.0g

Preparation of $NaHCO_3$ Solution: Add $NaHCO_3$ to O_2-free distilled/deionized water and bring volume to 10.0mL. Mix thoroughly. Filter sterilize. Gas with 100% CO_2 for 15 min.

Substrate Solution:
Composition per 100mL:

Sugar	10.0g

Preparation of Substrate Solution: Add sugar to O_2-free distilled/deionized water. Mix thoroughly. Gas with 100% N_2 for 15 min. Autoclave for 15 min at 15 psi pressure–121°C. Cool to 45–50°C.

Preparation of Medium: Add 167.0mL of solution A, 167.0mL of solution B, and 1.0mL of resazurin solution to distilled/deionized water and bring volume to 733.0mL. Mix thoroughly. Gently heat and bring to boiling. Continue boiling until resazurin turns colorless indicating reduction. Bring volume back to 733.0mL (some evaporation will have occurred) with O_2-free distilled/deionized water. Cool to 45–50°C under O_2-free 100% CO_2. Anaerobically add rumen fluid. Anaerobically distribute into tubes in 10.0mL volumes. Cap with butyl rubber stoppers. Place tubes in a press. Autoclave for 15 min at 15 psi pressure–121°C. Cool to 25°C. Immediately prior to inoculation, aseptically and anaerobically add 0.1mL of sterile cysteine·HCl solution, 0.5mL of sterile $NaHCO_3$ solution and 0.1mL of substrate solution per 10.0mL of medium in each tube.

Use: For the cultivation of *Bacteroides* species from rumens.

HY Agar for *Flavobacterium*

Composition per liter:

Agar	8.0g
Glutamic acid	5.0g
K_2HPO_4	0.1g
$MgSO_4·7H_2O$	0.1g

pH 7.3 ± 0.2 at 25°C

Preparation of Medium: Add components to distilled/deionized water and bring volume to 1.0L. Glutamic acid may be replaced by 1.0g folic acid if desired. Mix thoroughly. Gently heat and bring to boiling. Distribute into tubes or flasks. Autoclave for 15 min at 15 psi pressure–121°C.

Use: For the cultivation of *Flavobacterium* species.

HY Medium for *Flavobacterium*

Composition per liter:

Glutamic acid	5.0g
K_2HPO_4	0.1g
$MgSO_4·7H_2O$	0.1g

pH 7.3 ± 0.2 at 25°C

Preparation of Medium: Add components to distilled/deionized water and bring volume to 1.0L. Glutamic acid may be replaced by 1.0g folic acid if desired. Mix thoroughly. Gently heat and bring to boiling. Distribute into tubes or flasks. Autoclave for 15 min at 15 psi pressure–121°C.

Use: For the cultivation of *Flavobacterium* species.

Hydrogen–Oxidizing Bacteria Medium

Composition per 1020mL:

Solution I	1.0L
Solution II	10.0mL
Solution III	10.0mL

Solution I:
Composition per liter:

$Na_2HPO_4·12H_2O$	9.0g
KH_2PO_4	1.5g
NH_4Cl	1.0g
$MgSO_4·7H_2O$	0.2g
Trace elements solution	1.0mL

Preparation of Solution I: Add components to distilled/deionized water and bring volume to 1.0L. Mix thoroughly. Gently heat until dissolved. Autoclave for 15 min at 15 psi pressure–121°C. Cool to 25°C.

Trace Elements Solution:
Composition per liter:

H_3BO_3	0.3g
$CoCl_2·6H_2O$	0.2g
$ZnSO_4·7H_2O$	0.1g
$MnCl_2·4H_2O$	0.03g
$NaMoO_4·2H_2O$	0.03g
$NiCl_2·6H_2O$	0.02g
$CuCl_2·2H_2O$	0.01g

Preparation of Trace Elements Solution: Add components to distilled/deionized water and bring volume to 1.0L. Mix thoroughly.

Solution II:
Composition per 100mL:

$CaCl_2·2H_2O$	0.1g
Ferric ammonium citrate	0.05g

Preparation of Solution II: Add components to distilled/deionized water and bring volume to

100.0mL. Mix thoroughly. Autoclave for 15 min at 15 psi pressure–121°C. Cool to 25°C.

Solution III:

Composition per 100mL:

$NaHCO_3$ 5.0g

Preparation of Solution III: Add $NaHCO_3$ to distilled/deionized water and bring volume to 100.0mL. Mix thoroughly. Filter sterilize.

Preparation of Medium: Aseptically combine 1.0L of cooled, sterile solution I, 10.0mL of cooled, sterile solution II and 10.0mL of sterile solution III. Mix thoroughly. Aseptically distribute into sterile tubes or flasks.

Use: For the cultivation of hydrogen–oxidizing bacteria.

Hyphomicrobium Enrichment Medium

Composition per 100mL:

KNO_3 .... 0.04g
$Na_2HPO_4 \cdot 7H_2O$ 0.02g
$MgSO_4 \cdot 7H_2O$ 0.48mg
$FeCl_3 \cdot 7H_2O$ 0.02mg
$MnCl_2 \cdot 4H_2O$ 0.01mg

pH 7.2 ± 0.2 at 25°C

Preparation of Medium: Add components to distilled/deionized water and bring volume to 1.0L. Mix thoroughly. Adjust pH to 7.2. Distribute into tubes or flasks. Autoclave for 15 min at 15 psi pressure–121°C.

Use: For the cultivation and enrichment of *Hyphomicrobium* species.

Hyphomicrobium Medium

Composition per liter:

Agar 15.0g
Na_2HPO_4 .... 2.13g
KH_2PO_4 .... 1.36g
$MgSO_4 \cdot 7H_2O$ 0.2g
$CaCl_2 \cdot 2H_2O$ 9.95mg
$FeSO_4 \cdot 7H_2O$ 5.0mg
$MnSO_4 \cdot 4H_2O$ 2.5mg
$Na_2MoO_4 \cdot 2H_2O$ 2.5mg
Urea solution 30.0mL
Methanol 4.0mL

Urea Solution:

Composition per 100mL:

Urea 20.0g

Preparation of Urea Solution: Add urea to distilled/deionized water and bring volume to 100.0mL. Mix thoroughly. Filter sterilize.

Preparation of Medium: Filter sterilize methanol. Add components, except urea solution and methanol, to distilled/deionized water and bring volume to 966.0mL. Mix thoroughly. Gently heat and bring to boiling. Autoclave for 15 min at 15 psi pressure–121°C. Cool to 45–50°C. Aseptically add sterile urea solution and sterile methanol. Mix thoroughly. Aseptically distribute into sterile tubes or bottles.

Use: For the cultivation of *Hyphomicrobium* species.

Hyphomicrobium Medium

Composition per liter:

Noble agar 18.0g
Na_2HPO_4 .... 2.15g
KH_2PO_4 .... 1.36g
$(NH_4)_2SO_4$ 0.5g
$MgSO_4 \cdot 7H_2O$ 0.2g
Trace element solution 5.0mL
Methylamine·HCl solution 20.0mL

pH 7.1 ± 0.1 at 25°C

Trace Element Solution:

Composition per 100mL:

$CuCl_2$ 0.15g
$FeSO_4 \cdot 7H_2O$ 0.1g
$Na_2MoO_4 \cdot 2H_2O$ 0.05g
$MnSO_4 \cdot H_2O$ 0.035g

Preparation of Trace Element Solution: Add components to distilled/deionized water and bring volume to 100.0mL. Mix thoroughly.

Methylamine·HCl Solution:

Composition per 20mL:

Methylamine·HCl 3.38g

Preparation of Methylamine·HCl Solution: Add methylamine·HCl to distilled/deionized water and bring volume to 20.0mL. Mix thoroughly. Filter sterilize.

Preparation of Medium: Add components, except methylamine·HCl solution, to distilled/deionized water and bring volume to 980.0mL. Mix thoroughly. Gently heat and bring to boiling. Autoclave for 15 min at 15 psi pressure–121°C. Cool to 45–50°C. Aseptically add sterile methylamine·HCl solution. Mix thoroughly. Adjust pH to 7.1, if necessary. Pour into sterile Petri dishes or distribute into sterile tubes.

Use: For the cultivation and maintenance of *Hyphomicrobium aestuarii, H. facilis, H. hollandicum, H. vulgare,* and *H. zavarzinii*.

Hyphomicrobium Medium 337a

Composition per liter:

KH_2PO_4	1.3g
Na_2HPO_4	1.13g
$(NH_4)_2SO_4$	0.50g
$MgSO_4 \cdot 7H_2O$	0.20g
$CaCl_2 \cdot 2H_2O$	3.09mg
$FeSO_4 \cdot 7H_2O$	2.0mg
$Na_2MoO_4 \cdot 2H_2O$	1.0mg
$MnSO_4 \cdot 4H_2O$	0.88mg

pH 7.2–7.5 at 25°C

Preparation of Medium: Add components to distilled/deionized water and bring volume to 1.0L. Mix thoroughly. Distribute into tubes or flasks. Autoclave for 15 min at 15 psi pressure–121°C.

Use: For the enrichment and cultivation of *Hyphomicrobium* species.

Hyphomonas Enrichment Medium

Composition per liter:

Peptone	0.05g
Yeast extract	0.05g

Preparation of Medium: Add components to distilled/deionized water and bring volume to 1.0L. Mix thoroughly. Distribute into tubes or flasks. Autoclave for 15 min at 15 psi pressure–121°C.

Use: For the isolation and cultivation of *Hyphomonas* species.

Hypoxanthine Agar

Composition per 1100mL:

Agar	15.0g
Beef extract	3.0g
Peptone	5.0g
Hypoxanthine solution	5.0g

pH 7.0 ± 0.1 at 25°C

Hypoxanthine Solution:
Composition per 100mL:

Hypoxanthine	5.0g

Preparation of Hypoxanthine Solution: Add hypoxanthine to distilled/deionized water and bring volume to 100.0mL. Mix thoroughly. Filter sterilize.

Preparation of Medium: Add components, except hypoxanthine solution, to distilled/deionized water and bring volume to 900.0mL. Mix thoroughly. Autoclave for 15 min at 15 psi pressure–121°C. Cool to 45–50°C. Aseptically add 100.0mL of sterile hypoxanthine solution. Mix thoroughly. Pour into sterile 15 mm × 100 mm Petri dishes in 25.0 mL volumes.

Use: For the cultivation and differentiation of bacteria based on hypoxanthine hydrolysis. Bacteria that hydrolyze hypoxanthine, such as *Streptomyces griseus*, appear with a clear zone under and around the colonies. *Nocardia asteroides* does not hydrolyze hypoxanthine.

IFO Agar

Composition per liter:

Agar	20.0g
$(NH_4)_2HPO_4$	3.0g
NaCl	1.0g
$MgSO_4 \cdot 7H_2O$	0.2g
$FeSO_4 \cdot 7H_2O$	10.0mg
$MnSO_4 \cdot 4\text{-}6H_2O$	5.0mg
Riboflavin	0.02mg
Calcium pantothenate	0.02mg
Pyridoxine·HCl	0.02mg
Nicotinic acid	0.02mg
p-Aminobenzoic acid	0.01mg
Thiamine·HCl	0.01mg
Biotin	1.0μg
Methanol	10.0mL

pH 7.0 ± 0.2 at 25°C

Preparation of Medium: Add components, except agar and methanol, to distilled/deionized water and bring volume to 490.0mL. Mix thoroughly. Autoclave for 15 min at 15 psi pressure–121°C. Cool to 45–50°C. In a separate flask, add agar to distilled/deionized water and bring volume to 500.0mL. Mix thoroughly. Gently heat and bring to boiling. Autoclave for 15 min at 15 psi pressure–121°C. Cool to 45–50°C. Aseptically combine the two sterile solutions. Aseptically add 10.0mL filter sterilized methanol. Mix thoroughly. Adjust pH to 7.0. Pour into sterile Petri dishes or distribute into sterile tubes.

Use: For the cultivation and maintenance of *Hyphomicrobium methylovorum*.

IFO Broth

Composition per liter:

$(NH_4)_2HPO_4$	3.0g
NaCl	1.0g
$MgSO_4 \cdot 7H_2O$	0.2g
$FeSO_4 \cdot 7H_2O$	10.0mg
$MnSO_4 \cdot 4\text{-}6H_2O$	5.0mg
Riboflavin	20.0μg
Calcium pantothenate	20.0μg
Pyridoxine·HCl	20.0μg
Nicotinic acid	20.0μg
p-Aminobenzoic acid	10.0μg
Thiamine·HCl	10.0μg

Biotin 1.0μg
Methanol 10.0mL

pH 7.0 ± 0.2 at 25°C

Preparation of Medium: Add components, except methanol, to distilled/deionized water and bring volume to 990.0mL. Mix thoroughly. Autoclave for 15 min at 15 psi pressure–121°C. Aseptically add 10.0mL filter sterilized methanol. Mix thoroughly. Adjust pH to 7.0. Aseptically distribute into sterile tubes or flasks.

Use: For the cultivation and maintenance of *Hyphomicrobium methylovorum.*

Ilyobacter Agar

Composition per liter:

NaCl 20.0g
Agar 15.0g
$MgCl_2 \cdot 6H_2O$ 3.0g
KCl 0.5g
NH_4Cl 0.25g
KH_2PO_4 .... 0.2g
$CaCl_2 \cdot 2H_2O$ 0.15g
Resazurin 1.0mg
Sodium sulfide solution 10.0mL
Sodium L-tartrate solution 10.0mL
$NaHCO_3$ solution 10.0mL
Trace element solution SL-7 .... 1.0mL

pH 7.2 ± 0.2 at 25°C

Sodium Sulfide Solution:

Composition per 100mL:

$Na_2S \cdot 9H_2O$ 3.6g

Preparation of Sodium Sulfide Solution: Add $Na_2S \cdot 9H_2O$ to distilled/deionized water and bring volume to 100.0mL. Mix thoroughly. Autoclave for 15 min at 15 psi pressure–121°C under N_2. Maintain under 100% N_2.

Sodium L-Tartrate Solution:

Composition per 10mL:

Sodium L-tartrate 2.0g

Preparation of Sodium L-Tartrate Solution: Add sodium L-tartrate to distilled/deionized water and bring volume to 10.0mL. Mix thoroughly. Filter sterilize. Maintain under 80% N_2 + 20% CO_2.

$NaHCO_3$ Solution:

Composition per 10mL:

$NaHCO_3$ 2.5g

Preparation of $NaHCO_3$ Solution: Add $NaHCO_3$ to distilled/deionized water and bring volume to 10.0mL. Mix thoroughly. Filter sterilize. Maintain under 80% N_2 + 20% CO_2.

Trace Elements Solution SL-7:

Composition per liter:

$FeCl_2 \cdot 4H_2O$ 1.5g
$CoCl_2 \cdot 6H_2O$ 0.19g
$MnCl_2 \cdot 4H_2O$ 0.1g
$ZnCl_2$ 0.07g
H_3BO_3 .... 0.062g
$Na_2MoO_4 \cdot 2H_2O$ 0.036g
$NiCl_2 \cdot 6H_2O$ 0.024g
$CuCl_2 \cdot 2H_2O$ 0.017g
HCl (25% solution) 10.0mL

Preparation of Trace Elements Solution SL-7: Add the $FeCl_2 \cdot 4H_2O$ to the HCl. Add distilled/deionized water and bring volume to 1.0L. Add remaining components. Mix thoroughly. Filter sterilize. Maintain under 80% N_2 + 20% CO_2.

Preparation of Medium: Add components—except agar, sodium sulfide solution, sodium L-tartrate solution, $NaHCO_3$ solution, and trace element solution SL-7 solution—to distilled/deionized water and bring volume to 469.0mL. Mix thoroughly. Gently heat and bring to boiling. Autoclave for 15 min at 15 psi pressure–121°C. Cool to 45–50°C. Maintain under 80% N_2 + 20% CO_2. Aseptically add 10.0mL L-tartrate solution, 10.0mL $NaHCO_3$ solution, and 1.0mL trace element solution SL-7 solution under 80% N_2 + 20% CO_2. Mix thoroughly. In a separate flask add agar to distilled/deionized water and bring volume to 500.0mL. Mix thoroughly. Gently heat and bring to boiling. Autoclave for 15 min at 15 psi pressure–121°C. Cool to 45–50°C. Combine sterile agar and sterile basal medium. Adjust pH to 7.2. Aseptically add 10.0mL sodium sulfide solution. Pour into sterile Petri dishes or distribute into sterile tubes. Maintain under 80% N_2 + 20% CO_2.

Use: For the cultivation and maintenance of *Ilyobacter tartaricus.*

Ilyobacter Broth

Composition per liter:

NaCl 20.0g
$MgCl_2 \cdot 6H_2O$ 3.0g
KCl 0.5g
NH_4Cl 0.25g
KH_2PO_4 .... 0.2g
$CaCl_2 \cdot 2H_2O$ 0.15g
Resazurin 1.0mg
Sodium sulfide solution 10.0mL
Sodium L-tartrate solution 10.0mL
$NaHCO_3$ solution 10.0mL
Trace element solution SL-7 .... 1.0mL

pH 7.2 ± 0.2 at 25°C

Sodium Sulfide Solution:

Composition per 100mL:

$Na_2S \cdot 9H_2O$ 3.6g

Preparation of Sodium Sulfide Solution: Add $Na_2S \cdot 9H_2O$ to distilled/deionized water and bring volume to 100.0mL. Mix thoroughly. Autoclave for 15 min at 15 psi pressure–121°C under 100% N_2. Maintain under 100% N_2.

Sodium L-Tartrate Solution:

Composition per 10mL:

Sodium L-tartrate 2.0g

Preparation of Sodium L-Tartrate Solution: Add sodium L-tartrate to distilled/deionized water and bring volume to 10.0mL. Mix thoroughly. Filter sterilize. Maintain under 80% N_2 + 20% CO_2.

$NaHCO_3$ Solution:

Composition per 10mL:

$NaHCO_3$ 2.5g

Preparation of $NaHCO_3$ Solution: Add $NaHCO_3$ to distilled/deionized water and bring volume to 10.0mL. Mix thoroughly. Filter sterilize. Maintain under 80% N_2 + 20% CO_2.

Trace Elements Solution SL-7:

Composition per liter:

$FeCl_2 \cdot 4H_2O$ 1.5g
$CoCl_2 \cdot 6H_2O$ 0.19g
$MnCl_2 \cdot 4H_2O$ 0.1g
$ZnCl_2$ 0.07g
H_3BO_3 .... 0.062g
$Na_2MoO_4 \cdot 2H_2O$ 0.036g
$NiCl_2 \cdot 6H_2O$ 0.024g
$CuCl_2 \cdot 2H_2O$ 0.017g
HCl (25% solution) 10.0mL

Preparation of Trace Elements Solution SL-7: Add the $FeCl_2 \cdot 4H_2O$ to the HCl. Add distilled/deionized water and bring volume to 1.0L. Add remaining components. Mix thoroughly. Filter sterilize. Maintain under 80% N_2 + 20% CO_2.

Preparation of Medium: Add components—except sodium sulfide solution, sodium L-tartrate solution, $NaHCO_3$ solution, and trace element solution SL-7 solution—to distilled/deionized water and bring volume to 969.0mL. Mix thoroughly. Gently heat and bring to boiling. Autoclave for 15 min at 15 psi pressure–121°C. Cool to 45–50°C. Maintain under 80% N_2 + 20% CO_2. Aseptically add 10.0mL L-tartrate solution, 10.0mL $NaHCO_3$ solution, and 1.0mL trace element solution SL-7 solution under 80% N_2 + 20% CO_2. Mix thoroughly. Aseptically distribute into sterile tubes or flasks under 80% N_2 + 20% CO_2. Adjust pH to 7.2. At time of inoculation add sodium sulfide solution to a final concentration of 0.1%.

Use: For the cultivation and maintenance of *Ilyobacter tartaricus*.

Ilyobacter Medium

Composition per liter:

Crotonic acid 1.7g
NaCl 1.0g
Yeast extract 1.0g
$Na_2HPO_4 \cdot 12H_2O$ 0.7g
KCl 0.5g
$MgCl_2 \cdot 6H_2O$ 0.4g
NH_4Cl 0.3g
Na_2SO_4 .... 0.1g
Sodium sulfide solution 10.0mL
$CaCl_2 \cdot 2H_2O$ (1.0%) 1.0mL
$FeCl_3$ (0.5%) 1.0mL
Modified SL-7 trace elements solution 1.0mL
Resazurin (0.1%) 1.0mL
Selenite/tungstate solution 1.0mL

pH 6.8–7.2 at 25°C

Sodium Sulfide Solution:

Composition per 100mL:

$Na_2S \cdot 9H_2O$ 3.6g

Preparation of Sodium Sulfide Solution: Add $Na_2S \cdot 9H_2O$ to distilled/deionized water and bring volume to 100.0mL. Mix thoroughly. Autoclave for 15 min at 15 psi pressure–121°C under N_2. Maintain under 100% N_2.

Modified Sl-7 Trace Elements Solution:

Composition per liter:

$CoCl_2 \cdot 6H_2O$ 0.2g
$MnCl_2 \cdot 4H_2O$ 0.1g
$ZnCl_2$ 0.07g
H_3BO_3 .... 0.06g
$Na_2MoO_4 \cdot 2H_2O$ 0.04g
$CuCl_2 \cdot 2H_2O$ 0.02g
$NiCl_2 \cdot 6H_2O$ 0.02g
HCl (1*N*) 3.0mL

Preparation of Modified SL-7 Trace Elements Solution: Add components to distilled/deionized water and bring volume to 1.0L. Mix thoroughly. Filter sterilize. Maintain under 80% N_2 + 20% CO_2.

Selenite/Tungstate Solution:

Composition per liter:

NaOH 0.5g
$Na_2WO_4 \cdot 2H_2O$ 4.0mg
$Na_2SeO_3 \cdot 5HO$ 3.0mg

Preparation of Selenite/Tungstate Solution: Add components to distilled/deionized water and bring volume to 1.0L. Mix thoroughly. Filter sterilize. Maintain under 80% N_2 + 20% CO_2.

Preparation of Medium: Add components—except sodium sulfide solution, modified SL-7 trace elements solution, and selenite/tungstate solution—to distilled/deionized water and bring volume to 969.0mL. Mix thoroughly. Gently heat and bring to boiling. Autoclave for 15 min at 15 psi pressure–121°C. Adjust pH to 5.5. Cool to 45–50°C under 100% N_2. Maintain under 100% N_2. Aseptically add 1.0mL sterile modified SL-7 trace elements solution and 1.0mL sterile selenite/tungstate solution under 100% N_2. Mix thoroughly. Aseptically distribute into sterile tubes or flasks under 100% N_2. At time of inoculation add sodium sulfide solution to a final concentration of 1.0% sodium sulfide solution. Maintain under 100% N_2.

Use: For the cultivation and maintenance of *Ilyobacter delafieldii.*

Imhoff's Medium, Modified

Composition per liter:

NaCl	30.0g
$NaHCO_3$	3.0g
KH_2PO_4	1.0g
NH_4Cl	1.0g
Sodium acetate	1.0g
Na_2SO_4	0.7g
$MgCl_2 \cdot 6H_2O$	0.5g
Sodium ascorbate	0.5g
$CaCl_2 \cdot 2H_2O$	0.1g
Yeast extract	0.1g
Sodium sulfide solution	10.0mL
SLA trace elements solution	1.0mL
VA vitamin solution	1.0mL

pH 6.9–7.0 at 25°C

Sodium Sulfide Solution:

Composition per 100mL:

$Na_2S \cdot 9H_2O$	2.0g

Preparation of Sodium Sulfide Solution: Add $Na_2S \cdot 9H_2O$ to distilled/deionized water and bring volume to 100.0mL. Mix thoroughly. Autoclave for 15 min at 15 psi pressure–121°C under N_2. Maintain under 100% N_2.

SLA Trace Elements Solution:

Composition per liter:

$FeCl_2 \cdot 4H_2O$	1.8g
H_3BO_3	0.5g
$CoCl_2 \cdot 6H_2O$	0.25g
$ZnCl_2$	0.1g
$MnCl_2 \cdot 4H_2O$	0.07g
$Na_2MoO_4 \cdot 2H_2O$	0.03g
$CuCl_2 \cdot 2H_2O$	0.01g
$Na_2SeO_3 \cdot 5H_2O$	0.01g
$NiCl_2 \cdot 6H_2O$	0.01g

Preparation of SLA Trace Elements Solution: Add components to distilled/deionized water and bring volume to 1.0L. Mix thoroughly. Adjust pH to pH 2–3. Filter sterilize.

VA Vitamin Solution:

Composition per 500 mL:

Nicotinamide	0.17g
Thiamine·HCl	0.15g
p-Aminobenzoic acid	0.1g
Biotin	0.05g
Calcium pantothenate	0.05g
Pyridoxine·2HCl	0.05g
Cyanocobalamin	0.02g

Preparation of VA Vitamin Solution: Add components to distilled/deionized water and bring volume to 500.0mL. Mix thoroughly. Filter sterilize.

Preparation of Medium: Add components—except sodium sulfide solution, SLA trace elements solution, and VA vitamin solution—to distilled/deionized water and bring volume to 988.0mL. Mix thoroughly. Autoclave for 15 min at 15 psi pressure–121°C. Cool to 25°C. Aseptically add 1.0mL sterile SLA trace elements solution and 1.0mL sterile VA vitamin solution. Aseptically add 10.0mL sterile sodium sulfide solution. Mix thoroughly. Aseptically distribute into sterile tubes or flasks.

Use: For the cultivation and maintenance of *Rhodobacter adriaticus* and *R. sulfidophilus.*

Imidazole Utilization Medium

Composition per liter:

Imidazole	5.0g
KH_2PO_4	0.5g
$MgSO_4 \cdot 7H_2O$	0.5g
$CaCl_2$	3.0mg
$FeSO_4 \cdot 7H_2O$	3.0mg
Molybdenum solution	1.0mL
Trace element solution	1.0mL

pH 6.0 ± 0.2 at 25°C

Molybdenum Solution:

Composition per 18mL:

$Na_2MoO_4 \cdot 2H_2O$	0.5mg

Preparation of Molybdenum Solution: Add components to distilled/deionized water and bring volume to 18.0mL. Mix thoroughly. Filter sterilize.

Trace Element Solution:

Composition per 18mL:

H_3BO_3	11.0mg
$MnCl_2 \cdot 4H_2O$	7.0mg
$Al_2(SO_4)_3 \cdot 18\ H_2O$	1.94mg
$Co(NO_3)_2 \cdot 6H_2O$	1.0mg

$CuSO_4 \cdot 5H_2O$	1.0mg
$NiSO_4 \cdot 6H_2O$	1.0mg
$ZnSO_4 \cdot H_2O$	0.62mg
KBr	0.5mg
KI	0.5mg
LiCl	0.5mg
$SnCl_2 \cdot 2H_2O$	0.5mg

Preparation of Trace Element Solution: Add components to distilled/deionized water and bring volume to 18.0mL. Mix thoroughly. Filter sterilize.

Preparation of Medium: Add components, except molybdenum solution and trace elements solution, to distilled/deionized water and bring volume to 998.0mL. Mix thoroughly. Distribute into tubes or flasks. Autoclave for 15 min at 15 psi pressure–121°C. Cool to 25°C. Aseptically add 1.0mL molybdenum solution and 1.0mL trace element solution. Mix thoroughly. Adjust pH to 6.0 with phosphoric acid. Mix thoroughly. Aseptically distribute into sterile tubes or flasks.

Use: For the cultivation and maintenance of *Pseudomonas* species.

Indole Medium

Composition per 200mL:

K_2HPO_4	3.13g
L-Tryptophan	1.0g
NaCl	1.0g
KH_2PO_4	0.27g

pH 7.2 ± 0.2 at 25°C

Preparation of Medium: Add components to distilled/deionized water and bring volume to 200.0mL. Mix thoroughly. Distribute into tubes or flasks. Autoclave for 15 min at 15 psi pressure–121°C.

Use: For the differentiation of microorganisms by means of the indole production from tryptophan test.

Indole Medium

Composition per liter:

Pancreatic digest of casein	20.0g

pH 7.3 ± 0.2 at 25°C

Preparation of Medium: Add pancreatic digest of casein to distilled/deionized water and bring volume to 1.0L. Mix thoroughly. Distribute into tubes or flasks. Autoclave for 15 min at 15 psi pressure–121°C.

Use: For the differentiation of microorganisms by means of the indole test.

Indole Nitrite Medium (Trypticase™ Nitrate Broth)

Composition per liter:

Pancreatic digest of casein	20.0g
Na_2HPO_4	2.0g
Agar	1.0g
Glucose	1.0g
KNO_3	1.0g

pH 7.2 ± 0.2 at 25°C

Source: This medium is available as a premixed powder from BBL Microbiology Systems.

Preparation of Medium: Add components to distilled/deionized water and bring volume to 1.0L. Mix thoroughly. Gently heat and bring to boiling with frequent agitation. Distribute into tubes or flasks. Autoclave for 15 min at 15 psi pressure–121°C.

Use: For the identification of microorganisms by means of the nitrate reduction and indole tests.

Infusion Broth

Composition per liter:

Pancreatic digest of casein	13.0g
NaCl	5.0g
Yeast extract	5.0g
Heart muscle, solids from infusion	2.0g

pH 7.4 ± 0.2 at 25°C

Source: This medium is available as a premixed powder from BBL Microbiology Systems.

Preparation of Medium: Add components to distilled/deionized water and bring volume to 1.0L. Mix thoroughly. Distribute into tubes or flasks. Autoclave for 15 min at 15 psi pressure–121°C.

Use: For the cultivation of a wide variety of microorganisms.

Inorganic Salts Maltose Medium

Composition per liter:

Yeast extract	4.0g
Peptone	2.0g
Inorganic salts solution	980.0mL
Maltose solution	20.0mL

pH 7.5 ± 0.2 at 25°C

Inorganic Salts Solution:

Composition per liter:

$MgSO_4 \cdot 7H_2O$	49.37g
NaCl	43.8g
$CaCl_2 \cdot 2H_2O$	1.29g

Preparation of Inorganic Salts Solution: Add NaCl first and then other components to distilled/deionized water and bring volume to 1.0L. Mix thoroughly.

Maltose Solution:

Composition per 100mL:

Maltose	25.0g

Preparation of Maltose Solution: Add maltose to distilled/deionized water and bring volume to 100.0mL. Mix thoroughly. Filter sterilize.

Preparation of Medium: Add components, except maltose solution, to inorganic salts solution and bring volume to 980.0mL. Mix thoroughly. Adjust pH to 7.5 with KOH. Autoclave for 15 min at 15 psi pressure–121°C. Cool to 25°C. Aseptically add 20.0mL sterile maltose solution. Mix thoroughly. Aseptically distribute into sterile tubes or flasks.

Use: For the cultivation and maintenance of *Spirochaeta halophila*.

Inositol Brilliant Green Bile Salts Agar (IBB Agar) (*Pleisomonas* Differential Agar)

Composition per liter:

Agar	15.0g
meso-Inositol	10.0g
Proteose peptone	10.0g
Bile salts no. 3	8.5g
Meat extract	5.0g
NaCL	5.0g
Neutral red (2% solution)	1.25mL
Brilliant green (0.1% solution)	0.33mL

pH 7.2 ± 0.1 at 25°C

Preparation of Medium: Add components to distilled/deionized water and bring volume to 1.0L. Mix thoroughly. Gently heat and bring to boiling. Distribute into tubes or flasks. Autoclave for 15 min at 15 psi pressure–121°C. Pour into sterile Petri dishes or leave in tubes.

Use: For isolation of *Aeromonas* and *Plesiomonas* species.

Inositol Urea Caffeic Acid Medium

Composition per liter:

Agar solution	900.0mL
Base solution	100.0mL

Agar Solution:

Composition per 900mL:

Agar	15.0g

Preparation of Agar Solution: Add agar to distilled/deionized water and bring volume to 900.0mL. Mix thoroughly. Gently heat and bring to boiling. Autoclave for 15 min at 15 psi pressure–121°C. Cool to 45–50°C.

Base Solution:

Composition per 100mL:

Inositol	10.0g
Urea	5.0g
KH_2PO_4	1.0g
$MgSO_4 \cdot 7H_2O$	0.5g
Caffeic acid	0.2g
NaCl	0.1g
$CaCl_2 \cdot 2H_2O$	0.1g
Gentamicin sulfate	0.04g
H_3BO_3	0.5mg
$ZnSO_4 \cdot 7H_2O$	0.4mg
$MnSO_4 \cdot 4H_2O$	0.4mg
Thiamine·HCl	0.4mg
Pyroxidine·HCl	0.4mg
Niacin	0.4mg
Calcium pantothenate	0.4mg
p-Aminobenzoic acid	0.2mg
Riboflavin	0.2mg
$FeCl_3$	0.2mg
$Na_2MoO_4 \cdot 4H_2O$	0.2mg
KI	0.1mg
$CuSO_4 \cdot 5H_2O$	0.04mg
Folic acid	2.0μg
Biotin	2.0μg
Ferric citrate solution (1% solution)	1.0mL

Preparation of Base Solution: Add components, except urea, to distilled/deionized water and bring volume to 100.0mL. Mix thoroughly. Gently heat just until components are dissolved. Cool to 75–80°C. Add urea. Mix thoroughly. Do not heat after addition of urea. Do not autoclave. Filter sterilize.

Preparation of Medium: Aseptically combine the cooled, sterile agar solution with the sterile base solution. Mix thoroughly. Pour into sterile Petri dishes.

Use: For the selective isolation and differentiation of *Cryptococcus* species based on inositol and urea utilization and pigment production from caffeic acid. On this medium, only *Cryptococcus* species utilize inositol as sole carbon source and urea as sole nitrogen source. *Cryptococcus neoformans* appears as dark brown colonies. Other *Cryptococcus* species are unpigmented.

Iron Agar, Lyngby (Lyngby Iron Agar)

Composition per liter:

Peptone	20.0g
Agar	12.0g

NaCl	5.0g
Beef extract	3.0g
Yeast extract	3.0g
L-Cysteine	0.6g
Ferric citrate	0.3g
$Na_2S_2O_3$	0.3g

pH 7.4 ± 0.2 at 25°C

Source: This medium is available as a premixed powder from Oxoid Unipath.

Preparation of Medium: Add components to distilled/deionized water and bring volume to 1.0L. Mix thoroughly. Gently heat and bring to boiling. Distribute into tubes or flasks. Autoclave for 15 min at 15 psi pressure–121°C. Pour into sterile Petri dishes or leave in tubes.

Use: For the cultivation and enumeration of H_2S-producing bacteria and total counts of heterotrophic bacteria from fish and fish products.

Iron Bacteria Isolation Medium

Composition per liter:

Agar	10.0g
$(NH_4)_2SO_4$	0.5g
Glucose	0.15g
$CaCO_3$	0.1g
K_2HPO_4	0.05g
$MgSO_4 \cdot 7H_2O$	0.05g
KCl	0.05g
$Ca(NO_3)_2$	0.01g
Vitamin solution	10.0mL

Vitamin Solution:

Composition per 10mL:

Thiamine	0.4mg
Cyanocobalamin	0.01mg

Preparation of Vitamin Solution: Add components to distilled/deionized water and bring volume to 10.0mL. Mix thoroughly. Filter sterilize.

Preparation of Medium: Add components, except vitamin solution, to distilled/deionized water and bring volume to 990.0mL. Mix thoroughly. Gently heat and bring to boiling. Autoclave for 15 min at 15 psi pressure–121°C. Cool to 45–50°C. Aseptically add 10.0mL vitamin solution. Mix thoroughly. Pour into sterile Petri dishes or distribute into sterile tubes.

Use: For the isolation of iron bacteria.

Iron Milk Medium

Composition per liter:

Iron filings	1.0g
Whole milk	1.0L

pH 6.8 ± 0.2 at 25°C

Preparation of Medium: Add iron filings, which may be small balls of steel wool, to whole milk and bring volume to 1.0L. Mix thoroughly. Distribute into tubes or flasks. Autoclave for 15 min at 15 psi pressure–121°C.

Use: For the cultivation of lactic acid bacteria. For the cultivation and differentiation of *Clostridium* species. The medium turns black if H_2S is produced. The medium turns red if acid is produced from milk carbohydrate fermentation. Acid and gas production is characteristic of *C. perfringens* and *C. butyricum*.

Iron–Oxidizing Medium

Composition per liter:

$(NH_4)_2SO_4$	3.0g
K_2HPO_4	0.50g
$MgSO_4 \cdot 7H_2O$	0.50g
KCl	0.10g
$Ca(NO_3)_2$	0.01g
$FeSO_4 \cdot 7H_2O$ solution	300.0mL
H_2SO_4 (10*N*)	1.0mL

pH 3.0–3.6 at 25°C

$FeSO_4 \cdot 7H_2O$ Solution:

Composition per 300mL:

$FeSO_4 \cdot 7H_2O$	44.22g

Preparation of Medium: Add $FeSO_4 \cdot 7H_2O$ to distilled/deionized water and bring volume to 300.0mL. Mix thoroughly. Autoclave for 15 min at 15 psi pressure–121°C. Cool to 25°C.

Preparation of Medium: Add components, except $FeSO_4 \cdot 7H_2O$ solution, to distilled/deionized water and bring volume to 700.0mL. Mix thoroughly. Gently heat and bring to boiling. Autoclave for 15 min at 15 psi pressure–121°C. Cool to 25°C. Aseptically add 300.0mL sterile $FeSO_4 \cdot 7H_2O$ solution. Mix thoroughly. Aseptically distribute into sterile tubes or flasks.

Use: For enumeration, isolation, and cultivation of iron and sulfur bacteria, such as *Thiobacillus ferrooxidans*.

Iron Sulfite Agar

Composition per liter:

Agar	12.0g
Pancreatic digest of casein	10.0g
Ferric citrate	0.5g
$Na_2S \cdot 9H_2O$	0.5g

pH 7.1 ± 0.2 at 25°C

Source: This medium is available as a premixed powder from Oxoid Unipath.

Preparation of Medium: Add components to distilled/deionized water and bring volume to 1.0L. Mix thoroughly. Gently heat and bring to boiling. Distrib-

ute into tubes or flasks. Autoclave for 15 min at 15 psi pressure–121°C. Mix thoroughly. Pour into sterile Petri dishes or leave in tubes.

Use: For the detection of thermophilic anaerobic organisms.

ISM Agar

Composition per liter:

$MgSO_4 \cdot 7H_2O$	49.2g
NaCl	43.5g
Agar	7.5g
Yeast exract	4.0g
Peptone	2.0g
$CaCl_2 \cdot 2H_2O$	1.5g
Maltose solution	100.0mL

Maltose Solution:

Composition per 100mL:

Maltose	5.0g

Preparation of Maltose Solution: Add maltose to distilled/deionized water and bring volume to 100.0mL. Mix thoroughly. Filter sterilize.

Preparation of Medium: Add components, except maltose solution, to distilled/deionized water and bring volume to 900.0mL. Mix thoroughly. Gently heat and bring to boiling. Autoclave for 15 min at 15 psi pressure–121°C. Cool to 45–50°C. Aseptically add sterile maltose solution. Mix thoroughly. Pour into sterile Petri dishes or distribute into sterile tubes.

Use: For the cultivation and maintenance of *Spirochaeta halophila.*

ISM Broth

Composition per liter:

$MgSO_4 \cdot 7H_2O$	49.2g
NaCl	43.5g
Yeast exract	4.0g
Peptone	2.0g
$CaCl_2 \cdot 2H_2O$	1.5g
Maltose solution	100.0mL

Maltose Solution:

Composition per 100mL:

Maltose	5.0g

Preparation of Maltose Solution: Add maltose to distilled/deionized water and bring volume to 100.0mL. Mix thoroughly. Filter sterilize.

Preparation of Medium: Add components, except maltose solution, to distilled/deionized water and bring volume to 900.0mL. Mix thoroughly. Gently heat and bring to boiling. Autoclave for 15 min at 15 psi pressure–121°C. Cool to 45–50°C. Aseptically add sterile maltose solution. Mix thoroughly. Aseptically distribute into sterile tubes or flasks.

Use: For the cultivation of *Spirochaeta halophila.*

Isoleucine Hydroxamate Medium

Composition per liter:

Agar	15.0g
K_2HPO_4	7.0g
Glucose	5.0g
KH_2PO_4	3.0g
L-Isoleucine hydroxamate	1.0g
$(NH_4)_2SO_4$	1.0g

pH 7.0 ± 0.2 at 25°C

Preparation of Medium: Add components to distilled/deionized water and bring volume to 1.0L. Mix thoroughly. Gently heat and bring to boiling. Distribute into tubes or flasks. Autoclave for 15 min at 15 psi pressure–121°C. Pour into sterile Petri dishes or leave in tubes.

Use: For the cultivation and maintenance of *Serratia marcescens.*

IsoVitaleX® Enrichment

Composition per liter:

Glucose	100.0g
L-Cysteine·HCl	25.9g
L-Glutamine	10.0g
Adenine	1.0g
Thiamine pyrophosphate	0.1g
Vitamin B_{12}	0.1g
Guanine·HCl	0.03g
$Fe(NO_3)_3 \cdot 9H_2O$	0.02g
p-Aminobenzoic acid	0.013g
Thiamine·HCl	0.003g

Preparation of IsoVitaleX® Enrichment: Add components to distilled/deionized water and bring volume to 1.0L. Mix thoroughly. Filter sterilize.

Use: As a nutrient supplement for the isolation and cultivation of fastidious microorganisms.

ISP Medium 1 (International *Streptomyces* Project Medium 1) (Tryptone Yeast Extract Broth)

Composition per liter:

Pancreatic digest of casein	5.0g
Yeast extract	3.0g

pH 7.0–7.2 at 25°C

Preparation of Medium: Add components to distilled/deionized water and bring volume to 1.0L. Mix

thoroughly. Distribute into tubes or flasks. Autoclave for 15 min at 15 psi pressure–121°C.

Use: For cultivation of *Streptomyces* species according to the International *Streptomyces* Project.

ISP Medium 2 (International *Streptomyces* Project Medium 2) (Yeast Extract Malt Extract Agar)

Composition per liter:

Agar	20.0g
Malt extract	10.0g
Yeast extract	4.0g
Glucose	4.0g

pH 7.3 ± 0.2 at 25°C

Preparation of Medium: Add components to distilled/deionized water and bring volume to 1.0L. Mix thoroughly. Gently heat and bring to boiling. Distribute into tubes or flasks. Autoclave for 15 min at 15 psi pressure–121°C. Pour into sterile Petri dishes or leave in tubes.

Use: For cultivation of *Streptomyces* species according to the International *Streptomyces* Project.

ISP Medium 3 (International *Streptomyces* Project Medium 3) (Oatmeal Agar)

Composition per liter:

Oatmeal	20.0g
Agar	18.0g
Trace salts solution	1.0mL

Source: This medium is available as a premixed powder from Difco Laboratories.

Trace Salts Solution:
Composition per 100mL:

$FeSO_4 \cdot 7H_2O$	0.1g
$MnCl_2 \cdot 4H_2O$	0.1g
$ZnSO_4 \cdot 7H_2O$	0.1g

Preparation of Trace Salts Solution: Add components to distilled/deionized water and bring the volume to 100.0mL. Mix thoroughly. Filter sterilize.

Preparation of Medium: Add oatmeal to distilled/deionized water and bring volume to 1.0L. Mix thoroughly. Gently heat and bring to boiling. Steam for 20 min. Filter through cheesecloth. Add agar. Add sufficient distilled/deionized water to bring volume to 999.0mL. Gently heat and bring to boiling. Mix thoroughly. Distribute into tubes or flasks. Autoclave for 15 min at 15 psi pressure–121°C. Cool to 45–50°C. Aseptically add 1.0mL sterile trace salts solution. Mix thoroughly. Pour into sterile Petri dishes or distribute into sterile tubes.

Use: For cultivation of *Streptomyces* species according to the International *Streptomyces* Project.

ISP Medium 4 (International *Streptomyces* Project Medium 4) (Inorganic Salts Starch Agar)

Composition per liter:

Agar	20.0g
Soluble starch	10.0g
$CaCO_3$	2.0g
$(NH_4)_2SO_4$	2.0g
K_2HPO_4	1.0g
$MgSO_4 \cdot 7H_2O$	1.0g
NaCl	1.0g
$FeSO_4 \cdot 7H_2O$	1.0mg
$MnCl_2 \cdot 7H_2O$	1.0mg
$ZnSO_4 \cdot 7H_2O$	1.0mg

pH 7.2 ± 0.2 at 25°C

Source: This medium is available as a premixed powder from Difco Laboratories.

Preparation of Medium: Add components to distilled/deionized water and bring volume to 1.0L. Mix thoroughly. Gently heat and bring to boiling with frequent agitation. Distribute into tubes or flasks. Autoclave for 15 min at 15 psi pressure–121°C. Pour into sterile Petri dishes or leave in tubes.

Use: For characterizing *Streptomyces* species. For the cultivation and maintenance of *Actinomadura fastidiosa, Actinomadura roseoviolacea, Actinomadura* species, *Actinoplanes* species, *Amycolatopsis mediterranei, Kitasatosporia grisea, Kitasatosporia papulosa, Saccharomonospora internatus, Saccharopolyspora hirsuta, Streptomyces* species, *Streptosporangium* species, and *Streptoverticillium* species.

ISP Medium 4 with Glucose (International *Streptomyces* Project Medium 4 with Glucose)

Composition per liter:

Agar	20.0g
Glucose	20.0g
Soluble starch	10.0g
$CaCO_3$	2.0g
$(NH_4)_2SO_4$	2.0g
K_2HPO_4	1.0g

$MgSO_4 \cdot 7H_2O$	1.0g
NaCl	1.0g
$FeSO_4 \cdot 7H_2O$	1.0mg
$MnCl_2 \cdot 7H_2O$	1.0mg
$ZnSO_4 \cdot 7H_2O$	1.0mg

pH 7.2 ± 0.2 at 25°C

Preparation of Medium: Add components to distilled/deionized water and bring volume to 1.0L. Mix thoroughly. Gently heat and bring to boiling with frequent agitation. Distribute into tubes or flasks. Autoclave for 15 min at 15 psi pressure–121°C. Pour into sterile Petri dishes with swirling or leave in tubes.

Use: For the cultivation and maintenance of *Streptomyces purpureus*.

ISP Medium 4 with Yeast Extract (International *Streptomyces* Project Medium 4 with Yeast Extract)

Composition per liter:

Agar	20.0g
Soluble starch	10.0g
$CaCO_3$	2.0g
$(NH_4)_2SO_4$	2.0g
K_2HPO_4	1.0g
$MgSO_4 \cdot 7H_2O$	1.0g
NaCl	1.0g
Yeast extract	1.0g
$FeSO_4 \cdot 7H_2O$	1.0mg
$MnCl_2 \cdot 7H_2O$	1.0mg
$ZnSO_4 \cdot 7H_2O$	1.0mg

pH 7.2 ± 0.2 at 25°C

Preparation of Medium: Add components to distilled/deionized water and bring volume to 1.0L. Mix thoroughly. Gently heat and bring to boiling with frequent agitation. Distribute into tubes or flasks. Autoclave for 15 min at 15 psi pressure–121°C. Pour into sterile Petri dishes with swirling or leave in tubes.

Use: For the cultivation and maintenance of *Thermomonospora mesouviformis*.

ISP Medium 5 (International *Streptomyces* Project Medium 5) (Glycerol Asparagine Agar)

Composition per liter:

Agar	20.0g
Glycerol	10.0g
L-Asparagine	1.0g
K_2HPO_4	1.0g
Trace salts solution	1.0mL

pH 7.4 ± 0.2 at 25°C

Trace Salts Solution:

Composition per 100mL:

$FeSO_4 \cdot 7H_2O$	0.1g
$MnCl_2 \cdot 4H_2O$	0.1g
$ZnSO_4 \cdot 7H_2O$	0.1g

Preparation of Trace Salts Solution: Add components to distilled/deionized water and bring the volume to 100.0mL. Mix thoroughly. Filter sterilize.

Preparation of Medium: Add components, except trace salts solution, to distilled/deionized water and bring volume to 999.0mL. Mix thoroughly. Gently heat and bring to boiling. Autoclave for 15 min at 15 psi pressure–121°C. Cool to 45–50°C. Aseptically add 1.0mL sterile trace salts solution. Mix thoroughly. Pour into sterile Petri dishes or distribute into sterile tubes.

Use: For the cultivation and maintenance of the *Pseudonocardia* species and *Streptomyces peucetius*.

ISP Medium 6 (International *Streptomyces* Project Medium 6) (Peptone Yeast Extract Iron Agar)

Composition per liter:

Agar	15.0g
Peptone	15.0g
Proteose peptone	5.0g
K_2HPO_4	1.0g
Yeast extract	1.0g
Ferric ammonium citrate	0.5g
$Na_2S_2O_3$	0.08g

Preparation of Medium: Add components to distilled/deionized water and bring volume to 1.0L. Mix thoroughly. Gently heat and bring to boiling. Distribute into tubes or flasks. Autoclave for 15 min at 15 psi pressure–121°C. Pour into sterile Petri dishes or leave in tubes.

Use: For the cultivation and maintenance of *Streptomyces* species.

ISP Medium 9 (International Streptomyces Project Medium 9)

Composition per liter:

$K_2HPO_4 \cdot 3H_2O$	5.65g
$(NH_4)_2SO_4$	2.64g

KH_2PO_4	2.38g
$MgSO_4 \cdot 7H_2O$	1.0g
Carbohydrate solution	100.0mL
Pridham and Gottlieb trace salts	1.0mL

pH 6.8–7.0 at 25°C

Carbohydrate Solution:

Composition per 100mL:

Carbohydrate	10.0g

Preparation of Carbohydrate Solution: Add carbohydrate to distilled/deionized water and bring volume to 100.0mL.Use glucose, arabinose, sucrose, xylose, inositol, mannitol, fructose, rhamnose, raffinose or cellulose. Mix thoroughly. Filter sterilize.

Pridham And Gottlieb Trace Salts:

Composition per 100mL:

$MnCl_2 \cdot 7H_2O$	0.79g
$CuSO_4 \cdot 5H_2O$	0.64g
$ZnSO_4 \cdot 7H_2O$	0.15g
$FeSO_4 \cdot 7H_2O$	0.11g

Preparation of Pridham And Gottlieb Trace Salts: Add components to distilled/deionized water and bring volume to 100.0mL. Mix thoroughly.

Preparation of Medium: Add components, except carbohydrate solution, to distilled/deionized water and bring volume to 900.0mL. Mix thoroughly. Gently heat and bring to boiling with frequent agitation. Autoclave for 15 min at 15 psi pressure–121°C. Cool to 45–50°C. Aseptically add sterile carbohydrate solution. Mix thoroughly. Aseptically distribute into sterile tubes or flasks.

Use: For the cultivation and differentiation of *Streptomyces purpureus* and other *Streptomyces* species based on carbohydrate utilization.

J Agar

Composition per liter:

Agar	20.0g
Yeast extract	15.0g
Pancreatic digest of casein	5.0g
K_2HPO_4	3.0g
Glucose solution	10.0mL

pH 7.3–7.5 at 25°C

Glucose Solution:

Composition per 10mL:

Glucose	2.0g

Preparation of Glucose Solution: Add glucose to distilled/deionized water and bring volume to 10.0mL. Mix thoroughly. Filter sterilize.

Preparation of Medium: Add components, except glucose solution, to distilled/deionized water and bring volume to 990.0mL. Mix thoroughly. Gently heat and bring to boiling. Autoclave for 15 min at 15 psi pressure–121°C. Cool to 45–50°C. Aseptically add sterile glucose solution. Mix thoroughly. Pour into sterile Petri dishes.

Use: For the cultivation of *Bacillus* species and *Sporolactobacillus* species.

J Broth

Composition per liter:

Yeast extract	15.0g
Pancreatic digest of casein	5.0g

pH 7.3–7.5 at 25°C

Preparation of Medium: Add components to distilled/deionized water and bring volume to 1.0L. Mix thoroughly. Adjust pH to 7.3–7.5. Distribute into tubes or flasks. Autoclave for 20 min at 15 psi pressure–121°C.

Use: For the cultivation of *Bacillus* species and *Sporolactobacillus* species for performing the Voges-Proskauer test.

JB Medium with Glucose

Composition per liter:

Yeast extract	15.0g
Pancreatic digest of casein	5.0g
K_2HPO_4	3.0g
Glucose	2.0g

pH 7.3–7.5 at 25°C

Preparation of Medium: Add components to distilled/deionized water and bring volume to 1.0L. Mix thoroughly. Gently heat and bring to boiling. Distribute into tubes or flasks. Autoclave for 15 min at 15 psi pressure–121°C. Pour into sterile Petri dishes or leave in tubes.

Use: For the cultivation and maintenance of *Bacillus popilliae*.

JD1 Medium

Composition per liter:

Beef heart, solids from infusion	25.0g
Agar	15.0g
Peptone	5.0g
NaCl	2.5g
Bovine albumin	0.5g
Hemin chloride	0.04g

pH 7.8 ± 0.2 at 25°C

Preparation of Medium: Add components to distilled/deionized water and bring volume to 1.0L. Mix thoroughly. Gently heat and bring to boiling. Distrib-

ute into tubes or flasks. Autoclave for 15 min at 15 psi pressure–121°C. Pour into sterile Petri dishes or leave in tubes.

Use: For the isolation and cultivation of PD-ALS (Pierce's disease-almond leaf scorch) bacteria.

Jordan's Tartrate Agar

Composition per liter:

Agar	15.0g
Pancreatic digest of casein	10.0g
Sodium potassium tartrate	10.0g
NaCl	5.0g
Phenol red	0.024g

pH 7.7 ± 0.3 at 25°C

Source: This medium is available as a prepared medium in tubes from BBL Microbiology Systems.

Preparation of Medium: Add components to distilled/deionized water and bring volume to 1.0L. Mix thoroughly. Gently heat and bring to boiling. Adjust pH to 7.7. Distribute into tubes. Autoclave for 15 min at 15 psi pressure–121°C.

Use: For differentiation and identification of members of the Enterobacteriaceae, especially *Salmonella* species, based upon the ability to utilize tartrate. Utilization of tartrate turns the medium yellow. *S. enteritidis* utilizes tartrate. *S. paratyphi* A does not utilize tartrate.

K101 *Flexibacter* Medium

Composition per liter:

Agar	10.0g
Casamino acids	1.0g
Glucose	1.0g
Tris(hydroxymethyl)aminomethane buffer	1.0g
$CaCl_2$	0.1g
KNO_3	0.1g
$MgSO_4 \cdot 7H_2O$	0.1g
Sodium glycerophosphate	0.1g
Thiamine·HCl	1.0mg
Cyanocobalamin	1.0μg
Trace elements solution HO-LE	1.0mL

pH 7.5 ± 0.2 at 25°C

Trace Element Solution HO-LE:

Composition per liter:

H_3BO_3	2.85g
$MnCl_2 \cdot 4H_2O$	1.8g
Sodium tartrate	1.77g
$FeSO_4 \cdot 7H_2O$	1.36g
$CoCl_2 \cdot 6H_2O$	0.04g
$CuCl_2.2H_2O$	0.027g
$Na_2MoO_4 \cdot 2H_2O$	0.025g
$ZnCl_2$	0.020g

Preparation of Trace Element Solution HO-LE: Add components to distilled/deionized water and bring volume to 1.0L. Mix thoroughly. Filter sterilize.

Preparation of Medium: Add components, except trace elements solution HO-LE, to distilled/deionized water and bring volume to 999.0mL. Mix thoroughly. Gently heat and bring to boiling. Autoclave for 15 min at 15 psi pressure–121°C. Cool to 45–50°C. Aseptically add 1.0mL trace elements solution HO-LE. Mix thoroughly. Pour into sterile Petri dishes or distribute into sterile tubes.

Use: For the cultivation and maintenance of *Cytophaga* species, *Flexibacter* species, *Herpetosiphon geysericola*, and *Myxococcus fulvus*.

KCN Broth

Composition per liter:

Na_2HPO_4	5.64g
NaCl	5.0g
Peptone	3.0g
KH_2PO_4	0.225g
KCN (0.5% solution)	15.0mL

pH 7.6 ± 0.2 at 25°C

Caution: Cyanide is toxic.

Preparation of Medium: Add components, except KCN solution, to distilled/deionized water and bring volume to 985.0mL. Mix thoroughly. Autoclave for 15 min at 15 psi pressure–121°C. Cool to 25°C. Aseptically add KCN solution. Mix thoroughly. Aseptically distribute into sterile tubes. Stopper immediately.

Use: For the differentiation of Enterobacteriaceae based upon growth in the presence of potassium cyanide.

Kelly Medium, Nonselective Modified

Composition per 1430mL:

HEPES buffer (*N*-2-Hydroxyethylpiperazine-*N*-2-ethanesulfonic acid)	6.0g
Proteose peptone No. 2	5.0g
D-Glucose	3.0g
$NaHCO_3$	2.2g
Pancreatic digest of casein	1.0g
Yeast, autolyzed	1.0g
Sodium pyruvate	0.8g
Sodium citrate	0.7g
N-Acetylglucosamine	0.4g
$MgCl_2 \cdot 6H_2O$	0.3g
Gelatin solution	200.0mL
Bovine serum albumin	143.0mL
CMRL-1066 medium with glutamine, 10X	100.0mL

Rabbit serum, heat inactivated	86.0mL
Hemin solution	1.0mL

pH 7.2 ± 0.2 at 25°C

CMRL-1066 Medium with Glutamine, 10X:

Composition per liter:

NaCl	6.8g
$NaHCO_3$	2.2g
D-Glucose	1.0g
KCl	0.4g
L-Cysteine·HCl·H_2O	0.26g
$CaCl_2$, anhydrous	0.2g
$MgSO_4 \cdot 7H_2O$	0.2g
$NaH_2PO_4 \cdot H_2O$	0.14g
L-Glutamine	0.1g
Sodium acetate·$3H_2O$	0.083g
L-Glutamic acid	0.075g
L-Arginine·HCl	0.070g
L-Lysine·HCl	0.070g
L-Leucine	0.060g
Glycine	0.050g
Ascorbic acid	0.050g
L-Proline	0.040g
L-Tyrosine	0.040g
L-Aspartic acid	0.030g
L-Threonine	0.030g
L-Alanine	0.025g
L-Phenylalanine	0.025g
L-Serine	0.025g
L-Valine	0.025g
L-Cystine	0.020g
L-Histidine·HCl·H_2O	0.020g
L-Isoleucine	0.020g
Phenol red	0.020g
L-Methionine	0.015g
Deoxyadenosine	0.010g
Deoxycytidine	0.010g
Deoxyguanosine	0.010g
Glutathione, reduced	0.010g
Thymidine	0.010g
Hydroxy-L-proline	0.010g
L-Tryptophan	0.010g
Nicotinamide adenine dinucleotide	7.0mg
Tween™ 80	5.0mg
Sodium glucuronate·H_2O	4.2mg
Coenzyme A	2.5mg
Cocarboxylase	1.0mg
Flavin adenine dinucleotide	1.0mg
Nicotinamide adenine dinucleotide phosphate	1.0mg
Uridine triphosphate	1.0mg
Choline chloride	0.50mg
Cholesterol	0.20mg
5-Methyldeoxycytidine	0.10mg
Inositol	0.05mg
p-Aminobenzoic acid	0.05mg
Niacin	0.025mg
Niacinamide	0.025mg
Pyridoxine	0.025mg
Pyridoxal·HCl	0.025mg
Biotin	0.01mg
D-Calcium pantothenate	0.01mg
Folic acid	0.01mg
Riboflavin	0.01mg
Thiamine·HCl	0.01mg

pH 7.2 ± 0.2 at 25°C

Source: This solution is available as a premixed powder from GIBCO BRL.

Preparation of CMRL-1066 Medium with Glutamine, 10X: Add components to distilled/deionized water and bring volume to 1.0L. Mix thoroughly. Adjust pH to 7.2. Filter sterilize.

Gelatin Solution:

Composition per 200mL:

Gelatin	14.0g

Preparation of Gelatin Solution: Add gelatin to distilled/deionized water and bring volume to 1.0L. Mix thoroughly. Gently heat and bring to boiling. Autoclave for 15 min at 15 psi pressure–121°C. Cool to 50°C.

Hemin Solution:

Composition per 100mL:

Hemin	1.0g
NaOH (1*N* solution)	20.0mL

Preparation of Hemin Solution: Add hemin to 20.0mL of 1*N* NaOH solution. Mix thoroughly. Bring volume to 100.0mL with distilled/deionized water.

Bovine Serum Albumin Solution:

Composition per 200mL:

Bovine serum albumin	70.0g

Preparation of Bovine Serum Albumin Solution: Add bovine serum albumin to distilled/deionized water and bring volume to 200.0mL. Filter sterilize.

Preparation of Medium: Add components, except gelatin solution, bovine serum albumin solution, and rabbit serum to distilled/deionized water and bring volume to 1001.0mL. Mix thoroughly. Bring pH to 7.6 with 5*N* NaOH. Filter sterilize. Aseptically add 200.0mL sterile gelatin solution, 143.0mL sterile bovine serum albumin and 86.0mL sterile heat-inactivated rabbit serum. Mix thoroughly. Aseptically dispense into sterile tubes or flasks.

Use: For isolation of *Borrelia burgdorferi* and other spirochetes.

Kelly Medium, Selective Modified

Composition per liter:

Bovine serum albumin fraction V	50.0g
HEPES buffer (*N*-2-Hydroxyethylpiperazine-*N*-2-ethanesulfonic acid)	6.0g
Glucose	5.0g
Neopeptone	5.0g
$NaHCO_3$	2.2g
Sodium pyruvate	0.8g
Sodium citrate	0.7g
N-Acetylglucosamine	0.4g
Kanamycin	8.0mg
5-Fluorouracil	2.3mg
Gelatin solution	200.0mL
CMRL-1066 medium with glutamine, 10X	100.0mL
Rabbit serum, partially hemolyzed	70.0mL

pH 7.7 ± 0.2 at 25°C

Gelatin Solution:

Composition per 200mL:

Gelatin	14.0g

Preparation of Gelatin Solution: Add gelatin to distilled/deionized water and bring volume to 1.0L. Mix thoroughly. Gently heat and bring to boiling. Autoclave for 15 min at 15 psi pressure–121°C. Cool to 50°C.

CMRL-1066 Medium with Glutamine, 10X:

Composition per liter:

NaCl	6.8g
$NaHCO_3$	2.2g
D-Glucose	1.0g
KCl	0.4g
L-Cysteine·HCl·H_2O	0.26g
$CaCl_2$, anhydrous	0.2g
$MgSO_4 \cdot 7H_2O$	0.2g
$NaH_2PO_4 \cdot H_2O$	0.14g
L-Glutamine	0.1g
Sodium acetate·$3H_2O$	0.083g
L-Glutamic acid	0.075g
L-Arginine·HCl	0.070g
L-Lysine·HCl	0.070g
L-Leucine	0.060g
Glycine	0.050g
Ascorbic acid	0.050g
L-Proline	0.040g
L-Tyrosine	0.040g
L-Aspartic acid	0.030g
L-Threonine	0.030g
L-Alanine	0.025g
L-Phenylalanine	0.025g
L-Serine	0.025g
L-Valine	0.025g
L-Cystine	0.020g
L-Histidine·HCl·H_2O	0.020g
L-Isoleucine	0.020g
Phenol red	0.020g
L-Methionine	0.015g
Deoxyadenosine	0.010g
Deoxycytidine	0.010g
Deoxyguanosine	0.010g
Glutathione, reduced	0.010g
Thymidine	0.010g
Hydroxy-L-proline	0.010g
L-Tryptophan	0.010g
Nicotinamide adenine dinucleotide	7.0mg
Tween™ 80	5.0mg
Sodium glucuronate·H_2O	4.2mg
Coenzyme A	2.5mg
Cocarboxylase	1.0mg
Flavin adenine dinucleotide	1.0mg
Nicotinamide adenine dinucleotide phosphate	1.0mg
Uridine triphosphate	1.0mg
Choline chloride	0.50mg
Cholesterol	0.20mg
5-Methyldeoxycytidine	0.10mg
Inositol	0.05mg
p-Aminobenzoic acid	0.05mg
Niacin	0.025mg
Niacinamide	0.025mg
Pyridoxine	0.025mg
Pyridoxal·HCl	0.025mg
Biotin	0.01mg
D-Calcium pantothenate	0.01mg
Folic acid	0.01mg
Riboflavin	0.01mg
Thiamine·HCl	0.01mg

pH 7.2 ± 0.2 at 25°C

Source: This solution is available as a premixed powder from GIBCO BRL.

Preparation of CMRL-1066 Medium with Glutamine, 10X: Add components to distilled/deionized water and bring volume to 1.0L. Mix thoroughly. Adjust pH to 7.2. Filter sterilize.

Preparation of Medium: Add components, except gelatin solution, partially hemolyzed rabbit serum solution, kanamycin, and 5-fluorouracil to distilled/deionized water and bring volume to 1.0mL. Mix thoroughly. Bring pH to 7.6 with 5*N* NaOH. Filter sterilize. Aseptically add 200.0mL sterile gelatin solution, 70.0mL of partially hemolyzed rabbit serum, 8.0mg kanamycin, and 230.0mg 5-fluorouracil. Mix thoroughly. Aseptically distribute into sterile tubes or flasks.

Use: For the isolation of *Borrelia burgdorferi*.

Kerosene Mineral Salts Medium

Composition per liter:

KH_2PO_4	1.0g
K_2HPO_4	1.0g
NH_4NO_3	1.0g
$MgSO_4 \cdot 7H_2O$	0.2g
$CaCl_2$	0.02g
$FeCl_3$	0.05g
Kerosene	20.0mL

pH 6.9–7.0 at 25°C

Preparation of Medium: Add components, except kerosene, to distilled/deionized water and bring volume to 1.0L. Mix thoroughly. Adjust pH to 6.9–7.0 with dilute NaOH. Distribute into tubes in 10.0mL volumes or flasks in 100.0mL volumes. Autoclave for 15 min at 15 psi pressure–121°C. Overlay tubes with 0.2mL of kerosene per tube. Overlay flasks with 2.0mL of kerosene per flask.

Use: For the cultivation and maintenance of *Pseudomonas aeruginosa.*

Ketogluconate Broth

Composition per liter:

Potassium gluconate	20.0g
KH_2PO_4	5.4g
KNO_3	2.0g

pH 6.5 ± 0.2 at 25°C

Preparation of Medium: Add components to distilled/deionized water and bring volume to 1.0L. Mix thoroughly. Filter sterilize. Aseptically distribute into sterile tubes.

Use: For use in identifying bacteria that can utilize 2-ketogluconate.

Ketolactonate Broth

Composition per liter:

Agar	20.0g
Lactose	10.0g
Yeast extract	10.0g

Preparation of Medium: Add components to distilled/deionized water and bring volume to 1.0L. Mix thoroughly. Gently heat and bring to boiling. Distribute into tubes or flasks. Autoclave for 15 min at 15 psi pressure–121°C. Pour into sterile Petri dishes or leave in tubes.

Use: For use in identification of agrobacteria and other bacteria based upon utilization of 3-ketogluconate. After incubation Benedicts solution is added to the plates. Yellow zones around colonies indicate positive utilization of 3-ketogluconate.

KF *Streptococcus* Agar

Composition per liter:

Agar	20.0g
Maltose	20.0g
Proteose peptone	10.0g
Sodium glycerophosphate	10.0g
Yeast extract	10.0g
NaCl	5.5g
Lactose	1.0g
NaN_3	0.4g
Bromcresol Purple	0.015g
2,3,5-Triphenyltetrazolium chloride solution	10.0mL

pH 7.2 ± 0.2 at 25°C

Source: This medium is available as a premixed powder from Difco Laboratories and Oxoid Unipath.

2,3,5-Triphenyltetrazolium Chloride Solution:
Composition per 10mL:

2,3,5-Triphenyltetrazolium chloride	0.1g

Preparation of 2,3,5-Triphenyltetrazolium Chloride Solution: Add 2,3,5-triphenyltetrazolium chloride to distilled/deionized water and bring volume to 10.0mL. Mix thoroughly. Filter sterilize.

Caution: Sodium azide is toxic. Azides also react with metals and disposal must be highly diluted.

Preparation of Medium: Add components, except 2,3,5-triphenyltetrazolium chloride solution, to distilled/deionized water and bring volume to 990.0mL. Mix thoroughly. Gently heat and bring to boiling. Autoclave for 15 min at 15 psi pressure–121°C. Cool to 45–50°C. Aseptically add 2,3,5-triphenyltetrazolium chloride solution. Mix thoroughly. Pour into sterile Petri dishes or distribute into sterile tubes.

Use: For the isolation and enumeration of enterococci.

KF *Streptococcus* Agar

Composition per liter:

Agar	20.0g
Maltose	20.0g
Sodium glycerophosphate	10.0g
Yeast extract	10.0g
NaCl	5.0g
Pancreatic digest of casein	5.0g
Peptic digest of animal tissue	5.0g
Lactose	1.0g
NaN_3	0.4g
2,3,5-Triphenyltetrazolium chloride solution	10.0mL

pH 7.2 ± 0.2 at 25°C

Source: This medium is available as a premixed powder from BBL Microbiology Systems.

2,3,5-Triphenyltetrazolium Chloride Solution:
Composition per 10mL:

2,3,5-Triphenyltetrazolium chloride	0.1g

Preparation of 2,3,5-Triphenyltetrazolium Chloride Solution: Add 2,3,5-triphenyltetrazolium chloride to distilled/deionized water and bring volume to 10.0mL. Mix thoroughly. Filter sterilize.

Caution: Sodium azide is toxic. Azides also react with metals and disposal must be highly diluted.

Preparation of Medium: Add components, except 2,3,5-triphenyltetrazolium chloride solution, to distilled/deionized water and bring volume to 990.0mL. Mix thoroughly. Gently heat and bring to boiling. Autoclave for 15 min at 15 psi pressure–121°C. Cool to 45–50°C. Aseptically add 2,3,5-triphenyltetrazolium chloride solution. Mix thoroughly. Pour into sterile Petri dishes or distribute into sterile tubes.

Use: For the selective cultivation and enumeration of fecal streptococci.

KF *Streptococcus* Broth

Composition per liter:

Maltose	20.0g
Sodium glycerophosphate	10.0g
Yeast extract	10.0g
NaCl	5.0g
Pancreatic digest of casein	5.0g
Peptic digest of animal tissue	5.0g
Lactose	1.0g
Na_2CO_3	0.636g
NaN_3	0.4g
Phenol red	0.018g
2,3,5-Triphenyltetrazolium chloride solution	10.0mL

pH 7.2 ± 0.2 at 25°C

Source: This medium is available as a premixed powder from BBL Microbiology Systems and Difco Laboratories.

Caution: Sodium azide is toxic. Azides also react with metals and disposal must be highly diluted.

Preparation of Medium: Add components, except 2,3,5-triphenyltetrazolium chloride solution, to distilled/deionized water and bring volume to 990.0mL. Mix thoroughly. Gently heat and bring to boiling. Autoclave for 15 min at 15 psi pressure–121°C. Cool to 45–50°C. Aseptically add 2,3,5-triphenyltetrazolium chloride solution. Mix thoroughly. Aseptically distribute into sterile tubes or flasks.

Use: For the selective cultivation of fecal streptococci.

King's Medium A

Composition per liter:

Proteose peptone	20.0g
Agar	15.0g
Glycerol	10.0g
K_2SO_4	10.0g
$MgCl_2·6H_2O$	3.5g

pH 7.2–7.4 ± 0.2 at 25°C

Preparation of Medium: Add components to distilled/deionized water and bring volume to 1.0L. Mix thoroughly. Gently heat and bring to boiling. Distribute into tubes or flasks. Autoclave for 15 min at 15 psi pressure–121°C. Pour into sterile Petri dishes or leave in tubes.

Use: For the nonselective isolation, cultivation and pigment production of *Pseudomonas*.

King's Medium B

Composition per liter:

Agar	20.0g
Proteose peptone No. 3	20.0g
K_2HPO_4, anhydrous	1.5g
$MgSO_4·7H_2O$	1.5g
Glycerol	15.0mL

pH 7.2 ± 0.2 at 25°C

Preparation of Medium: Add components to distilled/deionized water and bring volume to 1.0L. Mix thoroughly. Gently heat and bring to boiling. Distribute into tubes or flasks. Autoclave for 15 min at 15 psi pressure–121°C. Pour into sterile Petri dishes or leave in tubes.

Use: For the nonselective isolation, cultivation and pigment production of *Pseudomonas* species.

King's Medium B

Composition per liter:

Proteose peptone No.3	20.0g
Agar	15.0g
K_2HPO_4	1.5g
$MgSO_4·7H_2O$	1.5g
Glycerol	10.0mL

pH 7.2 ± 0.2 at 25°C

Preparation of Medium: Add components to distilled/deionized water and bring volume to 1.0L. Mix thoroughly. Gently heat and bring to boiling. Distribute into tubes or flasks. Autoclave for 15 min at 15 psi pressure–121°C. Pour into sterile Petri dishes or leave in tubes.

Use: For the cultivation and maintenance of *Pseudomonas glumae*.

Kleb Agar (m-Kleb Agar)

Composition per liter:

Agar	15.0g
Proteose peptone No. 3	10.0g
NaCl	5.0g
Adonitol	5.0g
Beef extract	1.0g
Aniline blue	0.1g
Sodium lauryl sulfate	0.1g
Phenol red	0.025g
Ethanol (95% solution)	20.0mL
Carbenicillin solution	10.0mL

pH 7.4 ± 0.2 at 25°C

Carbenicillin Solution:

Composition per 10mL:

Carbenicillin	0.05g

Preparation of Carbenicillin Solution: Add carbenicillin to distilled/deionized water and bring volume to 10.0mL. Mix thoroughly. Filter sterilize.

Preparation of Medium: Add components, except ethanol and carbenicillin solution, to distilled/deionized water and bring volume to 970.0mL. Mix thoroughly. Gently heat and bring to boiling. Autoclave for 15 min at 15 psi pressure–121°C. Cool to 45–50°C. Aseptically add 20.0mL ethanol and 10.0mL carbenicillin solution. Mix thoroughly. Pour into sterile Petri dishes or distribute into sterile tubes.

Use: For the enumeration of bacteria from waters.

Klebsiella Medium (m-*Klebsiella* Medium)

Composition per 1041mL:

Agar	15.0g
Adonitol	4.0g
2× salt solution	500.0mL
Uric acid solution	200.0mL
Phenol Red solution	10.0mL
Sodium taurocholate solution	30.0mL
Carbenicillin solution	1.0mL

2X Salt Solution:

Composition per liter:

KCl	8.0g
K_2HPO_4	3.0g
NaCl	2.0g
KH_2PO_4	1.0g
$MgSO_4 \cdot 7H_2O$	0.2g

Preparation of 2X Salt Solution: Add components to distilled/deionized water and bring volume to 1.0L. Mix thoroughly.

Uric Acid Solution:

Composition per 200mL:

Uric acid	0.3g

Preparation of Uric Acid Solution: Dissolve uric acid in a small volume of 1*N* NaOH. Bring volume to 200.0mL with distilled/deionized water. Adjust pH to 7.1 with 1*N* HCl. Filter sterilize.

Phenol Red Solution:

Composition per 10mL:

Phenol Red	0.1g

Preparation of Phenol Red Solution: Add phenol red to sterile distilled/deionized water and bring volume to 10.0mL. Mix thoroughly.

Sodium Taurocholate Solution:

Composition per 30mL:

Sodium taurocholate	0.4g

Preparation of Sodium Taurocholate Solution: Add sodium taurocholate to sterile distilled/deionized water and bring volume to 30.0mL. Mix thoroughly.

Carbenicillin Solution:

Composition per 1mL:

Carbenicillin	5.0mg

Preparation of Carbenicillin Solution: Add carbenicillin to distilled/deionized water and bring volume to 1.0mL. Mix thoroughly. Filter sterilize.

Preparation of Medium: Add adonitol and agar to 500.0mL of 2× salt solution. Bring volume to 800.0mL with distilled/deionized water. Mix thoroughly. Gently heat and bring to boiling. Autoclave for 15 min at 15 psi pressure–121°C. Cool to 45–50°C. Aseptically add 200.0mL of uric acid solution, 30.0mL of sodium taurocholate solution, 10.0mL of phenol red solution, and 1.0mL of carbenicillin solution. Mix thoroughly. Pour into sterile Petri dishes or distribute into sterile tubes.

Use: For the enumeration of *Klebsiella* species by the membrane filter method.

Klebs-Loeffler Virulence Agar

See: K-L Virulence Agar

Kligler Iron Agar

Composition per liter:

Peptone	20.0g
Agar	12.0g
Lactose	10.0g
NaCl	5.0g
Beef extract	3.0g

Yeast extract	3.0g
Glucose	1.0g
Ferric citrate	0.3g
$Na_2S_2O_3$	0.3g
Phenol red	0.05g

pH 7.4 ± 0.2 at 25°C

Source: This medium is available as a premixed powder from Difco Laboratories and Oxoid Unipath.

Preparation of Medium: Add components to distilled/deionized water and bring volume to 1.0L. Mix thoroughly. Gently heat and bring to boiling. Distribute into tubes. Autoclave for 15 min at 15 psi pressure–121°C. Pour into sterile Petri dishes or leave in tubes.

Use: For the differentiation and identification of Enterobacteriaceae based upon sugar fermentation and hydrogen sulfide production. Sugar fermentation is indicated by the medium turning yellow. H_2S production results in the medium turning black.

Kligler Iron Agar

Composition per liter:

Agar	15.0g
Lactose	10.0g
Pancreatic digest of casein	10.0g
Peptic digest of animal tissue	10.0g
NaCl	5.0g
Glucose	1.0g
Ferric ammonium citrate	0.5g
$Na_2S_2O_3$	0.5g
Phenol red	0.025g

pH 7.4 ± 0.2 at 25°C

Source: This medium is available as a premixed powder from BBL Microbiology Systems.

Preparation of Medium: Add components to distilled/deionized water and bring volume to 1.0L. Mix thoroughly. Gently heat and bring to boiling. Distribute into tubes or flasks. Autoclave for 15 min at 15 psi pressure–121°C. Pour into sterile Petri dishes or leave in tubes.

Use: For the differentiation and identification of Enterobacteriaceae based upon sugar fermentation and hydrogen sulfide production. Sugar fermentation is indicated by the medium turning yellow. H_2S production results in the medium turning black.

Kligler Iron Agar (FDA M71)

Composition per liter:

Agar	15.0g
Lactose	20.0g
Pancreatic digest of casein	10.0g
Peptic digest of animal tissue	10.0g
NaCl	5.0g
Glucose	1.0g
Ferric ammonium citrate	0.5g
$Na_2S_2O_3$	0.5g
Phenol red	0.025g

pH 7.4 ± 0.2 at 25°C

Preparation of Medium: Add components to distilled/deionized water and bring volume to 1.0L. Mix thoroughly. Gently heat and bring to boiling. Distribute into tubes or flasks. Autoclave for 15 min at 15 psi pressure–121°C. Pour into sterile Petri dishes or leave in tubes.

Use: For the differentiation and identification of Enterobacteriaceae based upon sugar fermentation and hydrogen sulfide production. Sugar fermentation is indicated by the medium turning yellow. H_2S production results in the medium turning black.

Knisely Medium for *Bacillus anthracis*

Composition per liter:

Beef heart, solids from infusion	500.0g
Agar	15.0g
Pancreatic digest of casein	10.0g
NaCl	5.0g
EDTA	200.0mg
Lysozyme	40.0mg
Thallous acetate	40.0mg
Polymyxin	30,000U

Preparation of Medium: Add components, except EDTA, lysozyme, thallous acetate, and polymyxin, to distilled/deionized water and bring volume to 1.0mL. Mix thoroughly. Gently heat and bring to boiling. Adjust pH to 7.3. Autoclave for 15 min at 15 psi pressure–121°C. Cool to 45–50°C. Aseptically add sterile EDTA, lysozyme, thallous acetate, and polymyxin. Mix thoroughly. Pour into sterile Petri dishes or distribute into sterile tubes.

Use: For the cultivation and maintenance of *Bacillus anthracis*.

Koch's K1 Medium

Composition per liter:

Glucose	1.8g
Peptone	0.6g
Yeast extract	0.4g

Preparation of Medium: Add components to distilled/deionized water and bring volume to 1.0L. Mix thoroughly. Distribute into tubes or flasks. Autoclave for 15 min at 15 psi pressure–121°C.

Use: For the cultivation of a variety of fungi.

Koser Citrate Medium

Composition per liter:

Sodium citrate	3.0g
$NaNH_4HPO_4 \cdot 4H_2O$	1.5g
KH_2PO_4	1.0g
$MgSO_4 \cdot 7H_2O$	0.2g

pH 6.7 ± 0.2 at 25°C

Source: This medium is available as a premixed powder from Difco Laboratories.

Preparation of Medium: Add components to distilled/deionized water and bring volume to 1.0L. Mix thoroughly. Gently heat and bring to boiling. Distribute into tubes or flasks. Autoclave for 15 min at 15 psi pressure–121°C. Pour into sterile Petri dishes or leave in tubes.

Use: For the differentiation of *Escherichia coli* and *Enterobacter aerogenes* based oncitrate utilization.

Kosmachev's Medium

Composition per liter:

Agar	15.0g
$CaCO_3$	4.0g
KNO_3	1.0g
$(NH_4)_2SO_4$	1.0g
Na_2HPO_4	1.0g
$MgSO_4 \cdot 7H_2O$	0.5g
$FeSO_4 \cdot 7H_2O$	0.01g
Yeast autolysate (30% solution)	15.0mL

Preparation of Medium: Add components to distilled/deionized water and bring volume to 1.0L. Mix thoroughly. Gently heat and bring to boiling. Distribute into tubes or flasks. Autoclave for 15 min at 15 psi pressure–121°C. Pour into sterile Petri dishes or leave in tubes.

Use: For the isolation and cultivation of *Actinomadura* species, *Actinopolyspora* species, *Excellospora* species and *Microspora* species.

L and F Basal Salts, Modified with Heptadecane

Composition per liter:

NH_4Cl	2.0g
Na_2HPO_4	0.21g
$MgSO_4 \cdot 7H_2O$	0.2g
NaH_2PO_4	0.09g
KCl	0.04g
$CaCl_2$	0.015g
$FeSO_4 \cdot 7H_2O$	1.0mg
$ZnSO_4 \cdot 7H_2O$	0.070mg
H_3BO_3	0.010mg
$MnSO_4 \cdot 5H_2O$	0.010mg
MoO_3	0.010mg
$CuSO_4 \cdot 5H_2O$	5.0μg
Heptadecane	2.0mL

pH 7.2 ± 0.2 at 25°C

Preparation of Medium: Add components, except heptadecane, to distilled/deionized water and bring volume to 1.0L. Mix thoroughly. Gently heat and bring to boiling. Distribute equally into four 250.0mL volumes. Autoclave for 15 min at 15 psi pressure–121°C. Cool to 60°C. Filter sterilize heptadecane. To one 250.0mL fraction of basal salts, aseptically add 0.5mL of sterile heptadecane. Pour mixture into a sterile blender. Homogenize slowly to mix heptadecane with basal salts and not to create excess bubbles. Rapidly distribute medium to sterile screw-capped tubes. Chill tubes quickly in an ice pack or in the refrigerator. Allow tubes to solidify in a slanted position.

Use: For the cultivation and maintenance of *Thermoleophilum album* and *T. minutum*.

L Medium (ATCC Medium 167)

Composition per liter:

Agar	20.0g
$NaNO_3$	2.0g
Na_2HPO_4	0.21g
$MgSO_4 \cdot 7H_2O$	0.2g
NaH_2PO_4	0.09g
KCl	0.04g
$CaCl_2$	0.015g
$FeSO_4 \cdot 7H_2O$	1.0mg
Salts solution	1.0mL

Salts Solution:

Composition per 100mL:

$ZnSO_4 \cdot 7H_2O$	7.0mg
H_3BO_3	1.0mg
$MnSO_4 \cdot 5H_2O$	1.0mg
MoO_3	1.0mg
$CuSO_4 \cdot 5H_2O$	0.5mg

Preparation of Salts Solution: Add components to distilled/deionized water and bring volume to 100.0mL. Mix thoroughly.

Preparation of Medium: Add components to distilled/deionized water and bring volume to 1.0L. Mix thoroughly. Gently heat and bring to boiling. Distribute into tubes or flasks. Autoclave for 15 min at 15 psi pressure–121°C. Pour into sterile Petri dishes or leave in tubes.

Use: For the cultivation and maintenance of *Methylococcus capsulatus* and *Pseudomonas methanica*.

L Medium (ATCC Medium 1154)

Composition per liter:

Pancreatic digest of casein	10.0g
NaCl	5.0g
Yeast extract	5.0g

pH 7.0 ± 0.2 at 25°C

Preparation of Medium: Add components to distilled/deionized water and bring volume to 1.0L. Mix thoroughly. Adjust pH to 7.0. Distribute into tubes or flasks. Autoclave for 25 min at 15 psi pressure–121°C.

Use: For the cultivation and maintenance of *Escherichia coli.*

L Medium with Methanol

Composition per liter:

Agar	20.0g
$NaNO_3$	2.0g
Na_2HPO_4	0.21g
$MgSO_4 \cdot 7H_2O$	0.2g
NaH_2PO_4	0.09g
KCl	0.04g
$CaCl_2$	0.015g
$FeSO_4 \cdot 7H_2O$	1.0mg
Methanol	20.0mL
Salts solution	1.0mL

Salts Solution:

Composition per 100mL:

$ZnSO_4 \cdot 7H_2O$	7.0mg
H_3BO_3	1.0mg
$MnSO_4 \cdot 5H_2O$	1.0mg
MoO_3	1.0mg
$CuSO_4 \cdot 5H_2O$	0.5mg

Preparation of Salts Solution: Add components to distilled/deionized water and bring volume to 100.0mL. Mix thoroughly.

Preparation of Medium: Add components, except methanol, to distilled/deionized water and bring volume to 980.0mL. Mix thoroughly. Gently heat and bring to boiling. Autoclave for 15 min at 15 psi pressure–121°C. Cool to 45–50°C. Filter sterilize methanol. Aseptically add 20.0mL of sterile methanol to cooled sterile basal medium. Mix thoroughly. Pour into sterile Petri dishes or distribute into sterile tubes.

Use: For the cultivation and maintenance of *Methylobacillus glycogenes.*

L15 Medium, Modified Leibovitz

Composition per liter:

NaCl	8.0g
DL-Threonine	0.6g
Sodium pyruvate	0.6g
DL-Alanine	0.5g
L-Arginine, free base	0.5g
KCl	0.4g
L-Asparagine·H_2O	0.3g
L-Histidine, free base	0.3g
L-Glutamine	0.3g
L-Isoleucine	0.3g
L-Phenylalanine	0.3g
L-Tyrosine	0.3g
DL-Methionine	0.2g
DL-Valine	0.2g
Glycine	0.2g
L-Serine	0.2g
Na_2HPO_4, anhydrous	0.2g
$CaCl_2$, anhydrous	0.1g
L-Cysteine, free base	0.1g
L-Leucine·HCl	0.1g
$MgCl_2$, anhydrous	0.094g
D-Galactose	0.090g
KH_2PO_4	0.060g
L-Tryptophan	0.020g
Phenol red	0.010g
i-Inositol	2.0mg
Choline chloride	1.0mg
D-Calcium pantothenate	1.0mg
Folic acid	1.0mg
Nicotinamide	1.0mg
Pyridoxine·HCl	1.0mg
Thiamine monophosphate·$2H_2O$	1.0mg
Riboflavin-5-phosphate	0.1mg

pH 7.5 ± 0.2 at 25°C

Preparation of Medium: Add components to distilled/deionized water and bring volume to 1.0L. Mix thoroughly. Filter sterilize. Store at 5°C.

Use: For the cultivation of oysters used for the growth of enteroviruses.

Lab-Lemco Agar

Composition per liter:

Agar	15.0g
Peptone	5.0g
Lab-lemco meat extract	3.0g

pH 7.4 ± 0.2 at 25°C

Preparation of Medium: Add components to distilled/deionized water and bring volume to 1.0L. Mix thoroughly. Gently heat and bring to boiling. Distrib-

ute into tubes or flasks. Autoclave for 15 min at 15 psi pressure–121°C. Pour into sterile Petri dishes or leave in tubes.

Use: For the cultivation and maintenance of a variety of heterotrophic microorganisms.

Lab-Lemco Broth

Composition per liter:

Peptone	5.0g
Lab-lemco meat extract	3.0g

pH 7.4 ± 0.2 at 25°C

Preparation of Medium: Add components to distilled/deionized water and bring volume to 1.0L. Mix thoroughly. Distribute into tubes or flasks. Autoclave for 15 min at 15 psi pressure–121°C.

Use: For the cultivation of a variety of heterotrophic microorganisms, including microorganisms from wastewater.

Lactate Agar

Composition per liter:

Yeast extract	3.0g
K_2HPO_4	2.80g
Agar	2.0g
Peptone	2.0g
KH_2PO_4	0.52g
Sodium lactate (60% solution)	10.0mL

pH 7.2 ± 0.2 at 25°C

Preparation of Medium: Add components to distilled/deionized water and bring volume to 1.0L. Mix thoroughly. Gently heat and bring to boiling. Adjust pH to 7.2. Distribute into tubes or flasks. Autoclave for 15 min at 15 psi pressure–121°C. Pour into sterile Petri dishes or leave in tubes.

Use: For the cultivation and maintenance of *Serpens flexibilis*.

Lactate Broth

Composition per liter:

Yeast extract	3.0g
K_2HPO_4	2.80g
Peptone	2.0g
KH_2PO_4	0.52g
Sodium lactate (60% solution)	10.0mL

pH 7.2 ± 0.2 at 25°C

Preparation of Medium: Add components to distilled/deionized water and bring volume to 1.0L. Mix thoroughly. Gently heat and bring to boiling. Adjust pH to 7.2. Distribute into tubes or flasks. Autoclave for 15 min at 15 psi pressure–121°C.

Use: For the cultivation and maintenance of *Serpens flexibilis*.

Lactose Broth

Composition per liter:

Lactose	5.0g
Pancreatic digest of gelatin	5.0g
Beef extract	3.0g

pH 6.9 ± 0.2 at 25°C

Source: This medium is available as a premixed powder from BBL Microbiology Systems, Difco Laboratories, and Oxoid Unipath.

Preparation of Medium: Add components to distilled/deionized water and bring volume to 1.0L. Mix thoroughly. Distribute into tubes containing an inverted Durham tube in 10.0mL volumes. Autoclave for 12 min at 15 psi pressure–121°C. Cool broth quickly to 25°C. For testing water samples with 10.0mL volumes, prepare medium double strength.

Use: For detection of lactose-fermenting, Gram-negative coliforms, as a pre-enrichment broth for *Salmonella* species and in the study of lactose fermentation of bacteria in general.

Lauryl Sulfate Broth (m-Lauryl Sulfate Broth)

Composition per liter:

Peptone	39.0g
Lactose	30.0g
Yeast extract	6.0g
Sodium lauryl sulfate	1.0g
Phenol red	0.2g

pH 7.4 ± 0.2 at 25°C

Source: This medium is available as a premixed powder from Oxoid Unipath.

Preparation of Medium: Add components to distilled/deionized water and bring volume to 1.0L. Mix thoroughly. Distribute into bottles or flasks. Autoclave for 15 min at 15 psi pressure–121°C.

Use: For the cultivation and enumeration of coliform bacteria, especially *Escherichia coli*, in water by the membrane filter method.

Lauryl Sulfate Broth (Lauryl Tryptose Broth)

Composition per liter:

Pancreatic digest of casein	20.0g
Lactose	5.0g
NaCl	5.0g
K_2HPO_4	2.75g

KH_2PO_4	2.75g
Sodium lauryl sulfate	0.1g

pH 6.8 ± 0.2 at 25°C

Source: This medium is available as a premixed powder from BBL Microbiology Systems and Difco Laboratories.

Preparation of Medium: Add components to distilled/deionized water and bring volume to 1.0L. Mix thoroughly. Distribute into tubes containing an inverted Durham tube in 10.0mL volumes. Autoclave for 12 min at 15 psi pressure–121°C. Cool broth quickly to 25°C. For testing water samples with 10mL volumes, prepare medium double strength.

Use: For the detection of coliform bacteria in a variety of specimens. Also, for the enumeration of coliform bacteria by the multiple-tube fermentation technique.

Lauryl Sulfate Broth with MUG

Composition per liter:

Pancreatic digest of casein	20.0g
Lactose	5.0g
NaCl	5.0g
K_2HPO_4	2.75g
KH_2PO_4	2.75g
Sodium lauryl sulfate	0.1g
4-Methylumbellferyl-β-D-glucuronide (MUG)	0.05g

pH 6.8 ± 0.2 at 25°C

Source: This medium is available as a premixed powder from BBL Microbiology Systems.

Preparation of Medium: Add components to distilled/deionized water and bring volume to 1.0L. Mix thoroughly. Distribute into tubes containing an inverted Durham tube in 10.0mL volumes. Autoclave for 12 min at 15 psi pressure–121°C. Cool broth quickly to 25°C. For testing water samples with 10.0mL volumes, prepare medium double strength.

Use: For the detection of *Escherichia coli* in water and food samples by a fluorogenic procedure.

Lauryl Tryptose Mannitol Broth with Tryptophan

Composition per liter:

Pancreatic digest of casein	20.0g
Lactose	5.0g
NaCl	5.0g
K_2HPO_4	2.75g
KH_2PO_4	2.75g
Sodium lauryl sulfate	0.1g
L-Tryptophan	0.2g

pH 6.8 ± 0.2 at 25°C

Source: This medium is available as a premixed powder from Oxoid Unipath.

Preparation of Medium: Add components to distilled/deionized water and bring volume to 1.0L. Mix thoroughly. Distribute into tubes containing an inverted Durham tube in 10.0mL volumes. Autoclave for 10 min at 10 psi pressure–115°C. Cool broth quickly to 25°C.

Use: For the detection of *Escherichia coli* in water samples.

LAVMm2 Medium

Composition per liter:

Lactalbumin hydrolysate	10.0g
Sodium acetate	5.0g
$MgCl_2 \cdot 6H_2O$	20.3mg
Nitrilotriacetic acid	19.1mg
$CaCl_2$	11.1mg
$FeSO_4$	0.152mg
Thiamine·HCl	0.05mg
Cupric acetate	0.04mg
Biotin	0.02mg

pH 8.0–8.1 at 25°C

Preparation of Medium: Add components to distilled/deionized water and bring volume to 1.0L. Mix thoroughly. Adjust pH to 7.5 with Na_2CO_3. Distribute into tubes or flasks. Autoclave for 15 min at 15 psi pressure–121°C. The pH should be 8.0–8.1 after autoclaving.

Use: For the cultivation of *Caryophanon latum*.

Lead Acetate Agar

Composition per liter:

Agar	15.0g
Peptone	15.0g
Proteose peptone	5.0g
Glucose	1.0g
Lead acetate	0.2g
$Na_2S_2O_3$	0.08g

pH 6.6 ± 0.2 at 25°C

Preparation of Medium: Add components to distilled/deionized water and bring volume to 1.0L. Mix thoroughly. Gently heat and bring to boiling. Distribute into tubes or flasks. Autoclave for 15 min at 15 psi pressure–121°C. Pour into sterile Petri dishes or leave in tubes. Allow tubes to cool in a slanted position.

Use: For the cultivation and differentiation of Gram-negative coliform bacteria based on H_2S production. Bacteria that produce H_2S turn the medium brown.

Lecithin Lipase Anaerobic Agar

Composition per liter:

Pancreatic digest of casein	40.0g
Agar	25.0g
Yeast extract	5.0g
Glucose	2.0g
NaCl	2.0g
KH_2PO_4	1.0g
$Na_2HPO_4 \cdot 12H_2O$	5.0g
$MgSO_4 \cdot 7H_2O$	0.1g
Egg yolk emulsion	100.0mL

pH 7.6 ± 0.2 at 25°C

Egg Yolk Emulsion:
Composition:

Chicken egg yolks	11
Whole chicken egg	1

Preparation of Egg Yolk Emulsion: Soak eggs with 1:100 dilution of saturated mercuric chloride solution for 1 min. Crack eggs and separate yolks from whites. Mix egg yolks with 1 chicken egg. Filter sterilize.

Preparation of Medium: Add components, except egg yolk emulsion, to distilled/deionized water and bring volume to 900.0mL. Mix thoroughly. Gently heat and bring to boiling. Autoclave for 15 min at 15 psi pressure–121°C. Cool to 45–50°C. Aseptically add sterile egg yolk emulsion. Mix thoroughly. Pour into sterile Petri dishes or distribute into sterile tubes.

Use: For the isolation, cultivation and differentiation of *Clostridium* species based on lecithinase production and lipase production. Bacteria that produce lecithinase appear as colonies surrounded by a zone of insoluble precipitate. Bacteria that produce lipase appear as colonies with a pearly iridescent sheen.

Legionella Agar Base (*Legionella* Medium) (BCYEα Agar, Modified)

Composition per liter:

Agar	17.0g
Yeast extract	10.0g
ACES buffer (*N*-2-acetamido-2-aminoethane sulfonic acid)	6.0g
Charcoal, activated	1.5g
KOH	1.5g
α-Ketoglutarate	1.0g

pH 6.85–7.0 at 25°C

Source: This medium is available as a prepared medium from Difco Laboratories.

***Legionella* Agar Enrichment:**
Composition per 10mL:

L-Cysteine·HCl·H_2O	0.4g
$Fe_4(P_2O_7)_3$	0.25g

Preparation of *Legionella* Agar Enrichment: Add components to distilled/deionized water and bring volume to 10.0mL. Mix thoroughly. Filter sterilize.

Preparation of Medium: Add components, except *Legionella* agar enrichment, to distilled/deionized water and bring volume to 990.0mL. Mix thoroughly. Gently heat to boiling. Autoclave for 15 min at 15 psi pressure–121°C. Cool to 50° C. Add 10.0mL of sterile *Legionella* agar enrichment. Adjust pH to 6.9 at 50°C by adding 4.0–4.5mL of 1.0*N* KOH—this is a critical step. Mix thoroughly. Pour into sterile Petri dishes in 20.0mL volumes. Swirl medium while pouring to keep charcoal in suspension.

Use: For the preparation of *Legionella* agars. Also, for the isolation and cultivation of *Legionella* species from clinical and nonclinical materials.

Legionella Selective Agar

Composition per liter:

Agar	15.0g
ACES (2-[(2-amino-2-oxoethyl)-amino]ethane sulfonic acid) buffer	10.0g
Yeast extract	10.0g
Charcoal, activated	2.0g
α-Ketoglutarate	1.0g
L-Cysteine·HCl·H_2O solution	10.0mL
$Fe_4(P_2O_7)_3$ solution	10.0mL
Antibiotic solution	10.0mL

pH 6.85–7.0 at 25°C

Source: This medium is available as a prepared medium from BBL Microbiology Systems.

Cysteine·HCl·H_2O Solution:
Composition per 10mL:

L-Cysteine·HCl·H_2O	0.4g

Preparation of Cysteine·HCl·H_2O Solution: Add cysteine·HCl·H_2O to distilled/deionized water and bring volume to 10.0mL. Mix thoroughly. Filter sterilize.

$Fe_4(P_2O_7)_3$ Solution:
Composition per 10mL:

$Fe_4(P_2O_7)_3$	0.25g

Preparation of $Fe_4(P_2O_7)_3$ Solution: Add $Fe_4(P_2O_7)_3$ to distilled/deionized water and bring volume to 10.0mL. Mix thoroughly. Filter sterilize.

Antibiotic Solution:

Composition per 10mL:

Anisomycin	10.0mg
Colistin	3.75mg
Vancomycin	2.0mg

Preparation of Antibiotic Solution: Add components to distilled/deionized water and bring volume to 10.0mL. Mix thoroughly. Filter sterilize.

Preparation of Medium: Add components—except cysteine·HCl·H_2O, $Fe_4(P_2O_7)_3$, and antibiotic solutions—to distilled/deionized water and bring volume to 970.0mL. Mix thoroughly. Gently heat and bring to boiling. Autoclave for 15 min at 15 psi pressure–121°C. Cool to 45–50°C. Aseptically add sterile cysteine·HCl·H_2O, $Fe_4(P_2O_7)_3$, and antibiotic solutions. Mix thoroughly. Pour into sterile Petri dishes. Swirl medium while pouring to keep charcoal in suspension.

Use: *Legionella* Selective Agar is used in qualitative procedures for the isolation of *Legionella* species.

Legume Extract Agar

Composition per liter:

Alfalfa roots	35.0g
Agar	20.0g
Soybean meal	10.0g
Sucrose	10.0g
$CaCO_3$	5.0g
Glucose	5.0g
K_2HPO_4	1.0g
$MgSO_4·7H_2O$	0.2g
$CaCl_2$	0.1g
NaCl	0.1g
$FeCl_3$	1.0mg

Preparation of Medium: Wash the alfalfa roots well and cut them up. Add 10.0g of soybean meal. Add three times the volume of distilled/deionized water. Steam for 1 hr. Let stand at 25°C overnight. Bring volume to 1.0L with distilled/deionized water. Filter through paper pulp. To the filtrate, add the K_2HPO_4, $CaCl_2$, $MgSO_4·7H_2O$, NaCl, $FeCl_3$, and agar. Autoclave for 20 min at 15 psi pressure–121°C. Cool to 45–50°C. Add the $CaCO_3$, sucrose and glucose. Mix thoroughly. Distribute into tubes or flasks. Autoclave for 20 min at 10 psi pressure–115°C.

Use: For the cultivation of *Rhizobium* species.

Leifson Medium

Composition per liter:

Agar	15.0g
Pancreatic digest of casein	2.0g
$MgCl_2$	1.0g
Yeast extract	1.0g

pH 8.0 ± 0.2 at 25°C

Preparation of Medium: Add components to distilled/deionized water and bring volume to 1.0L. Mix thoroughly. Gently heat and bring to boiling. Adjust pH to 8.0. Distribute into tubes or flasks. Autoclave for 15 min at 15 psi pressure–121°C. Pour into sterile Petri dishes or leave in tubes.

Use: For the direct isolation and routine culturing of *Hyphomonas* species.

Leptospira Medium

Composition per liter:

$(NH_4)_2Fe(SO_4)_2·6H_2O$	6.0g
NaH_2PO_4	0.53g
L-Asparagine	0.5g
Glycerol	0.2g
Tween™ 60	0.2g
$MgSO_4·7H_2O$	0.15g
KH_2PO_4	0.069g
Tween™ 80	0.05g
EDTA	0.01g
$CaCO_3$	4.0mg
Thiamine·HCl	1.0mg
Vitamin B_{12}	1.0μg

pH 7.4–7.6 at 25°C

Preparation of Medium: Add components, except thiamine·HCl to distilled/deionized water and bring volume to 990.0mL. Mix thoroughly. Gently heat and bring to boiling. Autoclave for 15 min at 15 psi pressure–121°C. Aseptically add 1.0mg of thiamine·HCl. Aseptically distribute into sterile tubes or flasks.

Use: For the cultivation of *Leptospira* species.

Leptospira Medium, EMJH (*Leptospira* Medium, Ellinghausen–McCullough/ Johnson–Harris)

Composition per liter:

Na_2HPO_4	1.0g
NaCl	1.0g
KH_2PO_4	0.3g
NH_4Cl	0.25g
Thiamine	5.0mg
Rabbit serum	100.0mL

pH 7.5 ± 0.2 at 25°C

Source: This medium is available as a premixed powder from Difco Laboratories.

Preparation of Medium: Add components, except rabbit serum, to distilled/deionized water and bring volume to 900.0mL. Mix thoroughly. Gently heat and bring to boiling. Autoclave for 15 min at 15 psi pressure–121°C. Cool to 25°C. Aseptically add sterile rabbit serum. Mix thoroughly. Aseptically distribute into sterile tubes or flasks.

Use: For the cultivation and maintenance of *Leptospira* species.

Leptospira Medium, Modified

Composition per liter:

Agar....1.5g
NaCl....0.5g
Peptone....0.3g
Beef extract....0.2g
Hemin solution....2.5mL
Sterile rabbit serum....100.0mL

pH 7.3 ± 0.1 at 25°C

Hemin Solution:

Composition per 100mL:

Hemin....0.05g
NaOH (1*N* solution)....1.0mL

Preparation of Hemin Solution: Add hemin to 1.0mL of 1*N* NaOH solution. Mix thoroughly. Bring volume to 100.0mL with distilled/deionized water. Autoclave for 15 min at 15 psi pressure–121°C. Cool to 45–50°C.

Preparation of Medium: Add components, except hemin solution and rabbit serum, to distilled/deionized water and bring volume to 897.5mL. Mix thoroughly. Gently heat and bring to boiling. Adjust pH to 7.4. Autoclave for 15 min at 15 psi pressure–121°C. Cool to 45–50°C. Aseptically add 2.5mL of sterile hemin solution and 100.0mL of sterile rabbit serum. Mix thoroughly. The pH of the medium should be 7.3. Store at 4°C for 24 hr. Inactivate medium at 56°C for 60 min. Aseptically distribute into sterile tubes or flasks.

Use: For the cultivation and maintenance of *Leptospira biflexa, L. borgpetersenii, L. interrogans, L. meyeri, L. noguchii, L. santarosai,* and *L. weili.*

Leptospira Protein–Free Medium (*Leptospira* PF Medium)

Composition per liter:

TES (*N*-tris[hydroxymethyl]methyl-2-aminoethane sulfonic acid) buffer....1.2g
NaCl....0.9g
Sodium pyruvate....0.2g
CT-Tween™ 60....12.0mL
CT-Tween™ 40....3.0mL
$MgCl_2$-$CaCl_2$ solution....1.0mL
Cyanocobalamin (0.02% solution)....1.0mL
Glycerol (10% solution)....1.0mL
KH_2PO_4 (1% solution)....1.0mL
$MnSO_4 \cdot H_2O$ (0.1% solution)....1.0mL
$ZnSO_4$ (0.4% solution)....0.1mL

pH 7.6 ± 0.2 at 25°C

CT-Tween™ 60:

Composition per 200mL:

Charcoal, Norit A....40.0g
Tween™ 60....20.0g

Preparation of CT-Tween™ 60: Add Tween™ 60 to 200.0mL of distilled/deionized water. Mix thoroughly. While stirring, add charcoal. Stir mixture for 18 hr at 25°C. Allow charcoal to settle out of suspension for 18 hr at 4°C. Carefully decant the Tween™ solution off the sediment. Centrifuge the Tween™ solution at 10,000 × g for 1 hr. Decant supernatant solution. Pass Tween™ solution through a thin channel ultrafiltration XM 100 membrane. Store stock solution at –20°C.

CT-Tween™ 40:

Composition per 200mL:

Charcoal, Norit A....40.0g
Tween™ 40....20.0g

Preparation of CT-Tween™ 40: Add Tween™ 40 to 200.0mL of distilled/deionized water. Mix thoroughly. While stirring, add charcoal. Stir mixture for 18 hr at 25°C. Allow charcoal to settle out of suspension for 18 hr at 4°C. Carefully decant the Tween™ solution off the sediment. Centrifuge the Tween™ solution at 10,000 × g for 1 hr. Decant supernatant solution. Pass Tween™ solution through a thin channel ultrafiltration XM 100 membrane. Store stock solution at –20°C.

$MgCl_2$–$CaCl_2$ Solution:

Composition per 100mL:

$CaCl_2 \cdot 2H_2O$....1.5g
$MgCl_2 \cdot 6H_2O$....1.5g

Preparation of $MgCl_2$–$CaCl_2$ Solution: Add components to distilled/deionized water and bring volume to 100.0mL. Mix thoroughly.

Preparation of Medium: Add components to distilled/deionized water and bring volume to 1.0L. Mix thoroughly. Filter sterilize. Aseptically distribute into sterile tubes or flasks.

Use: For the cultivation of *Leptospira* species.

Leptothrix 2X PYG Medium

Composition per liter:

HEPES (*N*-2-hydroxyethyl piperazine-*N*′-2-ethanesulfonic acid) buffer....3.57g
$MgSO_4 \cdot 7H_2O$....0.6g
Glucose....0.5g

Peptone	0.5g
Yeast extract	0.5g
$CaCl_2 \cdot 2H_2O$	0.07g
$MnSO_4 \cdot H_2O$	0.017g

pH 7.3 ± 0.2 at 25°C

Preparation of Medium: Add components to distilled/deionized water and bring volume to 1.0L. Mix thoroughly. Adjust pH to 7.3. Distribute into tubes or flasks. Autoclave for 15 min at 15 psi pressure–121°C.

Use: For the cultivation and maintenance of *Leptothrix discophora.*

Leptothrix ochracea Medium

Composition per liter:

Agar	10.0g
Manganous acetate	0.1g
Manganese bicarbonate solution	100.0mL

Manganese Bicarbonate Solution:

Composition per 100mL:

$MnCO_3$	2.0g

Preparation of Manganese Bicarbonate Solution: Add $MnCO_3$ to distilled/deionized water and bring volume to 100.0mL. Mix thoroughly. Gas with 100% CO_2 for 20 min. Filter through Whatman #1 filter paper.

Preparation of Medium: Add components to distilled/deionized water and bring volume to 1.0L. Mix thoroughly. Gently heat and bring to boiling. Distribute into tubes or flasks. Autoclave for 15 min at 15 psi pressure–121°C. Pour into sterile Petri dishes or leave in tubes.

Use: For the cultivation of *Leptothrix ochracea.*

Leptothrix Strains Medium

Composition per liter:

Agar	7.5g
$MnCO_2$	2.0g
Beef extract	1.0g
$Fe(NH_4)_2(SO_4)_2$	0.15g
Sodium citrate	0.15g
Yeast extract	0.075g
Vitamin B_{12}	5.0μg

Preparation of Medium: Add components to distilled/deionized water and bring volume to 1.0L. Mix thoroughly. Distribute into tubes or flasks. Autoclave for 15 min at 15 psi pressure–121°C.

Use: For the isolation of *Leptothrix* species.

Leptothrix Strains Medium

Composition per liter:

Agar	12.0g
Peptone	5.0g
$MgSO_4 \cdot 7H_2O$	0.20g
Ferric ammonium citrate	0.15g
$CaCl_2$	0.05g
$FeCl_3 \cdot 6H_2O$	0.01g
$MnSO_4 \cdot H_2O$	0.01g

Preparation of Medium: Add components to distilled/deionized water and bring volume to 1.0L. Mix thoroughly. Gently heat and bring to boiling. Distribute into tubes or flasks. Autoclave for 15 min at 15 psi pressure–121°C. Pour into sterile Petri dishes or leave in tubes.

Use: For the isolation and cultivation of *Leptothrix* species.

Leptotrichia Medium

Composition per liter:

Pancreatic digest of casein	10.0g
NaCl	5.0g
Peptone	5.0g
Yeast extract	3.0g
Na_2HPO_4	2.5g
L-Cysteine·HCl·H_2O	0.5g
Horse serum	100.0mL
Tomato decoction	50.0mL

pH 7.2–7.4 at 25°C

Tomato Decoction:

Composition per 100mL:

Tomatoes	50.0mL

Preparation of Tomato Decoction: Mince fresh tomatoes and measure 50.0mL. Add 50.0mL of tap water. Mix thoroughly. Gently heat and bring to boiling. Continue boiling for 10 min. Filter through Whatman #1 filter paper. Autoclave filtrate for 15 min at 15 psi pressure–121°C.

Preparation of Medium: Add components, except horse serum and tomato decoction, to distilled/deionized water and bring volume to 850.0mL. Mix thoroughly. Gently heat and bring to boiling. Adjust pH to 7.2–7.4. Autoclave for 15 min at 15 psi pressure–121°C. Cool to 25°C. Aseptically add sterile horse serum and tomato decoction. Mix thoroughly. Aseptically distribute into sterile tubes or flasks.

Use: For the cultivation and maintenance of *Leptotrichia buccalis.*

Leucothrix Medium (ATCC Medium 429)

Composition per liter:

NaCl	11.7g
Monosodium glutamate	10.0g
$MgCl_2 \cdot 6H_2O$	5.35g
Na_2SO_4	2.0g
$CaCl_2 \cdot 2H_2O$	0.75g
Tris(hydroxymethyl)aminomethane buffer	0.5g
KCl	0.35g
Na_2HPO_4	0.05g

pH 7.6 ± 0.2 at 25°C

Preparation of Medium: Add components to distilled/deionized water and bring volume to 1.0L. Mix thoroughly. Distribute into tubes or flasks. Autoclave for 15 min at 15 psi pressure–121°C.

Use: For the cultivation and maintenance of *Leucothrix mucor*.

Leucothrix Medium (ATCC Medium 430)

Composition per liter:

NaCl	11.7g
$MgCl_2 \cdot 6H_2O$	5.35g
Na_2SO_4	2.0g
$CaCl_2 \cdot 2H_2O$	0.75g
Pancreatic digest of casein	0.5g
Yeast extract	0.5g
Tris(hydroxymethyl)aminomethane buffer	0.5g
KCl	0.35g
Na_2HPO_4	0.05g

pH 7.6 ± 0.2 at 25°C

Preparation of Medium: Add components to distilled/deionized water and bring volume to 1.0L. Mix thoroughly. Distribute into tubes or flasks. Autoclave for 15 min at 15 psi pressure–121°C.

Use: For the cultivation and maintenance of *Leucothrix mucor*.

Levine EMB Agar (Levine Eosin Methylene Blue Agar) (Eosin Methylene Blue Agar, Levine) (LEMB Agar)

Composition per liter:

Agar	15.0g
Lactose	10.0g
Peptone	10.0g
K_2HPO_4	2.0g
Eosin Y	0.4g
Methylene blue	0.065mg

pH 7.1 ± 0.2 at 25°C

Source: This medium is available as a premixed powder from BBL Microbiology Systems and Difco Laboratories.

Preparation of Medium: Add components to distilled/deionized water and bring volume to 1.0L. Mix thoroughly. Gently heat and bring to boiling. Distribute into tubes or flasks. Autoclave for 15 min at 15 psi pressure–121°C. Pour into sterile Petri dishes or leave in tubes.

Use: For the isolation, cultivation and differentiation of Gram-negative enteric bacteria based on lactose fermentation. Bacteria that ferment lactose, especially the coliform bacterium *Escherichia coli*, appear as colonies with a green metallic sheen or blue-black to brown color. Bacteria that do not ferment lactose appear as colorless or transparent light purple colonies.

LHET2 Medium

Composition per liter:

Solution A	500.0mL
Solution B	500.0mL

pH 2.5–3.0 at 25°C

Solution A:

Composition per 500mL:

$(NH_4)_2SO_4$	2.0g
K_2HPO_4	0.51g
$MgSO_4 \cdot 7H_2O$	0.5g
KCl	0.1g
Pancreatic digest of casein	0.06g
NaCl	0.02g
Papaic digest of soybean meal	0.01g

Preparation of Solution A: Add components to distilled/deionized water and bring volume to 500.0mL. Mix thoroughly. Gently heat and bring to boiling. Adjust pH to 2.5–3.0 with 1*N* H_2SO_4. Autoclave for 15 min at 15 psi pressure–121°C. Cool to 45–50°C.

Solution B:

Composition per 500mL:

Agar	12.0g
Glucose	1.0g

Preparation of Solution B: Add components to distilled/deionized water and bring volume to 500.0mL. Mix thoroughly. Gently heat and bring to boiling. Autoclave for 15 min at 15 psi pressure–121°C. Cool to 45–50°C.

Preparation of Medium: Aseptically combine sterile solution A and sterile solution B. Mix thoroughly. Pour into sterile Petri dishes or distribute into sterile tubes.

Use: For the cultivation and maintenance of *Acidiphilium cryptum.*

LHET2 Medium with Yeast Extract or Yeast Autolysate

Composition per liter:

Solution A	500.0mL
Solution B	500.0mL

pH 2.5–3.0 at 25°C

Solution A:

Composition per 500mL:

$(NH_4)_2SO_4$	2.0g
K_2HPO_4	0.51g
$MgSO_4 \cdot 7H_2O$	0.5g
KCl	0.1g
Yeast extract or yeast autolysate	0.1g
Pancreatic digest of casein	0.06g
NaCl	0.02g
Papaic digest of soybean meal	0.01g

Preparation of Solution A: Add components to distilled/deionized water and bring volume to 500.0mL. Mix thoroughly. Gently heat and bring to boiling. Adjust pH to 2.5–3.0 with 1*N* H_2SO_4. Autoclave for 15 min at 15 psi pressure–121°C. Cool to 45–50°C.

Solution B:

Composition per 500mL:

Agar	12.0g
Glucose	1.0g

Preparation of Solution B: Add components to distilled/deionized water and bring volume to 500.0mL. Mix thoroughly. Gently heat and bring to boiling. Autoclave for 15 min at 15 psi pressure–121°C. Cool to 45–50°C.

Preparation of Medium: Aseptically combine sterile solution A and sterile solution B. Mix thoroughly. Pour into sterile Petri dishes or distribute into sterile tubes.

Use: For the cultivation and maintenance of *Acidiphilium angustum, A. facilis,* and *A. rubrum.*

Lima Bean Agar

Composition per liter:

Lima beans, solids from infusion	62.5g
Agar	15.0g

pH 5.6 ± 0.2 at 25°C

Source: This medium is available as a premixed powder from Difco Laboratories.

Preparation of Medium: Add components to distilled/deionized water and bring volume to 1.0L. Mix thoroughly. Gently heat and bring to boiling. Distribute into tubes or flasks. Autoclave for 15 min at 15 psi pressure–121°C. Pour into sterile Petri dishes or leave in tubes.

Use: For the cultivation of a variety of phytopathological fungi and other fungi.

Lipovitellin Salt Mannitol Agar

Composition per liter:

NaCl	75.0g
Egg yolk	20.0g
Agar	15.0g
D-Mannitol	10.0g
Polypeptone™	10.0g
Beef extract	1.0g
Phenol red	0.025g

Preparation of Medium: Add components to distilled/deionized water and bring volume to 1.0L. Mix thoroughly. Gently heat and bring to boiling. Distribute into tubes or flasks. Autoclave for 15 min at 15 psi pressure–121°C. Pour into sterile Petri dishes or leave in tubes.

Use: For the detection of *Staphylococcus aureus* in swimming pool water based on lipovitellin-lipase activity and mannitol fermentation. *Staphylococcus aureus* and other bacteria with lipovitellin-lipase activity attack the egg yolk and appear as colonies surrounded by an opaque zone. Bacteria that ferment mannitol appear as colonies surrounded by a yellow zone.

Litmus Lactose Agar (LL Agar)

Composition per liter:

Agar	10.0g
Lactose	10.0g
Meat peptone	5.0g
Beef extract	3.0g
Litmus	1.0g

pH 7.0 ± 0.2 at 25°C

Preparation of Medium: Add components to distilled/deionized water and bring volume to 1.0L. Mix thoroughly. Gently heat and bring to boiling. Distribute into tubes or flasks. Autoclave for 15 min at 15 psi pressure–121°C. Pour into sterile Petri dishes or leave in tubes.

Use: For the maintenance of lactic acid bacteria and for the differentiation of several bacteria on the basis of lactose fermentation. Bacteria that ferment lactose appear as red colonies. Bacteria that do not ferment lactose appear as dark blue-purple colonies.

Litmus Lactose Agar with Crystal Violet (LLK Agar)

Composition per liter:

Agar	10.0g
Lactose	10.0g
Meat peptone	5.0g
Beef extract	3.0g
Litmus	1.0g
Crystal violet	5.0mg

pH 7.0 ± 0.2 at 25°C

Preparation of Medium: Add components to distilled/deionized water and bring volume to 1.0L. Mix thoroughly. Gently heat and bring to boiling. Distribute into tubes or flasks. Autoclave for 15 min at 15 psi pressure–121°C. Pour into sterile Petri dishes or leave in tubes.

Use: For the maintenance of lactic acid bacteria and for the differentiation of several bacteria on the basis of lactose fermentation. Bacteria that ferment lactose appear as red colonies. Bacteria that do not ferment lactose appear as dark blue-purple colonies.

Litmus Milk

Composition per liter:

Skim milk	100.0g
Azolitmin	0.5g
Na_2SO_3	0.5g

pH 6.5 ± 0.2 at 25°C

Source: This medium is available as a premixed powder from BBL Microbiology Systems, Difco Laboratories, and Oxoid Unipath.

Preparation of Medium: Add components to distilled/deionized water and bring volume to 1.0L. Mix thoroughly. Gently heat and bring to boiling. Distribute into tubes or flasks. Autoclave for 20 min at 10 psi pressure–115°C.

Use: For the maintenance of lactic acid bacteria and for the differentiation of several bacteria, especially *Clostridium* species, based on their action on milk. Bacteria that do not ferment carbohydrates, such as *Proteus vulgaris* or *Moraxella lacunata*, show no change in the litmus indicator. Bacteria that ferment lactose or glucose with the production of gas, such as *Clostridium perfringens*, turn the medium pink and frothy. Bacteria that proteolytically degrade lactalbumin turn the medium blue. Bacteria that coagulate casein form a curd or clot. Bacteria that peptonize casein, such as *Pseudomonas aeruginosa*, show a dissolution of the clot.

Liver Broth

Composition per liter:

Beef liver, fresh	453.0g
Pancreatic digest of casein	10.0g
K_2HPO_4	1.0g
Soluble starch	1.0g

pH 7.6 ± 0.2 at 25°C

Preparation of Medium: Remove the fat from fresh beef liver. Grind the liver. Add 1.0L of distilled/deionized water. Gently heat and bring to boiling. Continue boiling for 60 min. Adjust pH to 7.6. Filter through cheesecloth. Reserve meat. To filtrate, add pancreatic digest of casein, K_2HPO_4, and soluble starch. Bring volume to 1.0L with distilled/deionized water. Refilter solution. Add meat particles to test tubes to a depth of approximately 2 cm. Distribute broth into tubes with meat particles in 15.0mL volumes. Autoclave for 20 min at 15 psi pressure–121°C.

Use: For the isolation and cultivation of anaerobic microorganisms, especially *Clostridium botulinum*, from foods.

Lombard–Dowell Agar (LD Agar)

Agar	20.0g
Pancreatic digest of casein	5.0g
Yeast extract	5.0g
NaCl	2.5g
L-Cystine	0.4g
L-Tryptophan	0.2g
Na_2SO_3	0.1g
Hemin	10.0mg
NaOH (1*N* NaOH)	5.0mL
Vitamin K_1 solution	1.0mL

pH 7.5 ± 0.2 at 25°C

Vitamin K_1 Solution:

Composition per 100mL:

Vitamin K_1	1.0g
Ethanol	99.0mL

Preparation of Vitamin K_1 Solution: Add vitamin K_1 to 99.0mL of absolute ethanol. Mix thoroughly.

Preparation of Medium: Add hemin and cystine to 5.0mL of NaOH. Mix thoroughly. Add remaining components. Bring volume to 1.0L with distilled/deionized water. Mix thoroughly. Gently heat and bring to boiling. Distribute into tubes or flasks. Autoclave for 15 min at 15 psi pressure–121°C. Pour into sterile Petri dishes.

Use: For the cultivation and identification of a variety of obligate anaerobic bacteria. For the cultivation of *Bacteroides species, Fusobacterium species, Clostridium* species, and nonspore–forming Gram–positive anaerobes.

Lombard–Dowell Broth (LD Broth)

Composition per liter:

Pancreatic digest of casein	5.0g
Yeast extract	5.0g
Agar	0.7g
NaCl	2.5g
L-Tryptophan	0.2g
Na_2SO_3	0.1g
NaOH (1*N* NaOH)	5.0mL
Hemin solution	1.0mL
Vitamin K_1 solution	1.0mL

pH 7.5 ± 0.2 at 25°C

Hemin Solution:

Composition per 100mL:

Hemin	1.0g
NaOH (1*N* solution)	20.0mL

Preparation of Hemin Solution: Add hemin to 20.0mL of 1*N* NaOH solution. Mix thoroughly. Bring volume to 100.0mL with distilled/deionized water.

Vitamin K_1 Solution:

Composition per 100mL:

Vitamin K_1	1.0g
Ethanol	99.0mL

Preparation of Vitamin K_1 Solution: Add vitamin K_1 to 99.0mL of absolute ethanol. Mix thoroughly.

Preparation of Medium: Add tryptophan to 5.0mL of NaOH. Mix thoroughly. Add remaining components. Bring volume to 1.0L with distilled/deionized water. Mix thoroughly. Gently heat and bring to boiling. Adjust pH to 7.5. Distribute into screw-capped tubes in 7.0mL volumes. Autoclave for 15 min at 15 psi pressure–121°C. Cool tubes, with caps loose, under 85% N_2 + 10% H_2 + 5% CO_2. Tighten caps.

Use: For the cultivation of a wide variety of anaerobic bacteria.

Lombard–Dowell Egg Yolk Agar (LD Egg Yolk Agar) (Egg Yolk Agar, Lombard–Dowell)

Composition per 9100mL:

$Na_2HPO_4 \cdot 12H_2O$	5.0g
Glucose	2.0g
LD Agar	9000.0mL
Egg yolk emulsion	100.0mL
$MgSO_4 \cdot 7H_20$ (5% solution)	0.2mL

pH 7.5 ± 0.2 at 25°C

LD Agar:

Composition per liter:

Agar	20.0g
Pancreatic digest of casein	5.0g
Yeast extract	5.0g
NaCl	2.5g
L-Cystine	0.4g
L-Tryptophan	0.2g
Na_2SO_3	0.1g
Hemin	10.0mg
NaOH (1*N* NaOH)	5.0mL
Vitamin K_1 solution	1.0mL

Preparation of LD Agar: Add hemin and cystine to 5.0mL of NaOH. Mix thoroughly. Add remaining components. Mix thoroughly. Gently heat and bring to boiling.

Vitamin K_1 Solution:

Composition per 100mL:

Vitamin K_1	1.0g
Ethanol	99.0mL

Preparation of Vitamin K_1 Solution: Add vitamin K_1 to 99.0mL of absolute ethanol. Mix thoroughly.

Egg Yolk Emulsion:

Composition:

Chicken egg yolks	11
Whole chicken egg	1

Preparation of Egg Yolk Emulsion: Soak eggs with 1:100 dilution of saturated mercuric chloride solution for 1 min. Crack eggs and separate yolks from whites. Mix egg yolks with 1 chicken egg.

Preparation of Medium: Combine components, except egg yolk emulsion. Mix thoroughly. Autoclave for 15 min at 15 psi pressure–121°C. Cool to 45–50°C. Aseptically add 100.0mL of egg yolk emulsion. Mix thoroughly. Pour into sterile Petri dishes.

Use: For the cultivation of a wide variety of anaerobic bacteria. For the differentiation of anaerobic bacteria based on lecithinase production, lipase production and proteolytic ability. Bacteria that produce lecithinase appear as colonies surrounded by a zone of insoluble precipitate. Bacteria that produce lipase appear as colonies with a pearly iridescent sheen. Bacteria that produce proteolytic activity appear as colonies surrounded by a clear zone.

Lombard–Dowell Esculin Agar (LD Esculin Agar) (Esculin Agar, Lombard–Dowell)

Composition per liter:

Agar	20.0g
Pancreatic digest of casein	5.0g
Yeast extract	5.0g
NaCl	2.5g
Esculin	1.0g
Ferric citrate	0.5g
L-Cystine	0.4g
L-Tryptophan	0.2g
Hemin	10.0mg
NaOH (1*N* NaOH)	5.0mL
Vitamin K_1 solution	1.0mL

pH 7.5 ± 0.2 at 25°C

Vitamin K_1 Solution:

Composition per 100mL:

Vitamin K_1	1.0g
Ethanol	99.0mL

Preparation of Vitamin K_1 Solution: Add vitamin K_1 to 99.0mL of absolute ethanol. Mix thoroughly.

Preparation of Medium: Add hemin and cystine to 5.0mL of NaOH. Mix thoroughly. Add remaining components. Bring volume to 1.0L with distilled/deionized water. Mix thoroughly. Gently heat and bring to boiling. Distribute into tubes or flasks. Autoclave for 15 min at 15 psi pressure–121°C. Pour into sterile Petri dishes.

Use: For the cultivation of a wide variety of anaerobic bacteria. For the differentiation of anaerobic bacteria based on esculin hydrolysis, H_2S production and catalase production. Bacteria that hydrolyze esculin appear as colonies surrounded by a red-brown to dark brown zone. Bacteria that produce H_2S appear as black colonies.

Lombard–Dowell Gelatin Agar (LD Gelatin Agar)

Composition per liter:

Agar	20.0g
Pancreatic digest of casein	5.0g
Yeast extract	5.0g
Gelatin	4.0g
NaCl	2.5g
Glucose	1.0g
L-Cystine	0.4g
L-Tryptophan	0.2g
Na_2SO_3	0.1g
Hemin	10.0mg
NaOH (1*N* NaOH)	5.0mL
Vitamin K_1 solution	1.0mL

pH 7.5 ± 0.2 at 25°C

Vitamin K_1 Solution:

Composition per 100mL:

Vitamin K_1	1.0g
Ethanol	99.0mL

Preparation of Vitamin K_1 Solution: Add vitamin K_1 to 99.0mL of absolute ethanol. Mix thoroughly.

Preparation of Medium: Add hemin and cystine to 5.0mL of NaOH. Mix thoroughly. Add remaining components, except agar and gelatin. Bring volume to 750.0mL with distilled/deionized water. Mix thoroughly. Gently heat and bring to boiling. In a separate flask, add gelatin to 100.0mL of cold distilled/deionized water. Gently heat and bring to 70°C. Add gelatin solution to the 750.0mL of basal medium. Mix thoroughly. Add agar. Bring volume to 1.0L with distilled/deionized water. Autoclave for 15 min at 15 psi pressure–121°C. Pour into sterile Petri dishes.

Use: For the cultivation of a wide variety of anaerobic bacteria. For the differentiation of anaerobic bacteria based on gelatinase production. After incubation of plates, gelatinase activity is determined by addition of Frazier's reagent. Bacteria that hydrolyze gelatin appear as colonies surrounded by a clear zone.

Low Phosphate Buffered Basal Medium, Modified

Composition per 1030mL:

Pectin	4.0g
NH_4Cl	1.0g
Na_2HPO_4	0.72g
KH_2PO_4	0.3g
$MgCl_2 \cdot 6H_2O$	0.2g
Reducing agent	20.0mL
Yeast extract solution	10.0mL
Trace minerals	10.0mL
Vitamins	5.0mL
Resazurin (0.2% solution)	1.0mL
$FeSO_4 \cdot 7H_2O$ (2.5% solution)	0.03mL

pH 7.3 ± 0.1 at 25°C

Reducing Agent:

Composition per 20mL:

$Na_2S \cdot 9H_2O$	0.5g

Preparation of Reducing Agent: Add $Na_2S \cdot 9H_2O$ to distilled/deionized water and bring volume to

20.0mL. Mix thoroughly. Gas with 100% N_2 for 20 min. Cap with a rubber stopper. Autoclave for 45 min at 15 psi pressure–121°C. Use freshly prepared solution.

Yeast Extract Solution:
Composition per 10mL:

Yeast extract	1.0g

Preparation of Yeast Extract Solution: Add yeast extract to distilled/deionized water and bring volume to 10.0mL. Mix thoroughly. Autoclave for 45 min at 15 psi pressure–121°C. Cool to 25°C.

Trace Minerals:
Composition per liter:

Nitrilotriacetic acid	12.8g
NaCl	1.0g
$CoCl_2 \cdot 6H_2O$	0.16g
$CaCl_2 \cdot 2H_2O$	0.1g
$FeSO_4 \cdot 7H_2O$	0.1g
$MnCl_2 \cdot 4H_2O$	0.1g
$ZnCl_2$	0.1g
$CuCl_2$	0.02g
Na_2SeO_3	0.02g
H_3BO_3	0.01g
$Na_2MoO_4 \cdot 2H_2O$	0.01g
$NiSO_4 \cdot 6H_2O$	0.026g

Preparation of Trace Minerals: Add nitrilotriacetic acid to 500.0mL of distilled/deionized water. Dissolve by adjusting pH to 6.5 with KOH. Add remaining components. Add distilled/deionized water to 1.0L.

Wolfe's Vitamin Solution:
Composition per liter:

Pyridoxine·HCl	0.01g
Thiamine·HCl	5.0mg
Riboflavin	5.0mg
Nicotinic acid	5.0mg
Calcium pantothenate	5.0mg
p-Aminobenzoic acid	5.0mg
Thioctic acid	5.0mg
Biotin	2.0mg
Folic acid	2.0mg
Cyanocobalamin	0.1mg

Preparation of Wolfe's Vitamin Solution: Add components to distilled/deionized water and bring volume to 1.0L. Mix thoroughly.

Preparation of Medium: Add components, except yeast extract solution and reducing agent, to distilled/deionized water and bring volume to 1.0L. Mix thoroughly. Gently heat and bring to boiling. Cool under 90% N_2 + 10% CO_2. Anaerobically distribute into tubes in 6.0mL volumes. Autoclave for 45 min at 15 psi pressure–121°C. Aseptically add 0.06mL of sterile yeast extract solution to each tube. Mix thoroughly. Immediately prior to inoculation aseptically add 0.12mL of sterile reducing agent to each tube. Mix thoroughly.

Use: For the cultivation and maintenance of *Clostridium thermosulfurogenes*.

Lowenstein–Gruft Medium

Composition per 1600mL:

Potato starch	30.0g
Asparagine	3.6g
KH_2PO_4	2.4g
Magnesium citrate	0.6g
Malachite green	0.4g
$MgSO_4 \cdot 7H_2O$	0.24g
Nalidixic acid	0.056g
Ribonucleic acid	0.08mg
Homogenized whole egg	1.0L
Glycerol	12.0mL
Penicillin	80,000U

Homogenized Whole Egg:
Composition per liter:

Whole eggs	18–24

Preparation of Homogenized Whole Egg: Use fresh eggs, less than 1 week old. Scrub the shells with soap. Let stand in a soap solution for 30 min. Rinse in running water. Soak eggs in 70% ethanol for 15 min. Break the eggs into a sterile container. Homogenize by shaking. Filter through four layers of sterile cheesecloth into a sterile graduated cylinder. Measure out 1.0L.

Preparation of Medium: Add glycerol to 600.0mL of distilled/deionized water. Mix thoroughly. Add remaining components, except fresh egg mixture. Mix thoroughly. Gently heat while stirring and bring to boiling. Autoclave for 15 min at 15 psi pressure–121°C. Cool to 50°C. Aseptically add 1.0L of homogenized whole egg. Mix thoroughly. Distribute into sterile screw-capped tubes. Place tubes in a slanted position. Inspissate at 85°C (moist heat) for 45 min.

Use: For the cultivation and differentiation of *Mycobacterium* species. *M. tuberculosis* appears as granular, rough, dry colonies. *M. kansasii* appears as smooth to rough photochromogenic colonies. *M. gordonae* appears as smooth yellow-orange colonies. *M. avium* appears as smooth, colorless colonies. *M. smegmatis* appears as wrinkled, creamy white colonies.

Lowenstein–Jensen Medium

Composition per 1600mL:

Potato starch	30.0g
Asparagine	3.6g

KH_2PO_4 .... 2.4g
Magnesium citrate 0.6g
Malachite green 0.4g
$MgSO_4 \cdot 7H_2O$ 0.24g
Homogenized whole egg 1.0L
Glycerol 12.0mL

Source: Available as a prepared medium from BBL Microbiology Systems, Difco Laboratories and Oxoid Unipath.

Homogenized Whole Egg:
Composition per liter:
Whole eggs 18–24

Preparation of Homogenized Whole Egg: Use fresh eggs, less than 1 week old. Scrub the shells with soap. Let stand in a soap solution for 30 min. Rinse in running water. Soak eggs in 70% ethanol for 15 min. Break the eggs into a sterile container. Homogenize by shaking. Filter through four layers of sterile cheesecloth into a sterile graduated cylinder. Measure out 1.0L.

Preparation of Medium: Add glycerol to 600.0mL of distilled/deionized water. Mix thoroughly. Add remaining components, except fresh egg mixture. Mix thoroughly. Gently heat while stirring and bring to boiling. Autoclave for 15 min at 15 psi pressure–121°C. Cool to 50°C. Aseptically add 1.0L of homogenized whole egg. Mix thoroughly. Distribute into sterile screw-capped tubes. Place tubes in a slanted position. Inspissate at 85°C (moist heat) for 45 min.

Use: For the cultivation and differentiation of *Mycobacteirum* species. *M. tuberculosis* appears as granular, rough, dry colonies. *M. kansasii* appears as smooth to rough photochromogenic colonies. *M. gordonae* appears as smooth yellow-orange colonies. *M. avium* appears as smooth, colorless colonies. *M. smegmatis* appears as wrinkled, creamy white colonies. Also used for the cultivation and maintenance of *Gordona* species, *Nocardia* species, *Rhodococcus* species, and *Tsukamurella paurometabolum*.

Lowenstein–Jensen Medium with NaCl

Composition per 1600mL:
NaCl 80.0g
Potato starch 30.0g
Asparagine 3.6g
KH_2PO_4 .... 2.4g
Magnesium citrate 0.6g
Malachite green 0.4g
$MgSO_4 \cdot 7H_2O$ 0.24g
Homogenized whole egg 1.0L
Glycerol 12.0mL

Homogenized Whole Egg:
Composition per liter:
Whole eggs 18–24

Preparation of Homogenized Whole Egg: Use fresh eggs, less than 1 week old. Scrub the shells with soap. Let stand in a soap solution for 30 min. Rinse in running water. Soak eggs in 70% ethanol for 15 min. Break the eggs into a sterile container. Homogenize by shaking. Filter through four layers of sterile cheesecloth into a sterile graduated cylinder. Measure out 1.0L.

Preparation of Medium: Add glycerol to 600.0mL of distilled/deionized water. Mix thoroughly. Add remaining components, except fresh egg mixture. Mix thoroughly. Gently heat while stirring and bring to boiling. Autoclave for 15 min at 15 psi pressure–121°C. Cool to 50°C. Aseptically add 1.0L of homogenized whole egg. Mix thoroughly. Distribute into sterile screw-capped tubes. Place tubes in a slanted position. Inspissate at 85°C (moist heat) for 45 min.

Use: For the cultivation of *Mycobacterum smegmatis* and other salt-tolerant *Mycobacterum* species.

LPBM Acido–Thermophile Medium

Composition per liter:
Agar 20.0g
Cellulose 5.0g
KH_2PO_4 .... 1.0g
NH_4Cl 1.0g
Yeast extract 1.0g
Cellobiose 0.5g
$MgSO_4 \cdot 7H_2O$ 0.2g
$Na_2HPO_4 \cdot 7H_2O$ 0.1g
$CaCl_2 \cdot 2H_2O$ 0.02g
pH 5.2 ± 0.2 at 25°C

Preparation of Medium: Add components, except cellulose and cellobiose, to distilled/deionized water and bring volume to 1.0L. Mix thoroughly. Adjust pH to 5.2 with H_3PO_4. Add cellulose and cellobiose. Mix thoroughly. Gently heat and bring to boiling. Distribute into tubes or flasks. Autoclave for 15 min at 15 psi pressure–121°C. Pour into sterile Petri dishes or leave in tubes.

Use: For the cultivation and maintenance of *Acidothermus cellulolyticus*.

Luminous Medium

Composition per liter:
NaCl 30.0g
Agar 20.0g

NH_4Cl	5.0g
Pancreatic digest of casein	5.0g
Yeast extract	5.0g
K_2HPO_4	3.9g
KH_2PO_4	2.1g
$CaCO_3$	1.0g
$MgSO_4 \cdot 7H_2O$	1.0g
KCl	0.75g
Tris buffer (1*M* solution, pH 7.5)	50.0mL
Glycerol	3.0mL

pH 7.2 ± 0.2 at 25°C

Preparation of Medium: Add components to distilled/deionized water and bring volume to 1.0L. Mix thoroughly. Gently heat and bring to boiling. Distribute into tubes or flasks. Autoclave for 15 min at 15 psi pressure–121°C. Pour into sterile Petri dishes or leave in tubes.

Use: For the cultivation and maintenance of *Alteromonas hanedai, Photobacterium* species, *Shewanella hanedai,* and *Vibrio* species.

Lysine Arginine Iron Agar

Composition per liter:

Agar	15.0g
L-Arginine	10.0g
L-Lysine	10.0g
Peptone	5.0g
Yeast extract	3.0g
Glucose	1.0g
Ferric ammonium citrate	0.5g
Sodium thiosulfate	0.04g
Bromcresol purple	0.02g

pH 6.8 ± 0.2 at 25°C

Preparation of Medium: Add components to distilled/deionized water and bring volume to 1.0L. Mix thoroughly. Gently heat and bring to boiling. Adjust pH to 6.8. Distribute into screw-capped tubes in 5.0mL volumes. Autoclave for 12 min at 15 psi pressure–121°C. Allow tubes to cool in a slanted position.

Use: For the cultivation and differentiation of bacteria based on their ability to decarboxylate lysine, decarboxylate arginine and produce H_2S. Bacteria that decarboxylate lysine or arginine turn the medium purple. Bacteria that produce H_2S appear as black colonies.

Lysine Decarboxylase Broth, Falkow

Composition per liter:

Peptone	5.0g
L-Lysine	5.0g
Yeast extract	3.0g
Glucose	1.0g
Bromcresol purple	0.02g

pH 6.5–6.8 at 25°C

Preparation of Medium: Add components to distilled/deionized water and bring volume to 1.0L. Mix thoroughly. Gently heat and bring to boiling. Adjust pH to 6.5–6.8. Distribute into tubes in 5.0mL volumes. Autoclave for 15 min at 15 psi pressure–121°C.

Use: For the cultivation and differentiation of bacteria, especially *Salmonella,* based on their ability to decarboxylate lysine. Bacteria that decarboxylate lysine turn the medium turbid purple.

Lysine Decarboxylase Broth, Taylor Modification

Composition per liter:

L-Lysine	5.0g
Yeast extract	3.0g
Glucose	1.0g
Bromcresol purple	0.02g

pH 6.1 ± 0.2 at 25°C

Source: This medium is available as a premixed powder from Oxoid Unipath.

Preparation of Medium: Add components to distilled/deionized water and bring volume to 1.0L. Mix thoroughly. Gently heat and bring to boiling. Adjust pH to 6.1. Distribute into tubes in 5.0mL volumes. Autoclave for 15 min at 15 psi pressure–121°C.

Use: For the cultivation and differentiation of bacteria, especially *Salmonella,* based on their ability to decarboxylate lysine. Bacteria that decarboxylate lysine turn the medium turbid purple.

Lysine Decarboxylase Broth, Taylor Modification (Lysine Decarboxylase Broth)

Composition per liter:

L-Lysine	5.0g
Peptone	5.0g
Yeast extract	3.0g
Glucose	1.0g
Bromcresol purple	0.02g

pH 6.8 ± 0.2 at 25°C

Source: This medium is available as a premixed powder from Difco Laboratories.

Preparation of Medium: Add components to distilled/deionized water and bring volume to 1.0L. Mix thoroughly. Gently heat and bring to boiling. Adjust pH to 6.1. Distribute into tubes in 5.0mL volumes. Autoclave for 15 min at 15 psi pressure–121°C.

Use: For the cultivation and differentiation of bacteria, especially *Salmonella*, based on their ability to decarboxylate lysine. Bacteria that decarboxylate lysine turn the medium turbid purple.

Lysine Decarboxylase Medium

Composition per liter:

Glucose	0.5g
KH_2PO_4	0.5g
L-Lysine·HCl	0.5g

pH 4.6 ± 0.2 at 25°C

Preparation of Medium: Add components to distilled/deionized water and bring volume to 1.0L. Mix thoroughly. Gently heat and bring to boiling. Adjust pH to 4.6. Autoclave for 15 min at 15 psi pressure–121°C. Aseptically distribute into sterile tubes in 1.0mL volumes.

Use: For the cultivation and differentiation of Gram-negative nonfermentative bacteria based on their ability to decarboxylate lysine. Bacteria that decarboxylate lysine turn the medium turbid purple.

Lysine Iron Agar

Composition per liter:

Agar	13.5g
L-Lysine	10.0g
Pancreatic digest of gelatin	5.0g
Yeast extract	3.0g
Glucose	1.0g
Ferric ammonium citrate	0.5g
$Na_2S_2O_3 \cdot 5H_2O$	0.04g
Bromcresol purple	0.02g

pH 6.7 ± 0.2 at 25°C

Source: This medium is available as a premixed powder from BBL Microbiology Systems, Difco Laboratories and Oxoid Unipath.

Preparation of Medium: Add components to distilled/deionized water and bring volume to 1.0L. Mix thoroughly. Gently heat while stirring and bring to boiling. Distribute into tubes in 10.0mL volumes. Autoclave for 12 min at 15 psi pressure–121°C. Allow tubes to cool in a slanted position.

Use: For the cultivation and differentiation of members of the Enterobacteriaceae based on their ability to decarboxylate lysine and to form H_2S. Bacteria that decarboxylate lysine turn the medium purple. Bacteria that produce H_2S appear as black colonies.

M1 Medium

Composition per liter:

L-Leucine	2.0g
L-Alanine	1.0g
L-Isoleucine	1.0g
L-Phenylalanine	1.0g
L-Proline	1.0g
L-Tryptophane	1.0g
L-Asparagine	0.50g
L-Lysine	0.50g
L-Methionine	0.50g
L-Tyrosine	0.40g
L-Valine	0.20g
L-Serine	0.20g
$MgSO_4 \cdot 7H_2O$	0.20g
NaCl	0.20g
KH_2PO_4	0.14g
L-Arginine	0.10g
L-Cysteine	0.10g
L-Glycine	0.10g
L-Histidine	0.10g
L-Threonine	0.10g
$CaCl_2$	2.0mg
$FeCl_3 \cdot 6H_2O$	2.0mg
Tris(hydroxymethyl)aminomethane buffer (0.01*M* solution, pH 7.6)	1.0L

pH 7.6 ± 0.2 at 25°C

Preparation of Medium: Add solid components to 1.0L of Tris buffer. Mix thoroughly. Filter sterilize. Aseptically distribute into tubes or flasks.

Use: For the cultivation of *Myxococcus xanthus*.

M1A Medium

Composition per 1001mL:

Sorbitol	23.3g
Peptone	6.0g
Sucrose	3.3g
Pancreatic digest of casein	3.3g
Beef heart infusion	2.0g
Glucose	1.3g
Yeast extract	1.0g
Fructose	0.3g
Phenol red	20.0mg
Schneider's *Drosophila* medium	533.0mL
Fetal calf serum, heat inactivated	167.0mL
Fresh yeast extract solution	33.0mL
Penicillin solution	8.0mL

Schneider's *Drosophila* Medium:
Composition per liter:

$MgSO_4 \cdot 7H_2O$	3.7g
NaCl	2.1g
Yeast extract	2.0g
Trehalose	2.0g

D-Glucose	2.0g
L-Glutamine	1.8g
L-Lysine·HCl	1.7g
L-Proline	1.7g
KCl	1.6g
$Na_2HPO_4 \cdot 7H_2O$	1.3g
L-Glutamic acid	0.8g
L-Methionine	0.8g
$CaCl_2$, anhydrous	0.6g
KH_2PO_4	0.5g
β-Alanine	0.5g
L-Tyrosine	0.5g
L-Arginine	0.4g
L-Aspartic acid	0.4g
L-Histidine	0.4g
L-Threonine	0.4g
$NaHCO_3$	0.4g
Glycine	0.3g
L-Serine	0.3g
L-Valine	0.3g
L-Isoleucine	0.2g
L-Leucine	0.2g
L-Phenylalanine	0.2g
α-Ketoglutaric acid	0.2g
Fumaric acid	0.1g
Malic acid	0.1g
Succinic acid	0.1g
L-Cystine	0.1g
L-Tryptophan	0.1g
L-Cysteine	0.06g

Preparation of Schneider's *Drosophila* Medium: Add components to distilled/deionized water and bring volume to 1.0L. Mix thoroughly. Filter sterilize.

Penicillin Solution:
Composition per 10mL:

Penicillin	2,500,000U

Preparation of Penicillin Solution: Add penicillin to distilled/deionized water and bring volume to 10.0mL. Filter sterilize.

Fresh Yeast Extract Solution:
Composition per 100mL:

Baker's yeast, live, pressed, starch-free	25.0g

Preparation of Fresh Yeast Extract Solution: Add the live Baker's yeast to 100.0mL of distilled/deionized water. Autoclave for 90 min at 15 psi pressure–121°C. Allow to stand. Remove supernatant solution. Adjust pH to 6.6–6.8. Filter sterilize.

Preparation of Medium: Add components—except Schneider's *Drosophila* medium, fetal calf serum, fresh yeast extract solution, and penicillin solution— to distilled/deionized water and bring volume to 260.0mL. Mix thoroughly. Gently heat and bring to boiling. Autoclave for 15 min at 15 psi pressure–121°C. Cool to 45–50°C. Aseptically add 533.0mL of sterile Schneider's *Drosophila* medium, 167.0mL of sterile fetal calf serum, 33.0mL of sterile fresh yeast extract solution, and 8.0mL of sterile penicillin solution. Mix thoroughly. Pour into sterile Petri dishes or distribute into sterile tubes.

Use: For the isolation and cultivation of *Spiroplasma* species that cause corn stunt.

M3 Agar

Composition per 1020mL:

Agar	18.0g
Na_2HPO_4	0.732g
KH_2PO_4	0.466g
NaCl	0.29g
Sodium propionate	0.2g
$MgSO_4 \cdot 7H_2O$	0.1g
$CaCO_3$	0.02g
KNO_3	0.01g
$FeSO_4 \cdot 7H_2O$	0.2mg
$ZnSO_4 \cdot 7H_2O$	0.18mg
$MnSO_4 \cdot 4H_2O$	0.020mg
Cycloheximide solution	10.0mL
Thiamine·HCl solution	10.0mL

pH 7.0 ± 0.2 at 25°C

Cycloheximide Solution:
Composition per 10mL:

Cycloheximide	0.05g

Preparation of Cycloheximide Solution: Add cycloheximide to distilled/deionized water and bring volume to 10.0mL. Mix thoroughly. Filter sterilize.

Thiamine·HCl Solution:
Composition per 10mL:

Thiamine·HCl	4.0mg

Preparation of Thiamine·HCl Solution: Add thiamine·HCl to distilled/deionized water and bring volume to 10.0mL. Mix thoroughly. Filter sterilize.

Preparation of Medium: Add components, except cycloheximide solution and thiamine·HCl solution, to distilled/deionized water and bring volume to 980.0mL. Mix thoroughly. Gently heat and bring to boiling. Autoclave for 15 min at 15 psi pressure–121°C. Cool to 45–50°C. Aseptically add 10.0mL of sterile cycloheximide solution and 10.0mL of thiamine·HCl solution. Mix thoroughly. Pour into sterile Petri dishes or distribute into sterile tubes.

Use: For the selective isolation and cultivation of *Nocardia* species and *Rhodococcus* species.

M3 Agar Medium

Composition per liter:

Agar	18.0g
Na_2HPO_4	0.732g
KH_2PO_4	0.466g
NaCl	0.29g
Sodium propionate	0.20g
KNO_3	0.10g
$MgSO_4 \cdot 7H_2O$	0.10g
$CaCO_3$	0.02g
Thiamine·HCl	4.0mg
$FeSO_4 \cdot 7H_2O$	0.2mg
$ZnSO_4 \cdot 7H_2O$	0.18mg
$MnSO_4 \cdot 4H_2O$	0.02mg
Cycloheximide solution	10.0mL
Thiamine·HCl solution	10.0mL

pH 7.0 ± 0.2 at 25°C

Cycloheximide Solution:

Composition per 10mL:

Cycloheximide	0.04g

Preparation of Cycloheximide Solution: Add cycloheximide to distilled/deionized water and bring volume to 10.0mL. Mix thoroughly. Filter sterilize.

Thiamine·HCl Solution:

Composition per 10mL:

Thiamine·HCl	0.04g

Preparation of Thiamine·HCl Solution: Add thiamine·HCl to distilled/deionized water and bring volume to 10.0mL. Mix thoroughly. Filter sterilize.

Preparation of Medium: Add components, except cycloheximide solution and thiamine·HCl solution, to distilled/deionized water and bring volume to 980.0mL. Mix thoroughly. Gently heat and bring to boiling. Autoclave for 15 min at 15 psi pressure–121°C. Cool to 45–50°C. Aseptically add sterile cycloheximide solution and thiamine·HCl solution. Mix thoroughly. Pour into sterile Petri dishes or distribute into sterile tubes.

Use: For the cultivation of *Micromonospora* species.

M9 Medium

Composition per liter:

Na_2HPO_4	6.0g
KH_2PO_4	3.0g
NH_4Cl	1.0g
NaCl	0.5g
Glucose solution	10.0mL
$MgSO_4 \cdot 7H_2O$ solution	1.0mL
Thiamine·HCl solution	1.0mL
$CaCl_2$ solution	1.0mL

pH 7.0 ± 0.2 at 25°C

Glucose Solution:

Composition per 100mL:

D-Glucose	20.0g

Preparation of Glucose Solution: Add glucose to distilled/deionized water and bring volume to 1.0L. Mix thoroughly. Autoclave for 15 min at 15 psi pressure–121°C.

$MgSO_4 \cdot 7H_2O$ Solution:

Composition per liter:

$MgSO_4 \cdot 7H_2O$	246.5g

Preparation of $MgSO_4 \cdot 7H_2O$ Solution: Add $MgSO_4 \cdot 7H_2O$ to distilled/deionized water and bring volume to 1.0L. Mix thoroughly. Autoclave for 15 min at 15 psi pressure–121°C.

Thiamine·HCl Solution:

Composition per 10mL:

Thiamine·HCl	10.0mg

Preparation of Thiamine·HCl Solution: Add thiamine·HCl to distilled/deionized water and bring volume to 1.0L. Mix thoroughly. Filter sterilize.

$CaCl_2$ Solution:

Composition per liter:

$CaCl_2$ solution	14.7g

Preparation of $CaCl_2$ Solution: Add $CaCl_2$ solution to distilled/deionized water and bring volume to 1.0L. Mix thoroughly. Autoclave for 15 min at 15 psi pressure–121°C.

Preparation of Medium: Add components, except $MgSO_4 \cdot 7H_2O$ solution, glucose solution, thiamine·HCl solution, and $CaCl_2$ solution, to distilled/deionized water and bring volume to 987.0mL. Mix thoroughly. Adjust pH to 7.0. Autoclave for 15 min at 15 psi pressure–121°C. Cool to room temperature. Aseptically add sterile $MgSO_4 \cdot 7H_2O$ solution, sterile glucose solution, sterile thiamine·HCl solution, and sterile $CaCl_2$ solution. Mix thoroughly. Distribute into tubes or flasks.

Use: For the cultivation and maintenance of *Escherichia coli* and a variety of other bacteria.

M13 *Verrucomicrobium* Medium

Composition per liter:

Glucose	0.25g
Peptone	0.25g
Yeast extract	0.25g
Distilled water	670.0mL
Artificial seawater	250.0mL
Tris-HCl buffer, (0.1*M* solution, pH 7.5)	50.0mL

Modified Huntner's basal salts	20.0mL
Vitamin solution	10.0mL

pH 7.5 ± 0.2 at 25°C

Artificial Seawater:

Composition per liter:

NaCl	23.48g
$MgCl_2$	4.98g
Na_2SO_4	3.92g
$CaCl_2$	1.10g
KCl	0.66g
$NaHCO_3$	0.19g
H_3BO_3	0.026g
$SrCl_2$	0.024g
KBr	6.0mg
NaF	3.0mg

Preparation of Artificial Seawater: Add components to distilled/deionized water and bring volume to 1.0L. Mix thoroughly.

Modified Hutner's Basal Salts:

Composition per liter:

$MgSO_4 \cdot 7H_2O$	29.7g
Nitrilotriacetic acid	10.0g
$CaCl_2 \cdot 2H_2O$	3.34g
$FeSO_4 \cdot 7H_2O$	99.0mg
$(NH_4)_2MoO_4$	9.25mg
Metals "44"	50.0mL

Preparation of Modified Hutner's Basal Salts: Add nitrilotriacetic acid to 500.0mL of distilled/deionized water. Dissolve by adjusting pH to 6.5 with KOH. Add remaining components. Add distilled/deionized water to 1.0L.

Metals "44":

Composition per 100mL:

$ZnSO_4 \cdot 7H_2O$	1.1g
$FeSO_4 \cdot 7H_2O$	0.5g
EDTA	0.25g
$MnSO_4 \cdot 7H_2O$	0.154g
$CuSO_4 \cdot 5H_2O$	0.04g
$Co(NO_3)_2 \cdot 6H_2O$	0.025g
$Na_2B_4O_7 \cdot 10H_2O$	0.018g

Preparation of Metals "44": Add components to distilled/deionized water and bring volume to 100.0mL. Mix thoroughly.

Vitamin Solution:

Composition per liter:

D-Calcium pantothenate	5.0mg
Riboflavin	5.0mg
Thiamine·HCl	5.0mg
Biotin	2.0mg
Folic acid	2.0mg
Vitamin B_{12}	0.1mg

Preparation of Vitamin Solution: Add components to distilled/deionized water and bring volume to 1.0L. Mix thoroughly. Filter sterilize.

Preparation of Medium: Add components, except Modified Hutner's Basal Salts, to distilled/deionized water and bring volume to 980.0mL. Mix thoroughly. Autoclave for 15 min at 15 psi pressure–121°C. Cool to room temperature. Aseptically add 20.0mL of sterile Modified Hutner's basal salts. Mix thoroughly. Aseptically distribute into sterile tubes or flasks.

Use: For the cultivation and maintenance of *Verrucomicrobium spinosum.*

M14 Medium

Composition per liter:

Yeast extract	1.0g
D-Glucose	1.0g
Tris(hydroxymethyl)aminomethane	0.753g
Artificial seawater	250.0mL
Modified Hutner's basal salts	20.0mL

pH 7.5 ± 0.2 at 25°C

Artificial Seawater:

Composition per liter:

NaCl	23.48g
$MgCl_2$	4.98g
Na_2SO_4	3.92g
$CaCl_2$	1.10g
KCl	0.66g
$NaHCO_3$	0.19g
H_3BO_3	0.026g
$SrCl_2$	0.024g
KBr	6.0mg
NaF	3.0mg

Preparation of Artificial Seawater: Add components to distilled/deionized water and bring volume to 1.0L. Mix thoroughly.

Modified Hutner's Basal Salts:

Composition per liter:

$MgSO_4 \cdot 7H_2O$	29.7g
Nitrilotriacetic acid	10.0g
$CaCl_2 \cdot 2H_2O$	3.34g
$FeSO_4 \cdot 7H_2O$	99.0mg
$(NH_4)_2MoO_4$	9.25mg
Metals "44"	50.0mL

Preparation of Modified Hutner's Basal Salts: Add nitrilotriacetic acid to 500.0mL of distilled/deionized water. Dissolve by adjusting pH to 6.5 with KOH. Add remaining components. Add distilled/deionized water to 1.0L.

Metals "44":

Composition per 100mL:

$ZnSO_4 \cdot 7H_2O$	1.1g
$FeSO_4 \cdot 7H_2O$	0.5g
EDTA	0.25g
$MnSO_4 \cdot 7H_2O$	0.154g
$CuSO_4 \cdot 5H_2O$	0.04g
$Co(NO_3)_2 \cdot 6H_2O$	0.025g
$Na_2B_4O_7 \cdot 10H_2O$	0.018g

Preparation of Metals "44": Add components to distilled/deionized water and bring volume to 100.0mL. Mix thoroughly.

Preparation of Medium: Add components, except Modified Hutner's basal salts, to distilled/deionized water and bring volume to 980.0mL. Mix thoroughly. Autoclave for 15 min at 15 psi pressure–121°C. Cool to room temperature. Aseptically add 20.0mL of sterile Modified Hutner's Basal Salts. Mix thoroughly. Aseptically distribute into sterile tubes or flasks.

Use: For the cultivation and maintenance of *Pirellula marina*.

MacConkey Agar

Composition per liter:

Pancreatic digest of gelatin	17.0g
Agar	13.5g
Lactose	10.0g
NaCl	5.0g
Bile salts	1.5g
Pancreatic digest of casein	1.5g
Peptic digest of animal tissue	1.5g
Neutral red	0.03g
Crystal violet	1.0mg

pH 7.1 ± 0.2 at 25°C

Source: This medium is available as a premixed powder from BBL Microbiology Systems and Difco Laboratories.

Preparation of Medium: Add components to distilled/deionized water and bring volume to 1.0L. Mix thoroughly. Gently heat while stirring until boiling. Autoclave for 15 min at 15 psi pressure–121°C. Pour into sterile Petri dishes or distribute into sterile tubes.

Use: For the selective isolation, cultivation and differentiation of coliforms and enteric pathogens based on the ability to ferment lactose. Lactose-fermenting organisms appear as red to pink colonies. Lactose-nonfermenting organisms appear as colorless or transparent colonies.

MacConkey Agar

Composition per liter:

Peptone	20.0g
Agar	12.0g
Lactose	10.0g
Bile salts	5.0g
NaCl	5.0g
Neutral red	0.075g

pH 7.4 ± 0.2 at 25°C

Source: This medium is available as a premixed powder from Oxoid Unipath.

Preparation of Medium: Add components to distilled/deionized water and bring volume to 1.0L. Mix thoroughly. Gently heat while stirring until boiling. Autoclave for 15 min at 15 psi pressure–121°C. Pour into sterile Petri dishes or distribute into sterile tubes.

Use: For the selective isolation, cultivation and differentiation of coliforms and enteric pathogens based on the ability to ferment lactose. Lactose-fermenting organisms appear as red to pink colonies. Lactose-nonfermenting organisms appear as colorless or transparent colonies.

MacConkey Agar No. 2 (MacConkey II Agar)

Composition per liter:

Peptone	20.0g
Agar	15.0g
Lactose	10.0g
NaCl	5.0g
Bile salts No.2	1.5g
Neutral red	0.05g
Crystal violet	1.0mg

pH 7.2 ± 0.2 at 25°C

Source: This medium is available as a premixed powder from BBL Microbiology Systems and Oxoid Unipath.

Preparation of Medium: Add components to distilled/deionized water and bring volume to 1.0L. Mix thoroughly. Gently heat while stirring until boiling. Autoclave for 15 min at 15 psi pressure–121°C. Pour into sterile Petri dishes or distribute into sterile tubes.

Use: For the selective isolation, cultivation and differentiation of enteric pathogens, especially enterococci, in materials of sanitary importance.

MacConkey Agar without Crystal Violet

Composition per liter:

Agar	12.0g
Lactose	10.0g

Pancreatic digest of casein	10.0g
Peptic digest of animal tissue	10.0g
Bile salts	5.0g
NaCl	5.0g
Neutral red	0.05g

pH 7.4 ± 0.2 at 25°C

Source: This medium is available as a premixed powder from BBL Microbiology Systems and Difco Laboratories.

Preparation of Medium: Add components to distilled/deionized water and bring volume to 1.0L. Mix thoroughly. Gently heat while stirring until boiling. Autoclave for 15 min at 15 psi pressure–121°C. Pour into sterile Petri dishes or distribute into sterile tubes.

Use: For the detection of members of the *Enterobacteriaceae* and enterococci as well as some staphylococci. Used for the isolation and detection of coliforms and enteric pathogens from water and wastewater.

MacConkey Agar, CS

Composition per liter:

Peptone	17.0g
Agar	13.5g
Lactose	10.0g
NaCl	5.0g
Proteose peptone	3.0g
Bile salts	1.5g
Neutral red	0.03g
Crystal violet	1.0mg

pH 7.1 ± 0.2 at 25°C

Source: Available as a prepared medium from Difco Laboratories.

Preparation of Medium: Add components to distilled/deionized water and bring volume to 1.0L. Mix thoroughly. Gently heat while stirring until boiling. Autoclave for 15 min at 15 psi pressure–121°C. Pour into sterile Petri dishes or distribute into sterile tubes.

Use: For the cultivation and differentiation of lactose-fermenting and nonfermenting Gram-negative bacteria while also controlling the swarming of *Proteus* species, if present. Lactose-fermenting organisms appear as red to pink colonies. Lactose-nonfermenting organisms appear as colorless or transparent colonies.

MacConkey Broth

Composition per liter:

Pancreatic digest of gelatin	20.0g
Lactose	10.0g
Oxgall	5.0g
Bromcresol purple	0.02g

pH 7.3 ± 0.2 at 25°C

Source: This medium is available as a premixed powder from BBL Microbiology Systems and Difco Laboratories.

Preparation of Medium: Add components to distilled/deionized water and bring volume to 1.0L. If testing 10.0mL samples, prepare medium double strength. Mix thoroughly. Gently heat while stirring until boiling. Distribute into test tubes containing inverted Durham tubes. Autoclave for 15 min at 15 psi pressure–121°C.

Use: For the selective isolation and cultivation of coliforms in water.

MacConkey Broth

Composition per liter:

Peptone	20.0g
Lactose	10.0g
Bile salts	5.0g
NaCl	5.0g
Neutral red	0.075g

pH 7.4 ± 0.2 at 25°C

Source: This medium is available as a premixed powder from Oxoid Unipath.

Preparation of Medium: Add components to distilled/deionized water and bring volume to 1.0L. If testing 10.0mL samples, prepare medium double strength. Mix thoroughly. Gently heat while stirring until boiling. Distribute into test tubes containing inverted Durham tubes. Autoclave for 15 min at 15 psi pressure–121°C.

Use: For the selective isolation and cultivation of coliforms in water.

MacConkey Broth, Purple

Composition per liter:

Peptone	20.0g
Lactose	10.0g
Bile salts	5.0g
NaCl	5.0g
Bromcresol purple	0.01g

pH 7.4 ± 0.2 at 25°C

Source: This medium is available as a premixed powder or tablets from Oxoid Unipath.

Preparation of Medium: Add components to distilled/deionized water and bring volume to 1.0L. If testing 10.0mL samples, prepare medium double strength. Mix thoroughly. Gently heat while stirring until boiling. Distribute into test tubes containing inverted Durham tubes. Autoclave for 15 min at 15 psi pressure–121°C.

Use: For the selective isolation and cultivation of coliforms in water.

Magnetic *Spirillum* Growth Medium, Revised (MSGM, Revised)

Composition per liter:

Agar	1.3g
KH_2PO_4	0.68g
Tartaric acid	0.37g
Succinic acid	0.37g
$NaNO_3$	0.12g
Sodium acetate	0.05g
Ascorbic acid	0.035g
Wolfe's vitamin solution	10.0mL
Wolfe's mineral solution	5.0mL
Ferric quinate solution	2.0mL
Resazurin (0.1% solution)	0.45mL

pH 6.75 ± 0.2 at 25°C

Wolfe's Vitamin Solution:
Composition per liter:

Pyridoxine·HCl	10.0mg
Thiamine·HCl	5.0mg
Riboflavin	5.0mg
Nicotinic acid	5.0mg
Calcium pantothenate	5.0mg
p-Aminobenzoic acid	5.0mg
Thioctic acid	5.0mg
Biotin	2.0mg
Folic acid	2.0mg
Cyanocobalamin	100.0μg

Preparation of Wolfe's Vitamin Solution: Add components to distilled/deionized water and bring volume to 1.0L. Mix thoroughly.

Wolfe's Mineral Solution:
Composition per liter:

$MgSO_4 \cdot 7H_2O$	3.0g
Nitriloacetic acid	1.5g
NaCl	1.0g
$MnSO_4 \cdot H_2O$	0.5g
$FeSO_4 \cdot 7H_2O$	0.1g
$CoCl_2 \cdot 6H_2O$	0.1g
$CaCl_2$	0.1g
$ZnSO_4 \cdot 7H_2O$	0.1g
$CuSO_4 \cdot 5H_2O$	0.01g
$AlK(SO_4)_2 \cdot 12H_2O$	0.01g
H_3BO_3	0.01g
$Na_2MoO_4 \cdot 2H_2O$	0.01g

Preparation of Wolfe's Mineral Solution: Add nitrilotriacetic acid to 500.0mL of distilled/deionized water. Dissolve by adjusting pH to 6.5 with KOH. Add remaining components. Add distilled/deionized water to 1.0L.

Ferric Quinate Solution:
Composition per 100mL:

$FeCl_3$	0.27g
Quinic acid	0.19g

Preparation of Ferric Quinate Solution: Add components to distilled/deionized water and bring volume to 1.0L. Mix thoroughly. Autoclave for 15 min at 15 psi pressure–121°C.

Preparation of Medium: To 1.0L of distilled/deionized water add components in the following order: Wolfe's Vitamin solution, Wolfe's Mineral solution, ferric quinate solution, resazurin, KH_2PO_4, $NaNO_3$, ascorbic acid, tartaric acid, succinic acid, sodium acetate and agar. Mix thoroughly after each addition. Adjust pH to 6.75 with NaOH. Autoclave for 15 min at 15 psi pressure–121°C. Aseptically distribute into sterile screw-capped tubes. Fill tubes to capacity with medium. Use a heavy inoculum in each tube and do not introduce a head space of air. Screw down caps tightly.

Use: For the cultivation and maintenance of *Aquaspirillum magnetotacticum.*

Malonate Broth

Composition per liter:

Sodium malonate	3.0g
NaCl	2.0g
$(NH_4)_2SO_4$	2.0g
K_2HPO_4	0.6g
KH_2PO_4	0.4g
Bromthymol blue	0.025g

pH 6.7 ± 0.2 at 25°C

Source: This medium is available as a premixed powder from Difco Laboratories.

Preparation of Medium: Add components to distilled/deionized water and bring volume to 1.0L. Mix thoroughly. Distribute into tubes or flasks. Autoclave for 15 min at 15 psi pressure–121°C. Avoid introduction of carbon and nitrogen from other sources.

Use: For the cultivation and differentiation of coliforms and other enteric organisms, particularly *Enterobacter* and *Escherichia* based on their ability to utilize malonate as the sole carbon source and ammonium sulfate as the sole nitrogen source. Malonate-utilizing organisms turn the medium blue.

Malonate Broth, Ewing Modified

Composition per liter:

Sodium malonate	3.0g
NaCl	2.0g
$(NH_4)_2SO_4$	2.0g
Yeast extract	1.0g

Glucose 0.25g
K_2HPO_4 .... 0.6g
KH_2PO_4 .... 0.4g
Bromthymol blue 0.025g
pH 6.7 ± 0.2 at 25°C

Source: This medium is available as a premixed powder from BBL Microbiology Systems and Difco Laboratories.

Preparation of Medium: Add components to distilled/deionized water and bring volume to 1.0L. Mix thoroughly. Distribute into tubes or flasks. Autoclave for 15 min at 15 psi pressure–121°C.

Use: For the cultivation and differentiation of coliforms and other enteric organisms, particularly *Enterobacter* and *Escherichia* based on their ability to utilize malonate as a carbon source and ammonium sulfate as a nitrogen source. The small amount of yeast extract and glucose encourages the growth of some organisms that may be distressed or fail to respond. Malonate-utilizing organisms turn the medium blue.

Malt Agar

Composition per liter:
Malt extract 30.0g
Agar 15.0g
pH 5.5 ± 0.2 at 25°C

Source: This medium is available as a premixed powder from BBL Microbiology Systems and Difco Laboratories.

Preparation of Medium: Add components to distilled/deionized water and bring volume to 1.0L. Mix thoroughly. Gently heat while stirring until boiling. Distribute into tubes or flasks. Autoclave for 15 min at 15 psi pressure–118°C. Do not overheat or agar will not harden. Pour into sterile Petri dishes or distribute into sterile tubes.

Use: For the cultivation of yeasts and molds.

Malt and Peptone Medium

Composition per liter:
Agar 15.0g
Malt extract 10.0g
Peptone 5.0g
NaCl 1.0g
pH 6.5 ± 0.2 at 25°C

Preparation of Medium: Add components to distilled/deionized water and bring volume to 1.0L. Mix thoroughly. Distribute into tubes or flasks. Adjust pH to 6.5. Autoclave for 15 min at 15 psi pressure–121°C.

Use: For the cultivation and maintenance of *Flavobacterium* species.

Malt Extract Agar

Composition per liter:
Malt extract 30.0g
Agar 15.0g
Peptone 5.0g
pH 7.0 ± 0.2 at 25°C

Preparation of Medium: Add components to distilled/deionized water and bring volume to 1.0L. Mix thoroughly. Gently heat and bring to boiling. Distribute into tubes or flasks. Autoclave for 20 min at 10 psi pressure–115°C. Pour into sterile Petri dishes or leave in tubes.

Use: For the cultivation of *Xanthomonas* species.

Malt Extract Agar

Composition per liter:
Malt extract 30.0g
Agar 15.0g
Mycological peptone 5.0g
pH 5.4 ± 0.2 at 25°C

Source: This medium is available as a premixed powder from Oxoid Unipath.

Preparation of Medium: Add components to distilled/deionized water and bring volume to 1.0L. Mix thoroughly. Gently heat while stirring until boiling. Distribute into tubes or flasks. Autoclave for 10 min at 15 psi pressure–115°C. Do not overheat or agar will not harden. If a lower pH (3.5) is desired, cool medium to 55°C and aseptically add 100.0mL of sterile lactic acid. Pour into sterile Petri dishes or distribute into sterile tubes.

Use: For the detection, isolation and enumeration of yeasts and molds. The addition of lactic acid suppresses bacterial growth.

Malt Extract Agar

Composition per liter:
Malt extract 30.0g
Agar 20.0g
Chlortetracycline solution 10.0mL
pH 5.5 ± 0.2 at 25°C

Chlortetracycline Solution:
Composition per 10mL:
Chlortetracycline 0.04mg

Preparation of Chlortetracycline Solution: Add chlortetracycline to distilled/deionized water and bring volume to 10.0mL. Mix thoroughly. Filter sterilize.

Preparation of Medium: Add components, except chlortetracycline solution, to distilled/deionized water and bring volume to 990.0mL. Mix thoroughly. Gently heat and bring to boiling. Autoclave for 15 min at 15 psi pressure–121°C. Cool to 45–50°C. Aseptically add sterile chlortetracycline solution. Mix thoroughly. Pour into sterile Petri dishes in 20.0mL volumes.

Use: For the cultivation of yeasts and filamentous fungi (molds) from cosmetics.

Malt Extract Agar (ATCC Medium 109)

Composition per liter:

Agar....................15.0g
Maltose....................12.75g
Dextrin....................2.75g
Glycerol....................2.35g
Pancreatic digest of gelatin....................0.78g

pH 4.6 ± 0.2 at 25°C

Source: This medium is available as a premixed powder from BBL Microbiology Systems and Difco Laboratories.

Preparation of Medium: Add components to distilled/deionized water and bring volume to 1.0L. Mix thoroughly. Gently heat while stirring until boiling. Distribute into tubes or flasks. Autoclave for 15 min at 15 psi pressure–118°C. Do not overheat or agar will not harden. Pour into sterile Petri dishes or distribute into sterile tubes.

Use: For the cultivation and maintenance of yeasts, molds and *Flavobacterium lucecoloratum*.

Malt Extract Broth

Composition per liter:

Malt extract....................6.0g
Maltose....................1.8g
Glucose....................6.0g
Yeast extract....................1.2g

pH 4.7 ± 0.2 at 25°C

Source: This medium is available as a premixed powder from Difco Laboratories.

Preparation of Medium: Add components to distilled/deionized water and bring volume to 1.0L. Mix thoroughly. Distribute into tubes or flasks. Autoclave for 15 min at 15 psi pressure–121°C. Do not overheat—this results in darkening of the broth.

Use: For the isolation, cultivation and enumeration of yeast and filamentous fungi (mold).

Malt Extract Broth

Composition per liter:

Malt extract, purified solids....................15.0g

pH 4.7 ± 0.2 at 25°C

Source: This medium is available as a premixed powder from BBL Microbiology Systems.

Preparation of Medium: Add components to distilled/deionized water and bring volume to 1.0L. Mix thoroughly. Distribute into tubes or flasks. Autoclave for 15 min at 15 psi pressure–118°C. Do not overheat.

Use: For the cultivation of yeasts and molds.

Manganese Agar No. 1 (Mn Agar No. 1)

Composition per liter:

Agar....................10.0g
$MnCO_3$....................2.0g
Beef extract....................1.0g
$Fe(NH_4)_2(SO_4)_2$....................0.15g
Sodium citrate....................0.15g
Yeast extract....................0.075g
Cyanocobalamin solution....................10.0mL

Cyanocobalamin Solution:

Composition per 10mL:

Cyanocobalamin....................0.005mg

Preparation of Cyanocobalamin Solution: Add cyanocobalamin to distilled/deionized water and bring volume to 10.0mL. Mix thoroughly. Filter sterilize.

Preparation of Medium: Add components, except cyanocobalamin, to distilled/deionized water and bring volume to 990.0mL. Mix thoroughly. Autoclave for 15 min at 15 psi pressure–121°C. Cool to 45–50°C. Aseptically add 10.0mL of the sterile cyanocobalamin solution. Mix thoroughly. Pour into sterile Petri dishes or distribute into sterile tubes.

Use: For the isolation and cultivation of iron and sulfur bacteria. Also used to differentiate *Leptothrix* (*Sphaerotilus*) *discophorus* from *Sphaerotilus natans*.

Manganese Agar No. 2 (Mn Agar No. 2)

Composition per liter:

Agar....................15.0g
$MnSO_4 \cdot H_2O$....................10.0mg

Preparation of Medium: Add components to distilled/deionized water and bring volume to 1.0L. Mix thoroughly. Gently heat and bring to boiling. Distrib-

ute into tubes or flasks. Autoclave for 15 min at 15 psi pressure–121°C. Pour into sterile Petri dishes or leave in tubes. Use freshly prepared solution.

Use: For the enumeration, enrichment and isolation of iron and sulfur bacteria. For the isolation and cultivation of *Leptothrix* species from water.

Manganese Medium for *Pseudomonas* species

Composition per liter:

Noble agar	10.0g
$MnCO_3$	1.0g
$Fe(NH_4)_2(SO_4)_2 \cdot 6H_2O$	0.15g
Sodium citrate	0.15g
Yeast extract	0.075g
$Na_4P_2O_7 \cdot 10H_2O$	0.05g

pH 6.8 ± 0.2 at 25°C

Preparation of Medium: Add components to distilled/deionized water and bring volume to 1.0L. Mix thoroughly. Gently heat and bring to boiling. Distribute into tubes or flasks. Autoclave for 15 min at 15 psi pressure–121°C. Pour into sterile Petri dishes or leave in tubes.

Use: For the cultivation and maintenance of *Pseudomonas putida* and other *P.* pecies.

Manganese Nutrient Agar

Composition per liter:

Agar	15.0g
Peptone	5.0g
Meat extract	3.0g
$MnSO_4 \cdot H_2O$	5.0mg

Preparation of Medium: Add components to distilled/deionized water and bring volume to 1.0L. Mix thoroughly. Gently heat and bring to boiling. Distribute into tubes or flasks. Autoclave for 15 min at 15 psi pressure–121°C. Pour into sterile Petri dishes or leave in tubes.

Use: For the cultivation and obtaining sporulation of *Bacillus* species.

Marine Agar 2216

Composition per liter:

NaCl	19.45g
Agar	15.0g
$MgCl_2$	8.8g
Peptone	5.0g
Na_2SO_3	3.24g
$CaCl_2$	1.8g
Yeast extract	1.0g
KCl	0.55g
$NaHCO_3$	0.16g
Ferric citrate	0.1g
KBr	0.08g
$SrCl_2$	0.03g
H_3BO_3	0.02g
Na_2HPO_4	8.0mg
Na_2SiO_3	4.0mg
NaF	2.4mg
NH_4NO_3	1.6mg

pH 7.6 ± 0.2 at 25°C

Source: This medium is available as a premixed powder from Difco Laboratories.

Preparation of Medium: Add components to distilled/deionized water and bring volume to 1.0L. Mix thoroughly. Gently heat while stirring and bring to boiling. Distribute into tubes or flasks. Autoclave for 15 min at 15 psi pressure–121°C. Pour into sterile Petri dishes or leave in tubes.

Use: For the isolation, cultivation and maintenance of a wide variety of heterotrophic marine bacteria.

Marine Agar with κ- and λ-Carrageenan

Composition per 1070mL:

Solution A	1.0L
Solution B	60.0mL
Solution C	10.0mL

pH 7.2 ± 0.2 at 25°C

Solution A:

Composition per liter:

NaCl	25.0g
Agar	15.0g
$MgSO_4 \cdot 7H_2O$	5.0g
Casamino acids	2.5g
$NaNO_3$	2.0g
κ-Carrageenan	1.25g
λ-Carrageenan	1.25g
$CaCl_2 \cdot 2H_2O$	0.2g
KCl	0.1g

Preparation of Solution A: Add components to distilled/deionized water and bring volume to 1.0L. Mix thoroughly. Gently heat and bring to boiling. Autoclave for 15 min at 15 psi pressure–121°C.

Solution B:

Composition per 100mL:

$Na_2HPO_4 \cdot 2H_2O$	3.56g

Preparation of Solution B: Add component to distilled/deionized water and bring volume to 100.0mL. Mix thoroughly. Autoclave for 15 min at 15 psi pressure–121°C.

Solution C:

Composition per 100mL:

$FeSO_4 \cdot 7H_2O$	0.3g

Preparation of Solution C: Add component to distilled/deionized water and bring volume to 100.0mL. Mix thoroughly. Autoclave for 15 min at 15 psi pressure–121°C.

Preparation of Medium: Aseptically add 60.0mL of sterile solution B and 10.0mL of sterile solution C to 1.0L of sterile solution A. Mix thoroughly. Pour into sterile Petri dishes or distribute into sterile tubes.

Use: For the cultivation and maintenance of *Pseudomonas carrageenovora*.

Marine Broth 2216

Composition per liter:

NaCl	19.45g
$MgCl_2$	8.8g
Peptone	5.0g
Na_2SO_3	3.24g
$CaCl_2$	1.8g
Yeast extract	1.0g
KCl	0.55g
$NaHCO_3$	0.16g
Ferric citrate	0.1g
KBr	0.08g
$SrCl_2$	0.03g
H_3BO_3	0.02g
Na_2HPO_4	8.0mg
Na_2SiO_3	4.0mg
NaF	2.4mg
NH_4NO_3	1.6mg

pH 7.6 ± 0.2 at 25°C

Source: This medium is available as a premixed powder from Difco Laboratories.

Preparation of Medium: Add components to distilled/deionized water and bring volume to 1.0L. Mix thoroughly. Gently heat while stirring and bring to boiling. Distribute into tubes or flasks. Autoclave for 15 min at 15 psi pressure–121°C.

Use: For the isolation, cultivation and maintenance of a wide variety of heterotrophic marine bacteria.

Marine Broth with κ- and λ-Carrageenan

Composition per 1070mL:

Solution A	1.0L
Solution B	60.0mL
Solution C	10.0mL

pH 7.2 ± 0.2 at 25°C

Solution A:

Composition per liter:

NaCl	25.0g
$MgSO_4 \cdot 7H_2O$	5.0g
Casamino acids	2.5g
$NaNO_3$	2.0g
κ-Carrageenan	1.25g
λ-Carrageenan	1.25g
$CaCl_2 \cdot 2H_2O$	0.2g
KCl	0.1g

Preparation of Solution A: Add components to distilled/deionized water and bring volume to 1.0L. Mix thoroughly. Gently heat and bring to boiling. Autoclave for 15 min at 15 psi pressure–121°C.

Solution B:

Composition per 100mL:

$Na_2HPO_4 \cdot 2H_2O$	3.56g

Preparation of Solution B: Add components to distilled/deionized water and bring volume to 100.0mL. Mix thoroughly. Autoclave for 15 min at 15 psi pressure–121°C.

Solution C:

Composition per 100mL:

$FeSO_4 \cdot 7H_2O$	0.3g

Preparation of Solution C: Add components to distilled/deionized water and bring volume to 100.0mL. Mix thoroughly. Autoclave for 15 min at 15 psi pressure–121°C.

Preparation of Medium: Aseptically add 60.0mL of sterile solution B and 10.0mL of sterile solution C to 1.0L of sterile solution A. Mix thoroughly. Distribute into sterile tubes or flasks.

Use: For the cultivation and maintenance of *Pseudomonas carrageenovora*.

Marine Chlorobiaceae Medium 2

Composition per 1051mL:

Solution 1	950.0mL
$Na_2S \cdot 9H_2O$ solution	60.0mL
$NaHCO_3$ solution	40.0mL
Vitamin B_{12} solution	1.0mL

pH 6.8 ± 0.2 at 25°C

Solution 1:

Composition per 950mL:

NaCl	20.0g
$MgSO_4 \cdot 7H_2O$	3.0g
KH_2PO_4	1.0g
NH_4Cl	0.5g
$CaCl_2 \cdot 2H_2O$	0.05g
Trace element solution SL-8	1.0mL

Preparation of Solution 1: Add components to distilled/deionized water and bring volume to 950.0mL. Mix thoroughly. Autoclave for 15 min at 15 psi pressure–121°C. Cool to 45–50°C.

Trace Element Solution SL-8:

Composition per liter:

Disodium EDTA	5.2g
$FeCl_2 \cdot 4H_2O$	1.5g
$CoCl_2 \cdot 6H_2O$	0.19g
$MnCl_2 \cdot 4H_2O$	0.10g
$ZnCl_2$	0.07g
H_3BO_3	0.06g
$NaMoO_4 \cdot 2H_2O$	0.04g
$CuCl_2 \cdot 2H_2O$	0.02g
$NiCl_2 \cdot 6H_20$	0.02g

Preparation of Trace Element Solution SL-8: Add components to distilled/deionized water and bring volume to 1.0L. Mix thoroughly.

$Na_2S \cdot 9H_2O$ Solution:

Composition per 100mL:

$Na_2S \cdot 9H_2O$	5.0g

Preparation of $Na_2S \cdot 9H_2O$ Solution: Add $Na_2S \cdot 9H_2O$ to distilled/deionized water and bring volume to 100.0mL. Autoclave for 15 min at 15 psi pressure–121°C. Cool to 45–50°C.

$NaHCO_3$ Solution:

Composition per 100mL:

$NaHCO_3$	5.0g

Preparation of $NaHCO_3$ Solution: Add $NaHCO_3$ to distilled/deionized water and bring volume to 100.0mL. Mix thoroughly. Filter sterilize.

Vitamin B_{12} Solution:

Composition per 100mL:

Vitamin B_{12}	2.0mg

Preparation of Vitamin B_{12} Solution: Add vitamin B_{12} to distilled/deionized water and bring volume to 100.0mL. Mix thoroughly. Filter sterilize.

Preparation of Medium: To 950.0mL of cooled, sterile solution 1, aseptically add 60.0mL of sterile $Na_2S \cdot 9H_2O$ solution, 40.0mL of sterile $NaHCO_3$ solution, and 1.0mL of sterile vitamin B_{12} solution. Mix thoroughly. Adjust pH to 6.8 with sterile H_2SO_4 or Na_2CO_3. Aseptically distribute into sterile 50.0mL or 100.0mL bottles with metal screw caps and rubber seals. Completely fill bottles with medium except for a pea-sized air bubble.

Use: For the isolation and cultivation of marine members of the Chlorobiaceae.

Marine Chromatiaceae Medium 2

Composition per 1051mL:

Solution 1	950.0mL
$Na_2S \cdot 9H_2O$ solution	60.0mL
$NaHCO_3$ solution	40.0mL
Vitamin B_{12} solution	1.0mL

pH 7.3± 0.2 at 25°C

Solution 1:

Composition per 950mL:

NaCl	20.0g
$MgSO_4 \cdot 7H_2O$	3.0g
KH_2PO_4	1.0g
NH_4Cl	0.5g
$CaCl_2 \cdot 2H_2O$	0.05g
Trace element solution SL-8	1.0mL

Preparation of Solution 1: Add components to distilled/deionized water and bring volume to 950.0mL. Mix thoroughly. Autoclave for 15 min at 15 psi pressure–121°C. Cool to 45–50°C.

Trace Element Solution SL-8:

Composition per liter:

Disodium EDTA	5.2g
$FeCl_2 \cdot 4H_2O$	1.5g
$CoCl_2 \cdot 6H_2O$	0.19g
$MnCl_2 \cdot 4H_2O$	0.10g
$ZnCl_2$	0.07g
H_3BO_3	0.06g
$NaMoO_4 \cdot 2H_2O$	0.04g
$CuCl_2 \cdot 2H_2O$	0.02g
$NiCl_2 \cdot 6H_20$	0.02g

Preparation of Trace Element Solution SL-8: Add components to distilled/deionized water and bring volume to 1.0L. Mix thoroughly.

$Na_2S \cdot 9H_2O$ Solution:

Composition per 100mL:

$Na_2S \cdot 9H_2O$	5.0g

Preparation of $Na_2S \cdot 9H_2O$ Solution: Add $Na_2S \cdot 9H_2O$ to distilled/deionized water and bring volume to 100.0mL. Autoclave for 15 min at 15 psi pressure–121°C. Cool to 45–50°C.

$NaHCO_3$ Solution:

Composition per 100mL:

$NaHCO_3$	5.0g

Preparation of $NaHCO_3$ Solution: Add $NaHCO_3$ to distilled/deionized water and bring volume to 100.0mL. Mix thoroughly. Filter sterilize.

Vitamin B_{12} Solution:

Composition per 100mL:

Vitamin B_{12}	2.0mg

Preparation of Vitamin B_{12} Solution: Add vitamin B_{12} to distilled/deionized water and bring volume to 100.0mL. Mix thoroughly. Filter sterilize.

Preparation of Medium: To 950.0mL of cooled, sterile solution 1, aseptically add 60.0mL of sterile $Na_2S{\cdot}9H_2O$ solution, 40.0mL of sterile $NaHCO_3$ solution, and 1.0mL of sterile vitamin B_{12} solution. Mix thoroughly. Adjust pH to 7.3 with sterile H_2SO_4 or Na_2CO_3. Aseptically distribute into sterile 50.0mL or 100.0mL bottles with metal screw caps and rubber seals. Completely fill bottles with medium except for a pea-sized air bubble.

Use: For the isolation and cultivation of marine members of the Chromatiaceae.

Marine *Cytophaga* Agar

Composition per liter:

Agar	15.0g
Nutrient broth	8.0g
Yeast extract	5.0g
Salt solution	1.0L

Salt Solution:

Composition per liter:

NaCl	12.86g
$MgCl_2$	2.48g
KCl	0.75g
$CaCl_2$	0.56g
$Fe(SO_4)_2(NH_4)_2$	0.048g

Preparation of Salt Solution: Add components to distilled/deionized water and bring volume to 1.0L. Mix thoroughly.

Preparation of Medium: Add solid components to 1.0L of salt solution. Mix thoroughly. Gently heat while stirring and bring to boiling. Distribute into tubes or flasks. Autoclave for 15 min at 15 psi pressure–121°C. Pour into sterile Petri dishes or leave in tubes.

Use: For the cultivation and maintenance of *Cytophaga* species.

Marine Glucose Trypticase™ Yeast Extract Agar (MGTY Agar)

Composition per liter:

Agar	8.0g
Glucose	2.0g
Pancreatic digest of casein	1.0g
Yeast extract	1.0g
L-Cysteine·HCl·H_2O	0.5g
Seawater	750.0mL
Tris-HCl buffer (5.0 m*M*, pH 7.5)	50.0mL
Resazurin (0.1% solution)	1.0mL

pH 7.5 ± 0.2 at 25°C

Preparation of Medium: Add components to distilled/deionized water and bring volume to 1.0L. Mix thoroughly. Gently heat while stirring and bring to boiling. Distribute into tubes or flasks under 97% N_2 + 3% H_2. Cap with rubber stoppers and place tubes in a press. Autoclave for 15 min at 15 psi pressure–121°C with fast exhaust.

Use: For the cultivation and maintenance of *Spirochaeta isovalerica.*

Marine Glucose Trypticase™ Yeast Extract Broth (MGTY Broth)

Composition per liter:

Glucose	2.0g
Pancreatic digest of casein	1.0g
Yeast extract	1.0g
L-Cysteine·HCl·H_2O	0.5g
Seawater	750.0mL
Tris-HCl buffer (5.0 m*M*, pH 7.5)	50.0mL
Resazurin (0.1% solution)	1.0mL

pH 7.5 ± 0.2 at 25°C

Preparation of Medium: Add components to distilled/deionized water and bring volume to 1.0L. Mix thoroughly. Gently heat while stirring and bring to boiling. Distribute into tubes or flasks under 97% N_2 + 3% H_2. Cap with rubber stoppers and place tubes in a press. Autoclave for 15 min at 15 psi pressure–121°C with fast exhaust.

Use: For the cultivation and maintenance of *Spirochaeta isovalerica.*

Marine Methanol Medium

Composition per liter:

NaCl	20.0g
$(NH_4)_2SO_4$	2.0g
K_2HPO_4	2.0g
KH_2PO_4	1.0g
$MgSO_4{\cdot}7H_2O$	0.3g
Methanol	10.0mL
Vitamin B_{12} solution	10.0mL
Trace metals solution	1.0mL

pH 7.0 ± 0.2 at 25°C

Vitamin B_{12} Solution:

Composition per 100mL:

Vitamin B_{12}	10.0μg

Preparation of Vitamin B_{12} Solution: Add the vitamin B_{12} to distilled/deionized water and bring volume to 100.0mL. Adjust pH to 5. Autoclave for 15 min at 15 psi pressure–121°C.

Trace Metals Solution:

Composition per liter:

$ZnSO_4·7H_2O$	1.4g
$MnSO_4·H_2O$	0.84g
$FeSO_4·7H_2O$	0.28g
$CuSO_4·5H_2O$	0.25g
$Na_2MoO_4·2H_2O$	0.24g
$CoCl_2·6H_2O$	0.24g
$CaCl_2·2H_2O$	0.15g

Preparation of Trace Metals Solution: Add components to distilled/deionized water and bring volume to 1.0L. Mix thoroughly.

Preparation of Medium: Add components, except vitamin B_{12} solution and methanol to distilled/deionized water and bring volume to 980.0mL. Adjust pH to 7.0 with NaOH. Autoclave for 15 min at 15 psi pressure–121°C. Filter sterilize methanol. Aseptically add sterile vitamin B_{12} solution and filter-sterilized methanol. Distibute into sterile tubes or flasks.

Use: For the cultivation and maintenance of *Methylophaga thalassica.*

Marine Peptone Succinate Salts Medium (PSS Medium)

Composition per liter:

Peptone	10.0g
Succinic acid	1.0g
$(NH_4)_2SO_4$	1.0g
$MgSO_4·7H_2O$	1.0g
$FeCl_3·6H_2O$	2.0mg
$MnSO_4·H_2O$	2.0mg
Synthetic seawater	1.0L

pH 6.8 ± 0.2 at 25°C

Synthetic Seawater:

Composition per liter:

NaCl	27.5g
$MgCl_2$	5.0g
$MgSO_4·7H_2O$	2.0g
KCl	1.0g
$CaCl_2$	0.5g
$FeSO_4$	1.0mg

Preparation of Synthetic Seawater: Add components to distilled/deionized water and bring volume to 1.0L. Mix thoroughly.

Preparation of Medium: Add components to 1.0L of synthetic seawater. Mix thoroughly. Gently heat while stirring and bring to boiling. Adjust pH to 6.8 with KOH. Distribute into tubes or flasks. Autoclave for 15 min at 15 psi pressure–121°C.

Use: For the cultivation and maintenance of *Oceanospirillum beijerinckii* and *O. multiglobuliferum.*

Marine Peptone Yeast Medium with Magnesium Sulfate

Composition per liter:

NaCl	20.0g
Peptone	10.0g
$MgSO_4·7H_2O$	2.0g
$(NH_4)_2SO_4$	2.0g
Yeast extract	1.0g

pH 7.0 ± 0.2 at 25°C

Preparation of Medium: Add components to distilled/deionized water and bring volume to 1.0L. Mix thoroughly. Distribute into tubes or flasks. Autoclave for 15 min at 15 psi pressure–121°C.

Use: For the cultivation and maintenance of *Oceanospirillum pusillum.*

Marine *Pseudomonas* Medium

Composition per liter:

Agar	15.0g
Nutrient broth	8.0g
Yeast extract	5.0g
Salt solution	1.0L

Salt Solution:

Composition per liter:

NaCl	12.86g
$MgCl_2$	2.48g
KCl	0.75g
$CaCl_2$	0.56g
$Fe(SO_4)_2(NH_4)_2$	0.048g

Preparation of Salt Solution: Add components to distilled/deionized water and bring volume to 1.0L. Mix thoroughly.

Preparation of Medium: Add components to 1.0L of salt solution. Mix thoroughly. Gently heat and bring to boiling. Distribute into tubes or flasks. Autoclave for 15 min at 15 psi pressure–121°C. Pour into sterile Petri dishes or leave in tubes.

Use: For the cultivation and maintenance of *Alteromonas haloplanktis.*

Marine *Rhodococcus* Medium

Composition per liter:

Yeast extract	10.0g
Malt extract	4.0g

Glucose 4.0g
Seawater 750.0mL

Preparation of Medium: Add components to distilled/deionized water and bring volume to 1.0L. Mix thoroughly. Gently heat while stirring and bring to boiling. Distribute into tubes or flasks. Autoclave for 15 min at 15 psi pressure–121°C.

Use: For the cultivation and maintenance of *Rhodococcus marinonascens.*

Marine *Rhodopseudomonas* Medium

Composition per liter:

NaCl 30.4g
Yeast extract 1.0g
Disodium succinate 1.0g
KH_2PO_4 ....0.5g
$MgSO_4 \cdot 7H_2O$ 0.4g
NH_4Cl 0.4g
$CaCl_2 \cdot 2H_2O$ 0.05g
Ferric citrate (0.1% solution) 5.0mL
Trace elements SL-6 ....1.0mL
Ethanol 0.5mL

pH 6.8 ± 0.2 at 25°C

Trace Elements Solution SL-6:

Composition per liter:

H_3BO_3 ....0.3g
$CoCl_2 \cdot 6H_2O$ 0.2g
$ZnSO_4 \cdot 7H_2O$ 0.10g
$MnCl_2 \cdot 4H_2O$ 0.03g
$Na_2MoO_4 \cdot H_2O$ 0.03g
$NiCl_2 \cdot 6H_2O$ 0.02g
$CuCl_2 \cdot 2H_2O$ 0.01g

Preparation of Trace Elements Solution SL-6: Add components to distilled/deionized water and bring volume to 1.0L. Mix thoroughly. Adjust pH to 3.4.

Preparation of Medium: Add components to distilled/deionized water and bring volume to 1.0L. Mix thoroughly. Gently heat while stirring and bring to boiling. Distribute into tubes or flasks. Autoclave for 15 min at 15 psi pressure–121°C.

Use: For the cultivation and maintenance of *Rhodopseudomonas marina.*

Marine Salts Medium for *Sporosarcina halophila*

Composition per liter:

Marine salts mix 100.0g
Agar 20.0g
Yeast extract 10.0g
Proteose peptone No. 3 ....5.0g
Glucose 1.0g

pH 7.0 ± 0.2 at 25°C

Preparation of Medium: Add components to distilled/deionized water and bring volume to 1.0L. Mix thoroughly. Gently heat while stirring and bring to boiling. Distribute into tubes or flasks. Autoclave for 15 min at 15 psi pressure–121°C. Pour into sterile Petri dishes or leave in tubes.

Use: For the cultivation and maintenance of *Sporosarcina halophila.*

Marine Spirochete Medium

Composition per liter:

Cellobiose 2.0g
Peptone 2.0g
Yeast extract 1.0g
Sodium thioglycollate 1.0g
Seawater, charcoal-filtered 800.0mL

pH 7.5 ± 0.2 at 25°C

Preparation of Medium: Add components, except sodium thioglycollate, to glass distilled water and bring volume to 1.0L. Mix thoroughly. Bubble 100% N_2 into medium for 1.5 min. Add sodium thioglycollate. Adjust pH to 7.5 with 10*N* KOH. Distribute into tubes or flasks. Autoclave for 15 min at 15 psi pressure–121°C.

Use: For the cultivation and maintenance of *Spirochaeta bajacaliforniensis.*

Maximum Recovery Diluent

Composition per liter:

NaCl 8.5g
Peptone 1.0g

pH 7.0 ± 0.2 at 25°C

Source: This medium is available as a premixed powder from Oxoid Unipath.

Preparation of Medium: Add components to distilled/deionized water and bring volume to 1.0L. Mix thoroughly. Distribute into tubes or flasks. Autoclave for 15 min at 15 psi pressure–121°C.

Use: This diluent is a physiologically isotonic and protective medium for maximal recovery of microorganisms from a variety of sources.

MBM Acetate Medium (Mineral Base Medium with Acetate)

Composition per liter:

Agar 16.0g
NaCl 5.0g

K_2HPO_4 .... 1.0g
$NH_4H_2PO_4$ 1.0g
Sodium acetate·$3H_2O$ 1.0g
$MgSO_4·7H_2O$ 0.1g
Bromothymol blue 0.01g
pH 6.5 ± 0.2 at 25°C

Preparation of Medium: Add components to distilled/deionized water and bring volume to 1.0L. Mix thoroughly. Adjust pH to 6.5. Gently heat and bring to boiling. Distribute into screw-capped tubes in 3.0mL volumes. Autoclave for 15 min at 15 psi pressure–121°C. Allow tubes to cool in a slanted position.

Use: For determining the nutritional independence of bacteria. Bacteria that are nutritionally independent turn the medium blue.

MD 1 Medium

Composition per liter:
Pancreatic digest of casein 3.0g
$MgSO_4·7H_2O$ 2.0g
$CaCl_2$ 0.5g
Trace element solution 1.0mL
Vitamin B_{12} solution 1.0mL

Trace Element Solution:
Composition per liter:
EDTA 8.0g
$MnCl_2·4H_2O$ 0.1g
$CoCl_2$ 0.02g
KBr 0.02g
$ZnCl_2$ 0.02g
$CuSO_4$ 0.01g
H_3BO_3 .... 0.01g
$NaMoO_4·2H_2O$ 0.01g
$BaCl_2$ 5.0mg
LiCl 5.0mg
$SnCl_2·2H_2O$ 5.0mg

Preparation of Trace Element Solution: Add components to distilled/deionized water and bring volume to 1.0L. Mix thoroughly.

Vitamin B_{12} Solution:
Composition per 10mL:
Vitamin B_{12} 5.0mg

Preparation of Vitamin B_{12} Solution: Add vitamin B_{12} to distilled/deionized water and bring volume to 10.0mL. Mix thoroughly.

Preparation of Medium: Add components to distilled/deionized water and bring volume to 1.0L. Mix thoroughly. Distribute into tubes or flasks. Autoclave for 15 min at 15 psi pressure–121°C.

Use: For the cultivation of myxobacteria.

MED IIa

Composition per liter:
Tris buffer stock solution 10.0mL
$CaCl_2$ (5.0% solution) 10.0mL
$MgSO_4·7H_2O$ (3.33% solution) 1.0mL
pH 7.2 ± 0.2 at 25°C

Tris Buffer Stock Solution:
Composition per 500mL:
Tris(hydroxymethyl)aminomethane·HCl 35.01g
Tris(hydroxymethyl)aminomethane 3.35g

Preparation of Tris Buffer Stock Solution: Add components to distilled/deionized water and bring volume to 500.0mL. Mix thoroughly. Adjust pH to 7.2.

Preparation of Medium: Add components to distilled/deionized water and bring volume to 1.0L. Mix thoroughly. Distribute into tubes or flasks. Autoclave for 20 min at 15 psi pressure–121°C.

Use: For the cultivation and maintenance of *Vampirovibrio chlorellavorus*.

Medium 2A

Composition per liter:
Arginine 10.0g
NaCl 5.0g
Agar 4.0g
Peptone 1.0g
$K_2HPO_4·3H_2O$ 0.3g
Phenol red 0.01g
pH 7.2–7.4 at 25°C

Preparation of Medium: Add components to distilled/deionized water and bring volume to 1.0L. Mix thoroughly. Gently heat and bring to boiling. Distribute into tubes. Autoclave for 15 min at 15 psi pressure–121°C.

Use: For the cultivation and differentiation of *Pseudomonas* species based on their production of arginine dihydrolase activity.

Medium A

Composition per liter:
D-Glucose 20.0g
Agar 20.0g
Yeast extract 10.0g
Biotin 1.0mg
Calcium pantothenate 1.0mg
pH 7.3 ± 0.2 at 25°C

Preparation of Medium: Add components, except biotin and calcium pantothenate, to distilled/

deionized water and bring volume to 990.0mL. Mix thoroughly. Gently heat and bring to boiling. Autoclave for 15 min at 15 psi pressure–121°C. Cool to 45–50°C. Add biotin and calcium pantothenate to distilled deionized water and bring volume to 10.0mL. Mix thoroughly. Filter sterilize. Aseptically add the sterile biotin and calcium pantothenate solution to the cooled sterile basal medium. Mix thoroughly. Pour into sterile Petri dishes or distribute into sterile tubes.

Use: For the cultivation and maintenance of *Zymomonas mobilis*.

Medium AS4

Composition per liter:

Sucrose	80.0g
PPLO broth without crystal violet	500.0mL
Horse serum	200.0mL
Phenol red (0.5% solution)	5.0mL

pH 7.2 ± 0.2 at 25°C

PPLO Broth without Crystal Violet:

Composition per 500mL:

Beef heart (solids from infusion)	11.53g
Peptone	2.33g
NaCl	1.15g

Source: PPLO broth without crystal violet is available as a premixed powder from Difco Laboratories.

Preparation of PPLO Broth without Crystal Violet: Add components to distilled/deionized water and bring volume to 500.0mL.Mix thoroughly. Beef heart for infusion may be substituted; 100g of beef heart for infusion are equivalent to 500g of fresh heart tissue.

Preparation of Medium: Add components, except horse serum, to distilled/deionized water and bring volume to 800.0mL. Mix thoroughly. Adjust pH to 7.2. Autoclave for 10 min at 15 psi pressure–121°C. Cool to 45–50°C. Aseptically add 200.0mL of non-inactivated, sterile horse serum. Mix thoroughly. Aseptically distribute into sterile tubes or flasks.

Use: For the cultivation and maintenance of *Spiroplasma melliferum*.

Medium B for Sulfate Reducers (Postgate's Medium B for Sulfate Reducers)

Composition per liter:

Sodium lactate	3.5g
$MgSO_4 \cdot 7H_2O$	2.0g
NH_4Cl	1.0g
$CaSO_4$	1.0g
Yeast extract	1.0g
KH_2PO_4	0.5g
$FeSO_4 \cdot 7H_2O$	0.5g
Ascorbic acid	0.1g
Thioglycollic acid	0.1g

pH 7.0–7.5 at 25°C

Preparation of Medium: Add components, except ascorbic acid and thioglycollic acid, to tap water and bring volume to 1.0L. For marine bacteria, NaCl may be added or seawater used in place of tap water. Mix thoroughly. Adjust pH to 7.0–7.5. Thioglycollate and ascorbate should be added immediately prior to sterilization. Distribute into tubes or flasks. Autoclave for 15 min at 15 psi pressure–121°C.

Use: For isolation, cultivation and maintenance of *Desulfovibrio* species and *Desulfotomaculum* species. This medium turns black as a result of H_2S production due to bacterial growth.

Medium C for Sulfate Reducers (Postgate's Medium C for Sulfate Reducers)

Composition per liter:

Sodium lactate	6.0g
Na_2SO_4	4.5g
NH_4Cl	1.0g
Yeast extract	1.0g
KH_2PO_4	0.5g
Sodium citrate·$2H_2O$	0.3g
$CaCl_2 \cdot 6H_2O$	0.06g
$MgSO_4 \cdot 7H_2O$	0.06g
$FeSO_4 \cdot 7H_2O$	0.004g

pH 7.5 ± 0.2 at 25°C

Preparation of Medium: Add components to distilled/deionized water and bring volume to 1.0L. For marine bacteria, NaCl may be added or seawater used in place of distilled/deionized water. Mix thoroughly. Adjust pH to 7.5. Distribute into tubes or flasks. Autoclave for 15 min at 15 psi pressure–121°C.

Use: For detection, culturing and storage of *Desulfovibrio* species and many *Desulfotomaculum* species. This medium should be used when a clear culture medium is desired such as for chemostat culture. This medium may be cloudy after sterilization but usually clears on cooling. It turns black as a result of H_2S production due to bacterial growth.

Medium D2

Composition per liter:

Agar	15.0g
Glucose	10.0g

LiCl	5.0g
Pancreatic digest of casein	4.0g
Yeast extract	2.0g
Tris(hydroxymethyl)amino-methane·HCl buffer	1.2g
NH_4Cl	1.0g
$MgSO_4 \cdot 7H_2O$	0.3g
Polymyxin sulfate solution	10.0mL
NaN_3 solution	10.0mL

pH 6.9 ± 0.2 at 25°C

Polymyxin Sulfate Solution:
Composition per 10mL:

Polymyxin sulfate	0.04g

Preparation of Polymyxin Sulfate Solution: Add polymyxin sulfate to distilled/deionized water and bring volume to 10.0mL. Mix thoroughly. Filter sterilize. Use freshly prepared solution.

NaN_3 Solution:
Composition per 10mL:

NaN_3	2.0mg

Preparation of NaN_3 Solution: Add NaN_3 to distilled/deionized water and bring volume to 10.0mL. Mix thoroughly. Filter sterilize. Use freshly prepared solution.

Caution: Sodium azide is toxic. Azides also react with metals and disposal must be highly diluted.

Preparation of Medium: Add components, except polymyxin sulfate solution and NaN_3 solution, to distilled/deionized water and bring volume to 980.0mL. Mix thoroughly. Gently heat and bring to boiling. Autoclave for 15 min at 15 psi pressure–121°C. Cool to 45–50°C. Aseptically add sterile polymyxin sulfate solution and NaN_3 solution. Mix thoroughly. Pour into sterile Petri dishes or distribute into sterile tubes.

Use: For the selective isolation and cultivation of *Corynebacterium* species.

Medium D4

Composition per liter:

Agar	15.0g
Sucrose	10.0g
NH_4Cl	5.0g
Na_2HPO_4, anhydrous	2.3g
Pancreatic digest of casein	1.0g
Sodium dodecyl sulfate	0.6g
Glycerol	10.0mL

Preparation of Medium: Add components to distilled/deionized water and bring volume to 1.0L. Mix thoroughly. Gently heat and bring to boiling. Distribute into tubes or flasks. Autoclave for 15 min at 15 psi pressure–121°C. Pour into sterile Petri dishes or leave in tubes.

Use: For the selective isolation and cultivation of *Pseudomonas syringae.*

Medium D for Sulfate Reducers (Postgate's Medium D for Sulfate Reducers)

Composition per liter:

Sodium pyruvate	3.5g
$MgCl_2 \cdot 6H_2O$	1.6g
NH_4Cl	1.0g
Yeast extract	1.0g
KH_2PO_4	0.5g
$CaCl_2 \cdot 2H_2O$	0.1g
$FeSO_4 \cdot 7H_2O$	0.004g

pH 7.5 ± 0.2 at 25°C

Preparation of Medium: Add components to distilled/deionized water and bring volume to 1.0L. Malate or fumarate may also be used as a carbon source. For marine bacteria, NaCl may be added or sea water used in place of distilled/deionized water. Mix thoroughly. Adjust pH to 7.5. Filter sterilize. Aseptically distribute into sterile tubes or flasks.

Use: For cultivation of *Desulfovibrio* species and *Desulfotomaculum* species that can grow in the absence of sulfate.

Medium D for Sulfate Reducers (Postgate's Medium D for Sulfate Reducers)

Composition per liter:

$MgCl_2 \cdot 6H_2O$	1.6g
Choline chloride	1.0g
NH_4Cl	1.0g
Yeast extract	1.0g
KH_2PO_4	0.5g
$CaCl_2 \cdot 2H_2O$	0.1g
$FeSO_4 \cdot 7H_2O$	0.004g

pH 7.5 ± 0.2 at 25°C

Preparation of Medium: Add components to distilled/deionized water and bring volume to 1.0L. Malate or fumarate may also be used as a carbon source. For marine bacteria, NaCl may be added or sea water used in place of distilled/deionized water. Mix thoroughly. Adjust pH to 7.5. Filter sterilize. Aseptically distribute into sterile tubes or flasks.

Use: For cultivation of *Desulfovibrio* species and *Desulfotomaculum* species that can grow in the absence of sulfate.

Medium D for *Thermus*

Composition per liter:

Pancreatic digest of casein	1.0g
Yeast extract	1.0g
$NaNO_3$	0.7g
KNO_3	0.1g
$MgSO_4 \cdot 7H_2O$	0.1g
Na_2HPO_4	0.1g
Nitrilotriacetic acid	0.1g
$CaSO_4 \cdot 2H_2O$	0.06g
NaCl	8.0mg
$MnSO_4 \cdot H_2O$	2.2mg
$ZnSO_4 \cdot 7H_2O$	0.5mg
H_3BO_3	0.5mg
$FeCl_3$	0.28mg
$Na_2MoO_4 \cdot 2H_2O$	0.03mg
$CuSO_4$	0.02mg

pH 8.2 ± 0.2 at 25°C

Preparation of Medium: Add nitrilotriacetic acid to 500.0mL of distilled/deionized water. Dissolve by adjusting pH to 6.5 with KOH. Add remaining components. Readjust pH to 8.2 with H_2SO_4 or KOH. Add distilled/deionized water to 1.0L. Distribute into tubes or flasks. Autoclave for 15 min at 15 psi pressure–121°C.

Use: For the cultivation of *Thermus* species.

Medium E for *Bacillus*

Composition per liter:

NaCl	50.0g
K_2HPO_4	10.6g
Sucrose	10.0g
KH_2PO_4	5.3g
$(NH_4)_2SO_4$	1.0g
$MgSO_4$	0.25g
Trace salts solution	10.0mL

Trace Salts Solution:

Composition per liter:

$MnSO_4 \cdot H_2O$	3.0g
Disodium EDTA	1.0g
$FeSO_4 \cdot 7H_2O$	0.1g
$CaCl_2 \cdot 2H_2O$	0.1g
$CoCl_2 \cdot 6H_2O$	0.1g
$ZnSO_4 \cdot 7H_2O$	0.1g
$CuSO_4 \cdot 5H_2O$	0.01g
$AlK(SO_4)_2 \cdot 12H_2O$	0.01g
H_3BO_3	0.01g
$Na_2MoO_4 \cdot 2H_2O$	0.01g

Preparation of Trace Salts Solution: Add components to distilled/deionized water and bring volume to 1.0L. Mix thoroughly.

Preparation of Medium: Add components to distilled/deionized water and bring volume to 1.0L. Mix thoroughly. Autoclave for 15 min at 15 psi pressure–121°C. Aseptically distribute into sterile tubes or flasks.

Use: For the cultivation and maintenance of *Bacillus* species.

Medium E for Sulfate Reducers (Postgate's Medium E for Sulfate Reducers)

Composition per liter:

Agar	15.0g
Sodium lactate	3.5g
$MgCl_2 \cdot 6H_2O$	2.0g
NH_4Cl	1.0g
Na_2SO_4	1.0g
$CaCl_2 \cdot 2H_2O$	1.0g
Yeast extract	1.0g
KH_2PO_4	0.5g
Ascorbic acid	0.1g
Thioglycollic acid	0.1g
$FeSO_4 \cdot 7H_2O$	0.004g

pH 7.6 ± 0.2 at 25°C

Preparation of Medium: Add components, except ascorbic acid and thioglycollic acid, to tap water and bring volume to 1.0L. For marine bacteria, NaCl may be added or sea water used in place of tap water. Mix thoroughly. Gently heat and bring to boiling. Adjust pH to 7.6. Thioglycollate and ascorbate should be added immediately prior to sterilization. Distribute into screw-capped tubes or flasks. Autoclave for 15 min at 15 psi pressure–121°C.

Use: For the cultivation and enumeration of *Desulfovibrio* species and *Desulfotomaculum* species as black colonies in deep agar cultures. Also used for isolation of pure cultures of of *Desulfovibrio* species and *Desulfotomaculum* species.

Medium F

Composition per liter:

$MgSO_4 \cdot 7H_2O$	0.5g
$(NH_4)_2SO_4$	0.15g
KCl	0.05g
KH_2PO_4	0.05g
$Ca(NO_3)_2$	0.01g
$FeSO_4 \cdot 7H_2O$ solution	10.0mL

pH 3.5 ± 0.2 at 25°C

$FeSO_4 \cdot 7H_2O$ Solution:

Composition per 10mL:

$FeSO_4 \cdot 7H_2O$	1.0g

Preparation of $FeSO_4 \cdot 7H_2O$ Solution: Add the $FeSO_4 \cdot 7H_2O$ to distilled/deionized water and bring volume to 10.0mL. Mix thoroughly. Filter sterilize.

Preparation of Medium: Add components, except $FeSO_4 \cdot 7H_2O$ solution, to tap water and bring volume to 990.0mL. Mix thoroughly. Gently heat until dissolved. Adjust pH to 3.5. Autoclave for 15 min at 15 psi pressure–121°C. Cool to 45–50°C. Aseptically add 10.0mL of sterile $FeSO_4 \cdot 7H_2O$ solution. Mix thoroughly. Aseptically distribute into sterile tubes or flasks.

Use: For the cultivation of *Thiobacillus* species.

Medium F for Sulfate Reducers (Postgate's Medium F for Sulfate Reducers)

Composition per liter:

Agar	12.0g
Pancreatic digest of casein	10.0g
Sodium lactate	3.5g
Ferrous citrate	0.5g
Na_2SO_3	0.5g
$MgSO_4 \cdot 7H_2O$	0.2g

pH 7.1 ± 0.2 at 25°C

Preparation of Medium: Add components, except ascorbic acid and thioglycollic acid, to tap water and bring volume to 1.0L. For marine bacteria, NaCl may be added or seawater used in place of tap water. Mix thoroughly. Gently heat and bring to boiling. Adjust pH to 7.1. Thioglycollate and ascorbate should be added immediately prior to sterilization. Distribute into screw-capped tubes or flasks. Autoclave for 15 min at 15 psi pressure–121°C.

Use: For isolation and cultivation of *Desulfotomaculum nigrificans, Desulfovibrio* species and other *Desulfotomaculum* species especially in food. These bacteria form black colonies in deep agar cultures.

Medium G for Sulfate Reducers (Postgate's Medium G for Sulfate Reducers)

Composition per 1015.2mL:

Solution 1	970.0mL
Solution 4	30.0mL
Solution 8A, 8B, 8C, 8D or 8E	10.0mL
Solution 5	3.0mL
Solution 2	1.0mL
Solution 3	1.0mL
Solution 6	0.1mL
Solution 7	0.1mL

pH 7.2 ± 0.2 at 25°C

Solution 1:

Composition per 970mL:

Na_2SO_4	3.0g
NaCl	1.2
$MgCl_2 \cdot 6H_2O$	0.4g
KCl	0.3g
NH_4Cl	0.3g
KH_2PO_4	0.2g
$CaCl_2 \cdot 2H_2O$	0.15g

Preparation of Solution 1: Add components to distilled/deionized water and bring volume to 970.0mL. Mix thoroughly. Adjust pH to 7.2 with 2*N* HCl. Autoclave for 15 min at 15 psi pressure–121°C. Cool to 25°C.

Solution 2:

Composition per 10mL:

NaOH	5.0mg
Na_2SeO_3	0.03mg

Preparation of Solution 2: Add NaOH and Na_2SeO_3 to distilled/deionized water and bring volume to 10.0mL. Mix thoroughly. Autoclave for 15 min at 15 psi pressure–121°C. Cool to 25°C.

Solution 3:

Composition per liter:

$FeCl_2 \cdot 4H_2O$	1.5g
$CoCl_2 \cdot 6H_2O$	0.12g
$MnCl_2 \cdot 4H_2O$	0.1g
$ZnCl_2$	0.070g
H_3BO_3	0.060g
$NiCl_2 \cdot 6H_2O$	0.025g
$NaMoO_4 \cdot 2H_2O$	0.025g
$CuCl_2 \cdot 2H_2O$	0.015g

Preparation of Solution 3: Add components to distilled/deionized water and bring volume to 1.0L. Mix thoroughly. Autoclave for 15 min at 15 psi pressure–121°C. Cool to 25°C.

Solution 4:

Composition per 30mL:

$NaHCO_3$	2.55g

Preparation of Solution 4: Add $NaHCO_3$ to distilled/deionized water and bring volume to 30.0mL. Mix thoroughly. Gas with 100% CO_2 for 10–15 min. Filter sterilize.

Solution 5:

Composition per 3mL:

$Na_2S \cdot 9H_2O$	0.36g

Preparation of Solution 5: Add $Na_2S \cdot 9H_2O$ to distilled/deionized water and bring volume to 3.0mL. Mix thoroughly. Gas with 100% N_2 for 5–10 min. Cap tube with a rubber stopper. Autoclave for 15 min at 15 psi pressure–121°C. Cool to 25°C.

Solution 6:

Composition per 100mL:

Thiamine·HCl	0.010g
Cyanocobalamin	5.0mg
p-Aminobenzoic acid	5.0mg
Biotin	1.0mg

Preparation of Solution 6: Add components to distilled/deionized water and bring volume to 100.0mL. Mix thoroughly. Filter sterilize.

Solution 7:

Composition per 100mL:

Succinic acid	0.6g
Isobutyric acid	0.5g
Valeric acid	0.5g
2-Methylbutyric acid	0.5g
3-Methylbutyric acid	0.5g
Caproic acid	0.2g

Preparation of Solution 7: Add components to distilled/deionized water and bring volume to 100.0mL. Mix thoroughly. Adjust pH to 9.0 with NaOH. Autoclave for 15 min at 15 psi pressure–121°C. Cool to 25°C.

Solution 8A:

Composition per 100mL:

Sodium acetate·$3H_2O$	20.0g

Solution 8B:

Composition per 100mL:

Propionic acid	7.0g

Solution 8C:

Composition per 100mL:

n-Butyric acid	8.0g

Solution 8D:

Composition per 100mL:

Benzoic acid	5.0g

Solution 8E:

Composition per 100mL:

n-Palmitic acid	5.0g

Preparation of Solutions 8A-E: Add the appropriate amount of component to distilled/deionized water and bring volume to 100.0mL. Mix thoroughly. Adjust pH to 9.0 with NaOH. Autoclave for 15 min at 15 psi pressure–121°C. Cool to 25°C.

Preparation of Medium: To 970.0mL of cooled, sterile solution 1 aseptically add 1.0mL of sterile solution 2, 1.0mL of sterile solution 3, 30.0mL of sterile solution 4, 3.0mL of sterile solution 5, 0.1mL of sterile solution 6, 0.1mL of sterile solution 7, and 10.0mL of sterile solution 8A, 8B, 8C, 8D or 8E. Mix thoroughly. Aseptically distribute into sterile tubes or flasks.

Use: For the isolation and cultivation of *Desulfovibrio baarsii, Desulfovibrio sapovorans, Desulfobacter* species, *Desulfonema* species, *Desulfobulbus* species, and *Desulfotomaculum acetoxidans*.

Medium for Ammonia Oxidizers

Composition per liter:

$MgSO_4 \cdot 7H_2O$	0.20g
$(NH_4)_2SO_4$	0.13g
K_2HPO_4	0.09g
$CaCl_2 \cdot 2H_2O$	0.02g
Chelated iron	1.0mg
$MnCl_2 \cdot 4H_2O$	0.2mg
$Na_2MoO_4 \cdot 2H_2O$	0.1mg
$ZnSO_4 \cdot 7H_2O$	0.1mg
$CuSO_4 \cdot 5H_2O$	0.02mg
$CoCl_2 \cdot 6H_2O$	2.0μg

Preparation of Medium: Add components to distilled/deionized water and bring volume to 1.0L. Mix thoroughly. Distribute into tubes or flasks. Autoclave for 15 min at 15 psi pressure–121°C.

Use: For the isolation, cultivation and enrichment of ammonia-oxidizing bacteria from soil.

Medium for Ammonia Oxidizers

Composition per liter:

$(NH_4)_2SO_4$	2.0g
$MgSO_4 \cdot 7H_2O$	0.2g
$CaCl_2 \cdot 2H_2O$	0.02g
K_2HPO_4	0.02g
Chelated iron	1.0mg
$MnCl_2 \cdot 4H_2O$	0.2mg
$Na_2MoO_4 \cdot 2H_2O$	0.1mg
$ZnSO_4 \cdot 7H_2O$	0.1mg
$CuSO_4 \cdot 5H_2O$	0.02mg
$CoCl_2 \cdot 6H_2O$	2.0μg

Preparation of Medium: Add components to distilled/deionized water and bring volume to 1.0L. Mix thoroughly. Distribute into tubes or flasks. Autoclave for 15 min at 15 psi pressure–121°C.

Use: For the isolation, cultivation and enrichment of ammonia-oxidizing bacteria from soil.

Medium for Ammonia Oxidizers

Composition per liter:

K_2HPO_4	0.5g
$(NH_4)_2SO_4$	0.5g
Phenol red	0.5g
$MgSO_4 \cdot 7H_2O$	0.05g
$CaCl_2 \cdot 2H_2O$	0.02g

$NaCl$ 0.02g
$Na_2MoO_4 \cdot 2H_2O$ 2.4μg
Metals "44" 1.0mL

Metals "44":
Composition per 100mL:

$ZnSO_4 \cdot 7H_2O$ 1.1g
$FeSO_4 \cdot 7H_2O$ 0.5g
EDTA 0.25g
$MnSO_4 \cdot 7H_2O$ 0.154g
$CuSO_4 \cdot 5H_2O$ 0.04g
$Co(NO_3)_2 \cdot 6H_2O$ 0.025g
$Na_2B_4O_7 \cdot 10H_2O$ 0.018g

Preparation of Metals "44": Add a few drops of H_2SO_4 to distilled/deionized water to inhibit precipitate formation. Add components to acidified distilled/deionized water and bring volume to 100.0mL. Mix thoroughly.

Preparation of Medium: Add components to distilled/deionized water and bring volume to 1.0L. Mix thoroughly. Distribute into tubes or flasks. Autoclave for 15 min at 15 psi pressure–121°C.

Use: For the isolation, cultivation and enrichment of ammonia-oxidizing bacteria from soil.

Medium for Ammonia Oxidizers

Composition per liter:

$(NH_4)_2SO_4$ 0.5g
KH_2PO_4 .... 0.2g
$CaCl_2 \cdot 2H_2O$ 0.04g
$MgSO_4 \cdot 7H_2O$ 0.04g
Ferric citrate 0.5mg
Phenol red 0.5mg

Preparation of Medium: Add components to distilled/deionized water and bring volume to 1.0L. Mix thoroughly. Distribute into tubes or flasks. Autoclave for 15 min at 15 psi pressure–121°C.

Use: For the isolation, cultivation and enrichment of ammonia-oxidizing bacteria from soil.

Medium for Ammonia Oxidizers, Brackish

Composition per liter:

$CaCO_3$ 5.0g
NH_4Cl 0.5g
K_2HPO_4 .... 0.05g
Sea water 400.0mL

Preparation of Medium: Add components to distilled/deionized water and bring volume to 1.0L. Mix thoroughly. Distribute into tubes or flasks. Autoclave for 15 min at 15 psi pressure–121°C.

Use: For the isolation, cultivation and enrichment of ammonia-oxidizing bacteria from brackish specimens.

Medium for Ammonia Oxidizers, Marine

Composition per liter:

$(NH_4)_2SO_4$ 1.32g
$MgSO_4 \cdot 7H_2O$ 0.20g
Chelated iron 0.13g
K_2HPO_4 .... 0.11g
$CaCl_2 \cdot 2H_2O$ 0.02g
$ZnSO_4 \cdot 7H_2O$ 0.1mg
$CuSO_4 \cdot 5H_2O$ 0.02mg
$CoCl_2 \cdot 6H_2O$ 2.0μg
$MnCl_2 \cdot 4H_2O$ 2.0μg
$Na_2MoO_4 \cdot 2H_2O$ 1.0μg
Sea water 1.0L

Preparation of Medium: Combine components. Mix thoroughly. Distribute into tubes or flasks. Autoclave for 15 min at 15 psi pressure–121°C.

Use: For the isolation, cultivation and enrichment of marine ammonia-oxidizing bacteria.

Medium for Hydrocarbon–Degrading Bacteria

Composition per 1020mL:

NH_4Cl 0.5g
$MgSO_4 \cdot 7H_2O$ 0.5g
NaCl 0.4g
Hydrocarbon 20.0mL
KH_2PO_4 solution 0.5mL
$Na_2HPO_4 \cdot H_2O$ solution 0.5mL

KH_2PO_4 Solution:
Composition per 100mL:

KH_2PO_4 .... 10.0g

Preparation of KH_2PO_4 Solution: Add KH_2PO_4 to distilled/deionized water and bring volume to 100.0mL. Mix thoroughly. Autoclave for 15 min at 15 psi pressure–121°C. Cool to 25°C.

$Na_2HPO_4 \cdot H_2O$ Solution:
Composition per 100mL:

$Na_2HPO_4 \cdot H_2O$ 10.0g

Preparation of $Na_2HPO_4 \cdot H_2O$ Solution: Add $Na_2HPO_4 \cdot H_2O$ to distilled/deionized water and bring volume to 100.0mL. Mix thoroughly. Autoclave for 15 min at 15 psi pressure–121°C. Cool to 25°C.

Preparation of Medium: Add components—except hydrocarbon, KH_2PO_4 solution and $Na_2HPO_4 \cdot H_2O$ solution—to distilled/deionized water and bring volume to 999.0mL. Mix thoroughly. Gently heat and bring to boiling. Autoclave for 15 min at 15 psi pressure–121°C. Cool to 45–50°C. Aseptically add 0.5mL of sterile KH_2PO_4 solution and 0.5mL of the sterile $Na_2HPO_4 \cdot H_2O$ solution. Mix thoroughly. Aseptically distribute into sterile tubes in 10.0mL volumes. Add 0.2mL of sterile hydrocarbon to each tube.

Use: For the cultivation and enumeration of hydrocarbon-degrading bacteria in fresh water.

Medium for Hydrocarbon– Degrading Bacteria (Naphthalene Mineral Salts Medium)

Composition per liter:

K_2HPO_4	1.0g
$(NH_4)_2SO_4$	1.0g
$MgSO_4 \cdot 7H_2O$	0.3g
$CaCl_2$	0.1g
$FeSO_4 \cdot 7H_2O$	0.02g
Naphthalene	2.0mL

pH 7.0 ± 0.2 at 25°C

Preparation of Medium: Add components, except naphthalene, to distilled/deionized water and bring volume to 998.0mL. Mix thoroughly. Gently heat and bring to boiling. Autoclave for 15 min at 15 psi pressure–121°C. Cool to 45–50°C. Aseptically add 2.0mL of sterile naphthalene to 20.0mL of sterile basal salts. Ultrasonically homogenize the solution. Add the naphthalene–basal salts homogenate back to the remainder of the sterile basal salts medium. Mix thoroughly. Aseptically distribute into sterile tubes or flasks.

Use: For the cultivation and enrichment of hydrocarbon-degrading bacteria.

Medium for Nitrite Oxidizers

Composition per liter:

$KHCO_3$	1.5g
KH_2PO_4	0.5g
K_2HPO_4	0.5g
KNO_2	0.3g
$MgSO_4 \cdot 7H_2O$	0.2g
NaCl	0.2g
$CaCl_2 \cdot 2H_2O$	0.01g
$FeSO_4 \cdot 7H_2O$	0.01g

Preparation of Medium: Add components to distilled/deionized water and bring volume to 1.0L. Mix thoroughly. Distribute into tubes or flasks. Autoclave for 15 min at 15 psi pressure–121°C.

Use: For the isolation, cultivation and enrichment of nitrate-oxidizing bacteria.

Medium for Nitrite Oxidizers, Marine

Composition per liter:

$MgSO_4 \cdot 7H_2O$	0.1g
$NaNO_2$	0.07g
$CaCl_2 \cdot 2H_2O$	6.0mg
K_2HPO_4	1.74mg
Chelated iron	1.0mg
$MnCl_2 \cdot 4H_2O$	66.0μg
$Na_2MoO_4 \cdot 2H_2O$	30.0μg
$ZnSO_4 \cdot 7H_2O$	30.0μg
$CuSO_4 \cdot 5H_2O$	6.0μg
$CoCl_2 \cdot 6H_2O$	0.6μg
Seawater	700.0mL

Preparation of Medium: Add components to distilled/deionized water and bring volume to 1.0L. Mix thoroughly. Distribute into tubes or flasks. Autoclave for 15 min at 15 psi pressure–121°C.

Use: For the isolation, cultivation and enrichment of marine nitrate-oxidizing bacteria.

Medium for *Prosthecomicrobium* and *Ancalomicrobium*

Composition per liter:

Agar	15.0g
Peptone	0.1g
Hutner's modified salts solution	20.0mL
Vitamin solution	10.0mL

Hutner's Mineral Base:

Composition per liter:

$MgSO_4 \cdot 7H_2O$	29.7g
Nitrilotriacetic acid	10.0g
$CaCl_2 \cdot 2H_2O$	3.34g
$FeSO_4 \cdot 7H_2O$	0.1g
$(NH_4)_2MoO_4$	9.25mg
Metals "44"	50.0mL

Preparation of Hutner's Mineral Base: Add nitrilotriacetic acid to 500.0mL of distilled/deionized water. Dissolve by adjusting pH to 6.5 with KOH. Add remaining components. Add distilled/deionized water to 1.0L.

Metals "44":

Composition per 100mL:

$ZnSO_4 \cdot 7H_2O$	1.1g
$FeSO_4 \cdot 7H_2O$	0.5g
EDTA	0.25g

$MnSO_4 \cdot 7H_2O$	0.154g
$CuSO_4 \cdot 5H_2O$	0.04g
$Co(NO_3)_2 \cdot 6H_2O$	0.025g
$Na_2B_4O_7 \cdot 10H_2O$	0.018g

Preparation of Metals "44": Add components to distilled/deionized water and bring volume to 100.0mL. Mix thoroughly.

Vitamin Solution:

Composition per liter:

Pyridoxine·HCl	0.01g
Calcium pantothenate	5.0mg
Nicotinamide	5.0mg
Riboflavin	5.0mg
Thiamine HCl	5.0mg
Biotin	2.0mg
Folic acid	2.0mg
Vitamin B_{12}	0.1mg

Preparation of Vitamin Solution: Add components to distilled/deionized water and bring volume to 1.0L. Mix thoroughly. Filter sterilize.

Preparation of Medium: Add components, except vitamin solution, to distilled/deionized water and bring volume to 990.0mL. Mix thoroughly. Gently heat and bring to boiling. Autoclave for 15 min at 15 psi pressure–121°C. Cool to 45–50°C. Aseptically add sterile vitamin solution. Mix thoroughly. Pour into sterile Petri dishes or distribute into sterile tubes.

Use: For the isolation of *Prosthecomicrobium* species and *Ancalomicrobium* species.

Medium for *Prosthecomicrobium* and *Ancalomicrobium*

Composition per liter:

$(NH_4)_2SO_4$	0.25g
Glucose	0.25g
Na_2HPO_4	0.071g
Modified Hutner's basal salts	20.0mL
Vitamin solution	10.0mL

pH 7.2 ± 0.2 at 25°C

Modified Hutner's Basal Salts:

Composition per liter:

$MgSO_4 \cdot 7H_2O$	29.7g
Nitrilotriacetic acid	10.0g
$CaCl_2 \cdot 2H_2O$	3.34g
$FeSO_4 \cdot 7H_2O$	0.10g
$(NH_4)_2MoO_4$	9.25mg
Metals "44"	50.0mL

Preparation of Modified Hutner's Basal Salts: Add nitrilotriacetic acid to 500.0mL of distilled/deionized water. Dissolve by adjusting pH to 6.5 with KOH. Add remaining components. Readjust pH to 7.2 with H_2SO_4 or KOH. Add distilled/deionized water to 1.0L. Store at 5°C.

Metals "44":

Composition per 100mL:

$ZnSO_4 \cdot 7H_2O$	1.1g
$FeSO_4 \cdot 7H_2O$	0.5g
EDTA	0.25g
$MnSO_4 \cdot 7H_2O$	0.154g
$CuSO_4 \cdot 5H_2O$	0.04g
$Co(NO_3)_2 \cdot 6H_2O$	0.025g
$Na_2B_4O_7 \cdot 10H_2O$	0.018g

Preparation of Metals "44": Add a few drops of H_2SO_4 to distilled/deionized water to inhibit precipitate formation. Add components to acidified distilled/deionized water and bring volume to 100.0mL. Mix thoroughly.

Vitamin Solution:

Composition per liter:

Thiamine·HCl	5.0mg
D-Calcium pantothenate	5.0mg
Riboflavin	5.0mg
Biotin	2.0mg
Folic acid	2.0mg
Vitamin B_{12}	0.1mg

Preparation of Vitamin Solution: Add components to distilled/deionized water and bring volume to 1.0L. Mix thoroughly. Filter sterilize.

Preparation of Medium: Add components, except vitamin solution. to distilled deionized water and bring volume to 990.0mL. Mix thoroughly. Autoclave for 15 min at 15 psi pressure–121°C. Cool to room temperature. Aseptically add 10.0mL of sterile vitamin solution. Mix thoroughly. Aseptically distribute into sterile tubes or flasks.

Use: For the cultivation and maintenance of *Prosthecomicrobium enhydrum, Prosthecomicrobium pneumaticum,* and *Ancalomicrobium* species.

Medium for *Prosthecomicrobium* and *Ancalomicrobium* with Nicotinamide

Composition per liter:

$(NH_4)_2SO_4$	0.25g
Glucose	0.25g
Na_2HPO_4	0.071g
Modified Hutner's basal salts	20.0mL
Vitamin solution	10.0mL

pH 7.2 ± 0.2 at 25°C

Modified Hutner's Basal Salts:

Composition per liter:

$MgSO_4 \cdot 7H_2O$	29.7g

Nitrilotriacetic acid 10.0g
$CaCl_2 \cdot 2H_2O$ 3.34g
$FeSO_4 \cdot 7H_2O$ 0.10g
$(NH_4)_2MoO_4$ 9.25mg
Metals "44" 50.0mL

Preparation of Modified Hutner's Basal Salts: Add nitrilotriacetic acid to 500.0mL of distilled/deionized water. Dissolve by adjusting pH to 6.5 with KOH. Add remaining components. Readjust pH to 7.2 with H_2SO_4 or KOH. Add distilled/deionized water to 1.0L. Store at 5°C.

Metals "44":

Composition per 100mL:

$ZnSO_4 \cdot 7H_2O$ 1.1g
$FeSO_4 \cdot 7H_2O$ 0.5g
EDTA 0.25g
$MnSO_4 \cdot 7H_2O$ 0.154g
$CuSO_4 \cdot 5H_2O$ 0.04g
$Co(NO_3)_2 \cdot 6H_2O$ 0.025g
$Na_2B_4O_7 \cdot 10H_2O$ 0.018g

Preparation of Metals "44": Add a few drops of H_2SO_4 to distilled/deionized water to inhibit precipitate formation. Add components to acidified distilled/deionized water and bring volume to 100.0mL. Mix thoroughly.

Vitamin Solution:

Composition per liter:

Thiamine·HCl 5.0mg
D-Calcium pantothenate 5.0mg
Riboflavin 5.0mg
Nicotinamide 5.0mg
Biotin 2.0mg
Folic acid 2.0mg
Vitamin B_{12} 0.1mg

Preparation of Vitamin Solution: Add components to distilled/deionized water and bring volume to 1.0L. Mix thoroughly. Filter sterilize.

Preparation of Medium: Add components, except vitamin solution, to distilled/deionized water and bring volume to 990.0mL. Mix thoroughly. Autoclave for 15 min at 15 psi pressure–121°C. Cool to room temperature. Aseptically add 10.0mL of sterile vitamin solution. Mix thoroughly. Aseptically distribute into sterile tubes or flasks.

Use: For the cultivation and maintenance of *Ancalomicrobium adetum* and *Prosthecomicrobium* species.

Medium for *Prosthecomicrobium* and *Ancalomicrobium*, Modified

Composition per liter:

Agar 15.0g
Glucose 1.0g
$(NH_4)_2SO_4$ 0.25g
Peptone 0.15g
Yeast extract 0.15g
Modified Hutner's basal salts 20.0mL
Vitamin solution 10.0mL

Modified Hutner's Basal Salts:

Composition per liter:

$MgSO_4 \cdot 7H_2O$ 29.7g
Nitrilotriacetic acid 10.0g
$CaCl_2 \cdot 2H_2O$ 3.34g
$FeSO_4 \cdot 7H_2O$ 0.10g
$(NH_4)_2MoO_4$ 9.25mg
Metals "44" 50.0mL

Preparation of Modified Hutner's Basal Salts: Add nitrilotriacetic acid to 500.0mL of distilled/deionized water. Dissolve by adjusting pH to 6.5 with KOH. Add remaining components. Readjust pH to 7.2 with H_2SO_4 or KOH. Add distilled/deionized water to 1.0L. Store at 5°C.

Metals "44":

Composition per 100mL:

$ZnSO_4 \cdot 7H_2O$ 1.1g
$FeSO_4 \cdot 7H_2O$ 0.5g
EDTA 0.25g
$MnSO_4 \cdot 7H_2O$ 0.154g
$CuSO_4 \cdot 5H_2O$ 0.04g
$Co(NO_3)_2 \cdot 6H_2O$ 0.025g
$Na_2B_4O_7 \cdot 10H_2O$ 0.018g

Preparation of Metals "44": Add a few drops of H_2SO_4 to distilled/deionized water to inhibit precipitate formation. Add components to acidified distilled/deionized water and bring volume to 100.0mL. Mix thoroughly.

Vitamin Solution:

Composition per liter:

Thiamine·HCl 5.0mg
D-Calcium pantothenate 5.0mg
Riboflavin 5.0mg
Biotin 2.0mg
Folic acid 2.0mg
Vitamin B_{12} 0.1mg

Preparation of Vitamin Solution: Add components to distilled/deionized water and bring volume to 1.0L. Mix thoroughly. Filter sterilize.

Preparation of Medium: Add components, except vitamin solution, to distilled deionized water and bring volume to 990.0mL. Mix thoroughly. Autoclave for 15 min at 15 psi pressure–121°C. Cool to room temperature. Aseptically add 10.0mL of sterile vitamin solution. Mix thoroughly. Aseptically distribute into sterile tubes or flasks.

Use: For the cultivation and maintenance of *Ancalomicrobium adetum*, *Prosthecomicrobium hirschii*, and *Prosthecomicrobium* species.

Medium for Sulfate Reducers (ATCC Medium 1282)

Composition per 1050mL:

Modified Baar's medium for sulfate reducers	1020.0mL
Organic acid solution	10.0mL
Vitamin solution	10.0mL
Wolfe's mineral solution	10.0mL

pH 7.5 ± 0.2 at 25°C

Modified Baar's Medium for Sulfate Reducers:

Composition per 1020mL:

Component I	400.0mL
Component III	400.0mL
Component II	200.0mL
$Fe(NH_4)_2(SO_4)_2$ (5% solution)	20.0mL

Component I:

Composition per 400mL:

Sodium citrate	5.0g
$MgSO_4$	2.0g
$CaSO_4$	1.0g
NH_4Cl	1.0g

Preparation of Component I: Add components to distilled/deionized water and bring volume to 400.0mL. Mix thoroughly. Adjust pH to 7.5. Autoclave for 15 min at 15 psi pressure–121°C.

Component II:

Composition per 200mL:

K_2HPO_4	0.5g

Preparation of Component II: Add K_2HPO_4 to distilled/deionized water and bring volume to 200.0mL. Mix thoroughly. Adjust pH to 7.5. Autoclave for 15 min at 15 psi pressure–121°C.

Component III:

Composition per 400mL:

Sodium lactate	3.5g
Yeast extract	1.0g

Preparation of Component III: Add components to distilled/deionized water and bring volume to 400.0mL. Mix thoroughly. Adjust pH to 7.5. Autoclave for 15 min at 15 psi pressure–121°C.

Preparation of Modified Baar's Medium for Sulfate Reducers: Aseptically combine the three sterile solutions, except the $Fe(NH_4)_2(SO_4)_2$ solution. Mix thoroughly. Distribute 5.0mL volumes into tubes under 97% N_2 + 3% H_2. Add medium to tubes while still warm to exclude as much O_2 as possible. Prepare a 5% solution of ferrous ammonium sulfate, $Fe(NH_4)_2(SO_4)_2$. Sterilize by filtration. Add 0.2mL of sterile $Fe(NH_4)_2(SO_4)_2$ solution to 10.0mL of medium immediately prior to inoculation.

Organic Acid Solution:

Composition per 100mL:

Butyric acid	5.18mL
Caproic acid	2.4mL
Octanoic acid	1.25mL

Preparation of Organic Acid Solution: Add components to distilled/deionized water and bring volume to 75.0mL. Adjust pH to 7.0 with 5*N* NaOH. Bring volume to 100.0mL with distilled/deionized water. Filter sterilize.

Wolfe's Vitamin Solution:

Composition per liter:

Pyridoxine·HCl	10.0mg
Thiamine·HCl	5.0mg
Riboflavin	5.0mg
Nicotinic acid	5.0mg
Calcium pantothenate	5.0mg
p-Aminobenzoic acid	5.0mg
Thioctic acid	5.0mg
Biotin	2.0mg
Folic acid	2.0mg
Cyanocobalamin	100.0μg

Preparation of Wolfe's Vitamin Solution: Add components to distilled/deionized water and bring volume to 1.0L. Mix thoroughly. Filter sterilize.

Wolfe's Mineral Solution:

Composition per liter

$MgSO_4 \cdot 7H_2O$	3.0g
Nitriloacetic acid	1.5g
NaCl	1.0g
$MnSO_4 \cdot H_2O$	0.5g
$FeSO_4 \cdot 7H_2O$	0.1g
$CoCl_2 \cdot 6H_2O$	0.1g
$CaCl_2$	0.1g
$ZnSO_4 \cdot 7H_2O$	0.1g
$CuSO_4 \cdot 5H_2O$	0.01g
$AlK(SO_4)_2 \cdot 12H_2O$	0.01g
H_3BO_3	0.01g
$Na_2MoO_4 \cdot 2H_2O$	0.01g

Preparation of Wolfe's Mineral Solution: Add nitrilotriacetic acid to 500.0mL of distilled/deionized water. Dissolve by adjusting pH to 6.5 with KOH. Add remaining components. Add distilled/deionized water to 1.0L. Filter sterilize.

Preparation of Medium: To each test tube containing 10.0mL of modified Baar's medium for sulfate reducers aseptically add 0.1mL of sterile organic acid solution, 0.1mL of sterile Wolfe's vitamin solution and 0.1mL of sterile Wolfe's mineral solution immediately prior to inoculation.

Use: For the cultivation and maintenance of *Desulfotomaculum thermobenzoicum* and *Desulfovibrio sapovorans.*

Medium for Sulfate Reducers (Postgate's Medium for Sulfate Reducers) (ATCC Medium 1283)

Composition per liter:

Part A 869.0mL
Part C 100.0mL
Part D 10.0mL
Part E 10.0mL
Part F 10.0mL
Part B, trace element solution SL-7 .... 1.0mL

pH 7.7 ± 0.2 at 25°C

Part A:

Composition per 869mL:

Na_2SO_4 .... 3.0g
NaCl 1.0g
KCl 0.5g
$MgCl_2 \cdot 6H_2O$ 0.4g
NH_4Cl 0.3g
KH_2PO_4 .... 0.2g
$CaCl_2 \cdot 2H_2O$ 0.15g

Preparation of Part A: Add components to distilled/deionized water and bring volume to 869.0mL. Mix thoroughly. Prepare and autoclave part A under 90% N_2 + 10% CO_2. Autoclave for 15 min at 15 psi pressure–121°C. Cool to room temperature.

Part B, Trace element solution SL-7:

Composition per liter:

$FeCl_2 \cdot 4H_2O$ 1.5g
$CoCl_2 \cdot 6H_2O$ 0.19g
$MnCl_2 \cdot 4H_2O$ 0.10g
$ZnCl_2$ 0.07g
H_3BO_3 .... 0.06g
$Na_2MoO_4 \cdot 2H_2O$ 0.04g
$NiCl_2 \cdot 6H_2O$ 0.02g
$CuCl_2 \cdot 2H_2O$ 0.02g
HCl, 25% 10.0mL

Preparation of Part B: Add the $FeCl_2 \cdot 4H_2O$ to the HCl. Add distilled/deionized water and bring volume to 1.0L. Add remaining components. Mix thoroughly. Autoclave under 100% N_2 for 15 min at 15 psi pressure–121°C. Cool to room temperature.

Part C:

Composition per 100mL:

$NaHCO_3$ 5.0g

Preparation of Part C: Add the $NaHCO_3$ to distilled/deionized water and bring volume to 100.0mL. Mix thoroughly. Filter sterilize. Gas with 90% N_2 + 10% CO_2 to remove residual O_2.

Part D:

Composition per 10mL:

Sodium butyrate 0.7g
Sodium caproate 0.3g
Sodium octanoate 0.15g

Preparation of Part D: Add components to distilled/deionized water and bring volume to 10.0mL. Mix thoroughly. Autoclave under 100% N_2 for 15 min at 15 psi pressure–121°C. Cool to room temperature.

Part E:

Composition per 10mL:

Yeast extract 1.0g
Thiamine·HCl 100.0μg
p-Aminobenzoic acid 40.0μg
D(+)-Biotin 10.0μg

Preparation of Part E: Add components to distilled/deionized water and bring volume to 10.0mL. Mix thoroughly. Autoclave under 100% N_2 for 15 min at 15 psi pressure–121°C. Cool to room temperature.

Part F:

Composition per 10mL:

$Na_2S \cdot 9H_2O$ 0.4g

Preparation of Part F: Add $Na_2S \cdot 9H_2O$ to distilled/deionized water and bring volume to 10.0mL. Mix thoroughly. Autoclave under 100% N_2 for 15 min at 15 psi pressure–121°C. Cool to room temperature.

Preparation of Medium: To 869.0mL of sterile cooled Part A, aseptically add the remaining sterile solutions in the following order: Part B, Part C, Part D, Part E, and Part F. Mix thoroughly. Adjust pH to 7.7. Anaerobically distribute under 80% N_2 + 20% CO_2 into sterile tubes or flasks.

Use: For the cultivation and maintenance of *Desulfovibrio baarsii* and *Desulfovibrio sapovorans.*

Medium M71

Composition per liter:

Agar 20.0g
Peptone 10.0g
Glucose 5.0g
H_3BO_3 .... 1.0g
Pancreatic digest of casein 1.0g
Cycloheximide 0.05g
2,3,5-Triphenyltetrazolium·HCl solution 10.0mL

2,3,5-Triphenyltetrazolium·HCl Solution:

Composition per 10mL:

2,3,5-Triphenyltetrazolium·HCl 0.05g

Preparation of 2,3,5-Triphenyltetrazolium-HCl Solution: Add 2,3,5-triphenyltetrazolium·HCl to distilled/deionized water and bring volume to 10.0mL. Mix thoroughly. Autoclave for 15 min at 15 psi pressure–121°C.

Preparation of Medium: Add components, except 2,3,5-triphenyltetrazolium·HCl solution, to distilled/deionized water and bring volume to 990.0mL. Mix thoroughly. Gently heat and bring to boiling. Autoclave for 15 min at 15 psi pressure–121°C. Cool to 45–50°C. Aseptically add 10.0mL of sterile 2,3,5-triphenyltetrazolium·HCl solution. Mix thoroughly. Pour into sterile Petri dishes.

Use: For the selective isolation and cultivation of *Pseudomonas syringae*.

Medium N for Sulfate Reducers (Postgate's Medium N for Sulfate Reducers)

Composition per liter:

$(NH_4)_2SO_4$	7.0g
Sodium lactate	6.0g
NH_4Cl	1.0g
Yeast extract	1.0g
KH_2PO_4	0.5g
Sodium citrate·$2H_2O$	0.3g
$FeSO_4 \cdot 7H_2O$	0.1g
$CaCl_2 \cdot 6H_2O$	0.06g
$MgSO_4 \cdot 7H_2O$	0.06g

pH 7.5 ± 0.2 at 25°C

Preparation of Medium: Add components to distilled/deionized water and bring volume to 1.0L. For marine bacteria, NaCl may be added or seawater used in place of distilled/deionized water. Mix thoroughly. Adjust pH to 7.5. Distribute into tubes or flasks. Autoclave for 15 min at 15 psi pressure–121°C.

Use: For detection, culturing and storage of *Desulfovibrio* species and many *Desulfotomaculum* species. This medium should be used when a clear culture medium is desired such as for chemostat culture. This medium may be cloudy after sterilization but usually clears on cooling. It turns black as a result of H_2S production due to bacterial growth.

Medium R

Composition per liter:

$Na_2S_2O_3 \cdot 5H_2O$	5.0g
KNO_3	2.0g
$MgCl_2 \cdot 6H_2O$	0.5g
NH_4Cl	0.5g
KH_2PO_4 solution	10.0mL
$NaHCO_3$ solution	10.0mL
$FeSO_4 \cdot 7H_2O$ solution	10.0mL

pH 7.0 ± 0.2 at 25°C

KH_2PO_4 Solution:

Composition per 10mL:

KH_2PO_4	2.0g

Preparation of KH_2PO_4 Solution: Add KH_2PO_4 to distilled/deionized water and bring volume to 10.0mL. Mix thoroughly. Filter sterilize.

$NaHCO_3$ Solution:

Composition per 10mL:

$NaHCO_3$	1.0g

Preparation of $NaHCO_3$ Solution: Add the $NaHCO_3$ to distilled/deionized water and bring volume to 10.0mL. Mix thoroughly. Filter sterilize.

$FeSO_4 \cdot 7H_2O$ Solution:

Composition per 10mL:

$FeSO_4 \cdot 7H_2O$	10.0mg

Preparation of $FeSO_4 \cdot 7H_2O$ Solution: Add the $FeSO_4 \cdot 7H_2O$ to distilled/deionized water and bring volume to 10.0mL. Mix thoroughly. Filter sterilize.

Preparation of Medium: Add components—except KH_2PO_4 solution, $NaHCO_3$ solution, and $FeSO_4 \cdot 7H_2O$ solution—to tap water and bring volume to 970.0mL. Mix thoroughly. Gently heat until dissolved. Adjust pH to 7.0. Autoclave for 15 min at 15 psi pressure–121°C. Cool to 45–50°C. Aseptically add 10.0mL of sterile KH_2PO_4 solution, 10.0mL of $NaHCO_3$ solution, and 10.0mL of $FeSO_4 \cdot 7H_2O$ solution. Mix thoroughly. Aseptically distribute into sterile tubes or flasks.

Use: For the cultivation of *Thiobacillus denitrificans*.

Medium S

Composition per liter:

$Na_2S_2O_3 \cdot 5H_2O$	5.0g
$(NH_4)_2SO_4$	4.0g
KH_2PO_4	4.0g
$MgSO_4$	0.5g
$CaCl_2$	0.25g
$FeSO_4$	0.01g

Preparation of Medium: Add components to distilled/deionized water and bring volume to 1.0L. Mix thoroughly. Distribute into tubes or flasks. Autoclave for 15 min at 15 psi pressure–121°C.

Use: For the cultivation of *Thiobacillus* species.

Medium SP 4

Composition per liter:

Pancreatic digest of casein	11.0g
Peptone	5.3g
Glucose	5.0g
NaCl	0.875g
Beef extract	0.525g
Yeast extract	0.525g
Beef heart, solids from infusion	0.35g
Fetal bovine serum, heat inactivated	170.0mL
Yeast extract solution	100.0mL
CMRL 1066, 10× solution	50.0mL
Fresh yeast extract solution	35.0mL
Phenol red solution	20.0mL
Penicillin solution	10.0mL

pH 7.6 ± 0.2 at 25°C

Yeast Extract Solution:

Composition per 100mL:

Yeast extract	2.0g

Preparation of Yeast Extract Solution: Add yeast extract to distilled/deionized water and bring volume to 100.0mL. Mix thoroughly. Autoclave for 15 min at 15 psi pressure–121°C.

CMRL 1066, 10X Solution:

Composition per liter:

NaCl	6.8g
$NaHCO_3$	2.2g
D-Glucose	1.0g
KCl	0.4g
L-Cysteine·HCl·H_2O	0.26g
$CaCl_2$, anhydrous	0.2g
$MgSO_4 \cdot 7H_2O$	0.2g
$NaH_2PO_4 \cdot H_2O$	0.14g
L-Glutamine	0.1g
Sodium acetate·$3H_2O$	0.083g
L-Glutamic acid	0.075g
L-Arginine·HCl	0.070g
L-Lysine·HCl	0.070g
L-Leucine	0.060g
Glycine	0.050g
Ascorbic acid	0.050g
L-Proline	0.040g
L-Tyrosine	0.040g
L-Aspartic acid	0.030g
L-Threonine	0.030g
L-Alanine	0.025g
L-Phenylalanine	0.025g
L-Serine	0.025g
L-Valine	0.025g
L-Cystine	0.020g
L-Histidine·HCl·H_2O	0.020g
L-Isoleucine	0.020g
Phenol red	0.020g
L-Methionine	0.015g
Deoxyadenosine	0.010g
Deoxycytidine	0.010g
Deoxyguanosine	0.010g
Glutathione, reduced	0.010g
Thymidine	0.010g
Hydroxy-L-proline	0.010g
L-Tryptophan	0.010g
Nicotinamide adenine dinucleotide	7.0mg
Tween™ 80	5.0mg
Sodium glucoronate·H_2O	4.2mg
Coenzyme A	2.5mg
Cocarboxylase	1.0mg
Flavin adenine dinucleotide	1.0mg
Nicotinamide adenine dinucleotide phosphate	1.0mg
Uridine triphosphate	1.0mg
Choline chloride	0.50mg
Cholesterol	0.20mg
5-Methyldeoxycytidine	0.10mg
Inositol	0.05mg
p-Aminobenzoic acid	0.05mg
Niacin	0.025mg
Niacinamide	0.025mg
Pyridoxine	0.025mg
Pyridoxal·HCl	0.025mg
Biotin	0.01mg
D-Calcium pantothenate	0.01mg
Folic acid	0.01mg
Riboflavin	0.01mg
Thiamine·HCl	0.01mg

Source: CMRL 1066, 10× medium is available as a premixed powder from GIBCO BRL.

Preparation of CMRL 1066, 10X Solution: Add components to distilled/deionized water and bring volume to 1.0L. Mix thoroughly. Adjust pH to 7.2. Filter sterilize.

Fresh Yeast Extract Solution:

Composition per 100mL:

Baker's yeast, live, pressed, starch-free	25.0g

Preparation of Fresh Yeast Extract Solution: Add the live Baker's yeast to 100.0mL of distilled/deionized water. Autoclave for 90 min at 15 psi pressure–121°C. Allow to stand. Remove supernatant solution. Adjust pH to 6.6–6.8. Filter sterilize.

Phenol Red Solution:

Composition per 100mL:

Phenol red	0.01g

Preparation of Phenol Red Solution: Add phenol red to distilled/deionized water and bring volume to 10.0mL. Mix thoroughly. Filter sterilize.

Penicillin Solution:

Composition per 10mL:

Penicillin	1,000,000U

Preparation of Penicillin Solution: Add penicillin to distilled/deionized water and bring volume to 10.0mL. Filter sterilize.

Preparation of Medium: Add components—except fetal bovine serum, yeast extract solution, CMRL 1066, 10× solution, fresh yeast extract solution, phenol red solution, and penicillin solution—to distilled/deionized water and bring volume to 615.0mL. Mix thoroughly. Gently heat and bring to boiling. Autoclave for 15 min at 15 psi pressure–121°C. Cool to 45–50°C. Aseptically add 170.0mL of sterile fetal bovine serum, 100.0mL of sterile yeast extract solution, 50.0mL of sterile CMRL 1066, 10× solution, 35.0mL of sterile fresh yeast extract solution, 20.0mL of sterile phenol red solution, and 10.0mL of sterile penicillin solution. Mix thoroughly. Aseptically distribute into sterile tubes or flasks.

Use: For the isolation and cultivation of *Spiroplasma* species from ticks.

Medium VTY

Composition per 100mL:

Peptone	1.0g
Noble agar	0.7g
Yeast extract	0.5g
L-Cysteine·HCl·H_2O	0.1g
Salts A	20.0mL
Salts B	20.0mL
Glucose (10% solution)	5.0mL
$NaHCO_3$ (5% solution)	1.0mL
Hemin solution	1.0mL
Volatile fatty acid solution	0.31mL
Resazurin (0.1% solution)	0.1mL

pH 7.2 ± 0.2 at 25°C

Salts A:

Composition per liter:

$CaCl_2 \cdot 2H_2O$	0.6g
$MgSO_4$	0.45g

Preparation of Salts A: Add components to distilled/deionized water and bring volume to 1.0L. Mix thoroughly.

Salts B:

Composition per liter:

NaCl	4.5g
$(NH_4)_2SO_4$	4.5g
Potassium phosphate buffer (0.05*M*, pH 7.4)	1.0L

Preparation of Salts B: Add NaCl and $(NH_4)_2SO_4$ to 1.0L of 0.05M potassium phosphate buffer, pH 7.4. Mix thoroughly.

Hemin Solution:

Composition per liter:

Hemin	0.5g
NaOH (0.01*N* solution)	1.0mL

Preparation of Hemin Solution: Add hemin to 1.0mL of 0.01*N* NaOH solution. Mix thoroughly.

Volatile Fatty Acid Solution:

Composition per 31mL:

Acetic acid	17.0mL
Propionic acid	6.0mL
n-Butyric acid	4.0mL
n-Valeric acid	1.0mL
Isovaleric acid	1.0mL
Isobutyric acid	1.0mL
DL-α-methylbutyric acid	1.0mL

Preparation of Volatile Fatty Acid Solution: Combine components. Mix thoroughly.

Preparation of Medium: Add components, except glucose and $NaHCO_3$, to distilled/deionized water and bring volume to 94.0mL. Mix thoroughly. Adjust pH to 7.2. Gently heat and gas with 95% N_2 + 5% CO_2 until reduced. Anaerobically distribute into tubes or flasks. Cap with rubber stoppers. Autoclave for 20 min at 15 psi pressure–121°C. Cool to 50°C. Filter sterilize glucose solution and $NaHCO_3$ solution separately. Aseptically and anaerobically add sterile glucose solution and sterile $NaHCO_3$ solution to cooled, sterile basal medium.

Use: For the cultivation and maintenance of *Roseburia cecicola*.

Megasphaera Medium

Composition per liter:

Yeast extract	4.0g
K_2HPO_4	3.2g
KH_2PO_4	1.6g
Agar	1.0g
NH_4Cl	0.5g
Sodium thioglycollate	0.45g
$CaCl_2$	0.2g
$MgCl_2$	0.2g
Sodium lactate (60% solution)	16.0mL

pH 7.0 ± 0.2 at 25°C

Preparation of Medium: Add components to distilled/deionized water and bring volume to 1.0L. Mix thoroughly. Gently heat and bring to boiling. Distribute into tubes or flasks. Autoclave for 15 min at 15 psi pressure–121°C.

Use: For the cultivation and maintenance of *Megasphaera elsdenii*.

Melissococcus pluton Medium

Composition per liter:

Glucose	10.0g
Neopeptone	5.0g
Peptone	2.5g
Yeast extract	2.5g
Soluble starch	2.0g
Pancreatic digest of casein	2.0g
L-Cysteine·HCl·H_2O	0.25g
Phosphate buffer (1*M*, pH 6.7)	50.0mL

pH 7.2 ± 0.2 at 25°C

Preparation of Medium: Add components to distilled/deionized water and bring volume to 1.0L. Mix thoroughly. Adjust pH to 7.2. Gently heat and bring to boiling. Distribute into tubes or flasks that have been flushed with 90% N_2 + 10% CO_2. Cap with butyl rubber stoppers. Place tubes in a press. Autoclave for 15 min at 15 psi pressure–121°C.

Use: For the cultivation and maintenance of *Melisococcus pluton.*

Membrane Lauryl Sulfate Broth

Composition per liter:

Peptone	39.0g
Lactose	30.0g
Yeast extract	6.0g
Sodium lauryl sulfate	1.0g
Phenol red	0.2g

pH 7.4 ± 0.2 at 25°C

Preparation of Medium: Add components to distilled/deionized water and bring volume to 1.0L. Mix thoroughly. Distribute into tubes or flasks. Autoclave for 15 min at 15 psi pressure–121°C.

Use: For the enumeration of coliform organisms and *Escherichia coli* in water.

Meniscus glaucopis Agar

Composition per liter:

Agar	15.0g
$CaCO_3$	10.0g
Maltose	5.0g
Yeast extract	1.0g
KH_2PO_4	0.5g
NaCl	0.4g
NH_4Cl	0.4g
Sodium thioglycollate	0.3g
$MgSO_4 \cdot 7H_2O$	0.2g
$CaCl_2 \cdot H_2O$	0.01g
$FeSO_4 \cdot 7H_2O$	1.0mg
Resazurin (0.025% solution)	4.0mL
Trace elements solution SL-6	1.0mL
Vitamin solution	10.0mL

pH 7.3 ± 0.2 at 25°C

Trace Elements Solution SL-6:

Composition per liter:

H_3BO_3	0.30g
$CoCl_2 \cdot 6H_2O$	0.20g
$ZnSO_4 \cdot 7H_2O$	0.10g
$MnCl_2 \cdot 4H_2O$	0.03g
$Na_2MoO_4 \cdot H_2O$	0.03g
$NiCl_2 \cdot 6H_2O$	0.02g
$CuCl_2 \cdot 2H_2O$	0.01g

Preparation of Trace Elements Solution SL-6: Add components to distilled/deionized water and bring volume to 1.0L. Mix thoroughly. Adjust pH to 3.4.

Vitamin Solution:

Composition per liter:

Vitamin B_{12}	2.8mg
Thiamine·HCl	0.28mg

Preparation of Vitamin Solution: Add components to distilled/deionized water and bring volume to 10.0mL. Mix thoroughly. Filter sterilize.

Preparation of Medium: Add components, except vitamin solution, to distilled/deionized water and bring volume to 990.0mL. Mix thoroughly. Adjust pH to 7.3 with 10% Na_2CO_3. Gently heat and bring to boiling. Continue boiling until resazurin changes color. Cool to 50°C. Distribute into tubes in 7.0mL volumes under O_2-free 97% N_2 + 3% H_2. Cap with rubber stoppers under O_2-free 97% N_2 + 3% H_2. Place tubes in a press. Autoclave for 15 min at 15 psi pressure–121°C using fast exhaust. Cool to 50°C. Aseptically add 0.25mL of sterile vitamin solution to each tube.

Use: For the cultivation and maintenance of *Meniscus glaucopis.*

Meniscus glaucopis Broth

Composition per liter:

Maltose	5.0g
Agar	3.0g
Yeast extract	1.0g
KH_2PO_4	0.5g
NaCl	0.4g
NH_4Cl	0.4g
Sodium thioglycollate	0.3g
$MgSO_4 \cdot 7H_2O$	0.2g
$CaCl_2 \cdot H_2O$	0.01g
$FeSO_4 \cdot 7H_2O$	1.0mg
Resazurin (0.025% solution)	4.0mL

Trace elements solution SL-6 .... 1.0mL
Vitamin solution 10.0mL
pH 7.3 ± 0.2 at 25°C

Trace Elements Solution SL-6:
Composition per liter:

H_3BO_3	0.30g
$CoCl_2 \cdot 6H_2O$	0.20g
$ZnSO_4 \cdot 7H_2O$	0.10g
$MnCl_2 \cdot 4H_2O$	0.03g
$Na_2MoO_4 \cdot H_2O$	0.03g
$NiCl_2 \cdot 6H_2O$	0.02g
$CuCl_2 \cdot 2H_2O$	0.01g

Preparation of Trace Elements Solution SL-6: Add components to distilled/deionized water and bring volume to 1.0L. Mix thoroughly. Adjust pH to 3.4.

Vitamin Solution:
Composition per liter:

Vitamin B_{12}	2.8mg
Thiamine·HCl	0.28mg

Preparation of Vitamin Solution: Add components to distilled/deionized water and bring volume to 10.0mL. Mix thoroughly. Filter sterilize.

Preparation of Medium: Add components, except vitamin solution, to distilled/deionized water and bring volume to 990.0mL. Mix thoroughly. Adjust pH to 7.3 with 10% Na_2CO_3. Gently heat and bring to boiling. Continue boiling until resazurin changes color. Cool to 50°C. Distribute into tubes in 7.0mL volumes under O_2-free 97% N_2 + 3% H_2. Cap with rubber stoppers under O_2-free 97% N_2 + 3% H_2. Place tubes in a press. Autoclave for 15 min at 15 psi pressure–121°C using fast exhaust. Cool to 50°C. Aseptically add 0.25mL of sterile vitamin solution to each tube.

Use: For the cultivation and maintenance of *Meniscus glaucopis.*

Metallogenium Cultivation Broth

Composition per liter:

Gum arabic	20.0g
$MnCO_3$	0.5g

$MnCO_3$:
Composition per 100mL:

$MnCl_2$	20.0g
$NaHCO_3$ (25% solution)	25.0mL

Preparation of $MnCO_3$: Add $MnCl_2$ to distilled/deionized water and bring volume to 100.0mL. Mix thoroughly. Add $NaHCO_3$ solution. Filter through Whatman #1 filter paper. Save the $MnCO_3$ precipitate. Wash and store under distilled/deionized water.

Preparation of Medium: Add components to distilled/deionized water and bring volume to 1.0L. Mix thoroughly. Distribute into tubes or flasks. Autoclave for 15 min at 15 psi pressure–121°C.

Use: For the cultivation of *Metallogenium* species.

Metallogenium Cultivation Broth

Composition per liter:

Starch, hydrolyzed	20.0g
$MnCO_3$	0.5g

$MnCO_3$:
Composition per 100mL:

$MnCl_2$	20.0g
$NaHCO_3$ (25% solution)	25.0mL

Preparation of $MnCO_3$: Add $MnCl_2$ to distilled/deionized water and bring volume to 100.0mL. Mix thoroughly. Add $NaHCO_3$ solution. Filter through Whatman #1 filter paper. Save the $MnCO_3$ precipitate. Wash and store under distilled/deionized water.

Preparation of Medium: Hydrolyze starch with HCl. Add components to distilled/deionized water and bring volume to 1.0L. Mix thoroughly. Distribute into tubes or flasks. Autoclave for 15 min at 15 psi pressure–121°C.

Use: For the cultivation of *Metallogenium* species.

Metallogenium Isolation Agar

Composition per liter:

Agar	15.0g
Manganese acetate	0.1g

Preparation of Medium: Add components to distilled/deionized water and bring volume to 1.0L. Mix thoroughly. Gently heat and bring to boiling. Distribute into tubes or flasks. Autoclave for 15 min at 15 psi pressure–121°C. Pour into sterile Petri dishes or leave in tubes.

Use: For the isolation and cultivation of *Metallogenium* species.

Metallogenium Medium

Composition per liter:

$MnCO_3$	2.0g
Starch, hydrolyzed	1.0g
DNA	0.01g
Catalase	5.0mg
Mycoplasma broth base	100.0mL
Yeast extract, ultrafiltrate	100.0mL
Horse serum	10.0mL

***Mycoplasma* Broth Base:**
Composition per liter:

Pancreatic digest of casein	7.0g
NaCl	5.0g

Beef extract 3.0g
Yeast extract 3.0g
Beef heart, solids from infusion 2.0g

Preparation of *Mycoplasma* Broth Base: Add components to distilled/deionized water and bring volume to 1.0L. Mix thoroughly. Autoclave for 15 min at 15 psi pressure–121°C. Cool to 25°C.

$MnCO_3$:

Composition per 100mL:

$MnCl_2$ 20.0g
$NaHCO_3$ (25% solution) 25.0mL

Preparation of $MnCO_3$: Add $MnCl_2$ to distilled/deionized water and bring volume to 100.0mL. Mix thoroughly. Add $NaHCO_3$ solution. Filter through Whatman #1 filter paper. Save the $MnCO_3$ precipitate. Wash and store under distilled/deionized water.

Preparation of Medium: Add $MnCO_3$, hydrolyzed starch and DNA to 25.0mL of distilled/deionized water. Mix thoroughly. Autoclave for 15 min at 15 psi pressure–121°C. Cool to 45–50°C. Aseptically add 100.0mL of sterile *Mycoplasma* broth base, 100.0mL of ultrafiltrate of yeast extract, 10.0mL of horse serum and 5.0mg of catalase. Mix thoroughly. Aseptically distribute into sterile tubes or flasks.

Use: For the cultivation of *Metallogenium* species.

Methanobacillus Medium

Composition per liter:

KH_2PO_4 .... 9.0g
K_2HPO_4 .... 6.0g
NH_4Cl 5.0g
$MgCl_2$ 1.0g
$CaCl_2$ 0.01g
$FeSO_4 \cdot 7H_2O$ 0.01g
Ethanol 10.0mL

pH 7.4 ± 0.2 at 25°C

Preparation of Medium: Filter sterilize ethanol. Add components, except ethanol, to tap water and bring volume to 990.0mL. Mix thoroughly. Gently heat until dissolved. Autoclave for 20 min at 10psi pressure–115°C. Cool to 45–50°C. Aseptically add sterile ethanol. Mix thoroughly. Aseptically distribute into sterile tubes or flasks.

Use: For the selective isolation and cultivation of *Methanobacillus* species from mixed cultures.

Methanobacteria Medium

Composition per liter:

Mineral Solution 2 .... 50.0mL
Sodium carbonate solution 50.0mL
Mineral Solution 1 .... 25.0mL
Cysteine-sulfide reducing agent 20.0mL
Wolfe's mineral solution 10.0mL
Vitamin solution 10.0mL
Resazurin (0.025% solution) 4.0mL

pH 7.2 ± 0.2 at 25°C

Mineral Solution 1:

Composition per liter:

K_2HPO_4 .... 6.0g

Preparation of Medium: Add K_2HPO_4 to distilled/deionized water and bring volume to 1.0L. Mix thoroughly.

Mineral Solution 2:

Composition per liter:

NaCl 12.0g
KH_2PO_4 .... 6.0g
$(NH_4)_2SO_4$ 6.0g
$MgSO_4 \cdot 7H_2O$ 2.4g
$CaCl_2 \cdot 2H_2O$ 1.6g

Preparation of Mineral Solution 2: Add components to distilled/deionized water and bring volume to 1.0L. Mix thoroughly.

Sodium Carbonate Solution:

Composition per 100mL:

Na_2CO_3 .... 8.0g

Preparation of Sodium Carbonate Solution: Add Na_2CO_3 to distilled/deionized water and bring volume to 100.0mL. Mix thoroughly.

Cysteine-Sulfide Reducing Agent:

Composition per 20mL:

L-Cysteine·HCl·H_2O 0.3g
$Na_2S \cdot 9H_2O$ 0.3g

Preparation of Cysteine-Sulfide Reducing Agent: Add L-Cysteine·HCl·H_2O to 10.0mL of distilled/deionized water. Mix thoroughly. In a separate tube, add $Na_2S \cdot 9H_2O$ to 10.0mL of distilled/deionized water. Mix thoroughly. Gas both solutions with 100% N_2 and cap tubes. Autoclave both solutions for 15 min at 15 psi pressure–121°C using fast exhaust. Cool to 50°C. Aseptically combine the two solutions under 100% N_2.

Wolfe's Mineral Solution:

Composition per liter

$MgSO_4 \cdot 7H_2O$ 3.0g
Nitriloacetic acid 1.5g
NaCl 1.0g
$MnSO_4 \cdot H_2O$ 0.5g
$FeSO_4 \cdot 7H_2O$ 0.1g
$CoCl_2 \cdot 6H_2O$ 0.1g
$CaCl_2$ 0.1g
$ZnSO_4 \cdot 7H_2O$ 0.1g
$CuSO_4 \cdot 5H_2O$ 0.01g

$AlK(SO_4)_2 \cdot 12H_2O$	0.01g
H_3BO_3	0.01g
$Na_2MoO_4 \cdot 2H_2O$	0.01g

Preparation of Wolfe's Mineral Solution: Add nitrilotriacetic acid to 500.0mL of distilled/deionized water. Dissolve by adjusting pH to 6.5 with KOH. Add remaining components. Add distilled/deionized water and bring volume to 1.0L.

Wolfe's Vitamin Solution:
Composition per liter:

Pyridoxine·HCl	10.0mg
Thiamine·HCl	5.0mg
Riboflavin	5.0mg
Nicotinic acid	5.0mg
Calcium pantothenate	5.0mg
p-Aminobenzoic acid	5.0mg
Thioctic acid	5.0mg
Biotin	2.0mg
Folic acid	2.0mg
Cyanocobalamin	100.0μg

Preparation of Wolfe's Vitamin Solution: Add components to distilled/deionized water and bring volume to 1.0L. Mix thoroughly. Filter sterilize.

Preparation of Medium: Add components, except vitamin solution and cysteine-sulfide reducing agent, to distilled/deionized water and bring volume to 970.0mL. Mix thoroughly. Autoclave for 15 min at 15 psi pressure–121°C. Cool under 80% N_2 + 20% CO_2. Aseptically add the sterile vitamin solution and then the sterile cysteine-sulfide reducing agent. Adjust the pH to 7.2. Distribute aseptically and anaerobically into sterile tubes.

Use: For the cultivation and maintenance of *Acetogenium kivui, Methanobacterium formicicum, Methanobacterium thermoautotrophicum,* and *Methanobrevibacter arboriphilicus.*

Methanobacteria Medium with Glucose and Yeast Extract

Composition per liter:

Glucose	5.0g
Yeast extract	2.0g
Mineral solution 2	50.0mL
Sodium carbonate solution	50.0mL
Mineral solution 1	25.0mL
Cysteine-sulfide reducing agent	20.0mL
Wolfe's mineral solution	10.0mL
Vitamin solution	10.0mL
Resazurin (0.025% solution)	4.0mL

pH 7.2 ± 0.2 at 25°C

Mineral Solution 1:
Composition per liter:

K_2HPO_4	6.0g

Preparation of Mineral Solution 1: Add K_2HPO_4 to distilled/deionized water and bring volume to 1.0L. Mix thoroughly.

Mineral Solution 2:
Composition per liter:

NaCl	12.0g
KH_2PO_4	6.0g
$(NH_4)_2SO_4$	6.0g
$MgSO_4 \cdot 7H_2O$	2.4g
$CaCl_2 \cdot 2H_2O$	1.6g

Preparation of Mineral Solution 2: Add components to distilled/deionized water and bring volume to 1.0L. Mix thoroughly.

Sodium Carbonate Solution:
Composition per 100mL:

Na_2CO_3	8.0g

Preparation of Sodium Carbonate Solution: Add Na_2CO_3 to distilled/deionized water and bring volume to 100.0mL. Mix thoroughly.

Cysteine-Sulfide Reducing Agent:
Composition per 20mL:

L-Cysteine·HCl·H_2O	0.3g
$Na_2S \cdot 9H_2O$	0.3g

Preparation of Cysteine-Sulfide Reducing Agent: Add L-Cysteine·HCl·H_2O to 10.0mL of distilled/deionized water. Mix thoroughly. In a separate tube, add $Na_2S \cdot 9H_2O$ to 10.0mL of distilled/deionized water. Mix thoroughly. Gas both solutions with 100% N_2 and cap tubes. Autoclave both solutions for 15 min at 15 psi pressure–121°C using fast exhaust. Cool to 50°C. Aseptically combine the two solutions under 100% N_2.

Wolfe's Mineral Solution:
Composition per liter

$MgSO_4 \cdot 7H_2O$	3.0g
Nitriloacetic acid	1.5g
NaCl	1.0g
$MnSO_4 \cdot H_2O$	0.5g
$FeSO_4 \cdot 7H_2O$	0.1g
$CoCl_2 \cdot 6H_2O$	0.1g
$CaCl_2$	0.1g
$ZnSO_4 \cdot 7H_2O$	0.1g
$CuSO_4 \cdot 5H_2O$	0.01g
$AlK(SO_4)_2 \cdot 12H_2O$	0.01g
H_3BO_3	0.01g
$Na_2MoO_4 \cdot 2H_2O$	0.01g

Preparation of Wolfe's Mineral Solution: Add nitrilotriacetic acid to 500.0mL of distilled/deionized water. Dissolve by adjusting pH to 6.5 with KOH. Add remaining components. Add distilled/deionized water to 1.0L.

Wolfe's Vitamin Solution:

Composition per liter:

Pyridoxine·HCl	10.0mg
Thiamine·HCl	5.0mg
Riboflavin	5.0mg
Nicotinic acid	5.0mg
Calcium pantothenate	5.0mg
p-Aminobenzoic acid	5.0mg
Thioctic acid	5.0mg
Biotin	2.0mg
Folic acid	2.0mg
Cyanocobalamin	100.0μg

Preparation of Wolfe's Vitamin Solution: Add components to distilled/deionized water and bring volume to 1.0L. Mix thoroughly. Filter sterilize.

Preparation of Medium: Add components, except vitamin solution and cysteine-sulfide reducing agent, to distilled/deionized water and bring volume to 970.0mL. Mix thoroughly. Autoclave for 15 min at 15 psi pressure–121°C. Cool under 80% N_2 + 20% CO_2. Aseptically add the sterile vitamin solution and then the sterile cysteine-sulfide reducing agent. Adjust the pH to 7.2. Distribute aseptically and anaerobically into sterile tubes.

Use: For the cultivation and maintenance of *Clostridium saccharolyticum, C. thermoaceticum,* and *C. thermohydrosulfuricum.*

Methanobacteria Medium with Xylose, Yeast Extract and Tryptone

Composition per liter:

Pancreatic digest of casein	10.0g
Xylose	5.0g
Yeast extract	3.0g
Mineral solution 2	50.0mL
Sodium carbonate solution	50.0mL
Mineral solution 1	25.0mL
Cysteine-sulfide reducing agent	20.0mL
Wolfe's mineral solution	10.0mL
Vitamin solution	10.0mL
Resazurin (0.025% solution)	4.0mL

pH 7.2 ± 0.2 at 25°C

Mineral Solution 1:

Composition per liter:

K_2HPO_4	6.0g

Preparation of Medium: Add K_2HPO_4 to distilled/deionized water and bring volume to 1.0L. Mix thoroughly.

Mineral Solution 2:

Composition per liter:

NaCl	12.0g
KH_2PO_4	6.0g
$(NH_4)_2SO_4$	6.0g
$MgSO_4·7H_2O$	2.4g
$CaCl_2·2H_2O$	1.6g

Preparation of Mineral Solution 2: Add components to distilled/deionized water and bring volume to 1.0L. Mix thoroughly.

Sodium Carbonate Solution:

Composition per 100mL:

Na_2CO_3	8.0g

Preparation of Sodium Carbonate Solution: Add Na_2CO_3 to distilled/deionized water and bring volume to 100.0mL. Mix thoroughly.

Cysteine-Sulfide Reducing Agent:

Composition per 20mL:

L-Cysteine·HCl·H_2O	0.3g
$Na_2S·9H_2O$	0.3g

Preparation of Cysteine-Sulfide Reducing Agent: Add L-Cysteine·HCl·H_2O to 10.0mL of distilled/deionized water. Mix thoroughly. In a separate tube, add $Na_2S·9H_2O$ to 10.0mL of distilled/deionized water. Mix thoroughly. Gas both solutions with 100% N_2 and cap tubes. Autoclave both solutions for 15 min at 15 psi pressure–121°C using fast exhaust. Cool to 50°C. Aseptically combine the two solutions under 100% N_2.

Wolfe's Mineral Solution:

Composition per liter:

$MgSO_4·7H_2O$	3.0g
Nitriloacetic acid	1.5g
NaCl	1.0g
$MnSO_4·H_2O$	0.5g
$FeSO_4·7H_2O$	0.1g
$CoCl_2·6H_2O$	0.1g
$CaCl_2$	0.1g
$ZnSO_4·7H_2O$	0.1g
$CuSO_4·5H_2O$	0.01g
$AlK(SO_4)_2·12H_2O$	0.01g
H_3BO_3	0.01g
$Na_2MoO_4·2H_2O$	0.01g

Preparation of Wolfe's Mineral Solution: Add nitrilotriacetic acid to 500.0mL of distilled/deionized water. Dissolve by adjusting pH to 6.5 with KOH. Add remaining components. Add distilled/deionized water to 1.0L.

Wolfe's Vitamin Solution:
Composition per liter:

Pyridoxine·HCl	10.0mg
Thiamine·HCl	5.0mg
Riboflavin	5.0mg
Nicotinic acid	5.0mg
Calcium pantothenate	5.0mg
p-Aminobenzoic acid	5.0mg
Thioctic acid	5.0mg
Biotin	2.0mg
Folic acid	2.0mg
Cyanocobalamin	100.0μg

Preparation of Wolfe's Vitamin Solution: Add components to distilled/deionized water and bring volume to 1.0L. Mix thoroughly. Filter sterilize.

Preparation of Medium: Add components, except vitamin solution and cysteine-sulfide reducing agent, to distilled/deionized water and bring volume to 970.0mL. Mix thoroughly. Autoclave for 15 min at 15 psi pressure–121°C. Cool under 80% N_2 + 20% CO_2. Aseptically add the sterile vitamin solution and then the sterile cysteine-sulfide reducing agent. Adjust the pH to 7.2. Distribute aseptically and anaerobically into sterile tubes.

Use: For the cultivation and maintenance of *Thermobacteroides acetoethylicus.*

Methanobacteria Medium with Yeast Extract, Sodium Acetate and Methanol

Composition per liter:

Glucose	5.0g
Sodium acetate	4.1g
Yeast extract	2.0g
Mineral solution 2	50.0mL
Sodium carbonate solution	50.0mL
Mineral solution 1	25.0mL
Cysteine-sulfide reducing agent	20.0mL
Wolfe's mineral solution	10.0mL
Vitamin solution	10.0mL
Methanol	4.0mL
Resazurin (0.025% solution)	4.0mL

pH 7.2 ± 0.2 at 25°C

Mineral Solution 1:
Composition per liter:

K_2HPO_4	6.0g

Preparation of Mineral Solution 1: Add K_2HPO_4 to distilled/deionized water and bring volume to 1.0L. Mix thoroughly.

Mineral Solution 2:
Composition per liter:

NaCl	12.0g
KH_2PO_4	6.0g
$(NH_4)_2SO_4$	6.0g
$MgSO_4·7H_2O$	2.4g
$CaCl_2·2H_2O$	1.6g

Preparation of Mineral Solution 2: Add components to distilled/deionized water and bring volume to 1.0L. Mix thoroughly.

Sodium Carbonate Solution:
Composition per 100mL:

Na_2CO_3	8.0g

Preparation of Sodium Carbonate Solution: Add Na_2CO_3 to distilled/deionized water and bring volume to 100.0mL. Mix thoroughly.

Cysteine-Sulfide Reducing Agent:
Composition per 20mL:

L-Cysteine·HCl·H_2O	300.0mg
$Na_2S·9H_2O$	300.0mg

Preparation of Cysteine-Sulfide Reducing Agent: Add L-Cysteine·HCl·H_2O to 10.0mL of distilled/deionized water. Mix thoroughly. In a separate tube, add $Na_2S·9H_2O$ to 10.0mL of distilled/deionized water. Mix thoroughly. Gas both solutions with 100% N_2 and cap tubes. Autoclave both solutions for 15 min at 15 psi pressure–121°C using fast exhaust. Cool to 50°C. Aseptically combine the two solutions under 100% N_2.

Wolfe's Mineral Solution:
Composition per liter:

$MgSO_4·7H_2O$	3.0g
Nitriloacetic acid	1.5g
NaCl	1.0g
$MnSO_4·H_2O$	0.5g
$FeSO_4·7H_2O$	0.1g
$CoCl_2·6H_2O$	0.1g
$CaCl_2$	0.1g
$ZnSO_4·7H_2O$	0.1g
$CuSO_4·5H_2O$	0.01g
$AlK(SO_4)_2·12H_2O$	0.01g
H_3BO_3	0.01g
$Na_2MoO_4·2H_2O$	0.01g

Preparation of Wolfe's Mineral Solution: Add nitrilotriacetic acid to 500.0mL of distilled/deionized water. Dissolve by adjusting pH to 6.5 with KOH. Add remaining components. Add distilled/deionized water to 1.0L.

Wolfe's Vitamin Solution:
Composition per liter:

Pyridoxine·HCl	10.0mg
Thiamine·HCl	5.0mg
Riboflavin	5.0mg
Nicotinic acid	5.0mg
Calcium pantothenate	5.0mg

p-Aminobenzoic acid	5.0mg
Thioctic acid	5.0mg
Biotin	2.0mg
Folic acid	2.0mg
Cyanocobalamin	100.0μg

Preparation of Wolfe's Vitamin Solution: Add components to distilled/deionized water and bring volume to 1.0L. Mix thoroughly. Filter sterilize.

Preparation of Medium: Add components, except vitamin solution, cysteine-sulfide reducing agent and methanol, to distilled/deionized water and bring volume to 970.0mL. Mix thoroughly. Autoclave for 15 min at 15 psi pressure–121°C. Cool under 80% N_2 + 20% CO_2. Filter sterilize methanol. Aseptically add 4.0mL of sterile methanol to cooled, sterile basal medium. Aseptically add the sterile vitamin solution and then the sterile cysteine-sulfide reducing agent. Adjust the pH to 7.2. Distribute aseptically and anaerobically into sterile tubes.

Use: For the cultivation and maintenance of *Butyribacterium methylotrophicum.*

Methanobacterium alcaliphilum Medium

Composition per liter:

$NaHCO_3$	10.0g
Yeast extract	2.0g
Peptone	2.0g
NH_4Cl	1.0g
L-Cysteine·HCl·H_2O	0.5g
K_2HPO_4	0.4g
$MgCl_2·6H_2O$	0.1g
$CaCl_2$	0.02g
Resazurin	1.0mg
Salt solution	5.0mL

pH 8.4 ± 0.2 at 25°C

Salt Solution:
Composition per 100mL:

Sodium EDTA·$2H_2O$	0.10g
$CoCl_2·6H_2O$	0.03g
$MnCl_2·4H_2O$	0.02g
$ZnCl_2$	0.02g
$AlCl_3·6H_2O$	8.0mg
$CuCl_2·2H_2O$	4.0mg
$NiSO_4·6H_2O$	4.0mg
Na_2SeO_3	2.7mg
$FeSO_4·7H_2O$	2.0mg
H_3BO_3	2.0mg
$NaMoO_4·2H_2O$	2.0mg

Preparation of Salt Solution: Add components to distilled/deionized water and bring volume to 100.0mL. Mix thoroughly.

Preparation of Medium: Add components, except $NaHCO_3$, yeast extract, peptone and L-cysteine·HCl·H_2O, to distilled/deionized water and bring volume to 990.0mL. Gently heat and bring to boiling under O_2-free 100% N_2. Continue boiling until resazurin becomes pale, indicating partial reduction. Add the yeast extract, peptone and L-cysteine·HCl·H_2O and continue boiling under O_2-free 100% N_2 until resazurin becomes colorless, indicating complete reduction. Cool to room temperature under O_2-free 100% N_2. Add $NaHCO_3$ to 10.0mL of distilled/deionized water. Mix thoroughly. Gas with O_2-free 100% N_2 in a sealed tube. Add reduced $NaHCO_3$ solution to cooled reduced medium. Distribute anaerobically into tubes. Cap with butyl rubber stoppers and secure with closures. Autoclave for 15 min at 15 psi pressure–121°C with fast exhaust.

Use: For cultivation and maintenance of *Methanobacterium alcaliphilum.*

Methanobacterium Enrichment Medium

Composition per liter:

$CaCO_3$	100.0g
K_2HPO_4	5.0g
$(NH_4)_2SO_4$	0.3g
$MgSO_47H_2O$	0.1g
$FeSO_4·7H_2O$	0.02g
Na_2CO_3 solution	10.0mL
$Na_2S·9H_2O$ solution	10.0mL
Ethanol	10.0mL
Yeast autolysate	5.0mL

pH 7.2 ± 0.2 at 25°C

Na_2CO_3 Solution:
Composition per 10mL:

$NaHCO_3$	0.5g

Preparation of Na_2CO_3 Solution: Add Na_2CO_3 to distilled/deionized water and bring volume to 10.0mL. Mix thoroughly. Filter sterilize.

$Na_2S·9H_2O$ Solution:
Composition per 10mL:

$Na_2S·9H_2O$	0.1g

Preparation of $Na_2S·9H_2O$ Solution: Add $Na_2S·9H_2O$ to distilled/deionized water and bring volume to 10.0mL. Mix thoroughly. Filter sterilize.

Preparation of Medium: Filter sterilize ethanol. Add components—except ethanol, Na_2CO_3 solution, and $Na_2S·9H_2O$ solution—to distilled/deionized water and bring volume to 970.0mL. Mix thoroughly. Gently heat and bring to boiling. Autoclave for 15 min at 15 psi pressure–121°C. Cool to 45–50°C. Aseptically add sterile ethanol, Na_2CO_3 solution, and

$Na_2S·9H_2O$ solution. Mix thoroughly. Aseptically distribute into sterile tubes or flasks.

Use: For the cultivation and enrichment of *Methanobacterium* species.

Methanobacterium ruminantium Medium

Composition per liter:

$NaHCO_3$	6.0g
NaCl	2.0g
Cysteine·HCl·H_2O	1.0g
$K_2HPO_4·3H_2O$	1.0g
KH_2PO_4	1.0g
NH_4Cl	1.0g
$CaCl_2·2H_2O$	0.1g
$MgSO_4·7H_2O$	0.1g
Resazurin	1.0mg
Rumen fluid	300.0mL
$Na_2S·9H_2O$ solution	10.0mL

6.8 ± 0.2 at 25°C

$Na_2S·9H_2O$ Solution:

Composition per 10mL:

$Na_2S·9H_2O$	0.25g

Preparation of $Na_2S·9H_2O$ Solution: Add $Na_2S·9H_2O$ to distilled/deionized water and bring volume to 10.0mL. Mix thoroughly. Autoclave for 15 min at 15 psi pressure–121°C. Cool to 25°C.

Preparation of Medium: Prepare and distribute medium anaerobically under 80% H_2 + 20% CO_2. Add components, except rumen fluid and $Na_2S·9H_2O$ solution, to distilled/deionized water and bring volume to 690.0mL. Mix thoroughly. Gently heat and bring to boiling. Continue boiling until resazurin turns colorless, indicating reduction. Autoclave for 15 min at 15 psi pressure–121°C. Cool to 25°C. Aseptically add 10.0mL of sterile $Na_2S·9H_2O$ solution and 300.0mL of sterile rumen fluid. Mix thoroughly. Aseptically and anaerobically distribute into sterile tubes or flasks.

Use: For the cultivation of *Methanobacterium ruminantium.*

Methanobacterium thermoautotrophicum Medium, Taylor and Pirt

Composition per liter:

Na_2CO_3	4.0g
$(NH_4)_2SO_4$	3.0g
NaCl	1.2g
KH_2PO_4	0.6g
Cysteine·HCl·H_2O	0.5g
K_2HPO_4	0.3g
Nitrilotriacetic acid	0.03g
$CoCl_2$	0.02g
$CaCl_2$	0.01g
$FeSO_4$	0.01g
$MgSO_4$	0.01g
$MnSO_4$	0.01g
$ZnSO_4$	2.0mg
Resazurin	1.0mg
$AlK(SO_4)_2$	0.2mg
$CuSO_4$	0.2mg
H_3BO_3	0.2mg
Na_2MoO_4	0.2mg
$Na_2S·9H_2O$ solution	10.0mL

pH 7.2 ± 0.2 at 25°C

$Na_2S·9H_2O$ Solution:

Composition per 10mL:

$Na_2S·9H_2O$	0.5g

Preparation of $Na_2S·9H_2O$ Solution: Add $Na_2S·9H_2O$ to distilled/deionized water and bring volume to 10.0mL. Mix thoroughly. Autoclave for 15 min at 15 psi pressure–121°C. Cool to 25°C.

Preparation of Medium: Prepare and distribute medium anaerobically under 80% H_2 + 20% CO_2. Add components, except $Na_2S·9H_2O$ solution, to distilled/deionized water and bring volume to 990.0mL. Mix thoroughly. Gently heat and bring to boiling. Continue boiling until resazurin turns colorless, indicating reduction. Autoclave for 15 min at 15 psi pressure–121°C. Cool to 25°C. Aseptically add 10.0mL of sterile $Na_2S·9H_2O$ solution. Mix thoroughly. Aseptically and anaerobically distribute into sterile tubes or flasks.

Use: For the cultivation of *Methanobacterium thermoautotrophicum.*

Methanococcus vannielii Medium

Composition per 1020mL:

Solution A	500.0mL
Inorganic salts solution	500.0mL
$Na_2S·9H_2O$ solution	10.0mL
Na_2CO_3 solution	10.0mL

Solution A:

Composition per 500mL:

Sodium formate	10.0g
Phenol red	3.0mg
Methylene blue	2.0mg

Preparation of Solution A: Add components to distilled/deionized water and bring volume to 500.0mL. Mix thoroughly. Autoclave for 15 min at 15 psi pressure–121°C. Cool to 25°C.

Inorganic Salts Solution:
Composition per 500mL:

$K_2HPO_4 \cdot 3H_2O$	1.45g
NH_4Cl	1.0g
KH_2PO_4	0.75g
$MgCl_2 \cdot 6H_2O$	0.2g
Nitrilotriacetic acid	0.04g
$CaCl_2 \cdot 2H_2O$	0.02g
$FeCl_2 \cdot 4H_2O$	3.6mg
$CoCl_2 \cdot 6H_2O$	1.5mg
MnCl2·4H2O	0.9mg
$ZnCl_2$	0.9mg
H_3BO_2	0.17mg
$Na_2MoO_4 \cdot 2H_2O$	.09mg

Preparation of Inorganic Salts Solution: Add nitrilotriacetic acid to 250.0mL of distilled/deionized water. Dissolve by adjusting pH to 6.5 with KOH. Add remaining components. Readjust pH to 7.2 with H_2SO_4 or KOH. Add distilled/deionized water to 500.0mL. Filter sterilize.

$Na_2S \cdot 9H_2O$ Solution:
Composition per 10mL:

$Na_2S \cdot 9H_2O$	0.3g

Preparation of $Na_2S \cdot 9H_2O$ Solution: Add $Na_2S \cdot 9H_2O$ to distilled/deionized water and bring volume to 10.0mL. Mix thoroughly. Autoclave for 15 min at 15 psi pressure–121°C. Cool to 25°C.

Na_2CO_3 Solution:
Composition per 10mL:

Na_2CO_3	2.5g

Preparation of Na_2CO_3 Solution: Add Na_2CO_3 to distilled/deionized water and bring volume to 10.0mL. Mix thoroughly. Autoclave for 15 min at 15 psi pressure–121°C. Cool to 25°C.

Preparation of Medium: Prepare and distribute medium anaerobically under 80% N_2 + 20% CO_2. Aseptically and anaerobically combine 500.0mL of sterile inorganic salts solution, 500.0mL of sterile solution A, 10.0mL of sterile $Na_2S \cdot 9H_2O$ solution, and 10.0mL of sterile Na_2CO_3 solution. Mix thoroughly. Aseptically and anaerobically distribute into sterile tubes or flasks.

Use: For the isolation and cultivation of *Methanococcus vannielii* from marine mud.

Methanococcus vannielii Medium

Composition per liter:

Sodium formate	15.0g
K_2HPO_4	3.48g
$CoCl_2 \cdot 6H_2O$	2.38g
NH_4Cl	1.0g
Cysteine·HCl·H_2O	0.3g
$MgSO_4 \cdot 7H_2O$	0.2g
$CaCl_2 \cdot 2H_2O$	0.01g
$FeSO_4 \cdot 7H_2O$	0.01g
$MnSO_4 \cdot H_2O$	7.5mg
$Na_2MoO_4 \cdot 2H_2O$	7.5mg
Na_2SeO_3	1.7mg
$Na_2S \cdot 9H_2O$	10.0mL

$Na_2S \cdot 9H_2O$ Solution:
Composition per 10mL:

$Na_2S \cdot 9H_2O$	0.15g

Preparation of $Na_2S \cdot 9H_2O$ Solution: Add $Na_2S \cdot 9H_2O$ to distilled/deionized water and bring volume to 10.0mL. Mix thoroughly. Autoclave for 15 min at 15 psi pressure–121°C. Cool to 25°C.

Preparation of Medium: Prepare and distribute medium anaerobically under 100% N_2. Add components, except $Na_2S \cdot 9H_2O$ solution, to distilled/deionized water and bring volume to 990.0mL. Mix thoroughly. Gently heat and bring to boiling. Continue boiling until resazurin turns colorless, indicating reduction. Autoclave for 15 min at 15 psi pressure–121°C. Cool to 25°C. Aseptically add 10.0mL of sterile $Na_2S \cdot 9H_2O$ solution. Mix thoroughly. Aseptically and anaerobically distribute into sterile tubes or flasks.

Use: For the cultivation of *Methanococcus vannielii.*

Methanogen Enrichment Medium, Barker

Composition per liter:

$CaCO_3$	20.0g
NH_4Cl	1.0g
$K_2HPO_4 \cdot 3H_2O$	0.4g
$MgCl_2 \cdot 6H_2O$	0.1g
Methanol	20.0mL

pH 7.0 ± 0.2 at 25°C

Preparation of Medium: Add components, except methanol and $CaCO_3$, to distilled/deionized water and bring volume to 1.0L. Mix thoroughly. Gently heat and bring to boiling. Autoclave for 15 min at 15 psi pressure–121°C. Cool to 25°C. Aseptically add filter-sterilized methanol solution. Mix thoroughly. Add 1.0g of $CaCO_3$ to each of 50.0mL screw-capped bottles. Autoclave for 15 min at 15 psi pressure–121°C. Cool to 25°C. Fill each bottle to capacity with enrichment medium.

Use: For the cultivation of methanogenic bacteria.

Methanogen Medium

Composition per 106mL:

$CaCO_3$	10.0g
Calcium acetate	2.0g
NH_4Cl	0.1g
$K_2HPO_4 \cdot 3H_2O$	0.04g
$MgCl_2 \cdot 6H_2O$	0.01g
$Na_2S \cdot 9H_2O$ solution	3.0mL
Na_2CO_3 solution	3.0mL

$Na_2S \cdot 9H_2O$ Solution:

Composition per 10mL:

$Na_2S \cdot 9H_2O$	0.1g

Preparation of $Na_2S \cdot 9H_2O$ Solution: Add $Na_2S \cdot 9H_2O$ to distilled/deionized water and bring volume to 10.0mL. Mix thoroughly. Autoclave for 15 min at 15 psi pressure–121°C. Cool to 25°C.

Na_2CO_3 Solution:

Composition per 10mL:

Na_2CO_3	0.5g

Preparation of Na_2CO_3 Solution: Add Na_2CO_3 to distilled/deionized water and bring volume to 10.0mL. Mix thoroughly. Autoclave for 15 min at 15 psi pressure–121°C. Cool to 25°C.

Preparation of Medium: Prepare and distribute medium anaerobically under 100% N_2. Add components, except $Na_2S \cdot 9H_2O$ solution and Na_2CO_3 solution, to distilled/deionized water and bring volume to 100.0mL. Mix thoroughly. Autoclave for 15 min at 15 psi pressure–121°C. Cool to 25°C. Aseptically add 3.0mL of sterile $Na_2S \cdot 9H_2O$ solution and 3.0mL of sterile Na_2CO_3 solution. Mix thoroughly. Aseptically and anaerobically distribute into sterile tubes or flasks.

Use: For the cultivation and enrichment of acetate-utilizing methanogenic bacteria.

Methanogen Medium, Zeikus

Composition per 1010mL:

Inorganic salts solution	500.0mL
Vitamin solution	500.0mL
$Na_2S \cdot 9H_2O$ solution	10.0mL

pH 7.0 ± 0.2 at 25°C

Inorganic Salts Solution:

Composition per 500mL:

$K_2HPO_4 \cdot 3H_2O$	1.45g
NH_4Cl	1.0g
KH_2PO_4	0.75g
$MgCl_2 \cdot 6H_2O$	0.2g
Nitrilotriacetic acid	0.04g
$CaCl_2 \cdot 2H_2O$	0.02g
$FeCl_2 \cdot 4H_2O$	3.6mg
$CoCl_2 \cdot 6H_2O$	1.5mg
MnCl2·4H2O	0.9mg
$ZnCl_2$	0.9mg
H_3BO_2	0.17mg
$Na_2MoO_4 \cdot 2H_2O$	.09mg

Preparation of Inorganic Salts Solution: Add nitrilotriacetic acid to 250.0mL of distilled/deionized water. Dissolve by adjusting pH to 6.5 with KOH. Add remaining components. Readjust pH to 7.2 with H_2SO_4 or KOH. Add distilled/deionized water to 500.0mL. Filter sterilize.

Vitamin Solution:

Composition per 500mL:

Pyridoxine·HCl	1.0mg
p-Aminobenzoic acid	0.5mg
Ca-D-pantothenate	0.5mg
Nicotinic acid	0.5mg
Riboflavin	0.5mg
Thiamine·HCl	0.5mg
Thioctic acid	0.5mg
Biotin	0.2mg
Folic acid	0.2mg
Vitamin B_{12}	0.01mg

Preparation of Vitamin Solution: Add components to distilled/deionized water and bring volume to 500.0mL. Mix thoroughly. Filter sterilize.

$Na_2S \cdot 9H_2O$ Solution:

Composition per 10mL:

$Na_2S \cdot 9H_2O$	0.3g

Preparation of $Na_2S \cdot 9H_2O$ Solution: Add $Na_2S \cdot 9H_2O$ to distilled/deionized water and bring volume to 10.0mL. Mix thoroughly. Autoclave for 15 min at 15 psi pressure–121°C. Cool to 25°C.

Preparation of Medium: Prepare and distribute medium anaerobically under 95% N_2 + 5% CO_2. Aseptically and anaerobically combine 500.0mL of sterile inorganic salts solution, 500.0mL of sterile vitamin solution and 10.0mL of sterile $Na_2S \cdot 9H_2O$ solution. Mix thoroughly. Aseptically and anaerobically distribute into sterile tubes or flasks.

Use: For the cultivation of methanogenic bacteria.

Methanogenium Medium

Composition per liter:

NaCl	18.0g
$NaHCO_3$	5.0g
$MgSO_4 \cdot 7H_2O$	3.45g
$MgCl_2 \cdot 7H_2O$	2.75g
Yeast extract	2.0g
Pancreatic digest of casein	2.0g
Sodium acetate	1.0g
Resazurin	1.0g

L-Cysteine·HCl·H_2O	0.5g
$Na_2S·9H_2O$	0.5g
KCl	0.335g
NH_4Cl	0.25g
$CaCl_2·2H_2O$	0.14g
K_2HPO_4	0.14g
$Fe(NH_4)_2(SO_4)_2·7H_2O$	2.0mg
Trace elements solution SL-6	10.0mL
Wolfe's vitamin solution	10.0mL

pH 6.8 ± 0.2 at 25°C

Trace Elements Solution SL-6:

Composition per liter:

H_3BO_3	0.30g
$CoCl_2·6H_2O$	0.20g
$ZnSO_4·7H_2O$	0.10g
$MnCl_2·4H_2O$	0.03g
$Na_2MoO_4·H_2O$	0.03g
$NiCl_2·6H_2O$	0.02g
$CuCl_2·2H_2O$	0.01g

Preparation of Trace Elements Solution SL-6: Add components to distilled/deionized water and bring volume to 1.0L. Mix thoroughly. Adjust pH to 3.4.

Wolfe's Vitamin Solution:

Composition per liter:

Pyridoxine·HCl	10.0mg
Thiamine·HCl	5.0mg
Riboflavin	5.0mg
Nicotinic acid	5.0mg
Calcium pantothenate	5.0mg
p-Aminobenzoic acid	5.0mg
Thioctic acid	5.0mg
Biotin	2.0mg
Folic acid	2.0mg
Cyanocobalamin	100.0μg

Preparation of Wolfe's Vitamin Solution: Add components to distilled/deionized water and bring volume to 1.0L. Mix thoroughly. Filter sterilize.

Preparation of Medium: Prepare and dispense medium under 80% N_2 + 20% CO_2. Add components, except Wolfe's vitamin solution, to distilled/deionized water and bring volume to 990.0mL. Mix thoroughly. Adjust pH to 6.8. Autoclave for 15 min at 15 psi pressure–121°C. Cool under 80% N_2 + 20% CO_2. Aseptically add sterile Wolfe's vitamin solution. Aseptically and anaerobically distribute into sterile tubes or flasks.

Use: For the cultivation and maintenance of *Methanococcus frisius, Methanococcus maripaludis, Methanococcus thermolithotrophicus,* and *Methanoplanus limicola.*

Methanol Ammonium Salts Medium

Composition per liter:

$MgSO_4·7H_2O$	1.0g
NH_4Cl	0.5g
Na_2HPO_4	0.33g
KH_2PO_4	0.26g
$CaCl_2$	0.2g
Ferrous EDTA	5.0mg
$Na_2MoO_4·2H_2O$	2.0mg
$FeSO_4·7H_2O$	500.0μg
$ZnSO_4·7H_2O$	400.0μg
EDTA	250.0μg
$CoCl_2·6H_2O$	50.0μg
$MnCl_2·4H_2O$	20.0μg
H_3BO_4	15.0μg
$NiCl_2·6H_2O$	10.0μg
Methanol	5.0mL

pH 6.8 ± 0.2 at 25°C

Preparation of Medium: Add Na_2HPO_4 and KH_2PO_4 to distilled/deionized water and bring volume to 100.0mL. Mix thoroughly. In a separate container, add remaining components, except methanol, to distilled/deionized water and bring volume to 895.0mL. Mix thoroughly. Autoclave both solutions for 15 min at 15 psi pressure–121°C. Cool to room temperature. Filter sterilize methanol. Aseptically add the sterile phosphate solution and the sterile methanol to the cooled, sterile basal medium. Mix thoroughly. Aseptically distribute into sterile tubes or flasks.

Use: For the maintenance and cultivation of *Methylomonas methylotrophus.*

Methanol Medium (ATCC Medium 436)

Composition per liter:

Agar	15.0g
K_2HPO_4	7.0g
$(NH_4)_2SO_4$	3.0g
KH_2PO_4	2.0g
$MgSO_4·7H_2O$	0.5g
Yeast extract	0.2g
$FeSO_4·7H_2O$	0.01g
$MnSO_4·H_2O$	8.0mg
Biotin	0.2μg
Thiamine·HCl	0.2μg
Methanol	10.0mL

pH 7.0 ± 0.2 at 25°C

Preparation of Medium: Add components, except methanol, to distilled/deionized water and bring volume to 990.0mL. Mix thoroughly. Gently heat and bring to boiling. Autoclave for 15 min at 15 psi pressure–121°C. Cool to 50–55°C. Filter sterilize methanol. Aseptically add the sterile methanol to the cooled, sterile basal medium. Mix thoroughly. Aseptically distribute into sterile tubes or flasks.

Use: For the cultivation and maintenance of *Ancylobacter* species, *Methanomonas methylovora*, and *Methylobacterium* species.

Methanol Medium (ATCC Medium 1096)

Composition per liter:

NH_4NO_3	0.75g
$FeCl_3$	0.743g
Methanol	0.45g
$MgSO_4$	0.09g
KH_2PO_4	0.044g
$Na_2MoO_4 \cdot 2H_2O$	0.1mg

pH 7.0 ± 0.2 at 25°C

Preparation of Medium: Prepare and dispense medium under 97% N_2 + 3% H_2. Add components, except methanol, to distilled/deionized water and bring volume to 999.0mL. Mix thoroughly. Gently heat and bring to boiling. Autoclave for 15 min at 15 psi pressure–121°C. Cool to 50–55°C. Filter sterilize methanol. Aseptically add the sterile methanol to the cooled, sterile basal medium. Mix thoroughly. Aseptically distribute into sterile tubes or flasks.

Use: For the cultivation and maintenance of *Ancylobacter*, *Methylobacterium* species and *Methanomonas methylovora*.

Methanol Medium for *Achromobacter*

Composition per liter:

NH_4Cl	5.0g
KH_2PO_4	2.0g
NaCl	0.5g
$MgSO_4$	0.2g
Yeast extract	0.2g
$FeSO_4$	2.0mg
$MnCl_2$	2.0mg
Methanol	20.0mL

pH 7.0 ± 0.2 at 25°C

Preparation of Medium: Add components, except methanol, to distilled/deionized water and bring volume to 980.0mL. Mix thoroughly. Autoclave for 15 min at 15 psi pressure–121°C. Cool to 50–55°C. Filter sterilize methanol. Aseptically add the sterile methanol to the cooled, sterile basal medium. Mix thoroughly. Aseptically distribute into sterile tubes or flasks.

Use: For the cultivation and maintenance of *Achromobacter methanolophila*, *Methylobacterium rhodesianum*, *Pseudomonas insueta*, and *Pseudomonas polysaccharogenes*.

Methanol Medium with 1% Peptone

Composition per liter:

Agar	15.0g
Peptone	10.0g
K_2HPO_4	7.0g
$(NH_4)_2SO_4$	3.0g
KH_2PO_4	2.0g
$MgSO_4 \cdot 7H_2O$	0.5g
Yeast extract	0.2g
$FeSO_4 \cdot 7H_2O$	0.01g
$MnSO_4 \cdot H_2O$	8.0mg
Biotin	0.2μg
Thiamine·HCl	0.2μg
Methanol	10.0mL

pH 7.0 ± 0.2 at 25°C

Preparation of Medium: Add components, except methanol, to distilled/deionized water and bring volume to 990.0mL. Mix thoroughly. Autoclave for 15 min at 15 psi pressure–121°C. Cool to 50–55°C. Filter sterilize methanol. Aseptically add the sterile methanol to the cooled, sterile basal medium. Mix thoroughly. Aseptically distribute into sterile tubes or flasks.

Use: For the cultivation and maintenance of *Methylobacterium* species.

Methanol Mineral Salts Medium

Composition per liter:

Agar	20.0g
$(NH_4)_2SO_4$	2.0g
NH_4Cl	2.0g
$(NH_4)_2HPO_4$	2.0g
Yeast extract	2.0g
KH_2PO_4	1.0g
K_2HPO_4	1.0g
$MgSO_4 \cdot 7H_2O$	0.5g
$Fe_2SO_4 \cdot 7H_2O$	0.01g
$CaCl_2 \cdot 2H_2O$	0.01g
Methanol	10.0mL

pH 7.0 ± 0.2 at 25°C

Preparation of Medium: Add components, except methanol, to distilled/deionized water and bring volume to 990.0mL. Mix thoroughly. Gently heat and bring to boiling. Autoclave for 15 min at 15 psi pressure–121°C. Cool to 50–55°C. Filter sterilize methanol. Aseptically add the sterile methanol to the cooled, sterile basal medium. Mix thoroughly. Aseptically distribute into sterile Petri dishes or sterile tubes.

Use: For the cultivation and maintenance of *Pseudomonas viscogena*.

Methanolobus Medium

Composition per liter:

NaCl	18.0g
$NaHCO_3$	5.0g
$MgSO_4·7H_2O$	3.45g
$MgCl_2·6H_2O$	2.75g
L-Cysteine·HCl·H_2O	0.5g
$Na_2S·9H_2O$	0.5g
KCl	0.335g
NH_4Cl	0.25g
$CaCl_2·2H_2O$	0.14g
K_2HPO_4	0.14g
$Fe(NH_4)_2(SO_4)_2·6H_2O$	2.0mg
Resazurin	1.0mg
Wolfe's mineral solution	10.0mL
Wolfe's vitamin solution	10.0mL
Methanol	5.0mL

pH 6.5 ± 0.2 at 25°C

Wolfe's Mineral Solution:

Composition per liter:

$MgSO_4·7H_2O$	3.0g
Nitriloacetic acid	1.5g
NaCl	1.0g
$MnSO_4·H_2O$	0.5g
$FeSO_4·7H_2O$	0.1g
$CoCl_2·6H_2O$	0.1g
$CaCl_2$	0.1g
$ZnSO_4·7H_2O$	0.1g
$CuSO_4·5H_2O$	0.01g
$AlK(SO_4)_2·12H_2O$	0.01g
H_3BO_3	0.01g
$Na_2MoO_4·2H_2O$	0.01g

Preparation of Wolfe's Mineral Solution: Add nitrilotriacetic acid to 500.0mL of distilled/deionized water. Dissolve by adjusting pH to 6.5 with KOH. Add remaining components. Add distilled/deionized water to 1.0L.

Wolfe's Vitamin Solution:

Composition per liter:

Pyridoxine·HCl	10.0mg
Thiamine·HCl	5.0mg
Riboflavin	5.0mg
Nicotinic acid	5.0mg
Calcium pantothenate	5.0mg
p-Aminobenzoic acid	5.0mg
Thioctic acid	5.0mg
Biotin	2.0mg
Folic acid	2.0mg
Cyanocobalamin	100.0μg

Preparation of Wolfe's Vitamin Solution: Add components to distilled/deionized water and bring volume to 1.0L. Mix thoroughly. Filter sterilize.

Preparation of Medium: Prepare and dispense medium under 80% N_2 + 20% CO_2. Add components, except methanol and Wolfe's vitamin solution, to distilled/deionized water and bring volume to 985.0mL. Mix thoroughly. Autoclave for 15 min at 15 psi pressure–121°C. Cool under 80% N_2 + 20% CO_2. Aseptically add sterile Wolfe's vitamin solution and sterile methanol. Adjust pH to 6.5. Aseptically and anaerobically distribute into sterile tubes or flasks.

Use: For the cultivation and maintenance of *Methanolobus tindarius*.

Methanomicrobium Medium

Composition per liter:

$NaHCO_3$	2.0g
Yeast extract	1.0g
Pancreatic digest of casein	1.0g
NaCl	0.6g
L-Cysteine·HCl·H_2O	0.5g
$Na_2S·9H_2O$	0.5g
K_2HPO_4	0.3g
KH_2PO_4	0.3g
$(NH_4)_2SO_4$	0.3g
$MgSO_4·7H_2O$	0.13g
$CaCl_2·2H_2O$	8.0mg
$FeSO_4·7H_2O$	2.0mg
Rumen fluid, clarified	300.0mL
Fatty acid mixture	20.0mL
Wolfe's mineral solution	10.0mL
Wolfe's vitamin solution	10.0mL
Resazurin (0.1% solution)	1.0mL

pH 6.5 ± 0.2 at 25°C

Fatty Acid Mixture:

Composition per liter:

Valeric acid	0.7mL
Isovaleric acid	0.7mL
α-Methylbutyric acid	0.5mL
Isobutyric acid	0.5mL

Preparation of Fatty Acid Mixture: Add components to distilled/deionized water and bring volume to 1.0L. Mix thoroughly.

Wolfe's Mineral Solution:

Composition per liter:

$MgSO_4·7H_2O$	3.0g
Nitriloacetic acid	1.5g
NaCl	1.0g
$MnSO_4·H_2O$	0.5g
$FeSO_4·7H_2O$	0.1g
$CoCl_2·6H_2O$	0.1g

$CaCl_2$	0.1g
$ZnSO_4 \cdot 7H_2O$	0.1g
$CuSO_4 \cdot 5H_2O$	0.01g
$AlK(SO_4)_2 \cdot 12H_2O$	0.01g
H_3BO_3	0.01g
$Na_2MoO_4 \cdot 2H_2O$	0.01g

Preparation of Wolfe's Mineral Solution: Add nitrilotriacetic acid to 500.0mL of distilled/deionized water. Dissolve by adjusting pH to 6.5 with KOH. Add remaining components. Add distilled/deionized water to 1.0L.

Wolfe's Vitamin Solution:

Composition per liter:

Pyridoxine·HCl	10.0mg
Thiamine·HCl	5.0mg
Riboflavin	5.0mg
Nicotinic acid	5.0mg
Calcium pantothenate	5.0mg
p-Aminobenzoic acid	5.0mg
Thioctic acid	5.0mg
Biotin	2.0mg
Folic acid	2.0mg
Cyanocobalamin	100.0μg

Preparation of Wolfe's Vitamin Solution: Add components to distilled/deionized water and bring volume to 1.0L. Mix thoroughly.

Preparation of Medium: Prepare and dispense medium under 80% N_2 + 20% CO_2. Add components to distilled/deionized water and bring volume to 1.0L. Mix thoroughly. Adjust pH to 6.5. Distribute into tubes or flasks under 80% N_2 + 20% CO_2. Cap with rubber stoppers. Autoclave for 15 min at 15 psi pressure–121°C.

Use: For the cultivation and maintenance of *Methanomicrobium mobile.*

Methanomonas Autotrophic Medium

Composition per liter:

$NaNO_3$	2.0g
Na_2HPO_4	0.21g
$MgSO_4 \cdot 7H_2O$	0.2g
NaH_2PO_4	0.09g
KCl	0.04g
$CaCl_2$	0.015g
$FeSO_4 \cdot 7H_2O$	1.0mg
$ZnSO_4 \cdot 7H_2O$	0.3mg
$CuSO_4 \cdot 5H_2O$	0.2mg
H_3BO_3	0.06mg
$MnSO_4 \cdot H_2O$	0.03mg
MoO_3	0.015mg

Preparation of Medium: Add components to distilled/deionized water and bring volume to 1.0L. Mix thoroughly. Gently heat until dissolved. Distribute into tubes or flasks. Autoclave for 15 min at 15 psi pressure–121°C.

Use: For the autotrophic cultivation of *Methanomonas* species.

Methanosarcina acetovorans Medium

Composition per liter:

NaCl	23.4g
Agar	10.0g
$MgSO_4$	6.3g
Na_2CO_3	5.0g
Trimethylamine·HCl	3.0g
Yeast extract	1.0g
NH_4Cl	1.0g
KCl	0.8g
Na_2HPO_4	0.6g
L-Cysteine·HCl·H_2O	0.25g
$Na_2S \cdot 9H_2O$	0.25g
$CaCl_2 \cdot 2H_2O$	0.14g
Resazurin	1.0mg
Wolfe's mineral solution	10.0mL

pH 7.2 ± 0.2 at 25°C

Wolfe's Mineral Solution:

Composition per liter:

$MgSO_4 \cdot 7H_2O$	3.0g
Nitriloacetic acid	1.5g
NaCl	1.0g
$MnSO_4 \cdot H_2O$	0.5g
$FeSO_4 \cdot 7H_2O$	0.1g
$CoCl_2 \cdot 6H_2O$	0.1g
$CaCl_2$	0.1g
$ZnSO_4 \cdot 7H_2O$	0.1g
$CuSO_4 \cdot 5H_2O$	0.01g
$AlK(SO_4)_2 \cdot 12H_2O$	0.01g
H_3BO_3	0.01g
$Na_2MoO_4 \cdot 2H_2O$	0.01g

Preparation of Wolfe's Mineral Solution: Add nitrilotriacetic acid to 500.0mL of distilled/deionized water. Dissolve by adjusting pH to 6.5 with KOH. Add remaining components. Add distilled/deionized water to 1.0L.

Preparation of Medium: Add components, except $Na_2S \cdot 9H_2O$, to glass distilled water and bring volume to 990.0mL. Mix thoroughly. Methanol or methylamine·HCl may be substituted for the trimethylamine·HCl at a concentration of 50 mM. Heat gently and bring to boiling. Adjust pH to 7.2 with 6*N* HCl. Autoclave for 5 min at 10 psi pressure–115°C. Cool to 50°C under 80% N_2 + 20% CO_2. If a large precipitate is present add a small amount of HCl and mix thoroughly. Add $Na_2S·9H_2O$. Mix thoroughly. Distribute into tubes under 80% N_2 + 20% CO_2. Cap with butyl rubber stoppers. Autoclave for 15 min at 15 psi pressure–121°C. A precipitate will form but resolubilizes as the medium cools. Invert tubes as they are cooling to facilitate resolubilization. Allow tubes to cool in a slanted position.

Use: For the cultivation and maintenance of *Methanococcoides methylutens* and *Methanosarcina acetivorans.*

Methanosarcina Medium

Composition per liter:

Agar	20.0g
NaCl	2.25g
Yeast extract	2.00g
Pancreatic digest of casein	2.00g
NH_4Cl	0.50g
$MgSO_4·7H_2O$	0.50g
K_2HPO_4	0.35g
$CaCl_2·2H_2O$	0.25g
KH_2PO_4	0.23g
$FeSO_4·7H_2O$	2.0mg
$NaHCO_3$ solution	20.0mL
Cysteine-sulfide reducing agent	20.0mL
Wolfe's vitamin solution	10.0mL
Methanol	10.0mL
Resazurin (0.025% solution)	4.0mL
Trace elements SL-6	3.0mL

pH 6.8 ± 0.2 at 25°C

$NaHCO_3$ Solution:
Composition per 20mL:

$NaHCO_3$	850.0mg

Preparation of $NaHCO_3$ Solution: Add $NaHCO_3$ to distille/deionized water and bring volume to 20.0mL. Mix thoroughly. Filter sterilize. Gas with 100% CO_2 for 20 min.

Cysteine-Sulfide Reducing Agent:
Composition per 20mL:

L-Cysteine·HCl·H_2O	0.3g
$Na_2S·9H_2O$	0.3g

Preparation of Cysteine-Sulfide Reducing Agent: Add L-Cysteine·HCl·H_2O to 10.0mL of distilled/deionized water. Mix thoroughly. In a separate tube, add $Na_2S·9H_2O$ to 10.0mL of distilled/deionized water. Mix thoroughly. Gas both solutions with 100% N_2 and cap tubes. Autoclave both solutions for 15 min at 15 psi pressure–121°C using fast exhaust. Cool to 50°C. Aseptically combine the two solutions under 100% N_2.

Wolfe's Vitamin Solution:
Composition per liter:

Pyridoxine·HCl	10.0mg
Thiamine·HCl	5.0mg
Riboflavin	5.0mg
Nicotinic acid	5.0mg
Calcium pantothenate	5.0mg
p-Aminobenzoic acid	5.0mg
Thioctic acid	5.0mg
Biotin	2.0mg
Folic acid	2.0mg
Cyanocobalamin	100.0μg

Preparation of Wolfe's Vitamin Solution: Add components to distilled/deionized water and bring volume to 1.0L. Mix thoroughly. Filter sterilize.

Trace Elements Solution SL-6:
Composition per liter:

H_3BO_3	0.3g
$CoCl_2·6H_2O$	0.2g
$ZnSO_4·7H_2O$	0.10g
$MnCl_2·4H_2O$	0.03g
$Na_2MoO_4·H_2O$	0.03g
$NiCl_2·6H_2O$	0.02g
$CuCl_2·2H_2O$	0.01g

Preparation of Trace Elements Solution SL-6: Add components to distilled/deionized water and bring volume to 1.0L. Mix thoroughly. Adjust pH to 3.4.

Preparation of Medium: Add components, except $NaHCO_3$ solution, cysteine-sulfide reducing agent, Wolfe's vitamin solution and methanol, to distilled/deionized water and bring volume to 940.0mL. Mix thoroughly. Autoclave for 15 min at 15 psi pressure–121°C. Cool under 80% N_2 + 20% CO_2. Aseptically and anaerobically add the sterile $NaHCO_3$ solution, the sterile cysteine-sulfide reducing agent, the sterile Wolfe's vitamin solution and filter-sterilized methanol. Mix thoroughly. Adjust pH to 6.8. Aseptically and anaerobically distribute into sterile tubes or flasks.

Use: For the cultivation and maintenance of *Bifidobacterium asteroides, Methanosarcina barkeri, Methanosarcina* species, and *Methanosarcina vacuolata.*

Methanospirillum hungatei Medium

Composition per 100mL:

Na_2CO_3	0.4g
Sodium formate	0.2g
Pancreatic digest of casein	0.2g
Yeast extract	0.2g

NaCl	0.05g
L-Cysteine·HCl·H_2O	0.03g
K_2HPO_4	0.02g
KH_2PO_4	0.02g
$(NH_4)_2SO_4$	0.02g
$MgSO_4 \cdot 7H_2O$	9.0mg
$CaCl_2 \cdot 2H_2O$	6.0mg
Resazurin	0.1mg
$Na_2S \cdot 9H_2O$ solution	10.0mL
Vitamin solution	1.0mL
Trace metal solution	1.0mL

pH 7.0 ± 0.2 at 25°C

$Na_2S \cdot 9H_2O$ Solution:

Composition per 10mL:

$Na_2S \cdot 9H_2O$	0.03g

Preparation of $Na_2S \cdot 9H_2O$ Solution: Add $Na_2S \cdot 9H_2O$ to distilled/deionized water and bring volume to 10.0mL. Mix thoroughly. Autoclave for 15 min at 15 psi pressure–121°C. Cool to 25°C.

Vitamin Solution:

Composition per 100mL:

Pyridoxine·HCl	1.0mg
p-Aminobenzoic acid	0.5mg
Calcium-D-pantothenate	0.5mg
Nicotinic acid	0.5mg
Riboflavin	0.5mg
Thiamine·HCl	0.5mg
Thioctic acid	0.5mg
Biotin	0.2mg
Folic acid	0.2mg
Vitamin B_{12}	0.01mg

Preparation of Vitamin Solution: Add components to distilled/deionized water and bring volume to 1.0L. Mix thoroughly. Filter sterilize.

Trace Metal Solution:

Composition per liter:

$K_2HPO_4 \cdot 3H_2O$	9.0g
K_2HPO_4	6.0g
NH_4Cl	5.0g
$MgCl_2 \cdot 6H_2O$	1.0g
$CaCl_2 \cdot 2H_2O$	0.01g

Preparation of Trace Metal Solution: Add components to distilled/deionized water and bring volume to 1.0L. Mix thoroughly.

Preparation of Medium: Prepare and distribute medium anaerobically under 80% H_2 + 20% CO_2. Add components, except $Na_2S \cdot 9H_2O$ solution, to distilled/deionized water and bring volume to 90.0mL. Mix thoroughly. Gently heat and bring to boiling. Continue boiling until resazurin turns colorless, indicating reduction. Autoclave for 15 min at 15 psi pressure–121°C. Cool to 25°C. Aseptically add 10.0mL of sterile $Na_2S \cdot 9H_2O$ solution. Mix thoroughly. Aseptically and anaerobically distribute into sterile tubes or flasks.

Use: For the cultivation of *Methanospirillum hungatei.*

Methylamine Salts Medium

Composition per liter:

Agar	15.0g
Methylamine·HCl	6.75g
K_2HPO_4	2.12g
KH_2PO_4	1.0g
Solution A	5.0mL
Solution B	1.0mL

pH 7.0 ± 0.2 at 25°C

Solution A:

Composition per 100mL:

$MgSO_4 \cdot 7H_2O$	2.0g
$CaCl_2 \cdot 2H_2O$	0.2g
$FeSO_4 \cdot 7H_2O$	0.2g

Preparation of Solution A: Add components to distilled/deionized water and bring volume tc 100.0mL. Mix thoroughly.

Solution B:

Composition per 100mL:

$MnSO_4 \cdot 7H_2O$	0.05g
$Na_2MoO_4 \cdot 2H_2O$	0.05g

Preparation of Solution B: Add components to distilled/deionized water and bring volume to 100.0mL. Mix thoroughly.

Preparation of Medium: Add components to distilled/deionized water and bring volume to 1.0L. Mix thoroughly. Gently heat and bring to boiling. Distribute into tubes or flasks. Autoclave for 15 min at 15 psi pressure–121°C. Pour into sterile Petri dishes or leave in tubes.

Use: For the cultivation and maintenance of *Methylobacterium extorquens* and *Pseudomonas* species.

Methylene Blue Milk Medium (MBM Medium)

Composition per liter:

Skim milk, dehydrated	100.0g
Methylene blue	10.0g

pH 6.4 ± 0.2 at 25°C

Preparation of Medium: Add components to distilled/deionized water and bring volume to 1.0L. Mix thoroughly. Distribute into tubes or flasks. Autoclave for 20 min at 10 psi pressure–115°C.

Use: For cultivation and differentiation of group D streptococci (enterococci) from other *Streptococcus* species.

Methylococcus Medium

Composition per liter:

Agar	8.0g
$NaNO_3$ (20% solution)	10.0mL
L-F Salts solution	10.0mL
Sodium-potassium phosphate buffer	6.5mL

pH 7.1 ± 0.2 at 25°C

Sodium-Potassium Phosphate Buffer:
Composition per liter:

KH_2PO_4	136.0g
NaOH	28.8g

Preparation of Sodium-Potassium Phosphate Buffer: Add components to distilled/deionized water and bring volume to 1.0L. Mix thoroughly. Adjust pH to 7.1.

L-F Salts Solution:
Composition per liter:

$MgSO_4 \cdot 7H_2O$ (10% solution)	200.0mL
$CaCl_2 \cdot 2H_2O$ (10% solution)	20.0mL
$FeSO_4$ (10% solution)	10.0mL
$ZnSO_4 \cdot 7H_2O$ (1% solution)	4.9mL
H_3BO_3 (1% solution)	0.6mL
$MnSO_4 \cdot H_2O$ (1% solution)	0.27mL
$CuSO_4 \cdot 5H_2O$ (1% solution)	0.2mL

Preparation of L-F Salts Solution: Filter sterilize $FeSO_4$ solution immediately prior to use. Add all components to distilled/deionized water and bring volume to 1.0L. Mix thoroughly.

Preparation of Medium: Add components to distilled/deionized water and bring volume to 1.0L. Mix thoroughly. Adjust pH to 7.1. Autoclave for 15 min at 15 psi pressure–121°C. Pour into sterile Petri dishes or leave in tubes.

Use: For cultivation and maintenance of *Methylococcus* species.

Methylophaga Agar

Composition per 103mL:

Agar solution	50.0mL
Mineral base, 2X	50.0mL
Solution T	2.0mL
Vitamin B_{12} solution	1.0mL
Methanol	0.3mL

pH 7.3 ± 0.2 at 25°C

Agar Solution:
Composition per 500mL:

Agar	15.0g

Preparation of Agar Solution: Add agar to distilled/deionized water and bring volume to 500.0mL. Mix thoroughly. Autoclave for 15 min at 15 psi pressure–121°C. Cool to 50°C.

Mineral Base, 2X:
Composition per 500mL:

NaCl	24.0g
$MgCl_2 \cdot 6H_2O$	3.0g
$MgSO_4 \cdot 7H_2O$	2.0g
$CaCl_2 \cdot 2H_2O$	1.0g
KCl	0.5g
Bis-Tris buffer (bis[2-Hydroxyethyl]imino-tris[hydroxymethyl]-methane)	0.5g
Wolfe's mineral solution	10.0mL

Preparation of Mineral Base, 2X: Add components to distilled/deionized water and bring volume to 500.0mL. Mix thoroughly. Adjust pH to 7.3. Autoclave for 15 min at 15 psi pressure–121°C. Cool to 50°C.

Wolfe's Mineral Solution:
Composition per liter:

$MgSO_4 \cdot 7H_2O$	3.0g
Nitriloacetic acid	1.5g
NaCl	1.0g
$MnSO_4 \cdot H_2O$	0.5g
$FeSO_4 \cdot 7H_2O$	0.1g
$CoCl_2 \cdot 6H_2O$	0.1g
$CaCl_2$	0.1g
$ZnSO_4 \cdot 7H_2O$	0.1g
$CuSO_4 \cdot 5H_2O$	0.01g
$AlK(SO_4)_2 \cdot 12H_2O$	0.01g
H_3BO_3	0.01g
$Na_2MoO_4 \cdot 2H_2O$	0.01g

Preparation of Wolfe's Mineral Solution: Add nitrilotriacetic acid to 500.0mL of distilled/deionized water. Dissolve by adjusting pH to 6.5 with KOH. Add remaining components. Add distilled/deionized water to 1.0L.

Solution T:
Composition per 100mL:

NH_4Cl	10.0g
Bis-Tris buffer (bis[2-Hydroxyethyl]imino-tris[hydroxymethyl]-methane)	10.0g
KH_2PO_4	0.7g
Ferric ammonium citrate	0.3g

Preparation of Solution T: Add components to distilled/deionized water and bring volume to 100.0mL. Mix thoroughly. Adjust pH to 7.3. Autoclave for 15 min at 15 psi pressure–121°C.

Vitamin B_{12} Solution:
Composition per 10mL:

Vitamin B_{12}	1.0μg

Preparation of Vitamin B_{12} Solution: Add vitamin B_{12} to 10.0mL of distilled/deionized water. Mix thoroughly. Filter sterilize.

Preparation of Medium: Aseptically mix 50.0mL of the sterile agar solution with 50.0mL of the sterile mineral base, 2X. Aseptically combine the sterile solution T and sterile vitamin B_{12} solution with the sterile mineral base. Filter sterilize methanol and add to basal medium. Pour into sterile Petri dishes or distribute into sterile tubes.

Use: For the cultivation and maintenance of *Methylophaga marina.*

Methylophaga Broth

Composition per 103mL:

Mineral base	100.0mL
Solution T	2.0mL
Vitamin B_{12} solution	1.0mL
Methanol	0.3mL

pH 7.3 ± 0.2 at 25°C

Mineral Base:

Composition per liter:

NaCl	24.0g
$MgCl_2 \cdot 6H_2O$	3.0g
$MgSO_4 \cdot 7H_2O$	2.0g
$CaCl_2 \cdot 2H_2O$	1.0g
KCl	0.5g
Bis-Tris buffer (bis[2-Hydroxyethyl]imino-tris[hydroxymethyl]-methane)	0.5g
Wolfe's mineral solution	10.0mL

Preparation of Mineral Base: Add components to distilled/deionized water and bring volume to 1.0L. Mix thoroughly. Adjust pH to 7.3. Autoclave for 15 min at 15 psi pressure–121°C.

Wolfe's Mineral Solution:

Composition per liter:

$MgSO_4 \cdot 7H_2O$	3.0g
Nitriloacetic acid	1.5g
NaCl	1.0g
$MnSO_4 \cdot H_2O$	0.5g
$FeSO_4 \cdot 7H_2O$	0.1g
$CoCl_2 \cdot 6H_2O$	0.1g
$CaCl_2$	0.1g
$ZnSO_4 \cdot 7H_2O$	0.1g
$CuSO_4 \cdot 5H_2O$	0.01g
$AlK(SO_4)_2 \cdot 12H_2O$	0.01g
H_3BO_3	0.01g
$Na_2MoO_4 \cdot 2H_2O$	0.01g

Preparation of Wolfe's Mineral Solution: Add nitrilotriacetic acid to 500.0mL of distilled/deionized water. Dissolve by adjusting pH to 6.5 with KOH. Add remaining components. Add distilled/deionized water to 1.0L.

Solution T:

Composition per 100mL:

NH_4Cl	10.0g
Bis-Tris buffer (bis[2-Hydroxyethyl]imino-tris[hydroxymethyl]-methane)	10.0g
KH_2PO_4	0.7g
Ferric ammonium citrate	0.3g

Preparation of Solution T: Add components to distilled/deionized water and bring volume to 100.0mL. Mix thoroughly. Adjust pH to 7.3. Autoclave for 15 min at 15 psi pressure–121°C.

Vitamin B_{12} Solution:

Composition per 10mL:

Vitamin B_{12}	1.0μg

Preparation of Vitamin B_{12} Solution: Add vitamin B_{12} to 10.0mL of distilled/deionized water. Mix thoroughly. Filter sterilize.

Preparation of Medium: Aseptically combine the sterile solution T and sterile vitamin B_{12} solution with the sterile mineral base. Filter sterilize methanol and add to basal medium. Aseptically distribute into tubes or flasks.

Use: For the cultivation and maintenance of *Methylophaga marina.*

MH IH Agar

Composition per liter:

Solution A	490.0mL
Solution B	490.0mL
Supplement solution	20.0mL

pH 6.9 ± 0.2 at 25°C

Solution A:

Composition per 490mL:

Beef infusion	300.0g
Acid hydrolysate of casein	17.5g
Agar	17.0g
Starch	1.5g

Preparation of Solution A: Add components to distilled/deionized water and bring volume to 490.0mL. Mix thoroughly. Gently heat and bring to boiling. Autoclave for 15 min at 15 psi pressure–121°C. Cool to 45–50°C.

Solution B:

Composition per 490mL:

Hemoglobin	10.0g

Preparation of Solution B: Add hemoglobin to distilled/deionized water and bring volume to 490.0mL. Mix thoroughly. Gently heat and bring to boiling. Autoclave for 15 min at 15 psi pressure–121°C. Cool to 45–50°C.

Supplement Solution:

Composition per liter:

Glucose	100.0g
L-Cysteine·HCl	25.9g

L-Glutamine	10.0g
L-Cystine	1.1g
Adenine	1.0g
Nicotinamide adenine dinucleotide	0.25g
Vitamin B_{12}	0.1g
Thiamine pyrophosphate	0.1g
Guanine·HCl	0.03g
$Fe(NO_3)_3 \cdot 6H_2O$	0.02g
p-Aminobenzoic acid	0.013g
Thiamine·HCl	3.0mg

Source: The supplement solution IsoVitaleX® enrichment is available from BBL Microbiology Laboratories. This enrichment may be replaced by supplement VX from Difco Laboratories.

Preparation of Supplement Solution: Add components to distilled/deionized water and bring volume to 1.0L. Mix thoroughly. Filter sterilize.

Preparation of Medium: Aseptically combine cooled, sterile solution A and cooled, sterile solution B. Mix thoroughly. Adjust pH to 6.9 with sterile 1*N* HCl or sterile 1*N* KOH. Aseptically add 20.0mL of sterile supplement solution. Pour into sterile Petri dishes or distribute into sterile tubes.

Use: For the cultivation and differentiation of *Legionella* species.

MH Medium

Composition per liter:

NaCl	60.7g
Agar	20.0g
$MgCl_2 \cdot 6H_2O$	15.0g
Yeast extract	10.0g
$MgSO_4 \cdot 7H_2O$	7.4g
Proteose peptone No. 3	5.0g
KCl	1.5g
Glucose	1.0g
$CaCl_2$	0.27g
$NaHCO_3$	0.45g
NaBr	0.19g

Preparation of Medium: Add components to distilled/deionized water and bring volume to 1.0L. Mix thoroughly. Gently heat and bring to boiling. Distribute into tubes or flasks. Autoclave for 15 min at 15 psi pressure–121°C. Pour into sterile Petri dishes or leave in tubes.

Use: For the cultivation and maintenance of *Deleya salina* and *Volcaniella eurihalina*.

MH Salts

Composition per liter:

NaCl	120.5g
$MgCl_2 \cdot 6H_2O$	22.4g
Agar	20.0g
$MgSO_4$	14.4g
Yeast extract	10.0g
Proteose peptone No. 3	5.0g
KCl	3.0g
Glucose	1.0g
$CaCl_2$	0.54g
$NaHCO_3$	0.09g
NaBr	0.039g

pH 7.5 ± 0.2 at 25°C

Preparation of Medium: Add components to distilled/deionized water and bring volume to 1.0L. Mix thoroughly. Adjust pH to 7.5. Gently heat and bring to boiling. Distribute into tubes or flasks. Autoclave for 15 min at 15 psi pressure–121°C. Pour into sterile Petri dishes or leave in tubes.

Use: For the cultivation and maintenance of *Bacillus halophilus*.

Microbacterium Medium

Composition per liter:

Glucose	10.0g
KH_2PO_4	5.0g
K_2HPO_4	5.0g
Potassium aspartate	5.0g
$(NH_4)_2SO_4$	2.0g
$MgSO_4 \cdot 7H_2O$	0.5g
Calcium pantothenate	0.2g
β-Mercaptopurine	0.1g
$FeSO_4 \cdot 7H_2O$	0.01g
Thiamine·HCl	0.01g
Biotin	0.1mg

pH 7.0 ± 0.2 at 25°C

Preparation of Medium: Add components to distilled/deionized water and bring volume to 1.0L. Mix thoroughly. Distribute into tubes or flasks. Autoclave for 10 min at 15 psi pressure–121°C.

Use: For the cultivation and maintenance of *Microbacterium* species.

Microbial Content Test Agar

Composition per liter:

Agar	15.0g
Pancreatic digest of casein	15.0g
NaCl	5.0g
Tween™ 80	5.0g
Enzymatic hydrolysate of soybean meal	5.0g
Lecithin	0.7g

pH 7.3 ± 0.2 at 25°C

Source: This medium is available as a premixed powder from Difco Laboratories.

Preparation of Medium: Add components to distilled/deionized water and bring volume to 1.0L. Mix thoroughly. Gently heat and bring to boiling. Boil for 1–2 min. Distribute into tubes or flasks. Autoclave for 15 min at 15 psi pressure–121°C. Pour into sterile Petri dishes or leave in tubes.

Use: For use in the microbial content test of water soluble cosmetic products. Also used for determining the efficiency of sanitization of containers, equipment and environmental surfaces.

Micrococcus–Sarcina Medium

Composition per liter:

Agar	16.0g
Pancreatic digest of casein	5.0g
Sodium succinate·$6H_2O$	2.0g
Starch	2.0g
Yeast autolysate	1.0g
Sodium citrate·$2H_2O$	0.5g
Sodium acetate·$3H_2O$	0.3g
K_2HPO_4	0.2g

pH 7.0 ± 0.2 at 25°C

Preparation of Medium: Add components to distilled/deionized water and bring volume to 1.0L. Mix thoroughly. Gently heat and bring to boiling. Distribute into tubes or flasks. Autoclave for 15 min at 15 psi pressure–121°C. Pour into sterile Petri dishes or leave in tubes.

Use: For the cultivation and maintenance of *Micrococcus luteus* and *Sarcina* species.

Microcyclus eburneus Medium

Composition per liter:

K_2HPO_4	7.0g
$(NH_4)SO_4$	3.0g
KH_2PO_4	2.0g
$MgSO_4 \cdot 7H_2O$	0.5g
Yeast extract	0.2g
Thiamine·HCl	0.2mg
Biotin	0.02mg
$FeSO_4 \cdot 7H_2O$	2.0μg
$MnSO_4 \cdot 4H_2O$	2.0μg

Preparation of Medium: Add components to distilled/deionized water and bring volume to 1.0L. Mix thoroughly. Distribute into tubes or flasks. Autoclave for 15 min at 15 psi pressure–121°C.

Use: For the cultivation of *Microcyclus eburneus.*

Microcylus major Medium

Composition per liter:

Glucose	1.0g
Peptone	1.0g
KNO_3	0.1g
K_2HPO_4	0.07g
$MgSO_4 \cdot 7H_2O$	0.03g
Trace element solution	1.0mL

Trace Element Solution:

Composition per liter:

Disodium EDTA	10.0g
$FeSO_4 \cdot 7H_20$	9.3g
$NaBO_3 \cdot 4H_2O$	2.6g
$MnCl_2 \cdot 4H_2O$	1.8g
$CaCl_2$	1.2g
$(NH_4)_6Mo_7O_{24} \cdot 4H_2O$	1.0g
$ZnSO_4 \cdot 7H_2O$	0.2g
$CuSO_4 \cdot 5H_2O$	0.08g
$Co(NO_3)_2 \cdot H_2O$	0.02g

Preparation of Trace Element Solution: Add components to distilled/deionized water and bring volume to 1.0L. Mix thoroughly.

Preparation of Medium: Add components to distilled/deionized water and bring volume to 1.0L. Mix thoroughly. Distribute into tubes or flasks. Autoclave for 15 min at 15 psi pressure–121°C.

Use: For the cultivation of *Microcyclus major.*

Microcyclus marinus Medium

Composition per liter:

NaCl	23.5g
$MgCl_2$	5.0g
Na_2SO_4	4.0g
$CaCl_2 \cdot 2H_2O$	1.5g
KCl	0.7g
$NaHCO_3$	0.2g

Preparation of Medium: Add components to distilled/deionized water and bring volume to 1.0L. Mix thoroughly. Distribute into tubes or flasks. Autoclave for 15 min at 15 psi pressure–121°C.

Use: For the cultivation of *Microcyclus marinus.*

Microcyclus Medium

Composition per liter:

Agar	15.0g
Glucose	5.0g
Peptone	5.0g
Yeast extract	5.0g

pH 6.8 ± 0.2 at 25°C

Preparation of Medium: Add components to distilled/deionized water and bring volume to 1.0L. Mix thoroughly. Gently heat and bring to boiling. Distribute into tubes or flasks. Autoclave for 15 min at 15 psi pressure–121°C. Pour into sterile Petri dishes or leave in tubes.

Use: For the cultivation and maintenance of *Flectobacillus major* and *Microcyclus* species.

Microcyclus–Spirosoma Medium

Composition per liter:

Agar	15.0g
Glucose	1.0g
Peptone	1.0g
Yeast extract	1.0g

pH 6.8–7.0 at 25°C

Preparation of Medium: Add components to distilled/deionized water and bring volume to 1.0L. Mix thoroughly. Gently heat and bring to boiling. Distribute into tubes or flasks. Autoclave for 15 min at 15 psi pressure–121°C. Pour into sterile Petri dishes or leave in tubes.

Use: For the cultivation and maintenance of *Spirosoma linguale* and *Microcyclus* species.

MIL Medium (Motility Indole Lysine Medium)

Composition per liter:

Peptone	10.0g
Pancreatic digest of casein	10.0g
L-Lysine·HCl	10.0g
Yeast extract	3.0g
Agar	2.0g
Dextrose	1.0g
Ferric ammonium citrate	0.5g
Bromcresol purple	0.02g

pH 6.6 ± 0.2 at 25°C

Source: Available as a premixed powder and prepared medium from Difco Laboratories.

Preparation of Medium: Add components to distilled/deionized water and bring volume to 1.0L. Mix thoroughly. Gently heat and bring to boiling. Distribute into tubes in 5.0mL volumes. Autoclave for 15 min at 15 psi pressure–121°C.

Use: For the cultivation and differentiation of members of the Enterobacteriaceae on the basis of motility, lysine decarboxylase activity, lysine deaminase activity and indole production.

Mineral Base E for Autotrophic Growth

Composition per liter:

Noble agar	15.0g
K_2HPO_4	1.2g
KH_2PO_4	0.624g
$(NH_4)_2SO_4$	0.5g
NaCl	0.1g
$CaCl_2 \cdot 6H_2O$ solution	10.0mL
$MgSO_4 \cdot 7H_2O$ solution	10.0mL
Mineral solution	1.0mL
p-Aminobenzoic acid solution	1.0mL

$CaCl_2 \cdot 6H_2O$ Solution:

Composition per liter:

$CaCl_2 \cdot 6H_2O$	5.0g

Preparation of $CaCl_2 \cdot 6H_2O$ Solution: Add $CaCl_2 \cdot 6H_2O$ to distilled/deionized water and bring volume to 1.0L. Mix thoroughly. Autoclave for 15 min at 15 psi pressure–121°C.

$MgSO_4 \cdot 7H_2O$ Solution:

Composition per liter:

$MgSO_4 \cdot 7H_2O$	20.0g

Preparation of $MgSO_4 \cdot 7H_2O$ Solution: Add $MgSO_4 \cdot 7H_2O$ to distilled/deionized water and bring volume to 1.0L. Mix thoroughly. Autoclave for 15 min at 15 psi pressure–121°C.

***p*-Aminobenzoic Acid Solution:**

Composition per 10mL:

p-Aminobenzoic acid	100.0mg

Preparation of *p*-Aminobenzoic Acid Solution: Add *p*-aminobenzoic acid to distilled/deionized water and bring volume to 10.0mL. Mix thoroughly. Autoclave for 15 min at 15 psi pressure–121°C.

Mineral Solution:

Composition per 100mL:

Disodium EDTA	1.58g
ZnSO4·7H2O	0.7g
$MnSO_4 \cdot 4H_2O$	0.18g
$FeSO_4 \cdot 7H_2O$	0.16g
$CoCl_2 \cdot 6H_2O$	0.052g
$Na_2MoO_4 \cdot 2H_2O$	0.047g
$CuSO_4 \cdot 5H_2O$	0.047g

Preparation of Medium: Add components, except $CaCl_2 \cdot 6H_2O$ solution, $MgSO_4 \cdot 7H_2O$ solution, and *p*-aminobenzoic acid solution to distilled/deionized water and bring volume to 979.0mL. Mix thoroughly. Autoclave for 15 min at 15 psi pressure–121°C. Cool to 50°C. Aseptically add in the following order: 10.0mL of sterile $CaCl_2 \cdot 6H_2O$ solution,

10.0mL of sterile $MgSO_4 \cdot 7H_2O$ solution, and 1.0mL of sterile *p*-aminobenzoic acid solution. Mix thoroughly. Aseptically distribute into sterile tubes or flasks. Incubate inoculated tubes in 50% CO_2.

Use: For the autotrophic cultivation and maintenance of *Pseudomonas thermocarboxydovorans.*

Mineral Base E for Heterotrophic Growth

Composition per liter:

Noble agar	15.0g
K_2HPO_4	1.2g
KH_2PO_4	0.624g
$(NH_4)_2SO_4$	0.5g
NaCl	0.1g
$CaCl_2 \cdot 6H_2O$ solution	10.0mL
$MgSO_4 \cdot 7H_2O$ solution	10.0mL
Sodium pyruvate solution	10.0mL
Mineral solution	1.0mL
p-Aminobenzoic acid solution	1.0mL

$CaCl_2 \cdot 6H_2O$ Solution:
Composition per liter:

$CaCl_2 \cdot 6H_2O$	5.0g

Preparation of $CaCl_2 \cdot 6H_2O$ Solution: Add $CaCl_2 \cdot 6H_2O$ to distilled/deionized water and bring volume to 1.0L. Mix thoroughly. Autoclave for 15 min at 15 psi pressure–121°C.

$MgSO_4 \cdot 7H_2O$ Solution:
Composition per liter:

$MgSO_4 \cdot 7H_2O$	20.0g

Preparation of $MgSO_4 \cdot 7H_2O$ Solution: Add $MgSO_4 \cdot 7H_2O$ to distilled/deionized water and bring volume to 1.0L. Mix thoroughly. Autoclave for 15 min at 15 psi pressure–121°C.

Sodium Pyruvate Solution:
Composition per 10mL:

Sodium pyruvate	2.0g

Preparation of Sodium Pyruvate Solution: Add sodium pyruvate to distilled/deionized water and bring volume to 10.0mL. Mix thoroughly. Filter sterilize.

***p*-Aminobenzoic Acid Solution:**
Composition per 10mL:

p-Aminobenzoic acid	100.0mg

Preparation of *p*-Aminobenzoic Acid Solution: Add *p*-aminobenzoic acid to distilled/deionized water and bring volume to 10.0mL. Mix thoroughly. Autoclave for 15 min at 15 psi pressure–121°C.

Mineral Solution:
Composition per 100mL:

Disodium EDTA	1.58g
ZnSO4·7H2O	0.7g
$MnSO_4 \cdot 4H_2O$	0.18g
$FeSO_4 \cdot 7H_2O$	0.16g
$CoCl_2 \cdot 6H_2O$	0.052g
$Na_2MoO_4 \cdot 2H_2O$	0.047g
$CuSO_4 \cdot 5H_2O$	0.047g

Preparation of Medium: Add components, except $CaCl_2 \cdot 6H_2O$ solution, $MgSO_4 \cdot 7H_2O$ solution, sodium pyruvate solution, and *p*-aminobenzoic acid solution to distilled/deionized water and bring volume to 969.0mL. Mix thoroughly. Autoclave for 15 min at 15 psi pressure–121°C. Cool to 45–50°C. Aseptically add in the following order: 10.0mL of the sterile $CaCl_2 \cdot 6H_2O$ solution, 10.0mL of the sterile $MgSO_4 \cdot 7H_2O$ solution, 10.0mL of sterile sodium pyruvate solution and 1.0mL of sterile *p*-aminobenzoic acid solution. Mix thoroughly. Aseptically distribute into sterile tubes or flasks.

Use: For the heterotrophic cultivation and maintenance of *Pseudomonas thermocarboxydovorans.*

Mineral Medium

Composition per liter:

Yeast extract	2.0g
Mineral Base 5X	200.0mL
Trace element solution SL-6	1.0mL
Thiamine·HCl	3.0μg
Biotin	0.2μg

pH 6.8 ± 0.2 at 25°C

Mineral Base 5X:
Composition per liter:

NaCl	5.0g
NH_4Cl	2.0g
KH_2PO_4	1.35g
$MgSO_4 \cdot 7H_2O$	1.0g
K_2HPO_4	0.87g
$CaCl_2$	0.05g
$FeCl_3 \cdot 6H_2O$	1.25mg

Preparation of Mineral Base 5X: Add components to distilled/deionized water and bring volume to 1.0L. Mix thoroughly.

Trace Elements Solution SL-6:
Composition per liter:

H_3BO_3	0.3g
$CoCl_2 \cdot 6H_2O$	0.2g
$ZnSO_4 \cdot 7H_2O$	0.10g
$MnCl_2 \cdot 4H_2O$	0.03g
$Na_2MoO_4 \cdot H_2O$	0.03g
$NiCl_2 \cdot 6H_2O$	0.02g
$CuCl_2 \cdot 2H_2O$	0.01g

Preparation of Trace Elements Solution SL-6: Add components to distilled/deionized water and bring volume to 1.0L. Mix thoroughly. Adjust pH to 3.4.

Preparation of Medium: Add components to distilled/deionized water and bring volume to 1.0L. Mix thoroughly. Adjust pH to 6.8. Distribute into tubes or flasks. Autoclave for 15 min at 15 psi pressure–121°C.

Use: For the cultivation of the *Arthrobacter* species.

Mineral Medium A

Composition per liter:

$(NH_4)_2SO_4$	1.0g
K_2HPO_4	1.0g

Preparation of Medium: Add components to tap water and bring volume to 1.0L. Mix thoroughly. Distribute into tubes or flasks. Autoclave for 15 min at 15 psi pressure–121°C.

Use: For the cultivation of *Saccharobacterium ovale.*

Mineral Medium for Hydrogen Bacteria

Composition per liter:

Agar	15.0g
$Na_2HPO_4 \cdot 2H_2O$	2.9g
KH_2PO_4	2.3g
NH_4Cl	1.0g
$MgSO_4 \cdot 7H_2O$	0.5g
$NaHCO_3$	0.5g
$CaCl_2 \cdot 2H_2O$	0.01g
Ferric ammonium citrate solution	20.0mL

Ferric Ammonium Citrate Solution:
Composition per 20mL:

Ferric ammonium citrate	0.05g

Preparation of Ferric Ammonium Citrate Solution: Add ferric ammonium citrate to 20.0mL of distilled/deionized water. Filter sterilize.

Preparation of Medium: Add components, except ferric ammonium citrate solution, to distilled/deionized water and bring volume to 980.0mL. Mix thoroughly. Gently heat and bring to boiling. Autoclave for 15 min at 15 psi pressure–121°C. Cool to 50°C. Aseptically add the sterile ferric ammonium citrate solution. Mix thoroughly. Pour into sterile Petri dishes or distribute into sterile tubes. Incubate inoculated medium at 30°C under 60% H_2 + 25% N_2 + 10% CO_2 + 5% O_2.

Use: For the cultivation and maintenance of *Alcaligenes eutrophus, Hydrogenophaga flava,* and *H. pseudoflava.*

Mineral Medium with Dichlorobenzoate

Composition per liter:

Na_2HPO_4	2.78g
KH_2PO_4	2.78g
$(NH_4)_2SO_4$	1.0g
Hutner's mineral base	20.0mL
2,4-Dichlorobenzoate solution	10.0mL

pH 6.8 ± 0.2 at 25°C

Hutner's Mineral Base:
Composition per liter:

$MgSO_4 \cdot 7H_2O$	29.7g
Nitrilotriacetic acid	10.0g
$CaCl_2 \cdot 2H_2O$	3.34g
$FeSO_4 \cdot 7H_2O$	99.0mg
$(NH_4)_2MoO_4$	9.25mg
Metals "44"	50.0mL

Preparation of Hutner's Mineral Base: Add nitrilotriacetic acid to 500.0mL of distilled/deionized water. Dissolve by adjusting pH to 6.5 with KOH. Add remaining components. Readjust pH to 7.2 with H_2SO_4 or KOH. Add distilled/deionized water to 1.0L. Store at 5°C.

Metals "44":
Composition per 100mL:

$ZnSO_4 \cdot 7H_2O$	1.1g
$FeSO_4 \cdot 7H_2O$	0.5g
EDTA	0.25g
$MnSO_4 \cdot 7H_2O$	0.154g
$CuSO_4 \cdot 5H_2O$	0.04g
$Co(NO_3)_2 \cdot 6H_2O$	0.025g
$Na_2B_4O_7 \cdot 10H_2O$	0.018g

Preparation of Metals "44": Add a few drops of H_2SO_4 to distilled/deionized water to inhibit precipitate formation. Add components to acidified distilled/deionized water and bring volume to 100.0mL. Mix thoroughly.

2,4-Dichlorobenzoate Solution:
Composition per 10mL:

2,4-Dichlorobenzoate	5.0mg

Preparation of 2,4-Dichlorobenzoate Solution: Add 2,4-dichlorobenzoate to 10.0mL of distilled/deionized water. Mix thoroughly. Filter sterilize.

Preparation of Medium: Add components, except 2,4-dichlorobenzoate solution, to distilled/deionized water and bring volume to 990.0mL. Mix thoroughly. Adjust pH to 6.8 with 1*N* KOH. Autoclave for 15 min at 15 psi pressure–121°C. Cool to 45–50°C. Aseptically add the sterile 2,4-dichlorobenzoate solution. Mix thoroughly. Distribute into sterile tubes or flasks.

Use: For the cultivation and maintenance of *Actinomyces viscosus*.

Mineral Medium with Glucose

Composition per liter:

Agar	20.0g
Na_2HPO_4	4.8g
KH_2PO_4	4.4g
NH_4Cl	1.0g
$MgSO_4 \cdot 7H_2O$	0.5g
Solution B	10.0mL
Solution A	5.0mL

pH 6.8 ± 0.2 at 25°C

Solution A:

Composition per 100mL:

Ferric ammonium citrate	1.0g
$CaCl_2$	0.1g

Preparation of Solution A: Add ferric ammonium citrate and $CaCl_2$ to distilled/deionized water and bring volume to 100.0mL. Mix thoroughly. Filter sterilize.

Solution B:

Composition per 100mL:

Glucose	10.0g

Preparation of Solution B: Add glucose to distilled/deionized water and bring volume to 100.0mL. Mix thoroughly. Filter sterilize.

Preparation of Medium: Add components, except solution A and solution B, to distilled/deionized water and bring volume to 985.0mL. Mix thoroughly. Gently heat and bring to boiling. Autoclave for 15 min at 15 psi pressure–121°C. Cool to 50°C. Aseptically add the sterile solution A and solution B. Mix thoroughly. Pour into sterile Petri dishes or distribute into sterile tubes.

Use: For the cultivation and maintenance of *Alcaligenes latus*.

Mineral Medium with Phenol

Composition per liter:

Phenol	1.0g
K_2HPO_4	1.0g
NH_4NO_3	1.0g
$(NH_4)_2SO_4$	0.5g
$MgSO_4$	0.5g
KH_2PO_4	0.5g
NaCl	0.5g
$CaCl_2$	0.02g
$FeSO_4$	0.02g
Wolfe's mineral solution	10.0mL

Wolfe's Mineral Solution:

Composition per liter

$MgSO_4 \cdot 7H_2O$	3.0g
Nitriloacetic acid	1.5g
NaCl	1.0g
$MnSO_4 \cdot H_2O$	0.5g
$FeSO_4 \cdot 7H_2O$	0.1g
$CoCl_2 \cdot 6H_2O$	0.1g
$CaCl_2$	0.1g
$ZnSO_4 \cdot 7H_2O$	0.1g
$CuSO_4 \cdot 5H_2O$	0.01g
$AlK(SO_4)_2 \cdot 12H_2O$	0.01g
H_3BO_3	0.01g
$Na_2MoO_4 \cdot 2H_2O$	0.01g

Preparation of Wolfe's Mineral Solution: Add nitrilotriacetic acid to 500.0mL of distilled/deionized water. Dissolve by adjusting pH to 6.5 with KOH. Add remaining components. Add distilled/deionized water to 1.0L.

Preparation of Medium: Add components, except phenol, to distilled/deionized water and bring volume to 1.0L. Mix thoroughly. Autoclave for 15 min at 15 psi pressure–121°C. Aseptically add the phenol. Mix thoroughly. Distribute into sterile tubes or flasks.

Use: For the cultivation and maintenance of *Pseudomonas putida*.

Mineral Medium with Santonin

Composition per liter:

K_2HPO_4	6.3g
α-Santonin	4.0g
KH_2PO_4	1.82g
NH_4NO_3	1.0g
$CaCl_2 \cdot 2H_2O$	0.75g
$MgSO_4 \cdot 7H_2O$	0.1g
$FeSO_4 \cdot 7H_2O$	0.06g
$MnSO_4$ (anhydrous)	600.0μg
$Na_2MoO_4 \cdot 2H_2O$	600.0μg

pH 7.0 ± 0.2 at 25°C

Preparation of Medium: Add components to distilled/deionized water and bring volume to 1.0L. Mix thoroughly. Autoclave for 15 min at 15 psi pressure–121°C. Distribute into sterile tubes or flasks.

Use: For the cultivation and maintenance of *Pseudomonas* species.

Mineral Salts Agar

Composition per liter:

Agar	15.0g
$NaNO_3$	2.0g

K_2HPO_4	1.2g
$MgSO_4$	0.5g
KCl	0.5g
KH_2PO_4	0.14g
Yeast extract	0.02g
$Fe_2(SO_4)_3 \cdot H_2O$	0.01g

pH 7.2 ± 0.2 at 25°C

Preparation of Medium: Add components to distilled/deionized water and bring volume to 1.0L. Mix thoroughly. Adjust pH to 7.2. Gently heat and bring to boiling. Distribute into tubes. Autoclave for 15 min at 15 psi pressure–121°C. Allow tubes to cool in a slanted position. Add a strip of sterile filter paper onto cooled slant. Inoculate organisms on filter paper.

Use: For the cultivation and maintenance of *Cytophaga aurantiaca* and *Sporocytophaga myxococcoides.*

Mineral Salts Enrichment Medium

Composition per liter:

KH_2PO_4	1.36g
$(NH_4)_2SO_4$	0.5g
$MgSO_4 \cdot 7H_2O$	0.2g
$CaCl_2 \cdot 2H_2O$	0.01g
$FeSO_4 \cdot 7H_2O$	5.0mg
$MnSO_4 \cdot 7H_2O$	2.5mg
$Na_2MoO_4 \cdot 2H_2O$	2.5mg
Na_2HPO_4	2.13mg

pH 7.2 ± 0.2 at 25°C

Preparation of Medium: Add components to distilled/deionized water and bring volume to 1.0L. Mix thoroughly. Distribute into tubes or flasks. Autoclave for 15 min at 15 psi pressure–121°C.

Use: For the enrichment and cultivation of *Caulobacter* species.

Mineral Salts for Thermophiles (L Salts for Thermophiles)

Composition per liter:

$NaNO_3$	0.25g
NH_4Cl	0.25g
Na_2HPO_4	0.21g
$MgSO_4 \cdot 7H_2O$	0.20g
NaH_2PO_4	0.09g
KCl	0.04g
$CaCl_2$	0.02g
$FeSO_4$	1.0mg
Trace mineral solution	10.0mL
n-Heptadecane	1.0mL

Trace Mineral Solution:
Composition per liter:

$ZnSO_4 \cdot 7H_2O$	7.0mg
H_3BO_4	1.0mg
MoO_3	1.0mg
$CuSO_4 \cdot 5H_2O$	500.0μg
$CoSO_4 \cdot 7H_2O$	18.0μg
$MnSO_4 \cdot 5H_2O$	7.0μg

Preparation of Trace Mineral Solution: Add components to distilled/deionized water and bring volume to 1.0L. Mix thoroughly.

Preparation of Medium: Add components, except *n*-heptadecane, to distilled/deionized water and bring volume to 1.0L. Mix thoroughly. Autoclave for 15 min at 15 psi pressure–121°C. Aseptically add the *n*-heptadecane. Mix thoroughly. Distribute into sterile tubes or flasks.

Use: For the cultivation and maintenance of *Bacillus thermoleovorans.*

Mineral Salts Medium

Composition per liter:

Na_2HPO_4	4.0g
KH_2PO_4	1.5g
NH_4Cl	1.0g
$MgSO_4 \cdot 7H_2O$	0.2g
Ferric ammonium citrate	5.0mg
Modified Hoagland trace element solution	1.0mL

pH 7.0 ± 0.2 at 25°C

Modified Hoagland Trace Element Solution:
Composition per 3.6 liters:

H_3BO_3	11.0g
$MnCl_2 \cdot 4H_2O$	7.0g
$AlCl_3$	1.0g
$CoCl_2$	1.0g
$CuCl_2$	1.0g
KI	1.0g
$NiCl_2$	1.0g
$ZnCl_2$	1.0g
$BaCl_2$	0.5g
KBr	0.5g
LiCl	0.5g
Na_2MoO_4	0.5g
$SeCl_4$	0.5g
$SnCl_2 \cdot 2H_2O$	0.5g
$NaVO_3 \cdot H_2O$	0.1g

Preparation of Modified Hoagland Trace Element Solution: Prepare each component as a separate solution. Dissolve each salt in approximately 100.0mL of distilled/deionized water. Adjust the pH of each solution to below 7.0. Combine all the salt solutions and bring the volume to 3.6L with distilled/

deionized water. Adjust the pH to 3–4. A yellow precipitate may form after mixing. After a few days, it will turn into a fine white precipitate. Mix the solution thoroughly before using.

Preparation of Medium: Add components to distilled/deionized water and bring volume to 1.0L. Mix thoroughly. Distribute into tubes or flasks. Autoclave for 15 min at 15 psi pressure–121°C.

Use: For the cultivation and maintenance of *Rhodococcus rhodochrous.*

Mineral Salts Medium with Methanol

Composition per liter:

$NaNH_4HPO_4{\cdot}4H_2O$	1.74g
$NaH_2PO_4{\cdot}H_2O$	0.54g
$MgSO_4{\cdot}7H_2O$	0.2g
KCl	0.04g
$FeSO_4{\cdot}7H_2O$	5.0mg
Methanol	5.0mL
Trace mineral solution	1.0mL

pH 7.2 ± 0.2 at 25°C

Trace Mineral Solution:

Composition per liter:

H_3BO_3	2.86g
$MnCl_2{\cdot}4H_2O$	1.81g
$ZnSO_4{\cdot}7H_2O$	0.22g
$CuSO_4{\cdot}5H_2O$	0.08g
$CoCl_2{\cdot}6H_2O$	0.06g
$Na_2MoO_4{\cdot}2H_2O$	25.0mg

Preparation of Trace Mineral Solution: Add components to distilled/deionized water and bring volume to 1.0L. Mix thoroughly.

Preparation of Medium: Add components, except methanol, to distilled/deionized water and bring volume to 1.0L. Mix thoroughly. Distribute into tubes or flasks. Autoclave for 15 min at 15 psi pressure–121°C. Cool to 50°C. Filter sterilize methanol. Aseptically add sterile methanol to cooled, sterile basal medium.

Use: For the cultivation and maintenance of *Rhodococcus rhodochrous.*

Mineral Salts Medium with Methanol and Yeast Extract

Composition per liter:

$NaNH_4HPO_4{\cdot}4H_2O$	1.74g
$NaH_2PO_4{\cdot}H_2O$	0.54g
$MgSO_4{\cdot}7H_2O$	0.2g
Yeast extract	0.2g
KCl	0.04g
$FeSO_4{\cdot}7H_2O$	5.0mg
Methanol	5.0mL
Trace mineral solution	1.0mL

pH 7.2 ± 0.2 at 25°C

Trace Mineral Solution:

Composition per liter:

H_3BO_3	2.86g
$MnCl_2{\cdot}4H_2O$	1.81g
$ZnSO_4{\cdot}7H_2O$	0.22g
$CuSO_4{\cdot}5H_2O$	0.08g
$CoCl_2{\cdot}6H_2O$	0.06g
$Na_2MoO_4{\cdot}2H_2O$	25.0mg

Preparation of Trace Mineral Solution: Add components to distilled/deionized water and bring volume to 1.0L. Mix thoroughly.

Preparation of Medium: Add components, except methanol, to distilled/deionized water and bring volume to 1.0L. Mix thoroughly. Autoclave for 15 min at 15 psi pressure–121°C. Cool to 50°C. Filter sterilize methanol. Aseptically add sterile methanol to cooled, sterile basal medium. Aseptically distribute into sterile tubes or flasks.

Use: For the cultivation and maintenance of *Pseudomonas* species.

Mineral Salt Peptonized Milk Agar (SPMA)

Composition per liter:

Agar	15.0g
Milk, peptonized	1.0g
Mineral solution	100.0mL

Mineral Solution:

Composition per 100mL:

$MgSO_4{\cdot}7H_2O$	0.50g
$CaCl_2$	0.25g
K_2HPO_4	0.25g
$(NH_4)_2SO_4$	0.10g
$FeCl_3{\cdot}6H_2O$	0.01g
$MnCl_2$	0.1mg

Preparation of Mineral Solution: Add components to distilled/deionized water and bring volume to 100.0mL. Mix thoroughly. Filter sterilize.

Preparation of Medium: Add components, except mineral solution, to distilled/deionized water and bring volume to 900.0mL. Mix thoroughly. Gently heat and bring to boiling. Autoclave for 15 min at 15 psi pressure–121°C. Cool to 45–50°C. Aseptically add 100.0mL of sterile mineral solution. Mix thoroughly. Pour into sterile Petri dishes or distribute into sterile tubes.

Use: For the cultivation of freshwater *Myxobacterium* species.

Mineral Salts with Butane

Composition per liter of tap water:

$(NH_4)_2HPO_4$	8.0g
$Na_2HPO_4 \cdot 12H_2O$	2.5g
KH_2PO_4	2.0g
$MgSO_4 \cdot 7H_2O$	0.5g
Yeast extract	100.0mg
$CaCl_2 \cdot 2H_2O$	60.0mg
$FeSO_4 \cdot 7H_2O$	30.0mg
$MnCl_2 \cdot 4H_2O$	60.0μg
$CuSO_4 \cdot 5H_2O$	15.0μg

pH 7.1 ± 0.2 at 25°C

Preparation of Medium: Add components to distilled/deionized water and bring volume to 1.0L. Mix thoroughly. Distribute into tubes or flasks. Autoclave for 15 min at 15 psi pressure–121°C. Incubate inoculated medium in 88% air + 7% *n*-butane + 5% CO_2.

Use: For the cultivation and maintenance of *Pseudomonas butanovora*.

Minerals Modified Medium

Composition per liter:

Lactose	20.0g
Sodium glutamate	12.7g
NH_4Cl	5.0g
K_2HPO_4	1.80g
Sodium formate	0.5g
$MgSO_4 \cdot 7H_2O$	0.2g
L-Aspartic acid	0.048g
L-Cystine	0.04g
L-Arginine	0.04g
Ferric ammonium citrate	0.020g
$CaCl_2 \cdot 2H_2O$	0.020g
Bromcresol Purple	0.020g
Thiamine	2.0mg
Nicotinic acid	2.0mg
Pantothenic acid	2.0mg

pH 6.7 ± 0.2 at 25°C

Source: This medium is available as a premixed powder from Oxoid Unipath.

Preparation of Medium: Add NH_4Cl to distilled/deionized water and bring volume to 800.0mL. Add remaining components and bring volume to 1.0L. Mix thoroughly. Adjust pH to 6.7. Distribute into tubes or flasks. Autoclave for 10 min at 10 psi pressure–116°C. Check pH after autoclaving. This medium is double-strength.

Use: For the enumeration of coliform bacteria in water.

Minimal Medium for Denitrifying Bacteria

Composition per liter:

Solution A	980.0mL
Solution B	10.0mL
Solution C	10.0mL

Solution A:

Composition per 980mL:

KNO_3	5.0g
Carbon source	4.0g
$(NH_4)_2SO_4$	1.0g
$K_2HPO_4 \cdot 3H_2O$	0.87g
KH_2PO_4	0.54g

Preparation of Solution A: Add components to distilled/deionized water and bring volume to 1.0L. Mix thoroughly. Autoclave for 15 min at 15 psi pressure–121°C. Cool to 25°C.

Solution B:

Composition per 100mL:

$MgSO_4 \cdot 7H_2O$	2.0g

Preparation of Solution B: Add $MgSO_4 \cdot 7H_2O$ to distilled/deionized water and bring volume to 100.0mL. Mix thoroughly. Autoclave for 15 min at 15 psi pressure–121°C. Cool to 25°C.

Solution C:

Composition per 100mL:

$CaCl_2 \cdot 2H_2O$	0.2g
$FeSO_4 \cdot 7H_2O$	0.1g
$MnSO_4 \cdot H_2O$	0.05g
$CuSO_4 \cdot 5H_2O$	0.01g
$Na_2MoO_4 \cdot 2H_2O$	0.01g
HCl (0.1*N* solution)	100.0mL

Preparation of Solution C: Combine components. Mix thoroughly. Autoclave for 15 min at 15 psi pressure–121°C. Cool to 25°C.

Preparation of Medium: Aseptically combine 980.0mL of cooled sterile solution A, 10.0mL of cooled sterile solution B, and 10.0mL of cooled sterile solution C. Mix thoroughly. Aseptically distribute into sterile tubes or flasks.

Use: For the isolation and cultivation of denitrifying bacteria.

Mist Agar

Composition per liter:

Cow-manure, dry	50.0g
Agar	15.0g

Preparation of Medium: Add cow manure to 1.0L of tap water. Boil for 1 hr. Filter through cheesecloth. Filter through paper. Add agar to filtrate and bring volume to 1.0L with tap water. Gently heat and bring to boiling. Distribute into tubes or flasks. Autoclave for 15 min at 15 psi pressure–121°C. Pour into sterile Petri dishes or leave in tubes.

Use: For the cultivation and maintenance of *Streptomyces fragmentosporus.*

ML Medium (Minimal Lactate Medium)

Composition per liter:

Sodium lactate	5.0g
$MgSO_4 \cdot 7H_2O$	2.0g
NH_4Cl	1.0g
Na_2SO_4	1.0g
Yeast extract	1.0g
K_2HPO_4	0.5g
Cysteine	0.5g
$CaCl_2 \cdot 6H_2O$	0.1g
Resazurin	1.0mg
$NaHCO_3$ solution	25.0mL
$FeSO_4 \cdot 7H_2O$ solution	25.0mL

pH 6.8 ± 0.2 at 25°C

$NaHCO_3$ Solution:

Composition per 25mL:

$NaHCO_3$	4.0g

Preparation of $NaHCO_3$ Solution: Add $NaHCO_3$ to distilled/deionized water and bring volume to 25.0mL. Mix thoroughly. Filter sterilize. Gas with O_2-free 97% N_2 + 3% H_2. Cap with a rubber stopper.

$FeSO_4 \cdot 7H_2O$ Solution:

Composition per 25mL:

$FeSO_4 \cdot 7H_2O$	4.0mg

Preparation of $FeSO_4 \cdot 7H_2O$ Solution: Add $FeSO_4 \cdot 7H_2O$ to distilled/deionized water and bring volume to 25.0mL. Mix thoroughly. Filter sterilize. Gas with O_2-free 97% N_2 + 3% H_2. Cap with a rubber stopper.

Preparation of Medium: Add components, except $NaHCO_3$ solution and $FeSO_4 \cdot 7H_2O$ solution, to distilled/deionized water and bring volume to 1.0L. Gently heat and bring to boiling under O_2-free 97% N_2 + 3% H_2. Adjust pH to 6.8. Continue boiling until rezasurin becomes colorless, indicating reduction. Distribute anaerobically under O_2-free 97% N_2 + 3% H_2 into tubes in 10.0mL volumes. Cap with rubber stoppers. Place tubes in a press. Autoclave for 15 min at 15 psi pressure–121°C. Cool to room temperature. Prior to inoculation, add 0.25mL of sterile $NaHCO_3$ solution and 0.25mL of sterile $FeSO_4 \cdot 7H_2O$ solution to each test tube containing 10.0mL of sterile basal medium.

Use: For the cultivation and maintenance of *Desulfovibrio* species.

MMS Medium for *Thermotoga neapolitana*

Composition per liter:

NaCl	6.93g
Starch	5.0g
$MgSO_4 \cdot 7H_2O$	1.75g
$MgCl_2 \cdot 6H_2O$	1.38g
KH_2PO_4	0.5g
$Na_2S \cdot 9H_2O$	0.5g
$CaCl_2$	0.38g
KCl	0.16g
NaBr	25.0mg
H_3BO_3	7.5mg
$SrCl_2 \cdot 6H_2O$	3.8mg
$(NH_4)_2Ni(SO_4)_2$	2.0mg
Resazurin	1.0mg
KI	0.025mg
Wolfe's mineral solution	15.0mL

pH 6.5 ± 0.2 at 25°C

Wolfe's Mineral Solution:

Composition per liter:

$MgSO_4 \cdot 7H_2O$	3.0g
Nitriloacetic acid	1.5g
NaCl	1.0g
$MnSO_4 \cdot H_2O$	0.5g
$FeSO_4 \cdot 7H_2O$	0.1g
$CoCl_2 \cdot 6H_2O$	0.1g
$CaCl_2$	0.1g
$ZnSO_4 \cdot 7H_2O$	0.1g
$CuSO_4 \cdot 5H_2O$	0.01g
$AlK(SO_4)_2 \cdot 12H_2O$	0.01g
H_3BO_3	0.01g
$Na_2MoO_4 \cdot 2H_2O$	0.01g

Preparation of Wolfe's Mineral Solution: Add nitrilotriacetic acid to 500.0mL of distilled/deionized water. Dissolve by adjusting pH to 6.5 with KOH. Add remaining components. Add distilled/deionized water to 1.0L.

Preparation of Medium: Prepare and dispense medium under 80% N_2 and 20% CO_2. Add components to distilled/deionized water and bring volume to 1.0L. Mix thoroughly. Adjust pH to 6.5 with H_2SO_4. Distribute into tubes or flasks. Autoclave for 15 min at 15 psi pressure–121°C.

Use: For the cultivation and maintenance of *Thermotoga neapolitana.*

MN Marine Medium

Composition per liter:

Noble agar	10.0g
$NaNO_3$	0.75g
$MgSO_4 \cdot 7H_2O$	0.04g
$CaCl_2 \cdot 2H_2O$	0.02g
$K_2HPO_4 \cdot 3H_2O$	0.02g
Na_2CO_3	0.02g
Citric acid	3.0mg
Ferric ammonium citrate	3.0mg
Disodium potassium EDTA	0.5mg
Trace metals A-5	1.0mL

pH 8.5 ± 0.2 at 25°C

A-5 Trace Metal Mix:

Composition per liter:

H_3BO_3	2.86g
$MnCl_2 \cdot 4H_2O$	1.81g
$ZnSO_4 \cdot 7H_2O$	0.222g
$CuSO_4 \cdot 5H_2O$	0.079g
$Na_2MoO_4 \cdot 2H_2O$	0.039g
$Co(NO_3)_2 \cdot 6H_2O$	0.049g

Preparation of A-5 Trace Metal Mix: Add components to distilled/deionized water and bring volume to 1.0L. Mix thoroughly.

Preparation of Medium: Add components to 750mL of seawater and bring volume to lL with glass-distilled water. Mix thoroughly. Gently heat and bring to boiling. Autoclave for 15 min at 15 psi pressure–121°C. After autoclaving, adjust pH to 8.5 with KOH.

Use: For cultivation and maintenance of marine cyanobacteria.

MN Marine Medium with Vitamin B_{12}

Composition per liter:

Noble agar	10.0g
$NaNO_3$	0.75g
$MgSO_4 \cdot 7H_2O$	0.04g
$CaCl_2 \cdot 2H_2O$	0.02g
$K_2HPO_4 \cdot 3H_2O$	0.02g
Na_2CO_3	0.02g
Citric acid	3.0mg
Ferric ammonium citrate	3.0mg
Disodium potassium EDTA	0.5mg
Vitamin B_{12}	20.0μg
Trace metals A-5	1.0mL

pH 8.5 ± 0.2 at 25°C

A-5 Trace Metal Mix:

Composition per liter:

H_3BO_3	2.86g
$MnCl_2 \cdot 4H_2O$	1.81g
$ZnSO_4 \cdot 7H_2O$	0.222g
$CuSO_4 \cdot 5H_2O$	0.079g
$Na_2MoO_4 \cdot 2H_2O$	0.039g
$Co(NO_3)_2 \cdot 6H_2O$	0.049g

Preparation of A-5 Trace Metal Mix: Add components to distilled/deionized water and bring volume to 1.0L. Mix thoroughly.

Preparation of Medium: Add components to 750.0mL of seawater and bring volume to lL with glass-distilled water. Mix thoroughly. Gently heat and bring to boiling. Autoclave for 15 min at 15 psi pressure–121°C. After autoclaving, adjust pH to 8.5 with KOH.

Use: For the cultivation and maintenance of *Dermocarpa* species, *Dermocarpella* species, *Myxosarcina* species, *Phormidium* species, *Pleurocapsa* species, *Synechococcus* species, *Synechocystis* species, and *Xenococcus* species.

Modified Semi-Solid Rappaport Vassiliadis Medium (MSRV Medium)

Composition per liter:

$MgCl_2$, anhydrous	10.93g
NaCl	7.34g
Casein hydrolysate	4.59g
Tryptose	4.59g
Agar	2.7g
KH_2PO_4	1.47g
Malachite green oxalate	0.037g
Novobiocin	10.0mL

pH 5.2 ± 0.2 at 25°C

Source: This medium is available as a premixed powder from Oxoid Unipath.

Novobiocin Solution:

Composition per 10mL:

Novobiocin	0.02g

Preparation of Novobiocin Solution: Add novobiocin to 10.0mL of distilled/deionized water. Mix thoroughly. Filter sterilize.

Preparation of Medium: Add components, except novobiocin solution, to distilled/deionized water and bring volume to 990.0mL. Mix thoroughly. Gently heat to boiling. Do not autoclave. Cool to 45–50°C. Aseptically add 10.0mL of sterile novobiocin solution. Mix thoroughly. Pour into sterile Petri dishes. Air dry plates for at least 1 hr.

Use: For the isolation and cultivation of motile *Salmonella* species from environmental samples.

Møller Decarboxylase Broth

Composition per liter:

Amino acid	10.0g
Peptic digest of animal tissue	5.0g
Beef extract	5.0g
Glucose	0.5g
Bromcresol purple	0.01g
Cresol red	5.0mg
Pyridoxal	5.0mg

pH 6.0 ± 0.2 at 25°C

Source: Available as a premixed powder from BBL Microbiology Systems and Difco Laboratories.

Preparation of Medium: Add components to distilled/deionized water and bring volume to 1.0L. Use L-lysine, L-arginine or L-ornithine. Mix thoroughly. Gently heat until dissolved. Distribute into screw-capped tubes in 5.0mL volumes. Autoclave for 15 min at 15 psi pressure–121°C. A slight precipitate may form in the ornithine broth.

Use: For the differentiation of Gram-negative enteric bacteria based on the production of arginine dihydrolase, lysine decarboxylase or ornithine decarboxylase.

Møller KCN Broth Base

Composition per liter:

Na_2HPO_4	5.64g
NaCl	5.0g
Pancreatic digest of casein	1.5g
Peptic digest of animal tissue	1.5g
KH_2PO_4	0.225g
KCN solution	0.15mL

pH 7.6 ± 0.2 at 25°C

Source: This medium is available as a premixed powder from BBL Microbiology Systems.

KCN Solution:

Composition per100mL:

KCN	0.5g

Preparation of KCN Solution: Add KCN to 100.0mL of cold distilled/deionized water. Mix thoroughly and cap. Do not mouth pipette.

Caution: Cyanide is toxic.

Preparation of Medium: Add components, except KCN solution, to distilled/deionized water and bring volume to 1.0L. Mix thoroughly. Autoclave for 15 min at 15 psi pressure–121°C. Cool to room temperature. Prior to use, add 0.15mL of KCN solution. Mix thoroughly. Aseptically distribute into sterile tubes.

Use: For the differentiation of Gram-negative enteric bacteria on the basis of their ability to grow in the presence of cyanide.

Motility GI Medium

Composition per liter:

Gelatin	53.4g
Heart infusion broth	25.0g
Agar	3.0g

pH 7.2 ± 0.2 at 25°C

Source: This medium is available as a premixed powder from Difco Laboratories.

Preparation of Medium: Add components to cold distilled/deionized water and bring to 1.0L. Mix thoroughly. Gently heat and bring to boiling. Distribute into tubes or flasks. Autoclave for 15 min at 15 psi pressure–121°C. Pour into sterile Petri dishes in 20.0mL volumes or leave in tubes.

Use: For demonstrating motility of microorganisms and for separating organisms in their motile phase.

Motility Indole Ornithine Medium (MIO Medium)

Composition per liter:

Pancreatic digest of gelatin	10.0g
Pancreatic digest of casein	9.5g
L-Ornithine·HCl	5.0g
Yeast extract	3.0g
Agar	2.0g
Glucose	1.5g
Bromcresol purple	0.02g

pH 6.6 ± 0.2 at 25°C

Source: This medium is available as a premixed powder from BBL Microbiology Systems and Difco Laboratories.

Preparation of Medium: Add components to distilled/deionized water and bring to 1.0L. Mix thoroughly. Gently heat and bring to boiling. Distribute into tubes or flasks. Autoclave for 15 min at 15 psi pressure–121°C.

Use: For the differentiation of Gram-negative enteric bacteria based on their motility, indole production and ornithine decarboxylase activity.

Motility Medium S

Composition per liter:

Beef heart solids from infusion	500.0g
Gelatin	30.0g
Enzymatic hydrolysate of protein	10.0g
NaCl	5.0g
K_2HPO_4	2.0g
KNO_3	2.0g
Agar	1.0g

2,3,5-triphenyltetrazolium chloride solution 10.0mL

pH 7.2 ± 0.2 at 25°C

Source: This medium is available as a premixed powder from Difco Laboratories.

2,3,5-Triphenyltetrazolium Chloride Solution:
Composition per 10mL:

2,3,5-triphenyltetrazolium chloride 0.1g

Preparation of 2,3,5-Triphenyltetrazolium Chloride Solution: Add 2,3,5-triphenyltetrazolium chloride to distilled/deionized water and bring volume to 10.0mL. Mix thoroughly. Filter sterilize.

Preparation of Medium: Add components, except 2,3,5-triphenyltetrazolium chloride solution, to distilled/deionized water and bring volume to 990.0mL. Mix thoroughly. Gently heat while stirring and bring to boiling. Autoclave for 15 min at 15 psi pressure–121°C. Cool to 60°C. Aseptically add 10.0mL of the sterile 2,3,5-triphenyltetrazolium chloride solution. Mix thoroughly. Aseptically distribute into sterile tubes. Keep at 4–8°C until used.

Use: For the determination of bacterial motility.

Motility Nitrate Agar

Composition per liter:

Beef heart, solids from infusion 100.0g
Tryptose 12.0g
Agar 3.0g
NaCl 1.0g
KNO_3 .... 1.0g
Glucose 0.5g

pH 7.4 ± 0.2 at 25°C

Preparation of Medium: Add components to distilled/deionized water and bring volume to 1.0L. Mix thoroughly. Gently heat and bring to boiling. Distribute into tubes in 4.0mL volumes. Autoclave for 15 min at 15 psi pressure–121°C.

Use: For the cultivation and observation of motility and nitrate reduction in a variety of Gram-negative bacteria.

Motility Sulfide Medium

Composition per liter:

Gelatin 80.0g
Proteose peptone 10.0g
NaCl 5.0g
Agar 4.0g
Beef extract 3.0g
Sodium citrate 2.0g
L-Cystine 0.2g
Ferrous ammonium citrate 0.2g

pH 7.3 ± 0.2 at 25°C

Source: This medium is available as a premixed powder from Difco Laboratories.

Preparation of Medium: Add components to distilled/deionized water and bring volume to 1.0L. Mix thoroughly. Gently heat while stirring and bring to boiling. Distribute into tubes in 4–5.0mL volumes. Autoclave for 15 min at 10 psi pressure–116°C.

Use: For the determination of bacterial motility and the ability of bacteria to produce H_2S from L-cystine. Used for the differentiation of Gram-negative bacteria of the Enterobacteriaceae.

Motility Test and Maintenance Medium

Composition per liter:

Agar 9.0g
Tryptose 8.0g
NaCl 5.0g
Pancreatic digest of gelatin 2.5g
Beef extract 1.5g

pH 7.2 ± 0.1 at 25°C

Preparation of Medium: Add components to distilled/deionized water and bring volume to 1.0L. Mix thoroughly. Gently heat and bring to boiling. Distribute into tubes in 7.0mL volumes. Autoclave for 15 min at 15 psi pressure–121°C. Cool to 45–50°C. Pass the cooled tubes into an anaerobic chamber containing 85% N_2 + 10% H_2 + 5% CO_2.

Use: For the cultivation, maintenance and observation of motility in a variety of anaerobic bacteria.

Motility Test and Maintenance Medium

Composition per liter:

Peptone 10.0g
NaCl 5.0g
Agar 4.0g
Beef extract 3.0g

pH 7.4 ± 0.1 at 25°C

Preparation of Medium: Add components to distilled/deionized water and bring volume to 1.0L. Mix thoroughly. Distribute into screw-capped tubes in 8.0mL volumes. Autoclave for 15 min at 15 psi pressure–121°C.

Use: For the cultivation, maintenance and observation of motility in members of the *Enterobacteriaceae*.

Motility Test and Maintenance Medium, Gilardi

Composition per liter:

Pancreatic digest of casein 10.0g
NaCl 5.0g

Agar	3.0g
Yeast extract	3.0g

pH 7.2 ± 0.1 at 25°C

Preparation of Medium: Add components to distilled/deionized water and bring volume to 1.0L. Mix thoroughly. Distribute into screw-capped tubes in 3.5mL volumes. Autoclave for 15 min at 15 psi pressure–121°C.

Use: For the cultivation, maintenance and observation of motility in nonfermenting Gram–negative bacteria.

Motility Test and Maintenance Medium, Tatum

Composition per liter:

Tryptose	8.0g
NaCl	5.0g
Agar	4.0g
Pancreatic digest of gelatin	2.5g
Beef extract	1.5g

pH 6.9 ± 0.2 at 25°C

Preparation of Medium: Add components to distilled/deionized water and bring volume to 1.0L. Mix thoroughly. Distribute into screw-capped tubes in 8.0mL volumes. Autoclave for 15 min at 15 psi pressure–121°C.

Use: For the cultivation, maintenance and observation of motility in nonfermenting Gram–negative bacteria.

Motility Test Medium

Composition per liter:

Pancreatic digest of gelatin	10.0g
NaCl	5.0g
Agar	4.0g
Beef extract	3.0g

pH 7.3 ± 0.2 at 25°C

Source: This medium is available as a premixed powder from BBL Microbiology Systems.

2,3,5-Triphenyltetrazolium Chloride Solution:
Composition per 10mL:

2,3,5-triphenyltetrazolium chloride	0.1g

Preparation of 2,3,5-Triphenyltetrazolium Chloride Solution: Add 2,3,5-triphenyltetrazolium chloride to distilled/deionized water and bring volume to 10.0mL. Mix thoroughly. Filter sterilize.

Preparation of Medium: Add components to distilled/deionized water and bring volume to 995.0mL. Mix thoroughly. Gently heat while stirring and bring to boiling. Autoclave for 15 min at 15 psi pressure–121°C. Cool to 45–50°C. Aseptically add 5.0mL of sterile 2,3,5-triphenyltetrazolium chloride solution. Mix thoroughly. Aseptically distribute into sterile tubes.

Use: For the detection of motility of Gram-negative enteric bacteria.

Motility Test Medium

Composition per liter:

Tryptose	10.0g
NaCl	5.0g
Agar	5.0g

pH 7.2 ± 0.2 at 25°C

Source: This medium is available as a premixed powder from Difco Laboratories.

Preparation of Medium: Add components to distilled/deionized water and bring volume to 1.0L. Mix thoroughly. Gently heat while stirring and bring to boiling. Distribute into tubes in 4–5.0mL volumes. Autoclave for 15 min at 15 psi pressure–121°C. Cool tubes quickly in an upright position.

Use: For the determination of bacterial motility.

Motility Test Medium, Semisolid

Composition per liter:

Peptone	10.0g
NaCl	5.0g
Agar	4.0g
Beef extract	3.0g

pH 7.4 ± 0.2 at 25°C

Preparation of Medium: Add components to distilled/deionized water and bring volume to 1.0L. Mix thoroughly. Gently heat while stirring and bring to boiling. Distribute into screw-capped tubes in 8.0mL or 20.0mL volumes. Autoclave for 15 min at 15 psi pressure–121°C. Pour into sterile Petri dishes in 20.0mL volumes or leave in tubes.

Use: For the cultivation and observation of motility in a variety of bacteria, especially *Salmonella* species.

MP Agar

Composition per liter:

Agar	15.0g
Sodium acetate	0.1g
Basal medium	1.0L
Sodium sulfide solution	3.0mL

pH 7.0–7.5 at 25°C

Basal Medium:
Composition per liter:

$CaSO_4 \cdot 2H_2O$ (saturated solution)	20.0mL
$MgSO_4 \cdot 7H_2O$ (1% solution)	1.0mL

NH_4Cl (4% solution) 5.0mL
Trace elements solution 5.0mL
K_2HPO_4 (1% solution) 1.0mL

Preparation of Basal Medium: Add components to distilled/deionized water and bring volume to 1.0L. Mix thoroughly.

Trace Elements Solution:
Composition per liter:
$ZnSO_4 \cdot 7H_2O$ (0.1% solution) 10.0mL
$MnSO_4 \cdot 4H_2O$ (0.02% solution) 10.0mL
$CuSO_4 \cdot 5H_2O$ (0.00005% solution) 10.0mL
H_3BO_3 (0.1% solution) 10.0mL
$Co(NO_3)_2$ or
$CoCl_2 \cdot 6H_2O$ (0.01% solution) 10.0mL
$Na_2MoO_4 \cdot 2H_2O$ (0.01% solution) 10.0mL
Ferrous EDTA solution 20.0mL

Preparation of Trace Elements Solution: Add components to distilled/deionized water and bring volume to 1.0L. Mix thoroughly.

Ferrous EDTA Solution:
Composition per 100mL:
$FeSO_4 \cdot 7H_2O$ 7.0g
EDTA 2.0g
HCl, concentrated 1.0mL

Preparation of Ferrous EDTA Solution: Add components to distilled/deionized water and bring volume to 100.0mL. Mix thoroughly.

Sodium Sulfide Solution:
Composition per 10mL:
$Na_2S \cdot 9H_2O$ 1.0g

Preparation of Sodium Sulfide Solution: Add $Na_2S \cdot 9H_2O$ to distilled/deionized water and bring volume to 10.0mL. Mix thoroughly. Autoclave for 15 min at 15 psi pressure–121°C. Prepare freshly.

Preparation of Medium: Add sodium acetate and agar to 1.0L of basal medium. Mix thoroughly. Adjust pH to 7.0–7.5. Gently heat and bring to boiling. Autoclave for 15 min at 15 psi pressure–121°C. Cool to 45–50°C. Aseptically add 3.0mL of sterile sodium sulfide solution immediately prior to dispensing. Mix thoroughly. Pour into sterile Petri dishes or distribute into sterile screw-capped tubes.

Use: For the isolation and cultivation of *Beggiatoa* species and myxotrophic strains of *Thiothrix* species from water and environmental sources.

MP 5 Medium (Mineral Pectin 5 Medium)

Composition per liter:
Agar solution 500.0mL
Basal medium 250.0mL
Mineral solution 250.0mL
pH 5.0–6.0 at 25°C

Agar Solution:
Composition per 500mL:
Agar 15.0g

Preparation of Agar Solution: Add agar to distilled/deionized water and bring volume to 500.0mL. Mix thoroughly. Gently heat and bring to boiling. Autoclave for 15 min at 15 psi pressure–121°C. Cool to 45–50°C.

Basal Medium:
Composition per 250mL:
Na_2HPO_4 .... 6.0g
Pectin, citrus or apple 5.0g
KH_2PO_4 .... 4.0g
NH_4SO_4 .... 2.0g
Yeast extract 1.0g

Preparation of Basal Medium: Add components to distilled/deionized water and bring volume to 250.0mL. Mix thoroughly. Gently heat and bring to boiling.

Mineral Solution:
Composition per 250mL:
$FeSO_4$ (0.1% solution) 1.0mL
$MgSO_4 \cdot 7H_2O$ (20% solution) 1.0mL
$CaCl_2 \cdot 2H_2O$ (0.1% solution) 1.0mL
H_3BO_3 (0.001% solution) 1.0mL
$MnSO_4 \cdot H_2O$ (0.001% solution) 1.0mL
$ZnSO_4 \cdot 7H_2O$ (0.007% solution 1.0mL
$CuSO_4 \cdot 5H_2O$ (0.005% solution) 1.0mL
MoO_3 (0.001% solution) 1.0mL

Preparation of Mineral Solution: Add components to distilled/deionized water and bring volume to 250.0mL. Mix thoroughly.

Preparation of Medium: Combine 250.0mL of basal medium and 250.0mL of mineral solution. Mix thoroughly. Adjust pH to 5.0–6.0 with 1*N* HCl. Autoclave the basal medium-mineral solution and agar solution separately for 15 min at 15 psi pressure–121°C. Cool to 45–50°C. Aseptically combine the two sterile solutions. Mix thoroughly. Pour immediately into sterile Petri dishes to prevent hydrolysis of the agar.

Use: For the cultivation of microorganisms that produce polygalactanase.

MP 7 Medium (Mineral Pectin 7 Medium)

Composition per liter:
Basal medium 500.0mL
Mineral solution 500.0mL
pH 7.2 ± 0.2 at 25°C

Basal Medium:

Composition per 500mL:

Agar....15.0g
Na_2HPO_4....6.0g
Pectin, citrus or apple....5.0g
KH_2PO_4....4.0g
NH_4SO_4....2.0g
Yeast extract....1.0g

Preparation of Basal Medium: Add components to distilled/deionized water and bring volume to 500.0mL. Mix thoroughly. Gently heat and bring to boiling.

Mineral Solution:

Composition per 500mL:

$FeSO_4$ (0.1% solution)....1.0mL
$MgSO_4 \cdot 7H_2O$ (20% solution)....1.0mL
$CaCl_2 \cdot 2H_2O$ (0.1% solution)....1.0mL
H_3BO_3 (0.001% solution)....1.0mL
$MnSO_4 \cdot H_2O$ (0.001% solution)....1.0mL
$ZnSO_4 \cdot 7H_2O$ (0.007% solution....1.0mL
$CuSO_4 \cdot 5H_2O$ (0.005% solution....1.0mL
MoO_3 (0.001% solution)....1.0mL

Preparation of Mineral Solution: Add components to distilled/deionized water and bring volume to 500.0mL. Mix thoroughly.

Preparation of Medium: Combine 500.0mL of basal medium and 500.0mL of mineral solution. Mix thoroughly. Adjust pH to 7.2. Autoclave for 15 min at 15 psi pressure–121°C. Cool to 50°C. Pour into sterile Petri dishes.

Use: For the cultivation of microorganisms that produce pectate lyase.

MPH Agar (Milk Protein Hydrolysate Agar)

Composition per liter:

Agar....15.0g
Casein hydrolysate....9.0g
Glucose....1.0g

pH 7.0 ± 0.2 at 25°C

Source: This medium is available as a premixed powder from BBL Microbiology Systems.

Preparation of Medium: Add components to distilled/deionized water and bring volume to 1.0L. Mix thoroughly. Gently heat while stirring and bring to boiling. Autoclave for 15 min at 15 psi pressure–121°C. Aseptically distibute into sterile tubes. Cool to 43–45°C before using.

Use: For use in the enumeration of bacteria in water.

MPSS Broth

Composition per liter:

Peptone....5.0g
$MgSO_4 \cdot 7H_2O$....1.0g
$(NH_4)_2SO_4$....1.0g
Succinic acid....1.0g
$FeCl_3 \cdot 6H_2O$ (0.2% solution)....1.0mL
$MnSO_4 \cdot H_2O$ (0.2% solution)....1.0mL

pH 6.8 ± 0.2 at 25°C

Preparation of Medium: Add components to distilled/deionized water and bring volume to 1.0L. Mix thoroughly. Distribute into tubes or flasks. Autoclave for 15 min at 15 psi pressure–121°C.

Use: For the cultivation of *Spirillum volutans*.

MPY Agar (Maltose Peptone Yeast Extract Medium) (ATCC Medium 518)

Composition per liter:

Agar....10.0g
Maltose....2.0g
Peptone....2.0g
Yeast extract....1.0g
Potassium phosphate buffer (1*M*, pH 7.5)....10.0mL

pH 7.5 ± 0.2 at 25°C

Preparation of Medium: Add components, except potassium phosphate buffer, to distilled/deionized water and bring volume to 990.0mL. Mix thoroughly. Gently heat and bring to boiling. Autoclave for 15 min at 15 psi pressure–121°C. Cool to 45–50°C. Filter sterilize potassium phosphate bufffer. Aseptically add sterile potassium phosphate bufffer to sterile cooled basal medium. Distribute into sterile tubes or flasks.

Use: For the cultivation and maintenance of *Spirochaeta aurantia*.

MPY Agar (Malt Peptone Yeast Extract Agar) (ATCC Medium 582)

Composition per liter:

Agar....15.0g
Malt extract....5.0g
Xylose....2.0g
Fructose....2.0g
Lactose....2.0g

Peptone....1.0g
Yeast extract....1.0g
pH 7.0 ± 0.2 at 25°C

Preparation of Medium: Add components to distilled/deionized water and bring volume to 1.0L. Mix thoroughly. Gently heat and bring to boiling. Distribute into tubes or flasks. Autoclave for 15 min at 15 psi pressure–121°C. Pour into sterile Petri dishes or leave in tubes.

Use: For the cultivation and maintenance of *Streptomyces naniwaensis*.

MPY Broth (Maltose Peptone Yeast Extract Broth)

Composition per liter:
Maltose....2.0g
Peptone....2.0g
Yeast extract....1.0g
Potassium phosphate buffer (1*M*, pH 7.5)....10.0mL
pH 7.5 ± 0.2 at 25°C

Preparation of Medium: Add components, except potassium phosphate buffer, to distilled/deionized water and bring volume to 990.0mL. Mix thoroughly. Autoclave for 15 min at 15 psi pressure–121°C. Filter sterilize potassium phosphate bufffer. Aseptically add sterile potassium phosphate bufffer to sterile cooled basal medium. Distribute into sterile tubes or flasks.

Use: For the cultivation and maintenance of *Spirochaeta aurantia*.

MRVP Broth (Methyl Red–Voges–Proskauer Broth)

Composition per liter:
Glucose....5.0g
KH_2PO_4....5.0g
Pancreatic digest of casein....3.5g
Peptic digest of animal tissue....3.5g
pH 6.9 ± 0.2 at 25°C

Source: Available as a premixed powder from BBL Microbiology Systems and as a prepared medium from Difco Laboratories.

Preparation of Medium: Add components to distilled/deionized water and bring volume to 1.0L. Mix thoroughly. Distribute into tubes or flasks. Autoclave for 15 min at 15 psi pressure–121°C.

Use: For the differentiation of bacteria based on acid production (methyl red test) and acetoin production (Voges-Proskauer reaction).

MRVP Medium (Methyl Red Voges–Proskauer Medium)

Composition per liter:
Glucose....5.0g
Peptone....5.0g
Phosphate buffer....5.0g
pH 7.5 ± 0.2 at 25°C

Source: This medium is available as a premixed powder from Oxoid Unipath.

Preparation of Medium: Add components to distilled/deionized water and bring volume to 1.0L. Mix thoroughly. Distribute into tubes or flasks. Autoclave for 15 min at 15 psi pressure–121°C.

Use: For the differentiation of bacteria based on acid production (methyl red test) and acetoin production (Voges-Proskauer reaction).

MS 1 Agar

Composition per liter:
Agar....15.0g

Preparation of Medium: Add agar to 1.0L of natural seawater. Mix thoroughly. Gently heat and bring to boiling. Distribute into tubes or flasks. Autoclave for 15 min at 15 psi pressure–121°C. Pour into sterile Petri dishes or leave in tubes.

Use: For the isolation and cultivation of *Cytophaga* species, *Herpetosiphon* species, *Saprospira* species, and *Flexithrix* species.

MS 3 Agar

Composition per liter:
Agar....15.0g
$(NH_4)_2SO_4$....1.0g

Preparation of Medium: Add agar to 500.0mL of natural seawater. Mix thoroughly. Gently heat and bring to boiling. In a separate flask, add $(NH_4)_2SO_4$ to 500.0mL of natural seawater. Mix thoroughly. Autoclave both solutions separately for 15 min at 15 psi pressure–121°C. Aseptically combine the two sterile solutions. Pour into sterile Petri dishes or distribute into sterile tubes.

Use: For the isolation and cultivation of *Cytophaga* species, *Herpetosiphon* species, *Saprospira* species, and *Flexithrix* species.

MS 4 Agar

Composition per liter:

Agar	15.0g
Glucose	2.0g
$(NH_4)_2SO_4$	1.0g

Preparation of Medium: Add agar to 500.0mL of natural seawater. Mix thoroughly. Gently heat and bring to boiling. In a separate flask, add $(NH_4)_2SO_4$ to 250.0mL of natural seawater. Mix thoroughly. In a third flask add glucose to 250.0mL of natural seawater. Mix thoroughly. Autoclave the three solutions separately for 15 min at 15 psi pressure–121°C. Aseptically combine the three sterile solutions. Pour into sterile Petri dishes or distribute into sterile tubes.

Use: For the isolation and cultivation of *Cytophaga* species, *Herpetosiphon* species, *Saprospira* species, and *Flexithrix* species.

MS Agar

Composition per liter:

Agar	15.0g
Peptone	1.0g
Yeast extract	1.0g
Glucose	1.0g

Preparation of Medium: Add components to distilled/deionized water and bring volume to 1.0L. Mix thoroughly. Gently heat and bring to boiling. Distribute into tubes or flasks. Autoclave for 15 min at 15 psi pressure–121°C. Pour into sterile Petri dishes or leave in tubes.

Use: For the cultivation and maintenance of *Runella slithyformis.*

MS Medium for Methanogens

Composition per 340mL:

Agar	8.0g
$NaHCO_3$	2.4g
Cysteine-sulfide reducing agent	16.0mL
Mineral solution 1	15.0mL
Mineral solution 2	15.0mL
Sodium formate (20% solution)	6.0mL
Yeast extract -soybean casein solution	4.0mL
Sodium acetate (25% solution)	4.0mL
Wolfe's vitamin solution	4.0mL
Wolfe's mineral solution	4.0mL
$FeSO_4 \cdot 7H_2O$ (0.2% solution)	0.4mL
Resazurin (0.1% solution)	0.4mL

pH 7.0 ± 0.2 at 25°C

Cysteine-Sulfide Reducing Agent:

Composition per 400mL:

L-Cysteine·HCl·H_2O	5.0g
Na_2S (12.5% solution)	40.0mL
NaOH (1*N* solution)	30.0mL

Preparation of Cysteine-Sulfide Reducing Agent: Add distilled/deionized water to a 500.0mL round-bottom flask. Add freshly prepared NaOH solution. Gently heat and bring to boiling under 100% N_2. Remove gassing probe. Add cysteine·HCl·H_2O. Add freshly prepared Na_2S solution. Renew gassing for several minutes. Cap with rubber stoppers. Distribute into 8.0mL/18 mm Hungate tubes.

Mineral Solution 1:

Composition per liter:

K_2HPO_4	6.0g

Preparation of Mineral Solution 1: Add K_2HPO_4 to distilled/deionized water and bring volume to 1.0L. Mix thoroughly.

Mineral Solution 2:

Composition per liter:

NaCl	12.0g
KH_2PO_4	6.0g
$(NH_4)_2SO_4$	6.0g
$MgSO_4 \cdot 7H_2O$	2.6g
$CaCl_2 \cdot 2H_2O$	0.16g

Preparation of Mineral Solution 2: Add components to distilled/deionized water and bring volume to 1.0L. Mix thoroughly.

Yeast Extract–Soybean Casein Solution:

Composition per 100mL:

Yeast extract	20.0g
Pancreatic digest of casein	20.0g

Preparation of Yeast Extract–Soybean Casein Solution: Add components to distilled/deionized water and bring volume to 100.0mL. Mix thoroughly.

Wolfe's Mineral Solution:

Composition per liter:

$MgSO_4 \cdot 7H_2O$	3.0g
Nitriloacetic acid	1.5g
NaCl	1.0g
$MnSO_4 \cdot H_2O$	0.5g
$FeSO_4 \cdot 7H_2O$	0.1g
$CoCl_2 \cdot 6H_2O$	0.1g
$CaCl_2$	0.1g
$ZnSO_4 \cdot 7H_2O$	0.1g
$CuSO_4 \cdot 5H_2O$	0.01g
$AlK(SO_4)_2 \cdot 12H_2O$	0.01g
H_3BO_3	0.01g
$Na_2MoO_4 \cdot 2H_2O$	0.01g

Preparation of Wolfe's Mineral Solution: Add nitrilotriacetic acid to 500.0mL of distilled/deionized water. Dissolve by adjusting pH to 6.5 with KOH.

Add remaining components. Add distilled/deionized water to 1.0L.

Wolfe's Vitamin Solution:

Composition per liter:

Pyridoxine·HCl	10.0mg
Thiamine·HCl	5.0mg
Riboflavin	5.0mg
Nicotinic acid	5.0mg
Calcium pantothenate	5.0mg
p-Aminobenzoic acid	5.0mg
Thioctic acid	5.0mg
Biotin	2.0mg
Folic acid	2.0mg
Cyanocobalamin	100.0μg

Preparation of Wolfe's Vitamin Solution: Add components to distilled/deionized water and bring volume to 1.0L. Mix thoroughly.

Preparation of Medium: Add components to distilled/deionized water and bring volume to 408.0mL. Gently heat and bring to boiling under 80% N_2 + 20% CO_2. Continue boiling until resazurin turns colorless, indicating reduction. Adjust pH to 7.0. Anaerobically distribute into into tubes under 80% N_2 + 20% CO_2. Cap with rubber stoppers and aluminum crimp closures. Autoclave for 15 min at 15 psi pressure–121°C.

Use: For the cultivation and maintenance of *Methanobacterium thermoautotrophicum, Methanobacterium wolfei, Methanobrevibacter smithii, Methanogenium bourgense,* and *Methanogenium* species.

MSV AcS Agar

Composition per liter:

$Na_2S·9H_2O$	0.187g
Sodium acetate	0.15g
MSV Agar	1.0L

pH 7.2–7.5 at 25°C

MSV Agar:

Composition per liter:

Agar	12.0g
$(NH_4)_2SO_4$	0.5g
K_2HPO_4	0.11g
KH_2PO_4	0.085g
$MgSO_4·7H_2O$	0.05g
$CaCl_2·2H_20$	0.05g
EDTA	3.0mg
$FeCl_3·H_2O$	2.0mg
Vitamin mix	1.0mL

Preparation of MSV Agar: Add components to distilled/deionized water and bring volume to 1.0L. Mix thoroughly.

Vitamin Mix:

Composition per 100mL:

Calcium pantothenate	0.01g
Niacin	0.01g
Pyridoxine	0.01g
p-Aminobenzoic acid	0.01g
Cocarboxylase	0.01g
Inositol	0.01g
Thiamine	0.01g
Riboflavin	0.01g
Biotin	0.5mg
Cyanocobalamin	0.5mg
Folic acid	0.5mg

Preparation of Vitamin Mix: Add components to distilled/deionized water and bring volume to 100.0mL. Mix thoroughly.

Preparation of Medium: To 1.0L of MSV Agar add sodium acetate and $Na_2S·9H_2O$. Adjust pH to 7.2–7.5. Gently heat to boiling. Distribute into tubes or flasks. Autoclave for 15 min at 15 psi pressure–121°C. Pour into sterile Petri dishes or leave in tubes.

Use: For the isolation, cultivation and enrichment of heterotrophic strains of *Thiothrix* species from water and environmental sources.

MSV Agar

Composition per liter:

Agar	12.0g
$(NH_4)_2SO_4$	0.5g
K_2HPO_4	0.11g
KH_2PO_4	0.085g
$MgSO_4·7H_2O$	0.05g
$CaCl_2·2H_20$	0.05g
EDTA	3.0mg
$FeCl_3·H_2O$	2.0mg
Vitamin mix	1.0mL

pH 7.2–7.5 at 25°C

Vitamin Mix:

Composition per 100mL:

Calcium pantothenate	0.01g
Niacin	0.01g
Pyridoxine	0.01g
p-Aminobenzoic acid	0.01g
Cocarboxylase	0.01g
Inositol	0.01g
Thiamine	0.01g
Riboflavin	0.01g
Biotin	0.5mg
Cyanocobalamin	0.5mg
Folic acid	0.5mg

Preparation of Vitamin Mix: Add components to distilled/deionized water and bring volume to 100.0mL. Mix thoroughly.

Preparation of Medium: Add components to distilled/deionized water and bring volume to 1.0L. Mix thoroughly. Adjust pH to 7.2–7.5. Gently heat to boiling. Distribute into tubes or flasks. Autoclave for 15 min at 15 psi pressure–121°C. Pour into sterile Petri dishes or leave in tubes.

Use: For the isolation, cultivation and enrichment of heterotrophic strains of *Thiothrix* species from water and environmental sources.

MSV Broth

Composition per liter:

$(NH_4)_2SO_4$	0.5g
K_2HPO_4	0.11g
$MgSO_4 \cdot 7H_2O$	0.05g
$CaCl_2 \cdot 2H_20$	0.05g
KH_2PO_4	0.085g
EDTA	3.0mg
$FeCl_3 \cdot H_2O$	2.0mg
Vitamin mix	1.0mL

pH 7.2–7.5 at 25°C

Vitamin Mix:

Composition per 100mL:

Calcium pantothenate	0.01g
Niacin	0.01g
Pyridoxine	0.01g
p-Aminobenzoic acid	0.01g
Cocarboxylase	0.01g
Inositol	0.01g
Thiamine	0.01g
Riboflavin	0.01g
Biotin	0.5mg
Cyanocobalamin	0.5mg
Folic acid	0.5mg

Preparation of Vitamin Mix: Add components to distilled/deionized water and bring volume to 100.0mL. Mix thoroughly.

Preparation of Medium: Add components to distilled/deionized water and bring volume to 1.0L. Mix thoroughly. Adjust pH to 7.2–7.5. Distribute into tubes or flasks. Autoclave for 15 min at 15 psi pressure–121°C.

Use: For the isolation, cultivation and enrichment of heterotrophic strains of *Thiothrix* species from water and environmental sources.

MSV GS Agar

Composition per liter:

$Na_2S \cdot 9H_2O$	0.187g
Glucose	0.15g
MSV agar	1.0L

pH 7.2–7.5 at 25°C

MSV Agar:

Composition per liter:

Agar	12.0g
$(NH_4)_2SO_4$	0.5g
K_2HPO_4	0.11g
KH_2PO_4	0.085g
$MgSO_4 \cdot 7H_2O$	0.05g
$CaCl_2 \cdot 2H_20$	0.05g
EDTA	3.0mg
$FeCl_3 \cdot H_2O$	2.0mg
Vitamin mix	1.0mL

Preparation of MSV Agar: Add components to distilled/deionized water and bring volume to 1.0L. Mix thoroughly.

Vitamin Mix:

Composition per 100mL:

Calcium pantothenate	0.01g
Niacin	0.01g
Pyridoxine	0.01g
p-Aminobenzoic acid	0.01g
Cocarboxylase	0.01g
Inositol	0.01g
Thiamine	0.01g
Riboflavin	0.01g
Biotin	0.5mg
Cyanocobalamin	0.5mg
Folic acid	0.5mg

Preparation of Vitamin Mix: Add components to distilled/deionized water and bring volume to 100.0mL. Mix thoroughly.

Preparation of Medium: To 1.0L of MSV Agar add glucose and $Na_2S \cdot 9H_2O$. Adjust pH to 7.2–7.5. Gently heat to boiling. Distribute into tubes or flasks. Autoclave for 15 min at 15 psi pressure–121°C. Pour into sterile Petri dishes or leave in tubes.

Use: For the isolation, cultivation and enrichment of heterotrophic strains of *Thiothrix* species from water and environmental sources.

MSV I Agar

Composition per liter:

Glucose	0.15g
MSV agar	1.0L

pH 7.2–7.5 at 25°C

MSV Agar:

Composition per liter:

Agar	12.0g
$(NH_4)_2SO_4$	0.5g
K_2HPO_4	0.11g
KH_2PO_4	0.085g
$MgSO_4 \cdot 7H_2O$	0.05g

$CaCl_2 \cdot 2H_2O$	0.05g
EDTA	3.0mg
$FeCl_3 \cdot H_2O$	2.0mg
Vitamin mix	1.0mL

Preparation of MSV Agar: Add components to distilled/deionized water and bring volume to 1.0L. Mix thoroughly.

Vitamin Mix:

Composition per 100mL:

Calcium pantothenate	0.01g
Niacin	0.01g
Pyridoxine	0.01g
p-Aminobenzoic acid	0.01g
Cocarboxylase	0.01g
Inositol	0.01g
Thiamine	0.01g
Riboflavin	0.01g
Biotin	0.5mg
Cyanocobalamin	0.5mg
Folic acid	0.5mg

Preparation of Vitamin Mix: Add components to distilled/deionized water and bring volume to 100.0mL. Mix thoroughly.

Preparation of Medium: To 1.0L of MSV Agar add glucose. Adjust pH to 7.2–7.5. Gently heat to boiling. Distribute into tubes or flasks. Autoclave for 15 min at 15 psi pressure–121°C. Pour into sterile Petri dishes or leave in tubes.

Use: For the isolation, cultivation and enrichment of heterotrophic strains of *Thiothrix* species from water and environmental sources.

MSV LT Agar

Composition per liter:

Sodium lactate	0.5g
$Na_2S_2O_3$	0.5g
MSV agar	1.0L

pH 7.2–7.5 at 25°C

MSV Agar:

Composition per liter:

Agar	12.0g
$(NH_4)_2SO_4$	0.5g
K_2HPO_4	0.11g
$MgSO_4 \cdot 7H_2O$	0.05g
$CaCl_2 \cdot 2H_2O$	0.05g
KH_2PO_4	0.085g
EDTA	3.0mg
$FeCl_3 \cdot H_2O$	2.0mg
Vitamin mix	1.0mL

Preparation of MSV Agar: Add components to distilled/deionized water and bring volume to 1.0L. Mix thoroughly.

Vitamin Mix:

Composition per 100mL:

Calcium pantothenate	0.01g
Niacin	0.01g
Pyridoxine	0.01g
p-Aminobenzoic acid	0.01g
Cocarboxylase	0.01g
Inositol	0.01g
Thiamine	0.01g
Riboflavin	0.01g
Biotin	0.5mg
Cyanocobalamin	0.5mg
Folic acid	0.5mg

Preparation of Vitamin Mix: Add components to distilled/deionized water and bring volume to 100.0mL. Mix thoroughly.

Preparation of Medium: To 1.0L of MSV agar add sodium lactate and $Na_2S_2O_3$. Adjust pH to 7.2–7.5. Gently heat to boiling. Distribute into tubes or flasks. Autoclave for 15 min at 15 psi pressure–121°C. Pour into sterile Petri dishes or leave in tubes.

Use: For the isolation, cultivation and enrichment of heterotrophic strains of *Thiothrix* species from water and environmental sources.

MSV S Agar

Composition per liter:

$Na_2S \cdot 9H_2O$	0.187g
MSV agar	1.0L

pH 7.2–7.5 at 25°C

MSV Agar:

Composition per liter:

Agar	12.0g
$(NH_4)_2SO_4$	0.5g
K_2HPO_4	0.11g
KH_2PO_4	0.085g
$MgSO_4 \cdot 7H_2O$	0.05g
$CaCl_2 \cdot 2H_2O$	0.05g
EDTA	3.0mg
$FeCl_3 \cdot H_2O$	2.0mg
Vitamin mix	1.0mL

Preparation of MSV Agar: Add components to distilled/deionized water and bring volume to 1.0L. Mix thoroughly.

Vitamin Mix:

Composition per 100mL:

Calcium pantothenate	0.01g
Niacin	0.01g
Pyridoxine	0.01g
p-Aminobenzoic acid	0.01g
Cocarboxylase	0.01g
Inositol	0.01g
Thiamine	0.01g

Riboflavin	0.01g
Biotin	0.5mg
Cyanocobalamin	0.5mg
Folic acid	0.5mg

Preparation of Vitamin Mix: Add components to distilled/deionized water and bring volume to 100.0mL. Mix thoroughly.

Preparation of Medium: To 1.0L of MSV agar add $Na_2S \cdot 9H_2O$. Adjust pH to 7.2–7.5. Gently heat to boiling. Distribute into tubes or flasks. Autoclave for 15 min at 15 psi pressure–121°C. Pour into sterile Petri dishes or leave in tubes.

Use: For the isolation, cultivation and enrichment of heterotrophic strains of *Thiothrix* species from water and environmental sources.

MSV SS Agar

Composition per liter:

$Na_2S \cdot 9H_2O$	0.187g
Sucrose	0.15g
MSV agar	1.0L

pH 7.2–7.5 at 25°C

MSV Agar:

Composition per liter:

Agar	12.0g
$(NH_4)_2SO_4$	0.5g
K_2HPO_4	0.11g
KH_2PO_4	0.085g
$MgSO_4 \cdot 7H_2O$	0.05g
$CaCl_2 \cdot 2H_20$	0.05g
EDTA	3.0mg
$FeCl_3 \cdot H_2O$	2.0mg
Vitamin mix	1.0mL

Preparation of MSV Agar: Add components to distilled/deionized water and bring volume to 1.0L. Mix thoroughly.

Vitamin Mix:

Composition per 100mL:

Calcium pantothenate	0.01g
Niacin	0.01g
Pyridoxine	0.01g
p-Aminobenzoic acid	0.01g
Cocarboxylase	0.01g
Inositol	0.01g
Thiamine	0.01g
Riboflavin	0.01g
Biotin	0.5mg
Cyanocobalamin	0.5mg
Folic acid	0.5mg

Preparation of Vitamin Mix: Add components to distilled/deionized water and bring volume to 100.0mL. Mix thoroughly.

Preparation of Medium: To 1.0L of MSV agar add $Na_2S \cdot 9H_2O$ and sucrose. Adjust pH to 7.2–7.5. Gently heat to boiling. Distribute into tubes or flasks. Autoclave for 15 min at 15 psi pressure–121°C. Pour into sterile Petri dishes or leave in tubes.

Use: For the isolation, cultivation and enrichment of heterotrophic strains of *Thiothrix* species from water and environmental sources.

MSV SUC Agar

Composition per liter:

Sodium succinate	0.15g
MSV agar	1.0L

pH 7.2–7.5 at 25°C

MSV Agar:

Composition per liter:

Agar	12.0g
$(NH_4)_2SO_4$	0.5g
K_2HPO_4	0.11g
KH_2PO_4	0.085g
$MgSO_4 \cdot 7H_2O$	0.05g
$CaCl_2 \cdot 2H_20$	0.05g
EDTA	3.0mg
$FeCl_3 \cdot H_2O$	2.0mg
Vitamin mix	1.0mL

Preparation of MSV Agar: Add components to distilled/deionized water and bring volume to 1.0L. Mix thoroughly.

Vitamin Mix:

Composition per 100mL:

Calcium pantothenate	0.01g
Niacin	0.01g
Pyridoxine	0.01g
p-Aminobenzoic acid	0.01g
Cocarboxylase	0.01g
Inositol	0.01g
Thiamine	0.01g
Riboflavin	0.01g
Biotin	0.5mg
Cyanocobalamin	0.5mg
Folic acid	0.5mg

Preparation of Vitamin Mix: Add components to distilled/deionized water and bring volume to 100.0mL. Mix thoroughly.

Preparation of Medium: To 1.0L of MSV Agar add sodium succinate. Adjust pH to 7.2–7.5. Gently heat to boiling. Distribute into tubes or flasks. Autoclave for 15 min at 15 psi pressure–121°C. Pour into sterile Petri dishes or leave in tubes.

Use: For the isolation, cultivation and enrichment of heterotrophic strains of *Thiothrix* species from water and environmental sources.

MWY Medium (Wadowsky and Yee Medium, Modified)

Composition per liter:

Agar	13.0g
Yeast extract	10.0g
Glycine	3.0g
ACES buffer (2-[(2-Amino-2-oxoethyl)-amino]-ethane sulfonic acid)	2.0g
Charcoal, activated	2.0g
α-Ketoglutarate	0.2g
$Fe_4(P_2O_7)_3 \cdot 9H_2O$	0.05g
Bromcresol purple	0.01g
Bromcresol blue	0.01g
Antibiotic inhibitor	10.0mL
L-Cysteine·HCl·H_2O solution	10.0mL

pH 6.9 ± 0.2 at 25°C

Antibiotic Inhibitor:
Composition per 10mL:

Anisomycin	0.16g
Cefamandole	4.0mg
Vancomycin	1.0mg
Polymyxin B	130,000U

Preparation of Antibiotic Inhibitor: Add components to distilled/deionized water and bring volume to 10.0mL. Mix thoroughly. Filter sterilize.

L-Cysteine·HCl·H_2O Solution:
Composition per 10mL:

L-Cysteine·HCl·H_2O	0.08g

Preparation of L-Cysteine·HCl·H_2O Solution: Add L-Cysteine·HCl·H_2O to distilled/deionized water and bring volume to 10.0mL. Mix thoroughly. Filter sterilize.

Preparation of Medium: Add components, except cysteine and antibiotic inhibitor, to distilled/deionized water and bring volume to 980.0mL. Mix thoroughly. Adjust medium to pH 6.9 with 1*N* KOH. Heat gently and bring to boiling for 1 min. Autoclave for 15 min at 15 psi pressure–121°C. Cool to 50–55°C. Add 10.0mL of the sterile L-cysteine·HCl·H_2O solution and 10.0mL of the sterile antibiotic solution. Mix thoroughly. Pour into sterile Petri dishes with constant agitation to keep charcoal in suspension.

Use: For the selective isolation and cultivation of *Legionella pneumophila* and other *L.* species.

MY Agar

Composition per liter:

Agar	15.0g
Sodium acetate	0.1g
Yeast extract	0.1g
Pancreatic digest of gelatin	0.06g
Beef extract	0.04g
Basal medium	1.0L
Sodium sulfide solution	3.0mL

pH 7.0–7.5 at 25°C

Basal Medium:
Composition per liter:

$CaSO_4 \cdot 2H_2O$ (saturated solution)	20.0mL
NH_4Cl (4% solution)	5.0mL
Trace elements solution	5.0mL
K_2HPO_4 (1% solution)	1.0mL
$MgSO_4 \cdot 7H_2O$ (1% solution)	1.0mL

Preparation of Basal Medium: Add components to distilled/deionized water and bring volume to 1.0L. Mix thoroughly.

Trace Elements Solution:
Composition per liter:

Ferrous EDTA solution	20.0mL
$ZnSO_4 \cdot 7H_2O$ (0.1% solution)	10.0mL
$MnSO_4 \cdot 4H_2O$ (0.02% solution)	10.0mL
$CuSO_4 \cdot 5H_2O$ (0.00005% solution)	10.0mL
H_3BO_3 (0.1% solution)	10.0mL
$Co(NO_3)_2$ or $CoCl_2 \cdot 6H_2O$ (0.01% solution)	10.0mL
$Na_2MoO_4 \cdot 2H_2O$ (0.01% solution)	10.0mL

Preparation of Trace Elements Solution: Add components to distilled/deionized water and bring volume to 1.0L. Mix thoroughly.

Ferrous EDTA Solution:
Composition per 100mL:

$FeSO_4 \cdot 7H_2O$	7.0g
EDTA	2.0g
HCl, concentrated	1.0mL

Preparation of Ferrous EDTA Solution: Add components to distilled/deionized water and bring volume to 100.0mL. Mix thoroughly.

Sodium Sulfide Solution:
Composition per 10mL:

$Na_2S \cdot 9H_20$	1.0g

Preparation of Sodium Sulfide Solution: Add $Na_2S \cdot 9H_20$ to distilled/deionized water and bring volume to 10.0mL. Mix thoroughly. Autoclave for 15 min at 15 psi pressure–121°C. Freshly prepare.

Preparation of Medium: Add sodium acetate, nutrient broth powder, yeast extract and agar to 1.0L of basal medium. Mix thoroughly. Adjust pH to 7.0–7.5. Gently heat and bring to boiling. Autoclave for 15 min at 15 psi pressure–121°C. Cool to 45–50°C. Aseptically add 3.0mL of sterile sodium sulfide solution immediately prior to dispensing. Mix thoroughly. Pour into sterile Petri dishes or distribute into sterile screw-capped tubes.

Use: For the isolation and cultivation of *Beggiatoa* species and myxotrophic strains of *Thiothrix* species from water and environmental sources.

Mycobacterium Medium

Composition per liter:

Noble agar	15.0g
$(NH_4)_2SO_4$	1.0g
Na_2HPO_4	0.5g
KH_2PO_4	0.5g
$MgSO_4$	0.2g
$FeSO_4 \cdot 7H_2O$	5.0mg
$MnSO_4$	2.0mg
Liquid paraffin	5.0mL

Preparation of Medium: Add components, except agar, to distilled/deionized water and bring volume to 1.0L. Homogenize in a blender. Add agar. Gently heat and bring to boiling. Distribute into tubes or flasks. Autoclave for 15 min at 15 psi pressure–121°C. Pour into sterile Petri dishes or leave in tubes.

Use: For the cultivation and maintenance of *Mycobacterium paraffinicum.*

Mycobacterium Yeast Extract Medium

Composition per liter:

Agar	15.0g
Pancreatic digest of casein	5.0g
Yeast extract	2.5g
Glucose	1.0g

Preparation of Medium: Add components to distilled/deionized water and bring volume to 1.0L. Mix thoroughly. Gently heat and bring to boiling. Distribute into tubes or flasks. Autoclave for 15 min at 15 psi pressure–121°C. Pour into sterile Petri dishes or leave in tubes.

Use: For the cultivation and maintenance of *Mycobacterium* species and *Rhodococcus* species.

Mycological Broth

Composition per liter:

Glucose	40.0g
Enzymatic hydrolysate of soybean meal	10.0g

pH 7.0 ± 0.2 at 25°C

Source: This medium is available as a premixed powder from Difco Laboratories.

Preparation of Medium: Add components to distilled/deionized water and bring volume to 1.0L. Mix thoroughly. Gently heat and bring to boiling. Distribute into tubes or flasks. Autoclave for 15 min at 15 psi pressure–121°C.

Use: For the cultivation of fungi.

Mycological Broth with Low pH

Composition per liter:

Glucose	40.0g
Enzymatic hydrolysate of soybean meal	10.0g

pH 4.8 ± 0.2 at 25°C

Source: This medium is available as a premixed powder from Difco Laboratories.

Preparation of Medium: Add components to distilled/deionized water and bring volume to 1.0L. Mix thoroughly. Gently heat and bring to boiling. Adjust pH to 4.8. Distribute into tubes or flasks. Autoclave for 15 min at 15 psi pressure–121°C.

Use: For the cultivation of saprophytic species of yeasts and molds. It is also suitable for cultivation of aciduric bacteria.

Mycophil™ Agar

Composition per liter:

Agar	16.0g
Papaic digest of soybean meal	10.0g
Glucose	10.0g

pH 7.0 ± 0.2 at 25°C

Source: This medium is available as a premixed powder from BBL Microbiology Systems.

Preparation of Medium: Add components to distilled/deionized water and bring volume to 1.0L. Mix thoroughly. Gently heat and bring to boiling. Distribute into tubes or flasks. Autoclave for 15 min at 15 psi pressure–121°C. Pour into sterile Petri dishes or distribute into sterile tubes.

Use: For the cultivation, maintenance, and enumeration of fungi. For the demonstration of pigment production in fungal species. Also used for the cultivation and maintenance of *Bacillus* species.

Mycophil™ Agar with Low pH

Composition per liter:

Agar	18.0g
Papaic digest of soybean meal	10.0g
Glucose	10.0g

pH 4.7 ± 0.2 at 25°C

Source: This medium is available as a premixed powder from BBL Microbiology Systems.

Preparation of Medium: Add components to distilled/deionized water and bring volume to 1.0L. Mix thoroughly. Gently heat and bring to boiling. Distrib-

ute into tubes or flasks. Autoclave for 15 min at 15 psi pressure–121°C. Adjust pH to 4.7 by adding approximately 10.0mL of sterile 10% lactic acid. Mix thoroughly. Pour into sterile Petri dishes or distribute into sterile tubes.

Use: For the isolation and enumeration of yeasts and molds.

Mycophil™ Broth

Composition per liter:

Glucose	40.0g
Pancreatic digest of casein	5.0g
Peptic digest of animal tissue	5.0g

pH 7.0 ± 0.2 at 25°C

Source: This medium is available as a premixed powder from BBL Microbiology Systems.

Preparation of Medium: Add components to distilled/deionized water and bring volume to 1.0L. Mix thoroughly. Gently heat and bring to boiling. Distribute into tubes or flasks. Autoclave for 15 min at 15 psi pressure–118°C. Do not overheat.

Use: For the isolation and cultivation of a wide variety of fungi.

Mycoplasma Agar

Composition per liter:

Basal medium	700.0mL
Horse serum	200.0mL
Fresh yeast extract solution	100.0mL

pH 7.5-7.8 at 25°C

Basal Medium:

Composition per 700mL:

Sorbitol	50.0g
Beef heart (solids from infusion)	16.2g
Noble agar	13.0g
Peptone	3.26g
NaCl	1.62g
Fructose	1.0g
Glucose	1.0g
Sucrose	1.0g
Pancreatic digest of casein	1.0g

Preparation of Basal Medium: Add components to distilled/deionized water and bring volume to 700.0mL. Mix thoroughly. Adjust pH to 7.5-7.8. Autoclave for 15 min at 15 psi pressure–121°C. Cool to 50°C.

Fresh Yeast Extract Solution:

Composition per 100mL:

Live, pressed, starch-free, Baker's yeast	25.0g

Preparation of Fresh Yeast Extract Solution: Add the live Baker's yeast to 100.0mL of distilled/deionized water. Autoclave for 90 min at 15 psi pressure–121°C. Allow to stand. Remove supernatant solution. Adjust pH to 6.6–6.8.

Preparation of Medium: Filter sterilize horse serum and fresh yeast extract solution. Aseptically add to cooled, sterile basal medium. Mix thoroughly. Aseptically distribute into sterile tubes or flasks.

Use: For the cultivation and maintenance of *Mycoplasma mycoides, Spiroplasma apis, S. citri,* and *S. melliferum.*

Mycoplasma Agar

Composition per1004mL:

Noble agar	8.0g
Arginine	1.0g
Glucose	1.0g
L-Cysteine·HCl·H_2O	1.0g
Mycoplasma broth base	850.0mL
Horse serum (not inactivated)	100.0mL
Fresh yeast extract (25% solution)	50.0mL
Phenol red (1.0% solution)	2.0mL
DNA calf thymus solution	2.0mL

pH 7.8 ± 0.2 at 25°C

***Mycoplasma* Broth Base:**

Composition per 850mL:

Pancreatic digest of casein	7.0g
NaCl	5.0g
Beef extract	3.0g
Yeast extract	3.0g
Beef heart solids from infusion	2.0g

Preparation of *Mycoplasma* Broth Base: Add components to distilled/deionized water and bring volume to 850.0mL. Add the 8.0g of Noble agar. Mix thoroughly. Gently heat and bring to boiling. Autoclave for 15 min at 15 psi pressure–121°C. Cool to 50°C.

Fresh Yeast Extract Solution:

Composition per 100mL:

Live, pressed, starch-free, Baker's yeast	25.0g

Preparation of Fresh Yeast Extract Solution: Add the live Baker's yeast to 100.0mL of distilled/deionized water. Autoclave for 90 min at 15 psi pressure–121°C. Allow to stand. Remove supernatant solution. Adjust pH to 6.6–6.8.

DNA Calf Thymus Solution:

Composition per 10mL:

DNA calf thymus	1.0g

Preparation of DNA Calf Thymus Solution: Add DNA calf thymus to distilled/deionized water and bring volume to 10.0mL. Mix thoroughly. Filter sterilize.

Preparation of Medium: Combine components, except *Mycoplasma* Broth Base and DNA calf thymus solution, and mix thoroughly. Filter sterilize through a 0.2μm membrane. Add sterile solution to 850.0mL of cooled, sterile *Mycoplasma* broth base. Aseptically add 2.0mL of sterile DNA calf thymus solution. Mix thoroughly. Pour into sterile Petri dishes or distribute into sterile tubes.

Use: For the cultivation and maintenance of *Mycoplasma lipophilum* and *Mycoplasma* species.

Mycoplasma Agar (ATCC Medium 555)

Composition per 103mL:

Noble agar	0.7g
Hartley's digest broth	30.0mL
Pig serum	20.0mL
Enzymatic hydrolysate of lactalbumin	10.0mL
Hanks' balanced salt solution, 10X	4.0mL
Fresh yeast extract solution	2.0mL
Phenol red (0.25% solution)	1.0mL

pH 7.4 ± 0.2 at 25°C

Hartley's Digest Broth:
Composition per 10L:

Ox heart	3,000.0g
Pancreatin	50.0g
Na_2CO_3, anhydrous (0.8% solution)	5L
HCl, concentrated	80.0mL

pH 7.5 ± 0.2 at 25°C

Preparation of Hartley's Digest Broth: Finely mince the ox heart. Add the meat to 5.0L of distilled/deionized water. Gently heat and bring to 80°C. Add the 5.0L of Na_2CO_3 solution. Cool to 45°C. Add pancreatin and maintain at 45°C for 4 hr while stirring. Add the HCl and steam at 100°C for 30 min. Cool to room temperature. Adjust pH to 8.0 with 1*N* NaOH. Gently heat and bring to boiling. Continue boiling for 25 min. Filter while hot. Cool to room temperature. Adjust pH to 7.5. Autoclave for 15 min at 15 psi pressure–121°C.

Pig Serum:
Composition per 100mL:

Pig serum	100.0mL

Preparation of Pig Serum: Adjust pH of pig serum to 4.4 with sterile 1*N* HCl. Do not let pH go below 4.2. Let serum stand at 4°C for 18–20 hr. Adjust pH to 7.0 with sterile 1*N* NaOH. Centrifuge at 9,000 rpm for 20 min. Discard pellet. Filter supernatant solution through a 0.2 μm membrane. Store at –70°C.

Enzymatic Hydrolysate of Lactalbumin:
Composition per 100mL:

Enzymatic hydrolysate of lactalbumin	5.0g

Preparation of Enzymatic Hydrolysate of Lactalbumin: Add enzymatic hydrolysate of lactalbumin to 100.0mL of phosphate buffered saline, 1X, pH 7.0.

Phosphate Buffered Saline Solution, 1X:
Composition per liter:

NaCl	8.0g
$Na_2HPO_4 \cdot 7H_2O$	2.16g
KCl	0.2g
KH_2PO_4	0.2g
$MgCl_2 \cdot 6H_2O$	0.1g
$CaCl_2$	0.1g

Hanks' Balanced Salt Solution, 10X:
Composition per liter:

NaCl	80.0g
Glucose	10.0g
KCl	4.0g
$CaCl_2$	1.4g
$MgCl_2 \cdot 6H_2O$	1.0g
$MgSO_4 \cdot 7H_2O$	1.0g
$Na_2HPO_4 \cdot 7H_2O$	0.9g
KH_2PO_4	0.6g

Preparation of Hanks' Balanced Salt Solution, 10X: Add components to distilled/deionized water and bring volume to 1.0L. Mix thoroughly.

Preparation of Medium: Combine components, except agar, in the following order: Hanks' balanced salt solution, 10X, phenol red, Hartley's digest broth, pig serum, enzymatic hydrolysate of lactalbumin, and fresh yeast extract solution. Mix thoroughly. Adjust pH to 7.4 with 1*N* NaOH. Filter sterilize through a 0.2μm membrane. Add 0.7g Noble agar to 36.0mL of distilled/deionized water. Autoclave for 15 min at 15 psi pressure–121°C. Cool to 56°C. Warm basal medium to 56°C. Aseptically combine the two solutions. Pour into sterile Petri dishes or distribute into sterile tubes.

Use: For the cultivation of *Mycoplasma* species.

Mycoplasma Agar (ATCC Medium 1435)

Composition per 930mL:

Agar, noninhibitory to mycoplasmas	10.0g
Glucose	1.0g
Nicotinamide adenine dinucleotide	0.1g
PPLO broth without crystal violet	680.0mL
Swine serum (56°C, 30 min)	150.0mL
Yeast extract (25% w/v solution)	100.0mL

pH 7.8 ± 0.2 at 25°C

PPLO Broth without Crystal Violet:
Composition per 680mL:

Beef heart, infusion from	11.3g
Peptone	2.28g
NaCl	1.13g

Source: PPLO broth without crystal violet is available as a premixed powder from Difco Laboratories.

Preparation of PPLO Broth without Crystal Violet: Add components and agar to distilled/deionized water and bring volume to 680.0mL. Mix thoroughly. Gently heat and bring to boiling. Autoclave for 15 min at 15 psi pressure–121°C. Cool to 45–50°C.

Preparation of Medium: Mix glucose, nicotinamide adenine dinucleotide, swine serum, and fresh yeast extract solution. Mix thoroughly. Heat to 56°C. Add to cooled, sterile PPLO broth without crystal violet. Mix thoroughly. Pour into sterile Petri dishes or distribute into sterile tubes.

Use: For the cultivation and maintenance of *Mycoplasma anseris* and *Mycoplasma lipofaciens*.

Mycoplasma Agar Base (PPLO Agar Base)

Composition per 1300mL:

Agar	14.0g
Pancreatic digest of casein	7.0g
NaCl	5.0g
Beef extract	3.0g
Yeast extract	3.0g
Beef heart, infusion from (solids)	2.0g
Horse serum	260.0mL
Fresh yeast extract solution	65.0mL

pH 7.8 ± 0.2 at 25°C

Source: This medium is available as a premixed powder from BBL Microbiology Systems.

Fresh Yeast Extract Solution:
Composition per 100mL:

Baker's yeast, live, pressed, starch-free	25.0g

Preparation of Fresh Yeast Extract Solution: Add the live Baker's yeast to 100mL of distilled/deionized water. Autoclave for 90 min at 15 psi pressure–121°C. Allow to stand. Remove supernatant solution. Adjust pH to 6.6–6.8.

Preparation of Medium: Add components, except horse serum and special yeast extract, to distilled/deionized water and bring volume to 1.0L. Mix thoroughly. Gently heat and bring to boiling. Distribute into tubes or flasks. Autoclave for 15 min at 15 psi pressure–121°C. Cool to 50°C. To each 75.0mL of cooled, sterile basal medium add 20.0mL of sterile horse serum and 5.0mL of special yeast extract. Mix thoroughly. Pour into sterile Petri dishes or distribute into sterile tubes.

Use: For the preparation of media for cultivation of *Mycoplasma*.

Mycoplasma Agar with Increased Selectivity

Composition per 1300mL:

Agar	14.0g
Pancreatic digest of casein	7.0g
NaCl	5.0g
Beef extract	3.0g
Yeast extract	3.0g
Beef heart, solids from infusion	2.0g
Thallous acetate	0.7g
Penicillin	70,000U
Horse serum	260.0mL
Fresh yeast extract solution	65.0mL

pH 7.8 ± 0.2 at 25°C

Caution: Thallous acetate is a poison.

Fresh Yeast Extract Solution:
Composition per 100mL:

Baker's yeast, live, pressed, starch-free	25.0g

Preparation of Fresh Yeast Extract Solution: Add the live Baker's yeast to 100.0mL of distilled/deionized water. Autoclave for 90 min at 15 psi pressure–121°C. Allow to stand. Remove supernatant solution. Adjust pH to 6.6–6.8.

Preparation of Medium: Add components—except horse serum, special yeast extract, thallous acetate and penicillin—to distilled/deionized water and bring volume to 1.0L. Mix thoroughly. Gently heat and bring to boiling. Distribute into tubes or flasks. Autoclave for 15 min at 15 psi pressure–121°C. Cool to 50°C. To each of 70.0mL of cooled, sterile basal medium add 20.0mL of sterile horse serum, 10.0mL of special yeast extract, 50.0mg of thallous acetate and 5000U of penicillin. Mix thoroughly. Pour into sterile Petri dishes or distribute into sterile tubes.

Use: For the preparation of media for cultivation of *Mycoplasma* species.

Mycoplasma Agar with Supplement G

Composition per liter:

Agar	10.0g
Bacteriological peptone	10.0g
Beef extract	10.0g

NaCl	5.0g
Special mineral supplement, Oxoid Unipath	0.5g
Mycoplasma supplement G	250.0mL

pH 7.8 ± 0.2 at 25°C

Source: This medium is available as a premixed powder from Oxoid Unipath.

***Mycoplasma* Supplement G:**
Composition per 20mL:

Thallous acetate	25.0mg
Horse serum	20.0mL
Yeast extract (25% solution)	10.0mL
Penicillin	20,000U

Preparation of *Mycoplasma* Supplement G: Add components to distilled/deionized water and bring volume to 20.0mL. Mix thoroughly. Filter sterilize.

Caution: Thallous acetate is a poison.

Preparation of Medium: Add components, except *Mycoplasma* Supplement G, to distilled/deionized water and bring volume to 1.0L. Mix thoroughly. Gently heat and bring to boiling. Distribute into flasks in 80.0mL volumes. Autoclave for 15 min at 15 psi pressure–121°C. Cool to 50°C. Aseptically add 20.0mL of sterile *Mycoplasma* Supplement G to each 80.0mL of basal medium. Mix thoroughly.

Use: For the growth of *Mycoplasma* species.

Mycoplasma Agar with Supplement P

Composition per liter:

Agar	10.0g
Bacteriological peptone	10.0g
Beef extract	10.0g
NaCl	5.0g
Special mineral supplement, Oxoid Unipath	0.5g
Mycoplasma supplement P	250.0mL

pH 7.8 ± 0.2 at 25°C

Source: This medium is available as a premixed powder from Oxoid Unipath.

***Mycoplasma* Supplement P:**
Composition per 20mL:

Glucose	0.3g
Mycoplasma broth base	0.145g
Thallous acetate	8.0mg
Phenol red	1.2mg
Methylene blue chloride	0.3mg
Penicillin	12,000U
Horse serum	6.0mL
Yeast extract (25% solution)	3.0mL

Preparation of *Mycoplasma* Supplement P: Add components to distilled/deionized water and bring volume to 20.0mL. Mix thoroughly. Filter sterilize.

Caution: Thallous acetate is a poison.

Fresh Yeast Extract Solution:
Composition per 100mL:

Baker's yeast, live, pressed, starch-free	25.0g

Preparation of Fresh Yeast Extract Solution: Add the live Baker's yeast to 100.0mL of distilled/deionized water. Autoclave for 90 min at 15 psi pressure–121°C. Allow to stand. Remove supernatant solution. Adjust pH to 6.6–6.8.

Preparation of Medium: Add components, except *Mycoplasma* Supplement P, to distilled/deionized water and bring volume to 1.0L. Mix thoroughly. Gently heat and bring to boiling. Distribute into bottles in 1.0mL volumes. Autoclave for 15 min at 15 psi pressure–121°C. Cool to room temperature. Aseptically add 2.0mL of sterile *Mycoplasma* Supplement P to each bottle.

Use: For the growth of *Mycoplasma* species.

Mycoplasma Broth

Composition per liter:

Basal medium	700.0mL
Horse serum	200.0mL
Fresh yeast extract solution	100.0mL

pH 7.5–7.8 at 25°C

Basal Medium:
Composition per 700mL:

Sorbitol	50.0g
Beef heart, solids from infusion	16.2g
Peptone	3.26g
NaCl	1.62g
Fructose	1.0g
Glucose	1.0g
Sucrose	1.0g
Pancreatic digest of casein	1.0g

Preparation of Basal Medium: Add components to distilled/deionized water and bring volume to 700.0mL. Mix thoroughly. Adjust pH to 7.5–7.8. Autoclave for 15 min at 15 psi pressure–121°C. Cool to 50°C.

Fresh Yeast Extract Solution:
Composition per 100mL:

Baker's yeast, live, pressed, starch-free	25.0g

Preparation of Fresh Yeast Extract Solution: Add the live Baker's yeast to 100.0mL of distilled/deionized water. Autoclave for 90 min at 15 psi pressure–121°C. Allow to stand. Remove supernatant solution. Adjust pH to 6.6–6.8.

Preparation of Medium: Filter sterilize horse serum and fresh yeast extract solution. Aseptically add to cooled, sterile basal medium. Mix thoroughly. Aseptically distribute into sterile tubes or flasks.

Use: For the cultivation and maintenance of *Mycoplasma mycoides, Spiroplasma apis, S. citri,* and *S. melliferum.*

Mycoplasma Broth

Composition per 950mL:

Glucose	1.0g
Nicotinamide adenine dinucleotide	0.1g
PPLO broth without crystal violet	680.0mL
Swine serum (56°C, 30 min)	150.0mL
Fresh yeast extract solution	100.0mL
Phenol red (0.1% w/v solution)	20.0mL

pH 7.8 ± 0.2 at 25°C

PPLO Broth without Crystal Violet:
Composition per 680mL:

Beef heart, solids from infusion	11.3g
Peptone	2.28g
NaCl	1.13g

Source: PPLO broth without crystal violet is available as a premixed powder from Difco Laboratories.

Preparation of PPLO Broth without Crystal Violet: Add components to distilled/deionized water and bring volume to 900.0mL. Autoclave for 15 min at 15 psi pressure–121°C. Cool to 56°C.

Fresh Yeast Extract Solution:
Composition per 100mL:

Baker's yeast, live, pressed, starch-free	25.0g

Preparation of Fresh Yeast Extract Solution: Add the live Baker's yeast to 100.0mL of distilled/deionized water. Autoclave for 90 min at 15 psi pressure–121°C. Allow to stand. Remove supernatant solution. Adjust pH to 6.6–6.8.

Preparation of Medium: Mix glucose, nicotinamide adenine dinucleotide, swine serum, fresh yeast extract solution, and phenol red. Mix thoroughly. Heat to 56°C. Add to cooled, sterile PPLO broth without crystal violet. Mix thoroughly. Aseptically distribute into sterile tubes or flasks.

Use: For the cultivation and maintenance of *Mycoplasma anseris* and *M. lipofaciens.*

Mycoplasma Broth (ATCC Medium 555)

Composition per 103mL:

Hartley's digest broth	30.0mL
Pig serum	20.0mL
Enzymatic hydrolysate of lactalbumin	10.0mL
Hanks' balanced salt solution, 10X	4.0mL
Fresh yeast extract solution	2.0mL
Phenol red (0.25% solution)	1.0mL

pH 7.4 ± 0.2 at 25°C

Hartley's Digest Broth:
Composition per 10L:

Ox heart	3,000.0g
Pancreatin	50.0g
Na_2CO_3, anhydrous (0.8% solution)	5L
HCl, concentrated	80.0mL

pH 7.5 ± 0.2 at 25°C

Preparation of Hartley's Digest Broth: Finely mince the ox heart. Add the meat to 5.0L of distilled/deionized water. Gently heat and bring to 80°C. Add the 5.0L of Na_2CO_3 solution. Cool to 45°C. Add pancreatin and maintain at 45°C for 4 hr while stirring. Add the HCl and steam at 100°C for 30 min. Cool to room temperature. Adjust pH to 8.0 with 1*N* NaOH. Gently heat and bring to boiling. Continue boiling for 25 min. Filter while hot. Cool to room temperature. Adjust pH to 7.5. Autoclave for 15 min at 15 psi pressure–121°C.

Pig Serum:
Composition per 100mL:

Pig serum	100.0mL

Preparation of Pig Serum: Adjust pH of pig serum to 4.4 with sterile 1*N* HCl. Do not let pH go below 4.2. Let serum stand at 4°C for 18–20 hr. Adjust pH to 7.0 with sterile 1*N* NaOH. Centrifuge at 9,000 rpm for 20 min. Discard pellet. Filter supernatant solution through a 0.2μm membrane. Store at –70°C.

Enzymatic Hydrolysate of Lactalbumin:
Composition per 100mL:

Enzymatic hydrolysate of lactalbumin	5.0g

Preparation of Enzymatic Hydrolysate of Lactalbumin: Add enzymatic hydrolysate of lactalbumin to 100.0mL of phosphate buffered saline, 1X, pH 7.0.

Phosphate Buffered Saline Solution, 1X:
Composition per liter:

NaCl	8.0g
$Na_2HPO_4·7H_2O$	2.16g
KCl	0.2g
KH_2PO_4	0.2g
$MgCl_2·6H_2O$	0.1g
$CaCl_2$	0.1g

Hanks' Balanced Salt Solution, 10X:
Composition per liter:

NaCl	80.0g
Glucose	10.0g
KCl	4.0g
$CaCl_2$	1.4g
$MgCl_2·6H_2O$	1.0g
$MgSO_4·7H_2O$	1.0g
$Na_2HPO_4·7H_2O$	0.9g
KH_2PO_4	0.6g

Preparation of Hanks' Balanced Salt Solution, 10X: Add components to distilled/deionized water and bring volume to 1.0L. Mix thoroughly.

Preparation of Medium: Combine components in to following order: Hanks' balanced salt solution, 10X, phenol red, Hartley's digest broth, pig serum, enzymatic hydrolysate of lactalbumin, and fresh yeast extract solution. Mix thoroughly. Add 36.0mL of distilled/deionized water. Adjust pH to 7.4 with 1*N* NaOH. Filter sterilize through a 0.2μm membrane. Store at 4°C for up to 3 weeks.

Use: For the cultivation of *Mycoplasma* species.

Mycoplasma Broth Base (PPLO Broth Base without Crystal Violet)

Composition per liter:

Pancreatic digest of casein	7.0g
NaCl	5.0g
Beef extract	3.0g
Yeast extract	3.0g
Beef heart, solids from infusion	2.0g

pH 7.8 ± 0.2 at 25°C

Source: This medium is available as a premixed powder from BBL Microbiology Systems.

Preparation of Medium: Add components to distilled/deionized water and bring volume to 1.0L. Gently heat and bring to boiling. Mix thoroughly. Distribute into tubes or flasks. Autoclave for 15 min at 15 psi pressure–121°C.

Use: Use as a basal medium that should be enriched for the isolation and cultivation of *Mycoplasma* species.

Mycoplasma Broth, Supplemented

Composition per liter:

Pancreatic digest of casein	7.0g
NaCl	5.0g
Beef extract	3.0g
Yeast extract	3.0g
Beef heart, solids from infusion	2.0g
Horse serum	260.0mL
Fresh yeast extract solution	65.0mL

pH 7.8 ± 0.2 at 25°C

Fresh Yeast Extract Solution:
Composition per 100mL:

Baker's yeast, live, pressed, starch-free	25.0g

Preparation of Fresh Yeast Extract Solution: Add the live Baker's yeast to 100.0mL of distilled/deionized water. Autoclave for 90 min at 15 psi pressure–121°C. Allow to stand. Remove supernatant solution. Adjust pH to 6.6–6.8.

Preparation of Medium: Add components, except horse serum and special yeast extract, to distilled/deionized water and bring volume to 1.0L. Mix thoroughly. Gently heat and bring to boiling. Distribute into tubes or flasks. Autoclave for 15 min at 15 psi pressure–121°C. Cool to 50°C. To each of 75.0mL of cooled, sterile basal medium add 20.0mL of sterile horse serum and 5.0mL of special yeast extract. Mix thoroughly. Aseptically distribute into sterile tubes.

Use: For the isolation and cultivation of *Mycoplasma* species.

Mycoplasma Broth with Supplement G

Composition per liter:

Bacteriological peptone	10.0g
Beef extract	10.0g
NaCl	5.0g
Special mineral supplement, Oxoid Unipath	0.5g
Mycoplasma Supplement G	250.0mL

pH 7.8 ± 0.2 at 25°C

Source: This medium is available as a premixed powder from Oxoid Unipath.

***Mycoplasma* Supplement G:**
Composition per 20mL:

Thallous acetate	25.0mg
Horse serum	20.0mL
Yeast extract (25% solution)	10.0mL
Penicillin	20,000U

Preparation of *Mycoplasma* Supplement G: Add components to distilled/deionized water and bring volume to 20.0mL. Mix thoroughly. Filter sterilize.

Caution: Thallous acetate is a poison.

Preparation of Medium: Add components, except *Mycoplasma* Supplement G, to distilled/deionized water and bring volume to 1.0L. Mix thoroughly. Gently heat and bring to boiling. Distribute into flasks in 80.0mL volumes. Autoclave for 15 min at 15 psi pressure–121°C. Cool to 50°C. Aseptically add 20.0mL of sterile *Mycoplasma* Supplement G to each 80.0mL of basal medium. Mix thoroughly.

Use: For the growth of *Mycoplasma* species.

Mycoplasma Broth with Supplement P

Composition per liter:

Bacteriological peptone	10.0g
Beef extract	10.0g

NaCl	5.0g
Special mineral supplement, Oxoid Unipath	0.5g
Mycoplasma supplement P	250.0mL

pH 7.8 ± 0.2 at 25°C

Source: This medium is available as a premixed powder from Oxoid Unipath.

***Mycoplasma* Supplement P:**
Composition per 20mL:

Glucose	0.3g
Mycoplasma broth base	0.145g
Thallous acetate	8.0mg
Phenol red	1.2mg
Methylene blue chloride	0.3mg
Penicillin	12,000U
Horse serum	6.0mL
Yeast extract (25% solution)	3.0mL

Preparation of *Mycoplasma* Supplement P: Add components to distilled/deionized water and bring volume to 20.0mL. Mix thoroughly. Filter sterilize.

Caution: Thallous acetate is a poison.

Preparation of Medium: Add components, except *Mycoplasma* Supplement P, to distilled/deionized water and bring volume to 1.0L. Mix thoroughly. Gently heat and bring to boiling. Distribute into bottles in 1.0mL volumes. Autoclave for 15 min at 15 psi pressure–121°C. Cool to room temperature. Aseptically add 2.0mL of sterile *Mycoplasma* Supplement P to each bottle.

Use: For the cultivation of *Mycoplasma* species.

Mycosel™ Agar (Cycloheximide Chloramphenicol Agar)

Composition per liter:

Agar	15.5g
Papaic digest of soybean meal	10.0g
Glucose	10.0g
Cycloheximide	0.4g
Chloramphenicol	0.05g

pH 6.9 ± 0.2 at 25°C

Source: This medium is available as a premixed powder from BBL Microbiology Systems.

Preparation of Medium: Add components to distilled/deionized water and bring volume to 1.0L. Mix thoroughly. Gently heat while stirring and bring to boiling. Autoclave for 15 min at 14 psi pressure–118°C. Avoid overheating. Pour into sterile Petri dishes or distribute into sterile tubes.

Use: For the selective isolation of pathogenic fungi from specimens having a large flora of other fungi and bacteria.

Mysorens Medium

Composition per liter:

Peptone	10.0g
Meat extract	10.0g
Yeast extract	5.0g
NaCl	3.0g

Preparation of Medium: Add components to distilled/deionized water and bring volume to 1.0L. Mix thoroughly. Distribute into tubes or flasks. Autoclave for 15 min at 15 psi pressure–121°C.

Use: For the cultivation and maintenance of *Arthrobacter mysorens.*

Myxobacteria Medium

Composition per liter:

Agar	15.0g
Skim milk powder	5.0g
Yeast extract	0.5g

Preparation of Medium: Add components to distilled/deionized water and bring volume to 1.0L. Mix thoroughly. Do not adjust pH. Gently heat and bring to boiling. Distribute into tubes or flasks. Autoclave for 15 min at 15 psi pressure–121°C.

Use: For the cultivation and maintenance of *Archangium primigenium, Chondrococcus macrosporus* and *Myxococcus coralloides.*

Myxococcus Medium

Composition per liter:

Agar	12.0g
Pancreatic digest of casein	1.0g
Meat extract	1.0g
Glucose solution	50.0mL

pH 7.2 ± 0.2 at 25°C

Glucose Solution:
Composition per 50mL:

Glucose	1.0g

Preparation of Glucose Solution: Add glucose to distilled/deionized water and bring volume to 50.0mL. Mix thoroughly. Autoclave for 15 min at 15 psi pressure–121°C. Cool to 25°C.

Preparation of Medium: Add components, except glucose solution, to distilled/deionized water and bring volume to 950.0mL. Mix thoroughly. Adjust pH to 7.2. Gently heat and bring to boiling. Autoclave for 15 min at 15 psi pressure–121°C. Cool to

45–50°C. Aseptically add sterile glucose solution. Mix thoroughly. Pour into sterile Petri dishes or distribute into sterile tubes or bottles. Allow tubes or bottles to cool in a slanted position.

Use: For the cultivation of *Myxococcus* species.

Myxococcus xanthus Medium

Agar	20.0g
Pancratic digest of casein	10.0g
$MgSO_4 \cdot 7H_2O$	0.5g
K_2HPO_4	0.148g
KH_2PO_4	0.017g

pH 7.6 ± 0.2 at 25°C

Preparation of Medium: Add components to distilled/deionized water and bring volume to 1.0L. Mix thoroughly. Gently heat and bring to boiling. Distribute into tubes or flasks. Autoclave for 15 min at 15 psi pressure–121°C. Pour into sterile Petri dishes or leave in tubes.

Use: For the cultivation and maintenance of *Myxococcus xanthus*.

NANAT Agar (Nalidixic Acid Novobiocin Actidione Tellurite Agar)

Composition per liter:

Pancreatic digest of casein	17.0g
Agar	15.0g
NaCl	5.0g
Tween™ 80	5.0g
Papaic digest of soybean meal	3.0g
K_2HPO_4	2.5g
Glucose	2.5g
Yeast extract	1.0g
Tellurite solution	10.0mL
Antibiotic solution	10.0mL

pH 7.2 ± 0.2 at 25°C

Tellurite Solution:

Composition per 100mL:

K_2TeO_3	0.05g

Preparation of Tellurite Solution: Add K_2TeO_3 to distilled/deionized water and bring volume to 100.0mL. Mix thoroughly. Filter sterilize.

Antibiotic Solution:

Composition per 10mL:

Actidione (cycloheximide)	0.04g
Novobiocin	0.025g
Nalidixic acid	0.020g
Polymyxin B (optional)	0.030g

Preparation of Antibiotic Solution: Add components to distilled/deionized water and bring volume to 10.0mL. Mix thoroughly. Filter sterilize.

Caution: Potassium tellurite is toxic.

Preparation of Medium: Add components, except tellurite solution and antibiotic solution, to distilled/deionized water and bring volume to 980.0mL. Mix thoroughly. Gently heat and bring to boiling. Autoclave for 15 min at 15 psi pressure–121°C. Cool to 45–50°C. Aseptically add sterile tellurite solution and antibiotic solution. Mix thoroughly. Pour into sterile Petri dishes or distribute into sterile tubes.

Use: For the isolation and cultivation of *Rhodococcus (Corynebacterium) equi* from animal feces, especially from horses and swine. The addition of polymyxin B inhibits the growth of *Pseudomonas aeruginosa* which may interfere with the isolation of *R. equi*.

Nannocystis Agar

Composition per liter:

Agar	15.0g
$CaCl_2 \cdot 2H_2O$	1.0g

pH 7.2 ± 0.2 at 25°C

Preparation of Medium: Add components to distilled/deionized water and bring volume to 1.0L. Mix thoroughly. Gently heat and bring to boiling. Autoclave for 15 min at 15 psi pressure–121°C. Pour into sterile Petri dishes. After the agar has solidified, overlay the suface with 0.5mL of a suspension of dead (autoclaved) *Escherichia coli* cells.

Use: For the cultivation and maintenance of *Nannocystis* species.

Natronobacteria Medium

Composition per liter:

NaCl	200.0g
Agar	20.0g
Yeast extract	5.0g
Casamino acids	5.0g
KH_2PO_4	1.0g
KCl	1.0g
NH_4Cl	1.0g
Sodium glutamate	1.0g
$MgSO_4 \cdot 7H_2O$	0.24g
$CaSO_4 \cdot 2H_2O$	0.17g
Na_2CO_3 solution	100.0mL
Trace elements solution SL-6	1.0mL

pH 9.0 ± 0.2 at 25°C

Na_2CO_3 Solution:

Composition per 100mL:

Na_2CO_3	5.0g

Preparation of Na_2CO_3 Solution: Add Na_2CO_3 to distilled/deionized water and bring volume to 100.0mL. Mix thoroughly. Autoclave for 15 min at 15 psi pressure–121°C. Cool to 50°C.

Trace Elements Solution SL-6:
Composition per liter:

H_3BO_3	0.3g
$CoCl_2·6H_2O$	0.2g
$ZnSO_4·7H_2O$	0.10g
$MnCl_2·4H_2O$	0.03g
$Na_2MoO_4·H_2O$	0.03g
$NiCl_2·6H_2O$	0.02g
$CuCl_2·2H_2O$	0.01g

Preparation of Trace Elements Solution SL-6: Add components to distilled/deionized water and bring volume to 1.0L. Mix thoroughly. Adjust pH to 3.4.

Preparation of Medium: Add components, except Na_2CO_3 solution, to distilled/deionized water and bring volume to 900.0mL. Mix thoroughly. Gently heat and bring to boiling. Autoclave for 15 min at 15 psi pressure–121°C. Cool to 45–50°C. Aseptically add sterile Na_2CO_3 solution. Mix thoroughly. Adjust pH to 9.0, if necessary. Pour into sterile Petri dishes or distribute into sterile tubes.

Use: For the cultivation and maintenance of *Natronobacterium gregoryi, Natronobacterium magadii, Natronobacterium pharaonis* and *Natronococcus occultus*.

NBA Medium

Composition per liter:

Pancreatic digest of gelatin	5.0g
Casamino acids	5.0g
Beef extract	3.0g
Yeast extract	1.0g

pH 6.8 ± 0.2 at 25°C

Preparation of Medium: Add components to distilled/deionized water and bring volume to 1.0L. Mix thoroughly. Distribute into tubes or flasks. Autoclave for 15 min at 15 psi pressure–121°C.

Use: For the cultivation of *Bdellovibrio* species.

NBY Medium (Nutrient Broth Yeast Extract Medium) (ATCC Medium 763)

Composition per 940mL:

Agar	15.0g
Nutrient broth	8.0g
Yeast extract	2.0g
K_2HPO_4	2.0g
KH_2PO_4	0.5g
Glucose solution	50.0mL
$MgSO_4·7H_2O$ solution	50.0mL

Glucose Solution:
Composition per 50mL:

D-Glucose	5.0g

Preparation of Glucose Solution: Add glucose to distilled/deionized water and bring volume to 50.0mL. Mix thoroughly. Filter sterilize.

$MgSO_4$ Solution:
Composition per 50mL:

$MgSO_4·7H_2O$	0.25g

Preparation of $MgSO_4$ Solution: Add the solid $MgSO_4·7H_2O$ to distilled/deionized water and bring volume to 50.0mL. Mix thoroughly. Autoclave for 15 min at 15 psi pressure–121°C. Cool to 50°C.

Preparation of Medium: Add components, except glucose solution and $MgSO_4·7H_2O$ solution, to distilled/deionized water and bring volume to 900.0mL. Mix thoroughly. Gently heat and bring to boiling. Autoclave for 25 min at 15 psi pressure–121°C. Cool to 45–50°C. Aseptically add sterile glucose solution and $MgSO_4·7H_2O$ solution. Mix thoroughly. Pour into sterile Petri dishes or distribute into sterile tubes.

Use: For the cultivation and maintenance of *Bacillus sphaericus*.

NBY Medium (Nutrient Broth Yeast Extract Medium)

Composition per 950mL:

Nutrient broth, dehydrated	8.0g
Yeast extract	2.0g
K_2HPO_4	2.0g
KH_2PO_4	0.5g
Glucose solution	50.0mL
$MgSO_4·7H_2O$ (1*M* solution)	1.0mL

Glucose Solution:
Composition per 50mL:

D-Glucose	5.0g

Preparation of Glucose Solution: Add glucose to distilled/deionized water and bring volume to 50.0mL. Mix thoroughly. Filter sterilize.

Preparation of Medium: Add components, except glucose solution, to distilled/deionized water and bring volume to 950.0mL. Mix thoroughly. Gently heat and bring to boiling. Autoclave for 15 min at 15 psi pressure–121°C. Cool to 45–50°C.

Aseptically add sterile glucose solution. Mix thoroughly. Pour into sterile Petri dishes or distribute into sterile tubes.

Use: For the cultivation and maintenance of *Curtobacterium flaccumfaciens* and *Pseudomonas syringae.*

Neopeptone Glucose Agar

Composition per liter:

Agar	20.0g
Glucose	10.0g
Neopeptone	5.0g

pH 6.5 ± 0.2 at 25°C

Preparation of Medium: Add components to distilled/deionized water and bring volume to 1.0L. Mix thoroughly. Gently heat and bring to boiling. Distribute into tubes or flasks. Autoclave for 15 min at 15 psi pressure–121°C. Adjust pH to 6.5. Pour into sterile Petri dishes or leave in tubes.

Use: For the maintenance of stock cultures of a variety of microorganisms.

Neopeptone Glucose Rose Bengal Aureomycin® Agar

Composition per liter:

Agar	20.0g
Neopeptone	5.0g
Glucose	1.0g
Tetracycline solution	5.0mL
Rose bengal solution	3.5mL

pH 6.5 ± 0.2 at 25°C

Tetracycline Solution:

Composition per 150mL:

Tetracycline	1.0g

Preparation of Tetracycline Solution: Add tetracycline to distilled/deionized water and bring volume to 150.0mL. Mix thoroughly. Filter sterilize.

Rose Bengal Solution:

Composition per 100mL:

Rose bengal	1.0g

Preparation of Rose Bengal Solution: Add rose bengal to distilled/deionized water and bring volume to 100.0mL. Mix thoroughly. Filter sterilize.

Preparation of Medium: Add components, except tetracycline solution, to distilled/deionized water and bring volume to 995.0mL. Mix thoroughly. Gently heat and bring to boiling. Autoclave for 15 min at 15 psi pressure–121°C. Cool to 45–50°C. Aseptically add 5.0mL of sterile tetracycline solution. Mix thoroughly. Pour into sterile Petri dishes or distribute into sterile tubes.

Use: For the isolation and cultivation of a wide variety of fungal species.

Neurospora Medium

Composition per liter:

Agar	15.0g
Glucose	5.0g
Malt syrup, spray-dried	5.0g
Sucrose	5.0g
Yeast extract	2.5g
Vitamin solution	10.0mL
Casein, hydrolyzed	5.0mL

Vitamin Solution:

Composition per liter:

Ribonucleic acid, alkali hydrolyzed	0.50g
Inositol	0.40g
Choline	0.20g
Nicotinamide	0.20g
Pantothenic acid	0.20g
Thiamine	0.10g
p-Aminobenzoic acid	0.05g
Pyridoxine	0.05g
Riboflavin	0.05g
Folic acid	4.0μg

Preparation of Vitamin Solution: Add components to distilled/deionized water and bring volume to 1.0L. Mix thoroughly. Filter sterilize.

Preparation of Medium: Add components, except vitamin solution, to distilled/deionized water and bring volume to 990.0mL. Mix thoroughly. Gently heat and bring to boiling. Autoclave for 15 min at 15 psi pressure–121°C. Cool to 45–50°C. Aseptically add 10.0mL of sterile vitamin solution. Mix thoroughly. Pour into sterile Petri dishes or distribute into sterile tubes.

Use: For the cultivation of *Neurospora* species on complete medium.

Nitrate Agar

Composition per liter:

Agar	12.0g
Peptone	5.0g
Beef extract	3.0g
KNO_3	1.0g

pH 6.8 ± 0.2 at 25°C

Preparation of Medium: Add components to distilled/deionized water and bring volume to 1.0L. Mix thoroughly. Gently heat and bring to boiling. Distribute into tubes. Autoclave for 15 min at 15 psi pressure–121°C. Allow tubes to cool in a slanted position.

Use: For the differentiation of aerobic and facultative Gram-negative microorganisms based on their ability to reduce nitrate. Test for nitrates with sulfanilic acid and α-naphthylamine reagents. Bacteria that reduce nitrate to nitrite turn the reagents red or pink.

Nitrate Assimilation Medium, Auxanographic Method for Yeast Identification

Composition per liter:

Noble agar	20.00g
Glucose	10.0g
KH_2PO_4	1.0g
$MgSO_4 \cdot 7H_2O$	0.5g
NaCl	0.1g
$CaCl_2 \cdot 2H_2O$	0.1g
DL-Methionine	0.02g
DL-Tryptophan	0.02g
L-Histidine·HCl	0.01g
Inositol	2.0mg
H_3BO_3	0.5mg
$ZnSO_4 \cdot 7H_2O$	0.4mg
$MnSO_4 \cdot 4H_2O$	0.4mg
Thiamine·HCl	0.4mg
Pyridoxine	0.4mg
Niacin	0.4mg
Calcium pantothenate	0.4mg
p-Aminobenzoic acid	0.2mg
Riboflavin	0.2mg
$FeCl_3$	0.2mg
$Na_2MoO_4 \cdot 4H_2O$	0.2mg
KI	0.1mg
$CuSO_4 \cdot 5H_2O$	0.04mg
Folic acid	2.0μg
Biotin	2.0μg

pH 4.5 ± 0.2 at 25°C

Preparation of Medium: Add components to distilled/deionized water and bring volume to 1.0L. Mix thoroughly. Gently heat and bring to boiling. Distribute into screw-capped tubes in 20.0mL volumes. Autoclave for 15 min at 15 psi pressure–121°C.

Use: For nitrate assimilation tests by the auxanographic method.

Nitrate Broth (International Streptomyces Project Medium 8) (ISP Medium 8)

Composition per liter:

Peptone	5.0g
Beef extract	3.0g
KNO_3	1.0g

pH 7.0 ± 0.2 at 25°C

Source: This medium is available as a premixed powder from Difco Laboratories.

Preparation of Medium: Add components to distilled/deionized water and bring volume to 1.0L. Mix thoroughly. Distribute into tubes or flasks. Autoclave for 15 min at 15 psi pressure–121°C.

Use: For the differentiation of aerobic and facultative Gram-negative microorganisms based on their ability to reduce nitrate. Test for nitrates with sulfanilic acid and α-naphthylamine reagents. Bacteria that reduce nitrate to nitrite turn the reagents red or pink.

Nitrate Broth

Composition per liter:

Pancreatic digest of gelatin	20.0g
KNO_3	2.0g

pH 7.2 ± 0.2 at 25°C

Source: This medium is available as a premixed powder from BBL Microbiology Systems.

Preparation of Medium: Add components to distilled/deionized water and bring volume to 1.0L. Mix thoroughly. Distribute into tubes or flasks. Autoclave for 15 min at 15 psi pressure–121°C.

Use: For the differentiation of aerobic and facultative Gram-negative microorganisms based on their ability to reduce nitrate. Test for nitrates with sulfanilic acid and α-naphthylamine reagents. Bacteria that reduce nitrate to nitrite turn the reagents red or pink.

Nitrate Broth, Enriched

Composition per liter:

Pancreatic digest of casein	13.0g
NaCl	5.0g
Yeast extract	5.0g
Heart muscle, solids from infusion	2.0g
KNO_3	2.0g

pH 7.3 ± 0.2 at 25°C

Preparation of Medium: Add components to distilled/deionized water and bring volume to 1.0L. Mix thoroughly. Distribute into test tubes that contain an inverted Durham tube. Autoclave for 15 min at 15 psi pressure–121°C.

Use: For the differentiation of aerobic and facultative Gram-negative microorganisms based on their ability to reduce nitrate to nitrite or form N_2 gas. Test for nitrates with sulfanilic acid and α-naphthylamine reagents. Bacteria that reduce nitrate to nitrite turn the reagents red or pink.

Nitrate Liquid Medium

Composition per liter:

Solution A	500.0mL
Solution B	250.0mL
Solution C	250.0mL

Solution A:

Composition per 500mL:

Mannitol	10.0g
KNO_3	0.6g
$Na_2HPO_4 \cdot 12H_2O$	0.45g
Na_2SO_4	0.03g

Preparation of Solution A: Add components to distilled/deionized water and bring volume to 500.0mL. Mix thoroughly. Autoclave for 15 min at 15 psi pressure–121°C. Cool to 25°C.

Solution B:

Composition per 250mL:

$MgSO_4 \cdot 7H_2O$	0.12g
$CaCl_2 \cdot 6H_2O$	0.1g
$FeCl_3 \cdot 6H_2O$	0.01g

Preparation of Solution B: Add components to distilled/deionized water and bring volume to 250.0mL. Mix thoroughly. Autoclave for 15 min at 15 psi pressure–121°C. Cool to 25°C.

Solution C:

Composition per 250mL:

Calcium pantothenate	0.5mg
Thiamine·HCl	0.1mg
Biotin	0.5μg

Preparation of Solution C: Add components to distilled/deionized water and bring volume to 250.0mL. Mix thoroughly. Filter sterilize.

Preparation of Medium: Aseptically combine 500.0mL of cooled, sterile solution A, 250.0mL of cooled, sterile solution B, and 250.0mL of sterile solution C. Mix thoroughly. Aseptically distribute into sterile tubes or flasks.

Use: For the isolation and cultivation of *Rhizobium* species.

Nitrate Methanol Medium

Composition per liter:

$NaNO_3$	5.0g
K_2HPO_4	2.0g
NaCl	1.0g
$MgSO_4 \cdot 7H_2O$	0.02g
$Na_2MoO_4 \cdot H_2O$	1.0mg
Riboflavin	0.2mg
Calcium pantothenate	0.2mg
Pyridoxine·HCl	0.2mg
Nicotinic acid	0.2mg
Thiamine·HCl	0.1mg
p-Aminobenzoic acid	0.1mg
Biotin	0.01mg
Methanol	10.0mL

pH 7.0 ± 0.2 at 25°C

Preparation of Medium: Add components, except methanol, to distilled/deionized water and bring volume to 990.0mL. Mix thoroughly. Autoclave for 15 min at 15 psi pressure–121°C. Cool to 45–50°C. Filter sterilize methanol. Aseptically add sterile methanol to cooled sterile medium. Mix thoroughly. Aseptically distribute into sterile tubes or flasks.

Use: For the cultivation and maintenance of *Methylobacterium rhodinum.*

Nitrate Mineral Salts Medium (NMS Medium)

Composition per liter:

Noble agar	12.5g
$MgSO_4 \cdot 7H_2O$	1.0g
KNO_3	1.0g
$Na_2HPO_4 \cdot 12H_2O$	0.717g
KH_2PO_4	0.272g
$CaCl_2 \cdot 6H_2O$	0.20g
Ferric ammonium EDTA	4.0mg
Trace element solution	0.5mL

pH 6.8 ± 0.2 at 25°C

Trace Element Solution:

Composition per liter:

Disodium EDTA	0.5g
$FeSO_4 \cdot 7H_2O$	0.2g
H_3BO_3	0.030g
$CoCl_2 \cdot 6H_2O$	0.020g
$ZnSO_4 \cdot 7H_2O$	0.010g
$MnCl_2 \cdot 4H_2O$	3.0mg
$Na_2MoO_4 \cdot 2H_2O$	3.0mg
$NiCl_2 \cdot 6H_2O$	2.0mg
$CaCl_2 \cdot 2H_2O$	1.0mg

Preparation of Trace Element Solution: Add components to distilled/deionized water and bring volume to 1.0L. Mix thoroughly.

Preparation of Medium: Add components to distilled/deionized water and bring volume to 1.0L. Mix thoroughly. Gently heat and bring to boiling. Adjust pH to 6.8. Distribute into tubes or flasks. Autoclave for 15 min at 15 psi pressure–121°C. Pour into sterile Petri dishes or leave in tubes.

Use: For the cultivation and maintenance of *Methylobacterium* species, *Methylococcus capsulatus, Methylomonas agile* and *Methylomonas methanica.*

Nitrate Mineral Salts Medium with Methanol (NMS Medium with Methanol)

Composition per liter:

Noble agar	12.5g
$MgSO_4 \cdot 7H_2O$	1.0g
KNO_3	1.0g
$Na_2HPO_4 \cdot 12H_2O$	0.717g
KH_2PO_4	0.272g
$CaCl_2 \cdot 6H_2O$	0.20g
Ferric ammonium EDTA	4.0mg
Trace element solution	0.5mL
Methanol	1.0mL

pH 6.8 ± 0.2 at 25°C

Trace Element Solution:

Composition per liter:

Disodium EDTA	0.5g
$FeSO_4 \cdot 7H_2O$	0.2g
H_3BO_3	0.030g
$CoCl_2 \cdot 6H_2O$	0.020g
$ZnSO_4 \cdot 7H_2O$	0.010g
$MnCl_2 \cdot 4H_2O$	3.0mg
$Na_2MoO_4 \cdot 2H_2O$	3.0mg
$NiCl_2 \cdot 6H_2O$	2.0mg
$CaCl_2 \cdot 2H_2O$	1.0mg

Preparation of Trace Element Solution: Add components to distilled/deionized water and bring volume to 1.0L. Mix thoroughly.

Preparation of Medium: Add components, except methanol, to distilled/deionized water and bring volume to 999.0mL. Mix thoroughly. Gently heat and bring to boiling. Adjust pH to 6.8. Distribute into tubes or flasks. Autoclave for 15 min at 15 psi pressure–121°C. Cool to 45–50°C. Filter sterilize methanol. Aseptically add sterile methanol to cooled sterile medium. Mix thoroughly. Pour into sterile Petri dishes or leave in tubes.

Use: For the cultivation and maintenance of *Methylobacterium fujisawaense*, *Methylobacterium* species and *Methylomonas clara*.

Nitrate Reduction Broth

Composition per liter:

Pancreatic digest of gelatin	5.0g
Beef extract	3.0g
KNO_3	1.0g

pH 6.9 ± 0.2 at 25°C

Preparation of Medium: Add components to distilled/deionized water and bring volume to 1.0L. Mix thoroughly. Distribute into test tubes that contain an inverted Durham tube. Autoclave for 15 min at 15 psi pressure–121°C.

Use: For the differentiation of members of the Pseudomonadaceae based on their ability to reduce nitrate to nitrite or form N_2 gas. Test for nitrates with sulfanilic acid and α-naphthylamine reagents. Bacteria that reduce nitrate to nitrite turn the reagents red or pink.

Nitrate Reduction Broth

Composition per liter:

Pancreatic digest of casein	13.0g
NaCl	5.0g
Yeast extract	5.0g
Heart muscle, solids from infusion	2.0g
KNO_3 or $NaNO_3$	2.0g

pH 7.4 ± 0.2 at 25°C

Preparation of Medium: Add components to distilled/deionized water and bring volume to 1.0L. Mix thoroughly. Distribute into test tubes that contain an inverted Durham tube. Autoclave for 15 min at 15 psi pressure–121°C.

Use: For the differentiation of a variety of Gram-negative bacteria based on their ability to reduce nitrate to nitrite or form N_2 gas. Test for nitrates with sulfanilic acid and α-naphthylamine reagents. Bacteria that reduce nitrate to nitrite turn the reagents red or pink.

Nitrate Reduction Broth

Composition per liter:

Pancreatic digest of casein	13.0g
NaCl	5.0g
Yeast extract	5.0g
Heart muscle, solids from infusion	2.0g
KNO_3 or $NaNO_3$	2.0g

pH 7.4 ± 0.2 at 25°C

Preparation of Medium: Dispense in 3.0mL amounts into screw-cap tubes, add inverted vials and autoclave at 121°C for 15 min.

Use: For the differentiation of a variety of nonfermenting Gram-negative bacteria based on their ability to reduce nitrate to nitrite or form N_2 gas. Test for nitrates with sulfanilic acid and α-naphthylamine reagents. Bacteria that reduce nitrate to nitrite turn the reagents red or pink.

Nitrate Reduction Broth, Clark

Composition per liter:

Peptone	20.0g
KNO_3 or $NaNO_3$	2.0g

Preparation of Medium: Add components to distilled/deionized water and bring volume to 1.0L. Mix thoroughly. Distribute into test tubes that contain an inverted Durham tube. Autoclave for 15 min at 15 psi pressure–121°C.

Use: For the differentiation of a variety of Gram-negative bacteria based on their ability to reduce nitrate to nitrite or form N_2 gas. Test for nitrates with sulfanilic acid and α-naphthylamine reagents. Bacteria that reduce nitrate to nitrite turn the reagents red or pink.

Nitrate Reduction Broth for *Pseudomonas* and Related Genera

Composition per liter:

Peptone	5.0g
NaCl	5.0g
Yeast extract	2.0g
Beef extract	1.0g
$NaNO_3$	0.1g

pH 7.4 ± 0.2 at 25°C

Preparation of Medium: Add components to distilled/deionized water and bring volume to 1.0L. Mix thoroughly. Distribute into test tubes that contain an inverted Durham tube. Autoclave for 15 min at 15 psi pressure–121°C.

Use: For the differentiation of members of the Pseudomonadaceae based on their ability to reduce nitrate to nitrite or form N_2 gas. Test for nitrates with sulfanilic acid and α-naphthylamine reagents. Bacteria that reduce nitrate to nitrite turn the reagents red or pink.

Nitrobacter agilis Medium

Composition per liter:

$CaCO_3$	10.0g
NaCl	0.3g
Na_2CO_3	0.25g
KNO_2	0.17g
K_2HPO_4	0.14g
$MgSO_4·7H_2O$	0.14g
$FeSO_4·7H_2O$	0.03g
$MnSO_4·4H_2O$	0.01g
Biotin solution	10.0mL

Biotin Solution:

Composition per 10mL:

Biotin	0.15g

Preparation of Biotin Solution: Add biotin to distilled/deionized water and bring volume to 10.0mL. Mix thoroughly. Filter sterilize.

Preparation of Medium: Add Na_2CO_3 to distilled/deionized water and bring volume to 200.0mL. Mix thoroughly. In a separate flask add the remaining components, except the biotin solution, to distilled/deionized water and bring volume to 790.0mL. Autoclave the Na_2CO_3 solution and salts solution separately for 15 min at 15 psi pressure–121°C. Cool to 25°C. Aseptically combine the sterile Na_2CO_3 solution, sterile salts solution, and sterile biotin solution. Mix thoroughly. Aseptically distribute into sterile tubes or flasks.

Use: For the cultivation of *Nitrobacter agilis*.

Nitrobacter Medium 203

Composition per liter:

Solution A	0.5mL
Solution B	0.5mL
Solution C	1.0mL
Solution D	0.5mL
Solution E	0.5mL
Solution F	0.2mL

Solution A:

Composition per 100mL:

$CaCl_2$	2.0g

Preparation of Solution A: Add $CaCl_2$ to distilled/deionized water and bring volume to 100.0mL. Mix thoroughly.

Solution B:

Composition per 100mL:

$MgSO_4·7H_2O$	20.0g

Preparation of Solution B: Add $MgSO_4·7H_2O$ to distilled/deionized water and bring volume to 100.0mL. Mix thoroughly.

Solution C:

Composition per 100mL:

Chelated iron (Sequestrene)	0.1g

Preparation of Solution C: Add chelated iron to distilled/deionized water and bring volume to 100.0mL. Mix thoroughly.

Solution D:

Composition per liter:

$MnCl_2·4H_2O$	0.2g
$Na_2MoO_4·2H_2O$	0.1g
$ZnSO_4·7H_2O$	0.1g
$CuSO_4·5H_2O$	0.02g
$CoCl_2·6H_2O$	2.0mg

Preparation of Solution D: Add components to distilled/deionized water and bring volume to 1.0L. Mix thoroughly.

Solution E:

Composition per 100mL:

$NaNO_2$ 41.4g

Preparation of Solution E: Add $NaNO_2$ to distilled/deionized water and bring volume to 100.0mL. Mix thoroughly.

Solution F:

Composition per 100mL:

K_2HPO_4 .... 1.74g

Preparation of Solution F: Add K_2HPO_4 to distilled/deionized water and bring volume to 100.0mL. Mix thoroughly.

Preparation of Medium: Add the appropriate volumes of solutions A–F to distilled/deionized water and bring volume to 1.0L. Mix thoroughly. Distribute into tubes or flasks. Autoclave for 15 min at 15 psi pressure–121°C.

Use: For the cultivation and maintenance of *Nitrobacter* species and *Nitrobacter winogradskyi*.

Nitrobacter Medium 204

Composition per liter:

Seawater	700.0mL
Solution A	0.5mL
Solution B	0.5mL
Solution C	1.0mL
Solution D	0.5mL
Solution E	0.5mL
Solution F	0.2mL

Solution A:

Composition per 100mL:

$CaCl_2$ 2.0g

Preparation of Solution A: Add $CaCl_2$ to distilled/deionized water and bring volume to 100.0mL. Mix thoroughly.

Solution B:

Composition per 100mL:

$MgSO_4 \cdot 7H_2O$ 20.0g

Preparation of Solution B: Add $MgSO_4 \cdot 7H_2O$ to distilled/deionized water and bring volume to 100.0mL. Mix thoroughly.

Solution C:

Composition per 100mL:

Chelated iron (Sequestrene) 0.1g

Preparation of Solution C: Add chelated iron to distilled/deionized water and bring volume to 100.0mL. Mix thoroughly.

Solution D:

Composition per liter:

$MnCl_2 \cdot 4H_2O$	0.2g
$Na_2MoO_4 \cdot 2H_2O$	0.1g
$ZnSO_4 \cdot 7H_2O$	0.1g
$CuSO_4 \cdot 5H_2O$	0.02g
$CoCl_2 \cdot 6H_2O$	2.0mg

Preparation of Solution D: Add components to distilled/deionized water and bring volume to 1.0L. Mix thoroughly.

Solution E:

Composition per 100mL:

$NaNO_2$ 41.4g

Preparation of Solution E: Add $NaNO_2$ to distilled/deionized water and bring volume to 100.0mL. Mix thoroughly.

Solution F:

Composition per 100mL:

K_2HPO_4 .... 1.74g

Preparation of Solution F: Add K_2HPO_4 to distilled/deionized water and bring volume to 100.0mL. Mix thoroughly.

Preparation of Medium: Add the appropriate volumes of solutions A–F and seawater to distilled/deionized water and bring volume to 1.0L. Mix thoroughly. Distribute into tubes or flasks. Autoclave for 15 min at 15 psi pressure–121°C.

Use: For the cultivation and maintenance of *Nitrococcus mobilis*.

Nitrobacter Medium B

Composition per liter:

$NaNO_2$	1.0g
K_2HPO_4	0.5g
$MgSO_4$	0.5g
NaCl	0.3g
$Fe_2(SO_4)_3$	5.0mg
$MnSO_4$	2.0mg
Marble chips	as needed

pH 7.5 ± 0.2 at 25°C

Preparation of Medium: Add components, except marble chips, to distilled/deionized water and bring volume to 1.0L. Mix thoroughly. Autoclave for 15 min at 15 psi pressure–121°C. Cool to 25°C. Wash marble chips in distilled/deionized water. Put a few chips into test tubes. Autoclave for 60 min at 15 psi pressure–121°C. Cool to 25°C. Aseptically distribute cooled sterile medium into test tubes to cover marble chips.

Use: For the cultivation of *Nitrobacter* species.

Nitrogen-Fixing Hydrocarbon Oxidizers Medium

Composition per liter:

Na_2HPO_4	0.3g
KH_2PO_4	0.2g
$MgSO_4 \cdot 7H_2O$	0.1g
$FeSO_4 \cdot 7H_2O$	5.0mg
$Na_2MoO_4 \cdot 2H_2O$	2.0mg

Preparation of Medium: Add components to distilled/deionized water and bring volume to 1.0L. Mix thoroughly. Distribute into tubes or flasks. Autoclave for 15 min at 15 psi pressure–121°C.

Use: For the cultivation and enrichment of nitrogen-fixing hydrocarbon-oxidizing bacteria.

Nitrogen–Fixing Marine Medium

Composition per liter:

Noble agar	10.0g
$MgSO_4 \cdot 7H_2O$	0.04g
$CaCl_2 \cdot 2H_2O$	0.02g
$K_2HPO_4 \cdot 3H_2O$	0.02g
Na_2CO_3	0.02g
Citric acid	3.0mg
Ferric ammonium citrate	3.0mg
Disodium potassium EDTA	0.5mg
Seawater	750.0mL
Trace Metals A-5	1.0mL

pH 8.5 ± 0.2 at 25°C

Trace Metals A-5 Mix:

Composition per liter:

H_3BO_3	2.86g
$MnCl_2 \cdot 4H_2O$	1.81g
$ZnSO_4 \cdot 7H_2O$	0.222g
$CuSO_4 \cdot 5H_2O$	0.079g
$Na_2MoO_4 \cdot 2H_2O$	0.039g
$Co(NO_3)_2 \cdot 6H_2O$	0.050g

Preparation of Trace Metals A-5: Add components to distilled/deionized water and bring volume to 1.0L. Mix thoroughly.

Preparation of Medium: Add components to glass distilled water and bring volume to 1.0L. Mix thoroughly. Gently heat and bring to boiling. Autoclave for 15 min at 15 psi pressure–121°C. Adjust pH to 8.5 with KOH. Pour into sterile Petri dishes or distribute into sterile tubes.

Use: For the cultivation and maintenance of *Anabaena* species.

Nitrosococcus Medium

Composition per liter:

$(NH_4)_2SO_4$	1.32g
$MgSO_4 \cdot 7H_2O$	0.38g
$CaCl_2 \cdot 2H_2O$	0.020g
K_2HPO_4	8.7mg
Chelated iron	1.0mg
$MnCl_2 \cdot 4H_2O$	0.2mg
$Na_2MoO_4 \cdot 2H_2O$	0.1mg
$ZnSO_4 \cdot 7H_2O$	0.1mg
$CoCl_2 \cdot 6H_2O$	2.0μg
Phenol red (0.04% solution)	3.25mL

pH 7.5–7.8 at 25°C

Preparation of Medium: Add components to filtered seawater and bring volume to 1.0L. Mix thoroughly. Adjust pH to 7.5–7.8 with 1*N* HCl. Distribute into tubes. Autoclave for 15 min at 15 psi pressure–121°C.

Use: For the cultivation of *Nitrosococcus oceanus.*

Nitrosolobus Medium (ATCC Medium 438)

Composition per liter:

$(NH_4)_2SO_4$	1.65g
$MgSO_4 \cdot 7H_2O$	0.2g
K_2HPO_4	0.087g
$CaCl_2 \cdot 2H_2O$	0.02g
Phenol Red	5.0mg
Disodium EDTA	1.0mg
$MnCl_2 \cdot 4H_2O$	0.2mg
$Na_2MoO_4 \cdot 2H_2O$	0.1mg
$ZnSO_4 \cdot 7H_2O$	0.1mg
$CuSO_4 \cdot 5H_2O$	0.02mg
$CoCl_2 \cdot 6H_2O$	2.0μg

pH 7.5 ± 0.2 at 25°C

Preparation of Medium: Add components to distilled/deionized water and bring volume to 1.0L. Mix thoroughly. Adjust pH to 7.5 with 0.1*M* K_2CO_3. Distribute into tubes or flasks. Autoclave for 15 min at 15 psi pressure–121°C.

Use: For the cultivation and maintenance of *Nitrosolobus multiformis.*

Nitrosolobus Medium (ATCC Medium 929)

Composition per liter:

$(NH_4)_2SO_4$	1.32g
$MgSO_4 \cdot 7H_2O$	0.38g
K_2HPO_4	0.087g
$CaCl_2 \cdot 2H_2O$	0.020g
Chelated iron	1.0mg
$MnCl_2 \cdot 4H_2O$	0.2mg

$Na_2MoO_4 \cdot 2H_2O$	0.1mg
$ZnSO_4 \cdot 7H_2O$	0.1mg
$CoCl_2 \cdot 6H_2O$	2.0μg
Phenol red (0.5% solution)	0.25mL

pH 7.5 ± 0.2 at 25°C

Preparation of Medium: Add components to distilled/deionized water and bring volume to 1.0L. Mix thoroughly. Adjust pH to 7.5 with 0.1M K_2CO_3. Distribute into tubes or flasks. Autoclave for 15 min at 15 psi pressure–121°C.

Use: For the cultivation and maintenance of *Nitrosolobus multiformis.*

Nitrosomonas europaea Medium

Composition per liter:

$(NH_4)_2SO_4$	1.7g
$MgSO_4 \cdot 7H_2O$	0.2g
$CaCl_2 \cdot 2H_2O$	0.020g
K_2HPO_4	0.015g
Ferric EDTA	1.0mg
Trace elements solution	1.0mL

pH 7.5 ± 0.2 at 25°C

Trace Elements Solution:
Composition per 100mL:

$MnCl_2 \cdot 4H_2O$	0.020g
$Na_2MoO_4 \cdot 2H_2O$	0.010g
$ZnSO_4 \cdot 7H_2O$	0.010g
$CuSO_4 \cdot 5H_2O$	2.0mg
$CoCl_2 \cdot 6H_2O$	0.2mg

Preparation of Trace Elements Solution: Add components to distilled/deionized water and bring volume to 1.0L. Mix thoroughly.

Preparation of Medium: Add components to distilled/deionized water and bring volume to 1.0L. Mix thoroughly. Adjust pH to 7.5 with K_2CO_3. Distribute into tubes or flasks. Autoclave for 15 min at 15 psi pressure–121°C. After inoculation, maintain pH at 7.5–7.8 with sterile 50% K_2CO_3 solution.

Use: For the cultivation and maintenance of *Nitrosomonas europaea.*

Nitrosomonas Medium

Composition per liter:

$(NH_4)_2SO_4$	3.0g
K_2HPO_4	0.5g
$MgSO_4 \cdot 7H_2O$	0.05g
$CaCl_2 \cdot 2H_2O$	4.0mg
Cresol red (0.0005% solution)	25.0mL
Ferric EDTA solution	0.1mL

pH 8.2–8.4 at 25°C

Ferric EDTA Solution:
Composition per 100mL:

$FeSO_4 \cdot 7H_2O$	0.50g
Disodium EDTA	0.14g
H_2SO_4, concentrated	0.05mL

Preparation of Ferric EDTA Solution: Add components to distilled/deionized water and bring volume to 100.0mL. Mix thoroughly.

Preparation of Medium: Add $CaCl_2 \cdot 2H_2O$ and $MgSO_4 \cdot 7H_2O$ to distilled/deionized water and bring volume to 500.0mL. Mix thoroughly. In a separate flask, add remaining components to distilled/deionized water and bring volume to 500.0mL. Mix thoroughly. Autoclave both solutions separately for 15 min at 15 psi pressure–121°C. Cool to 25°C. Aseptically combine the two sterile solutions. Mix thoroughly. Aseptically distribute into sterile tubes or flasks. After inoculation, maintain pH at 8.2–8.4 with sterile 50% K_2CO_3 solution.

Use: For the cultivation and maintenance of *Nitrosomonas europaea.*

Nocardia Medium

Composition per liter:

Agar	20.0g
Peptone	10.0g
Beef extract	5.0g
NaCl	2.5g

Preparation of Medium: Add components to distilled/deionized water and bring volume to 1.0L. Mix thoroughly. Gently heat and bring to boiling. Distribute into tubes or flasks. Autoclave for 15 min at 15 psi pressure–121°C. Pour into sterile Petri dishes or leave in tubes.

Use: For the cultivation and maintenance of *Rhodococcus globerulus* and *Nocardia* species.

Nocardia Medium 1

Composition per 1010mL:

Agar	12.0g
Proteose peptone	10.0g
Veal infusion solids	10.0g
NaCl	3.0g
Na_2HPO_4	2.0g
Glucose	2.0g
Sodium acetate	1.0g
Adenine sulfate	0.01g
Guanine·HCl	0.01g
Uracil	0.01g
Xanthine	0.01g

Thiamine 0.02mg
Additives solution 10.0mL

pH 7.4 ± 0.2 at 25°C

Additives Solution:

Composition per 10mL:

Actidione (cycloheximide) 0.05mg
Mycostatin 0.05mg
Dimethylchlortetracycline·HCl 5.0μg

Preparation of Additives Solution: Add components to distilled/deionized water and bring volume to 10.0mL. Mix thoroughly. Filter sterilize.

Preparation of Medium: Add components, except additives solution, to distilled/deionized water and bring volume to 990.0mL. Mix thoroughly. Gently heat and bring to boiling. Autoclave for 15 min at 15 psi pressure–121°C. Cool to 45–50°C. Aseptically add sterile components. Mix thoroughly. Pour into sterile Petri dishes or distribute into sterile tubes.

Use: For the isolation and cultivation of *Nocardia.*

Nocardia Medium 2

Composition per 1010mL:

Agar 12.0g
Proteose peptone 10.0g
Veal infusion solids 10.0g
NaCl 3.0g
Na_2HPO_4 .... 2.0g
Glucose 2.0g
Sodium acetate 1.0g
Adenine sulfate 0.01g
Guanine·HCl 0.01g
Uracil 0.01g
Xanthine 0.01g
Thiamine 0.02mg
Additives solution 10.0mL

pH 7.4 ± 0.2 at 25°C

Additives Solution:

Composition per 10mL:

Actidione (cycloheximide) 0.05mg
Mycostatin 0.05mg
Methacycline HCl 0.01mg

Preparation of Additives Solution: Add components to distilled/deionized water and bring volume to 10.0mL. Mix thoroughly. Filter sterilize.

Preparation of Medium: Add components, except additives solution, to distilled/deionized water and bring volume to 990.0mL. Mix thoroughly. Gently heat and bring to boiling. Autoclave for 15 min at 15 psi pressure–121°C. Cool to 45–50°C. Aseptically add sterile components. Mix thoroughly. Pour into sterile Petri dishes or distribute into sterile tubes.

Use: For the isolation and cultivation of *Nocardia.*

Nocardia Medium 3

Composition per 1010mL:

Agar 12.0g
Proteose peptone 10.0g
Veal infusion solids 10.0g
NaCl 3.0g
Na_2HPO_4 .... 2.0g
Glucose 2.0g
Sodium acetate 1.0g
Adenine sulfate 0.01g
Guanine·HCl 0.01g
Uracil 0.01g
Xanthine 0.01g
Thiamine 0.02mg
Actidione 0.05mg
Mycostatin 0.05mg
Chlortetracycline HCl 0.045mg
Demethylchlortetracycline HCl 5.0μg
Additives solution 10.0mL

pH 7.4 ± 0.2 at 25°C

Additives Solution:

Composition per 10mL:

Actidione (cycloheximide) 0.05mg
Mycostatin 0.05mg
Dimethylchlortetracycline·HCl 5.0μg

Preparation of Additives Solution: Add components to distilled/deionized water and bring volume to 10.0mL. Mix thoroughly. Filter sterilize.

Preparation of Medium: Add components, except additives solution, to distilled/deionized water and bring volume to 990.0mL. Mix thoroughly. Gently heat and bring to boiling. Autoclave for 15 min at 15 psi pressure–121°C. Cool to 45–50°C. Aseptically add sterile components. Mix thoroughly. Pour into sterile Petri dishes or distribute into sterile tubes.

Use: For the isolation and cultivation of *Nocardia.*

Nocardia Medium 4

Composition per 1010mL:

Agar 12.0g
Proteose peptone 10.0g
Veal infusion solids 10.0g
NaCl 3.0g
Na_2HPO_4 .... 2.0g
Glucose 2.0g
Sodium acetate 1.0g
Adenine sulfate 0.01g
Guanine·HCl 0.01g
Uracil 0.01g
Xanthine 0.01g
Thiamine 0.02mg
Additives solution 10.0mL

pH 7.4 ± 0.2 at 25°C

Additives Solution:

Composition per 10mL:

Actidione (cycloheximide)	0.05mg
Mycostatin	0.05mg
Chlortetracycline·HCl	0.045mg
Methacycline·HCl	0.01mg

Preparation of Additives Solution: Add components to distilled/deionized water and bring volume to 10.0mL. Mix thoroughly. Filter sterilize.

Preparation of Medium: Add components, except additives solution, to distilled/deionized water and bring volume to 990.0mL. Mix thoroughly. Gently heat and bring to boiling. Autoclave for 15 min at 15 psi pressure–121°C. Cool to 45–50°C. Aseptically add sterile components. Mix thoroughly. Pour into sterile Petri dishes or distribute into sterile tubes.

Use: For the isolation and cultivation of *Nocardia* species.

NSMP, Modified

Composition per liter:

Casamino acids	5.0g
Glucose	2.0g
KH_2PO_4	0.86g
Sodium citrate	0.6g
K_2HPO_4	0.55g
$MgCl_2 \cdot 6H_2O$	0.43g
$CaCl_2$	0.1g
$MnCl_2 \cdot 4H_2O$	0.016g
$ZnCl_2$	7.0mg
$FeCl_3$	3.0mg

pH 6.5 ± 0.2 at 25°C

Preparation of Medium: Add components to distilled/deionized water and bring volume to 1.0L. Mix thoroughly. Distribute into tubes or flasks. Autoclave for 15 min at 15 psi pressure–121°C.

Use: For the cultivation and maintenance of *Bacillus thuringiensis*.

Nutrient Agar

Composition per liter:

Agar	15.0g
Peptone	5.0g
NaCl	5.0g
Yeast extract	2.0g
Beef extract	1.0g

pH 7.4 ± 0.2 at 25°C

Source: This medium is available as a premixed powder from Oxoid Unipath.

Preparation of Medium: Add components to distilled/deionized water and bring volume to 1.0L. Mix thoroughly. Gently heat and bring to boiling. Distribute into tubes or flasks. Autoclave for 15 min at 15 psi pressure–121°C. Pour into sterile Petri dishes or leave in tubes.

Use: For the cultivation and maintenance of a wide variety of microorganisms.

Nutrient Agar (ATCC Medium 3)

Composition per liter:

Agar	15.0g
Pancreatic digest of gelatin	5.0g
Beef extract	3.0g

pH 6.8 ± 0.2 at 25°C

Source: This medium is available as a premixed powder from BBL Microbiology Systems and Difco Laboratories.

Preparation of Medium: Add components to distilled/deionized water and bring volume to 1.0L. Mix thoroughly. Gently heat while stirring and bring to boiling. Distribute into tubes or flasks. Autoclave for 15 min at 15 psi pressure–121°C. Pour into sterile Petri dishes or leave in tubes.

Use: For the cultivation of a wide variety of bacteria and for the enumeration of organisms in water, sewage, feces and other materials.

Nutrient Agar, 1.5% (ATCC Medium 105)

Composition per liter:

Agar	15.0g
NaCl	8.0g
Pancreatic digest of gelatin	5.0g
Beef extract	3.0g

pH 7.3 ± 0.2 at 25°C

Source: This medium is available as a premixed powder from BBL Microbiology Systems and Difco Laboratories.

Preparation of Medium: Add components to distilled/deionized water and bring volume to 1.0L. Mix thoroughly. Gently heat while stirring and bring to boiling. Distribute into tubes or flasks. Autoclave for 15 min at 15 psi pressure–121°C. Pour into sterile Petri dishes or leave in tubes.

Use: For the cultivation and maintenance of a variety of nonfastidious bacteria.

Nutrient Agar, pH 6.0

Composition per liter:

Agar	15.0g
Peptone	5.0g
Beef extract	3.0g

pH 6.0 ± 0.2 at 25°C

Source: This medium is available as a premixed powder from Difco Laboratories.

Preparation of Medium: Add components to distilled/deionized water and bring volume to 1.0L. Mix thoroughly. Gently heat and bring to boiling. Adjust pH to 6.0. Distribute into tubes or flasks. Autoclave for 15 min at 15 psi pressure–121°C. Pour into sterile Petri dishes or leave in tubes.

Use: For the cultivation of microorganisms that prefer a slightly acid nutrient agar.

Nutrient Agar, pH 8.0

Composition per liter:

Agar	15.0g
Pancreatic digest of gelatin	5.0g
Beef extract	3.0g

pH 8.0 ± 0.2 at 25°C.

Preparation of Medium: Add components to distilled/deionized water and bring volume to 1.0L. Mix thoroughly. Gently heat and bring to boiling. Adjust pH to 8.0. Distribute into tubes or flasks. Autoclave for 15 min at 15 psi pressure–121°C. Pour into sterile Petri dishes or leave in tubes. Allow tubes to cool in a slanted position.

Use: For the cultivation and maintenance of *Bacillus alcalophilus*.

Nutrient Agar with 1% Methanol (ATCC Medium 620)

Composition per liter:

Agar	15.0g
Pancreatic digest of gelatin	5.0g
Beef extract	3.0g
Methanol	10.0mL

pH 6.8 ± 0.2 at 25°C

Source: Nutrient Agar is available as a premixed powder from Difco Laboratories.

Preparation of Medium: Filter sterilize methanol. Add components, except methanol, to distilled/deionized water and bring volume to 990.0mL. Mix thoroughly. Gently heat and bring to boiling. Autoclave for 15 min at 15 psi pressure–121°C. Cool to 45–50°C. Aseptically add sterile methanol. Mix thoroughly. Pour into sterile Petri dishes or distribute into sterile tubes.

Use: For the cultivation and maintenance of *Bacillus* species, *Methylomonas clara*, and *Pseudomonas methanolica*.

Nutrient Agar with 0.5% NaCl

Composition per liter:

Agar	15.0g
NaCl	5.0g
Pancreatic digest of gelatin	5.0g
Beef extract	3.0g

pH 6.8 ± 0.2 at 25°C

Source: Nutrient Agar is available as a premixed powder from Difco Laboratories.

Preparation of Medium: Add components to distilled/deionized water and bring volume to 1.0L. Mix thoroughly. Gently heat and bring to boiling. Distribute into tubes or flasks. Autoclave for 15 min at 15 psi pressure–121°C. Pour into sterile Petri dishes or leave in tubes.

Use: For the cultivation and maintenance of *Agrobacterium tumefaciens*, *Escherichia coli*, *Pseudomonas aeruginosa*, *Salmonella choleraesuis*, *Shigella dysenteriae*, *S. flexneri*, *Vibrio* species, and *Yersinia* species.

Nutrient Agar with 1.5% NaCl

Composition per liter:

NaCl	15.0g
Agar	15.0g
Pancreatic digest of gelatin	5.0g
Beef extract	3.0g

pH 6.8 ± 0.2 at 25°C

Source: Nutrient Agar is available as a premixed powder from Difco Laboratories.

Preparation of Medium: Add components to distilled/deionized water and bring volume to 1.0L. Mix thoroughly. Gently heat and bring to boiling. Distribute into tubes or flasks. Autoclave for 15 min at 15 psi pressure–121°C. Pour into sterile Petri dishes or leave in tubes.

Use: For the cultivation and maintenance of *Photobacterium leiognathi*, *Pseudomonas fluorescens*, and *Vibrio natriegens*.

Nutrient Agar with 3% NaCl

Composition per liter:

NaCl	30.0g
Agar	15.0g
Pancreatic digest of gelatin	5.0g
Beef extract	3.0g

pH 6.8 ± 0.2 at 25°C

Source: Nutrient Agar is available as a premixed powder from Difco Laboratories.

Preparation of Medium: Add components to distilled/deionized water and bring volume to 1.0L. Mix thoroughly. Gently heat and bring to boiling. Distribute into tubes or flasks. Autoclave for 15 min at 15 psi pressure–121°C. Pour into sterile Petri dishes or leave in tubes.

Use: For the cultivation and maintenance of *Bacillus* species, *Alteromonas nigrifaciens*, *Halococcus* species, *Planococcus citreus*, *Pseudomonas beijerinckii*, *Staphylococcus* species, *Streptococcus pyogenes* and *Vibrio* species.

Nutrient Agar with 10% NaCl

Composition per liter:

NaCl	100.0g
Agar	15.0g
Pancreatic digest of gelatin	5.0g
Beef extract	3.0g

pH 6.8 ± 0.2 at 25°C

Source: Nutrient Agar is available as a premixed powder from Difco Laboratories.

Preparation of Medium: Add components to distilled/deionized water and bring volume to 1.0L. Mix thoroughly. Gently heat and bring to boiling. Distribute into tubes or flasks. Autoclave for 15 min at 15 psi pressure–121°C. Pour into sterile Petri dishes or leave in tubes.

Use: For the cultivation and maintenance of *Paracoccus halodenitrificans* and *Micrococcus* species.

Nutrient Agar with 10% NaCl and Maltose

Composition per liter:

NaCl	100.0g
Agar	15.0g
Maltose	10.g
Pancreatic digest of gelatin	5.0g
Beef extract	3.0g

pH 6.8 ± 0.2 at 25°C

Source: Nutrient Agar is available as a premixed powder from Difco Laboratories.

Preparation of Medium: Add components to distilled/deionized water and bring volume to 1.0L. Mix thoroughly. Gently heat and bring to boiling. Distribute into tubes or flasks. Autoclave for 15 min at 15 psi pressure–121°C. Pour into sterile Petri dishes or leave in tubes.

Use: For the cultivation and maintenance of *Paracoccus halodenitrificans* and *Micrococcus* species.

Nutrient Agar with Dihydrostreptomycin

Composition per liter:

Agar	15.0g
Pancreatic digest of gelatin	5.0g
Beef extract	3.0g
Dihydrostreptomycin solution	10.0mL

pH 6.8 ± 0.2 at 25°C

Source: Nutrient Agar is available as a premixed powder from Difco Laboratories.

Dihydrostreptomycin Solution:

Composition per 10mL:

Dihydrostreptomycin	0.625g

Preparation of Dihydrostreptomycin Solution: Add dihydrostreptomycin to distilled/deionized water and bring volume to 10.0mL. Mix thoroughly. Filter sterilize.

Preparation of Medium: Add components, except dihydrostreptomycin solution, to distilled/deionized water and bring volume to 990.0mL. Mix thoroughly. Gently heat and bring to boiling. Autoclave for 15 min at 15 psi pressure–121°C. Cool to 45–50°C. Aseptically add sterile dihydrostreptomycin solution. Mix thoroughly. Pour into sterile Petri dishes or distribute into sterile tubes.

Use: For the cultivation and maintenance of *Escherichia coli*, *Micrococcus luteus*, *Shigella* species and *Vibrio cholerae*.

Nutrient Agar with Ethylene Glycol

Composition per liter:

Agar	15.0g
Pancreatic digest of gelatin	5.0g
Beef extract	3.0g
Ethylene glycol	2.0mL

pH 6.8 ± 0.2 at 25°C

Source: Nutrient Agar is available as a premixed powder from Difco Laboratories.

Preparation of Medium: Add components to distilled/deionized water and bring volume to 1.0L. Mix thoroughly. Gently heat and bring to boiling. Distribute into tubes or flasks. Autoclave for 15 min at 15 psi pressure–121°C. Pour into sterile Petri dishes or leave in tubes.

Use: For the cultivation and maintenance of *Pseudomonas putida.*

Nutrient Agar with Glucose

Composition per liter:

Agar	15.0g
Pancreatic digest of gelatin	5.0g
Beef extract	3.0g
Glucose	10.0g

pH 6.8 ± 0.2 at 25°C

Source: Nutrient Agar is available as a premixed powder from Difco Laboratories.

Preparation of Medium: Add components to distilled/deionized water and bring volume to 1.0L. Mix thoroughly. Gently heat and bring to boiling. Distribute into tubes or flasks. Autoclave for 15 min at 15 psi pressure–121°C. Pour into sterile Petri dishes or leave in tubes.

Use: For the cultivation and maintenance of *Amycolata saturnea, Arthrobacter* species, *Corynebacterium* species, *Curtobacterium flaccumfaciens, Deinococcus radiodurans, Escherichia coli, Hafnia alvei, Micrococcus aurantiacus, Myxomicrobium multiplex, Nocardia petroleophila, Nocardia* species, *Pseudomonas* species, *Rhodococcus rhodochrous, Streptomyces piedadensis* and *Xanthomonas* species.

Nutrient Agar with Horse Serum

Composition per liter:

Agar	15.0g
Pancreatic digest of gelatin	5.0g
Beef extract	3.0g
Horse serum	100.0mL

pH 6.8 ± 0.2 at 25°C

Source: Nutrient Agar is available as a premixed powder from Difco Laboratories.

Preparation of Medium: Add components, except horse serum, to distilled/deionized water and bring volume to 900.0mL. Mix thoroughly. Gently heat and bring to boiling. Autoclave for 15 min at 15 psi pressure–121°C. Cool to 45–50°C. Aseptically add sterile horse serum. Mix thoroughly. Pour into sterile Petri dishes or distribute into sterile tubes.

Use: For the cultivation and maintenance of *Alysiella filiformis* and *Simonsiella crassa.*

Nutrient Agar with Phytone

Composition per liter:

Agar	15.0g
Phytone	10.0g
Pancreatic digest of gelatin	5.0g
Beef extract	3.0g

pH 6.8 ± 0.2 at 25°C

Preparation of Medium: Add components to distilled/deionized water and bring volume to 1.0L. Mix thoroughly. Gently heat while stirring and bring to boiling. Distribute into tubes or flasks. Autoclave for 15 min at 15 psi pressure–121°C. Pour into sterile Petri dishes or leave in tubes.

Use: For the cultivation of a wide variety of bacteria.

Nutrient Agar with Potato Starch

Composition per liter:

Agar	15.0g
Potato starch	10.0g
Pancreatic digest of gelatin	5.0g
Beef extract	3.0g

pH 6.8 ± 0.2 at 25°C

Source: Nutrient Agar is available as a premixed powder from Difco Laboratories.

Preparation of Medium: Add components to distilled/deionized water and bring volume to 1.0L. Mix thoroughly. Gently heat while stirring and bring to boiling. Distribute into tubes or flasks. Autoclave for 15 min at 15 psi pressure–121°C. Pour into sterile Petri dishes or leave in tubes.

Use: For the cultivation and maintenance of *Bacillus polymyxa* and *B. subtilis.*

Nutrient Agar with Soil Extract

Composition per liter:

Agar	15.0g
Pancreatic digest of gelatin	5.0g
Beef extract	3.0g
Soil extract	250.0mL

pH 6.8 ± 0.2 at 25°C

Source: Nutrient Agar is available as a premixed powder from Difco Laboratories.

Soil Extract:

Composition per 300mL:

African violet soil 115.5g
Na_2CO_3 .... 0.3g

Preparation of Soil Extract: Add components to tap water and bring volume to 300.0mL. Autoclave for 60 min at 15 psi pressure–121°C. Filter through Whatman filter paper.

Preparation of Medium: Add components to distilled/deionized water and bring volume to 1.0L. Mix thoroughly. Gently heat and bring to boiling. Distribute into tubes or flasks. Autoclave for 15 min at 15 psi pressure–121°C. Pour into sterile Petri dishes.

Use: For the cultivation and maintenance of *Auerobacterium* species, *Bacillus* species, and *Saccharomonospora viridis*.

Nutrient Agar with Streptomycin

Composition per liter:

Agar 15.0g
Peptone 5.0g
Beef extract 3.0g
Streptomycin solution 10.0mL

pH 6.8 ± 0.2 at 25°C.

Streptomycin Solution:

Composition per 10mL:

Streptomycin 0.020g

Preparation of Streptomycin Solution: Add streptomycin to distilled/deionized water and bring volume to 10.0mL. Mix thoroughly. Filter sterilize.

Preparation of Medium: Add components, except streptomycin solution, to distilled/deionized water and bring volume to 990.0mL. Mix thoroughly. Gently heat and bring to boiling. Autoclave for 15 min at 15 psi pressure–121°C. Cool to 45–50°C. Aseptically add sterile streptomycin solution. Mix thoroughly. Pour into sterile Petri dishes or distribute into sterile tubes.

Use: For the cultivation and maintenance of *Corynebacterium glutamicum* and *C. herculis*.

Nutrient Agar with Sucrose

Composition per liter:

Sucrose 20.0g
Agar 15.0g
Pancreatic digest of gelatin 5.0g
Beef extract 3.0g

pH 6.8 ± 0.2 at 25°C

Source: Nutrient Agar is available as a premixed powder from Difco Laboratories.

Preparation of Medium: Add components to distilled/deionized water and bring volume to 1.0L. Mix thoroughly. Gently heat and bring to boiling. Distribute into tubes or flasks. Autoclave for 15 min at 15 psi pressure–121°C. Pour into sterile Petri dishes or leave in tubes.

Use: For the cultivation and maintenance of the *Pseudomonas* species.

Nutrient Agar with V-8™ Juice

Composition per liter:

Agar 15.0g
Pancreatic digest of gelatin 5.0g
Beef extract 3.0g
V-8™ Juice 200.0mL

pH 6.8 ± 0.2 at 25°C

Source: Nutrient Agar is available as a premixed powder from Difco Laboratories.

Preparation of Medium: Add components to distilled/deionized water and bring volume to 1.0L. Mix thoroughly. Gently heat and bring to boiling. Distribute into tubes or flasks. Autoclave for 15 min at 15 psi pressure–121°C. Pour into sterile Petri dishes or leave in tubes.

Use: For the cultivation and maintenance of *Pseudomonas tolaasii*.

Nutrient Broth

Composition per liter:

Peptone 5.0g
NaCl 5.0g
Yeast extract 2.0g
Beef extract 1.0g

pH 7.4 ± 0.2 at 25°C

Source: This medium is available as a premixed powder from Oxoid Unipath.

Preparation of Medium: Add components to distilled/deionized water and bring volume to 1.0L. Mix thoroughly. Distribute into tubes or flasks. Autoclave for 15 min at 15 psi pressure–121°C.

Use: For the cultivation of a wide variety of nonfastidious microorganisms.

Nutrient Broth

Composition per liter:

Pancreatic digest of gelatin 5.0g
Beef extract 3.0g

pH 6.9 ± 0.2 at 25°C

Source: This medium is available as a premixed powder from BBL Microbiology Systems and Difco Laboratories.

Preparation of Medium: Add components to distilled/deionized water and bring volume to 1.0L. Mix thoroughly. Distribute into tubes or flasks. Autoclave for 15 min at 15 psi pressure–121°C.

Use: For the cultivation of a wide variety of nonfastidious microorganisms.

Nutrient Broth Glycerol Medium

Composition per liter:

Glycerol	100.0g
Pancreatic digest of gelatin	5.0g
Beef extract	3.0g

pH 7.2–7.4 ± 0.2 at 25°C

Preparation of Medium: Add components to distilled/deionized water and bring volume to 1.0L. Mix thoroughly. Distribute into tubes or flasks. Autoclave for 15 min at 15 psi pressure–121°C.

Use: For the cultivation of *Pseudomonas* species.

Nutrient Broth, 1/2 Strength

Composition per liter:

Pancreatic digest of gelatin	2.5g
Beef extract	1.5g

pH 6.9 ± 0.2 at 25°C

Preparation of Medium: Add components to distilled/deionized water and bring volume to 1.0L. Mix thoroughly. Distribute into tubes or flasks. Autoclave for 15 min at 15 psi pressure–121°C.

Use: For the cultivation of *Cytophaga allerginae*.

Nutrient Broth, Diluted 1:100

Nutrient Broth:

Composition per liter:

Pancreatic digest of gelatin	5.0g
Beef extract	3.0g

pH 6.8 ± 0.2 at 25°C

Preparation of Nutrient Broth: Add components to distilled/deionized water and bring volume to 1.0L. Mix thoroughly.

Preparation of Medium: Add 10.0mL of nutrient broth to distilled/deionized water and bring volume to 1.0L. Mix thoroughly. Distribute into tubes or flasks. Autoclave for 15 min at 15 psi pressure–121°C.

Use: For the cultivation and maintenance of *Agromonas oligotrophica*.

Nutrient Broth No. 2

Composition per liter:

Beef extract	10.0g
Peptone	10.0g
NaCl	5.0g

pH 7.5 ± 0.2 at 25°C

Source: This medium is available as a premixed powder from Oxoid Unipath.

Preparation of Medium: Add components to distilled/deionized water and bring volume to 1.0L. Mix thoroughly. Distribute into tubes or flasks. Autoclave for 15 min at 15 psi pressure–121°C.

Use: For the cultivation of a variety of fastidious and nonfastidious microorganisms.

Nutrient Broth, Standard II

Composition per liter:

Special peptone	8.6g
NaCl	6.4g

pH 7.5 ± 0.1 at 37°C

Preparation of Medium: Add components to distilled/deionized water and bring volume to 1.0L. Mix thoroughly. Distribute into tubes or flasks. Autoclave for 15 min at 15 psi pressure–121°C.

Use: For the cultivation of a variety of fastidious and nonfastidious microorganisms.

Nutrient Broth with 6% NaCl

Composition per liter:

NaCl	60.0g
Pancreatic digest of gelatin	5.0g
Beef extract	3.0g

pH 6.8 ± 0.2 at 25°C

Preparation of Medium: Add components to distilled/deionized water and bring volume to 1.0L. Mix thoroughly. Distribute into tubes or flasks. Autoclave for 15 min at 15 psi pressure–121°C.

Use: For the cultivation of organisms in water, sewage, feces and other materials. For the cultivation and maintenance of *Paracoccus halodenitrificans*.

Nutrient Gelatin

Composition per liter:

Gelatin	120.0g
Pancreatic digest of gelatin	5.0g
Beef extract	3.0g

pH 6.8 ± 0.2 at 25°C

Source: This medium is available as a premixed powder from BBL Microbiology Systems and Difco Laboratories and Oxoid Unipath.

Preparation of Medium: Add components to distilled/deionized water and bring volume to 1.0L. Mix thoroughly. Gently heat while stirring to 50°C. Distribute into tubes. Autoclave for 15 min at 15 psi pressure–121°C.

Use: For the cultivation and differentiation of bacteria based on their ability to liquefy gelatin.

Nutrient Yeast Glucose Medium

Composition per liter:

Glucose	10.0g
Pancreatic digest of gelatin	5.0g
Yeast extract	5.0g
Beef extract	3.0g

pH 6.8 ± 0.2 at 25°C

Preparation of Medium: Add components to distilled/deionized water and bring volume to 1.0L. Mix thoroughly. Distribute into tubes or flasks. Autoclave for 15 min at 15 psi pressure–121°C.

Use: For the cultivation and maintenance of *Erwinia amylovora.*

NWRI Agar (HPC Agar)

Composition per liter:

Agar	15.0g
Peptone	3.0g
Soluble casein	0.5g
K_2HPO_4	0.2g
$MgSO_4$	0.05g
$FeCl_3$	1.0mg

pH 7.2 ± 0.2 at 25°C.

Preparation of Medium: Add components to distilled/deionized water and bring volume to 1.0L. Mix thoroughly. Gently heat and bring to boiling. Adjust pH to 7.2. Distribute into tubes or flasks. Autoclave for 15 min at 15 psi pressure–121°C. Pour into sterile Petri dishes.

Use: For estimation of the number of live heterotrophic bacteria in water using the heterotrophic plate count technique.

NY Medium (Nutrient Broth Yeast Extract Medium)

Composition per liter:

NaCl	8.0g
Peptone	5.0g
Yeast extract	5.0g
Beef extract	3.0g

pH 7.2–7.4 at 25°C

Preparation of Medium: Add components to distilled/deionized water and bring volume to 1.0L. Mix thoroughly. Distribute into tubes or flasks. Autoclave for 15 min at 15 psi pressure–121°C.

Use: For the cultivation of *Pseudomonas* species.

N–Z Amine A™ Medium

Composition per liter:

N-Z-Amine A™	5.0g
Beef extract	1.0g
Glycerol	80.0mL

Preparation of Medium: Add components to tap water and bring volume to 1.0L. Mix thoroughly. Distribute into tubes or flasks. Autoclave for 15 min at 15 psi pressure–121°C.

Use: For the cultivation and maintenance of *Nocardia brevicatena.*

N–Z Amine A™ Medium with Soluble Starch and Glucose

Composition per liter:

Soluble starch	20.0g
Agar	15.0g
Glucose	10.0g
Yeast extract	5.0g
N-Z-Amine A™	5.0g
$CaCO_3$	1.0g

Preparation of Medium: Add components to distilled/deionized water and bring volume to 1.0L. Mix thoroughly. Gently heat and bring to boiling. Distribute into tubes or flasks. Autoclave for 15 min at 15 psi pressure–121°C. Pour into sterile Petri dishes or leave in tubes.

Use: For the cultivation and maintenance of *Actinomadura* species, *Actinoplanes* species, *Amycolatopsis fastidiosa, Catenuloplanes japonicus, Dactylosporangium* species, *Geodermatophilus obscurus, Glycomyces* species, *Kitasatosporia mediocidica, Micromonospora* species, *Saccharomonospora caesia, Saccharothrix aerocolonigenes, Streptomyces* species, and *Streptosporangium* species.

N–Z Amine A™ Glycerol Agar

Composition per liter:

Agar	15.0g
N-Z-Amine A™	5.0g

Beef extract	1.0g
Glycerol	70.0mL

pH 6.5–7.0 at 25°C

Preparation of Medium: Add components to distilled/deionized water and bring volume to 1.0L. Mix thoroughly. Gently heat and bring to boiling. Distribute into tubes or flasks. Autoclave for 15 min at 15 psi pressure–121°C. Pour into sterile Petri dishes or leave in tubes.

Use: For the isolation and cultivation of *Actinomadura* species, *Actinopolyspora* species, *Excellospora* species and *Microspora* species.

NZY Agar

Composition per liter:

Agar	20.0g
N-Z- Amine A™	10.0g
NaCl	5.0g
Yeast extract	5.0g
$MgCl_2·6H_2O$	2.0g

Preparation of Medium: Add components to distilled/deionized water and bring volume to 1.0L. Mix thoroughly. Gently heat and bring to boiling. Distribute into tubes. Autoclave for 15 min at 15 psi pressure–121°C. Allow tubes to cool in a slanted position.

Use: For the cultivation and maintenance of a variety of microorganisms.

NZY Agar

Composition per liter:

Agar	11.0g
N-Z- Amine A™	10.0g
NaCl	5.0g
Yeast extract	5.0g
$MgCl_2·6H_2O$	2.0g

Preparation of Medium: Add components to distilled/deionized water and bring volume to 1.0L. Mix thoroughly. Gently heat and bring to boiling. Distribute into tubes or flasks. Autoclave for 15 min at 15 psi pressure–121°C. Pour into sterile Petri dishes.

Use: For the cultivation and enumeration of a variety of microorganisms.

NZY Broth

Composition per liter:

N-Z- Amine A™	10.0g
NaCl	5.0g
Yeast extract	5.0g
$MgCl_2·6H_2O$	2.0g

Preparation of Medium: Add components to distilled/deionized water and bring volume to 1.0L. Mix thoroughly. Gently heat and bring to boiling. Distribute into tubes. Autoclave for 15 min at 15 psi pressure–121°C.

Use: For the cultivation of a variety of microorganisms.

Oatmeal Agar (ATCC Medium 551)

Composition per liter:

Oatmeal	60.0g
Agar	12.5g

pH 6.0 ± 0.2 at 25°C

Source: This medium is available as a premixed powder from Difco Laboratories.

Preparation of Medium: Add components to distilled/deionized water and bring volume to 1.0L. Mix thoroughly. Gently heat and bring to boiling. Distribute into tubes or flasks. Autoclave for 15 min at 15 psi pressure–121°C. Pour into sterile Petri dishes or leave in tubes.

Use: Oatmeal agar is used for cultivation of fungi and actinomycetes, particularly for macrospore formation.

Ogawa Egg Medium

Composition:

Chicken eggs, whole	200.0mL
Sodium glutamate-KH_2PO_4 solution	100.0mL
Glycerol	6.0mL
Malachite green (2.0% solution)	6.0mL

pH 6.8 ± 0.2 at 25°C

Sodium Glutamate-KH_2PO_4 Solution:
Composition per 100mL:

Sodium glutamate	1.0g
KH_2PO_4	1.0g

Preparation of Sodium Glutamate-KH_2PO_4 Solution: Add components to distilled/deionized water and bring volume to 100.0mL. Mix thoroughly. Filter sterilize.

Preparation of Medium: Soak eggs with 1:100 dilution of saturated mercuric chloride solution for 1 min. Aseptically break eggs into a sterile graduated cylinder. Homogenize eggs. Add remaining components. Mix thoroughly. Aseptically distribute into sterile tubes in 10.0mL volumes. Inspissate at 90°C (moist heat) for 60 min.

Use: For the selective isolation and cultivation of *Nocardia* and *Rhodococci* species.

Oil Agar Medium

Composition per liter:

Agar, purified	20.0g
NaCl	10.0g
Oil powder	10.0g
NH_4NO_3	1.0g
$MgSO_4$	0.5g
Amphotericin B solution	10.0mL
K_2HPO_4 solution	7.0mL
KH_2PO_4 solution	3.0mL
$FeCl_3$	0.1mL

Oil Powder:

Composition per 10g:

Hydrocarbon	10.0g
Silica gel	10.0g
Diethyl ether	30.0mL

Preparation of Oil Powder: Add 10.0g of hydrocarbon to 30.0mL of diethyl ether. Mix thoroughly. Add 10.0g of silica gel. Allow ether to evaporate.

Amphotericin B Solution:

Composition per 10mL:

Amphotericin B	0.01g

Preparation of Amphotericin B Solution: Add amphotericin B to distilled/deionized water and bring volume to 10.0mL. Mix thoroughly. Filter sterilize.

K_2HPO_4 Solution:

Composition per 100mL:

K_2HPO_4	10.0g

Preparation of K_2HPO_4 Solution: Add K_2HPO_4 to distilled/deionized water and bring volume to 100.0mL. Mix thoroughly. Autoclave for 15 min at 15 psi pressure–121°C. Cool to 25°C.

KH_2PO_4 Solution:

Composition per 100mL:

KH_2PO_4	10.0g

Preparation of KH_2PO_4 Solution: Add KH_2PO_4 to distilled/deionized water and bring volume to 100.0mL. Mix thoroughly. Autoclave for 15 min at 15 psi pressure–121°C. Cool to 25°C.

Preparation of Medium: Add components—except amphotericin B solution, K_2HPO_4 solution, and KH_2PO_4 solution—to distilled/deionized water and bring volume to 980.0mL. Mix thoroughly. Gently heat and bring to boiling. Autoclave for 15 min at 15 psi pressure–121°C. Cool to 45–50°C. Aseptically add 10.0mL of sterile amphotericin B solution, 7.0mL of sterile K_2HPO_4 solution, and 3.0mL of sterile KH_2PO_4 solution. Mix thoroughly. Pour into sterile Petri dishes or distribute into sterile tubes.

Use: For the cultivation and enumeration of hydrocarbon-utilizing bacteria by direct plating of estuarine water and sediment samples.

Oleic Albumin Complex

Composition per liter:

Bovine albumin fraction V	50.0g
NaCl	8.5g
Oleic acid	0.6mL

Preparation of Medium: Add components to distilled/deionized water and bring volume to 1.0L. Mix thoroughly. Filter sterilize.

Use: For use in media employed for the cultivation of mycobacteria.

ONPG Broth

Composition per liter:

Peptone water	750.0mL
ONPG solution	250.0mL

pH 7.2–7.4 at 25°C

ONPG Solution:

Composition per 250mL:

ONPG (*o*-nitrophenyl-β-D-galactopyranoside)	1.5g
Sodium phosphate buffer (0.01*M*, pH 7.5)	250.0mL

Preparation of ONPG Solution: Add ONPG to 250.0mL of sodium phosphate buffer. Mix thoroughly. Filter sterilize.

Peptone Water:

Composition per 750mL:

Peptone	7.5g
NaCl	3.75g

Preparation of Peptone Water: Add components to distilled/deionized water and bring volume to 750.0mL. Mix thoroughly. Gently heat and bring to boiling. Adjust pH to 8.0–8.4. Continue boiling for 10 min. Filter through Whatman #1 filter paper. Readjust pH of filtrate to 7.2–7.4. Autoclave for 20 min at 10 psi pressure–115°C. Cool to 25°C.

Preparation of Medium: Aseptically combine the sterile ONPG solution with the cooled, sterile peptone water. Mix thoroughly. Aseptically distribute into tubes in 2.5–3.0.0mLvolumes. Store at 4°C for up to one month.

Use: For the differentiation of a variety of Gram-negative bacteria based on production of β-galactosidase. For the differentiation of lactose-delayed bacteria from lactose-negative bacteria. For the differentiation of *Pseudomonas cepacia* (positive) and *Pseudomonas maltophila* (positive) from other

Pseudomonas species (negative). Bacteria that produce β-galactosidase turn the medium yellow.

OR Indicator Agar (Oxidation–Reduction Indicator Agar)

Composition per liter:

Agar	15.0g
Sodium glycerol phosphate	10.0g
Sodium thioglycollate	1.7g
$CaCl_2 \cdot 2H_2O$	0.1g
Methylene blue	6.0mg

Preparation of Medium: Add components to distilled/deionized water and bring volume to 1.0L. Mix thoroughly. Gently heat and bring to boiling. Distribute into tubes or flasks. Autoclave for 15 min at 15 psi pressure–121°C. Pour into sterile Petri dishes or leave in tubes.

Use: For use as an indicator of oxygen free conditions in anaerobic culture chambers.

Organic Acid Medium KP (Organic Acid Medium, Kauffmann and Petersen)

Composition per liter:

Gelatin	10.0g
Bromthymol blue	0.024g
Organic acid solution	100.0mL

pH 7.4 ± 0.2 at 25°C

Organic Acid Solution:

Composition per 100mL:

Organic acid	10.0g

Preparation of Medium: Add organic acid to distilled/deionized water and bring volume to 100.0mL. Sodium potassium D-tartrate, sodium citrate or mucic acid may be used. Mix thoroughly

Preparation of Medium: Add components, except organic acid solution, to distilled/deionized water and bring volume to 900.0mL. Mix thoroughly. Gently heat and bring to boiling. Autoclave for 15 min at 15 psi pressure–121°C. Cool to 45–50°C. Aseptically add sterile organic acid solution. Mix thoroughly. Aseptically distribute into sterile tubes or flasks.

Use: For the cultivation and differentiation of members of the Enterobacteriaceae based on their ability to utilize different organic acids as carbon source. Bacteria that utilize tartrate, citrate or mucate turn the medium yellow.

Oxalate Medium

Basal Medium:

Composition per liter:

K_2HPO_4	4.4g
KH_2PO_4	3.4g
Potasssium oxalate	2.0g
$(NH_4)_2SO_4$	0.5g
$MgSO_4 \cdot 7H_2O$	0.2g
$FeCl_3$	0.015g
Phenol red (0.4% solution)	5.0mL
Mineral stock solution	1.0mL

Mineral Stock Solution:

Composition per liter:

$ZnSO_4 \cdot 7H_2O$	11.0g
$MnSO_4 \cdot H_2O$	5.0g
$Na_2MoO_4 \cdot 2H_2O$	2.0g
$CoSO_4$	0.05g
H_3BO_3	0.05g
$CuSO_4 \cdot 5H_2O$	7.0mg

Preparation of Mineral Stock Solution: Add components to distilled/deionized water and bring volume to 1.0L. Mix thoroughly.

Preparation of Medium: Add components to distilled/deionized water and bring volume to 1.0L. Mix thoroughly. Distribute into tubes or flasks. Autoclave for 15 min at 15 psi pressure–121°C.

Use: For the isolation and cultivation of oxalate–decomposing *Alcaligenes* species.

Oxalate Utilization Medium

Composition per liter:

Agar	12.0g
Potassium oxalate	1.0g
NaCl	1.0g
$(NH_4)_2HPO_4$	1.0g
KH_2PO_4	0.5g
$MgSO_4 \cdot 7H_2O$	0.2g
$CaCl_2$ solution	80.0mL

pH 7.0 ± 0.2 at 25°C

$CaCl_2$ Solution:

Composition per 100mL:

$CaCl_2$	1.47g

Preparation of $CaCl_2$ Solution: Add $CaCl_2$ to distilled/deionized water and bring volume to 10.0mL. Mix thoroughly. Gently heat until dissolved. Filter sterilize.

Preparation of Medium: Add components, except $CaCl_2$ solution, to distilled/deionized water and bring volume to 920.0mL. Mix thoroughly. Gently heat and bring to boiling. Distribute into flasks in 92.0mL volumes. Autoclave for 15 min at 15 psi pressure–121°C. Cool to 45–50°C. Aseptically add 8.0mL of sterile $CaCl_2$ solution to each flask. A fine precipitate of calcium oxalate will form. Mix thoroughly. Pour into sterile Petri dishes. Swirl flask while dispensing agar to disperse precipitate evenly.

Use: For the cultivation and differentiation of streptomycetes based on oxalate utilization. Bacteria that utilize oxalate turn the medium dark blue.

Oxidative Fermentative Glucose Medium, Semisolid (OF Glucose Medium, Semisolid)

Composition per liter:

Glucose	10.0g
NaCl	5.0g
Agar	2.0g
Pancreatic digest of casein	2.0g
K_2HPO_4	0.3g
Bromthymol blue dye	0.08g

pH 6.8 ± 0.2 at 25°C

Preparation of Medium: Add components to distilled/deionized water and bring volume to 1.0L. Mix thoroughly. Gently heat and bring to boiling. Distribute into tubes or flasks. Autoclave for 15 min at 15 psi pressure–121°C. Pour into sterile Petri dishes or leave in tubes.

Use: For differentiating Gram-negative bacteria based upon determining the oxidative and fermentative metabolism of glucose. Bacteria that ferment glucose turn the medium yellow.

Oxidation Fermentation Medium (OF Medium)

Composition per liter:

NaCl	5.0g
Agar	2.5g
Pancreatic digest of casein	2.0g
K_2HPO_4	0.3g
Bromthymol blue	0.03g
Carbohydrate solution	100.0mL

pH 6.8 ± 0.1 at 25°C

Source: This medium is available as a premixed powder from BBL Microbiology Systems.

Carbohydrate Solution:

Composition per 100mL:

Carbohydrate	10.0g

Preparation of Carbohydrate Solution: Add carbohydrate to distilled/deionized water and bring volume to 100.0mL. Mix thoroughly. Filter sterilize.

Preparation of Medium: Add components, except carbohydrate solution, to distilled/deionized water and bring volume to 900.0mL. Mix thoroughly. Gently heat and bring to boiling. Autoclave for 15 min at 15 psi pressure–121°C. Cool to 45–50°C. Aseptically add 100.0mL sterile carbohydrate solution. Mix thoroughly. Pour into sterile Petri dishes or distribute into sterile tubes.

Use: For differentiating Gram-negative bacteria based upon determining the oxidative and fermentative metabolism of carbohydrates.

Oxidation Fermentation Medium, Hugh–Leifson's (Hugh–Leifson's Oxidation Fermentation Medium)

Composition per liter:

NaCl	5.0g
Agar	3.0g
Peptone	2.0g
K_2HPO_4	0.3g
Carbohydrate solution	100.0mL
Bromthymol blue solution (0.2%)	15.0mL

pH 7.1 ± 0.2 at 25°C

Carbohydrate Solution:

Composition per 100mL:

Carbohydrate	10.0g

Preparation of Carbohydrate Solution: Add carbohydrate to distilled/deionized water and bring volume to 100.0mL. Mix thoroughly. Filter sterilize.

Preparation of Medium: Add components, except carbohydrate solution, to distilled/deionized water and bring volume to 900.0mL. Mix thoroughly. Gently heat and bring to boiling. Autoclave for 15 min at 15 psi pressure–121°C. Cool to 45–50°C. Aseptically add 100.0mL sterile carbohydrate solution. Mix thoroughly. Pour into sterile Petri dishes or distribute into sterile tubes.

Use: For differentiating Gram-negative bacteria, such as *Vibrio* species, based upon determining the oxidative and fermentative metabolism of carbohydrates. Bacteria that ferment the carbohydrate turn the medium yellow.

Oxidation Fermentation Medium, King's (King's OF Medium)

Composition per liter:

Base:

Agar	3.0g
Pancreatic digest of casein	2.0g

Carbohydrate solution 100.0mL
Phenol red (1.5% solution) 2.0mL
pH to 7.3 ± 0.2

Carbohydrate Solution:

Composition per 100mL:

Carbohydrate 10.0g

Preparation of Carbohydrate Solution: Add carbohydrate to distilled/deionized water and bring volume to 100.0mL. Mix thoroughly. Filter sterilize.

Preparation of Medium: Add components, except carbohydrate solution, to distilled/deionized water and bring volume to 900.0mL. Mix thoroughly. Gently heat and bring to boiling. Autoclave for 15 min at 15 psi pressure–121°C. Cool to 45–50°C. Aseptically add 100.0mL sterile carbohydrate solution. Mix thoroughly. Pour into sterile Petri dishes or distribute into sterile tubes.

Use: For differentiating bacteria based upon determining the oxidative and fermentative metabolism of carbohydrates. Bacteria that ferment the carbohydrate turn the medium yellow.

Oxidative–Fermentative Glucose Medium, Semisolid, with NaCl (OF Glucose Medium, Semisolid with NaCl)

Composition per liter:

NaCl 20.0g
Glucose 10.0g
Agar 2.0g
Pancreatic digest of casein 2.0g
K_2HPO_4 .... 0.3g
Bromthymol blue dye 0.08g
pH 6.8 ± 0.2 at 25°C

Preparation of Medium: Add components to distilled/deionized water and bring volume to 1.0L. Mix thoroughly. Gently heat and bring to boiling. Distribute into tubes or flasks. Autoclave for 15 min at 15 psi pressure–121°C. Pour into sterile Petri dishes or leave in tubes.

Use: For differentiating halophilic *Vibrio* species based upon determining the oxidative and fermentative metabolism of glucose. Bacteria that ferment glucose turn the medium yellow.

Oxidative–Fermentative Medium (OF Medium)

Composition per liter:

NaCl 5.0g
Agar 2.0g
Pancreatic digest of casein 2.0g
K_2HPO_4 .... 0.3g
Bromothymol blue 0.08g
Carbohydrate solution 100.0mL
pH 6.8 ± 0.1 at 25°C

Source: This medium is available as a premixed powder from Difco Laboratories and Oxoid Unipath.

Carbohydrate Solution:

Composition per 100mL:

Carbohydrate 10.0g

Preparation of Carbohydrate Solution: Add carbohydrate to distilled/deionized water and bring volume to 100.0mL. Mix thoroughly. Filter sterilize.

Preparation of Medium: Add components, except carbohydrate solution, to distilled/deionized water and bring volume to 900.0mL. Mix thoroughly. Gently heat and bring to boiling. Autoclave for 15 min at 15 psi pressure–121°C. Cool to 45–50°C. Aseptically add 100.0mL sterile carbohydrate solution. Mix thoroughly. Pour into sterile Petri dishes or distribute into sterile tubes.

Use: For differentiating bacteria based upon determining the oxidative and fermentative metabolism of carbohydrates. Bacteria that ferment the carbohydrate turn the medium yellow.

Oxidative–Fermentative Test Medium (OF Test Medium)

Composition per liter:

NaCl 5.0g
Agar 3.0g
Peptone 2.0g
K_2HPO_4 .... 0.3g
Bromthymol Blue 0.03g
Carbohydrate solution 100.0mL

Carbohydrate Solution:

Composition per 100mL:

Carbohydrate 10.0g

Preparation of Carbohydrate Solution: Add carbohydrate to distilled/deionized water and bring volume to 100.0mL. Mix thoroughly. Filter sterilize.

Preparation of Medium: Add components, except carbohydrate solution, to distilled/deionized water and bring volume to 1.0L. Mix thoroughly. Gently heat and bring to boiling. Distribute into tubes in 3.0mL volumes. Autoclave for 15 min at 15 psi pressure–121°C. Cool to 45–50°C. Aseptically add 0.3mL of sterile carbohydrate solution to each tube. Mix thoroughly.

Use: For the cultivation and differentiation of a variety of microorganisms based on their ability to ferment a specific carbohydrate. Bacteria that ferment the specific carbohydrate turn the medium yellow.

OZR Medium

Composition per liter:

Agar	15.0g
Pancreatic digest of casein	1.0g
Yeast extract	1.0g

Preparation of Medium: Add components to seawater and bring volume to 1.0L. Mix thoroughly. Gently heat and bring to boiling. Distribute into tubes or flasks. Autoclave for 15 min at 15 psi pressure–121°C. Pour into sterile Petri dishes or leave in tubes.

Use: For the cultivation and maintenance of *Leucothrix mucor.*

PA C Agar (mPA C Agar)

Composition per liter:

Agar	12.0g
L-Lysine·HCl	5.0g
NaCl	5.0g
$Na_2S_2O_3$	5.0g
Yeast extract	2.0g
$MgSO_4 \cdot 7H_2O$	1.5g
Lactose	1.25g
Sucrose	1.25g
Xylose	1.25g
Ferric ammonium citrate	0.80g
Phenol red	0.08g
Nalidixic acid	0.037g
Kanamycin	8.0mg

pH 7.2 ± 0.1 at 25°C

Source: This medium is available as a premixed powder BBL Microbiology Systems.

Preparation of Medium: Add components to distilled/deionized water and bring volume to 1.0L. Mix thoroughly. Gently heat and bring to boiling. Distribute into tubes or flasks. Autoclave for 15 min at 15 psi pressure–121°C. Pour into sterile Petri dishes or leave in tubes.

Use: For the selective recovery and enumeration of *Pseudomonas aeruginosa* from water samples.

Paraffin Agar

Composition per liter:

Agar	15.0g
K_2HPO_4	6.0g
NH_4NO_3	4.0g
KH_2PO_4	2.0g
Paraffin, liquid	1.0g
$ZnSO_4 \cdot 7H_2O$	0.049g
$MnCl_2 \cdot 4H_2O$	0.046g
$FeSO_4 \cdot 7H_2O$	5.4mg
$CuSO_4 \cdot 5H_2O$	2.5mg
$Na_2B_4O_7 \cdot 10H_2O$	0.94mg
$(NH_4)_6Mo_7O_{24} \cdot 4H_2O$	0.20mg

Preparation of Medium: Add components to distilled/deionized water and bring volume to 1.0L. Mix thoroughly. Gently heat and bring to boiling. Distribute into tubes or flasks. Autoclave for 15 min at 15 psi pressure–121°C. Pour into sterile Petri dishes or leave in tubes.

Use: For the selective isolation and cultivation of streptomycetes.

Paramecium Medium

Composition per liter:

Solution C	500.0mL
Solution A	10.0mL
Solution B	1.0mL

Solution A:

Composition per liter:

Thiamine·HCl	1.5g
Calcium pantothenate	1.0g
Nicotinamide	0.5g
Pyridoxal·HCl	0.5g
Riboflavin	0.5g
Folic acid	0.5g
α-Lipoic acid	0.1g
Biotin	0.1mg

Preparation of Solution A: Add components to distilled/deionized water and bring volume to 1.0L. Mix thoroughly. Distribute while stirring into screw-capped tubes in 10.0mL volumes. Store at −20°C. Thaw as needed.

Solution B:

Composition per 100mL:

TEM-4T (Hachmeister, Pittsburgh)	10.0g
Stigmasterol	0.5g
Ethanol, absolute	100.0mL

Preparation of Solution B: Add TEM-4T and stigmasterol to 100.0mL of hot ethanol. Mix thoroughly. Store at 4°C.

Solution C:

Composition per 500mL:

Proteose peptone	10.0g
Pancreatic digest of casein	5.0g
Ribonucleic acid	1.0g
$MgSO_4 \cdot 7H_2O$	0.5g

Preparation of Solution C: Add components to distilled/deionized water and bring volume to 500.0mL. Mix thoroughly.

Preparation of Medium: Combine 500.0mL of solution C, 10.0mL of solution A and 1.0mL of solution B. Mix thoroughly. Bring volume to 1.0L with distilled/deionized water. Adjust pH to 7.0–7.2 with 0.1*N* NaOH. Autoclave for 20 min at 15 psi pressure–121°C. Cool to 45–50°C. Aseptically distribute into sterile tubes or flasks.

Use: For the cultivation of *Paramecium* species to be used as host cells by bacterial symbionts.

PD2 Medium

Composition per liter:

Agar	15.0g
Pancreatic digest of casein	4.0g
Papaic digest of soybean meal	2.0g
K_2HPO_4	1.5g
Disodium succinate	1.0g
KH_2PO_4	1.0g
$MgSO_4 \cdot 7H_2O$	1.0g
Trisodium citrate	1.0g
Bovine serum albumin solution	10.0mL
Hemin chloride solution	10.0mL

pH 7.0 ± 0.2 at 25°C

Bovine Serum Albumin Solution:

Composition per 10mL:

Bovine serum albumin	2.0g

Preparation of Bovine Serum Albumin Solution: Add bovine serum albumin to distilled/deionized water and bring volume to 10.0mL. Mix thoroughly. Filter sterilize.

Hemin Chloride Solution:

Composition per 100mL:

Hemin chloride	0.1g
NaOH (0.05*N* solution)	100.0mL

Preparation of Hemin Chloride Solution: Add hemin chloride to 100.0mL of NaOH solution. Mix thoroughly.

Preparation of Medium: Add components, except bovine serum albumin solution, to distilled/deionized water and bring volume to 990.0mL. Mix thoroughly. Gently heat and bring to boiling. Adjust pH to 7.0. Autoclave for 15 min at 15 psi pressure–121°C. Cool to 45–50°C. Aseptically add sterile bovine serum albumin solution. Mix thoroughly. Pour into sterile Petri dishes or distribute into sterile tubes.

Use: For the isolation and cultivation of PD-ALS (Pierce's disease-almond leaf scorch) bacteria.

Pectin Agar

Composition per plate:

Base agar	15.0mL
Pectin gel	7.0mL

Base Agar:

Composition per liter:

Agar	15.0g
K_2HPO_4	5.0g
$CaCl_2 \cdot 2H_2O$	3.0g
NH_4Cl	1.0g
$MgSO_4 \cdot 7H_2O$	0.2g
Tris(hydroxymethyl)amino–methane buffer (1*M*, pH 8.0)	100.0mL
Trace element solution	1.0mL

Preparation of Base Agar: Add components to distilled/deionized water and bring volume to 1.0L. Mix thoroughly. Gently heat and bring to boiling. Autoclave for 15 min at 15 psi pressure–121°C. Cool to 45–50°C.

Trace Element Solution:

Composition per liter:

Disodium EDTA	8.0g
$MnCl_2 \cdot 4H_2O$	0.1g
$CoCl_2 \cdot 6H_2O$	0.02g
$ZnCl_2$	0.02g
KBr	0.02g
KI	0.02g
$CuSO_4$	0.01g
$Na_2MoO_4 \cdot 2H_2O$	0.01g
H_3BO_3	0.01g
LiCl	5.0mg
$SnCl_2 \cdot 2H_2O$	5.0mg

Preparation of Trace Element Solution: Add components to distilled/deionized water and bring volume to 1.0L. Mix thoroughly.

Pectin Gel:

Composition per liter:

Pectin, low esterified	20.0g

Preparation of Pectin Gel: Add pectin to 70°C distilled/deionized water and bring volume to 1.0L. Autoclave for 10 min at 7 psi pressure–110°C. Cool to 45–50°C.

Preparation of Medium: Pour cooled, sterile base agar into sterile Petri dishes in 15.0mL volumes. Allow agar to solidify. Overlay each plate with 7.0mL of cooled, sterile pectin gel. The pectin may take 5hr to form a gel.

Use: For the isolation and cultivation of *Cytophaga* species, *Herpetosiphon* species, *Saprospira* species, and *Flexithrix* species.

Pectin Medium

Composition per liter:

Pectin	30.0g
Yeast extract	5.0g
Bromthymol blue (0.1% solution)	1.0mL
$CaCl_2 \cdot 2H_2O$ (10.0% solution)	0.6mL

pH 7.3 ± 0.2 at 25°C

Preparation of Medium: Add 100.0mL of distilled/deionized water to a 2-liter flask and place on a magnetic stirrer with no heat. While stirring, slowly add $CaCl_2 \cdot 2H_2O$, bromthymol blue solution, yeast extract, and pectin. Add slowly to ensure each particle is wetted. Gently heat and bring to almost boiling. Adjust pH to 7.3 with 1N NaOH. Do not overshoot pH 7.3.

Use: For the isolation and presumptive identification of *Yersinia enterocolitica* and *Y. pseudotuberculosis* from other fermenting Gram–negative bacilli such as *Enterocolitica agglomerans* which is often confused with *Yersinia. E. agglomerans* does not produce pectinase. *Y. enterocolitica* is strongly positive for pectinase activity. *Y. pseudotuberculosis* is weakly positive for pectinase activity. *Yersinia pestis* is negative for pectinase activity. Also, use for the differentiation of *Klebsiella oxytoca* which is pectinase positive.

Peptone Cholic Acid Recovery

Composition per liter:

Meat extract	10.0g
Peptone	10.0g
Cholic acid	10.0g
NaCl	5.0g
NaOH	1.0g

pH 6.8 ± 0.2 at 25°C

Preparation of Medium: Add components to distilled/deionized water and bring volume to 1.0L. Mix thoroughly. Gently heat and bring to boiling. Adjust pH to 6.8. Distribute into tubes or flasks. Autoclave for 15 min at 15 psi pressure–121°C.

Use: For the cultivation and maintenance of *Arthrobacter* species and *Corynebacterium* species.

Peptone Recovery Broth

Composition per liter:

Meat extract	10.0g
Peptone	10.0g
NaCl	5.0g

pH 6.8 ± 0.2 at 25°C

Preparation of Medium: Add components to distilled/deionized water and bring volume to 1.0L. Mix thoroughly. Adjust pH to 6.8. Distribute into tubes or flasks. Autoclave for 15 min at 15 psi pressure–121°C.

Use: For the cultivation of *Brevibacterium* species.

Peptone Sodium Cholate

Composition per liter:

Meat extract	10.0g
Peptone	10.0g
NaCl	5.0g
Sodium cholate	5.0g

pH 6.8 ± 0.2 at 25°C

Preparation of Medium: Add components to distilled/deionized water and bring volume to 1.0L. Mix thoroughly. Adjust pH to 6.8. Distribute into tubes or flasks. Autoclave for 15 min at 15 psi pressure–121°C.

Use: For the cultivation and maintenance of *Anthrobacter* species.

Peptone Starch Carbonate Medium

Composition per liter:

Agar	15.0g
Soluble starch	10.0g
Peptone	5.0g
Yeast extract	5.0g
K_2HPO_4	1.0g
$MgSO_4 \cdot 7H_2O$	0.2g
Na_2CO_3 solution	100.0mL

Na_2CO_3 Solution:

Composition per 100mL:

Na_2CO_3	10.0g

Preparation of Na_2CO_3 Solution: Add Na_2CO_3 to distilled/deionized water and bring volume to 100.0mL. Mix thoroughly. Autoclave for 15 min at 15 psi pressure–121°C. Cool to 45–50°C.

Preparation of Medium: Add components, except Na_2CO_3 solution, to distilled/deionized water and bring volume to 990.0mL. Mix thoroughly. Gently heat and bring to boiling. Autoclave for 15 min at 15 psi pressure–121°C. Cool to 45–50°C. Aseptically add sterile Na_2CO_3 solution. Mix thoroughly. Pour into sterile Petri dishes or distribute into sterile tubes.

Use: For the cultivation and maintenance of *Bacillus alcalophilus* and other *Bacillus* species.

Peptone Succinate Agar

Composition per liter:

Peptone	5.0g
Succinic acid	1.68g

Agar....1.5g
$MgSO_4 \cdot 7H_2O$....1.0g
$(NH_4)_2SO_4$....1.0g
$FeCl_3 \cdot 6H_2O$....2.0mg
$MnSO_4 \cdot H_2O$....2.0mg

pH 7.0 ± 0.2 at 25°C

Preparation of Medium: Add components to distilled/deionized water and bring volume to 1.0L. Mix thoroughly. Gently heat and bring to boiling. Distribute into tubes or flasks. Autoclave for 15 min at 15 psi pressure–121°C. Pour into sterile Petri dishes or leave in tubes.

Use: For the cultivation and maintenance of *Aquaspirillum bengal, A. dispar,* and *Spirillum volutans.*

Peptone Succinate Agar in Seawater

Composition per liter:

Peptone....5.0g
Succinic acid....1.68g
Agar....1.5g
$MgSO_4 \cdot 7H_2O$....1.0g
$(NH_4)_2SO_4$....1.0g
$FeCl_3 \cdot 6H_2O$....2.0mg
$MnSO_4 \cdot H_2O$....2.0mg

pH 7.0 ± 0.2 at 25°C

Preparation of Medium: Add components to seawater and bring volume to 1.0L. Mix thoroughly. Gently heat and bring to boiling. Distribute into tubes or flasks. Autoclave for 15 min at 15 psi pressure–121°C. Pour into sterile Petri dishes or leave in tubes.

Use: For the cultivation and maintenance of *Oceanospirillum maris.*

Peptone Succinate Salts Broth (PSS Broth)

Composition per 100mL:

Peptone....1.0g
$MgSO_4 \cdot 7H_2O$....0.1g
$(NH_4)_2SO_4$....0.1g
Succinic acid....0.1g
$FeCl_3 \cdot 6H_2O$....0.2mg
$MnSO_4 \cdot H_2O$....0.2mg

pH 6.8 ± 0.2 at 25°C

Preparation of Medium: Add components to distilled/deionized water and bring volume to 1.0L. Mix thoroughly. Adjust pH to 6.8 with KOH. Distribute into tubes or flasks. Autoclave for 15 min at 15 psi pressure–121°C.

Use: For the cultivation of *Spirillum* species.

Peptone Succinate Salts in Seawater

Composition per liter:

Peptone....10.0g
$MgSO_4 \cdot 7H_2O$....1.0g
$(NH_4)_2SO_4$....1.0g
Succinic acid....1.0g
$FeCl_3 \cdot 6H_2O$....2.0mg
$MnSO_4 \cdot H_2O$....2.0mg
Synthetic seawater....1.0L

pH 6.8 ± 0.2 at 25°C

Synthetic Seawater:

Composition per liter:

NaCl....27.5g
$MgCl_2$....5.0g
$MgSO_4$....2.0g
KCl....1.0g
$CaCl_2$....0.5g
$FeSO_4$....1.0μg

Preparation of Synthetic Seawater: Add components to distilled/deionized water and bring volume to 1.0L. Mix thoroughly.

Preparation of Medium: Add solid components to synthetic seawater and bring volume to 1.0L. Mix thoroughly. Adjust pH to 6.8 with 2*N* KOH. Distribute into tubes or flasks. Autoclave for 15 min at 15 psi pressure–121°C.

Use: For the cultivation and maintenance of *Oceanospirillum maris.*

Peptone Succinate Salts Medium (PSS Medium)

Composition per liter:

Peptone....10.0g
$MgSO_4 \cdot 7H_2O$....1.0g
$(NH_4)_2SO_4$....1.0g
Succinic acid....1.0g
$FeCl_3 \cdot 6H_2O$....2.0mg
$MnSO_4 \cdot H_2O$....2.0mg

pH 6.8 ± 0.2 at 25°C

Preparation of Medium: Add solid components to seawater and bring volume to 1.0L. Mix thoroughly. Adjust pH to 6.8 with 2*N* KOH. Distribute into tubes or flasks. Autoclave for 15 min at 15 psi pressure–121°C.

Use: For the cultivation and maintenance of *Aquaspirillum anulus.*

Peptone Sucrose Broth

Composition per liter:

Sucrose....20.0g
Peptone....10.0g

Preparation of Medium: Add components to distilled/deionized water and bring volume to 1.0L. Mix thoroughly. Distribute into tubes or flasks. Autoclave for 15 min at 15 psi pressure–121°C.

Use: For the cultivation and maintenance of *Xanthomonas campestris*.

Peptone Water

Composition per liter:

Peptone	10.0g
NaCl	5.0g

pH 7.2 ± 0.2 at 25°C

Source: This medium is available as a premixed powder from Difco Laboratories and Oxoid Unipath.

Preparation of Medium: Add components to distilled/deionized water and bring volume to 1.0L. Mix thoroughly. Distribute into tubes or flasks. Autoclave for 15 min at 15 psi pressure–121°C.

Use: For the cultivation of nonfastidious microorganisms, for carbohydrate fermentation tests and for performing the indole test.

Peptone Water with Andrade's Indicator

Composition per liter:

Peptone	10.0g
NaCl	5.0g
Andrade's indicator	100.0mL
Carbohydrate solution	20.0mL

pH 7.4 ± 0.2 at 25°C

Source: This medium is available as a premixed powder from Oxoid Unipath.

Andrade's Indicator:

Composition per 100mL:

NaOH (1*N* solution)	16.0mL
Acid fuchsin	0.1 g

Caution: Acid fuchsin is a potential carcinogen and care must be taken to avoid inhalation of the powdered dye and contact with the skin.

Carbohydrate Solution:

Composition per 20mL:

Carbohydrate	5.0–10.0g

Preparation of Carbohydrate Solution: Add carbohydrate to distilled/deionized water and bring volume to 20.0mL. Mix thoroughly. Filter sterilize.

Preparation of Medium: Add components, except carbohydrate solution, to distilled/deionized water and bring volume to 980.0mL. Mix thoroughly. Adjust pH to 7.4 if necessary. Distribute into tubes containing an inverted Durham tube. Fill each tube with 9.8mL of medium. Autoclave for 15 min at 15 psi pressure–121°C. Aseptically add 0.2mL of sterile carbohydrate solution to each tube.

Use: For use in carbohydrate fermentation tests. Fermentation is determined by the production of acid—broth turns pink—and formation of gas—bubble trapped in Durham tube.

Peptone Yeast Extract Agar (ATCC Medium 526)

Composition per liter:

Agar	15.0g
Peptone	10.0g
Yeast extract	3.0g

Preparation of Medium: Add components to distilled/deionized water and bring volume to 1.0L. Mix thoroughly. Gently heat and bring to boiling. Distribute into tubes or flasks. Autoclave for 15 min at 15 psi pressure–121°C. Pour into sterile Petri dishes or leave in tubes.

Use: For the cultivation and maintenance of *Bdellovibrio bacteriovorus* and *B. stolpii*.

Peptone Yeast Extract Glucose Medium, Modified (MPYG Medium)

Composition per 950mL:

Peptone	10.0g
Yeast extract	10.0g
Glucose	5.0g
Cysteine·HCl·H_2O	0.5g
$(NH_4)_2SO_4$	0.5g
Salt solution	40.0mL
Vitamin K-heme solution	10.0mL
Resazurin (0.025% solution)	4.0mL
Volatile fatty acid solution	3.1mL

pH 7.0 ± 0.2 at 25°C

Salt Solution:

Composition per liter:

$NaHCO_3$	10.0g
NaCl	2.0g
K_2HPO_4	1.0g
KH_2PO_4	1.0g
$CaCl_2$ (anhydrous)	0.2g
$MgSO_4$	0.2g

Preparation of Salt Solution: Add $CaCl_2$ and $MgSO_4$ to 300.0mL of distilled/deionized water. Mix thoroughly until dissolved. Bring volume to

800.0mL with distilled/deionized water. Add remaining components while stirring. Bring volume to 1.0L. Mix thoroughly. Store at 4°C.

Vitamin K-Heme Solution:

Composition per liter:

Part A	100.0mL
Part B	1.0mL

Preparation of Vitamin K-Heme Solution: Aseptically add 1.0mL of sterile part B to 100.0mL of cooled sterile part A. Mix thoroughly.

Part A:

Composition per 100mL:

Hemin	50.0mg
NaOH (1*N* solution)	1.0mL

Preparation of Part A: Add hemin to NaOH solution and bring volume to 100.0mL with distilled/deionized water. Mix thoroughly. Autoclave for 15 min at 15 psi pressure–121°C. Cool to 45–50°C.

Part B:

Composition per 30mL:

Menadione (vitamin K_3)	100.0mg
Ethanol (95% solution)	30.0mL

Preparation of Part B: Add menadione to ethanol. Mix thoroughly. Filter sterilize.

Volatile Fatty Acid Solution:

Composition per 31mL:

Acetic acid	17.0mL
Propionic acid	6.0mL
n-Butyric acid	4.0mL
n-Valeric acid	1.0mL
Isovaleric acid	1.0mL
Isobutyric acid	1.0mL
DL-α-Methyl butyric acid	1.0mL

Preparation of Volatile Fatty Acid Solution: Combine components. Mix thoroughly.

Preparation of Medium: Add components—except vitamin K-heme solution, cysteine·HCl·H_2O and volatile fatty acid solution—to distilled/deionized water and bring volume to 936.9mL. Gently heat and bring to boiling under 97% N_2 + 3% H_2. Continue boiling until resazurin turns colorless indicating reduction. Cool to 45–50°C. Add vitamin K-heme solution, cysteine·HCl·H_2O and volatile fatty acid solution. Adjust pH to 7.0. Distribute into tubes under 97% N_2 + 3% H_2. Cap with rubber stoppers. Place tubes in a press. Autoclave for 15 min at 15 psi pressure–121°C with fast exhaust.

Use: For the cultivation and maintenance of *Acetivibrio ethanolgignens, Butyrivibrio fibrisolvens, Lachnospira multipara,* and *Succinivibrio dextrinosolvens.*

Peptonized Milk Agar (PMA Medium)

Composition per liter:

Agar	15.0g
Milk, peptonized	1.0g

Preparation of Medium: Add components to distilled/deionized water and bring volume to 1.0L. Mix thoroughly. Gently heat and bring to boiling. Distribute into tubes or flasks. Autoclave for 15 min at 15 psi pressure–121°C. Pour into sterile Petri dishes or leave in tubes.

Use: For the cultivation of freshwater *Myxobacterium* species.

Pfizer Selective *Enterococcus* Agar (PSE Agar)

Peptone C	17.0g
Agar	15.0g
Bile	10.0g
NaCl	5.0g
Yeast extract	5.0g
Peptone B	3.0g
Esculin	1.0g
Sodium citrate	1.0g
Ferric ammonium citrate	0.5g
NaN_3	0.25g

pH 7.1 ± 0.2 at 25°C

Caution: Sodium azide is toxic. Azides also react with metals and disposal must be highly diluted.

Preparation of Medium: Add components to distilled/deionized water and bring volume to 1.0L. Mix thoroughly. Gently heat and bring to boiling. Distribute into tubes or flasks. Autoclave for 15 min at 15 psi pressure–121°C. Pour into sterile Petri dishes or leave in tubes.

Use: For the selective isolation, cultivation and enumeration of *Enterococcus* species by the multiple tube technique.

PFS Medium (Peptone Fumarate Sulfate Medium)

Composition per liter:

Peptone	10.0g
Fumaric acid	2.0g
$(NH_4)_2SO_4$	1.0g
$MgSO_4 \cdot 7H_2O$	0.5g
$FeCl_3 \cdot 6H_2O$	0.2mg
$MnSO_4 \cdot H_2O$	0.2mg

pH 7.0 ± 0.2 at 25°C

Preparation of Medium: Add components to distilled/deionized water and bring volume to 1.0L. Mix thoroughly. Adjust pH to 7.0 with KOH. Distribute into tubes or flasks. Autoclave for 15 min at 15 psi pressure–121°C.

Use: For the cultivation and maintenance of *Aquaspirillum fasciculus*.

PGP Broth (Peptone Glycerol Phosphate Broth)

Composition per liter:

Peptone	5.0g
K_2HPO_4	2.0g
Glycerol	10.0mL

Preparation of Medium: Add components to distilled/deionized water and bring volume to 1.0L. Mix thoroughly. Distribute into tubes or flasks. Autoclave for 15 min at 15 psi pressure–121°C.

Use: For the cultivation and maintenance of *Serratia marcescens*.

PGS Agar (Peptone Glucose Salt Agar)

Composition per liter:

Agar	15.0g
Glucose	10.0g
NaCl	10.0g
Peptone	10.0g

pH 7.2 ± 0.2 at 25°C

Preparation of Medium: Add components to distilled/deionized water and bring volume to 1.0L. Mix thoroughly. Gently heat and bring to boiling. Distribute into tubes or flasks. Autoclave for 15 min at 15 psi pressure–121°C. Pour into sterile Petri dishes or leave in tubes.

Use: For the cultivation and maintenance of *Rhodococcus australis*.

PGY Agar (Peptone Glucose Yeast Extract Agar)

Composition per liter:

Agar	15.0g
Peptone	10.0g
Yeast extract	5.0g
Glucose	1.0g

Preparation of Medium: Add components to distilled/deionized water and bring volume to 1.0L. Mix thoroughly. Gently heat and bring to boiling. Distribute into tubes or flasks. Autoclave for 15 min at 15 psi pressure–121°C. Pour into sterile Petri dishes or leave in tubes.

Use: For the cultivation and maintenance of *Micrococcus luteus*.

Phenethyl Alcohol Agar (Phenylethanol Agar) (Phenylethyl Alcohol Agar)

Composition per liter:

Agar	15.0g
Pancreatic digest of casein	15.0g
NaCl	5.0g
Papaic digest of soybean meal	5.0g
β-Phenethyl alcohol	2.5g
Blood	50.0mL

pH 7.3 ± 0.2 at 25°C

Source: This medium is available as a premixed powder from BBL Microbiology Systems.

Preparation of Medium: Add components, except blood, to distilled/deionized water and bring volume to 950.0mL. Mix thoroughly. Gently heat and bring to boiling. Autoclave for 15 min at 13 psi pressure–118°C. Cool to 45–50°C. Aseptically add sterile defibrinated blood. Mix thoroughly. Pour into sterile Petri dishes or distribute into sterile tubes.

Use: For the selective isolation of Gram-positive bacteria, particularly Gram-positive cocci, from specimens with a mixed flora. Do not use for the observation of hemolytic reactions.

Phenol Red Agar

Composition per liter:

Agar	15.0g
Pancreatic digest of casein	10.0g
NaCl	5.0g
Phenol red	0.018g
Carbohydrate solution	20.0mL

pH 7.4 ± 0.2 at 25°C

Source: This medium is available as a premixed powder from BBL Microbiology Systems.

Carbohydrate Solution:
Composition per 20mL:

Carbohydrate	5.0–10.0g

Preparation of Carbohydrate Solution: Add carbohydrate to distilled/deionized water and bring volume to 20.0mL. Mix thoroughly. Filter sterilize.

Preparation of Medium: Add components, except carbohydrate solution, to distilled/deionized water and bring volume to 980.0mL. Mix thoroughly. Adjust pH to 7.4 if necessary. Autoclave for 15 min at 15 psi pressure–121°C. Cool to 45–50°C. Aseptically add 20.0mL of sterile carbohydrate solution. Pour into sterile Petri dishes or distribute into sterile tubes. Allow tubes to cool in a slanted position.

Use: For the determination of fermentation reactions. Bacteria that can ferment the added carbohydrate turn the medium yellow.

Phenol Red Agar

Composition per liter:

Agar	15.0g
Proteose peptone No. 3	10.0g
NaCl	5.0g
Beef extract	1.0g
Phenol red	0.025g
Carbohydrate solution	20.0mL

pH 7.4 ± 0.2 at 25°C

Source: This medium is available as a premixed powder from Difco Laboratories.

Carbohydrate Solution:
Composition per 20mL:

Carbohydrate	5.0–10.0g

Preparation of Carbohydrate Solution: Add carbohydrate to distilled/deionized water and bring volume to 20.0mL. Mix thoroughly. Filter sterilize.

Preparation of Medium: Add components, except carbohydrate solution, to distilled/deionized water and bring volume to 980.0mL. Mix thoroughly. Adjust pH to 7.4 if necessary. Autoclave for 15 min at 15 psi pressure–121°C. Cool to 45–50°C. Aseptically add 20.0mL of sterile carbohydrate solution. Pour into sterile Petri dishes or distribute into sterile tubes. Allow tubes to cool in a slanted position.

Use: For the determination of fermentation reactions. Bacteria that can ferment the added carbohydrate turn the medium yellow.

Phenol Red Broth

Composition per liter:

Pancreatic digest of casein	10.0g
NaCl	5.0g
Phenol red	0.018g
Carbohydrate solution	20.0mL

pH 7.4 ± 0.2 at 25°C

Source: This medium is available as a premixed powder from BBL Microbiology Systems.

Carbohydrate Solution:
Composition per 20mL:

Carbohydrate	5.0–10.0g

Preparation of Carbohydrate Solution: Add carbohydrate to distilled/deionized water and bring volume to 20.0mL. Mix thoroughly. Filter sterilize.

Preparation of Medium: Add components, except carbohydrate solution, to distilled/deionized water and bring volume to 980.0mL. Mix thoroughly. Adjust pH to 7.4 if necessary. Distribute into tubes containing an inverted Durham tube. Fill each tube with 9.8mL of medium. Autoclave for 15 min at 13 psi pressure–118°C. Cool to 45–50°C. Aseptically add 0.2mL of sterile carbohydrate solution to each tube.

Use: For the determination of fermentation reactions in the differentiation of microorganisms. Fermentation is determined by the production of acid—broth turns yellow—and formation of gas—bubble trapped in Durham tube.

Phenol Red Glucose Broth

Composition per liter:

Pancreatic digest of casein	10.0g
Glucose	5.0g
NaCl	5.0g
Phenol red	0.018g

pH 7.3 ± 0.2 at 25°C

Source: This medium is available as a premixed powder from BBL Microbiology Systems.

Preparation of Medium: Add components to distilled/deionized water and bring volume to 1.0L. Mix thoroughly. Adjust pH to 7.3 if necessary. Distribute into tubes containing an inverted Durham tube. Fill each tube with 10.0mL of medium. Autoclave for 15 min at 13 psi pressure–118°C.

Use: For the determination of the ability of a microorganism to ferment glucose. Fermentation is determined by the production of acid—broth turns yellow—and formation of gas—bubble trapped in Durham tube.

Phenol Red Lactose Agar

Composition per liter:

Agar	15.0g
Lactose	10.0g
Proteose peptone No. 3	10.0g
NaCl	5.0g
Beef extract	1.0g
Phenol red	25.0mg

pH 7.4 ± 0.2 at 25°C

Source: This medium is available as a premixed powder from Difco Laboratories.

Preparation of Medium: Add components to distilled/deionized water and bring volume to 1.0L. Mix thoroughly. Gently heat and bring to boiling. Distrib-

ute into tubes or flasks. Autoclave for 15 min at 13 psi pressure–118°C. Pour into sterile Petri dishes or leave in tubes. Allow tubes to cool in a slanted position.

Use: For the determination of the ability of a microorganism to ferment lactose. Fermentation is determined by the production of acid—medium turns yellow.

Phenol Red Lactose Broth

Composition per liter:

Pancreatic digest of casein	10.0g
Lactose	5.0g
NaCl	5.0g
Phenol red	0.018g

pH 7.3 ± 0.2 at 25°C

Source: This medium is available as a premixed powder from BBL Microbiology Systems.

Preparation of Medium: Add components to distilled/deionized water and bring volume to 1.0L. Mix thoroughly. Adjust pH to 7.3 if necessary. Distribute into tubes containing an inverted Durham tube. Fill each tube with 10.0mL of medium. Autoclave for 15 min at 13 psi pressure–118°C.

Use: For the determination of the ability of a microorganism to ferment lactose. Fermentation is determined by the production of acid—broth turns yellow—and formation of gas—bubble trapped in Durham tube.

Phenol Red Mannitol Agar

Composition per liter:

Agar	15.0g
Mannitol	10.0g
Proteose peptone No. 3	10.0g
NaCl	5.0g
Beef extract	1.0g
Phenol red	25.0mg

pH 7.4 ± 0.2 at 25°C

Source: This medium is available as a premixed powder from Difco Laboratories.

Preparation of Medium: Add components to distilled/deionized water and bring volume to 1.0L. Mix thoroughly. Gently heat and bring to boiling. Distribute into tubes or flasks. Autoclave for 15 min at 13 psi pressure–118°C. Pour into sterile Petri dishes or leave in tubes. Allow tubes to cool in a slanted position.

Use: For the determination of the ability of a microorganism to ferment mannitol. Fermentation is determined by the production of acid—medium turns yellow.

Phenol Red Mannitol Broth

Composition per liter:

Pancreatic digest of casein	10.0g
D-Mannitol	5.0g
NaCl	5.0g
Phenol red	0.018g

pH 7.3 ± 0.2 at 25°C

Source: This medium is available as a premixed powder from BBL Microbiology Systems.

Preparation of Medium: Add components to distilled/deionized water and bring volume to 1.0L. Mix thoroughly. Adjust pH to 7.3 if necessary. Distribute into tubes containing an inverted Durham tube. Fill each tube with 10.0mL of medium. Autoclave for 15 min at 13 psi pressure–118°C.

Use: For the determination of the ability of a microorganism to ferment mannitol. Fermentation is determined by the production of acid—broth turns yellow—and formation of gas—bubble trapped in Durham tube.

Phenol Red Sucrose Broth

Composition per liter:

Pancreatic digest of casein	10.0g
NaCl	5.0g
Sucrose	5.0g
Phenol red	0.018g

pH 7.3 ± 0.2 at 25°C

Source: This medium is available as a premixed powder from BBL Microbiology Systems.

Preparation of Medium: Add components to distilled/deionized water and bring volume to 1.0L. Mix thoroughly. Adjust pH to 7.3 if necessary. Distribute into tubes containing an inverted Durham tube. Fill each tube with 10.0mL of medium. Autoclave for 15 min at 13 psi pressure–118°C.

Use: For the determination of the ability of a microorganism to ferment sucrose. Fermentation is determined by the production of acid—broth turns yellow—and formation of gas—bubble trapped in Durham tube.

Phenylalanine Agar (Phenylalanine Deaminase Medium)

Composition per liter:

Agar	12.0g
NaCl	5.0g
Yeast extract	3.0g

DL-Phenylalanine 2.0g
Na_2HPO_4 .... 1.0g

pH 7.3 ± 0.2 at 25°C

Source: This medium is available as a premixed powder from BBL Microbiology Systems, and Difco Laboratories.

Preparation of Medium: Add components to distilled/deionized water and bring volume to 1.0L. Mix thoroughly. Gently heat while stirring and bring to boiling. Distribute into tubes or flasks. Autoclave for 10 min at 15 psi pressure–121°C. Pour into sterile Petri dishes or leave in tubes.

Use: For the differentiation of enteric Gram-negative bacilli on the basis of their ability to produce phenylpyruvic acid from phenylalanine. After appropriate incubation of bacteria, ferric chloride reagent is added on the agar. Formation of a green color in 1–5 min indicates the production of phenylpyruvic acid.

Phosphate Mineral Salts Medium with Octane

Composition per liter:

$(NH_4)_2HPO_4$ 10.0g
K_2HPO_4 .... 5.0g
Na_2SO_4 .... 0.5g
Octane 10.0mL

Preparation of Medium: Add components, except octane, to tap water and bring volume to 990.0mL. Mix thoroughly. Autoclave for 15 min at 15 psi pressure–121°C. Prior to inoculation, filter sterilize octane. Aseptically add sterile octane to sterile medium. Aseptically distribute into sterile tubes or flasks.

Use: For the cultivation and maintenance of *Pseudomonas oleovorans.*

Photobacterium Broth

Composition per liter:

NaCl 30.0g
Sodium glycerol phosphate 23.5g
Pancreatic digest of casein 5.0g
KH_2PO_4 .... 3.0g
Yeast extract 2.5g
$CaCO_3$ 1.0g
NH_4Cl 0.3g
$MgSO_4·7H_2O$ 0.3g
$FeCl_3$ 0.01g

Source: This medium is available as a premixed powder from Difco Laboratories.

Preparation of Medium: Add components to distilled/deionized water and bring volume to 1.0L. Mix thoroughly. Distribute into tubes or flasks to form a shallow layer of medium. Autoclave for 15 min at 15 psi pressure–121°C.

Use: For the cultivation and demonstration of luminescence by photobacteria. For the cultivation and maintenance of *Alteromonas hanedai, Photobacterium phosphoreum, Shewanella hanedai, Vibrio fischeri, V. harveyi,* and other *V.* species.

Photobacterium MPY Medium

Composition per liter:

NaCl 28.2g
$MgSO_4·7H_2O$ 6.9g
$MgCl_2·6H_2O$ 5.5g
Peptone 5.0g
Yeast extract 3.0g
$CaCl_2·2H_2O$ 1.5g
KCl 0.7g

pH 7.4 ± 0.2 at 25°C

Preparation of Medium: Add components to distilled/deionized water and bring volume to 1.0L. Mix thoroughly. Distribute into tubes or flasks. Autoclave for 15 min at 15 psi pressure–121°C.

Use: For the cultivation and maintenance of *Photobacterium leiognathi.*

Phthalic Acid Medium

Composition per liter:

Solution 1 .... 400.0mL
Solution 2 .... 400.0mL
Potassium hydrogen phthalate solution 200.0mL

pH 6.8 ± 0.2 at 25°C

Solution 1:

Composition per 400mL:

KH_2PO_4 .... 9.1g
$(NH_4)_2SO_4$ 1.2g

Preparation of Solution 1: Add components to distilled/deionized water and bring volume to 400.0mL. Mix thoroughly. Adjust pH to 6.8 with KOH. Autoclave for 15 min at 15 psi pressure–121°C. Cool to 25°C.

Solution 2:

Composition per 400mL:

$MgSO_4·7H_2O$ 0.4g
$FeSO_4·7H_2O$ 0.01g

Preparation of Solution 2: Add components to distilled/deionized water and bring volume to 400.0mL. Mix thoroughly. Adjust pH to 6.8 with KOH. Autoclave for 15 min at 15 psi pressure–121°C. Cool to 25°C.

Potassium Hydrogen Phthalate Solution:
Composition per 200mL:

Potassium hydrogen phthalate	1.0g

Preparation of Medium: Add component to distilled/deionized water and bring volume to 200.0mL. Mix thoroughly. Adjust pH to 6.8 with KOH. Autoclave for 15 min at 15 psi pressure–121°C. Cool to 25°C.

Preparation of Medium: Aseptically combine the three sterile solutions. Mix thoroughly. Aseptically distribute into sterile tubes or flasks.

Use: For the cultivation and maintenance of *Pseudomonas cepacia.*

Plant *Mycoplasma* Agar

Composition per 291.2mL:

Schneider's *Drosophila* medium	160.0mL
Solution 1	70.0mL
Fetal calf serum	50.0mL
Fresh yeast extract solution	10.0mL
Phenol red (0.5% solution)	1.2mL

pH 7.4 ± 0.2 at 25°C

Schneider's *Drosophila* Medium:
Composition per liter:

$MgSO_4 \cdot 7H_2O$	3.7g
NaCl	2.1g
Yeast extract	2.0g
Trehalose	2.0g
D-Glucose	2.0g
L-Glutamine	1.8g
L-Lysine·HCl	1.7g
L-Proline	1.7g
KCl	1.6g
$Na_2HPO_4 \cdot 7H_2O$	1.3g
L-Glutamic acid	0.8g
L-Methionine	0.8g
$CaCl_2$, anhydrous	0.6g
KH_2PO_4	0.5g
β-Alanine	0.5g
L-Tyrosine	0.5g
L-Arginine	0.4g
L-Aspartic acid	0.4g
L-Histidine	0.4g
L-Threonine	0.4g
$NaHCO_3$	0.4g
Glycine	0.3g
L-Serine	0.3g
L-Valine	0.3g
L-Isoleucine	0.2g
L-Leucine	0.2g
L-Phenylalanine	0.2g
α-Ketoglutaric acid	0.2g
Fumaric acid	0.1g
Malic acid	0.1g
Succinic acid	0.1g
L-Cystine	0.1g
L-Tryptophan	0.1g
L-Cysteine	0.06g

Preparation of Schneider's *Drosophila* Medium: Add components to distilled/deionized water and bring volume to 1.0L. Mix thoroughly. Filter sterilize.

Solution 1:
Composition per 70mL:

Sorbitol	7.0g
Noble agar	5.0g
Beef heart, solids from infusion	5.0g
Peptone	1.8g
Pancreatic digest of casein	1.0g
Sucrose	1.0g
NaCl	0.5g
D-Fructose	0.1g
D-Glucose	0.1g

Preparation of Solution 1: Add components to distilled/deionized water and bring volume to 70.0mL. Mix thoroughly. Adjust pH to 7.8 with 1*N* NaOH. Autoclave for 15 min at 15 psi pressure–121°C. Cool to 50°C.

Fresh Yeast Extract Solution:
Composition per 100mL:

Baker's yeast live, pressed, starch-free	25.0g

Preparation of Fresh Yeast Extract Solution: Add the live Baker's yeast to 100.0mL of distilled/deionized water. Autoclave for 90 min at 15 psi pressure–121°C. Allow to stand. Remove supernatant solution. Adjust pH to 6.6–6.8.

Preparation of Medium: Bring fetal calf serum and phenol red solution to 56°C. Rapidly bring Schneider's *Drosophila* Medium to 37°C. Rapidly combine the components. Mix thoroughly. Pour into sterile Petri dishes or distribute into sterile tubes or flasks.

Use: For the cultivation and maintenance of *Spiroplasma floricola, S. kunkelii, S. melliferum,* and *S.* species.

Plant *Mycoplasma* Broth

Composition per 291.2mL:

Schneider's *Drosophila* medium	160.0mL
Solution 1	70.0mL
Fetal calf serum	50.0mL
Fresh yeast extract solution	10.0mL
Phenol red (0.5% solution)	1.2mL

pH 7.4 ± 0.2 at 25°C

Schneider's *Drosophila* Medium:
Composition per liter:

$MgSO_4 \cdot 7H_2O$	3.7g
NaCl	2.1g
Trehalose	2.0g
Yeast extract	2.0g
D-Glucose	2.0g
L-Glutamine	1.8g
L-Lysine·HCl	1.7g
L-Proline	1.7g
KCl	1.6g
$Na_2HPO_4 \cdot 7H_2O$	1.3g
L-Glutamic acid	0.8g
L-Methionine	0.8g
$CaCl_2$, anhydrous	0.6g
KH_2PO_4	0.5g
β-Alanine	0.5g
L-Tyrosine	0.5g
L-Arginine	0.4g
L-Aspartic acid	0.4g
L-Histidine	0.4g
L-Threonine	0.4g
$NaHCO_3$	0.4g
Glycine	0.3g
L-Serine	0.3g
L-Valine	0.3g
L-Isoleucine	0.2g
L-Leucine	0.2g
L-Phenylalanine	0.2g
α-Ketoglutaric acid	0.2g
Fumaric acid	0.1g
Malic acid	0.1g
Succinic acid	0.1g
L-Cystine	0.1g
L-Tryptophan	0.1g
L-Cysteine	0.06g

Preparation of Schneider's *Drosophila* Medium: Add components to distilled/deionized water and bring volume to 1.0L. Mix thoroughly. Filter sterilize.

Solution 1:
Composition per 70mL:

Sorbitol	7.0g
Beef heart, solids from infusion	5.0g
Peptone	1.8g
Pancreatic digest of casein	1.0g
Sucrose	1.0g
NaCl	0.5g
D-Fructose	0.1g
D-Glucose	0.1g

Preparation of Solution 1: Add components to distilled/deionized water and bring volume to 70.0mL. Mix thoroughly. Adjust pH to 7.8 with 1*N* NaOH. Autoclave for 15 min at 15 psi pressure–121°C. Cool to 25°C.

Fresh Yeast Extract Solution:
Composition per 100mL:

Baker's yeast live, pressed, starch-free	25.0g

Preparation of Fresh Yeast Extract Solution: Add the live Baker's yeast to 100.0mL of distilled/deionized water. Autoclave for 90 min at 15 psi pressure–121°C. Allow to stand. Remove supernatant solution. Adjust pH to 6.6–6.8.

Preparation of Medium: Bring fetal calf serum and phenol red solution to 56°C. Rapidly bring Schneider's *Drosophila* medium to 37°C. Rapidly combine the components. Mix thoroughly. Aseptically distribute into sterile tubes or flasks.

Use: For the cultivation and maintenance of *Spiroplasma floricola*, *S. kunkelii*, *S. melliferum*, and *S.* species.

Plate Count Agar (Tryptone Glucose Yeast Agar)

Composition per liter:

Agar	9.0g
Pancreatic digest of casein	5.0g
Yeast extract	2.5g
Glucose	1.0g

pH 7.0 ± 0.2 at 25°C

Source: This medium is available as a premixed powder from Oxoid Unipath.

Preparation of Medium: Add components to distilled/deionized water and bring volume to 1.0L. Mix thoroughly. Gently heat while stirring and bring to boiling. Distribute into tubes or flasks. Autoclave for 15 min at 15 psi pressure–121°C. Pour into sterile Petri dishes or leave in tubes.

Use: For the enumeration of the number of live heterotrophic bacteria in water. For the cultivation and maintenance of *Brevibacterium casei*, *B. epedermidis*, and *Methylobacterium mesophilicum*.

Plate Count Agar (ATCC Medium 1048)

Composition per liter:

Agar	15.0g
Pancreatic digest of casein	5.0g
Yeast extract	2.5g
Glucose	1.0g

pH 7.0 ± 0.2 at 25°C

Source: This medium is available as a premixed powder from Difco Laboratories and Oxoid Unipath.

Preparation of Medium: Add components to distilled/deionized water and bring volume to 1.0L. Mix thoroughly. Gently heat and bring to boiling. Distrib-

ute into tubes or flasks. Autoclave for 15 min at 15 psi pressure–121°C. Pour into sterile Petri dishes or leave in tubes.

Use: For the enumeration of bacteria in water.

Plate Count Agar, Modified

Composition per liter:

Pancreatic digest of casein	20.0g
Yeast extract	20.0g
Agar	10.0g
Glucose	4.0g

pH 7.0 ± 0.1 at 25°C

Preparation of Medium: Add components to distilled/deionized water and bring volume to 1.0L. Mix thoroughly. Gently heat and bring to boiling. Distribute into bottles. Autoclave for 15 min at 15 psi pressure–121°C.

Use: For the cultivation and enumeration of microorganisms from food by the plate count method.

Plate Count Broth (m–Plate Count Broth)

Composition per liter:

Yeast extract	5.0g
Glucose	2.0g

pH 7.0 ± 0.2 at 25°C

Source: This medium is available as a premixed powder from Difco Laboratories.

Preparation of Medium: Add components to distilled/deionized water and bring volume to 1.0L. Mix thoroughly. Distribute into tubes or flasks. Autoclave for 15 min at 15 psi pressure–121°C.

Use: For the determination of bacterial counts by the membrane filter method.

PMY Medium

Composition per liter:

Agar	15.0g
Glucose	10.0g
NaCl	5.0g
Polypeptone™	5.0g
Beef extract	2.0g
Yeast extract	1.0g
$MgSO_4.7H_2O$	0.5g

pH 7.0 ± 0.2 at 25°C

Preparation of Medium: Add components to distilled/deionized water and bring volume to 1.0L. Mix thoroughly. Gently heat and bring to boiling. Distribute into tubes or flasks. Autoclave for 15 min at 15 psi pressure–121°C. Pour into sterile Petri dishes or leave in tubes.

Use: For the cultivation and maintenance of *Xanthomonas campestris*.

Poly-β-Hydroxybutyrate Medium (PHB Medium)

Composition per liter:

Part A	900.0mL
Part B	100.0mL

pH 7.2 ± 0.2 at 25°C

Part A:

Composition per 900mL:

$K_2HPO_4 \cdot 3H_2O$	0.6g
KH_2PO_4	0.2g
$MgSO_4 \cdot 7H_2O$	0.2g
$(NH_4)_2SO_4$	0.2g

Preparation of Medium: Add components to distilled/deionized water and bring volume to 900.0mL. Mix thoroughly. Adjust pH to 7.2. Autoclave for 15 min at 15 psi pressure–121°C. Cool to 25°C.

Part B:

Composition per 100mL:

Glucose	10.0g

Preparation of Medium: Add glucose to distilled/deionized water and bring volume to 100.0mL. Mix thoroughly. Autoclave for 15 min at 15 psi pressure–121°C. Cool to 25°C.

Preparation of Medium: Aseptically combine 900.0mL of cooled, sterile part A and 100.0mL of cooled, sterile part B. Mix thoroughly. Aseptically distribute into sterile tubes or flasks.

Use: For the cultivation and differentiation of *Pseudomonas* species based on their ability to produce intracellular poly-β-hydroxybutyrate. Production of poly-β-hydroxybutyrate is determined by staining cells with Sudan Black B.

Potassium Cyanide Broth

Composition per liter:

Na_2HPO_4	5.64g
NaCl	5.0g
Proteose peptone No. 3	3.0g
KH_2PO_4	0.225g
KCN solution	15.0mL

pH 7.6 ± 0.2 at 25°C

KCN Solution:

Composition per 100mL:

KCN	0.5g

Preparation of KCN Solution: Add KCN to distilled/deionized water and bring volume to 100.0mL. Mix thoroughly.

Caution: Cyanide is toxic.

Preparation of Medium: Add components, except KCN solution, to distilled/deionized water and bring volume to 985.0mL. Mix thoroughly. Gently heat and bring to boiling. Autoclave for 15 min at 15 psi pressure–121°C. Cool to 25°C. Aseptically add 15.0mL of KCN solution. Mix thoroughly. Distribute into sterile screw-capped tubes or flasks in 1.0–1.5mL volumes. Close caps tightly.

Use: For the cultivation and differentiation of urease negative Gram-negative enteric bacteria. *Salmonella* species and *Shigella* species are nonmotile in this medium. *Proteus* species are motile in this medium.

Potassium Tellurite Agar

Composition per liter:

Beef heart, solids from infusion	500.0g
Agar	15.0g
Tryptose	10.0g
NaCl	5.0g
K_2TeO_3 solution	20.0mL
Blood, defibrinated	50.0mL

pH 6.0 ± 0.2 at 25°C

K_2TeO_3 Solution:

Composition per 20mL:

K_2TeO_3	0.5g

Preparation of K_2TeO_3 Solution: Add K_2TeO_3 to distilled/deionized water and bring volume to 10.0mL. Mix thoroughly. Filter sterilize.

Caution: Potassium tellurite is toxic.

Preparation of Medium: Add components, except K_2TeO_3 solution, to distilled/deionized water and bring volume to 930.0mL. Mix thoroughly. Gently heat and bring to boiling. Autoclave for 15 min at 15 psi pressure–121°C. Cool to 45–50°C. Aseptically add sterile K_2TeO_3 solution and 50.0mL of blood. Rabbit or sheep blood may be used. Mix thoroughly. Pour into sterile Petri dishes or distribute into sterile tubes. Allow tubes to cool in a slanted position.

Use: For the cultivation and differentiation of *Enterococcus faecalis. E. faecalis* appears as black colonies.

Potato Carrot Medium

Composition per liter of tap water:

Potatoes (sliced with skin)	300.0g
Carrots (peeled and sliced)	25.0g
Agar	15.0g

Preparation of Medium: Slice potatoes with the skin on. Peel carrots and slice. Add potatoes and carrots to approximately 700.0mL of tap water. Gently heat and bring to boiling. Continue boiling for 20 min. Filter through cheesecloth. Bring volume of filtrate to 1.0L with distilled/deionized water. Add agar. Gently heat and bring to boiling. Distribute into tubes or flasks. Autoclave for 20 min at 15 psi pressure–121°C. Pour into sterile Petri dishes or leave in tubes.

Use: For the cultivation and maintenance of *Actinoplanes awajinensis, Actinoplanes nirasakinensis, Amorphosporangium auranticolor, Streptomyces flaveus,* and *Thermoactinomyces vulgaris.*

Potato Dextrose Agar

Composition per liter:

Glucose	20.0g
Agar	15.0g
Potato, infusion from	4.0g
Tartaric acid solution	14.0mL

pH 5.6 ± 0.2 at 25°C

Source: This medium is available as a premixed powder from BBL Microbiology Systems and Oxoid Unipath.

Tartaric Acid Solution:

Composition per 50mL:

Tartaric acid	5.0g

Preparation of Tartaric Acid Solution: Add tartaric acid to distilled/deionized water and bring volume to 50.0mL. Mix thoroughly. Filter sterilize.

Preparation of Medium: Add components to distilled/deionized water and bring volume to 986.0mL. Mix thoroughly. Gently heat and bring to boiling. Distribute into tubes or flasks. Autoclave for 15 min at 15 psi pressure–121°C. Cool to 45–50°C. Aseptically add 14.0mL of sterile tartaric acid solution. Mix thoroughly. If medium is to be used for the enumeration of yeasts and molds in butter, adjust pH to 3.5. Pour into sterile Petri dishes or distribute into sterile tubes.

Use: For the cultivation and enumeration of yeasts and molds.

Potato Dextrose Agar (PDA Agar)

Composition per liter:

Glucose	20.0g
Agar	15.0g
Potatoes, infusion from	1.0L

Potatoes, Infusion From:

Composition per liter:

Potatoes	300.0g

Preparation of Potatoes, Infusion From: Peel and dice potatoes. Add 500.0mL of distilled/deionized water. Gently heat and bring to boiling. Continue boiling for 30 min. Filter through cheesecloth. Bring volume of filtrate to 1.0L.

Preparation of Medium: To 1.0L of potato infusion, add glucose and agar. Mix thoroughly. Gently heat and bring to boiling. Distribute into tubes or flasks. Autoclave for 15 min at 15 psi pressure–121°C. Pour into sterile Petri dishes or leave in tubes.

Use: For the cultivation and maintenance of *Bacillus megaterium, B. subtilis, Pseudomonas lindbergii, P. syringae, Streptomyces testaceus,* and *Xanthomonas campestris.*

Potato Dextrose Broth

Composition per liter:

Potatoes, infusion from 200.0g
Glucose 20.0g

pH 5.1 ± 0.2 at 25°C

Source: This medium is available as a premixed powder from Difco Laboratories.

Potatoes, Infusion From:
Composition per 500mL:

Potatoes 300.0g

Preparation of Potatoes, Infusion From: Peel and dice potatoes. Add 500.0mL of distilled/deionized water. Gently heat and bring to boiling. Continue boiling for 30 min. Filter through cheesecloth.

Preparation of Medium: Add components to distilled/deionized water and bring volume to 1.0L. Mix thoroughly. Distribute into tubes or flasks. Autoclave for 15 min at 15 psi pressure–121°C.

Use: For the cultivation of a wide variety of yeasts and molds.

Potato Extract Agar

Composition per liter:

Agar..15.0g
Peptone..5.0g
NaCl...5.0g
Yeast extract..2.0g
Beef extract powder..1.0g
Potato extract 20.0mL

pH 7.4 ± 0.2 at 25°C

Potato Extract:
Composition per liter:

Potatoes 300.0g

Preparation of Potato Extract: Peel and dice potatoes. Add 500.0mL of distilled/deionized water. Gently heat and bring to boiling. Continue boiling for 30 min. Filter through cheesecloth.

Use: For the cultivation of a wide variety of yeasts and molds.

Potato Flakes Agar

Composition per liter:

Potato flakes 20.0g
Agar..15.0g
Glucose 10.0g

Preparation of Medium: Add components to distilled/deionized water and bring volume to 1.0L. Mix thoroughly. Gently heat and bring to boiling. Distribute into tubes or flasks. Autoclave for 15 min at 15 psi pressure–121°C. Pour into sterile Petri dishes or leave in tubes.

Use: For the cultivation and induction of sporulation in all fungi.

Potato Glucose Agar

Composition per liter:

Potato, infusion from 500.0g
Glucose 20.0g
Agar..15.0g

Preparation of Medium: Peel and slice potatoes thinly. Add 800.0mL distilled/deionized water immediately to potatoes to prevent oxidation. Gently heat and bring to 60°C. Maintain at 60°C for 60 min. Filter through cheesecloth. Adjust volume of filtrate to 1.0L with distilled/deionized water. Add agar. Gently heat and bring to boiling. Add glucose. Mix thoroughly. Distribute into tubes or flasks. Autoclave for 20 min at 10 psi pressure–115°C. Pour into sterile Petri dishes or leave in tubes.

Use: For the cultivation and maintenance of *Nocardia asteroides, Pseudomonas caryophylli, P. syringae, Rhodococcus* species, *Streptomyces nobilis, S. prasinosporus,* and *S.* species.

Potato Infusion Agar (ATCC Medium 421)

Composition per liter:

Potato 200.0g
Agar..15.0g

Preparation of Medium: Peel and finely dice potatoes. Add to 500.0mL of distilled/deionized water. Gently heat and bring to boiling. Continue boiling for 20 min. Filter through cheesecloth. Bring volume of filtrate to 1.0L with distilled/deionized water. Add agar.

Gently heat and bring to boiling. Distribute into tubes or flasks. Autoclave for 15 min at 15 psi pressure–121°C. Pour into sterile Petri dishes or leave in tubes.

Use: For the cultivation and maintenance of *Streptomyces fradiae.*

Potato Malt Agar

Composition per liter:

Potatoes, infusion from	200.0g
Sucrose	60.0g
Agar	20.0g
Malt extract	20.0g
Peptone	1.0g

pH 7.4 ± 0.2 at 25°C

Source: This medium is available as a premixed powder from Difco Laboratories.

Potatoes, Infusion From:

Composition per 500mL:

Potatoes	300.0g

Preparation of Potatoes, Infusion From: Peel and dice potatoes. Add 500.0mL of distilled/deionized water. Gently heat and bring to boiling. Continue boiling for 30 min. Filter through cheesecloth.

Preparation of Medium: Add components to distilled/deionized water and bring volume to 1.0L. Mix thoroughly. Gently heat and bring to boiling. Distribute into tubes or flasks. Autoclave for 15 min at 15 psi pressure–121°C. Pour into sterile Petri dishes or leave in tubes.

Use: For the cultivation and maintenance of fungi and other aciduric microorganisms.

Potato P–YE *Thermus* Medium

Composition per liter:

Agar	20.0g
Peptone	5.0g
Yeast extract	0.2g
Potatoes, infusion from	200.0mL

pH 7.8 ± 0.2 at 25°C

Potatoes, Infusion From:

Composition per 500mL:

Potatoes	300.0g

Preparation of Medium: Add components to distilled/deionized water and bring volume to 1.0L. Mix thoroughly. Gently heat and bring to boiling. Distribute into tubes or flasks. Autoclave for 15 min at 15 psi pressure–121°C. Pour into sterile Petri dishes or leave in tubes.

Use: For the cultivation and maintenance of *Thermus ruber*.

Powell and Errington's Medium

Composition per 1060mL:

Solution 1	50.0mL
Solution 3	50.0mL
Solution 4	10.0mL
Solution 2	5.0mL

pH 7.0 ± 0.2 at 25°C

Solution 1:

Composition per liter:

$(NH_4)_2HPO_4$	238.0g
K_2SO_4	70.0g
$NaH_2PO_4 \cdot 2H_2O$	31.0g

Preparation of Solution 1: Add components to distilled/deionized water and bring volume to 1.0L. Mix thoroughly.

Solution 2:

Composition per liter:

MgO	10.0g
$FeCl_3 \cdot 6H_2O$	5.4g
$CaCO_3$	2.0g
$ZnSO_4 \cdot 7H_2O$	1.44g
$MnSO_4 \cdot 4H_2O$	1.11g
$Na_2MoO_4 \cdot 2H_2O$	0.49g
$CoSO_4 \cdot 7H_2O$	0.28g
$CuSO_4 \cdot 5H_2O$	0.25g
H_3BO_4	0.062g
HCl, concentrated	50.0mL

Preparation of Solution 2: Add components to distilled/deionized water and bring volume to 1.0L. Mix thoroughly.

Solution 3:

Composition per 50mL:

Citric acid	4.2g
Glucose	3.6g
L-Glutamic acid	2.94g
Succinic acid	1.18g

Preparation of Solution 3: Add components to distilled/deionized water and bring volume to 50.0mL. Mix thoroughly. Filter sterilize.

Solution 4:

Composition per 10mL:

$Na_2S_2O_3 \cdot 5H_2O$	1.24g

Preparation of Solution 4: Add $Na_2S_2O_3 \cdot 5H_2O$ to distilled/deionized water and bring volume to 10.0mL. Mix thoroughly. Filter sterilize.

Preparation of Medium: Add 50.0mL of solution 1 and 5.0mL of solution 2. Mix thoroughly. Bring volume to 1.0L with distilled/deionized water. Autoclave for 15 min at 15 psi pressure–121°C. Cool to 25°C. Adjust pH to 7.0 with sterile NaOH. Aseptically add 50.0mL of solution 3 and 10.0mL of sterile solution 4. Mix thoroughly. Aseptically distribute into sterile tubes or flasks.

Use: For the cultivation of a variety of heterotrophic microorganisms.

PP Starch Medium

Composition per liter:

Polypeptone™	10.0g
Soluble starch	10.0g
K_2HPO_4	3.0g
$MgSO_4·7H_2O$	1.0g

Preparation of Medium: Add components to distilled/deionized water and bring volume to 1.0L. Mix thoroughly. Gently heat while stirring and bring to boiling. Distribute into tubes or flasks. Autoclave for 15 min at 15 psi pressure–121°C.

Use: For the cultivation and maintenance of *Bacillus mycoides*.

Presence–Absence Broth (P–A Broth)

Composition per liter:

Pancreatic digest of casein	10.0g
Lactose	7.5g
Pancreatic digest of gelatin	5.0g
Beef extract	3.0g
NaCl	2.5g
K_2HPO_4	1.375g
KH_2PO_4	1.375g
Sodium lauryl sulfate	0.05g
Bromcresol purple	8.5mg

pH 6.8 ± 0.2 at 25°C

Source: Available as a premixed powder from BBL Microbiology Systems and Difco Laboratories.

Preparation of Medium: Add components to distilled/deionized water and bring volume to 333.0mL. Mix thoroughly. Distribute into screw-capped 250.0mL milk dilution bottles in 50.0mL volumes. Autoclave for 15 min at 15 psi pressure–121°C.

Use: For the detection of coliform bacteria in water from treatment plants or distribution systems using the presence-absence coliform test.

Pril Xylose Ampicillin Agar (PXA Agar)

Composition per liter:

Agar	15.0g
Xylose	10.0g
Pancreatic digest of gelatin	5.0g
Beef extract	3.0g
Pril	0.2g
Ampicillin	0.03g
Phenol red	0.025g

pH 6.8 ± 0.2 at 25°C

Note: Pril is a quaternary ammonium detergent composed of a mixture of primary alkyl sulfate, alkyl-benzyl sulfonate and salts. It is available from Böhme Fettchemie GmbH, Düsseldorf, Germany.

Preparation of Medium: Add components to distilled/deionized water and bring volume to 1.0L. Mix thoroughly. Gently heat and bring to boiling. Distribute into tubes or flasks. Autoclave for 15 min at 15 psi pressure–121°C. Pour into sterile Petri dishes or leave in tubes.

Use: For the selective isolation and cultivation of *Aeromonas hydrophila*.

Propionispira Medium

Composition per liter:

Sodium lactate	4.0g
Yeast extract	1.0g
Mineral solution 2	50.0mL
Sodium carbonate solution	50.0mL
Mineral solution 1	25.0mL
Cysteine-sulfide reducing agent	20.0mL
Wolfe's mineral solution	10.0mL
Vitamin solution	10.0mL
Resazurin (0.025% solution)	4.0mL

pH 7.2 ± 0.2 at 25°C

Mineral Solution 1:

Composition per liter:

K_2HPO_4	6.0g

Preparation of Medium: Add K_2HPO_4 to distilled/deionized water and bring volume to 1.0L. Mix thoroughly.

Mineral Solution 2:

Composition per liter:

NaCl	12.0g
KH_2PO_4	6.0g
$(NH_4)_2SO_4$	6.0g
$MgSO_4·7H_2O$	2.4g
$CaCl_2·2H_2O$	1.6g

Preparation of Mineral Solution 2: Add components to distilled/deionized water and bring volume to 1.0L. Mix thoroughly.

Sodium Carbonate Solution:

Composition per 100mL:

Na_2CO_3	8.0g

Preparation of Sodium Carbonate Solution: Add Na_2CO_3 to distilled/deionized water and bring volume to 100.0mL Mix thoroughly.

Cysteine-Sulfide Reducing Agent:
Composition per 20mL:

L-Cysteine·HCl·H_2O	300.0mg
$Na_2S \cdot 9H_2O$	300.0mg

Preparation of Cysteine-Sulfide Reducing Agent: Add L-Cysteine·HCl·H_2O to 10.0mL of distilled/deionized water. Mix thoroughly. In a separate tube, add $Na_2S \cdot 9H_2O$ to 10.0mL of distilled/deionized water. Mix thoroughly. Gas both solutions with 100% N_2 and cap tubes. Autoclave both solutions for 15 min at 15 psi pressure–121°C using fast exhaust. Cool to 50°C. Aseptically combine the two solutions under 100% N_2.

Wolfe's Mineral Solution:
Composition per liter

$MgSO_4 \cdot 7H_2O$	3.0g
Nitrilotriacetic acid	1.5g
NaCl	1.0g
$MnSO_4 \cdot H_2O$	0.5g
$FeSO_4 \cdot 7H_2O$	0.1g
$CoCl_2 \cdot 6H_2O$	0.1g
$CaCl_2$	0.1g
$ZnSO_4 \cdot 7H_2O$	0.1g
$CuSO_4 \cdot 5H_2O$	0.01g
$AlK(SO_4)_2 \cdot 12H_2O$	0.01g
H_3BO_3	0.01g
$Na_2MoO_4 \cdot 2H_2O$	0.01g

Preparation of Wolfe's Mineral Solution: Add nitrilotriacetic acid to 500.0mL of distilled/deionized water. Dissolve by adjusting pH to 6.5 with KOH. Add remaining components. Add distilled/deionized water to 1.0L.

Wolfe's Vitamin Solution:
Composition per liter:

Pyridoxine·HCl	10.0mg
Thiamine·HCl	5.0mg
Riboflavin	5.0mg
Nicotinic acid	5.0mg
Calcium pantothenate	5.0mg
p-Aminobenzoic acid	5.0mg
Thioctic acid	5.0mg
Biotin	2.0mg
Folic acid	2.0mg
Cyanocobalamin	100.0μg

Preparation of Wolfe's Vitamin Solution: Add components to distilled/deionized water and bring volume to 1.0L. Mix thoroughly. Filter sterilize.

Preparation of Medium: Add components, except vitamin solution and cysteine-sulfide reducing agent, to distilled/deionized water and bring volume to 970.0mL. Mix thoroughly. Autoclave for 15 min at 15 psi pressure–121°C. Cool under 80% N_2 + 20% CO_2. Aseptically add the sterile vitamin solution and then the sterile cysteine-sulfide reducing agent. Adjust the pH to 7.2. Distribute aseptically and anaerobically into sterile tubes.

Use: For the cultivation and maintenance of *Propionispira arboris.*

Prosthecobacter Medium

Composition per liter:

Glucose	0.25g
$(NH_4)_2SO_4$	0.25g
Na_2HPO_4	0.071g
Hutner's mineral base	20.0mL

Hutner's Mineral Base:
Composition per liter:

$MgSO_4 \cdot 7H_2O$	29.7g
Nitrilotriacetic acid	10.0g
$CaCl_2 \cdot 2H_2O$	3.34g
$FeSO_4 \cdot 7H_2O$	0.10g
$(NH_4)_2MoO_4$	9.25mg
Metals "44"	50.0mL

Preparation of Hutner's Mineral Base: Add nitrilotriacetic acid to 500.0mL of distilled/deionized water. Dissolve by adjusting pH to 6.5 with KOH. Add remaining components. Readjust pH to 7.2 with H_2SO_4 or KOH. Add distilled/deionized water to 1.0L. Store at 5°C.

Metals "44":
Composition per 100mL:

$ZnSO_4 \cdot 7H_2O$	1.1g
$FeSO_4 \cdot 7H_2O$	0.5g
EDTA	0.25g
$MnSO_4 \cdot 7H_2O$	0.154g
$CuSO_4 \cdot 5H_2O$	0.04g
$Co(NO_3)_2 \cdot 6H_2O$	0.025g
$Na_2B_4O_7 \cdot 10H_2O$	0.018g

Preparation of Metals "44": Add a few drops of H_2SO_4 to distilled/deionized water to inhibit precipitate formation. Add components to acidified distilled/deionized water and bring volume to 100.0mL. Mix thoroughly.

Preparation of Medium: Add components to distilled/deionized water and bring volume to 1.0L. Mix thoroughly. Distribute into tubes or flasks. Autoclave for 15 min at 15 psi pressure–121°C.

Use: For the cultivation of *Prosthecobacter fusiformis.*

Proteose Yeast Extract Medium

Composition per liter:

Proteose peptone	20.0g
Glucose	10.0g
Yeast extract	5.0g

Preparation of Medium: Add components to distilled/deionized water and bring volume to 1.0L. Mix thoroughly. Distribute into tubes or flasks. Autoclave for 15 min at 15 psi pressure–121°C.

Use: For the cultivation of a variety of bacteria.

Provasoli Medium

Composition per liter:

NaCl	11.75g
$MgCl_2 \cdot 6H_2O$	5.35g
Na_2SO_4	2.0g
$CaCl_2 \cdot 2H_2O$	0.75g
Tris(hydroxymethyl)aminomethane	0.5g
KCl	0.35g
Na_2HPO_4	0.05g

pH 7.6 ± 0.2 at 25°C

Preparation of Medium: Add components to distilled/deionized water and bring volume to 1.0L. Mix thoroughly. Distribute into tubes or flasks. Autoclave for 15 min at 15 psi pressure–121°C.

Use: For the isolation and cultivation of *Leucothrix* species from marine habitats.

Pseudomonas aeruginosa Agar (*PA* Agar) (m–*PA* Agar) (m–*Pseudomonas aeruginosa* Agar)

Composition per liter:

Agar	15.0g
$Na_2S_2O_3$	6.8g
L-Lysine·HCl	5.0g
NaCl	5.0g
Xylose	2.5g
Yeast extract	2.0g
Lactose	1.25g
Sucrose	1.25g
Ferric ammonium citrate	0.8g
Sulfapyridine	0.176g
Cycloheximide	0.15g
Phenol red	0.08g
Nalidixic acid	0.037g
Kanamycin	8.5mg

pH 7.1 ± 0.2 at 25°C

Preparation of Medium: Add components—except sulfapyridine, cycloheximide, nalidixic acid, and kanamycin—to distilled/deionized water and bring volume to 1.0L. Mix thoroughly. Adjust pH to 6.5. Autoclave for 15 min at 15 psi pressure–121°C. Cool to 55–60°C. Readjust pH to 7.1. Aseptically add the sulfapyridine, cycloheximide, nalidixic acid, and kanamycin. Mix thoroughly. Pour into 50mm × 12mm Petri dishes in 3.0mL volumes.

Use: For the cultivation and estimation of numbers of *Pseudomonas aeruginosa* in water by the membrane filter method.

Pseudomonas Agar F

Composition per liter:

Proteose peptone No. 3	20.0g
Agar	15.0g
Glycerol	10.0g
Pancreatic digest of casein	10.0g
K_2HPO_4	1.5g
$MgSO_4 \cdot 7H_2O$	0.73g

pH 7.0 ± 0.2 at 25°C

Preparation of Medium: Add components to distilled/deionized water and bring volume to 1.0L. Mix thoroughly. Gently heat and bring to boiling. Distribute into tubes or flasks. Autoclave for 15 min at 15 psi pressure–121°C. Pour into sterile Petri dishes or leave in tubes.

Use: For the cultivation and observation of fluorescein production in *Pseudomonas* species.

Pseudomonas Agar F

Composition per liter:

Agar	15.0g
Glycerol	10.0g
Proteose peptone No. 3	10.0g
Pancreatic digest of casein	10.0g
K_2HPO_4	1.5g
$MgSO_4 \cdot 7H_2O$	1.5g

pH 7.0 ± 0.2 at 25°C

Source: This medium is available as a premixed powder from Difco Laboratories.

Preparation of Medium: Add components to distilled/deionized water and bring volume to 1.0L. Mix thoroughly. Gently heat and bring to boiling. Distribute into tubes or flasks. Autoclave for 15 min at 15 psi pressure–121°C. Pour into sterile Petri dishes or leave in tubes.

Use: For the isolation, cultivation and differentiation of *Pseudomonas aeruginosa* on the basis of pigment production.

Pseudomonas Agar P

Composition per liter:

Proteose peptone No. 3	20.0g
Agar	15.0g
Glycerol	10.0g

K_2HPO_4	10.0g
$MgCl_2 \cdot 6H_2O$	1.4g

pH 7.0 ± 0.2 at 25°C

Source: This medium is available as a premixed powder from Difco Laboratories.

Preparation of Medium: Add components to distilled/deionized water and bring volume to 1.0L. Mix thoroughly. Gently heat and bring to boiling. Distribute into tubes or flasks. Autoclave for 15 min at 15 psi pressure–121°C. Pour into sterile Petri dishes or leave in tubes.

Use: For the isolation, cultivation and differentiation of *Pseudomonas aeruginosa* on the basis of pigmentation.

Pseudomonas Basal Mineral Medium

Composition per liter:

K_2HPO_4	12.5g
KH_2PO_4	3.8g
$(NH_4)_2SO_4$	1.0g
$MgSO_4 \cdot 7H_2O$	0.1g
Carbon source (0.8*M* solution)	100.0mL
Trace element solution	5.0mL

pH 7.2 ± 0.2 at 25°C

Trace Element Solution:

Composition per liter:

H_3BO_3	0.232g
$ZnSO_4 \cdot 7H_2O$	0.174g
$FeSO_4(NH_4)_2SO_4 \cdot 6H_2O$	0.116g
$CoSO_4 \cdot 7H_2O$	0.096g
$(NH_4)_6Mo_7O_{24} \cdot 4H_2O$	0.022g
$CuSO_4 \cdot 5H_2O$	8.0mg
$MnSO_4 \cdot 4H_2O$	8.0mg

Preparation of Trace Element Solution: Add components to distilled/deionized water and bring volume to 1.0L. Mix thoroughly.

Carbon Source:

Composition per 100mL:

Glucose	14.4g

Preparation of Carbon Source: Add glucose to distilled/deionized water and bring volume to 100.0mL. Mix thoroughly. Filter sterilize. Other carbon sources may replace glucose. Prepare 0.8*M* carbon source solution.

Preparation of Medium: Add components, except carbon source, to distilled/deionized water and bring volume to 900.0mL. Mix thoroughly. Gently heat and bring to boiling. Autoclave for 15 min at 15 psi pressure–121°C. Cool to 45–50°C. Aseptically add 100.0mL of sterile carbon source. Mix thoroughly. Aseptically distribute into sterile tubes or flasks.

Use: For the cultivation and differentiation of *Pseudomonas* species based on their ability to grow on different carbon sources.

Pseudomonas bathycetes Medium

Composition per liter:

NaCl	24.0g
Proteose peptone	10.0g
$MgSO_4 \cdot 7H_2O$	7.0g
$MgCl_2$	5.3g
Yeast extract	3.0g
KCl	0.7g

pH 7.2–7.4 at 25°C

Preparation of Medium: Add components to distilled/deionized water and bring volume to 1.0L. Mix thoroughly. Distribute into tubes or flasks. Autoclave for 15 min at 15 psi pressure–121°C.

Use: For the cultivation and maintenance of *Alteromonas haloplanktis, Alteromonas nigrifaciens, Pseudomonas bathycetes,* and *Pseudomonas elongata.*

Pseudomonas CFC Agar

Composition per liter:

Pancreatic digest of gelatin	16.0g
Agar	11.0g
Pancreatic digest of casein	10.0g
K_2SO_4	10.0g
$MgCl_2 \cdot 6H_2O$	1.4g
CFC selective supplement	10.0mL
Glycerol	10.0mL

pH 7.1 ± 0.2 at 25°C

Source: This medium is available as a premixed powder from Oxoid Unipath.

CFC Selective Supplement:

Composition per 10mL:

Cephaloridine	0.05g
Fucidin	0.01g
Cetrimide	0.01g

Preparation of CFC Selective Supplement: Add components to distilled/deionized water and bring volume to 10.0mL. Mix thoroughly. Filter sterilize.

Preparation of Medium: Add components, except CFC selective supplement, to distilled/deionized water and bring volume to 990.0mL. Mix thoroughly. Gently heat and bring to boiling. Autoclave for 15 min at 15 psi pressure–121°C. Cool to 45–50°C. Aseptically add sterile CFC selective supplement. Mix thoroughly. Pour into sterile Petri dishes or distribute into sterile tubes.

Use: For the selective isolation and cultivation of *Pseudomonas* species.

Pseudomonas CN Agar

Composition per liter:

Pancreatic digest of gelatin 16.0g
Agar 11.0g
Pancreatic digest of casein 10.0g
K_2SO_4 .... 10.0g
$MgCl_2 \cdot 6H_2O$ 1.4g
CN selective supplement 10.0mL
Glycerol 10.0mL

CN Selective Supplement:

Cetrimide 0.1g
Sodium nalidixate 7.5mg

pH 7.1 ± 0.2 at 25°C

Source: This medium is available as a premixed powder from Oxoid Unipath.

Preparation of Medium: Add components, except CN selective supplement, to distilled/deionized water and bring volume to 990.0mL. Mix thoroughly. Gently heat and bring to boiling. Autoclave for 15 min at 15 psi pressure–121°C. Cool to 45–50°C. Aseptically add sterile CN selective supplement. Mix thoroughly. Pour into sterile Petri dishes or distribute into sterile tubes.

Use: For the selective isolation and cultivation of *Pseudomonas* species.

Pseudomonas denitrificans Medium

Composition per liter:

Agar 15.0g
Glucose 10.0g
Yeast extract 5.0g
$FeCl_3$ solution 20.0mL

$FeCl_3$ Solution:

Composition per 100mL:

$FeCl_3$ 0.03g

Preparation of $FeCl_3$ Solution: Add $FeCl_3$ to distilled/deionized water and bring volume to 100.0mL. Mix thoroughly. Filter sterilize.

Preparation of Medium: Add components, except $FeCl_3$ solution, to distilled/deionized water and bring volume to 980.0mL. Mix thoroughly. Gently heat and bring to boiling. Autoclave for 15 min at 15 psi pressure–121°C. Cool to 45–50°C. Aseptically add 20.0mL of sterile $FeCl_3$ solution. Mix thoroughly. Pour into sterile Petri dishes or distribute into sterile tubes.

Use: For the cultivation and maintenance of *Pseudomonas* species.

Pseudomonas Denitrification Medium

Composition per liter:

Glycerol 10.0g
KNO_3 .... 10.0g
Yeast extract 3.0g
$(NH_4)_2SO_4$ 1.5g
Agar 1.0g
$K_2HPO_4 \cdot 3H_2O$ 0.8g
$MgSO_4 \cdot 7H_2O$ 0.5g
KH_2PO_4 .... 0.2g
$CaCl_2$ 0.1g

pH 7.2 ± 0.2 at 25°C

Preparation of Medium: Add components to distilled/deionized water and bring volume to 1.0L. Mix thoroughly. Distribute into tubes in 10.0mL volumes. Autoclave for 15 min at 15 psi pressure–121°C.

Use: For the cultivation and differentiation of *Pseudomonas* species based on their ability to produce pyocin and other fluorescent pigments during denitrification.

Pseudomonas Isolation Agar

Composition per liter:

Peptone 20.0g
Agar 13.6g
K_2SO_4 .... 10.0g
$MgCl_2 \cdot 6H_2O$ 1.4g
Irgasan® (triclosan) 0.025g
Glycerol 20.0mL

pH 7.0 ± 0.2 at 25°C

Source: This medium is available as a premixed powder from BBL Microbiology Systems and Difco Laboratories.

Preparation of Medium: Add components to distilled/deionized water and bring volume to 1.0L. Mix thoroughly. Gently heat and bring to boiling. Distribute into tubes or flasks. Autoclave for 15 min at 15 psi pressure–121°C. Pour into sterile Petri dishes or leave in tubes.

Use: For the isolation and cultivation of *Pseudomonas* species.

Pseudomonas Medium (ATCC Medium 179)

Composition per 1020mL:

Solution 1 .... 1.0L
Solution 2 .... 5.0mL
Solution 3 .... 15.0mL

pH 6.8 ± 0.2 at 25°C

Solution 1:

Composition per liter:

Agar	20.0g
K_2HPO_4	2.56g
KH_2PO_4	2.08g
NH_4Cl	1.0g
$MgSO_4 \cdot 7H_2O$	0.5g

Preparation of Solution 1: Add components to distilled/deionized water and bring volume to 1.0L. Mix thoroughly. Gently heat and bring to boiling. Adjust pH to 6.8. Distribute into tubes or flasks. Autoclave for 15 min at 15 psi pressure–121°C. Cool to 45–50°C.

Solution 2:

Composition per 100mL:

Ferric ammonium citrate	1.0g
$CaCl_2$	0.1g

Preparation of Solution 2: Add components to distilled/deionized water and bring volume to 100.0mL. Mix thoroughly. Filter sterilize.

Solution 3:

Composition per 100mL:

Succinic acid	11.8g

Preparation of Solution 3: Add succinic acid to distilled/deionized water and bring volume to 100.0mL. Mix thoroughly. Adjust pH to 6.0 with NaOH. Filter sterilize.

Preparation of Medium: To 1.0L of cooled sterile solution 1, aseptically add 5.0mL of sterile solution 2 and 15.0mL of sterile solution 3. Mix thoroughly. Pour into sterile Petri dishes or distribute into sterile tubes.

Use: For the cultivation and maintenance of *Pseudomonas lemoignei* and *Pseudomonas putida*.

Pseudomonas Medium A

Composition per liter:

Peptone	20.0g
Agar	15.0g
Glycerol	10.0g
K_2SO_4	10.0g
$MgCl_2$	1.4g

pH 7.2 ± 0.2 at 25°C

Preparation of Medium: Add components to distilled/deionized water and bring volume to 1.0L. Mix thoroughly. Gently heat and bring to boiling. Distribute into tubes or flasks. Autoclave for 10 min at 10 psi pressure–115°C. Pour into sterile Petri dishes or leave in tubes.

Use: For the cultivation and production of pyocyanin by *Pseudomonas* species.

Pseudomonas Medium B

Composition per liter:

Peptone	20.0g
Agar	15.0g
Glycerol	10.0g
$MgSO_4 \cdot 7H_2O$	1.5g
K_2HPO_4 solution	100.0mL

pH 7.2 ± 0.2 at 25°C

K_2HPO_4 Solution:

Composition per 100mL:

K_2HPO_4	1.5g

Preparation of K_2HPO_4 Solution: Add K_2HPO_4 to distilled/deionized water and bring volume to 100.0mL. Mix thoroughly. Autoclave for 15 min at 15 psi pressure–121°C. Cool to 45–50°C.

Preparation of Medium: Add components, except K_2HPO_4 solution, to distilled/deionized water and bring volume to 900.0mL. Mix thoroughly. Gently heat and bring to boiling. Autoclave for 15 min at 15 psi pressure–121°C. Cool to 45–50°C. Aseptically add 100.0mL of sterile K_2HPO_4 solution. Mix thoroughly. Pour into sterile Petri dishes or distribute into sterile tubes.

Use: For the cultivation and observation of fluorescin production by *Pseudomonas* species.

Pseudomonas saccharophila Medium

Composition per 1015mL:

Agar	20.0g
Na_2HPO_4	4.8g
KH_2PO_4	4.4g
NH_4Cl	1.0g
$MgSO_4 \cdot 7H_2O$	0.5g
Solution A	5.0mL
Solution B	10.0mL

Solution A:

Composition per 100mL:

Ferric ammonium citrate	1.0g
$CaCl_2$	0.1g

Preparation of Solution A: Add components to distilled/deionized water and bring volume to 100.0mL. Mix thoroughly. Filter sterilize.

Solution B:

Composition per 100mL:

Sucrose	10.0g

Preparation of Solution B: Add sucrose to distilled/deionized water and bring volume to 100.0mL. Mix thoroughly. Filter sterilize.

Preparation of Medium: Add components, except solution A and solution B, to distilled/deionized water and bring volume to 1.0L. Mix thoroughly. Gently heat and bring to boiling. Autoclave for 15 min at 15 psi pressure–121°C. Cool to 45–50°C. Aseptically add sterile solution A and sterile solution B. Mix thoroughly. Pour into sterile Petri dishes or distribute into sterile tubes.

Use: For the cultivation and maintenance of *Pseudomonas saccharophila* and other *Pseudomonas* species.

Pseudomonas solanacearum Medium

Composition per liter:

Agar	17.0g
Peptone	10.0g
Glucose	5.0g
Pancreatic digest of casein	1.0g

Preparation of Medium: Add components to distilled/deionized water and bring volume to 1.0L. Mix thoroughly. Gently heat and bring to boiling. Distribute into tubes or flasks. Autoclave for 15 min at 15 psi pressure–121°C. Pour into sterile Petri dishes or leave in tubes.

Use: For the cultivation and maintenance of *Pseudomonas solanacearum*.

Pseudomonas syringae Selective Medium

Composition per liter:

Agar	15.0g
L-Proline	5.0g
$MgSO_4 \cdot 7H_2O$	0.2g
K_2HPO_4	0.08g
KH_2PO_4	0.02g
$MnSO_4 \cdot 4H_2O$ solution	10.0mL

pH 6.8 ± 0.2 at 25°C

$MnSO_4 \cdot 4H_2O$ Solution:
Composition per 10mL:

$MnSO_4 \cdot 4H_2O$	2.1g

Preparation of $MnSO_4 \cdot 4H_2O$ Solution: Add $MnSO_4 \cdot 4H_2O$ to distilled/deionized water and bring volume to 10.0mL. Mix thoroughly. Autoclave for 15 min at 15 psi pressure–121°C.

Preparation of Medium: Add components, except $MnSO_4 \cdot 4H_2O$ solution, to distilled/deionized water and bring volume to 990.0mL. Mix thoroughly. Gently heat and bring to boiling. Adjust pH to 6.8. Autoclave for 10 min at 10 psi pressure–115°C. Cool to 45–50°C. Aseptically add sterile $MnSO_4 \cdot 4H_2O$ solution. Mix thoroughly. Pour into sterile Petri dishes.

Use: For the selective isolation and cultivation of *Pseudomonas syringae*.

PT Agar

Composition per liter:

Agar	15.0g
Pancreatic digest of casein	4.0g
Yeast extract	4.0g
$MgSO_4 \cdot 7H_2O$	2.0g
$CaCl_2 \cdot 2H_2O$	1.0g

pH 7.2 ± 0.2 at 25°C

Preparation of Medium: Add components to distilled/deionized water and bring volume to 1.0L. Mix thoroughly. Gently heat and bring to boiling. Distribute into tubes or flasks. Autoclave for 15 min at 15 psi pressure–121°C. Pour into sterile Petri dishes or leave in tubes.

Use: For the cultivation of myxobacteria.

Purple Agar

Composition per liter:

Agar	15.0g
Proteose peptone No. 3	10.0g
NaCl	5.0g
Beef extract	1.0g
Bromcresol purple	0.02g
Carbohydrate solution	20.0mL

pH 6.8 ± 0.2 at 25°C

Source: This medium is available as a premixed powder from Difco Laboratories.

Carbohydrate Solution:
Composition per 20mL:

Carbohydrate	10.0g

Preparation of Carbohydrate Solution: Add carbohydrate to distilled/deionized water and bring volume to 20.0mL. For expensive carbohydrates, 5.0g may be used instead of 10.0g. Mix thoroughly. Filter sterilize.

Preparation of Medium: Add components, except carbohydrate solution, to distilled/deionized water and bring volume to 980.0mL. Mix thoroughly. Gently heat and bring to boiling. Distribute into tubes in 9.8mL volumes. Autoclave for 15 min at 15 psi pressure–121°C. Cool to 45–50°C. Aseptically add 0.2mL of sterile carbohydrate solution to each tube. Mix thoroughly. Allow tubes to cool in a slanted position.

Use: For the preparation of carbohydrate media used in fermentation studies for the identification of bacteria, especially members of the Enterobacteriaceae. Bacteria that can ferment the carbohydrate turn the medium yellow.

Purple Broth (Purple Carbohydrate Broth)

Composition per liter:

Proteose peptone No. 3	10.0g
NaCl	5.0g
Beef extract	1.0g
Bromcresol purple	0.015g
Carbohydrate solution	20.0mL

pH 6.8 ± 0.2 at 25°C

Source: This medium is available as a premixed powder from Difco Laboratories.

Carbohydrate Solution:
Composition per 20mL:

Carbohydrate	10.0g

Preparation of Carbohydrate Solution: Add carbohydrate to distilled/deionized water and bring volume to 20.0mL. For expensive carbohydrates, 5.0g may be used instead of 10.0g. Mix thoroughly. Filter sterilize.

Preparation of Medium: Add components, except carbohydrate solution, to distilled/deionized water and bring volume to 980.0mL. Mix thoroughly. Gently heat and bring to boiling. Distribute into tubes in 9.8mL volumes. Autoclave for 15 min at 15 psi pressure–121°C. Cool to 25°C. Aseptically add 0.2mL of sterile carbohydrate solution to each tube. Mix thoroughly.

Use: For the preparation of carbohydrate media used in fermentation studies for the identification of bacteria, especially members of the Enterobacteriaceae. Bacteria that can ferment the carbohydrate turn the medium yellow.

Purple Broth

Composition per liter:

Pancreatic digest of gelatin	10.0g
NaCl	5.0g
Bromcresol purple	0.02g
Carbohydrate solution	20.0mL

pH 6.8 ± 0.2 at 25°C

Source: This medium is available as a premixed powder from BBL Microbiology Systems.

Carbohydrate Solution:
Composition per 20mL:

Carbohydrate	10.0g

Preparation of Carbohydrate Solution: Add carbohydrate to distilled/deionized water and bring volume to 20.0mL. For expensive carbohydrates, 5.0g may be used instead of 10.0g. Mix thoroughly. Filter sterilize.

Preparation of Medium: Add components, except carbohydrate solution, to distilled/deionized water and bring volume to 980.0mL. Mix thoroughly. Adjust pH to 7.4 if necessary. Distribute into tubes containing an inverted Durham tube. Fill each tube with 9.8mL of medium. Autoclave for 15 min at 15 psi pressure–121°C. Aseptically add 0.2mL of sterile carbohydrate solution to each tube.

Use: For the preparation of liquid fermentation media. Bacteria that can ferment the carbohydrate turn the medium yellow.

Purple Lactose Agar

Composition per liter:

Agar	10.0g
Lactose	10.0g
Peptone	5.0g
Beef extract	3.0g
Bromcresol purple	0.025g

pH 6.8 ± 0.1 at 25°C

Source: This medium is available as a premixed powder from Difco Laboratories.

Preparation of Medium: Add components to distilled/deionized water and bring volume to 1.0L. Mix thoroughly. Gently heat and bring to boiling. Distribute into tubes or flasks. Autoclave for 15 min at 15 psi pressure–121°C. Pour into sterile Petri dishes or leave in tubes. Allow tubes to cool in a slanted position.

Use: For the detection and differentiation of members of the Enterobacteriaceae. Bacteria that can ferment lactose turn the medium yellow.

Purple Serum Agar Base

Composition per liter:

Agar	20.0g
Lactose	20.0g
Peptone	20.0g
NaCl	5.0g
Bromcresol purple	0.030g
Phenol Eed	0.024g

pH 7.6 ± 0.2 at 25°C

Preparation of Medium: Add components to distilled/deionized water and bring volume to 1.0L. Mix thoroughly. Gently heat and bring to boiling. Distribute into tubes or flasks. Autoclave for 15 min at 15 psi pressure–121°C. Pour into sterile Petri dishes or leave in tubes.

Use: For the cultivation and differentiation of Gram-negative bacteria isolated from the urinary tract. Bacteria that can ferment lactose turn the medium yellow.

PY CMC Medium (Peptone Yeast Extract Carboxymethyl Cellulose Medium)

Composition per liter:

Agar	15.0g
Carboxymethyl cellulose	10.0g
NaCl	5.0g
Polypeptone™	5.0g
Yeast extract	5.0g
$MgSO_4 \cdot 7H_2O$	2.0g
KH_2PO_4	1.0g
Na_2CO_3 solution	100.0mL

pH 9.5 ± 0.2 at 25°C

Na_2CO_3 Solution:

Composition per 100mL:

Na_2CO_3	10.0g

Preparation of Na_2CO_3 Solution: Add Na_2CO_3 to distilled/deionized water and bring volume to 100.0mL. Mix thoroughly. Autoclave for 15 min at 15 psi pressure–121°C. Cool to 45–50°C.

Preparation of Medium: Add components, except Na_2CO_3 solution, to distilled/deionized water and bring volume to 900.0mL. Mix thoroughly. Gently heat and bring to boiling. Autoclave for 15 min at 15 psi pressure–121°C. Cool to 45–50°C. Aseptically add sterile Na_2CO_3 solution. Mix thoroughly. Adjust pH to 9.5 if necessary. Pour into sterile Petri dishes or distribute into sterile tubes.

Use: For the cultivation and maintenance of alkalophilic *Bacillus* species.

PY Medium with Fructose (Peptone Yeast Extract Medium with Fructose)

Composition per liter:

Fructose	10.0g
Yeast extract	10.0g
Peptone	5.0g
Pancreatic digest of casein	5.0g
$Cysteine \cdot HCl \cdot H_2O$	0.5g
Salts solution	40.0mL
Hemin solution	10.0mL
Resazurin solution	4.0mL
Vitamin K_1 solution	0.2mL

pH 6.9 ± 0.2 at 25°C

Salts Solution:

Composition per liter:

$NaHCO_3$	10.0g
NaCl	2.0g
K_2HPO_4	1.0g
KH_2PO_4	1.0g
$CaCl_2$, anhydrous	0.2g
$MgSO_4$	0.2g

Preparation of Salts Solution: Add $CaCl_2$ and $MgSO_4$ to 300.0mL of distilled/deionized water. Mix thoroughly until dissolved. Bring volume to 800.0mL with distilled/deionized water. Add remaining components while stirring. Bring volume to 1.0L. Mix thoroughly. Store at 4°C.

Hemin Solution:

Composition per 100mL:

Hemin	0.050g
NaOH (1*N* solution)	1.0mL

Preparation of Hemin Solution: Add hemin to NaOH solution. Mix thoroughly. Adjust volume to 100.0mL with distilled/deionized water. Autoclave for 15 min at 15 psi pressure–121°C. Cool to 45–50°C.

Resazurin Solution:

Composition per 44mL:

Resazurin	0.044g

Preparation of Resazurin Solution: Add resazurin to distilled/deionized water and bring volume to 44.0mL. Mix thoroughly.

Vitamin K_1 Solution:

Composition per 30mL:

Vitamin K_1	0.15g
Ethanol (95% solution)	30.0mL

Preparation of Vitamin K_1 Solution: Add vitamin K_1 to ethanol. Mix thoroughly. Store in brown bottle and keep under refrigeration. Discard after one month.

Preparation of Medium: Add components—except $cysteine \cdot HCl \cdot H_2O$, hemin solution, and vitamin K_1 solution—to distilled/deionized water and bring volume to 989.8mL. Mix thoroughly. Gently heat and bring to boiling under 80% N_2 + 10% CO_2 + 10% H_2. Continue boiling until resazurin turns colorless indicating reduction. Cool to 45–50°C. Add the $cysteine \cdot HCl \cdot H_2O$, hemin solution, and vitamin K_1 solution. Adjust pH to 6.9, if necessary. Anaerobically distribute into tubes under 80% N_2 + 10% CO_2 + 10% H_2. Cap the tubes with rubber stoppers. Place tubes in a press. Autoclave for 15 min at 15 psi pressure–121°C with fast exhaust.

Use: For the cultivation and maintenance of *Megasphaera cerevisiae*.

PY Medium with Glucose (Peptone Yeast Extract Medium with Glucose) (PYG Medium)

Composition per liter:

Glucose	10.0g
Yeast extract	10.0g
Peptone	5.0g
Pancreatic digest of casein	5.0g
Cysteine·HCl·H_2O	0.5g
Salts solution	40.0mL
Hemin solution	10.0mL
Resazurin solution	4.0mL
Vitamin K_1 solution	0.2mL

pH 6.9 ± 0.2 at 25°C

Salts Solution:

Composition per liter:

$NaHCO_3$	10.0g
NaCl	2.0g
K_2HPO_4	1.0g
KH_2PO_4	1.0g
$CaCl_2$, anhydrous	0.2g
$MgSO_4$	0.2g

Preparation of Salts Solution: Add $CaCl_2$ and $MgSO_4$ to 300.0mL of distilled/deionized water. Mix thoroughly until dissolved. Bring volume to 800.0mL with distilled/deionized water. Add remaining components while stirring. Bring volume to 1.0L. Mix thoroughly. Store at 4°C.

Hemin Solution:

Composition per 100mL:

Hemin	0.050g
NaOH (1*N* solution)	1.0mL

Preparation of Hemin Solution: Add hemin to NaOH solution. Mix thoroughly. Adjust volume to 100.0mL with distilled/deionized water. Autoclave for 15 min at 15 psi pressure–121°C. Cool to 45–50°C.

Resazurin Solution:

Composition per 44mL:

Resazurin	0.044g

Preparation of Resazurin Solution: Add resazurin to distilled/deionized water and bring volume to 44.0mL. Mix thoroughly.

Vitamin K_1 Solution:

Composition per 30mL:

Vitamin K_1	0.15g
Ethanol (95% solution)	30.0mL

Preparation of Vitamin K_1 Solution: Add vitamin K_1 to ethanol. Mix thoroughly. Store in brown bottle and keep under refrigeration. Discard after one month.

Preparation of Medium: Add components—except cysteine·HCl·H_2O, hemin solution, and vitamin K_1 solution—to distilled/deionized water and bring volume to 989.8mL. Mix thoroughly. Gently heat and bring to boiling under 80% N_2 + 10% CO_2 + 10% H_2. Continue boiling until resazurin turns colorless indicating reduction. Cool to 45–50°C. Add the cysteine·HCl·H_2O, hemin solution, and vitamin K_1 solution. Adjust pH to 6.9, if necessary. Anaerobically distribute into tubes under 80% N_2 + 10% CO_2 + 10% H_2. Cap the tubes with rubber stoppers. Place tubes in a press. Autoclave for 15 min at 15 psi pressure–121°C with fast exhaust.

Use: For the cultivation and maintenance of *Clostridium* species.

PYEX Glucose Salt Medium (Peptone Yeast Extract Glucose Salt Medium)

Composition per liter:

Peptone	10.0g
NaCl	5.0g
Yeast extract	5.0g
Glucose	1.0g

Preparation of Medium: Add components to distilled/deionized water and bring volume to 1.0L. Mix thoroughly. Distribute into tubes or flasks. Autoclave for 15 min at 15 psi pressure–121°C.

Use: For the cultivation and maintenance of *Micrococcus luteus*.

PYG Broth (Peptone Yeast Extract Glucose Broth)

Composition per liter:

Peptone	20.0g
D-Glucose	10.0g
Yeast extract	10.0g
Cysteine·HCl·H_2O	0.5g
VPI salt solution	40.0mL
Resazurin solution	4.0mL

pH 7.2 ± 0.2 at 25°C

Resazurin Solution:

Composition per 44mL:

Resazurin	0.044g

Preparation of Resazurin Solution: Add resazurin to distilled/deionized water and bring volume to 44.0mL. Mix thoroughly.

VPI Salt Solution:

Composition per 40mL:

$CaCl_2$	0.2g
$MgSO_4$	0.2g
K_2HPO_4	1.0g
KH_2PO_4	1.0g

Preparation of VPI Salt Solution: Add $CaCl_2$ and $MgSO_4$ to 300.0mL of distilled/deionized water. Mix thoroughly until dissolved. Bring volume to 800.0mL with distilled/deionized water. Add remaining components while stirring. Bring volume to 1.0L. Mix thoroughly. Store at 4°C.

Preparation of Medium: Add components to distilled/deionized water and bring volume to 1.0L. Mix thoroughly. Distribute into screw-capped tubes in 7.0mL volumes. Autoclave for 15 min at 15 psi pressure–121°C. Cool to 45–50°C under 100% N_2.

Use: For the cultivation of a wide variety of anaerobic bacteria.

PYG Medium (Peptone Yeast Extract Glucose Medium)

Composition per liter:

Proteose peptone	20.0g
Glucose	18.0g
Yeast extract	2.0g
Sodium citrate·$2H_2O$	1.0g
$MgSO_4 \cdot 7H_2O$	0.98g
$Na_2HPO_4 \cdot 7H_2O$	0.355g
KH_2PO_4	0.34g
$CaCl_2$	0.059g
$Fe(NH_4)_2(SO_4)_2 \cdot 6H_2O$	0.02g

pH 6.5 ± 0.2 at 25°C

Preparation of Medium: Add components, except $CaCl_2$, to distilled/deionized water and bring volume to 900.0mL. Mix thoroughly until dissolved. Add $CaCl_2$. Mix thoroughly. Bring volume to 1.0L with distilled/deionized water. Distribute into screw-capped tubes in 5.0mL volumes. Autoclave for 15 min at 15 psi pressure–121°C.

Use: For the cultivation of *Acanthamoeba* species.

PYG Medium (Peptone Yeast Extract Glucose Medium)

Composition per liter:

Agar	15.0g
Glucose	0.25g
Peptone	0.25g
Yeast extract	0.25g
Hutner's modified salt solution	20.0mL
Vitamin solution	10.0mL

Hutner's Modified Salts Solution:

Composition per liter:

$MgSO_4 \cdot 7H_2O$	29.7g
Nitrilotriacetic acid	10.0g
$CaCl_2 \cdot 2H_2O$	3.34g
$FeSO_4 \cdot 7H_2O$	0.10g
Metals "44"	50.0mL

Preparation of Hutner's Modified Salts Solution: Add nitrilotriacetic acid to 500.0mL of distilled/deionized water. Dissolve by adjusting pH to 6.5 with KOH. Add remaining components. Add distilled/deionized water to 1.0L.

Metals "44":

Composition per 100mL:

$ZnSO_4 \cdot 7H_2O$	1.1g
$FeSO_4 \cdot 7H_2O$	0.5g
EDTA	0.25g
$MnSO_4 \cdot 7H_2O$	0.154g
$CuSO_4 \cdot 5H_2O$	0.04g
$Co(NO_3)_2 \cdot 6H_2O$	0.025g
$Na_2B_4O_7 \cdot 10H_2O$	0.018g

Preparation of Metals "44": Add components to distilled/deionized water and bring volume to 100.0mL. Mix thoroughly.

Vitamin Solution:

Composition per liter:

Pyridoxine HCl	0.01g
Calcium pantothenate	5.0mg
Nicotinamide	5.0mg
Riboflavin	5.0mg
Thiamine HCl	5.0mg
Biotin	2.0mg
Folic acid	2.0mg
Vitamin B_{12}	0.1mg

Preparation of Vitamin Solution: Add components to distilled/deionized water and bring volume to 1.0L. Mix thoroughly. Filter sterilize.

Preparation of Medium: Add components, except vitamin solution, to distilled/deionized water and bring volume to 990.0mL. Mix thoroughly. Gently heat and bring to boiling. Autoclave for 15 min at 15 psi pressure–121°C. Cool to 45–50°C. Aseptically add 10.0mL of sterile vitamin solution. Mix thoroughly. Pour into sterile Petri dishes or distribute into sterile tubes.

Use: For the isolation and cultivation of *Pasteuria ramosa*.

PYG Medium for *Spirillum* (Peptone Yeast Extract Glucose Medium for *Spirillum*)

Composition per liter:

Agar	15.0g
Peptone	10.0g
Yeast extract	5.0g
Glucose	3.0g

pH 7.2 ± 0.2 at 25°C

Glucose Solution:

Composition per 10mL:

D-Glucose	3.0g

Preparation of Glucose Solution: Add glucose to distilled/deionized water and bring volume to 10.0mL. Mix thoroughly. Filter sterilize.

Preparation of Medium: Add components, except glucose solution, to distilled/deionized water and bring volume to 990.0mL. Mix thoroughly. Gently heat and bring to boiling. Autoclave for 15 min at 15 psi pressure–121°C. Cool to 45–50°C. Aseptically add sterile glucose solution. Mix thoroughly. Pour into sterile Petri dishes or distribute into sterile tubes.

Use: For the cultivation and maintenance of *Spirillum pleomorphum.*

PYGV Marine Medium (Peptone Yeast Extract Glucose Vitamin Marine Medium)

Composition per liter:

Agar	15.0g
Peptone	0.25g
Yeast extract	0.25g
Mineral salt solution	20.0mL
Glucose solution	10.0mL
Vitamin solution	5.0mL

pH 7.5 ± 0.2 at 25°C

Mineral Salt Solution:

Composition per liter:

$MgSO_4.7H_2O$	29.7g
Nitrilotriacetic acid	10.0g
$CaCl_2 \cdot 2H_2O$	3.34g
$FeSO_4 \cdot 7H_2O$	0.099g
$Na_2MoO_4 \cdot 2H_2O$	0.013g
Metals "44"	50.0mL

Preparation of Mineral Salt Solution: Add nitrilotriacetic acid to 500.0mL of distilled/deionized water. Dissolve by adjusting pH to 6.5 with KOH. Add remaining components. Add distilled/deionized water to 1.0L. Readjust pH to 7.2.

Metals "44":

Composition per 100mL:

$ZnSO_4 \cdot 7H_2O$	1.1g
$FeSO_4 \cdot 7H_2O$	0.5g
EDTA	0.25g
$MnSO_4 \cdot 7H_2O$	0.154g
$CuSO_4 \cdot 5H_2O$	0.04g
$Co(NO_3)_2 \cdot 6H_2O$	0.025g
$Na_2B_4O_7 \cdot 10H_2O$	0.018g

Preparation of Metals "44": Add a few drops of H_2SO_4 to distilled/deionized water to inhibit precipitate formation. Add components to acidified distilled/deionized water and bring volume to 100.0mL. Mix thoroughly.

Glucose Solution:

Composition per 100mL:

D-Glucose	2.5g

Preparation of Glucose Solution: Add glucose to distilled/deionized water and bring volume to 100.0mL. Mix thoroughly. Filter sterilize.

Vitamin Solution:

Composition per liter:

Pyridoxine·HCl	0.02g
p-Aminobenzoic acid	0.01g
Calcium D-pantothenate	0.01g
Nicotinamide	0.01g
Riboflavin	0.01g
Thiamine·HCl	0.01g
Biotin	4.0mg
Folic acid	4.0mg
Cyanocobalamin	0.2mg

Preparation of Vitamin Solution: Add components to distilled/deionized water and bring volume to 1.0L. Mix thoroughly. Filter sterilize.

Preparation of Medium: Add components, except glucose solution and vitamin solution, to seawater and bring volume to 985.0mL. Mix thoroughly. Gently heat and bring to boiling. Autoclave for 15 min at 15 psi pressure–121°C. Cool to 45–50°C. Aseptically add 10.0mL of sterile glucose solution and 5.0mL of sterile vitamin solution. Mix thoroughly. Adjust pH to 7.5 with sterile KOH, if necessary. Pour into sterile Petri dishes or distribute into sterile tubes.

Use: For the cultivation and maintenance of *Planctomyces brasiliensis.*

PYGV Medium

Composition per liter:

Agar	15.0g
Peptone	0.25g
Yeast extract	0.25g
Mineral solution	20.0mL

Glucose solution 10.0mL
Vitamin solution 5.0mL
pH 7.5 ± 0.2 at 25°C

Mineral Solution:
Composition per liter:
$MgSO_4 \cdot 7H_20$ 29.7g
$NaMoO_4 \cdot 2H_2O$ 12.67g
Nitrilotriacetic acid 10.0g
$CaCl_2 \cdot 2H_2O$ 3.34g
$FeSO_4 \cdot 7H_2O$ 0.1g
Metals "44" solution 50.0mL

Preparation of Mineral Solution: Add nitrilotriacetic acid to 500.0mL of distilled/deionized water. Dissolve by adjusting pH to 6.5 with KOH. Add remaining components. Readjust pH to 7.2 with H_2SO_4 or KOH. Add distilled/deionized water to 1.0L. Store at 5°C.

Metals "44":
Composition per 100mL:
$ZnSO_4 \cdot 7H_2O$ 1.1g
$FeSO_4 \cdot 7H_2O$ 0.5g
EDTA 0.25g
$MnSO_4 \cdot 7H_2O$ 0.154g
$CuSO_4 \cdot 5H_2O$ 0.04g
$Co(NO_3)_2 \cdot 6H_2O$ 0.025g
$Na_2B_4O_7 \cdot 10H_2O$ 0.018g

Preparation of Metals "44": Add components to distilled/deionized water and bring volume to 100.0mL. Mix thoroughly.

Glucose Solution:
Composition per 100mL:
D-Glucose 2.5g

Preparation of Glucose Solution: Add D-glucose to distilled/deionized water and bring volume to 100.0mL. Mix thoroughly. Filter sterilize.

Vitamin Solution:
Composition per liter:
Pyridoxin·HCl 0.02g
p-Aminobenzoic acid 0.01g
Ca-panthothenate 0.01g
Nicotinamide 0.01g
Riboflavin 0.01g
Thiamine·HCl 0.01g
Biotin 4.0mg
Folic acid 4.0mg
Vitamin B_{12} 0.2mg

Preparation of Vitamin Solution: Add components to distilled/deionized water and bring volume to 1.0L.

Preparation of Medium: Add components, except glucose solution and vitamin solution, to distilled/deionized water and bring volume to 985.0mL. Mix thoroughly. Gently heat and bring to boiling. Autoclave for 20 min at 15 psi pressure–121°C. Cool to 60°C. Aseptically add 10.0mL of sterile glucose solution and 5.0mL of sterile vitamin solution. Mix thoroughly. Pour into sterile Petri dishes or distribute into sterile tubes.

Use: For the enrichment of *Stella* species from polluted waters.

PYGV Medium (Peptone Yeast Extract Glucose Vitamin Medium)

Composition per liter:
Agar 15.0g
Peptone 0.25g
Yeast extract 0.25g
Mineral salt solution 20.0mL
Glucose solution 10.0mL
Vitamin solution 5.0mL
pH 7.5 ± 0.2 at 25°C

Mineral Salt Solution:
Composition per liter:
$MgSO_4 \cdot 7H_2O$ 29.7g
Nitrilotriacetic acid 10.0g
$CaCl_2 \cdot 2H_2O$ 3.34g
$FeSO_4 \cdot 7H_2O$ 99.0mg
$Na_2MoO_4 \cdot 2H_2O$ 12.67mg
Metals "44" 50.0mL

Preparation of Mineral Salt Solution: Add nitrilotriacetic acid to 500.0mL of distilled/deionized water. Dissolve by adjusting pH to 6.5 with KOH. Add remaining components. Add distilled/deionized water to 1.0L. Readjust pH to 7.2.

Metals "44":
Composition per 100mL:
$ZnSO_4 \cdot 7H_2O$ 1.1g
$FeSO_4 \cdot 7H_2O$ 0.5g
EDTA 0.25g
$MnSO_4 \cdot 7H_2O$ 0.154g
$CuSO_4 \cdot 5H_2O$ 0.04g
$Co(NO_3)_2 \cdot 6H_2O$ 0.025g
$Na_2B_4O_7 \cdot 10H_2O$ 0.018g

Preparation of Metals "44": Add a few drops of H_2SO_4 to distilled/deionized water to inhibit precipitate formation. Add components to acidified distilled/deionized water and bring volume to 100.0mL. Mix thoroughly.

Glucose Solution:
Composition per 100mL:
D-Glucose 2.5g

Preparation of Glucose Solution: Add glucose to distilled/deionized water and bring volume to 100.0mL. Mix thoroughly. Filter sterilize.

Vitamin Solution:

Composition per liter:

Pyridoxine·HCl	0.02g
p-Aminobenzoic acid	0.01g
Calcium D-pantothenate	0.01g
Nicotinamide	0.01g
Riboflavin	0.01g
Thiamine·HCl	0.01g
Biotin	4.0mg
Folic acid	4.0mg
Cyanocobalamin	0.2mg

Preparation of Vitamin Solution: Add components to distilled/deionized water and bring volume to 1.0L. Mix thoroughly. Filter sterilize.

Preparation of Medium: Add components, except glucose solution and vitamin solution, to distilled/deionized water and bring volume to 985.0mL. Mix thoroughly. Gently heat and bring to boiling. Autoclave for 15 min at 15 psi pressure–121°C. Cool to 45–50°C. Aseptically add 10.0mL of sterile glucose solution and 5.0mL of sterile vitamin solution. Mix thoroughly. Adjust pH to 7.5 with sterile KOH, if necessary. Pour into sterile Petri dishes or distribute into sterile tubes.

Use: For the cultivation and maintenance of *Blastobacter aggregatus, B. capsulatus, B. denitrificans,* and *Planctomyces limnophilus.*

Pyrazinamide Medium

Composition per liter:

Agar	15.0g
Na_2HPO_4	2.5g
Sodium pyruvate	2.0g
L-Asparagine	2.0g
KH_2PO_4	1.0g
Pancreatic digest of casein	0.5g
Tween™ 80	0.2g
Pyrazinamide	0.1g
$CaCl_2 \cdot 2H_2O$	0.5mg
$CuSO_4 \cdot 5H_2O$	0.1mg
$ZnSO_4 \cdot 7H_2O$	0.1mg
Ferric ammonium citrate	0.05g
$MgSO_4 \cdot 7H_2O$	0.01g

pH 6.6 ± 0.2 at 25°C

Preparation of Medium: Combine components. Mix thoroughly. Gently heat and bring to boiling. Distribute into tubes in 5.0mL volumes. Autoclave for 15 min at 15 psi pressure–121°C. Allow tubes to cool in a slanted position.

Use: For cultivation and differentiation of *Corynebacterium* species and related organisms. Bacteria that produce pyrazinamidase turn the medium pink.

Pyridoxine Assay Medium

Composition per liter:

Sucrose	30.0g
Ammonium tartrate	10.0g
KH_2PO_4	5.0g
Sodium dihydrogen citrate	4.0g
$MgSO_4 \cdot 7H_2O$	1.0g
$CaCl_2 \cdot 2H_2O$	0.2g
NaCl	0.2g
Choline chloride	0.01g
$FeCl_3$	0.01g
Thiamine·HCl	0.01g
$ZnSO_4 \cdot 7H_2O$	4.0mg
Nicotinic acid	2.0mg
Calcium pantothenate	1.0mg
Riboflavin	1.0mg
p-Aminobenzoic acid	200μg
Biotin	8μg

pH 4.5 ± 0.2 at 25°C

Source: Available as a prepared medium from Difco Laboratories.

Preparation of Medium: Add components to distilled/deionized water and bring volume to 1.0L. Mix thoroughly. Gently heat and bring to boiling. Continue boiling for 2–3 min. Distribute into tubes in 5.0mL volumes. Add standard solutions and test solutions to each tube. Bring volume of each tube to 10.0mL with distilled/deionized water. Autoclave for 15 min at 15 psi pressure–121°C.

Use: For the microbiological assay of pyridoxine using *Neurospora sitophila* as the test organism.

Pyrrolidone Agar

Composition per liter:

Noble agar	21.0g
K_2HPO_4	5.65g
KH_2PO_4	2.95g
$MgSO_4 \cdot 7H_2O$	1.0g
Pyrrolidone carboxylic acid solution	30.0mL
NaOH solution	30.0mL
Trace metals	6.3mL

Pyrrolidone Carboxylic Acid Solution:

Composition per 300mL:

Pyrrolidone carboxylic acid	50.0g

Preparation of Pyrrolidone Carboxylic Acid Solution: Add pyrrolidone carboxylic acid to distilled/deionized water and bring volume to 300.0mL. Mix thoroughly. Filter sterilize.

NaOH Solution:

Composition per 100mL:

NaOH	5.0g

Preparation of Resazurin Solution: Add NaOH to distilled/deionized water and bring volume to 100.0mL. Mix thoroughly. Filter sterilize.

Trace Metals:

Composition per 100mL:

$FeSO_4 \cdot 7H_2O$	0.18g
$MnCl_2 \cdot 2H_2O$	0.13g
$CuSO_4 \cdot 5H_2O$	0.10g
$ZnSO_4 \cdot 7H_2O$	0.02g

Preparation of Trace Metals: Add a few drops of H_2SO_4 to distilled/deionized water to inhibit precipitate formation. Add components to acidified distilled/deionized water and bring volume to 100.0mL. Mix thoroughly.

Preparation of Medium: Add components, except pyrrolidone carboxylic acid solution and NaOH solution, to distilled/deionized water and bring volume to 940.0mL. Mix thoroughly. Gently heat and bring to boiling. Autoclave for 15 min at 15 psi pressure–121°C. Cool to 45–50°C. Aseptically add 30.0mL of sterile pyrrolidone carboxylic acid solution and 30.0mL of sterile NaOH solution. Mix thoroughly. Pour into sterile Petri dishes or distribute into sterile tubes.

Use: For the cultivation and maintenance of *Pseudomonas fluorescens*.

Quinolinic Acid Medium

Composition per liter:

Quinolinic acid	1.5g
K_2HPO_4	1.1g
NH_4NO_3	1.0g
KH_2PO_4	0.5g
$MgSO_4 \cdot 7H_2O$	0.25g

Preparation: Add quinolinic acid to distilled/deionized water and bring volume to 900.0mL. Mix thoroughly. Bring pH to 7.0 with NaOH. Add other components. Bring volume to 1.0L. Mix thoroughly. Distribute into tubes or flasks. Autoclave for 15 min at 15 psi pressure–121°C.

Use: For the cultivation of microorganisms that can utilize quinolinic acid as sole carbon source.

R2A Agar

Composition per liter:

Agar	15.0g
Yeast extract	0.5g
Acid hydrolysate of casein	0.5g
Glucose	0.5g
Soluble starch	0.5g
K_2HPO_4	0.3g
Sodium pyruvate	0.3g
Pancreatic digest of casein	0.25g
Peptic digest of animal tissue	0.25g
$MgSO_4$, anhydrous	0.024g

pH 7.2 ± 0.2 at 25°C

Source: Available as a premixed powder from BBL Microbiology Systems and Difco Laboratories.

Preparation of Medium: Add components to distilled/deionized water and bring volume to 1.0L. Mix thoroughly. Gently heat with mixing and bring to boiling. Distribute into tubes or flasks. Autoclave for 15 min at 15 psi pressure–121°C. Do not overheat. Pour into sterile Petri dishes or leave in tubes.

Use: For use in standard methods for pour plate, spread plate and membrane filter analysis to enumerate heterotrophic bacteria from potable waters.

R2YE Medium

Composition per 1062.2mL:

Thiostrepton	50.0mg
Basal solution	800.0mL
TES (*N*-tris[hydroxymethyl] methyl-2-amino–ethane-sulfonic acid) buffer	100.0mL
$CaCl_2 \cdot 2H_2O$ solution	80.2mL
Yeast extract solution	50.0mL
L-Proline solution	15.0mL
KH_2PO_4 solution	10.0mL
NaOH solution	5.0mL
Trace element solution	2.0mL

pH 7.2 ± 0.2 at 25°C

Basal Solution:

Composition per 800mL:

Sucrose	103.0g
$MgCl_2 \cdot 6H_2O$	10.12g
D-Glucose	10.0g
K_2SO_4	0.25g
Casamino acids	0.1g

Preparation of Basal Solution: Add components to distilled/deionized water and bring volume to 800.0mL. Mix thoroughly. Autoclave for 15 min at 15 psi pressure–121°C. Cool to 25°C.

TES Buffer:

Composition per liter:

TES (*N*-tris[hydroxymethyl] methyl-2-amino–ethane-sulfonic acid) buffer	57.3g

Preparation of TES Buffer: Add TES to distilled/deionized water and bring volume to 1.0L. Mix thoroughly. Adjust pH to 7.2. Filter sterilize.

$CaCl_2 \cdot 2H_2O$ Solution:

Composition per 100mL:

$CaCl_2 \cdot 2H_2O$	3.68g

Preparation of $CaCl_2 \cdot 2H_2O$ Solution: Add the $CaCl_2 \cdot 2H_2O$ to distilled/deionized water and bring volume to 100.0mL. Mix thoroughly. Filter sterilize.

Yeast Extract Solution:

Composition per 100mL:

Yeast extract	10.0g

Preparation of Yeast Extract Solution: Add yeast extract to distilled/deionized water and bring volume to 100.0mL. Mix thoroughly. Filter sterilize.

L-Proline Solution:

Composition per 100mL:

L-Proline	20.0g

Preparation of L-Proline Solution: Add the proline to distilled/deionized water and bring volume to 100.0mL. Mix thoroughly. Filter sterilize.

KH_2PO_4 Solution:

Composition per 100mL:

KH_2PO_4	0.5g

Preparation of KH_2PO_4 Solution Solution: Add the KH_2PO_4 to distilled/deionized water and bring volume to 100.0mL. Mix thoroughly. Filter sterilize.

NaOH Solution:

Composition per 100mL:

NaOH	40.0g

Preparation of NaOH Solution: Add the NaOH to distilled/deionized water and bring volume to 100.0mL. Mix thoroughly. Filter sterilize.

Trace Element Solution:

Composition per liter:

$FeCl_3 \cdot 6H_2O$	0.2g
$ZnCl_2$	0.04g
$CuCl_2 \cdot 2H_2O$	0.01g
$MnCl_2 \cdot 4H_2O$	0.01g
$Na_2B_4O_7 \cdot 10H_2O$	0.01g
$(NH_4)_6MoO_7O_{24} \cdot 4H_2O$	0.01g

Preparation of Trace Element Solution: Add components to distilled/deionized water and bring volume to 1.0L. Mix thoroughly. Filter sterilize.

Preparation of Medium: To 800.0mL of cooled, sterile basal solution aseptically add the remaining components. Mix thoroughly. Aseptically distribute into sterile tubes or flasks.

Use: For the cultivation and maintenance of *Streptomyces lividans.*

R3A Agar

Composition per liter:

Agar	15.0g
Yeast extract	1.0g
Acid hydrolysate of casein	1.0g
Glucose	1.0g
Soluble starch	1.0g
K_2HPO_4	0.6g
Sodium pyruvate	0.6g
Pancreatic digest of casein	0.5g
Peptic digest of animal tissue	0.5g
$MgSO_4$, anhydrous	0.048g

pH 7.2 ± 0.2 at 25°C

Source: This medium is available as a premixed powder from BBL Microbiology Systems.

Preparation of Medium: Add components to distilled/deionized water and bring volume to 1.0L. Mix thoroughly. Gently heat with mixing and bring to boiling. Distribute into tubes or flasks. Autoclave for 15 min at 15 psi pressure–121°C. Do not overheat. Pour into sterile Petri dishes or leave in tubes.

Use: For the cultivation and maintenance of heterotrophic bacteria from potable waters.

R8AH Medium

Composition per liter:

Malic acid	2.5g
$(NH_4)_2SO_4$	1.25g
Yeast extract	1.0g
K_2HPO_4	0.9g
KH_2PO_4	0.6g
$MgSO_4 \cdot 7H_2O$	0.2g
$CaCl_2 \cdot 2H_2O$	0.07g
EDTA	0.02g
Ferric citrate	0.01g
Vitamin solution	7.5mL
Trace element solution	1.0mL

pH 6.9 ± 0.2 at 25°C

Trace Element Solution:

Composition per 100mL:

Ferric citrate	0.3g
EDTA	0.05g
$CaCl_2 \cdot 2H_2O$	0.02g
$MnSO_4 \cdot H_2O$	2.0mg
$(NH_4)_6Mo_7O_{24} \cdot 4H_2O$	2.0mg
H_3BO_3	1.0mg
$CuSO_4 \cdot 5H_2O$	1.0mg
$ZnSO_4$	1.0mg

Preparation of Trace Element Solution: Add components to distilled/deionized water and bring volume to 100.0mL. Mix thoroughly.

Vitamin Solution:

Composition per liter:

Thiamine·HCl	0.4g
Nicotinic acid	0.2g
Nicotinamide	0.2g
Biotin	8.0mg

Preparation of Vitamin Solution: Add components to distilled/deionized water and bring volume to 1.0L. Mix thoroughly.

Preparation of Medium: Add malic acid to approximately 500.0mL of distilled/deionized water. Adjust pH to 6.9 with NaOH. Add remaining components. Bring volume to 1.0L with distilled/deionized water. Mix thoroughly. Adjust pH to 6.9. Distribute into tubes or flasks. Autoclave for 15 min at 15 psi pressure–121°C.

Use: For the cultivation and maintenance of *Rhodobacter sphaeroides, Rhodocyclus tenuis, Rhodopseudomonas rutila, Rhodospirillum photometricum,* and *Rhodospirillum rubrum.*

Rabbit Dung Agar

Composition per liter:

Rabbit dung	20.0g
Agar	15.0g

pH 7.2 ± 0.2 at 25°C

Preparation of Medium: Add rabbit dung to 1.0L of distilled/deionized water. Gently heat and bring to boiling. Continue boiling for 20 min. Filter through Whatman #1 filter paper. Bring volume of filtrate to 1.0L with distilled/deionized water. Add agar. Adjust pH to 7.2. Distribute into tubes or flasks. Autoclave for 15 min at 15 psi pressure–121°C. Pour into sterile Petri dishes or leave in tubes.

Use: For the cultivation of myxobacteria.

Raper *Achyla* Medium No. 1

Composition per liter:

Agar	20.0g
Lentil (hot water extract)	10.0g
Starch, soluble	3.0g
Peptone	1.0g
$CaCl_2$	1.0μg
$FeCl_3$	1.0μg
KH_2PO_4	1.0μg
$MgSO_4$	1.0μg
$ZnSO_4$	1.0μg

Preparation of Medium: Add components to distilled/deionized water and bring volume to 1.0L. Mix thoroughly. Gently heat and bring to boiling. Distribute into tubes or flasks. Autoclave for 15 min at 15 psi pressure–121°C. Pour into sterile Petri dishes or leave in tubes.

Use: For the cultivation of *Achyla* species.

Raper *Achyla* Medium No. 2

Composition per liter:

Agar	20.0g
Starch, soluble	3.0g
Inositol	1.0g
Peptone	1.0g

Preparation of Medium: Add components to distilled/deionized water and bring volume to 1.0L. Mix thoroughly. Gently heat and bring to boiling. Distribute into tubes or flasks. Autoclave for 15 min at 15 psi pressure–121°C. Pour into sterile Petri dishes or leave in tubes.

Use: For the cultivation of *Achyla* species.

Rappaport–Vassiliadis Enrichment Broth (RV Enrichment Broth)

Composition per 1110mL:

NaCl	8.0g
Papaic digest of soybean meal	5.0g
KH_2PO_4	1.6g
Magnesium chloride solution	100.0mL
Malachite green solution	10.0mL

pH 5.2 ± 0.2 at 25°C

Source: This medium is available as a premixed powder from Oxoid Unipath.

Magnesium Chloride Solution:

Composition per 100mL:

$MgCl_2·6H_2O$	40.0g

Preparation of Magnesium Chloride Solution: Add $MgCl_2·6H_2O$ to distilled/deionized water and bring volume to 100.0mL. Mix thoroughly. Autoclave for 15 min at 15 psi pressure–121°C. Cool to 45–50°C.

Malachite Green Solution:

Composition per 10.0mL:

Malachite green oxalate	0.04g

Preparation of Malachite Green Solution: Add malachite green to distilled/deionized water and bring volume to 10.0mL. Mix thoroughly. Autoclave for 15 min at 15 psi pressure–121°C. Cool to 45–50°C.

Preparation of Medium: Add components to distilled/deionized water and bring volume to 1.0L. Mix thoroughly. Distribute into tubes in 10.0mL volumes. Autoclave for 15 min at 10 psi pressure–115°C.

Use: For the isolation and cultivation of *Salmonella* species from environmental specimens.

Rappaport–Vassiliadis R10 Broth

Composition per liter:

$MgCl_2$, anhydrous	13.4g
NaCl	7.2g
Papaic digest of soybean meal	4.54g
KH_2PO_4	1.45g
Malachite green oxalate	0.036g

pH 5.1 ± 0.2 at 25°C

Source: This medium is available as a premixed powder from Difco Laboratories.

Preparation of Medium: Add components to distilled/deionized water and bring volume to 1.0L. Mix thoroughly. Distribute into screw-capped tubes in 10.0mL volumes. Autoclave for 15 min at 10 psi pressure–116°C.

Use: For the isolation and cultivation of *Salmonella* species from environmental specimens.

Rappaport–Vassiliadis Soya Peptone Broth (RVS Broth)

Composition per liter:

$MgCl_2$, anhydrous	13.58g
NaCl	7.2g
Papaic digest of soybean meal	4.5g
KH_2PO_4	1.26g
K_2HPO_4	0.18g
Malachite green	0.036g

pH 5.2 ± 0.2 at 25°C

Source: Available as a premixed powder from Oxoid Unipath.

Preparation of Medium: Add components to distilled/deionized water and bring volume to 1.0L. Mix thoroughly. Distribute into screw-capped tubes in 10.0mL volumes. Autoclave for 15 min at 10 psi pressure–115°C.

Use: For the isolation and cultivation of *Salmonella* species from environmental specimens.

RGCA Medium

Composition per 300.3mL:

Rumen fluid	120.0mL
Solution IV	65.0mL
Mineral solution I	45.0mL
Mineral solution II	45.0mL
Na_2CO_3 solution	20.0mL
Cysteine·HCl·H_2O solution	5.0mL
Solution III	0.3mL

pH 6.6 ± 0.2 at 25°C

Mineral Solution I:

Composition per 100mL:

K_2HPO_4	0.3g

Preparation of Mineral Solution I: Add K_2HPO_4 to distilled/deionized water and bring volume to 100.0mL. Mix thoroughly.

Mineral Solution II:

Composition per 100mL:

$(NH_4)_2SO_4$	0.6g
NaCl	0.6g
K_2HPO_4	0.3g
$MgSO_4$	0.06g
$CaCl_2$	0.06g

Preparation of Mineral Solution II: Add K_2HPO_4 to distilled/deionized water and bring volume to 100.0mL. Mix thoroughly.

Solution III:

Composition per 10mL:

Resazurin	0.01g

Preparation of Solution III: Add resazurin to 10.0mL of distilled/deionized water. Mix thoroughly.

Solution IV:

Composition per 65mL:

Agar	4.5g
Glucose	0.6g
Cellobiose	0.6g

Preparation of Solution IV: Add components to distilled/deionized water and bring volume to 65.0mL. Mix thoroughly.

Cysteine·HCl·H_2O Solution:

Composition per 100mL:

Cysteine·HCl·H_2O	3.0g

Preparation of Cysteine·HCl·H_2O Solution: Add cysteine·HCl·H_2O to distilled/deionized water and bring volume to 100.0mL. Mix thoroughly. Filter sterilize.

Na_2CO_3 Solution:

Composition per 100mL:

Na_2CO_3	6.0g

Preparation of Na_2CO_3 Solution: Add Na_2CO_3 to distilled/deionized water and bring volume to 100.0mL. Mix thoroughly. Filter sterilize.

Rumen Fluid:

Composition per 120mL:

Rumen fluid	120.0mL

Preparation of Rumen Fluid: Filter rumen contents, obtained from a cow on an alfalfa-hay concentrate ration, through two layers of cheesecloth to remove larger particles. Store under CO_2 in quart milk bottles in the refrigerator. Much of the particulate matter settles out. Use the supernatant fluid.

Preparation of Medium: Combine 45.0mL of mineral solution I, 45.0mL of mineral solution II, 0.3mL of solution III and 65.0mL of solution IV in a 500mL flask. Gently heat and bring to boiling. Add 120.0mL of rumen fluid. Gently heat and bring to boiling under 100% CO_2. Cap with a rubber stopper and wire the stopper secure. Autoclave for 20 min at 15 psi pressure–121°C. Cool to 45–50°C. Remove stopper and gas with 100% CO_2 to eliminate O_2. Aseptically add 5.0mL of sterile cysteine·HCl·H_2O solution and 20.0mL of sterile Na_2CO_3 solution. Mix thoroughly. Aseptically and anaerobically distribute into tubes under 100% CO_2 in 6.0mL volumes. Cap with rubber stoppers.

Use: For the cultivation and maintenance of *Ruminococcus albus, R. flavefaciens,* and *Succinimonas amylolytica.*

Rhizobium BIII Defined Agar

Composition per liter:

Agar	13.0g
Mannitol	10.0g
Sodium glutamate	1.10g
K_2HPO_4	0.23g
$MgSO_4·7H_2O$	0.10g
Trace element stock	1.0mL
Vitamin stock	1.0mL

pH 7.0 ± 0.2 at 25°C

Trace Element Stock:

Composition per liter:

Nitrilotriacetic acid	7.0g
$CaCl_2·2H_2O$	6.62g
H_3BO_3	0.145g
$FeSO_4·7H_2O$	0.125g
Na_2MoO_4	0.125g
$ZnSO_4·7H_2O$	0.108g
$CoSO_4·7H_2O$	0.070g
$CuSO_4·5H_2O$	5.0mg
$MnCl_2·4H_2O$	4.3mg

Preparation of Trace Element Stock: Add components to 500.0mL of distilled/deionized water in the following order: $CaCl_2·2H_2O$, H_3BO_3, $FeSO_4·7H_2O$, $CoSO_4·7H_2O$, $CuSO_4·5H_2O$, $MnCl_2·4H_2O$, $ZnSO_4·7H_2O$, and Na_2MoO_4. Adjust pH to 5.0. Add nitrilotriacetic acid. Bring volume to 1.0L with distilled/deionized water.

Vitamin Stock:

Composition per liter:

Inositol	0.12g
p-Aminobenzoic acid	0.02g
Biotin	0.02g
Calcium pantothenate	0.02g
Nicotinic acid	0.02g
Pyridoxine·HCl	0.02g
Riboflavin	0.02g
Thiamine·HCl	0.02g
Sodium phosphate buffer (50.0mM solution, pH 7.0)	1.0L

Preparation of Vitamin Stock: Combine components. Mix thoroughly. Filter sterilize. Store at 4°C in the dark.

Preparation of Medium: Add components, except vitamin stock, to distilled/deionized water and bring volume to 999.0mL. Mix thoroughly. Gently heat and bring to boiling. Autoclave for 15 min at 15 psi pressure–121°C. Cool to 45–50°C. Aseptically add 1.0mL of sterile vitamin stock. Mix thoroughly. Pour into sterile Petri dishes or distribute into sterile tubes.

Use: For the isolation and cultivation of *Rhizobium* species from root nodules.

Rhizobium BIII Defined Broth

Composition per liter:

Mannitol	10.0g
Sodium glutamate	1.10g
K_2HPO_4	0.23g
$MgSO_4·7H_2O$	0.10g
Trace element stock	1.0mL
Vitamin stock	1.0mL

pH 7.0 ± 0.2 at 25°C

Trace Element Stock:

Composition per liter:

Nitrilotriacetic acid	7.0g
$CaCl_2·2H_2O$	6.62g
H_3BO_3	0.145g
$FeSO_4·7H_2O$	0.125g
Na_2MoO_4	0.125g
$ZnSO_4·7H_2O$	0.108g
$CoSO_4·7H_2O$	0.070g
$CuSO_4·5H_2O$	5.0mg
$MnCl_2·4H_2O$	4.3mg

Preparation of Trace Element Stock: Add components to 500.0mL of distilled/deionized water in the following order: $CaCl_2 \cdot 2H_2O$, H_3BO_3, $FeSO_4 \cdot 7H_2O$, $CoSO_4 \cdot 7H_2O$, $CuSO_4 \cdot 5H_2O$, $MnCl_2 \cdot 4H_2O$, $ZnSO_4 \cdot 7H_2O$, and Na_2MoO_4. Adjust pH to 5.0. Add nitrilotriacetic acid. Bring volume to 1.0L with distilled/deionized water.

Vitamin Stock:

Composition per liter:

Inositol	0.12g
p-Aminobenzoic acid	0.02g
Biotin	0.02g
Calcium pantothenate	0.02g
Nicotinic acid	0.02g
Pyridoxine·HCl	0.02g
Riboflavin	0.02g
Thiamine·HCl	0.02g
Sodium phosphate buffer (50.0m*M* solution, pH 7.0)	1.0L

Preparation of Vitamin Stock: Combine components. Mix thoroughly. Filter sterilize. Store at 4°C in the dark.

Preparation of Medium: Add components, except vitamin stock, to distilled/deionized water and bring volume to 999.0mL. Mix thoroughly. Autoclave for 15 min at 15 psi pressure–121°C. Cool to 25°C. Aseptically add 1.0mL of sterile vitamin stock. Mix thoroughly. Aseptically distribute into sterile tubes or flasks.

Use: For the isolation and cultivation of *Rhizobium* species.

Rhizobium Medium 1

Composition per liter:

Agar	15.0g
Yeast extract	10.0g
K_2HPO_4	0.5g
$MgSO_4 \cdot 7H_2O$	0.2g
NaCl	0.2g
$FeCl_3 \cdot 6H_2O$	0.002g

pH 7.2 ± 0.2 at 25°C

Preparation of Medium: Add components, except agar, to distilled/deionized water and bring volume to 1.0L. Mix thoroughly. Adjust pH to 7.2. Add agar. Gently heat and bring to boiling. Distribute into tubes or flasks. Autoclave for 15 min at 15 psi pressure–121°C. Pour into sterile Petri dishes or leave in tubes.

Use: For the cultivation of members of the Rhizobiaceae.

Rhizobium Medium 2

Composition per liter:

Agar	15.0g
Glycerol	4.6g
$CaSO_4$	1.3g
K_2HPO_4	1.0g
L-Arabinose	1.0g
Yeast extract	1,0g
KNO_3	0.7g
$MgSO_4 \cdot 7H_2O$	0.36g
$FeCl_3 \cdot 6H_2O$	4.0mg

pH 7.2 ± 0.2 at 25°C

Preparation of Medium: Add components, except agar, to distilled/deionized water and bring volume to 1.0L. Mix thoroughly. Adjust pH to 7.2. Add agar. Gently heat and bring to boiling. Distribute into tubes or flasks. Autoclave for 15 min at 15 psi pressure–121°C. Pour into sterile Petri dishes or leave in tubes.

Use: For the cultivation of members of the Rhizobiaceae.

Rhizobium X Medium

Composition per liter:

Agar	15.0g
Mannitol	10.0g
Yeast extract	1.0g
Soil extract	200.0mL

pH 7.2 ± 0.2 at 25°C

Soil Extract:

Composition per 200mL:

African violet soil	77.0g
Na_2CO_3	0.2g

Preparation of Soil Extract: Add components to tap water and bring volume to 200.0mL. Autoclave for 60 min at 15 psi pressure–121°C. Filter through Whatman #1 filter paper.

Preparation of Medium: Add components to distilled/deionized water and bring volume to 1.0L. Mix thoroughly. Gently heat and bring to boiling. Distribute into tubes or flasks. Autoclave for 15 min at 15 psi pressure–121°C. Pour into sterile Petri dishes or leave in tubes.

Use: For the cultivation and maintenance of *Bradyrhizobium japonicum*, *Rhizobium* species, and *Sinorhizobium xinjiangensis*.

Rhizobium X Medium with Thiram

Composition per liter:

Agar	15.0g
Mannitol	10.0g
Yeast extract	1.0g
Soil extract	200.0mL
Thiram solution	10.0mL

pH 7.2 ± 0.2 at 25°C

Thiram Solution:

Composition per 10mL:

Thiram	1.0mg
Ethanol, absolute	10.0mL

Preparation of Thiram Solution: Add thiram to 10.0mL of absolute ethanol. Mix thoroughly. Filter sterilize.

Soil Extract:

Composition per 200mL:

African violet soil	77.0g
Na_2CO_3	0.2g

Preparation of Soil Extract: Add components to tap water and bring volume to 200.0mL.

Preparation of Medium: Add components, except thiram solution, to distilled/deionized water and bring volume to 990.0mL. Mix thoroughly. Gently heat and bring to boiling. Autoclave for 15 min at 15 psi pressure–121°C. Cool to 50°C. Aseptically add 10.0mL of sterile thiram solution. Pour into sterile Petri dishes or distribute into sterile tubes.

Use: For the cultivation and maintenance of *Bradyrhizobium japonicum, Rhizobium* species, and *Sinorhizobium xinjiangensis.*

Rhizoctonia Isolation Medium

Composition per liter:

Agar	20.0g
K_2HPO_4	1.0g
KCl	0.5g
$MgSO_4 \cdot 7H_2O$	0.5g
$NaNO_2$	0.2g
$FeSO_4 \cdot 7H_2O$	0.01g
Dexon® solution	10.0mL
Antibiotic solution	10.0mL
Gallic acid solution	10.0mL

Antibiotic Solution:

Composition per 10mL:

Chloramphenicol	0.05g
Streptomycin	0.05g

Preparation of Antibiotic Solution: Add components to distilled/deionized water and bring volume to 10.0mL. Mix thoroughly. Filter sterilize.

Dexon® Solution:

Composition per 10mL:

Dexon® (Chemagro®) wettable powder	0.09g

Preparation of Dexon® Solution: Add Dexon to distilled/deionized water and bring volume to 10.0mL. Mix thoroughly. Filter sterilize.

Gallic Acid Solution:

Composition per 10mL:

Gallic acid	0.4g

Preparation of Gallic Acid Solution: Add gallic acid to distilled/deionized water and bring volume to 10.0mL. Mix thoroughly. Filter sterilize.

Preparation of Medium: Add components—except Dexon® solution, antibiotic solution, and gallic acid solution—to distilled/deionized water and bring volume to 970.0mL. Mix thoroughly. Gently heat and bring to boiling. Autoclave for 15 min at 15 psi pressure–121°C. Cool to 45–50°C. Aseptically add sterile Dexon® solution, sterile antibiotic solution, and sterile gallic acid solution. Mix thoroughly. Pour into sterile Petri dishes or distribute into sterile tubes.

Use: For the isolation and cultivation of *Rhizoctonia* species.

Rhizomonas Medium

Composition per liter:

Noble agar	11.0g
Pancreatic digest of casein	5.0g
Glucose	2.5g
K_2HPO_4	1.0g
$MgSO_4 \cdot 7H_2O$	0.5g
KNO_3	0.5g
$Ca(NO_3)_2 \cdot 4H_2O$	0.06g

pH 7.2 ± 0.2 at 25°C

Preparation of Medium: Add components to distilled/deionized water and bring volume to 1.0L. Mix thoroughly. Gently heat and bring to boiling. Adjust pH to 7.2. Distribute into tubes or flasks. Autoclave for 15 min at 15 psi pressure–121°C. Pour into sterile Petri dishes or leave in tubes.

Use: For the cultivation and maintenance of *Rhizomonas suberifaciens.*

Rhodobacter veldkampii Medium

Composition per 127mL:

Solution 1	76.2mL
Solution 2 + Solution 3	44.8mL
Solution 4	6.0mL

Solution 1:

Composition per 2.5 liters:

$CaCl_2$ 2.0g

Preparation of Solution 1: Add $CaCl_2$ to distilled/deionized water and bring volume to 2.5L. Distribute in 80.0mL volumes into 127.0mL screw-capped bottles. Autoclave for 15 min at 15 psi pressure–121°C.

Solution 2:

Composition per 100mL:

Sodium ascorbate 2.4g
Sodium acetate 1.0g
KCl 1.0g
KH_2PO_4 .... 1.0g
$MgCl_2 \cdot 6H_2O$ 0.8g
NH_4Cl 0.8g
Heavy metal solution 50.0mL
Vitamin solution 15.0mL
Vitamin B_{12} solution 3.0mL

Preparation of Solution 2: Add components to distilled/deionized water and bring volume to 100.0mL. Mix thoroughly.

Heavy Metal Solution:

Composition per liter:

Ethylenediamine tetraacetate (EDTA) 1.5g
$FeSO_4 \cdot 7H_2O$ 0.2g
$ZnSO_4 \cdot 7H_2O$ 0.1g
$MnCl_2 \cdot 4H_2O$ 0.02g
Modified Hoagland trace element solution 6.0mL

Preparation of Heavy Metal Solution: Dissolve EDTA in 800.0mL of distilled/deionized water. Add remaining components. Bring volume to 1.0L with distilled/deionized water. Mix thoroughly.

Modified Hoagland Trace Element Solution:

Composition per 3.6L:

H_3BO_3 .... 11.0g
$MnCl_2 \cdot 4H_2O$ 7.0g
$AlCl_3$ 1.0g
$CoCl_2$ 1.0g
$CuCl_2$ 1.0g
KI 1.0g
$NiCl_2$ 1.0g
$ZnCl_2$ 1.0g
$BaCl_2$ 0.5g
KBr 0.5g
LiCl 0.5g
Na_2MoO_4 .... 0.5g
$SeCl_4$ 0.5g
$SnCl_2 \cdot 2H_2O$ 0.5g
$NaVO_3 \cdot H_2O$ 0.1g

Preparation of Modified Hoagland Trace Element Solution: Prepare each component as a separate solution. Dissolve each salt in approximately 100.0mL of distilled/deionized water. Adjust the pH of each solution to below 7.0. Combine all the salt solutions and bring the volume to 3.6L with distilled/deionized water. Adjust the pH to 3–4. A yellow precipitate may form after mixing. After a few days, it will turn into a fine white precipitate. Mix the solution thoroughly before using.

Vitamin Solution:

Composition per 100mL:

Pyridoxamine·2HCl 5.0mg
Nicotinic acid 2.0mg
Thiamine 1.0mg
Pantothenic acid 0.5mg
Biotin 0.2mg
p-Aminobenzoic acid 0.1mg

Preparation of Vitamin Solution: Add components to distilled/deionized water and bring volume to 100.0mL. Mix thoroughly.

Vitamin B_{12} Solution:

Composition per 100mL:

Vitamin B_{12} (cyanocobalamin) 2.0mg

Preparation of Vitamin B_{12} Solution: Add vitamin B_{12} to distilled/deionized water and bring volume to 100.0mL. Mix thoroughly.

Solution 3:

Composition per 900mL:

$NaHCO_3$ 4.5g

Preparation of Solution 3: Add $NaHCO_3$ to distilled/deionized water and bring volume to 900.0mL. Mix thoroughly. Bubble 100% CO_2 through the solution for 30 min. After CO_2 saturation of Solution 3, add Solution 2 and immediately filter the mixture through a Seitz filter (or a Millipore) using positive CO_2 pressure to push the liquid through.

Solution 4:

Composition per 200mL:

$Na_2S \cdot 9H_2O$ 3.0g

Preparation of Solution 4: Add $Na_2S \cdot 9H_2O$ to distilled/deionized water and bring volume to 200.0mL. Add a magnetic stir bar to the flask. Autoclave for 15 min at 15 psi pressure–121°C. On a magnetic stirrer, slowly add 2.0mL of sterile 2*M* H_2SO_4. This partially neutralizes the solution. The solution should turn yellow. H_2S gas will be liberated—neutralization and distribution of the solution should be done as rapidly as possible under adequate ventilation.

Preparation of Medium: To the 80.0mL of sterile solution 1 in screw-capped bottles, add combined solutions 2 and 3 immediately after filtration and fill bottles to capacity. Mix thoroughly. Aseptically remove 6.0mL of the medium from the bottles and replace it with 6.0mL of neutralized solution 4. Let stand for 24 hr. The medium should form a fine white precipitate before using. To inoculate, remove 6.0mL of the completed medium from the bottles and replace it with 6.0mL of inoculum.

Use: For the cultivation and maintenance of *Rhodobacter veldkampii*.

Rhodopila globiformis Medium

Composition per liter:

Mannitol	1.5g
Sodium gluconate	0.56g
KH_2PO_4	0.4g
NaCl	0.4g
$MgCl_2 \cdot 6H_2O$	0.4g
NH_4Cl	0.4g
$Na_2S_2O_3 \cdot 5H_2O$	0.2g
$CaCl_2 \cdot 2H_2O$	0.05g
Ferric citrate	5.0mg
VA vitamins	1.0mL
Trace elements solution SL-6	1.0mL

pH 4.9 ± 0.2 at 25°C

VA Vitamins:

Composition per 500mL:

Nicotinamide	0.175g
Thiamine·HCl	0.15g
p-Aminobenzoic acid	0.1g
Biotin	0.05g
Pyridoxine·2HCl	0.05g
Calcium pantothenate	0.05g
Cyanocobalamin	0.025g

Preparation of VA Vitamins: Add components to distilled/deionized water and bring volume to 500.0mL. Mix thoroughly.

Trace Elements Solution SL-6:

Composition per liter:

H_3BO_3	0.3g
$CoCl_2 \cdot 6H_2O$	0.2g
$ZnSO_4 \cdot 7H_2O$	0.10g
$MnCl_2 \cdot 4H_2O$	0.03g
$Na_2MoO_4 \cdot H_2O$	0.03g
$NiCl_2 \cdot 6H_2O$	0.02g
$CuCl_2 \cdot 2H_2O$	0.01g

Preparation of Trace Elements Solution SL-6: Add components to distilled/deionized water and bring volume to 1.0L. Mix thoroughly. Adjust pH to 3.4.

Preparation of Medium: Add components to distilled/deionized water and bring volume to 1.0L. Mix thoroughly. Adjust pH to 4.9. Distribute into tubes or flasks. Autoclave for 15 min at 15 psi pressure–121°C.

Use: For the cultivation and maintenance of *Rhodopila globiformis*.

Rhodopseudomonas blastica Medium

Composition per liter:

Sodium pyruvate	1.5g
Sodium hydrogen malate	1.5g
Yeast extract	1.0g
NH_4Cl	0.5g
$MgSO_4 \cdot 7H_2O$	0.4g
NaCl	0.4g
$CaCl_2 \cdot 2H_2O$	0.05g
Sodium phosphate buffer (0.1*M*, pH 6.8)	50.0mL

pH 6.8 ± 0.2 at 25°C

Preparation of Medium: Add components—except sodium pyruvate solution, sodium hydrogen malate solution and sodium phosphate buffer—to distilled/deionized water and bring volume to 950.0mL. Mix thoroughly. Gently heat and bring to boiling. Adjust pH to 6.8 with KOH. Autoclave for 15 min at 15 psi pressure–121°C. Cool to 45–50°C. Filter sterilize the sodium pyruvate solution, sodium hydrogen malate solution and sodium phosphate buffer solution. Aseptically add 1.5g of sterile sodium pyruvate solution, 1.5g of sodium hydrogen malate solution and 50.0mL of sodium phosphate buffer solution to cooled basal medium. Mix thoroughly. Pour into sterile Petri dishes or distribute into sterile tubes.

Use: For the cultivation and maintenance of *Rhodopseudomonas blastica* and other *Rhodopseudomonas* species.

Rhodopseudomonas Medium (ATCC Medium 543)

Composition per liter:

Sodium succinate	2.5g
$(NH_4)_2SO_4$	1.25g
K_2HPO_4	0.9g
KH_2PO_4	0.6g
Yeast extract	0.5g
$MgSO_4 \cdot 7H_2O$	0.2g
$CaCl_2$	0.07g
Ferric citrate	3.0mg
Ethylenediamine tetraacetate (EDTA)	2.0mg

pH 7.0 ± 0.2 at 25°C

Preparation of Medium: Add components to distilled/deionized water and bring volume to 1.0L. Mix thoroughly. Distribute into tubes or flasks. Autoclave for 15 min at 15 psi pressure–121°C.

Use: For the cultivation and maintenance of *Rhodopseudomonas* species.

Rhodopseudomonas Medium (ATCC Medium 650)

Composition per liter:

Sodium succinate	1.5g
KH_2PO_4	1.0g
NH_4Cl	0.5g
$MgSO_4 \cdot 7H_2O$	0.4g
NaCl	0.4g
$CaCl_2 \cdot 2H_2O$	0.05g
Trace metals solution	10.0mL

pH 5.6–6.0 at 25°C

Trace Metals Solution:

Composition per 100mL:

Ferric citrate	0.3g
Ethylenediamine tetraacetic acid (EDTA)	0.05g
$CaCl_2 \cdot 2H_2O$	0.02g
$MnSO_4 \cdot H_2O$	0.002g
$(NH_4)_6Mo_7O_{24} \cdot 4H_2O$	0.002g
H_3BO_3	0.001g
$CuSO_4 \cdot 5H_2O$	0.001g
$ZnSO_4$	0.001g

Preparation of Trace Metals Solution: Add components to distilled/deionized water and bring volume to 100.0mL. Mix thoroughly. Filter sterilize.

Preparation of Medium: Add components, except trace metals solution, to distilled/deionized water and bring volume to 990.0mL. Mix thoroughly. Autoclave for 15 min at 15 psi pressure–121°C. Cool to 25°C. Aseptically add 10.0mL of trace metals solution. Mix thoroughly. Aseptically distribute into sterile tubes or flasks.

Use: For the cultivation and maintenance of *Rhodopseudomonas viridis, R. acidophila,* and other *R.* species.

Rhodospirillaceae Enrichment Medium

Composition per liter:

Dicarboxylic acid substrate	1.0g
KH_2PO_4	0.5g
NaCl	0.4g
NH_4Cl	0.4g
$MgSO_4 \cdot 7H_2O$	0.2g
Yeast extract	0.2g
$CaCl_2 \cdot 2H_2O$	0.05g
Ferric citrate solution	5.0mL
Trace element solution SL-7	1.0mL
Vitamin B_{12} solution	1.0mL

pH 6.8 ± 0.2 at 25°C

Ferric Citrate Solution:

Composition per 100mL:

Ferric citrate	0.1g

Preparation of Ferric Citrate Solution: Add ferric citrate to distilled/deionized water and bring volume to 100.0mL. Mix thoroughly.

Trace Element Solution SL-7:

Composition per liter:

$CoCl_2 \cdot 6H_2O$	0.2g
$MnCl_2 \cdot 4H_2O$	0.1g
$ZnCl_2$	0.07g
H_3BO_3	0.06g
$NaMoO_4 \cdot 2H_2O$	0.04g
$CuCl_2 \cdot 2H_2O$	0.02g
$NiCl_2 \cdot 6H_2O$	0.02g
HCl (25% solution)	1.0mL

Preparation of Trace Element Solution SL-7: Add components to distilled/deionized water and bring volume to 1.0L. Mix thoroughly.

Vitamin B_{12} Solution:

Composition per 100mL:

Vitamin B_{12}	1.0mg

Preparation of Vitamin B_{12} Solution: Add vitamin B_{12} to distilled/deionized water and bring volume to 100.0mL. Mix thoroughly.

Preparation of Medium: Add components to distilled/deionized water and bring volume to 1.0L. Succinic acid or glutaric acid may be used for the dicarboxylic acid substrate. Mix thoroughly. Adjust pH to 6.8. Distribute into tubes or flasks. Autoclave for 15 min at 15 psi pressure–121°C.

Use: For the enrichment and isolation of members of the Rhodospirillaceae.

Rhodospirillaceae Enrichment Medium

Composition per liter:

Fatty acid substrate	1.0g
KH_2PO_4	0.5g
NaCl	0.4g
NH_4Cl	0.4g
$MgSO_4 \cdot 7H_2O$	0.2g
Yeast extract	0.2g

$CaCl_2 \cdot 2H_2O$ 0.05g
$NaHCO_3$ solution 40.0mL
Ferric citrate solution 5.0mL
Trace element solution SL-7 .... 1.0mL
Vitamin B_{12} solution 1.0mL
pH 7.3 ± 0.2 at 25°C

Ferric Citrate Solution:
Composition per 100mL:
Ferric citrate 0.1g

Preparation of Ferric Citrate Solution: Add ferric citrate to distilled/deionized water and bring volume to 100.0mL. Mix thoroughly.

Trace Element Solution SL-7:
Composition per liter:
$CoCl_2 \cdot 6H_2O$ 0.2g
$MnCl_2 \cdot 4H_2O$ 0.1g
$ZnCl_2$ 0.07g
H_3BO_3 .... 0.06g
$NaMoO_4 \cdot 2H_2O$ 0.04g
$CuCl_2 \cdot 2H_2O$ 0.02g
$NiCl_2 \cdot 6H_2O$ 0.02g
HCl (25% solution) 1.0mL

Preparation of Trace Element Solution SL-7: Add components to distilled/deionized water and bring volume to 1.0L. Mix thoroughly.

Vitamin B_{12} Solution:
Composition per 100mL:
Vitamin B_{12} 1.0mg

Preparation of Vitamin B_{12} Solution: Add vitamin B_{12} to distilled/deionized water and bring volume to 100.0mL. Mix thoroughly.

$NaHCO_3$ Solution:
Composition per 100mL:
$NaHCO_3$ 5.0g

Preparation of $NaHCO_3$ Solution: Add $NaHCO_3$ to distilled/deionized water and bring volume to 100.0mL. Mix thoroughly. Filter sterilize.

Preparation of Medium: Add components, except $NaHCO_3$ solution, to distilled/deionized water and bring volume to 1.0L. Acetate, propionate, or butyrate salts may be used for the fatty acid substrate. Mix thoroughly. Adjust pH to 7.3. Distribute into flasks in 50.0mL volumes. Autoclave for 15 min at 15 psi pressure–121°C. Cool to 25°C. Immediately prior to inoculation, aseptically add 2.0mL of sterile $NaHCO_3$ solution to each flask containing 50.0mL of medium.

Use: For the enrichment and isolation of members of the Rhodospirillaceae.

Rhodospirillaceae Enrichment Medium

Composition per liter:
Fatty acid or dicarboxylic acid substrate 1.0g
KH_2PO_4 .... 0.5g
NaCl 0.4g
NH_4Cl 0.4g
$MgSO_4 \cdot 7H_2O$ 0.2g
Yeast extract 0.2g
$CaCl_2 \cdot 2H_2O$ 0.05g
Ferric citrate solution 5.0mL
Trace element solution SL-7 .... 1.0mL
Vitamin B_{12} solution 1.0mL
pH 5.2–5.5 at 25°C

Ferric Citrate Solution:
Composition per 100mL:
Ferric citrate 0.1g

Preparation of Ferric Citrate Solution: Add ferric citrate to distilled/deionized water and bring volume to 100.0mL. Mix thoroughly.

Trace Element Solution SL-7:
Composition per liter:
$CoCl_2 \cdot 6H_2O$ 0.2g
$MnCl_2 \cdot 4H_2O$ 0.1g
$ZnCl_2$ 0.07g
H_3BO_3 .... 0.06g
$NaMoO_4 \cdot 2H_2O$ 0.04g
$CuCl_2 \cdot 2H_2O$ 0.02g
$NiCl_2 \cdot 6H_2O$ 0.02g
HCl (25% solution) 1.0mL

Preparation of Trace Element Solution SL-7: Add components to distilled/deionized water and bring volume to 1.0L. Mix thoroughly.

Vitamin B_{12} Solution:
Composition per 100mL:
Vitamin B_{12} 1.0mg

Preparation of Vitamin B_{12} Solution: Add vitamin B_{12} to distilled/deionized water and bring volume to 100.0mL. Mix thoroughly.

Preparation of Medium: Add components to distilled/deionized water and bring volume to 1.0L. Acetate, propionate, or butyrate salts may be used for the fatty acid substrate. Succinic acid or glutaric acid may be used for the dicarboxylic acid substrate. Lactate or ethanol may be used as an alternate substrate. Mix thoroughly. Adjust pH to 5.2–5.5. Distribute into tubes or flasks. Autoclave for 15 min at 15 psi pressure–121°C.

Use: For the enrichment and isolation of *Rhodopseudomonas acidophila* and *Rhodomicrobium vannielii*.

Rhodospirillum Medium (ATCC Medium 1308)

Composition per liter:

Yeast extract	1.0g
Disodium succinate	1.0g
KH_2PO_4	0.5g
Sodium ascorbate	0.5g
$MgSO_4 \cdot 7H_2O$	0.4g
NaCl	0.4g
NH_4Cl	0.4g
$CaCl_2 \cdot 2H_2O$	0.05g
Ferric citrate (0.1% solution)	5.0mL
Trace elements solution SL-6	1.0mL
Ethanol	0.5mL

pH 6.0 ± 0.2 at 25°C

Trace Elements Solution SL-6:

Composition per liter:

H_3BO_3	0.3g
$CoCl_2 \cdot 6H_2O$	0.2g
$ZnSO_4 \cdot 7H_2O$	0.10g
$MnCl_2 \cdot 4H_2O$	0.03g
$Na_2MoO_4 \cdot H_2O$	0.03g
$NiCl_2 \cdot 6H_2O$	0.02g
$CuCl_2 \cdot 2H_2O$	0.01g

Preparation of Trace Elements Solution SL-6: Add components to distilled/deionized water and bring volume to 1.0L. Mix thoroughly. Adjust pH to 3.4.

Preparation of Medium: Add components to distilled/deionized water and bring volume to 1.0L. Mix thoroughly. Adjust pH to 6.0. Distribute into tubes or flasks. Autoclave for 15 min at 15 psi pressure–121°C.

Use: For the cultivation and maintenance of *Rhodospirillum* species.

Rhodospirillum Medium (ATCC Medium 1408)

Composition per liter:

NaCl	100.0g
$MgCl_2 \cdot 6H_2O$	3.5g
Yeast extract	1.5g
Peptone	1.5g
Sodium malate	1.4g
KH_2PO_4	0.3g
SLA trace elements	1.0mL

pH 7.0 ± 0.2 at 25°C

SLA Trace Elements:

Composition per liter:

$FeCl_2 \cdot 4H_2O$	1.8g
H_3BO_3	0.5g
$CoCl_2 \cdot 6H_2O$	0.25g
$ZnCl_2$	0.1g
$MnCl_2 \cdot 4H_2O$	0.07g
$Na_2MoO_4 \cdot 2H_2O$	0.03g
$NiCl_2 \cdot 6H_2O$	0.01g
$CuCl_2 \cdot 2H_2O$	0.01g
$Na_2SeO_3 \cdot 5H_2O$	0.01g

Preparation of SLA Trace Elements: Add components to distilled/deionized water and bring volume to 1.0L. Mix thoroughly. Adjust pH to 2–3.

Preparation of Medium: Adjust medium for final pH 7.0. Sterilize by autoclaving at 121°C, 15 min.

Use: For the cultivation of *Rhodospirillum* species.

Rice Extract Agar

Composition per liter:

Agar	20.0g
White rice, solids from extract	5.0g
Polysorbate 80	10.0mL

pH 6.6 ± 0.2 at 25°C

Source: This medium is available as a premixed powder from BBL Microbiology Systems.

Preparation of Medium: Add components, except polysorbate 80, to distilled/deionized water and bring volume to 990.0mL. Mix thoroughly. Gently heat and bring to boiling. Add polysorbate 80. Mix thoroughly. Distribute into tubes or flasks. Autoclave for 15 min at 15 psi pressure–121°C. Pour into sterile Petri dishes.

Use: For the cultivation and differentiation of *Candida albicans* and *C. stellatoidea* from other *Candida* species based on chlamydospore formation.

Rice Extract Agar

Composition per liter:

Agar	20.0g
White rice, solids from extract	20.0g

pH 7.1 ± 0.2 at 25°C

Source: This medium is available as a premixed powder from Difco Laboratories.

Preparation of Medium: Add components, except polysorbate 80, to distilled/deionized water and bring volume to 990.0mL. Mix thoroughly. Gently heat and bring to boiling. Add polysorbate 80. Mix thoroughly. Distribute into tubes or flasks. Autoclave for 15 min at 15 psi pressure–121°C. Pour into sterile Petri dishes.

Use: For the cultivation and differentiation of *Candida albicans* and *Candida stellatoidea* from other *Candida* species based on chlamydospore formation.

Rila Marine Medium

Composition per liter:

Agar	15.0g
Peptone	0.5g
Yeast extract	0.5g
Pancreatic digest of casein	0.5g
Marine salts mixture	800.0mL

pH 7.6-8.0 at 25°C

Preparation of Medium: Add components to distilled/deionized water and bring volume to 1.0L. Mix thoroughly. Gently heat and bring to boiling. Autoclave for 15 min at 15 psi pressure–121°C. Adjust pH to 7.6–8.0. Pour into sterile Petri dishes or distribute into sterile tubes.

Use: For the cultivation and maintenance of *Alteromonas denitrificans.*

Rimler–Shotts Medium (RS Medium)

Composition per liter:

Agar	13.5g
$Na_2S_2O_3 \cdot 5H_2O$	6.8g
L-Ornithine·HCl	6.5g
NaCl	5.0g
L-Lysine·HCl	5.0g
Maltose	3.5g
Yeast extract	3.0g
Sodium deoxycholate	1.0g
Ferric ammonium citrate	0.8g
L-Cysteine·HCl	0.3g
Bromthymol blue	0.03g
Novobiocin solution	10.0mL

pH 7.0 ± 0.2 at 25°C

Novobiocin Solution:

Composition per 10mL:

Novobiocin	5.0mg

Preparation of Novobiocin Solution: Add novobiocin to distilled/deionized water and bring volume to 10.0mL. Mix thoroughly. Filter sterilize.

Preparation of Medium: Add components, except novobiocin solution, to distilled/deionized water and bring volume to 990.0mL. Mix thoroughly. Gently heat and bring to boiling. Autoclave for 15 min at 15 psi pressure–121°C. Cool to 45–50°C. Aseptically add sterile components. Mix thoroughly. Pour into sterile Petri dishes or distribute into sterile tubes.

Use: For the selective isolation, cultivation and presumptive identification of *Aeromonas hydrophila* and other Gram-negative bacteria based on their ability to decarboxylate lysine and ornithine, ferment maltose and produce H_2S. Maltose-fermenting bacteria appear as yellow colonies. Bacteria that produce lysine or ornithine decarboxylase turn the medium greenish-yellow to yellow. Bacteria that produce H_2S appear as colonies with black centers.

RIOT Agar (Rice Infusion Oxgall Tween™ 80 Agar)

Composition per 1010mL:

Agar	10.0g
Oxgall	10.0g
Rice extract	1.0L
Tween™ 80	10.0mL

pH 7.3 ± 0.2 at 25°C

Rice Extract:

Composition per liter:

Cream of rice cereal	10.0g

Preparation of Rice Extract: Add cream of rice cereal to 1.0L of boiling tap water. Mix thoroughly. Filter quickly through cheesecloth. Bring volume of filtrate to 1.0L with tap water.

Preparation of Medium: Combine components. Mix thoroughly. Gently heat and bring to boiling. Distribute into tubes or flasks. Autoclave for 15 min at 15 psi pressure–121°C. Pour into sterile Petri dishes or leave in tubes.

Use: For the cultivation and differentiation of *Candida albicans* and *C. stellatoidea* from other *Candida* species based on chlamydospore formation.

Rippey–Cabelli Agar (RC Agar)

Composition per liter:

Agar	15.0g
Meat peptone	5.0g
Trehalose	5.0g
NaCl	3.0g
KCl	2.0g
Yeast extract	2.0g
Bromthymol blue	0.44g
$MgSO_4 \cdot 7H_2O$	0.2g
$FeCl_3 \cdot 6H_2O$	0.1g
Sodium deoxycholate	0.1g
Ampicillin solution	10.0mL
Ethanol	10.0mL

pH 8.0 ± 0.2 at 25°C

Ampicillin Solution:

Composition per 10mL:

Ampicillin	0.02g

Preparation of Ampicillin Solution: Add ampicillin to distilled/deionized water and bring volume to 10.0mL. Mix thoroughly. Filter sterilize.

Preparation of Medium: Add components—except sodium deoxycholate, ampicillin solution and ethanol—to distilled/deionized water and bring volume to 990.0mL. Mix thoroughly. Gently heat and bring to boiling. Autoclave for 15 min at 15 psi pressure–121°C. Cool to 45–50°C. Aseptically add sodium deoxycholate, 10.0mL of sterile ampicillin solution and 10.0mL of ethanol. Mix thoroughly. Pour into sterile Petri dishes or distribute into sterile tubes.

Use: For the isolation, cultivation and differentiation of *Aeromonas* species and *Plesiomonas* species from water samples using the membrane filter method. This medium differentiates bacteria on the basis of trehalose fermentation. Bacteria that ferment trehalose turn the medium yellow.

RM Medium

Composition per liter:

Glucose	20.0g
Agar	15.0g
Yeast extract	10.0g
KH_2PO_4	2.0g
Solution 1	250.0mL
Solution 2	250.0mL
Solution 3	250.0mL
Solution 4	250.0mL

pH 6.0 ± 0.2 at 25°C

Solution 1:

Composition per 250mL:

Glucose	20.0g

Preparation of Solution 1: Add glucose to distilled/deionized water and bring volume to 250.0mL. Mix thoroughly. Autoclave for 15 min at 15 psi pressure–121°C. Cool to 45–50°C.

Solution 2:

Composition per 250mL:

Agar	15.0g

Preparation of Solution 2: Add agar to distilled/deionized water and bring volume to 250.0mL. Mix thoroughly. Autoclave for 15 min at 15 psi pressure–121°C. Cool to 45–50°C.

Solution 3:

Composition per 250mL:

Yeast extract	10.0g

Preparation of Solution 3: Add yeast extract to distilled/deionized water and bring volume to 250.0mL. Mix thoroughly. Autoclave for 15 min at 15 psi pressure–121°C. Cool to 45–50°C.

Solution 4:

Composition per 250mL:

KH_2PO_4	2.0g

Preparation of Solution 4: Add KH_2PO_4 to distilled/deionized water and bring volume to 250.0mL. Mix thoroughly. Autoclave for 15 min at 15 psi pressure–121°C. Cool to 45–50°C.

Preparation of Medium: Aseptically combine the four sterile solutions. Mix thoroughly. Adjust pH to 6.0. Pour into sterile Petri dishes or distribute into sterile tubes.

Use: For the cultivation and maintenance of *Zymomonas mobilis*.

Rose Bengal Chloramphenicol Agar

Composition per liter:

Agar	15.0g
Glucose	10.0g
Papaic digest of soybean meal	5.0g
KH_2PO_4	1.0g
$MgSO_4 \cdot 7H_2O$	0.5g
Rose bengal	0.05g
Chloramphenicol solution	10.0mL

pH 7.0 ± 0.2 at 25°C

Source: This medium is available as a premixed powder from Difco Laboratories and Oxoid Unipath.

Chloramphenicol Solution:

Composition per 10mL:

Chloramphenicol	0.10g

Preparation of Chloramphenicol Solution: Add chloramphenicol to distilled/deionized water and bring volume to 10.0mL. Mix thoroughly. Filter sterilize.

Preparation of Medium: Add components, except chloramphenicol solution, to distilled/deionized water and bring volume to 990.0mL. Mix thoroughly. Gently heat and bring to boiling. Autoclave for 15 min at 15 psi pressure–121°C. Cool to 45°C. Aseptically add sterile chloramphenicol solution. Mix thoroughly. Pour into sterile Petri dishes or distribute into sterile tubes.

Use: For the selective isolation, cultivation and enumeration of yeasts and molds from environmental specimens.

Rumen Fluid Cellobiose Agar (RFC Agar)

Composition per 10mL:

Rumen fluid cellobiose base medium 8.9mL
$NaHCO_3$–rifampin solution 1.0mL
Cellobiose solution 0.1mL

Rumen Fluid Cellobiose Base Medium:

Composition per 89mL:

Noble agar 0.7g
Cysteine·HCl·H_2O 0.1g
Clarified rumen fluid 30.0mL
Salts solution A 20.0mL
Salts solution B 20.0mL
Resazurin (0.1% solution) 0.1mL

pH 6.7–7.0 at 25°C

Preparation of Rumen Fluid Cellobiose Base Medium: Add components to distilled/deionized water and bring volume to 89.0mL. Mix thoroughly. Gently heat and bring to boiling. Continue boiling until resazurin turns colorless indicating reduction. Anaerobically distribute into tubes in 8.9mL volumes under 100% CO_2. Cap tubes with rubber stoppers. Autoclave for 15 min at 15 psi pressure–121°C. Cool to 25°C.

Salts Solution A:

Composition per liter:

$CaCl_2$ 0.45g
$MgSO_4$ 0.45g

Preparation of Salts Solution A: Add components to distilled/deionized water and bring volume to 1.0L. Mix thoroughly.

Salts Solution B:

Composition per liter:

NaCl 4.5g
$(NH_4)_2SO_4$ 4.5g
KH_2PO_4 .... 2.25g
K_2HPO_4 .... 2.25g

Preparation of Salts Solution B: Add components to distilled/deionized water and bring volume to 1.0L. Mix thoroughly.

$NaHCO_3$–Rifampin Solution:

Composition per 10mL:

$NaHCO_3$ 0.5g
Rifampin 0.1mg

Preparation of $NaHCO_3$–Rifampin Solution: Add components to distilled/deionized water and bring volume to 10.0mL. Mix thoroughly. Filter sterilize.

Cellobiose Solution:

Composition per 10mL:

Cellobiose 1.0g

Preparation of Cellobiose Solution: Add cellobiose to distilled/deionized water and bring volume to 10.0mL. Mix thoroughly. Filter sterilize.

Preparation of Medium: To each tube containing 8.9mL of sterile rumen fluid cellobiose base medium, aseptically add 1.0mL of sterile $NaHCO_3$–rifampin solution and 0.1mL of sterile cellobiose solution. Mix thoroughly.

Use: For the selective isolation of rumen treponemes.

Ruminococcus pasteuri Medium

Composition per liter:

$NaHCO_3$ 2.5g
Sodium tartrate 2.0g
NaCl 1.0g
KCl 0.5g
$MgCl_2·6H_2O$ 0.4g
$Na_2S·9H_2O$ 0.36g
NH_4Cl 0.25g
KH_2PO_4 .... 0.2g
$CaCl_2·2H_2O$ 0.15g
Resazurin 1.0mg
Trace elements solution SL-7 .... 1.0mL

pH 7.2 ± 0.2 at 25°C

Trace Elements Solution SL-7:

Composition per liter:

$FeCl_2·4H_2O$ 1.5g
$CoCl_2·6H_2O$ 0.19g
$MnCl_2·4H_2O$ 0.1g
$ZnCl_2$ 0.07g
H_3BO_3 .... 0.062g
$Na_2MoO_4·2H_2O$ 0.036g
$NiCl_2·6H_2O$ 0.024g
$CuCl_2·2H_2O$ 0.017g
HCl (25% solution) 10.0mL

Preparation of Trace Elements Solution SL-7: Add the $FeCl_2·4H_2O$ to the HCl. Add distilled/deionized water and bring volume to 1.0L. Add remaining components. Mix thoroughly. Autoclave for 15 min at 15 psi pressure–121°C under 100% N_2. Cool to room temperature.

$NaHCO_3$ Solution:

Composition per 10mL:

$NaHCO_3$ 2.5g

Preparation of $NaHCO_3$ Solution: Add the $NaHCO_3$ to distilled/deionized water and bring volume to 10.0mL. Mix thoroughly. Filter sterilize.

$Na_2S·9H_2O$ Solution:
Composition per 10mL:

$Na_2S·9H_2O$	0.36g

Preparation of $Na_2S·9H_2O$ Solution: Add $Na_2S·9H_2O$ to distilled/deionized water and bring volume to 10.0mL. Mix thoroughly. Autoclave for 15 min at 15 psi pressure–121°C under 100% N_2.

Preparation of Medium: Add components—except $NaHCO_3$ solution, $Na_2S·9H_2O$ solution, and trace elements solution SL-7—to distilled/deionized water and bring volume to 999.0mL. Mix thoroughly. Adjust pH to 7.2. Gently heat and bring to boiling under 80% N_2 + 20% CO_2. Distribute into tubes in 9.8mL volumes under 80% N_2 + 20% CO_2. Cool to 25°C. Aseptically add 0.1mL of sterile $NaHCO_3$ solution and 0.01mL of sterile trace elements solution SL-7 to each tube. Mix thoroughly. Immediately prior to inoculation aseptically add 0.1mL of sterile $Na_2S·9H_2O$ solution to each tube.

Use: For the cultivation and maintenance of *Ruminococcus pasteuri.*

Russell Double Sugar Agar

Composition per liter:

Agar	15.0g
Proteose peptone No. 3	12.0g
Lactose	10.0g
NaCl	5.0g
Beef extract	1.0g
Glucose	1.0g
Phenol red	0.025g

pH 7.5 ± 0.2 at 25°C

Preparation of Medium: Add components to distilled/deionized water and bring volume to 1.0L. Mix thoroughly. Gently heat and bring to boiling. Distribute into tubes. Autoclave for 15 min at 15 psi pressure–121°C. Allow tubes to cool in a slanted position.

Use: For the identification of Gram-negative enteric bacilli based on their fermentation of glucose and lactose. Bacteria that ferment both glucose and lactose produce a yellow slant and yellow butt. Bacteria that ferment glucose but do not ferment lactose produce a red slant and a yellow butt. Bacteria that ferment neither glucose nor lactose produce an unchanged pink-orange color.

S Salts

Composition per liter:

Dibenzothiophene	5.00g
NH_4Cl	0.50g
KH_2PO_4	0.25g
$MgCl_2·6H_2O$	0.25g

pH 6.5–7.0 at 25°C

Preparation of Medium: Add components to distilled/deionized water and bring volume to 1.0L. Mix thoroughly. Adjust pH to 6.5–7.0 with KOH. Distribute into tubes or flasks. Autoclave for 15 min at 15 psi pressure–121°C.

Use: For the cultivation of *Bacillus sulfasportare.*

S6 Medium for Thiobacilli

Composition per liter:

Agar	15.0g
$Na_2S_2O_3$	10.0g
KH_2PO_4	11.8g
Na_2HPO_4	1.2g
$MgSO_4·7H_2O$	0.1g
$(NH_4)_2SO_4$	0.1g
$CaCl_2$	0.03g
$FeCl_3$	0.02g
$MnSO_4$	0.02g

Preparation of Medium: Add components to distilled/deionized water and bring volume to 1.0L. Mix thoroughly. Gently heat and bring to boiling. Distribute into tubes or flasks. Autoclave for 15 min at 15 psi pressure–121°C. Pour into sterile Petri dishes or leave in tubes.

Use: For the cultivation and maintenance of *Thiobacillus denitrificans* and *T. thioparus.*

S8 Medium for Thiobacilli

Composition per liter:

Agar	15.0g
KH_2PO_4	11.8g
$Na_2S_2O_3$	10.0g
KNO_3	5.0g
Na_2HPO_4	1.2g
$NaHCO_3$	0.5g
$MgSO_4·7H_2O$	0.1g
$(NH_4)_2SO_4$	0.1g
$CaCl_2$	0.03g
$FeCl_3$	0.02g
$MnSO_4$	0.02g

Preparation of Medium: Add components to distilled/deionized water and bring volume to 1.0L. Mix thoroughly. Gently heat and bring to boiling. Distribute into tubes or flasks. Autoclave for 15 min at 15 psi pressure–121°C. Pour into sterile Petri dishes or leave in tubes.

Use: For the cultivation and maintenance of *Thiobacillus neapolitanus.*

S Medium

Composition per liter:

Glycogen	3.0g
$MgSO_4 \cdot 7H_2O$	2.0g
L-Leucine	1.0g
L-Tyrosine	0.60g
L-Asparagine	0.50g
L-Isoleucine	0.50g
L-Proline	0.50g
L-Lysine	0.25g
KH_2PO_4	0.13g
Djenkolic acid	0.10g
L-Arginine	0.10g
L-Serine	0.10g
L-Threonine	0.10g
L-Valine	0.10g
L-Alanine	0.05g
L-Glycine	0.05g
L-Histidine	0.05g
L-Methionine	0.05g
L-Tryptophan	0.05g

pH 7.6 ± 0.2 at 25°C

Preparation of Medium: Add components to distilled/deionized water and bring volume to 1.0L. Mix thoroughly. Filter sterilize. Aseptically distribute into tubes or flasks.

Use: For the cultivation of *Myxococcus xanthus.*

Sabouraud Glucose Agar, Emmons

Composition per liter:

Glucose	20.0g
Agar	17.0g
Pancreatic digest of casein	5.0g
Peptic digest of animal tissue	5.0g

pH 6.9 ± 0.2 at 25°C

Source: This medium is available as a premixed powder from BBL Microbiology Systems.

Preparation of Medium: Add components to distilled/deionized water and bring volume to 1.0L. Mix thoroughly. Gently heat and bring to boiling. Distribute into tubes or flasks. Autoclave for 15 min at 13 psi pressure–118°C. Pour into sterile Petri dishes or leave in tubes.

Use: For the cultivation of yeast and filamentous fungi.

Sabouraud Maltose Broth

Composition per liter:

Maltose	40.0g
Neopeptone	10.0g

pH 5.6 ± 0.2 at 25°C

Source: This medium is available as a premixed powder from Difco Laboratories.

Preparation of Medium: Add components to distilled/deionized water and bring volume to 1.0L. Mix thoroughly. Distribute into tubes or flasks. Autoclave for 15 min at 15 psi pressure–121°C. Avoid overheating.

Use: For the cultivation of a variety of fungi.

Sabouraud Medium, Fluid

Composition per liter:

Glucose	20.0g
Pancreatic digest of casein	5.0g
Peptamin	5.0g

pH 5.7 ± 0.2at 25°C

Source: This medium is available as a premixed powder from Difco Laboratories and Oxoid Unipath.

Preparation of Medium: Add components to distilled/deionized water and bring volume to 1.0L. Mix thoroughly. Distribute into tubes or flasks. Autoclave for 15 min at 15 psi pressure–121°C. Avoid overheating.

Use: For isolation and cultivation of yeasts and molds.

Saccharolytic Clostridia Medium

Composition per liter:

Pancreatic digest of casein	10.0g
Yeast extract	6.0g
Sodium thioglycollate	0.5g
Carbohydrate solution	100.0mL
Potassium phosphate (1*M* solution, pH 7.5)	30.0mL
$MgSO_4$ (1*M* solution)	1.0mL
Solution M	0.5mL
$FeSO_4$ solution	0.2mL

pH 7.0–7.2 at 25°C

Carbohydrate Solution:

Composition per 100mL:

Glucose or sucrose	20.0g

Preparation of Carbohydrate Solution: Add carbohydrate to distilled/deionized water and bring volume to 100.0mL. Mix thoroughly. Filter sterilize.

Solution M:

Composition per liter:

$CaCl_2$	3.33g

$MnCl_2 \cdot 4H_2O$ 1.98g
$Na_2MoO_4 \cdot 2H_2O$ 1.21g
$CoCl_2 \cdot 6H_2O$ 1.19g

Preparation of Solution M: Add components to distilled/deionized water and bring volume to 1.0L. Mix thoroughly.

$FeSO_4 \cdot 7H_2O$ Solution:
Composition per 10mL:
$FeSO_4 \cdot 7H_2O$ 55.6g
H_2SO_4 (0.1*M* solution) 10.0mL

Preparation of $FeSO_4 \cdot 7H_2O$ Solution: Add the $FeSO_4 \cdot 7H_2O$ to 10.0mL of H_2SO_4 solution. Mix thoroughly. Filter sterilize.

Preparation of Medium: Add components, except carbohydrate solution, to distilled/deionized water and bring volume to 900.0mL. Mix thoroughly. Gently heat and bring to boiling. Autoclave for 20 min at 15 psi pressure–121°C. Cool to 45–50°C. Aseptically add sterile carbohydrate solution. Mix thoroughly. Aseptically distribute into sterile tubes.

Use: For the cultivation of saccharolytic *Clostridium* species.

Saccharolytic Clostridia Medium

Composition per liter:
Sodium thioglycollate 1.0g
K_2HPO_4 .... 0.8g
KH_2PO_4 .... 0.2g
$MgSO_4 \cdot 7H_2O$ 0.2g
NaCl 0.2g
$Na_2MoO_4 \cdot 2H_2O$ 0.025g
Yeast extract 0.010g
$FeSO_4 \cdot 7H_2O$ 0.010g
$MnSO_4 \cdot 4H_2O$ 0.010g
CaCl2 .... 0.010g
Carbohydrate solution 100.0mL
Soil extract 10.0mL
Trace element solution 1.0mL
pH 7.2 ± 0.2 at 25°C

Carbohydrate Solution:
Composition per 100mL:
Glucose or sucrose 10.0g

Preparation of Carbohydrate Solution: Add glucose or sucrose to distilled/deionized water and bring volume to 100.0mL. Mix thoroughly. Filter sterilize.

Soil Extract:
Composition per 200mL:
Garden soil, neutral 100.0g

Preparation of Soil Extract: Add garden soil to 100.0mL of tap water. Gently heat and bring to 130°C for 60 min. Cool to 45°C. Filter through Whatman #1 filter paper. Autoclave for 15 min at 15 psi pressure–121°C. Cool to 45–50°C.

Trace Element Solution:
Composition per liter:
$Na_2B_4O_7 \cdot 10H_2O$ 0.05g
$CoNO_3 \cdot 6H_2O$ 0.05g
$CdSO_4 \cdot 2H_2O$ 0.05g
$CuSO_4 \cdot 5H_2O$ 0.05g
$ZnSO_4 \cdot 7H_2O$ 0.05g
$MnSO_4 \cdot H_2O$ 0.05g

Preparation of Trace Element Solution: Add components to distilled/deionized water and bring volume to 1.0L. Mix thoroughly.

Preparation of Medium: Add components, except sodium thioglycollate and carbohydrate solution, to distilled/deionized water and bring volume to 900.0mL. Mix thoroughly. Gently heat and bring to boiling. Add sodium thioglycollate. Mix thoroughly. Distribute 9.5mL into test tubes that contain inverted Durham tubes. Autoclave for 15 min at 15 psi pressure–121°C. Cool to 45–50°C. Aseptically add 0.5mL of sterile carbohydrate solution to each tube. Mix thoroughly.

Use: For the isolation of N_2-fixing, saccharolytic *Clostridium* species.

Salmonella Medium

Composition per liter:
Pancreatic digest of casein 10.0g
NaCl 5.0g
pH 7.4 ± 0.2 at 25°C

Preparation of Medium: Add components to distilled/deionized water and bring volume to 1.0L. Mix thoroughly. Distribute into tubes or flasks. Autoclave for 15 min at 15 psi pressure–121°C.

Use: For the cultivation and maintenance of *Escherichia coli* and *Salmonella choleraesuis*.

Salmonella Shigella Agar (SS Agar)

Composition per liter:
Agar 13.5g
Lactose 10.0g
Bile salts 8.5g
$Na_2S_2O_3$ 8.5g
Sodium citrate 8.5g
Beef extract 5.0g
Pancreatic digest of casein 2.5g
Peptic digest of animal tissue 2.5g
Ferric citrate 1.0g

Neutral red 0.025g
Brilliant green 0.330mg

pH 7.0 ± 0.2 at 25°C

Source: This medium is available as a premixed powder from BBL Microbiology Systems, Difco Laboratories and Oxoid Unipath.

Preparation of Medium: Add components to distilled/deionized water and bring volume to 1.0L. Mix thoroughly. Gently heat while stirring and bring to boiling. Do not autoclave. Cool to 45–50°C. Pour into sterile Petri dishes in 20.0mL volumes. Allow the surface of the plates to dry before inoculation.

Use: For the selective isolation and differentiation of pathogenic enteric bacilli, especially those belonging to the genus *Salmonella*. This medium is not recommended for the primary isolation of *Shigella* species. Lactose-fermenting bacteria such as *Escherichia coli* or *Klebsiella pneumoniae* appear as small pink or red colonies. Lactose-nonfermenting bacteria—such as *Salmonella* species, *Proteus* species and *Shigella* species—appear as colorless colonies. Production of H_2S by *Salmonella* species turns the center of the colonies black.

Salmonella Shigella Agar, Modified (SS Agar, Modified)

Composition per liter:

Agar..12.0g
Lactose..10.0g
Sodium citrate...10.0g
$Na_2S_2O_3$ 8.5g
Bile salts...5.5g
Beef extract...5.0g
Peptone...5.0g
Ferric citrate..1.0g
Neutral red 0.025g
Brilliant green 0.330mg

pH 7.3 ± 0.2 at 25°C

Source: This medium is available as a premixed powder from Oxoid Unipath.

Preparation of Medium: Add components to distilled/deionized water and bring volume to 1.0L. Mix thoroughly. Gently heat while stirring and bring to boiling. Do not autoclave. Cool to 45–50°C. Pour into sterile Petri dishes in 20.0mL volumes. Allow the surface of the plates to dry before inoculation.

Use: For the selective isolation and differentiation of pathogenic enteric bacilli, especially those belonging to the genus *Salmonella*. This medium provides better growth of *Shigella* species. Lactose-fermenting bacteria such as *Escherichia coli* or *Klebsiella pneumoniae* appear as small pink or red colonies. Lactose-nonfermenting bacteria—such as *Salmonella* species, *Proteus* species and *Shigella* species—appear as colorless colonies. Production of H_2S by *Salmonella* species turns the center of the colonies black.

Salt Broth, Modified

Composition per liter:

NaCl..65.0g
Enzymatic digest of animal tissue.....................10.0g
Heart digest..10.0g
Glucose..1.0g
Bromcresol purple.....................................0.016g

pH 7.2 ± 0.2 at 25°C

Source: Available as a prepared medium from BBL Microbiology Systems.

Preparation of Medium: Add components to distilled/deionized water and bring volume to 1.0L. Mix thoroughly. Distribute into tubes or flasks. Autoclave for 15 min at 15 psi pressure–121°C.

Use: For the cultivation and differentiation of the enterococcal group D streptococci from nonenterococcal group D streptococci based on salt tolerance.

Salt Colistin Broth

Composition per liter:

NaCl..20.0g
Peptone..10.0g
Yeast extract...3.0g
Colistin solution.......................................10.0mL

pH 7.4 ± 0.2 at 25°C

Colistin Solution:

Composition per 10mL:

Colistin methane sulfonate.......................500,000U

Preparation of Colistin Solution: Add colistin methane sulfonate to distilled/deionized water and bring volume to 10.0mL. Mix thoroughly. Filter sterilize.

Preparation of Medium: Add components, except colistin solution, to distilled/deionized water and bring volume to 990.0mL. Mix thoroughly. Gently heat until dissolved. Autoclave for 15 min at 15 psi pressure–121°C. Cool to 25°C. Aseptically add sterile colistin solution. Mix thoroughly. Aseptically distribute into sterile tubes or flasks.

Use: For the cultivation of halophilic *Vibrio* species.

Salt Medium

Composition per liter:

NaCl..58.4g
Agar..15.0g

Proteose peptone	5.0g
Pancreatic digest of casein	5.0g

pH 6.9 ± 0.2 at 25°C

Preparation of Medium: Add components to distilled/deionized water and bring volume to 1.0L. Mix thoroughly. Gently heat and bring to boiling. Distribute into tubes or flasks. Autoclave for 15 min at 15 psi pressure–121°C. Pour into sterile Petri dishes or leave in tubes.

Use: For the cultivation and maintenance of *Marinococcus halophilus*.

Salt Tolerance Medium

Composition per liter:

Beef heart, solids from infusion	500.0g
NaCl	65.0g
Tryptose	10.0g

pH 7.4 ± 0.2 at 25°C

Preparation of Medium: Add components to distilled/deionized water and bring volume to 1.0L. Mix thoroughly. Distribute into tubes or flasks. Autoclave for 15 min at 15 psi pressure–121°C.

Use: For testing the salt tolerance of a variety of microorganisms.

Salt Tolerance Medium, Gilardi

Composition per liter:

NaCl	65.0g
Pancreatic digest of casein	15.0g
Agar	15.0g
Papaic digest of soybean meal	5.0g

pH 7.3 ± 0.2 at 25°C

Preparation of Medium: Add components to distilled/deionized water and bring volume to 1.0L. Mix thoroughly. Gently heat and bring to boiling. Distribute into tubes or flasks. Autoclave for 15 min at 15 psi pressure–121°C. Do not overheat. Pour into sterile Petri dishes or leave in tubes.

Use: For the cultivation and maintenance of salt-tolerant, nonfermenting Gram-negative bacteria. For the differentiation of nonfermenting Gram-negative bacteria based on salt tolerance.

Salt Tolerance Medium, Tatum

Composition per liter:

NaCl	65.0g
Peptone	5.0g
Yeast extract	2.0g
Beef extract	1.0g

pH 7.4 ± 0.2 at 25°C

Preparation of Medium: Add components to distilled/deionized water and bring volume to 1.0L. Mix thoroughly. Distribute into tubes or flasks. Autoclave for 15 min at 15 psi pressure–121°C.

Use: For the cultivation of salt-tolerant, nonfermenting Gram-negative bacteria. For the differentiation of nonfermenting Gram-negative bacteria based on salt tolerance.

SAP 1 Agar

Composition per liter:

Agar	15.0g
Pancreatic digest of casein	5.0g
Yeast extract	5.0g
Artificial seawater	1.0L

pH 7.2 ± 0.2 at 25°C

Artificial Seawater:
Composition per liter:

NaCl	24.7g
$MgSO_4 \cdot 7H_2O$	6.3g
$MgCl_2 \cdot 6H_2O$	4.6g
$CaCl_2$	1.0g
KCl	0.7g
$NaHCO_3$	0.2g

Preparation of Artificial Seawater: Add components to distilled/deionized water and bring volume to 1.0L. Mix thoroughly.

Preparation of Medium: Add solid components to 1.0L of artificial seawater. Mix thoroughly. Gently heat and bring to boiling. Distribute into tubes or flasks. Autoclave for 15 min at 15 psi pressure–121°C. Pour into sterile Petri dishes or leave in tubes.

Use: For the isolation and cultivation of *Cytophaga* species, *Herpetosiphon* species, *Saprospira* species, and *Flexithrix* species.

SAP 2 Agar

Composition per liter:

Agar	15.0g
Pancreatic digest of casein	1.0g
Yeast extract	1.0g
Artificial seawater	1.0L

pH 7.2 ± 0.2 at 25°C

Artificial Seawater:
Composition per liter:

NaCl	24.7g
$MgSO_4 \cdot 7H_2O$	6.3g
$MgCl_2 \cdot 6H_2O$	4.6g
$CaCl_2$	1.0g
KCl	0.7g
$NaHCO_3$	0.2g

Preparation of Artificial Seawater: Add components to distilled/deionized water and bring volume to 1.0L. Mix thoroughly.

Preparation of Medium: Add solid components to 1.0L of artificial seawater. Mix thoroughly. Gently heat and bring to boiling. Distribute into tubes or flasks. Autoclave for 15 min at 15 psi pressure–121°C. Pour into sterile Petri dishes or leave in tubes.

Use: For the isolation and cultivation of *Cytophaga* species, *Herpetosiphon* species, *Saprospira* species, and *Flexithrix* species.

Saprospira grandis Medium

Composition per 1010mL:

Pancreatic digest of casein	5.0g
Yeast extract	5.0g
$Ca(NO_3)_2 \cdot 4H_2O$	0.1g
K_2HPO_4	0.02g
Seawater, filtered	1.0L
Trace elements	10.0mL

pH 7.0 ± 0.2 at 25°C

Trace Elements:

Composition per liter:

$FeSO_4 \cdot 7H_2O$	0.5mg
$ZnSO_4 \cdot 7H_2O$	0.3mg
H_3BO_3	0.1mg
$CoCl_2 \cdot 6H_2O$	0.1mg
$CuSO_4 \cdot 5H_2O$	0.1mg
$MnSO_4 \cdot 4H_2O$	0.1mg
$Na_2MoO_4 \cdot 2H_2O$	0.1mg

Preparation of Trace Elements: Add components to distilled/deionized water and bring volume to 1.0L. Mix thoroughly.

Preparation of Medium: Combine components. Mix thoroughly. Adjust pH to 7.0. Filter sterilize.

Use: For the cultivation of *Saprospira grandis.*

Sarcina maxima Medium

Composition per liter:

Glucose	10.0g
Peptone	10.0g
Yeast extract	5.0g
Cysteine·HCl solution	10.0mL

pH 6.0 ± 0.2 at 25°C

Cysteine·HCl Solution:

Composition per 10mL:

Cysteine·HCl	0.5g

Preparation of Cysteine·HCl Solution: Add cysteine·HCl to distilled/deionized water and bring volume to 10.0mL. Mix thoroughly. Filter sterilize.

Preparation of Medium: Add components, except cysteine·HCl solution, to distilled/deionized water and bring volume to 990.0mL. Mix thoroughly. Gently heat and bring to boiling. Autoclave for 15 min at 15 psi pressure–121°C. Cool to 25°C. Aseptically add sterile cysteine·HCl solution. Mix thoroughly. Aseptically distribute into sterile tubes or flasks.

Use: For the cultivation of *Sarcina maxima.*

Sarcina ventriculi Growth Medium

Composition per liter:

Glucose	30.0g
Peptone	5.0g
Yeast extract	5.0g

pH 6.0 ± 0.2 at 25°C

Preparation of Medium: Add components to distilled/deionized water and bring volume to 1.0L. Mix thoroughly. Distribute into tubes in 10.0mL volumes. Autoclave for 20 min at 15 psi pressure–121°C.

Use: For the cultivation and maintenance of *Sarcina maxima* and *Sarcina ventriculi.*

SC Agar

Composition per liter:

Agar	15.0g
Papaic digest of soybean meal	8.0g
Corn meal (solids from infusion)	2.0g
K_2HPO_4	1.0g
KH_2PO_4	1.0g
$MgSO_4 \cdot 7H_2O$	0.2g
Hemin solution	15.0mL
Bovine serum albumin, fraction V	10.0mL
Cysteine·H_2O solution	10.0mL
Glucose solution	1.0mL

pH 6.6 ± 0.2 at 25°C

Hemin Solution:

Composition per 100mL:

Hemin	0.1g
NaOH (0.05*N* solution)	100.0mL

Preparation of Hemin Solution: Add hemin to NaOH solution. Mix thoroughly.

Bovine Serum Albumin, Fraction V Solution:

Composition per 10mL:

Bovine serum albumin, fraction V	2.0g

Preparation of Bovine Serum Albumin, Fraction V Solution: Add bovine serum albumin to distilled/deionized water and bring volume to 10.0mL. Mix thoroughly. Filter sterilize.

Cysteine·H_2O Solution:

Composition per 10mL:

Cysteine·H_2O	1.0g

Preparation of Cysteine·H_2O Solution: Add cysteine·H_2O to distilled/deionized water and bring volume to 10.0mL. Mix thoroughly. Filter sterilize.

Glucose Solution:

Composition per 10mL:

D-Glucose	5.0g

Preparation of Glucose Solution: Add glucose to distilled/deionized water and bring volume to 10.0mL. Mix thoroughly. Filter sterilize.

Preparation of Medium: Add components—except bovine serum albumin, cysteine·H_2O solution, and glucose solution—to distilled/deionized water and bring volume to 979.0mL. Mix thoroughly. Adjust pH to 6.6 with NaOH. Gently heat and bring to boiling. Autoclave for 15 min at 15 psi pressure–121°C. Cool to 45–50°C. Aseptically add 10.0mL of sterile bovine serum albumin, 10.0mL of sterile cysteine·H_2O solution, and 1.0mL of sterile glucose solution. Mix thoroughly. Pour into sterile Petri dishes or distribute into sterile tubes.

Use: For the cultivation and maintenance of *Clavibacter xyli.*

SC Broth

Composition per liter:

Papaic digest of soybean meal	8.0g
KH_2PO_4	1.5g
K_2HPO_4	0.5g
$MgSO_4·7H_2O$	0.2g
Hemin solution	15.0mL
Bovine serum albumin, fraction V	10.0mL
Cysteine·H_2O solution	10.0mL
Glucose solution	1.0mL

pH 6.6 ± 0.2 at 25°C

Hemin Solution:

Composition per 100mL:

Hemin	0.1g
NaOH (0.05*N* solution)	100.0mL

Preparation of Hemin Solution: Add hemin to NaOH solution. Mix thoroughly.

Bovine Serum Albumin, Fraction V Solution:

Composition per 10mL:

Bovine serum albumin, fraction V	2.0g

Preparation of Bovine Serum Albumin, Fraction V Solution: Add bovine serum albumin to distilled/deionized water and bring volume to 10.0mL. Mix thoroughly. Filter sterilize.

Cysteine·H_2O Solution:

Composition per 10mL:

Cysteine·H_2O	1.0g

Preparation of Cysteine·H_2O Solution: Add cysteine·H_2O to distilled/deionized water and bring volume to 10.0mL. Mix thoroughly. Filter sterilize.

Glucose Solution:

Composition per 10mL:

D-Glucose	5.0g

Preparation of Glucose Solution: Add glucose to distilled/deionized water and bring volume to 10.0mL. Mix thoroughly. Filter sterilize.

Preparation of Medium: Add components—except bovine serum albumin, cysteine·H_2O solution, and glucose solution—to distilled/deionized water and bring volume to 979.0mL. Mix thoroughly. Adjust pH to 6.6 with NaOH. Gently heat and bring to boiling. Autoclave for 15 min at 15 psi pressure–121°C. Cool to 45–50°C. Aseptically add 10.0mL of sterile bovine serum albumin, 10.0mL of sterile cysteine·H_2O solution, and 1.0mL of sterile glucose solution. Mix thoroughly. Aseptically distribute into sterile tubes or flasks.

Use: For the cultivation of *Clavibacter xyli.*

SC Medium

Composition per 1021mL:

Agar	15.0g
Papaic digest of soybean meal	8.0g
Corn meal, solids from infusion	2.0g
Tween™ 80	1.0g
K_2HPO_4	1.0g
KH_2PO_4	1.0g
$MgSO_4·7H_2O$	0.2g
Hemin chloride solution	15.0mL
Bovine serum albumin solution	10.0mL
Cysteine solution	10.0mL
Glucose solution	1.0mL

pH 6.6 at 25°C

Hemin Chloride Solution:

Composition per 100mL:

Hemin chloride	0.1g
NaOH (0.05*N* solution)	100.0mL

Preparation of Hemin Chloride Solution: Add hemin chloride to 100.0mL of NaOH solution. Mix thoroughly.

Bovine Serum Albumin Solution:

Composition per 10mL:

Bovine serum albumin	2.0g

Preparation of Bovine Serum Albumin Solution: Add bovine serum albumin to distilled/deionized water and bring volume to 10.0mL. Mix thoroughly. Filter sterilize.

Cysteine Solution:
Composition per 10mL:

Cysteine, free base	1.0g

Preparation of Cysteine Solution: Add cysteine to distilled/deionized water and bring volume to 10.0mL. Mix thoroughly. Filter sterilize.

Glucose Solution:
Composition per 10mL:

Glucose	5.0g

Preparation of Glucose Solution: Add glucose to distilled/deionized water and bring volume to 10.0mL. Mix thoroughly. Autoclave for 15 min at 15 psi pressure–121°C. Cool to 25°C.

Preparation of Medium: Add components—except bovine serum albumin solution, cysteine solution, and glucose solution—to distilled/deionized water and bring volume to 1.0L. Mix thoroughly. Gently heat and bring to boiling. Autoclave for 15 min at 15 psi pressure–121°C. Cool to 45–50°C. Aseptically add 10.0mL of sterile bovine serum albumin solution, 10.0mL of sterile cysteine solution, and 1.0mL of sterile glucose solution. Mix thoroughly. Pour into sterile Petri dishes or distribute into sterile tubes.

Use: For the isolation and cultivation of coryneform bacteria that cause ratoon stunting disease of sugarcane.

Schaedler Agar (Schaedler Anaerobic Agar)

Composition per liter:

Agar	13.5g
Glucose	5.83g
Pancreatic digest of casein	5.7g
Proteose peptone No. 3	5.0g
Yeast extract	5.0g
Tris(hydroxymethyl)aminomethane buffer	3.0g
NaCl	1.65g
Papaic digest of soybean meal	1.0g
K_2HPO_4	0.83g
L-Cystine	0.4g
Hemin	0.01g

pH 7.6 ± 0.2 at 25°C

Source: This medium is available as a premixed powder from Difco Laboratories and Oxoid Unipath.

Preparation of Medium: Add components to distilled/deionized water and bring volume to 1.0L. Mix thoroughly. Gently heat and bring to boiling. Distribute into tubes or flasks. Autoclave for 15 min at 15 psi pressure–121°C. Pour into sterile Petri dishes or leave in tubes.

Use: For the isolation, cultivation and enumeration of anaerobic and aerobic microorganisms.

Schaedler Agar

Composition per liter:

Agar	13.5g
Pancreatic digest of casein	8.2g
Glucose	5.8g
Yeast extract	5.0g
Tris(hydroxymethyl)aminomethane buffer	3.0g
Peptic digest of animal tissue	2.5g
NaCl	1.7g
Papaic digest of soybean meal	1.0g
K_2HPO_4	0.8g
L-Cystine	0.4g
Hemin	0.01g

pH 7.6 ± 0.2 at 25°C

Source: This medium is available as a premixed powder from BBL Microbiology Systems.

Preparation of Medium: Add components to distilled/deionized water and bring volume to 1.0L. Mix thoroughly. Gently heat and bring to boiling. Distribute into tubes or flasks. Autoclave for 15 min at 15 psi pressure–121°C. Pour into sterile Petri dishes or leave in tubes.

Use: For the isolation, cultivation and enumeration of anaerobic and aerobic microorganisms.

Schaedler Broth (Schaedler Anaerobic Broth)

Composition per liter:

Pancreatic digest of casein	8.2g
Glucose	5.8g
Yeast extract	5.0g
Tris(hydroxymethyl)aminomethane buffer	3.0g
Peptic digest of animal tissue	2.5g
NaCl	1.7g
Papaic digest of soybean meal	1.0g
K_2HPO_4	0.8g
L-Cystine	0.4g
Hemin	0.01g

pH 7.6 ± 0.2 at 25°C

Source: This medium is available as a premixed powder from BBL Microbiology Systems, Difco Laboratories and Oxoid Unipath.

Preparation of Medium: Add components to distilled/deionized water and bring volume to 1.0L. Mix thoroughly. Distribute into tubes or flasks. Autoclave for 15 min at 15 psi pressure–121°C.

Use: For the cultivation and maintenance of *Eubacterium combesii, Eubacterium contortum* and a variety of other anaerobic bacteria.

SCY Medium (Maintenance SCY Medium)

Composition per liter:

Agar	10.0g
Sucrose	1.0g
Pancreatic digest of casein	0.92g
Yeast extract	0.25g
NaCl	0.05g
Papaic digest of soybean meal	0.03g
K_2HPO_4	0.025g
Thiamine	0.4mg
Cyanocobalamin	0.01mg

pH 7.3 ± 0.2 at 25°C

Preparation of Medium: Add components to distilled/deionized water and bring volume to 1.0L. Mix thoroughly. Filter sterilize.

Use: For the cultivation and maintenance of iron and sulfur bacteria.

SCY Medium (Maintenance SCY Medium)

Composition per liter:

Solution A	1.0L
Solution B	200.0mL

Solution A:

Composition per liter:

Agar	10.0g
Pancreatic digest of casein	0.92g
NaCl	0.05g
Papaic digest of soybean meal	0.03g
K_2HPO_4	0.025g

pH 7.0 ± 0.2 at 25°C

Preparation of Solution A: Add components to distilled/deionized water and bring volume to 1.0L. Mix thoroughly. Gently heat and bring to boiling. Distribute into tubes in 10.0mL volumes. Autoclave for 15 min at 15 psi pressure–121°C. Allow tubes to cool in a slanted position.

Solution B:

Composition per 200mL:

Sucrose	2.0g
Yeast extract	0.5g
Thiamine	0.8mg
Vitamin B_{12}	0.02mg

pH 8.5 ± 0.2 at 25°C

Preparation of Solution B: Add components to slightly alkaline tap water, pH 8.5, and bring volume to 200.0mL. Mix thoroughly. Filter sterilize.

Preparation of Medium: Inoculate bacteria onto prepared slants of solution A. After inoculation of tubes, aseptically add 2.0mL of sterile solution B on top of each slant.

Use: For the cultivation and maintenance of iron bacteria. For the cultivation and maintenance of *Haliscomenobacter hydrossis*.

Seawater Agar

Composition per 250mL:

Agar	20.0g
Beef extract	10.0g
Peptone	10.0g
Seawater	750.0mL

pH 7.2 ± 0.2 at 25°C

Preparation of Medium: Add components to tap water and bring volume to 1.0L. Mix thoroughly. Gently heat and bring to boiling. Distribute into tubes or flasks. Autoclave for 15 min at 15 psi pressure–121°C. Pour into sterile Petri dishes or leave in tubes.

Use: For the selective isolation and cultivation of *Planococcus* species.

Seawater Agar (SWA)

Composition per liter:

Agar	15.0g
Peptone	5.0g
Yeast extract	5.0g
Beef extract	3.0g
Seawater, synthetic	1.0L

pH 7.5 ± 0.2 at 25°C

Seawater, Synthetic:

Composition per liter:

NaCl	27.0g
$MgSO_4 \cdot 7H_2O$	7.0g
Tris(hydroxymethyl)aminomethane buffer	2.0g
KCl	0.6g
$CaCl_2$	0.3g

Preparation of Seawater, Synthetic: Add components to distilled/deionized water and bring volume to 1.0L. Mix thoroughly.

Preparation of Medium: Combine components. Mix thoroughly. Gently heat and bring to boiling. Distribute into tubes or flasks. Autoclave for 15 min at 15 psi pressure–121°C. Pour into sterile Petri dishes or leave in tubes.

Use: For the isolation and cultivation of halophilic microorganisms from foods, such as *Pseudomonas* species and *Vibrio* species from fish.

Seawater Agar (SWA)

Composition per liter:

Agar....................15.0g
Peptone....................5.0g
Yeast extract....................5.0g
Beef extract....................3.0g
Seawater, synthetic....................1.0L

pH 7.5 ± 0.2 at 25°C

Seawater, Synthetic:

Composition per liter:

NaCl....................24.0g
$MgSO_4·7H_2O$....................7.0g
$MgCl_2·6H_2O$....................5.3g
KCl....................0.7g
$CaCl_2$....................0.1g

Preparation of Seawater, Synthetic: Add components to distilled/deionized water and bring volume to 1.0L. Mix thoroughly. Adjust pH to 7.5.

Preparation of Medium: Combine components. Mix thoroughly. Gently heat and bring to boiling. Distribute into tubes or flasks. Autoclave for 15 min at 15 psi pressure–121°C. Pour into sterile Petri dishes or leave in tubes.

Use: For the isolation and cultivation of halophilic microorganisms from foods, such as *Pseudomonas* species and *Vibrio* species from fish.

Seawater Agar Medium

Composition per liter:

Agar....................15.0g
Beef extract....................10.0g
Peptone....................10.0g
Seawater, aged....................750.0mL

pH 7.2–7.3 at 25°C

Preparation of Medium: Add components to distilled/deionized water and bring volume to 1.0L. Mix thoroughly. Gently heat and bring to boiling. Distribute into tubes or flasks. Autoclave for 15 min at 15 psi pressure–121°C. Pour into sterile Petri dishes or leave in tubes.

Use: For the isolation and cultivation of marine *Flavobacterium* species.

Seawater Complete Medium

Composition per liter:

Pancreatic digest of casein....................5.0g
Yeast extract....................3.0g
Seawater....................750.0mL
Glycerol....................3.0mL

Preparation of Medium: Add components to distilled/deionized water and bring volume to 1.0L. Mix thoroughly. Distribute into tubes or flasks. Autoclave for 15 min at 15 psi pressure–121°C.

Use: For the cultivation and maintenance of *Vibrio fischeri*.

Seawater Medium

Composition per liter:

Agar....................15.0g
Peptone....................5.0g
Beef extract....................2.0g
KNO_3....................0.5g
Seawater, aged....................1.0L

pH 7.8 ± 0.2 at 25°C

Preparation of Medium: Combine components. Mix thoroughly. Gently heat and bring to boiling. Distribute into tubes or flasks. Autoclave for 15 min at 15 psi pressure–121°C. Pour into sterile Petri dishes or leave in tubes.

Use: For the cultivation of halophilic bacteria.

Seawater Medium

Composition per liter:

Agar....................15.0g
Peptone....................5.0g
Yeast extract....................1.0g
$FeSO_4$....................0.2g

Preparation of Medium: Add components to 1.0L of seawater. Mix thoroughly. Gently heat and bring to boiling. Distribute into tubes or flasks. Autoclave for 15 min at 15 psi pressure–121°C. Pour into sterile Petri dishes or leave in tubes.

Use: For the cultivation and maintenance of *Cyclobacterium marinus*.

Seawater Yeast Extract Agar

Composition per liter:

Marine salts mix....................37.9g
Agar....................15.0g
Proteose peptone....................10.0g
Yeast extract....................3.0g

pH 7.2–7.4 at 25°C

Preparation of Medium: Add components to distilled/deionized water and bring volume to 1.0L. Mix thoroughly. Gently heat and bring to boiling. Distribute into tubes or flasks. Autoclave for 15 min at 15 psi pressure–121°C. Pour into sterile Petri dishes or leave in tubes.

Use: For the cultivation and maintenance of *Alteromonas* species, *Caulobacter halobacteroides, Caulobacter maris, Cytophaga marinoflava* and *Cytophaga salmonicolor.*

Seawater Yeast Extract Broth, Modified

Composition per liter:

NaCl	23.4g
$MgSO_4 \cdot 7H_2O$	6.9g
Peptone	1.0g
Yeast extract	1.0g
KCl	0.75g

Preparation of Medium: Add components to distilled/deionized water and bring volume to 1.0L. Mix thoroughly. Distribute into tubes or flasks. Autoclave for 15 min at 15 psi pressure–121°C.

Use: For the cultivation and maintenance of the *Proteus* species and the *Vibrio* species.

Seawater Yeast Extract Peptone Medium

Composition per liter:

Agar	15.0g
Peptone	5.0g
Yeast extract	3.0g
Seawater, aged and filtered	750.0mL

pH 7.3 ± 0.2 at 25°C

Preparation of Medium: Add components, except agar, to distilled/deionized water and bring volume to 1.0L. Mix thoroughly. Adjust pH to 7.8. Gently heat and bring to boiling. Continue boiling for 3–5 min. Filter through Whatman filter paper. Adjust pH to 7.3. Add agar. Gently heat and bring to boiling. Distribute into tubes or flasks. Autoclave for 15 min at 15 psi pressure–121°C. Pour into sterile Petri dishes or leave in tubes.

Use: For the cultivation of *Planococcus kocurii.*

Selenite Broth (Selenite Broth, Lactose) (Selenite F Enrichment Medium) (Sodium Biselenite Medium) (Sodium Hydrogen Selenite Medium)

Composition per liter:

Na_2HPO_4	10.0g
Pancreatic digest of casein	5.0g
Lactose	4.0g
$NaHSeO_3 \cdot 5H_2O$	4.0g

pH 7.0 ± 0.2 at 25°C

Source: This medium is available as a premixed powder from Difco Laboratories and a prepared media from Oxoid Unipath.

Caution: Sodium biselenite is toxic and a potential teratogen and care must be taken to avoid inhalation of the powdered dye, contact with the skin, or ingestion, especially in pregnant laboratory workers.

Preparation of Medium: Add components to distilled/deionized water and bring volume to 1.0L. Mix thoroughly. Gently heat and bring to boiling. Do not autoclave. Distribute into sterile tubes in 10.0mL volumes.

Use: For the isolation and enrichment of *Salmonella* species from clinical specimens and food products.

Selenite Broth Base, Mannitol

Composition per liter:

Na_2HPO_4	10.0g
Peptone	5.0g
Mannitol	4.0g
$NaHSeO_3 \cdot 5H_2O$	4.0g

pH 7.1 ± 0.2 at 25°C

Source: This medium is available as a premixed powder from Oxoid Unipath.

Caution: Sodium selenite is toxic and a potential teratogen and care must be taken to avoid inhalation of the powdered dye, contact with the skin, or ingestion, especially in pregnant laboratory workers.

Preparation of Medium: Add components to distilled/deionized water and bring volume to 1.0L. Mix thoroughly. Gently heat. Do not autoclave. Distribute into sterile tubes in 10.0mL volumes. Sterilize for 10 min at 0 psi pressure–100°C.

Use: For the isolation and cultivation of *Salmonella typhi* and *Salmonella paratyphi B.*

Selenite Cystine Broth

Composition per liter:

Na_2HPO_4	10.0g
Pancreatic digest of casein	5.0g
Lactose	4.0g
$Na_2SeO_3 \cdot 5H_2O$	4.0g
L-Cystine	0.02g

pH 7.0 ± 0.2 at 25°C

Source: This medium is available as a premixed powder from BBL Microbiology Systems, Difco Laboratories and Oxoid Unipath.

Caution: Sodium selenite is toxic and a potential teratogen and care must be taken to avoid inhalation of the powdered dye, contact with the skin, or ingestion, especially in pregnant laboratory workers.

Preparation of Medium: Add components to distilled/deionized water and bring volume to 1.0L. Mix thoroughly. Gently heat. Do not autoclave. Distribute into sterile tubes in 10.0mL volumes. Sterilize for 15 min at 0 psi pressure–100°C.

Use: For the isolation and cultivation of *Salmonella* species from various specimens.

Selenite F Broth

Composition per liter:

KH_2PO_4	7.0g
Pancreatic digest of casein	5.0g
Lactose	4.0g
$Na_2SeO_3 \cdot 5H_2O$	4.0g
Na_2HPO_4	3.0g

pH 7.0 ± 0.2 at 25°C

Source: This medium is available as a premixed powder from BBL Microbiology Systems.

Caution: Sodium selenite is toxic and a potential teratogen and care must be taken to avoid inhalation of the powdered dye, contact with the skin, or ingestion, especially in pregnant laboratory workers.

Preparation of Medium: Add components to distilled/deionized water and bring volume to 1.0L. Mix thoroughly. Gently heat. Do not autoclave. Distribute into sterile tubes in 10.0mL volumes. Sterilize for 30 min at 0 psi pressure–100°C.

Use: For the isolation and cultivation of *Salmonella* species from various specimens.

Selenomonas acidaminophila Medium

Composition per liter:

Disodium β-glycerophosphate	19.0g
Beef extract	5.0g
Lactose	5.0g
Papaic digest of soybean meal	5.0g
Sodium glutamate	3.4g
Pancreatic digest of casein	2.5g
Peptic digest of animal tissue	2.5g
Yeast extract	2.5g
Ascorbic acid	0.5g
$MgSO_4 \cdot 7H_2O$	0.25g

pH 7.15 ± 0.05 at 25°C

Preparation of Medium: Add components to distilled/deionized water and bring volume to 1.0L. Mix thoroughly. Distribute into tubes or flasks. Autoclave for 15 min at 15 psi pressure–121°C.

Use: For the cultivation and maintenance of *Selenomonas acidaminophila.*

Selenomonas Selective Medium (SS Medium)

Composition per 100mL:

Pancreatic digest of casein	0.5g
Mannitol	0.2g
$FeSO_4 \cdot 7H_2O$	0.1g
Sodium acetate	0.1g
Yeast extract	0.1g
L-Cysteine·HCl	0.08g
Mineral solution S	4.0mL
Sodium carbonate (8% solution)	2.5mL
n-Valeric acid	0.05mL

pH 5.9–6.1 at 25°C

Mineral Solution S:

Composition per liter:

KH_2PO_4	12.0g
NaCl	12.0g
$(NH_4)_2SO_4$	6.0g
$MgSO_4 \cdot 7H_2O$	2.5g
$CaCl_2 \cdot 2H_2O$	1.6g

Preparation of Mineral Solution S: Add components to distilled/deionized water and bring volume to 1.0L. Mix thoroughly.

Preparation of Medium: Add components to distilled/deionized water and bring volume to 100.0mL. Mix thoroughly. Filter sterilize. Aseptically distribute into sterile tubes or flasks.

Use: For the isolation and cultivation of *Selenomonas* species.

Sellers Agar (Sellers Differential Agar)

Composition per 1015mL:

Pancreatic digest of gelatin	20.0g
Agar	13.5g

D-Mannitol 2.0g
NaCl 2.0g
$MgSO_4 \cdot 7H_2O$ 1.5g
K_2HPO_4 .... 1.0g
L-Arginine 1.0g
$NaNO_3$ 1.0g
Yeast extract 1.0g
$NaNO_3$ 0.35g
Bromthymol blue 0.04g
Phenol red 8.0mg
Glucose solution 15.0mL
pH 6.7 ± 0.2 at 25°C

Source: This medium is available as a premixed powder from BBL Microbiology Systems.

Glucose Solution:

Composition per 10mL:

D-Glucose 5.0g

Preparation of Glucose Solution: Add glucose to distilled/deionized water and bring volume to 10.0mL. Mix thoroughly. Filter sterilize.

Preparation of Medium: Add components, except glucose solution, to distilled/deionized water and bring volume to 1.0L. Mix thoroughly. Gently heat and bring to boiling. Distribute into tubes in 10.0mL volumes. Autoclave for 15 min at 15 psi pressure–121°C. Allow tubes to cool in a slanted position to form a 3 in. slant with a 1.5 in. butt. Immediately prior to inoculation aseptically add 0.15mL of sterile glucose solution to each tube. Let the glucose solution run down the side of the tube opposite the slant.

Use: For the cultivation and differentiation of nonfermentative Gram-negative bacilli, especially *Pseudomonas aeruginosa, Herellea vaginicola (Acinetobacter calcoaceticus) Mima polymorpha (Acinetobacter lwoffii), Alcaligenes faecalis* and *Bacterium anitratum (Acinetobacter calcoaceticus).*

Semisolid Medium for Motility

Composition per liter:

Biosate 5.0g
Polypeptone™ 5.0g
NaCl 5.0g
Agar 4.0g
Myosate 1.5g
Triphenyltetrazolium chloride solution 2.5mL
pH 6.9–7.1 at 25°C

Triphenyltetrazolium Chloride Solution:

Composition per 10mL:

Triphenyltetrazolium chloride 0.1g
Ethanol (95% solution) 10.0mL

Preparation of Triphenyltetrazolium Chloride Solution: Add triphenyltetrazolium chloride to 10.0mL of ethanol. Mix thoroughly.

Preparation of Medium: Add components, except triphenyltetrazolium chloride solution, to distilled/deionized water and bring volume to 997.5mL. Mix thoroughly. Gently heat and bring to boiling. Add 2.5mL of triphenyltetrazolium chloride solution. Mix thoroughly. Distribute into tubes in 10.0mL volumes. Autoclave for 15 min at 15 psi pressure–121°C.

Use: For the differentiation of bacteria based on motility.

Semisolid Pectin Agar

Composition per liter:

Pectin 30.0g
Yeast extract 5.0g
Agar 3.0g
Bromthymol blue (0.1% solution) 1.0mL
$CaCl_2 \cdot 2H_2O$ (10% solution 0.6mL
pH 7.3 ± 0.2 at 25°C

Preparation of Medium: Add approximately 100.0mL of distilled/deionized water to a 2-L flask. Place on a magnetic stirrer without heat. Slowly add the $CaCl_2 \cdot 2H_2O$ solution, bromthymol blue solution, yeast extract, and pectin while stirring. Mix thoroughly to ensure uniform wetting of the particles. Add agar. Gently heat and bring to almost boiling. Adjust pH to 7.3 with 1*N* NaOH if necessary. Do not overshoot pH above 7.3. Distribute into tubes or flasks. Autoclave for 15 min at 15 psi pressure–121°C.

Use: For the isolation and cultivation of bacteria such as *Erwinia* species and some *Klebsiella* species based on their ability to degrade pectin.

Serratia Differential Medium (SD Medium)

Composition per 102mL:

Solution A 92.0mL
Solution B 10.0mL
pH 6.7 ± 0.2 at 25°C

Solution A:

Composition per 92mL:

Yeast extract 1.0g
L-Ornithine 1.0g
NaCl 0.5g
Agar 0.4g
Irgasan inhibitor 1.0mL
Indicator solution 1.0mL

Preparation of Solution A: Add components to distilled/deionized water and bring volume to 92.0mL. Mix thoroughly. Adjust pH to 6.7 with 1*N* NaOH.

Irgasan Inhibitor:
Composition per 100mL:

Irgasan-DP-300 (4,2′, 4′-trichloro-2-hydroxydiphenylether)	0.1g
NaOH (1*N* solution)	10.0mL

Preparation of Irgasan Inhibitor: Add irgasan to 10.0mL of NaOH solution. Mix thoroughly. Gently heat to dissolve. Bring volume to 100.0mL with distilled/deionized water.

Indicator Solution:
Composition per 100mL:

Bromthymol blue	0.2g
Phenol red	0.1g

Preparation of Indicator Solution: Add components to 50.0mL of distilled/deionized water. Mix thoroughly for 1 hr. Bring volume to 100.0mL with distilled/deionized water.

Solution B:
Composition per 10mL:

L-Arabinose	1.0g

Preparation of Solution B: Add arabinose to distilled/deionized water and bring volume to 10.0mL. Mix thoroughly.

Preparation of Medium: Combine 92.0mL of solution A with 10.0mL of solution B. Mix thoroughly. Distribute into tubes. Autoclave for 15 min at 15 psi pressure–121°C. Allow tubes to cool in an upright position.

Use: For the cultivation and differentiation of *Serratia* species based on fermentation of arabinose and production of ornithine decarboxylase. *S. marcescens* changes the medium to purple throughout the tube. *S. liquefaciens* changes the medium to a band of purple at the top of the tube with a green/yellow butt. *S. rubidaea* changes the medium to yellow throughout the tube.

Serratia Hd–MHr

Composition per liter:

Agar	15.0g
K_2HPO_4	7.0g
Glucose	5.0g
KH_2PO_4	3.0g
2-Methyl-DL-histidine·2HCl	1.0g
$(NH_4)_2SO_4$	1.0g
$MgSO_4·7H_2O$	0.5g

Preparation of Medium: Add components to distilled/deionized water and bring volume to 1.0L. Mix thoroughly. Gently heat and bring to boiling. Distribute into tubes or flasks. Autoclave for 15 min at 15 psi pressure–121°C. Pour into sterile Petri dishes or leave in tubes.

Use: For the cultivation and maintenance of *Serratia marcescens.*

Serratia Medium (ATCC Medium 181)

Composition per liter:

Agar	20.0g
Pancreatic digest of casein	5.0g
Yeast extract	5.0g
Glucose	1.0g
K_2HPO_4	1.0g

pH 7.0 ± 0.2 at 25°C

Preparation of Medium: Add components to distilled/deionized water and bring volume to 1.0L. Mix thoroughly. Gently heat and bring to boiling. Distribute into tubes or flasks. Autoclave for 15 min at 15 psi pressure–121°C. Pour into sterile Petri dishes or leave in tubes.

Use: For the cultivation and maintenance of *Serratia marcescens.*

Serratia Medium (ATCC Medium 1399)

Composition per liter:

Agar	15.0g
K_2HPO_4	7.0g
Glucose	5.0g
KH_2PO_4	3.0g
Casein hydrolysate	1.0g
$(NH_4)_2SO_4$	1.0g
Yeast extract	1.0g
$MgSO_4·7H_2O$	0.1g

pH 7.0 ± 0.2 at 25°C

Preparation of Medium: Add components to distilled/deionized water and bring volume to 1.0L. Mix thoroughly. Gently heat and bring to boiling. Distribute into tubes or flasks. Autoclave for 15 min at 15 psi pressure–121°C. Pour into sterile Petri dishes or leave in tubes.

Use: For the cultivation and maintenance of *Serratia marcescens.*

SG Agar

Composition per liter:

Agar	15.0g
Pancreatic digest of casein	15.0g

$CaCl_2 \cdot 2H_2O$	2.0g
$MgSO_4 \cdot 7H_2O$	1.0g

pH 7.0 ± 0.2 at 25°C

Preparation of Medium: Add components to distilled/deionized water and bring volume to 1.0L. Mix thoroughly. Gently heat and bring to boiling. Distribute into tubes or flasks. Autoclave for 15 min at 15 psi pressure–121°C. Pour into sterile Petri dishes or leave in tubes.

Use: For the cultivation of myxobacteria.

Sierra Medium

Composition per liter:

Agar	15.0g
Peptone	10.0g
NaCl	5.0g
$CaCl_2 \cdot H_2O$	0.1g
Tween™ 80	10.0mL

pH 7.4 ± 0.2 at 25°C

Preparation of Medium: Add components, except Tween™ 80, to distilled/deionized water and bring volume to 990.0mL. Mix thoroughly. Gently heat and bring to boiling. Autoclave for 15 min at 15 psi pressure–121°C. Cool to 45–50°C. Separately autoclave Tween™ 80 for 15 min at 15 psi pressure–121°C. Cool to 45–50°C. Aseptically add 10.0mL of sterile Tween™ 80. Mix thoroughly. Pour into sterile Petri dishes.

Use: For the differentiation of bacteria based on lipase activity. Bacteria with lipase activity form colonies surrounded by a white precipitate.

SIM Medium

Composition per liter:

Peptone	30.0g
Agar	3.0g
Beef extract	3.0g
Peptonized iron (Difco)	0.2g
$Na_2S_2O_3 \cdot 5H_2O$	0.025g

pH 7.3 ± 0.2 at 25°C

Source: This medium is available as a premixed powder from Difco Laboratories.

Preparation of Medium: Add components to distilled/deionized water and bring volume to 1.0L. Mix thoroughly. Gently heat and bring to boiling. Distribute into tubes in 15.0mL volumes. Autoclave for 15 min at 15 psi pressure–121°C. Allow tubes to cool in an upright position.

Use: For the differentiation of members of the Enterobacteriaceae, based on H_2S production, indole production and motility.

SIM Medium

Composition per liter:

Pancreatic digest of casein	20.0g
Peptic digest of animal tissue	6.1g
Agar	3.5g
$Fe(NH_4)_2(SO_4)_2 \cdot 6H_2O$	0.2g
$Na_2S_2O_3 \cdot 5H_2O$	0.2g

pH 7.3 ± 0.2 at 25°C

Source: This medium is available as a premixed powder from BBL Microbiology Systems and Oxoid Unipath.

Preparation of Medium: Add components to distilled/deionized water and bring volume to 1.0L. Mix thoroughly. Gently heat and bring to boiling. Distribute into tubes in 15.0mL volumes. Autoclave for 15 min at 15 psi pressure–121°C. Allow tubes to cool in an upright position.

Use: For the differentiation of members of the Enterobacteriaceae, based on H_2S production, indole production and motility.

Simmons' Citrate Agar (Citrate Agar)

Composition per liter:

Agar	15.0g
NaCl	5.0g
Sodium citrate	2.0g
K_2HPO_4	1.0g
$(NH_4)H_2PO_4$	1.0g
$MgSO_4 \cdot 7H_2O$	0.2g
Bromthymol blue	0.08g

pH 6.9 ± 0.2 at 25°C

Source: This medium is available as a premixed powder from BBL Microbiology Systems, Difco Laboratories and Oxoid Unipath.

Preparation of Medium: Add components to distilled/deionized water and bring volume to 1.0L. Mix thoroughly. Gently heat while stirring and bring to boiling. Distribute into tubes or flasks. Autoclave for 15 min at 15 psi pressure–121°C. Pour into sterile Petri dishes or leave in tubes.

Use: For the differentiation of Gram-negative bacteria on the basis of citrate utilization. Bacteria which can utilize citrate as the sole carbon source turn the medium blue.

Singh's Medium, Modified

Composition per liter:

NaCl	8.75g
Lactalbumin hydrolysate	8.13g
Yeast extract	6.25g
D-Glucose	5.00g
$CaCl_2 \cdot 2H_2O$	0.25g
KCl	0.25g
$NaH_2PO_4 \cdot H_2O$	0.25g
$NaHCO_3$	0.15g
$MgCl_2 \cdot 6H_2O$	0.13g
Phenol red	0.01g
Fetal bovine serum (heat-inactivated at 56°C, 30 min)	200.0mL

pH 7.0 ± 0.2 at 25°C

Preparation of Medium: Add components to distilled/deionized water and bring volume to 1.0L. Mix thoroughly. Adjust pH to 7.0 with NaOH, if necessary. Filter sterilize. Aseptically distribute into sterile tubes or flasks.

Use: For the cultivation and maintenance of *Spiroplasma* species.

SJ Agar

Composition per liter:

Agar	15.0g
K_2HPO_4	1.0g
KCl	0.5g
$MgSO_4 \cdot 7H_2O$	0.5g
$NaNO_3$	0.5g
$FeSO_4 \cdot 7H_2O$	0.01g
Glucose solution	100.0mL

pH 7.2 ± 0.2 at 25°C

Glucose Solution:

Composition per 100mL:

D-Glucose	1.0g

Preparation of Glucose Solution: Add D-glucose to distilled/deionized water and bring volume to 100.0mL. Mix thoroughly. Autoclave for 15 min at 15 psi pressure–121°C. Cool to 25°C.

Preparation of Medium: Add components, except glucose solution, to distilled/deionized water and bring volume to 900.0mL. Mix thoroughly. Gently heat and bring to boiling. Autoclave for 15 min at 15 psi pressure–121°C. Cool to 45–50°C. Aseptically add sterile glucose solution. Mix thoroughly. Pour into sterile Petri dishes or distribute into sterile tubes.

Use: For the isolation and cultivation of *Cytophaga* species, *Herpetosiphon* species, *Saprospira* species, and *Flexithrix* species.

Skim Milk Acetate Medium

Composition per liter:

Agar	15.0g
Skim milk powder	5.0g
Yeast extract	0.5g
Sodium acetate	0.2g

Preparation of Medium: Add components to distilled/deionized water and bring volume to 1.0L. Mix thoroughly. Gently heat and bring to boiling. Distribute into tubes or flasks. Autoclave for 15 min at 15 psi pressure–121°C. Pour into sterile Petri dishes or leave in tubes.

Use: For the cultivation and maintenance of *Cytophaga johnsonae*.

Skim Milk Agar (Milk Agar) (ATCC Medium 377)

Composition per liter:

Agar	15.0g
Skim milk	8.0g

Preparation of Medium: Add components to distilled/deionized water and bring volume to 1.0L. Mix thoroughly. Gently heat and bring to boiling. Distribute into tubes or flasks. Autoclave for 15 min at 15 psi pressure–121°C. Pour into sterile Petri dishes or leave in tubes.

Use: For the cultivation and maintenance of *Herpetosiphon aurantiacus*.

Skim Milk Agar

Composition per 1100mL:

Agar	15.0g
Pancretic digest of casein	5.0g
Yeast extract	2.5g
Glucose	1.0g
Skim milk solution	100.0mL

pH 7.0 ± 0.1 at 25°C

Preparation of Skim Milk Solution: Add skim milk solids to distilled/deionized water and bring volume to 100.0mL. Mix thoroughly. Autoclave for 15 min at 15 psi pressure–121°C. Cool to 45–50°C.

Preparation of Medium: Add components, except skim milk solution, to distilled/deionized water and bring volume to 1.0L. Mix thoroughly. Gently heat and bring to boiling. Distribute into tubes or flasks. Autoclave for 15 min at 15 psi pressure–121°C. Cool to 45–50°C. Aseptically add 100.0mL of cooled, sterile skim milk solution. Mix thoroughly.

Pour into sterile Petri dishes or aseptically distribute into sterile tubes.

Use: For the cultivation and differentiation of bacteria based on proteolytic activity.

Slanetz and Bartley Medium

Composition per liter:

Tryptose	20.0g
Agar	10.0g
Yeast extract	5.0g
Na_2HPO_4 $2H_2O$	4.0g
Glucose	2.0g
NaN_3	0.4g
Tetrazolium chloride	0.1g

pH 7.2 ± 0.2 at 25°C

Source: This medium is available as a premixed powder from Oxoid Unipath.

Preparation of Medium: Add components to distilled/deionized water and bring volume to 1.0L. Mix thoroughly. Gently heat and bring to boiling. Distribute into tubes or flasks. Autoclave for 15 min at 15 psi pressure–121°C. Pour into sterile Petri dishes.

Use: For the detection and enumeration of enterococci by the membrane filter method.

Sludge Medium for Methanobacteria

Composition per liter:

$NaHCO_3$	4.0g
Sodium formate	2.0g
Sodium acetate	1.0g
Yeast extract	1.0g
Cysteine·HCl·H_2O	0.5g
KH_2PO_4	0.5g
$Na_2S·9H_2O$	0.5g
$MgSO_4·7H_2O$	0.4g
NaCl	0.4g
NH_4Cl	0.4g
$CaCl_2·2H_2O$	0.05g
$FeSO_4·7H_2O$	2.0mg
Resazurin	1.0mg
Sludge fluid	50.0mL
Fatty acid mixture	20.0mL
Trace elements SL-6	1.0mL

pH 6.7–7.0 at 25°C

Sludge Fluid:

Composition per 100mL:

Sludge	100.0mL
Yeast extract	0.4g

Preparation of Sludge Fluid: Add yeast extract to a concentration of 0.4% to sludge taken from an anaerobic digester. Gas with 100% N_2 for 5 min. Incubate at 37°C for 24 hr. Centrifuge at 13,000 × *g*. Remove the supernatant fluid. Gas with 100% N_2 for 5 min. Autoclave for 15 min at 15 psi pressure–121°C. Store at 25°C protected from light.

Fatty Acid Mixture:

Composition per 20mL:

α-Methylbutyric acid	0.5g
Isobutyric acid	0.5g
Isovaleric acid	0.5g
Valeric acid	0.5g

Preparation of Fatty Acid Mixture: Add components to distilled/deionized water and bring volume to 20.0mL. Mix thoroughly. Adjust pH to 7.5 with concentrated NaOH.

Trace Elements Solution SL-6:

Composition per liter:

H_3BO_3	0.3g
$CoCl_2·6H_2O$	0.2g
$ZnSO_4·7H_2O$	0.10g
$MnCl_2·4H_2O$	0.03g
$Na_2MoO_4·H_2O$	0.03g
$NiCl_2·6H_2O$	0.02g
$CuCl_2·2H_2O$	0.01g

Preparation of Trace Elements Solution SL-6: Add components to distilled/deionized water and bring volume to 1.0L. Mix thoroughly. Adjust pH to 3.4.

Preparation of Medium: Prepare and dispense medium under 80% N_2 + 20% CO_2. Add components to distilled/deionized water and bring volume to 1.0L. Mix thoroughly. Adjust pH to 6.7–7.0. Distribute anaerobically into tubes or bottles with aluminum seals. Autoclave for 15 min at 15 psi pressure–121°C with fast exhaust.

Use: For the cultivation and maintenance of *Methanobacterium uliginosum* and *Methanobrevibacter ruminantium.*

Sludge Medium for Methanobacteria, pH 7.9

Composition per liter:

$NaHCO_3$	4.0g
Sodium formate	2.0g
Sodium acetate	1.0g
Yeast extract	1.0g
Cysteine·HCl·H_2O	0.5g
KH_2PO_4	0.5g
$Na_2S·9H_2O$	0.5g
$MgSO_4·7H_2O$	0.4g

NaCl	0.4g
NH_4Cl	0.4g
$CaCl_2 \cdot 2H_2O$	0.05g
$FeSO_4 \cdot 7H_2O$	2.0mg
Resazurin	1.0mg
Sludge fluid	50.0mL
Fatty acid mixture	20.0mL
Trace elements SL-6	1.0mL

pH 7.9 ± 0.2 at 25°C

Sludge Fluid:

Composition per 100mL:

Sludge	100.0mL
Yeast extract	0.4g

Preparation of Sludge Fluid: Add yeast extract to a concentration of 0.4% to sludge taken from an anaerobic digester. Gas with 100% N_2 for 5 min. Incubate at 37°C for 24 hr. Centrifuge at 13,000 × *g*. Remove the supernatant fluid. Gas with 100% N_2 for 5 min. Autoclave for 15 min at 15 psi pressure–121°C. Store at 25°C protected from light.

Fatty Acid Mixture:

Composition per 20mL:

α-Methylbutyric acid	0.5g
Isobutyric acid	0.5g
Isovaleric acid	0.5g
Valeric acid	0.5g

Preparation of Fatty Acid Mixture: Add components to distilled/deionized water and bring volume to 20.0mL. Mix thoroughly. Adjust pH to 7.5 with concentrated NaOH.

Trace Elements Solution SL-6:

Composition per liter:

H_3BO_3	0.3g
$CoCl_2 \cdot 6H_2O$	0.2g
$ZnSO_4 \cdot 7H_2O$	0.10g
$MnCl_2 \cdot 4H_2O$	0.03g
$Na_2MoO_4 \cdot H_2O$	0.03g
$NiCl_2 \cdot 6H_2O$	0.02g
$CuCl_2.2H_2O$	0.01g

Preparation of Trace Elements Solution SL-6: Add components to distilled/deionized water and bring volume to 1.0L. Mix thoroughly. Adjust pH to 3.4.

Preparation of Medium: Prepare and dispense medium under 80% N_2 + 20% CO_2. Add components to distilled/deionized water and bring volume to 1.0L. Mix thoroughly. Adjust pH to 7.9. Distribute anaerobically into tubes or bottles with aluminum seals. Autoclave for 15 min at 15 psi pressure–121°C with fast exhaust.

Use: For the cultivation and maintenance of *Methanobacterium alcaliphilum* and *M. thermoalcalip*.

SM Selective Medium

Composition per liter:

Agar	15.0g
Mannitol	2.5g
L-Glutamic acid	1.0g
$MgSO_4 \cdot 7H_2O$	0.16g
2,3,5-Triphenyltetrazolium·HCl solution	10.0mL
Antibiotic solution	10.0mL
KH_2PO_4 (0.2m*M* solution)	1.0mL
Metals solution	0.05mL

pH 7.2 ± 0.2 at 25°C

Antibiotic Solution:

Composition per 10mL:

Bacitracin	0.05g
Cycloheximide	0.05g
Tyrothricin	0.02g
Captan	0.01g
Vancomycin	0.01g
Chloromycetin	5.0μg
Penicillin G	1.0μg

Preparation of Antibiotic Solution: Add components to distilled/deionized water and bring volume to 10.0mL. Mix thoroughly. Filter sterilize.

Metals Solution:

Composition per 100mL:

$ZnSO_4 \cdot 7H_2O$	1.1g
$MnSO_4 \cdot H_2O$	0.62g
$Fe(NH_4)_2(SO_4)_2 \cdot 6H_2O$	0.18g
$CuSO_4 \cdot 5H_2O$	0.029g
$CaSO_4 \cdot 5H_2O$	0.029g
H_3PO_3	0.011g
KI	0.013mg

Preparation of Metals Solution: Add components to distilled/deionized water and bring volume to 100.0mL. Mix thoroughly.

Preparation of Medium: Add components, except antibiotic solution, to distilled/deionized water and bring volume to 990.0mL. Mix thoroughly. Gently heat and bring to boiling. Adjust pH to 7.2 with KOH. Autoclave for 15 min at 15 psi pressure–121°C. Cool to 45–50°C. Aseptically add sterile antibiotic solution. Mix thoroughly. Pour into sterile Petri dishes or distribute into sterile tubes.

Use: For the isolation and cultivation of *Pseudomonas solanacearum*.

SMC Medium

Composition per liter:

Sorbitol	70.0g
Pancreatic digest of casein	17.0g
NaCl	5.0g

Beef extract	3.0g
Yeast extract	3.0g
Beef heart, solids from infusion	2.0g
Horse serum	200.0mL
Yeast extract solution	100.0mL
Phenol red solution	20.0mL
Sucrose solution	20.0mL
L-Arginine·HCl solution	10.0mL
Fructose solution	2.0mL
Glucose solution	2.0mL

pH 7.5 ± 0.2 at 25°C

Yeast Extract Solution:

Composition per 100mL:

Yeast extract	25.0g

Preparation of Yeast Extract Solution: Add yeast extract to distilled/deionized water and bring volume to 100.0mL. Mix thoroughly. Autoclave for 15 min at 15 psi pressure–121°C. Cool to 45–50°C.

Phenol Red Solution:

Composition per 100mL:

Phenol Red	0.01g

Preparation of Phenol Red Solution: Add phenol red to distilled/deionized water and bring volume to 100.0mL. Mix thoroughly. Autoclave for 15 min at 15 psi pressure–121°C. Cool to 45–50°C.

Sucrose Solution:

Composition per 20mL:

Sucrose	10.0g

Preparation of Sucrose Solution: Add sucrose to distilled/deionized water and bring volume to 20.0mL. Mix thoroughly. Autoclave for 15 min at 15 psi pressure–121°C. Cool to 45–50°C.

L-Arginine·HCl Solution:

Composition per 10mL:

L-Arginine·HCl	4.2g

Preparation of L-Arginine·HCl Solution: Add arginine·HCl to distilled/deionized water and bring volume to 10.0mL. Mix thoroughly. Autoclave for 15 min at 15 psi pressure–121°C. Cool to 45–50°C.

Fructose Solution:

Composition per 10mL:

Fructose	5.0g

Preparation of Fructose Solution: Add fructose to distilled/deionized water and bring volume to 10.0mL. Mix thoroughly. Autoclave for 15 min at 15 psi pressure–121°C. Cool to 45–50°C.

Glucose Solution:

Composition per 10mL:

Glucose	5.0g

Preparation of Glucose Solution: Add glucose to distilled/deionized water and bring volume to 10.0mL. Mix thoroughly. Autoclave for 15 min at 15 psi pressure–121°C. Cool to 45–50°C.

Preparation of Medium: Add components—except horse serum, yeast extract solution, phenol red solution, sucrose solution, L-arginine·HCl solution, fructose solution, and glucose solution—to distilled/deionized water and bring volume to 646.0mL. Mix thoroughly. Gently heat and bring to boiling. Autoclave for 15 min at 15 psi pressure–121°C. Cool to 45–50°C. Aseptically add 200.0mL of sterile horse serum, 100.0mL of sterile yeast extract solution, 20.0mL of sterile phenol red solution, 20.0mL of sterile sucrose solution, 10.0mL of sterile L-arginine·HCl solution, 2.0mL of sterile fructose solution, and 2.0mL of sterile glucose solution. Mix thoroughly. Aseptically distribute into sterile tubes or flasks.

Use: For the cultivation and maintenance of *Spiroplasma citri*.

SMC, Modified

Composition per liter:

Sorbitol	70.0g
Pancreatic digest of casein	17.0g
NaCl	5.0g
Beef extract	3.0g
Yeast extract	3.0g
Beef heart, solids from infusion	2.0g
Solution 1	100.0mL
Solution 3	20.0mL
Solution 2	10.0mL
NaOH (1*N* solution)	6.0mL

pH 7.7–7.8 at 25°C

Solution 1:

Composition per 100mL:

Sucrose	10.0g
Yeast extract	2.0g
Fructose	1.0g
Glucose	1.0g
Phenol red	0.02g

Preparation of Solution 1: Add components to distilled/deionized water and bring volume to 100.0mL. Mix thoroughly. Filter sterilize.

Solution 2:

Composition per 10mL:

Bovine serum albumin, fraction V	0.1g

Preparation of Solution 2: Add bovine serum albumin to distilled/deionized water and bring volume to 10.0mL. Mix thoroughly. Filter sterilize.

Solution 3:

Composition per 20mL:

Horse serum 20.0mL

Preparation of Solution 3: Inactivate horse serum at 56°C for 30 min. Filter sterilize.

Preparation of Medium: Add components—except solution 1, solution 2, and solution 3—to distilled/deionized water and bring volume to 870.0mL. Autoclave for 15 min at 15 psi pressure–121°C. Cool to 45–50°C. Aseptically add 100.0mL of sterile solution 1, 20.0mL of sterile solution 3, and 10.0mL of sterile solution 1. Mix thoroughly. Adjust pH to 7.7–7.8. Aseptically distribute into sterile tubes or flasks.

Use: For the cultivation and maintenance of *Spiroplasma citri*.

Sodium Chloride Broth, 6.5

Composition per liter:

Beef heart, solids from infusion 500.0g
NaCl 65.0g
Tryptose 10.0g

pH 7.4 ± 0.2 at 25°C

Preparation of Medium: Add components to distilled/deionized water and bring volume to 1.0L. Mix thoroughly. Distribute into tubes or flasks. Autoclave for 15 min at 15 psi pressure–121°C.

Use: For the cultivation of enterococci and other salt-tolerant organisms. Use for the differentiation of microorganisms based on salt tolerance.

Sodium Chloride Sucrose Medium 900 (Sodium Chloride SUC Medium 900)

Composition per liter:

Sucrose 97.3g
Pancreatic digest of gelatin 14.5g
NaCl 14.3g
Agar 13.3g
Brain heart, solids from infusion 6.0g
Peptic digest of animal tissue 6.0g
Yeast extract 5.0g
Glucose 3.0g
Na_2HPO_4 .... 2.5g
$MgSO_4$ 0.25g
Horse serum (γ-globulin-free, inactivated 30 min at 56°C) 100.0mL
Carbenicillin solution 10.0mL

pH 7.4 ± 0.2 at 25°C

Carbenicillin Solution:

Composition per 10mL:

Carbenicillin 5.0g

Preparation of Carbenicillin Solution: Add carbenicillin to distilled/deionized water and bring volume to 10.0mL. Mix thoroughly. Filter sterilize.

Preparation of Medium: Add components, except carbenicillin solution and horse serum, to distilled/deionized water and bring volume to 890.0mL. Mix thoroughly. Gently heat and bring to boiling. Autoclave for 15 min at 15 psi pressure–121°C. Cool to 45–50°C. Aseptically add carbenicillin solution and horse serum. Mix thoroughly. Pour into sterile Petri dishes or distribute into sterile tubes.

Use: For the cultivation of *Pseudomonas aeruginosa*.

Soft Agar Gelatin Overlay

Composition per plate:

Base agar 15.0mL
Soft agar gelatin overlay 2.5mL

pH 7.0 ± 0.2 at 25°C

Base Agar:

Composition per liter:

Agar 15.0g
Peptone 5.0g
NaCl 5.0g
Beef extract 3.0g
$MnSO_4 \cdot H_2O$ 0.05g

Preparation of Base Agar: Add components to distilled/deionized water and bring volume to 1.0L. Mix thoroughly. Gently heat and bring to boiling. Autoclave for 15 min at 15 psi pressure–121°C. Cool to 45–50°C.

Soft Agar Gelatin Overlay:

Composition per liter:

Gelatin 15.0g
Agar 8.0g
Peptone 5.0g
NaCl 5.0g
Beef extract 3.0g
$MnSO_4 \cdot H_2O$ 0.05g

Preparation of Soft Agar Gelatin Overlay: Add components to distilled/deionized water and bring volume to 1.0L. Mix thoroughly. Gently heat and bring to boiling. Autoclave for 15 min at 15 psi pressure–121°C. Cool to 45–50°C.

Preparation of Medium: Aseptically pour cooled, sterile base agar into sterile Petri dishes in 15.0mL volumes. Allow agar to solidify. Inoculate plates with samples. Overlay each plate with 2.5mL of soft agar gelatin overlay.

Use: For the cultivation and differentiation of microorganisms based on proteolytic activity.

Soil Extract Agar

Composition per liter:

Agar	20.0g
Glucose	1.0g
K_2HPO_4	1.0g
Peptone	1.0g
Yeast extract	1.0g
Soil extract	400.0mL
Cycloheximide solution	10.0mL

pH 6.6 ± 0.2 at 25°C

Soil Extract:

Composition per liter:

Garden soil, neutral	1.0Kg

Preparation of Soil Extract: Add garden soil to 1.0L of tap water. Autoclave for 20 min at 15 psi pressure–121°C. Filter through Whatman filter paper. Bring volume to 1.0L with tap water.

Cycloheximide Solution:

Composition per 10mL:

Cycloheximide	0.04g

Preparation of Cycloheximide Solution: Add cycloheximide to distilled/deionized water and bring volume to 10.0mL. Mix thoroughly. Filter sterilize.

Preparation of Medium: Add components, except cycloheximide solution, to distilled/deionized water and bring volume to 990.0mL. Mix thoroughly. Gently heat and bring to boiling. Autoclave for 15 min at 15 psi pressure–121°C. Cool to 45–50°C. Aseptically add sterile cycloheximide solution. Mix thoroughly. Pour into sterile Petri dishes or distribute into sterile tubes.

Use: For the isolation and cultivation of *Arthrobacter* species.

Soil Extract Agar

Composition per liter:

Soil	500.0g
Agar	15.0g
Glucose	2.0g
Yeast extract	1.0g
KH_2PO_4	0.5g

Preparation of Medium: Add 500.0g of garden soil to 1.0L of tap water. Autoclave for 3 hours at 15 psi pressure–121°C. Filter through Whatman No. 2 filter paper. Add remaining components to filtrate. Bring volume to 1.0L with tap water. Gently heat and bring to boiling. Distribute into tubes in 7.0mL volumes. Autoclave for 15 min at 15 psi pressure–121°C. Allow tubes to cool in a slanted position.

Use: For the cultivation and identification of *Histoplasma capsulatum, Blastomyces dermatitidis,* and *Bacillus* species based on formation of typical conidia.

Soil Extract Glucose Yeast Extract Agar

Composition per liter:

Agar	15.0g
Glucose	2.0g
Yeast extract	1.0g
Soil extract	250.0mL

pH 6.8 ± 0.2 at 25°C

Soil Extract:

Composition per liter:

Garden soil	500.0g

Preparation of Soil Extract: Add 500.0g of garden soil to 1.0L of tap water. Autoclave for 1 hr at 15 psi pressure–121°C. Filter through Whatman No. 2 filter paper.

Preparation of Medium: Add components to distilled/deionized water and bring volume to 1.0L. Mix thoroughly. Gently heat and bring to boiling. Distribute into tubes or flasks. Autoclave for 15 min at 15 psi pressure–121°C. Pour into sterile Petri dishes or leave in tubes.

Use: For the cultivation and maintenance of *Streptomyces rectus*.

Soil Extract Medium

Composition per liter:

Agar	15.0g
Pancreatic digest of gelatin	5.0g
Beef extract	3.0g
Soil extract	250.0mL

pH 6.8 ± 0.2 at 25°C

Soil Extract:

Composition per liter:

Garden soil	500.0g

Preparation of Soil Extract: Add 500.0g of garden soil to 1.0L of tap water. Autoclave for 1 hour at 15 psi pressure–121°C. Filter through Whatman No. 2 filter paper.

Preparation of Medium: Add components to distilled/deionized water and bring volume to 1.0L. Mix thoroughly. Gently heat and bring to boiling. Distribute into tubes or flasks. Autoclave for 15 min at 15 psi pressure–121°C. Pour into sterile Petri dishes or leave in tubes.

Use: For the cultivation and maintenance of *Streptomyces rectus.*

Soil Extract Peptone Beef Extract Medium

Composition per liter:

Agar	15.0g
Peptone	5.0g
Beef extract	3.0g
Soil extract	1.0L

pH 7.0 ± 0.2 at 25°C

Soil Extract:

Composition per liter:

Garden soil	400.0g

Preparation of Soil Extract: Add garden soil to 1.0L of tap water. Autoclave for 1 hour at 15 psi pressure–121°C. Filter through cheesecloth and Whatman No.2 filter paper. Autoclave filtrate again for 20 min at 15 psi pressure–121°C. Filter through Whatman No.2 filter paper.

Preparation of Medium: Add agar, peptone, and beef extract to 1.0L of soil extract. Mix thoroughly. Gently heat and bring to boiling. Distribute into tubes or flasks. Autoclave for 15 min at 15 psi pressure–121°C. Pour into sterile Petri dishes or leave in tubes.

Use: For the cultivation and maintenance of *Oerskovia turbata* and *O. xanthineolytica.*

Soil Extract Potato Extract Medium

Composition per 510mL:

Malt extract	10.0g
Yeast extract	4.0g
Potato extract	250.0mL
Soil extract	250.0mL

pH 7.0 ± 0.2 at 25°C

Soil Extract:

Composition per liter:

Garden soil	400.0g

Preparation of Soil Extract: Add garden soil to 1.0L of tap water. Autoclave for 45 min at 15 psi pressure–121°C. Filter through cheesecloth.

Potato Extract:

Composition per liter:

Potatoes	400.0g

Preparation of Potato Extract: Peel and dice potatoes. Add 500.0mL of distilled/deionized water. Gently heat and bring to boiling. Continue boiling for 15 min. Filter through cheesecloth. Bring volume to 1.0L with distilled/deionized water.

Preparation of Medium: Combine components. Mix thoroughly. Distribute into tubes or flasks. Autoclave for 15 min at 15 psi pressure–121°C.

Use: For the cultivation and maintenance of *Saccharopolyspora rectivirgula.*

Sorangium Medium

Composition per liter:

Agar	10.0g
KNO_3	1.0g
K_2HPO_4	1.0g
$MgSO_4$	0.2g
$CaCl_2$	0.1g
$FeCl_3$	0.02g

Preparation of Medium: Add components to tap water and bring volume to 1.0L. Mix thoroughly. Gently heat and bring to boiling. Distribute into tubes or flasks. Autoclave for 15 min at 15 psi pressure–121°C. Pour into sterile Petri dishes or leave in tubes. Allow tubes to cool in a slanted position. Aseptically add a sterile strip (4.5cm × 1.0cm) of Whatman No. 1 filter paper to the surface of each slant or 4–6 sterile strips of filter paper to the surface of each agar plate.

Use: For the cultivation and maintenance of *Polyangium cellulosum.*

Sorbitol Agar

Composition per liter:

Agar	20.0g
Peptone	10.0g
NaCl	2.0g
Yeast extract	2.0g
Sorbitol solution	50.0mL

pH 7.0 ± 0.2 at 25°C

Sorbitol Solution:

Composition per 50mL:

Sorbitol	5.0g

Preparation of Sorbitol Solution: Add sorbitol to distilled/deionized water and bring volume to 50.0mL. Mix thoroughly. Filter sterilize.

Preparation of Medium: Add components, except sorbitol solution, to distilled/deionized water and bring volume to 950.0mL. Mix thoroughly. Adjust pH to 7.0. Gently heat and bring to boiling. Autoclave for 15 min at 15 psi pressure–121°C. Cool to 45–50°C. Aseptically add sterile sorbitol solution. Mix thoroughly. Pour into sterile Petri dishes or distribute into sterile tubes.

Use: For the cultivation and maintenance of *Pseudomonas* species.

Sorbitol MacConkey Agar (MacConkey Agar with Sorbitol)

Composition per liter:

Peptone	20.0g
Agar	15.0g
Sorbitol	10.0g
NaCl	5.0g
Bile salts No.3	1.5g
Neutral red	0.03g
Crystal violet	1.0mg

pH 7.1 ± 0.2 at 25°C

Source: This medium is available as a premixed powder from Difco Laboratories and Oxoid Unipath.

Preparation of Medium: Add components to distilled/deionized water and bring volume to 1.0L. Mix thoroughly. Gently heat and bring to boiling. Distribute into tubes or flasks. Autoclave for 15 min at 15 psi pressure–121°C. Pour into sterile Petri dishes or leave in tubes.

Use: For the isolation and cultivation of pathogenic *Escherichia coli.*

SOT Medium

Composition per liter:

$NaHCO_3$	16.8g
$NaNO_3$	2.5g
K_2SO_4	1.0g
NaCl	1.0g
K_2HPO_4	0.5g
$MgSO_4 \cdot 7H_2O$	0.2g
Disodium EDTA·$2H_2O$	0.08g
$CaCl_2 \cdot 2H_2O$	0.04g
$FeSO_4 \cdot 7H_2O$	0.01g
Trace metal mix A5	1.0mL
Trace metal mix B6, modified	1.0mL

pH 9.0 ± 0.2 at 25°C

Trace Metal Mix A5:

Composition per liter:

H_3BO_3	2.86g
$MnCl_2 \cdot 4H_2O$	1.81g
$Na_2MoO_4 \cdot 2H_2O$	0.39g
$ZnSO_4 \cdot 7H_2O$	0.222g
$CuSO_4 \cdot 5H_2O$	0.079g
$Co(NO_3)_2 \cdot 6H_2O$	0.049g

Preparation of Trace Metal Mix A5: Add components to distilled/deionized water and bring volume to 1.0L. Mix thoroughly.

Trace Metal Mix B6, Modified:

Composition per liter:

NH_4NO_3	0.23g
$K_2Cr_2(SO_4)_4 \cdot 24H_2O$	0.096g
$NiSO_4 \cdot 7H_2O$	0.048g
$Ti_2(SO_4)_3$	0.040g
$Na_2WO_4 \cdot 2H_2O$	0.018g

Preparation of Trace Metal Mix B6, Modified: Add components to distilled/deionized water and bring volume to 1.0L. Mix thoroughly.

Preparation of Medium: Add components to distilled/deionized water and bring volume to 1.0L. Mix thoroughly. Adjust pH to 9.0. Distribute into tubes or flasks. Autoclave for 15 min at 15 psi pressure–121°C.

Use: For the cultivation and maintenance of *Spirulina maxima* and *S. platensis.*

Soybean Agar

Composition per liter:

White soybeans	100.0g
Agar	15.0g

Preparation of Medium: Add soybeans to 1.0L of distilled/deionized water. Soak overnight. Autoclave for 60 min at 15 psi pressure–121°C. Filter through cheesecloth. Measure volume of filtrate. Add agar to a concentration of 1.5%. Gently heat and bring to boiling. Distribute into tubes or flasks. Autoclave for 15 min at 15 psi pressure–121°C. Pour into sterile Petri dishes or leave in tubes.

Use: For the cultivation and maintenance of *Bacillus subtilis* and *Pseudomonas syringae.*

Soybean Extract, M-1

Composition per liter:

Soybeans	50.0g
Soluble starch	15.0g
$(NH_4)_2HPO_3$	10.0g
KCl	0.2g
$MgSO_4 \cdot 7H_2O$	0.2g

pH 7.0 ± 0.2 at 25°C

Preparation of Medium: Add soybeans to 1.0L of distilled/deionized water. Soak overnight. Add 2.0g NaOH. Adjust pH to 7.0 with HCl. Autoclave for 60 min at 0 psi pressure–100°C. Filter through cheesecloth. Bring volume of filtrate to 1.0L with distilled/deionized water. Add remaining components. Mix thoroughly. Distribute into tubes or flasks. Autoclave for 15 min at 15 psi pressure–121°C.

Use: For the cultivation and maintenance of *Bacillus subtilis.*

SP Agar

Composition per liter:

Agar	15.0g
Pancreatic digest of casein	2.5g
Galactose	1.0g
Raffinose	1.0g
Sucrose	1.0g
$MgSO_4 \cdot 7H_2O$	0.50g
K_2HPO_4	0.25g
Vitamin solution	2.5mL

Vitamin Solution:

Composition per liter:

Inositol	1.0g
Calcium pantothenate	0.2g
Choline hydrochloride	0.2g
Thiamine	0.1g
Nicotinamide	0.75g
Pyridoxin	0.75g
Riboflavin	0.75g
p-Aminobenzoic acid	5.0mg
Folic acid	1.0mg
Biotin	0.05mg
Vitamin B_{12}	0.05mg
Ethanol	1.0L

Preparation of Vitamin Solution: Add solid components to 1.0L of ethanol. Mix thoroughly.

Preparation of Medium: Add components to distilled/deionized water and bring volume to 1.0L. Mix thoroughly. Gently heat and bring to boiling. Distribute into tubes or flasks. Autoclave for 15 min at 15 psi pressure–121°C. Pour into sterile Petri dishes or leave in tubes.

Use: For the cultivation of myxobacteria.

SP 2 Agar

Composition per liter:

Agar	15.0g
Pancreatic digest of casein	3.0g
Yeast extract	1.0g
Sodium acetate	0.02g
Artificial seawater	1.0L

pH 7.2 ± 0.2 at 25°C

Artificial Seawater:

Composition per liter:

NaCl	24.7g
$MgSO_4 \cdot 7H_2O$	6.3g
$MgCl_2 \cdot 6H_2O$	4.6g
$CaCl_2$	1.0g
KCl	0.7g
$NaHCO_3$	0.2g

Preparation of Artificial Seawater: Add components to distilled/deionized water and bring volume to 1.0L. Mix thoroughly.

Preparation of Medium: Add solid components to 1.0L of artificial seawater. Mix thoroughly. Gently heat and bring to boiling. Distribute into tubes or flasks. Autoclave for 15 min at 15 psi pressure–121°C. Pour into sterile Petri dishes or leave in tubes.

Use: For the isolation and cultivation of *Cytophaga* species, *Herpetosiphon* species, *Saprospira* species, and *Flexithrix* species.

SP 6 Agar

Composition per liter:

Agar	15.0g
Pancreatic digest of casein	3.0g
Yeast extract	1.0g
Artificial seawater	1.0L

pH 7.2 ± 0.2 at 25°C

Artificial Seawater:

Composition per liter:

NaCl	24.7g
$MgSO_4 \cdot 7H_2O$	6.3g
$MgCl_2 \cdot 6H_2O$	4.6g
$CaCl_2$	1.0g
KCl	0.7g
$NaHCO_3$	0.2g

Preparation of Artificial Seawater: Add components to distilled/deionized water and bring volume to 1.0L. Mix thoroughly.

Preparation of Medium: Add solid components to 1.0L of artificial seawater. Mix thoroughly. Gently heat and bring to boiling. Distribute into tubes or flasks. Autoclave for 15 min at 15 psi pressure–121°C. Pour into sterile Petri dishes or leave in tubes.

Use: For the isolation and cultivation of *Cytophaga* species, *Herpetosiphon* species, *Saprospira* species, and *Flexithrix* species.

SP5 Broth

Composition per liter:

Pancreatic digest of casein	9.0g
Yeast extract	1.0g
Artificial seawater	1.0L

pH 7.2 ± 0.2 at 25°C

Artificial Seawater:

Composition per liter:

NaCl	24.7g
$MgSO_4 \cdot 7H_2O$	6.3g
$MgCl_2 \cdot 6H_2O$	4.6g
$CaCl_2$	1.0g

KCl	0.7g
$NaHCO_3$	0.2g

Preparation of Artificial Seawater: Add components to distilled/deionized water and bring volume to 1.0L. Mix thoroughly.

Preparation of Medium: Add solid components to 1.0L of artificial seawater. Mix thoroughly. Gently heat and bring to boiling. Distribute into tubes or flasks. Autoclave for 15 min at 15 psi pressure–121°C.

Use: For the isolation and cultivation of *Cytophaga* species, *Herpetosiphon* species, *Saprospira* species, and *Flexithrix* species.

SP Medium

Composition per liter:

Agar	15.0g
Soluble starch	5.0g
Pancreatic digest of casein	2.5g
Galactose	1.0g
Raffinose	1.0g
Sucrose	1.0g
$MgSO_4 \cdot 7H_2O$	0.5g
K_2HPO_4	0.25g

Preparation of Medium: Add components to distilled/deionized water and bring volume to 1.0L. Mix thoroughly. Gently heat and bring to boiling. Distribute into tubes or flasks. Autoclave for 15 min at 15 psi pressure–121°C. Pour into sterile Petri dishes or leave in tubes.

Use: For the cultivation and maintenance of *Archangium gephyra, Cystobacter fuscus, Melittangium lichenicola, Myxococcus* species, *Polyangium brachy- sporum, Stigmatella aurantiaca* and *Stigmatella erecta.*

SP4 Medium

Composition per liter:

Base solution	615.0mL
Fetal calf serum (inactivated at 56°C, 1 hr)	170.0mL
Yeast extract (2% solution)	100.0mL
CMRL 1066, 10X with glutamine	50.0mL
Fresh yeast extract solution	35.0mL
Phenol red (0.1% solution)	20.0mL
Penicillin solution	10.0mL

pH 7.0–7.4 ± 0.2 at 25°C

Base Solution:

Composition per 615mL:

Pancreatic digest of casein	11.2g
Noble agar	8.0g
Pancreatic digest of gelatin	5.3g
Glucose	5.0g
NaCl	0.875g
Beef extract	0.525g
Yeast extract	0.525g
Beef heart, solids from infusion	0.35g

Preparation of Base Solution: Add components to distilled/deionized water and bring volume to 615.0mL. Mix thoroughly. Adjust pH to 7.5. Gently heat and bring to boiling. Autoclave for 15 min at 15 psi pressure–121°C. Cool to 45–50°C.

CMRL 1066, 10X With Glutamine:

Composition per liter:

NaCl	6.8g
$NaHCO_3$	2.2g
D-Glucose	1.0g
KCl	0.4g
L-Cysteine·HCl·H_2O	0.26g
$CaCl_2$, anhydrous	0.2g
$MgSO_4 \cdot 7H_2O$	0.2g
$NaH_2PO_4 \cdot H_2O$	0.14g
L-Glutamine	0.1g
Sodium acetate·$3H_2O$	0.083g
L-Glutamic acid	0.075g
L-Arginine·HCl	0.070g
L-Lysine·HCl	0.070g
L-Leucine	0.060g
Glycine	0.050g
Ascorbic acid	0.050g
L-Proline	0.040g
L-Tyrosine	0.040g
L-Aspartic acid	0.030g
L-Threonine	0.030g
L-Alanine	0.025g
L-Phenylalanine	0.025g
L-Serine	0.025g
L-Valine	0.025g
L-Cystine	0.020g
L-Histidine·HCl·H_2O	0.020g
L-Isoleucine	0.020g
Phenol red	0.020g
L-Methionine	0.015g
Deoxyadenosine	0.010g
Deoxycytidine	0.010g
Deoxyguanosine	0.010g
Glutathione, reduced	0.010g
Thymidine	0.010g
Hydroxy-L-proline	0.010g
L-Tryptophan	0.010g
Nicotinamide adenine dinucleotide	7.0mg
Tween™ 80	5.0mg
Sodium glucoronate·H_2O	4.2mg
Coenzyme A	2.5mg
Cocarboxylase	1.0mg
Flavin adenine dinucleotide	1.0mg

Nicotinamide adenine
dinucleotide phosphate 1.0mg
Uridine triphosphate.......... 1.0mg
Choline chloride.......... 0.50mg
Cholesterol 0.20mg
5-Methyldeoxycytidine.......... 0.10mg
Inositol 0.05mg
p-Aminobenzoic acid.......... 0.05mg
Niacin.......... 0.025mg
Niacinamide 0.025mg
Pyridoxine.......... 0.025mg
Pyridoxal·HCl 0.025mg
Biotin 0.01mg
D-Calcium pantothenate 0.01mg
Folic acid.......... 0.01mg
Riboflavin.......... 0.01mg
Thiamine·HCl.......... 0.01mg

Preparation of CMRL 1066, 10X With Glutamine: Add components to distilled/deionized water and bring volume to 1.0L. Mix thoroughly. Adjust pH to 7.2. Filter sterilize.

Fresh Yeast Extract Solution:

Composition per 100mL:

Baker's yeast, live, pressed, starch-free, ... 25.0g

Preparation of Fresh Yeast Extract Solution: Add the live Baker's yeast to 100.0mL of distilled/deionized water. Autoclave for 90 min at 15 psi pressure–121°C. Allow to stand. Remove supernatant solution. Adjust pH to 6.6–6.8.

Penicillin Solution:

Composition per 10mL:

Penicillin G 1,000,000U

Preparation of Penicillin Solution: Add penicillin to distilled/deionized water and bring volume to 10.0mL. Mix thoroughly. Filter sterilize.

Preparation of Medium: To 615.0mL of cooled sterile base solution, aseptically add 170.0mL of sterile inactivated fetal calf serum, 100.0mL of sterile yeast extract, 50.0mL of sterile CMRL 1066, 10X with glutamine, 35.0mL of sterile fresh yeast extract solution, 20.0mL of phenol red solution, and 10.0mL of sterile penicillin solution. Mix thoroughly. Aseptically distribute into sterile tubes. Allow tubes to cool in a slanted position.

Use: For the cultivation of tick-derived *Mycoplasma* (*Spiroplasma*). Used for the enhanced recovery of *Mycoplasma pneumoniae, M. alvi*, and *M. hyopneumoniae.*

Sphaerotilus and *Leptothrix* Enrichment Medium

Composition per liter:
Glucose 1.0g
Peptone.......... 1.0g
$MgSO_4 \cdot 7H_2O$ 0.2g
$FeCl_3 \cdot 6H_2O$ 0.1g
$CaCl_2 \cdot 2H_2O$.......... 0.05g
pH 7.0 ± 0.2 at 25°C

Preparation of Medium: Add components to distilled/deionized water and bring volume to 1.0L. Mix thoroughly. Distribute into tubes or flasks. Autoclave for 15 min at 15 psi pressure–121°C.

Use: For the enrichment and cultivation of *Sphaerotilus* species and *Leptothrix* species.

Sphaerotilus CGYA Medium

Composition per liter:
Glycerol.......... 10.0g
Pancreatic digest of casein 5.0g
Yeast extract.......... 1.0g

Preparation of Medium: Add components to distilled/deionized water and bring volume to 1.0L. Mix thoroughly. Distribute into tubes or flasks. Autoclave for 15 min at 15 psi pressure–121°C.

Use: For the cultivation and maintenance of *Sphaerotilus natans* and *Sphaerotilus* species.

Sphaerotilus Defined Medium

Composition per liter:
Agar.......... 15.0g
Glycerol.......... 5.0g
Glutamic acid 0.9g
$FeSO_4 \cdot 7H_2O$.......... 0.5g
$MgSO_4 \cdot 7H_2O$ 0.1g
$CaCl_2 \cdot 2H_2O$.......... 0.03g
$ZnSO_4 \cdot 7H_2O$ 0.03g
Phosphate solution 100.0mL
pH 7.0 ± 0.2 at 25°C

Phosphate Solution:

Composition per 500mL:
K_2HPO_4.......... 5.7g
KH_2PO_4.......... 2.3g

Preparation of Phosphate Solution: Add components to distilled/deionized water and bring volume to 500.0mL. Mix thoroughly. Gently heat until dissolved. Autoclave for 15 min at 15 psi pressure–121°C.

Preparation of Medium: Add components, except phosphate solution, to distilled/deionized water and bring volume to 900.0mL. Mix thoroughly. Gently heat and bring to boiling. Autoclave for 10 min at 15 psi pressure–121°C. Cool to 45–50°C. Aseptically add 100.0mL of sterile phosphate solution. Mix thoroughly. Pour into sterile Petri dishes or distribute into sterile tubes.

Use: For the cultivation of *Sphaerotilus* species.

Sphaerotilus discophorus Medium

Composition per liter:

Agar	12.0g
Peptone	5.0g
$MgSO_4 \cdot 7H_2O$	0.2g
$CaCl_2$	0.05g
$MnSO_4 \cdot H_2O$	0.05g
Ferric solution	100.0mL

pH 7.0 ± 0.2 at 25°C

Ferric Solution:

Composition per 100mL:

Ferric ammonium citrate	0.5g
$FeCl_3 \cdot 6H_2O$	0.01g

Preparation of Ferric Solution: Add components to distilled/deionized water and bring volume to 100.0mL. Mix thoroughly. Filter sterilize.

Preparation of Medium: Add components, except ferric solution, to tap water and bring volume to 900.0mL. Mix thoroughly. Gently heat and bring to boiling. Autoclave for 15 min at 15 psi pressure–121°C. Cool to 45–50°C. Aseptically add sterile ferric solution. Mix thoroughly. Pour into sterile Petri dishes or distribute into sterile tubes.

Use: For the cultivation of *Sphaerotilus discophorus.*

Sphaerotilus Isolation Medium

Composition per liter:

Agar	15.0g
Glycerol	10.0g
Pancreatic digest of casein	5.0g
Yeast extract	1.0g

pH 7.0 ± 0.2 at 25°C

Preparation of Medium: Add components to distilled/deionized water and bring volume to 1.0L. Mix thoroughly. Gently heat and bring to boiling. Distribute into tubes or flasks. Autoclave for 15 min at 15 psi pressure–121°C. Pour into sterile Petri dishes or leave in tubes.

Use: For the isolation and cultivation of *Sphaerotilus* species.

Sphaerotilus natans Enrichment Medium

Composition per liter:

Sodium lactate	0.1g
$Na_2HPO_4 \cdot 7H_2O$	0.034g
$CaCl_2$	0.027g
$MgSO_4 \cdot 7H_2O$	0.023g
K_2HPO_4	0.022g
KH_2PO_4	8.5mg
NH_4Cl	1.7mg
$FeCl_3 \cdot 6H_2O$	0.25mg

pH 7.1–7.2 at 25°C

Preparation of Medium: Add components to distilled/deionized water and bring volume to 1.0L. Mix thoroughly. Distribute into tubes or flasks. Autoclave for 15 min at 15 psi pressure–121°C.

Use: For the enrichment and cultivation of *Sphaerotilus natans.*

Sphaerotilus natans Isolation Agar

Composition per liter:

Agar	15.0g
Meat extract	0.5g

Preparation of Medium: Add components to tap water and bring volume to 1.0L. Mix thoroughly. Gently heat and bring to boiling. Distribute into tubes or flasks. Autoclave for 15 min at 15 psi pressure–121°C. Pour into sterile Petri dishes or leave in tubes.

Use: For the isolation and cultivation of *Sphaerotilus natans.*

Sphaerotilus natans Isolation Agar

Composition per liter:

Agar	15.0g
Casein hydrolysate	1.5g

Preparation of Medium: Add components to tap water and bring volume to 1.0L. Mix thoroughly. Gently heat and bring to boiling. Distribute into tubes or flasks. Autoclave for 15 min at 15 psi pressure–121°C. Pour into sterile Petri dishes or leave in tubes.

Use: For the isolation and cultivation of *Sphaerotilus natans.*

Spirillum gracile Medium

Composition per liter:

Agar	15.0g
Peptone	5.0g
Yeast extract	0.5g

K_2HPO_4 .... 0.1g
Tween™ 80 .... 0.02g

pH 7.2 ± 0.2 at 25°C

Preparation of Medium: Add components to tap water and bring volume to 1.0L. Mix thoroughly. Gently heat and bring to boiling. Distribute into tubes or flasks. Autoclave for 15 min at 15 psi pressure–121°C. Pour into sterile Petri dishes or leave in tubes.

Use: For the cultivation and maintenance of *Aquaspirillum gracile.*

Spirillum lipoferum Medium

Composition per liter:

Sodium malate 5.0g
Agar 3.5g
KH_2PO_4 .... 0.4g
$MgSO_4 \cdot 7H_2O$ 0.2g
K_2HPO_4 .... 0.1g
NaCl 0.1g
$CaCl_2$ 0.02g
$FeCl_3$ 0.01g
$NaMoO_4 \cdot 2H_2O$ 2.0mg
Bromthymol blue solution 5.0mL

pH 6.8 ± 0.2 at 25°C

Bromthymol Blue Solution:

Composition per 10mL:

Bromthymol blue 0.5g
Ethanol 10.0mL

Preparation of Bromthymol Blue Solution: Add bromthymol blue to 10.0mL of ethanol. Mix thoroughly.

Preparation of Medium: Add components to distilled/deionized water and bring volume to 1.0L. Mix thoroughly. Gently heat and bring to boiling. Distribute into tubes or flasks. Autoclave for 15 min at 15 psi pressure–121°C.

Use: For the isolation and cultivation of *Spirillum leptoferum.*

Spirillum lipoferum Medium

Composition per liter:

Malic acid 5.0g
NaOH 4.7g
Agar 1.75g
KH_2PO_4 .... 0.4g
$MgSO_4 \cdot 7H_2O$ 0.2g
K_2HPO_4 .... 0.1g
NaCl 0.1g
$CaCl_2$ 0.02g
$FeCl_3$ 0.01g
$NaMoO_4 \cdot 2H_2O$ 2.0mg
Bromthymol blue solution 5.0mL

pH 6.8 ± 0.2 at 25°C

Bromthymol Blue Solution:

Composition per 10mL:

Bromthymol blue 0.5g
Ethanol 10.0mL

Preparation of Bromthymol Blue Solution: Add bromthymol blue to 10.0mL of ethanol. Mix thoroughly.

Preparation of Medium: Add components to distilled/deionized water and bring volume to 1.0L. Mix thoroughly. Gently heat and bring to boiling. Distribute into tubes or flasks. Autoclave for 15 min at 15 psi pressure–121°C.

Use: For the isolation and cultivation of *Spirillum leptoferum.*

Spirillum lipoferum Medium

Composition per liter:

Malic acid 5.0g
KOH 4.0g
Agar 1.75g
$FeSO_4 \cdot 7H_2O$ 0.5g
K_2HPO_4 .... 0.5g
$MgSO_4 \cdot 7H_2O$ 0.2g
NaCl 0.1g
$CaCl_2$ 0.02g
$MnSO_4 \cdot H_2O$ 0.01g
$NaMoO_4 \cdot 2H_2O$ 2.0mg
Bromthymol blue solution 5.0mL

pH 6.8 ± 0.2 at 25°C

Bromthymol Blue Solution:

Composition per 10mL:

Bromthymol blue 0.5g
Ethanol 10.0mL

Preparation of Bromthymol Blue Solution: Add bromthymol blue to 10.0mL of ethanol. Mix thoroughly.

Preparation of Medium: Add components to distilled/deionized water and bring volume to 1.0L. Mix thoroughly. Gently heat and bring to boiling. Distribute into tubes or flasks. Autoclave for 15 min at 15 psi pressure–121°C.

Use: For the isolation and cultivation of *Spirillum leptoferum.*

Spirillum Medium

Composition per liter:

Calcium lactate	10.0g
Peptone	5.0g
Beef extract	3.0g
Yeast extract	3.0g

pH 7.0 ± 0.2 at 25°C

Preparation of Medium: Add components to distilled/deionized water and bring volume to 1.0L. Mix thoroughly. Adjust pH to 7.0. Distribute into tubes or flasks. Autoclave for 20 min at 11 psi pressure–116°C. A precipitate will form during autoclaving.

Use: For the cultivation of *Spirillum* species.

Spirillum Nitrogen–Fixing Medium

Composition per liter:

Sodium malate	5.0g
KH_2PO_4	0.4g
$MgSO_4 \cdot 7H_2O$	0.2g
K_2HPO_4	0.1g
NaCl	0.1g
Yeast extract	0.05g
$CaCl_2$	0.02g
$FeCl_3$	0.01g
$NaMoO_4 \cdot 2H_2O$	2.0mg

pH 7.2-7.4 ± 0.2 at 25°C

Preparation of Medium: Add components to distilled/deionized water and bring volume to 1.0L. Mix thoroughly. Distribute into tubes or flasks. Autoclave for 15 min at 15 psi pressure–121°C.

Use: For the cultivation and maintenance of *Azospirillum brasilense*, *A. lipoferum* and *Herbaspirillum seropedicae*.

Spirillum volutans Defined Medium

Composition per liter:

BES (*N,N*-bis[2-hydroxyethyl]-2-aminoethane sulfonic acid) buffer	1.07g
$MgSO_4 \cdot 7H_2O$	1.0g
$(NH_4)_2SO_4$	1.0g
Succinic acid	1.0g
L-Histidine	0.2g
L-Isoleucine	0.2g
L-Methionine	0.2g
L-Threonine	0.2g
NaCl	0.085g
L-Cystine	0.025g
K_2HPO_4	0.02g
$FeCl_3 \cdot 6H_2O$	3.0mg
DL-Norepinephrine	2.0mg
$MnSO_4 \cdot H_2O$	2.0mg
$CaCO_3$	1.0mg
$ZnSO_4 \cdot 7H_2O$	0.72mg
$Na_2MoO_4 \cdot 2H_2O$	0.245mg
$CoSO_4 \cdot 7H_2O$	0.14mg
$CuSO_4 \cdot 5H_2O$	0.13mg
H_2BO_3	0.031mg

pH 6.8 ± 0.2 at 25°C

Preparation of Medium: Add components to distilled/deionized water and bring volume to 1.0L. Mix thoroughly. Adjust pH to 6.8. Distribute into tubes or flasks. Autoclave for 15 min at 15 psi pressure–121°C.

Use: For the cultivation of *Spirillum volutans.*

Spirit Blue Agar

Composition per liter:

Agar	20.0g
Pancreatic digest of casein	10.0g
Yeast extract	5.0g
Spirit blue	0.15g
Lipoidal emulsion	30.0mL

pH 6.8 ± 0.2 at 25°C

Source: This medium is available as a premixed powder from Difco Laboratories.

Lipoidal Emulsion:
Composition per 500mL:

Tween™ 80	1.0mL
Cottonseed oil or olive oil	100.0mL

Preparation of Lipoidal Emulsion: Add Tween™ 80 to 400.0mL of warm distilled/deionized water. Mix thoroughly. Add 100.0mL of cottonseed or olive oil. Emulsify in a blender. Autoclave for 15 min at 15 psi pressure–121°C. Cool to 45–50°C.

Preparation of Medium: Add components, except lipoidal emulsion, to distilled/deionized water and bring volume to 970.0mL. Mix thoroughly. Gently heat and bring to boiling. Autoclave for 15 min at 15 psi pressure–121°C. Cool to 45–50°C. Aseptically add 30.0mL sterile lipoidal emulsion. Mix thoroughly. Pour into sterile Petri dishes while shaking flask to keep emulsion dispersed.

Use: For the detection, enumeration and study of lipolytic microorganisms.

Spirochaeta aurantia Growth Medium

Composition per liter:

Yeast extract	4.0g
Maltose	2.0g

Peptone....2.0g
Potassium phosphate
buffer (0.1*M* solution, pH 7.0)....100.0mL
pH 7.2 ± 0.2 at 25°C

Preparation of Medium: Filter sterilize potassium phosphate buffer. Add components, except potassium phosphate buffer, to distilled/deionized water and bring volume to 900.0mL. Mix thoroughly. Gently heat and bring to boiling. Autoclave for 15 min at 15 psi pressure–121°C. Cool to 45–50°C. Aseptically add sterile potassium phosphate buffer. Mix thoroughly. Adjust pH to 7.2. Aseptically distribute into sterile tubes or flasks.

Use: For the cultivation of *Spirochetes aurantia.*

Spirochaeta aurantia Isolation Medium

Composition per liter:
Peptone....1.0g
Yeast extract....1.0g
Hay extract....500.0mL
pH 6.5 ± 0.2 at 25°C

Hay Extract:
Composition per liter:
Hay, dried....5.0g

Preparation of Hay Extract: Add hay to distilled/deionized water and bring volume to 1.0L. Mix thoroughly. Gently heat and bring to boiling. Continue boiling for 10 min. Filter through Whatman #1 filter paper.

Preparation of Medium: Add components to distilled/deionized water and bring volume to 1.0L. Mix thoroughly. Distribute into tubes or flasks. Autoclave for 15 min at 15 psi pressure–121°C.

Use: For the isolation and cultivation of *Spirochaeta aurantia.*

Spirochaeta halophila Medium

Composition per liter:
Glucose salts solution....970.0mL
Yeast extract peptone solution....30.0mL

Glucose Salts Solution:
Composition per liter:
NaCl....49.3g
$MgSO_4·7H_2O$....49.2g
$CaCl_2·2H_2O$....5.9g
Glucose solution....100.0mL
Sulfide solution....100.0mL

Preparation of Glucose Salts Solution: Add components, except glucose solution and sulfide solution, to distilled/deionized water and bring volume to 800.0mL. Mix thoroughly. Autoclave for 15 min at 15 psi pressure–121°C. Cool to 25°C. Aseptically add sterile glucose solution and sulfide solution. Mix thoroughly.

Sulfide Solution:
Composition per 100mL:
$Na_2S·9H_2O$....0.5g

Preparation of Sulfide Solution: Add $Na_2S·9H_2O$ to distilled/deionized water and bring volume to 100.0mL. Mix thoroughly. Autoclave for 15 min at 15 psi pressure–121°C. Cool to 25°C.

Glucose Solution:
Composition per 100mL:
Glucose....5.0g

Preparation of Glucose Solution: Add glucose to distilled/deionized water and bring volume to 100.0mL. Mix thoroughly. Filter sterilize.

Yeast Extract Peptone Solution:
Composition per 30mL:
Yeast exract....4.0g
Peptone....2.0g

Preparation of Yeast Extract Peptone Solution: Add components to distilled/deionized water and bring volume to 30.0mL. Mix thoroughly. Autoclave for 15 min at 15 psi pressure–121°C. Cool to 25°C.

Preparation of Medium: Aseptically combine 30.0mL yeast extract peptone solution with 970.0mL glucose slats solution. Mix thoroughly. Aseptically distribute into sterile tubes or flasks.

Use: For the isolation and cultivation of *Spirochaeta halophila.*

Spirochaeta litoralis Medium

Composition per liter:
Pancreatic digest of casein....3.0g
NaCl....2.0g
Yeast extract....0.5g
Glucose solution....2.0mL
Potasssium phosphate
buffer (1*M*, pH 7.4)....2.0mL
Sulfide solution....0.5mL
Salts solution....0.2mL
pH 7.3 ± 0.2 at 25°C

Glucose Solution:
Composition per 100mL:
D-Glucose....25.0g

Preparation of Glucose Solution: Add D-glucose to distilled/deionized water and bring volume to 100.0mL. Mix thoroughly. Filter sterilize.

Sulfide Solution:

Composition per 100mL:

$Na_2S \cdot 9H_2O$	10.0g

Preparation of Sulfide Solution: Add $Na_2S \cdot 9H_2O$ to distilled/deionized water and bring volume to 100.0mL. Autoclave for 15 min at 15 psi pressure–121°C. Cool to 25°C.

Salts Solution:

Composition per 100mL:

$MgSO_4 \cdot 7H_2O$	12.5g
$CaCl_2 \cdot 2H_2O$	3.75g
EDTA	1.0g
$FeSO_4 \cdot 7H_2O$	0.5g
Trace elements solution	25.0mL

Preparation of Salts Solution: Add components to distilled/deionized water and bring volume to 100.0mL. Mix thoroughly.

Trace Elements Solution:

Composition per 1800mL:

H_3BO_3	5.5g
$MnCl_2 \cdot 4H_2O$	3.5g
$AlCl_3 \cdot 6H_2O$	0.50g
$CoCl_2 \cdot 6H_2O$	0.50g
$CuCl_2 \cdot 2H_2O$	0.50g
$NiCl_2 \cdot 6H_2O$	0.50g
$ZnCl_2$	0.50g
KI	0.25g
LiCl	0.25g
$Na_2MoO_4 \cdot 2H_2O$	0.25g
$BaCl_2 \cdot 2H_2O$	0.15g
$SnCl_2 \cdot 2H_2O$	0.15g
$NaVO_3$	0.05g

Preparation of Trace Elements Solution: Add components to distilled/deionized water and bring volume to 1800.0mL. Mix thoroughly. Adjust pH to 3–4 with HCl.

Preparation of Medium: Add components, except glucose solution and sulfide solution, to distilled/deionized water and bring volume to 997.5mL. Mix thoroughly. Gently heat and bring to boiling. Autoclave for 15 min at 15 psi pressure–121°C. Cool to 45–50°C. Aseptically add sterile glucose solution and sulfide solution. Mix thoroughly. Aseptically distribute into sterile tubes or flasks.

Use: For the isolation of *Spirochaeta litoralis* from marine habitats.

Spirochaeta stenostrepta Medium

Composition per liter:

Glucose	5.0g
Peptone	2.0g
Yeast extract	0.3g
Vitamin B_{12}	0.01mg
Salts solution	100.0mL
Phosphate solution	15.0mL
Sulfide solution	10.0mL

pH 7.0 ± 0.2 at 25°C

Phosphate Solution:

Composition per liter:

KH_2PO_4	30.0g
K_2HPO_4	70.0g

Preparation of Phosphate Solution: Add components to distilled/deionized water and bring volume to 1.0L. Mix thoroughly. Filter sterilize.

Salts Solution:

Composition per liter:

$MgSO_4 \cdot 7H_2O$	2.0g
$CaCl_2 \cdot 2H_2O$	0.75g
EDTA	0.2g
$FeSO_4 \cdot 7H_2O$	0.1g
Trace elements solution	5.0mL

Preparation of Salts Solution: Add EDTA to approximately 800.0mL of distilled/deionized water. Gently heat until dissolved. Adjust pH to 7.0 with 2.5% KOH. Add the remaining components. Mix thoroughly. Bring volume to 1.0L with distilled/deionized water.

Trace Elements Solution:

Composition per 1800mL:

H_3BO_3	5.5g
$MnCl_2 \cdot 4H_2O$	3.5g
$AlCl_3 \cdot 6H_2O$	0.50g
$CoCl_2 \cdot 6H_2O$	0.50g
$CuCl_2 \cdot 2H_2O$	0.50g
$NiCl_2 \cdot 6H_2O$	0.50g
$ZnCl_2$	0.50g
KI	0.25g
LiCl	0.25g
$Na_2MoO_4 \cdot 2H_2O$	0.25g
$BaCl_2 \cdot 2H_2O$	0.15g
$SnCl_2 \cdot 2H_2O$	0.15g
$NaVO_3$	0.05g

Preparation of Trace Elements Solution: Add components to distilled/deionized water and bring volume to 1800.0mL. Mix thoroughly. Adjust pH to 3–4 with HCl.

Sulfide Solution:

Composition per 100mL:

$Na_2S \cdot 9H_2O$	2.0g

Preparation of Sulfide Solution: Add $Na_2S \cdot 9H_2O$ to distilled/deionized water and bring volume to 100.0mL. Autoclave for 15 min at 15 psi pressure–121°C. Cool to 25°C. Prepare solution freshly.

Preparation of Medium: Add components, except sulfide solution, to distilled/deionized water and bring volume to 990.0mL. Mix thoroughly. Gently heat and bring to boiling. Autoclave for 15 min at 15 psi pressure–121°C. Cool to 45–50°C. Aseptically add 10.0mL of sterile sulfide solution. Mix thoroughly. Aseptically distribute into sterile tubes or flasks.

Use: For the isolation of *Spirochaeta stenostrepta.*

Spirochaeta zuelzerae Medium

Composition per liter:

Solution 1	480.0mL
Solution 2	480.0mL
Solution 3	20.0mL
Solution 4	20.0mL

pH 7.2 ± 0.2 at 25°C

Solution 1:

Composition per 480mL:

KH_2PO_4	0.75g
Cysteine·HCl·H_2O	0.5g
$NaH_2PO_4 \cdot H_2O$	0.25g

Preparation of Solution 1: Add components to distilled/deionized water and bring volume to 480.0mL. Mix thoroughly. Adjust pH to 7.2 with 5.0% KOH. Autoclave for 15 min at 15 psi pressure–121°C. Cool to 45–50°C.

Solution 2:

Composition per 480mL:

NH_4Cl	1.0g
$MgSO_4 \cdot 7H_2O$	0.5g
Yeast extract	0.2g
$CaCl_2$	0.02g
Resazurin	1.0mg
$FeCl_3 \cdot 6H_2O$ solution (0.25g/L)	10.0mL
Trace element solution	2.0mL

Preparation of Solution 2: Add components to distilled/deionized water and bring volume to 480.0mL. Mix thoroughly. Autoclave for 15 min at 15 psi pressure–121°C. Cool to 45–50°C.

Trace Element Solution:

Composition per 100mL:

$Na_2MoO_4 \cdot 2H_2O$	0.075g
H_3BO_3	0.056g
$ZnSO_4 \cdot 7H_2O$	0.044g
$CoCl_2 \cdot 6H_2O$	0.020g
$CuSO_4 \cdot 5H_2O$	2.0mg
$MnCl_2$	2.0mg

Preparation of Trace Element Solution: Add components to distilled/deionized water and bring volume to 100.0mL. Mix thoroughly.

Solution 3:

Composition per 20mL:

$NaHCO_3$	1.0g

Preparation of Solution 3: Add $NaHCO_3$ to distilled/deionized water and bring volume to 20.0mL. Mix thoroughly. Filter sterilize under pressure.

Solution 4:

Composition per 20mL:

Glucose	2.0g

Preparation of Solution 4: Add glucose to distilled/deionized water and bring volume to 20.0mL. Mix thoroughly. Filter sterilize.

Preparation of Medium: Aseptically add 480.0 mL of sterile Solution 1 to 480.0mL of sterile Solution 2 under 80% N_2 + 20% CO_2. While gassing, add 20.0mL of sterile solution 3 and 20.0mL of sterile solution 4. Mix thoroughly. Adjust pH to 7.2. Aseptically and anaerobically distribute into tubes. Cap with rubber stoppers.

Use: For the cultivation and maintenance of *Spirochaeta zuelzerae.*

Spirochete Enrichment Medium

Composition per liter:

Agar	10.0g
Beef extract	1.0g
Peptone	1.0g
Yeast extract	1.0g
Seawater	500.0mL

Preparation of Medium: Add components to distilled/deionized water and bring volume to 1.0L. Mix thoroughly. Gently heat and bring to boiling. Distribute into tubes or flasks. Autoclave for 15 min at 15 psi pressure–121°C. Pour into sterile Petri dishes.

Use: For the isolation of spirochetes from muds. A well is cut into the agar plate and filled with mud samples. Spirochetes migrate out of the mud into the agar surrounding the well.

Spirochete Medium (ATCC Medium 164)

Composition per liter:

Agar	15.0g
KH_2PO_4	1.0g
NH_4Cl	1.0g

Yeast extract	1.0g
$MgSO_4$	0.5g
$CaCl_2$	0.04g
$FeCl_3 \cdot 6H_2O$	1.25mg
$NaHCO_3$ solution	20.0mL
Glucose solution	10.0mL
$Na_2S \cdot 9H_2O$ solution	5.0mL

Glucose Solution:

Composition per 100mL:

Glucose	10.0g

Preparation of Glucose Solution: Add glucose to distilled/deionized water and bring volume to 100.0mL. Mix thoroughly. Filter sterilize.

$NaHCO_3$ Solution:

Composition per 100mL:

$NaHCO_3$	5.0g

Preparation of $NaHCO_3$ Solution: Add the $NaHCO_3$ to distilled/deionized water and bring volume to 100.0mL. Mix thoroughly. Filter sterilize.

$Na_2S \cdot 9H_2O$ Solution:

Composition per 100mL:

$Na_2S \cdot 9H_2O$	10.0g

Preparation of $Na_2S \cdot 9H_2O$ Solution: Add $Na_2S \cdot 9H_2O$ to distilled/deionized water and bring volume to 100.0mL. Mix thoroughly. Autoclave for 15 min at 15 psi pressure–121°C.

Preparation of Medium: Add components—except $NaHCO_3$ solution, glucose solution and $Na_2S \cdot 9H_2O$ solution—to distilled/deionized water and bring volume to 965.0mL. Autoclave for 15 min at 15 psi pressure–121°C. Cool to 50°C. Aseptically add 20.0mL of sterile $NaHCO_3$ solution, 10.0mL of sterile glucose solution and 5.0mL of sterile $Na_2S \cdot 9H_2O$ solution. Mix thoroughly. Pour into sterile Petri dishes or distribute into sterile tubes.

Use: For the cultivation of spirochetes.

Spirochete Medium (ATCC Medium 1712)

Composition per liter:

Tris(hydroxymethyl)aminomethane buffer	7.52g
Pancreatic digest of casein	1.0g
Yeast extract	1.0g
Cysteine $HCl \cdot 2H_2O$	0.5g
Resazurin	1.0mg
Seawater	750.0mL
Glucose solution	20.0mL

pH 7.2 ± 0.2 at 25°C

Glucose Solution:

Composition per 20mL:

Glucose	2.0g

Preparation of Glucose Solution: Add glucose to distilled/deionized water and bring volume to 20.0mL. Mix thoroughly. Filter sterilize.

Preparation of Medium: Prepare and dispense medium under 100% N_2. Add components, except glucose solution, to distilled/deionized water and bring volume to 980.0mL. Mix thoroughly. Adjust pH to 7.5. Autoclave for 15 min at 15 psi pressure–121°C. Cool to 50°C. Aseptically add sterile glucose solution. Mix thoroughly. Aseptically distribute into sterile tubes or flasks.

Use: For the cultivation and maintenance of *Spirochaeta litoralis*.

Spirolate Broth

Composition per liter:

Pancreatic digest of casein	15.0g
Glucose	5.0g
Yeast extract	5.0g
NaCl	2.5g
L-Cysteine·$HCl \cdot H_2O$	1.0g
Sodium thioglycollate	0.5g
Palmitic acid	0.05g
Stearic acid	0.05g
Oleic acid	0.05g
Linoleic acid	0.05g
Serum	100.0mL

pH 7.1 ± 0.2 at 25°C

Source: This medium is available as a premixed powder from BBL Microbiology Systems.

Preparation of Medium: Add components, except serum, to distilled/deionized water and bring volume to 900.0mL. Mix thoroughly. Distribute into screw-capped tubes in 20.0mL volumes. Autoclave for 15 min at 15 psi pressure–121°C. Cool to 25°C. Aseptically add 2.0mL of serum to each tube. Heat-inactivated sheep, rabbit or bovine serum may be used. Tighten caps. Mix thoroughly.

Use: For the cultivation of *Treponema phagedenis* and other spirochetes.

Spiroplasma Agar MID

Composition per 291.2mL:

Schneider's *Drosophila* medium	160.0mL
Solution 1	80.0mL
Fetal calf serum	50.0mL
Phenol red (0.5% solution)	1.2mL

pH 7.4 ± 0.2 at 25°C

Schneider's *Drosophila* Medium:
Composition per liter:

$MgSO_4 \cdot 7H_2O$	3.7g
NaCl	2.1g
Yeast extract	2.0g
Trehalose	2.0g
D-Glucose	2.0g
L-Glutamine	1.8g
L-Lysine·HCl	1.7g
L-Proline	1.7g
KCl	1.6g
$Na_2HPO_4 \cdot 7H_2O$	1.3g
L-Glutamic acid	0.8g
L-Methionine	0.8g
$CaCl_2$, anhydrous	0.6g
KH_2PO_4	0.5g
β-Alanine	0.5g
L-Tyrosine	0.5g
L-Arginine	0.4g
L-Aspartic acid	0.4g
L-Histidine	0.4g
L-Threonine	0.4g
$NaHCO_3$	0.4g
Glycine	0.3g
L-Serine	0.3g
L-Valine	0.3g
L-Isoleucine	0.2g
L-Leucine	0.2g
L-Phenylalanine	0.2g
α-Ketoglutaric acid	0.2g
Fumaric acid	0.1g
Malic acid	0.1g
Succinic acid	0.1g
L-Cystine	0.1g
L-Tryptophan	0.1g
L-Cysteine	0.06g

Preparation of Schneider's *Drosophila* Medium: Add components to 1.0L of distilled/deionized water. Mix thoroughly. Filter sterilize.

Solution 1:
Composition per 80mL:

Sorbitol	7.0g
Noble agar	5.0g
Beef heart (solids from infusion)	5.0g
Peptone	1.8g
Sucrose	1.0g
Pancreatic digest of casein	1.0g
NaCl	0.5g
D-Fructose	0.1g
D-Glucose	0.1g

Preparation of Solution 1: Add components to distilled/deionized water and bring volume to 80.0mL. Mix thoroughly. Adjust pH to 7.8 with 1*N* NaOH. Autoclave for 15 min at 15 psi pressure–121°C. Cool to 50°C.

Preparation of Medium: Bring fetal calf serum and phenol red solution to 56°C. Rapidly bring Schneider's *Drosophila* Medium to 37°C. Rapidly combine the components. Mix thoroughly. Pour into sterile Petri dishes or distribute into sterile tubes or flasks.

Use: For the cultivation and maintenance of *Spiroplasma kunkelii* and *Spiroplasma* species.

Spiroplasma Broth MID

Composition per 291.2mL:

Schneider's *Drosophila* Medium	160.0mL
Solution 1	80.0mL
Fetal calf serum	50.0mL
Phenol red (0.5% solution)	1.2mL

pH 7.4 ± 0.2 at 25°C

Schneider's Drosophila Medium:
Composition per liter:

$MgSO_4 \cdot 7H_2O$	3.7g
NaCl	2.1g
Yeast extract	2.0g
Trehalose	2.0g
D-Glucose	2.0g
L-Glutamine	1.8g
L-Lysine·HCl	1.7g
L-Proline	1.7g
KCl	1.6g
$Na_2HPO_4 \cdot 7H_2O$	1.3g
L-Glutamic acid	0.8g
L-Methionine	0.8g
$CaCl_2$, anhydrous	0.6g
KH_2PO_4	0.5g
β-Alanine	0.5g
L-Tyrosine	0.5g
L-Arginine	0.4g
L-Aspartic acid	0.4g
L-Histidine	0.4g
L-Threonine	0.4g
$NaHCO_3$	0.4g
Glycine	0.3g
L-Serine	0.3g
L-Valine	0.3g
L-Isoleucine	0.2g
L-Leucine	0.2g
L-Phenylalanine	0.2g
α-Ketoglutaric acid	0.2g
Fumaric acid	0.1g
Malic acid	0.1g
Succinic acid	0.1g
L-Cystine	0.1g

L-Tryptophan	0.1g
L-Cysteine	0.06g

Preparation of Schneider's *Drosophila* Medium: Add components to distilled/deionized water and bring volume to 1.0L. Mix thoroughly. Filter sterilize.

Solution 1:

Composition per 80mL:

Sorbitol	7.0g
Beef heart (solids from infusion)	5.0g
Peptone	1.8g
Sucrose	1.0g
Pancreatic digest of casein	1.0g
NaCl	0.5g
D-Fructose	0.1g
D-Glucose	0.1g

Preparation of Solution 1: Add components to distilled/deionized water and bring volume to 80.0mL. Mix thoroughly. Adjust pH to 7.8 with 1*N* NaOH. Autoclave for 15 min at 15 psi pressure–121°C. Cool to 25°C.

Preparation of Medium: Bring fetal calf serum and phenol red solution to 56°C. Rapidly bring Schneider's *Drosophila* Medium to 37°C. Rapidly combine the components. Mix thoroughly. Aseptically distribute into sterile tubes or flasks.

Use: For the cultivation and maintenance of *Spiroplasma kunkelii* and *Spiroplasma* species.

Spiroplasma Medium

Composition per liter:

Sucrose	80.0g
Beef heart (solids from infusion)	34.7g
Peptone	6.9g
NaCl	3.5g
Horse serum, heat-inactivated	100.0mL

pH 7.2 ± 0.2 at 25°C

Preparation of Medium: Add components, except horse serum, to distilled/deionized water and bring volume to 900.0mL. Mix thoroughly. Autoclave for 15 min at 15 psi pressure–121°C. Cool to 25°C. Aseptically add horse serum. Mix thoroughly. Aseptically distribute into sterile tubes or flasks.

Use: For the cultivation and maintenance of *Spiroplasma* species.

Spiroplasma Medium with 25 mg/L Phenol Red

Composition per liter:

Sucrose	80.0g
Beef heart (solids from infusion)	34.7g
Peptone	6.9g
NaCl	3.5g
Phenol red	25.0mg
Horse serum, heat-inactivated	100.0mL

pH 7.2 ± 0.2 at 25°C

Preparation of Medium: Add components, except horse serum, to distilled/deionized water and bring volume to 900.0mL. Mix thoroughly. Gently heat and bring to boiling. Autoclave for 15 min at 15 psi pressure–121°C. Cool to 45–50°C. Aseptically add heat-inactivated horse serum. Mix thoroughly. Aseptically distribute into sterile tubes or flasks.

Use: For the cultivation and maintenance of *Spiroplasma floricola.*

Spizizen Potato Agar

Composition per liter:

Potatoes	200.0g
Agar	15.0g
$MnSO_4$	5.0mg

pH 6.8 ± 0.2 at 25°C

Preparation of Medium: Peel and dice potatoes. Add potatoes to 1.0L of tap water. Gently heat and bring to boiling. Continue boiling for 30 min. Filter through cheesecloth. Add $MnSO_4$ to filtrate and bring volume to 1.0L with tap water. Mix thoroughly. Adjust pH to 6.8. Add agar. Gently heat and bring to boiling. Distribute into tubes or flasks. Autoclave for 15 min at 15 psi pressure–121°C. Pour into sterile Petri dishes or leave in tubes.

Use: For the cultivation and maintenance of *Bacillus amyloliquefaciens.*

Spore Strip Broth

Composition per liter:

Spore strip broth	9.0g

Preparation of Medium: Add 9.0g of spore strip broth powder (a mixture of glucose, buffer salts, growth factors and bromthymol blue) to distilled/deionized water and bring volume to 1.0L. Mix thoroughly. Distribute into tubes or flasks. Autoclave for 15 min at 15 psi pressure–121°C.

Use: For the recovery of spores of *Bacillus stearothermophilus* on spore strips used to determine the sterilization efficiency of autoclaves.

Sporomusa Medium

Composition per 877mL:

NaCl	2.25g
Pancreatic digest of casein	2.0g

Yeast extract	2.0g
$MgSO_4 \cdot 7H_2O$	0.5g
NH_4Cl	0.5g
K_2HPO_4	0.35g
KH_2PO_4	0.23g
$CaCl_2 \cdot 2H_2O$	0.025g
$FeSO_4 \cdot 7H_2O$	2.0mg
Resazurin	1.0mg
$NaHSeO_3$	15.0µg
$NaHCO_3$ solution	50.0mL
Glycine betaine solution	50.0mL
Wolfe's vitamin solution	10.0mL
Cysteine·HCl·H_2O solution	10.0mL
Trace elements solution SL-6	3.0mL

pH 7.0–7.2 at 25°C

$NaHCO_3$ Solution:

Composition per 50mL:

$NaHCO_3$	4.0g

Preparation of $NaHCO_3$ Solution: Add the $NaHCO_3$ to distilled/deionized water and bring volume to 50.0mL. Mix thoroughly. Gas under 80% N_2 + 20% CO_2 for 20 min.

Glycine Betaine Solution:

Composition per 50mL:

Glycine betaine	5.0g

Preparation of Glycine Betaine Solution: Add the glycine betaine to distilled/deionized water and bring volume to 50.0mL. Mix thoroughly. Filter sterilize. Aseptically gas under 80% N_2 + 20% CO_2.

Wolfe's Vitamin Solution:

Composition per liter:

Pyridoxine·HCl	0.01g
Thiamine·HCl	5.0mg
Riboflavin	5.0mg
Nicotinic acid	5.0mg
Calcium pantothenate	5.0mg
p-Aminobenzoic acid	5.0mg
Thioctic acid	5.0mg
Biotin	2.0mg
Folic acid	2.0mg
Cyanocobalamin	0.1mg

Preparation of Wolfe's Vitamin Solution: Add components to distilled/deionized water and bring volume to 1.0L. Mix thoroughly.

Cysteine·HCl·H_2O Solution:

Composition per 10mL:

Cysteine·HCl·H_2O	0.3g

Preparation of Cysteine·HCl·H_2O Solution: Add cysteine·HCl·H_2O to distilled/deionized water and bring volume to 10.0mL. Mix thoroughly. Filter sterilize. Aseptically gas under 100% N_2.

Trace Elements Solution SL-6:

Composition per liter:

H_3BO_3	0.3g
$CoCl_2 \cdot 6H_2O$	0.2g
$ZnSO_4 \cdot 7H_2O$	0.10g
$MnCl_2 \cdot 4H_2O$	0.03g
$Na_2MoO_4 \cdot H_2O$	0.03g
$NiCl_2 \cdot 6H_2O$	0.02g
$CuCl_2 . 2H_2O$	0.01g

Preparation of Trace Elements Solution SL-6: Add components to distilled/deionized water and bring volume to 1.0L. Mix thoroughly. Adjust pH to 3.4.

Preparation of Medium: Add components—except $NaHCO_3$ solution, glycine betaine solution, and cysteine·HCl·H_2O solution—to distilled/deionized water and bring volume to 890.0mL. Mix thoroughly. Gently heat and bring to boiling. Continue boiling for 5 min. Cool rapidly to 25°C under 80% N_2 + 20% CO_2. Add 50.0mL of $NaHCO_3$ solution. Mix thoroughly. Autoclave anaerobically for 15 min at 15 psi pressure–121°C. Cool to 25°C. Aseptically add sterile glycine betaine solution. Mix thoroughly. Immediately prior to inoculation aseptically and anaerobically add cysteine·HCl·H_2O solution.

Use: For the cultivation and maintenance of *Sporomusa ovata* and *S. sphaeroides*.

Sporosarcina ureae

Composition per liter:

L-Asparagine·H_2O or L-glutamine	30.0g
KCl	3.4g
NaCl	2.92g
K_2HPO_4	0.25g
$(NH_4)_2SO_4$	0.2g
$MgSO_4 \cdot 7H_2O$	0.05g
$FeSO_4 \cdot 7H_2O$	2.5mg
$MnCl_2 \cdot 4H_2O$	0.25mg
Biotin solution	10.0mL
Cysteine solution	10.0mL
$(NH_4)_2SO_4$ solution	10.0mL

pH 8.7 ± 0.2 at 25°C

Biotin Solution:

Composition per 10mL:

D-Biotin	1.0mg

Preparation of Biotin Solution: Add biotin to distilled/deionized water and bring volume to 10.0mL. Mix thoroughly. Filter sterilize.

Cysteine Solution:

Composition per 10mL:

Cysteine	0.04g

Preparation of Cysteine Solution: Add cysteine to distilled/deionized water and bring volume to 10.0mL. Mix thoroughly. Filter sterilize.

$(NH_4)_2SO_4$ Solution:
Composition per 10mL:

$(NH_4)_2SO_4$	0.2g

Preparation of $(NH_4)_2SO_4$ Solution: Add $(NH_4)_2SO_4$ to distilled/deionized water and bring volume to 10.0mL. Mix thoroughly. Filter sterilize.

Preparation of Medium: Add components—except biotin solution, cysteine solution, and $(NH_4)_2SO_4$ solution—to distilled/deionized water and bring volume to 970.0mL. Mix thoroughly. Adjust pH to 8.7 with 1*N* NaOH. Autoclave for 15 min at 15 psi pressure–121°C. Cool to 45–50°C. Aseptically add sterile biotin solution, cysteine solution, and $(NH_4)_2SO_4$ solution. Mix thoroughly. Aseptically distribute into sterile tubes.

Use: For the cultivation of *Sporosarcina ureae.*

Sporosarcina ureae Medium

Composition per liter:

Agar	30.0g
Glucose	4.0g
$(NH_4)_2SO_4$	4.0g
Malt extract	3.0g
Peptone	3.0g
Yeast extract	2.0g
K_2HPO_4	1.0g
$MgSO_4$	0.8g
$CaCl_2$	0.1g
$MnSO_4 \cdot H_2O$	0.1g
$CuSO_4 \cdot 5H_2O$	0.01g
$ZnSO_4$	0.01g
$FeSO_4 \cdot 7H_2O$	1.0mg

Preparation of Medium: Add components to 1.0L of distilled/deionized water. Mix thoroughly. Gently heat and bring to boiling. Distribute into tubes or flasks. Autoclave for 15 min at 15 psi pressure–121°C. Pour into sterile Petri dishes or leave in tubes.

Use: For the cultivation and induction of sporulation of *Sporosarcina ureae.*

Sporulation Agar (m–Sporulation Agar)

Composition per liter:

Agar	15.0g
Glucose	10.0g
Tryptose	2.0g
Beef extract	1.0g
Yeast extract	1.0g
$FeSO_4$	1.0μg

pH 7.2 ± 0.2 at 25°C

Preparation of Medium: Add components to distilled/deionized water and bring volume to 1.0L. Mix thoroughly. Gently heat and bring to boiling. Distribute into tubes or flasks. Autoclave for 15 min at 15 psi pressure–121°C. Pour into sterile Petri dishes or leave in tubes.

Use: For the cultivation and sporulation of *Streptomyces, Streptoverticillium,* and *Thermoactinomyces* species. For the identification of sporulating bacteria by the membrane filter method.

Sporulation Broth

Composition per liter:

Glucose	3.3g
Tryptose	0.66g
Beef extract	0.33g
Yeast extract	0.33g
$FeSO_4$	0.33μg

pH 7.2 ± 0.2 at 25°C

Preparation of Medium: Add components to distilled/deionized water and bring volume to 1.0L. Mix thoroughly. Distribute into tubes or flasks. Autoclave for 15 min at 15 psi pressure–121°C.

Use: For the cultivation and sporulation of *Streptomyces, Streptoverticillium,* and *Thermoactinomyces* species.

SS Deoxycholate Agar (*Salmonella–Shigella* Deoxycholate Agar) (SSDC)

Composition per liter:

Agar	13.5g
Lactose	10.0g
Sodium desoxycholate	10.0g
Bile salts	8.5g
$Na_2S_2O_3$	8.5g
Sodium citrate	8.5g
Beef extract	5.0g
Pancreatic digest of casein	2.5g
Peptic digest of animal tissue	2.5g
$CaCl_2 \cdot 2H_2O$	1.0g
Ferric citrate	1.0g
Neutral red	0.025g
Brilliant green	0.330mg

pH 7.0 ± 0.2 at 25°C

Preparation of Medium: Add components to distilled/deionized water and bring volume to 1.0L. Mix thoroughly. Gently heat while stirring and bring to boiling. Do not autoclave. Cool to 45–50°C. Pour into sterile Petri dishes in 20.0mL volumes. Allow the surface of the plates to dry before inoculation.

Use: For the isolation and cultivation of *Yersinia enterocolitica* from foods.

SSL Agar

Composition per liter:

Agar	2.5g
$CaCl_2 \cdot 2H_2O$	1.0g
Gelatin	1.0g
KNO_3	1.0g
$MgSO_4 \cdot 7H_2O$	1.0g
NaCl	1.0g
Pancreatic digest of casein	1.0g
Yeast extract	1.0g
Sodium glycerophosphate	0.1g
Cyanocobalamin	1.0μg
Trace element solution	1.0mL

pH 7.5 ± 0.2 at 25°C

Trace Element Solution:

Composition per liter:

Disodium EDTA	8.0g
$MnCl_2 \cdot 4H_2O$	0.1g
$CoCl_2 \cdot 6H_2O$	0.02g
KBr	0.02g
KI	0.02g
$ZnCl_2$	0.02g
$CuSO_4$	0.01g
H_3BO_3	0.01g
$Na_2MoO_4 \cdot 2H_2O$	0.01g
LiCl	5.0mg
$SnCl_2 \cdot 2H_2O$	5.0mg

Preparation of Trace Element Solution: Add components to distilled/deionized water and bring volume to 1.0L. Mix thoroughly.

Preparation of Medium: Add components to distilled/deionized water and bring volume to 1.0L. Mix thoroughly. Gently heat and bring to boiling. Distribute into tubes or flasks. Autoclave for 15 min at 15 psi pressure–121°C. Pour into sterile Petri dishes or leave in tubes.

Use: For the isolation and cultivation of *Cytophaga* species, *Herpetosiphon* species, *Saprospira* species, and *Flexithrix* species.

ST Holding Medium (m–ST Holding Medium)

Composition per liter:

KH_2PO_4	3.0g
Tris(hydroxymethyl)aminomethane buffer	3.0g
Sulfanilamide	1.5g
$NaH_2PO_4 \cdot H_2O$	0.1g
Ethanol (95% solution)	10.0mL

pH 8.6 ± 0.2 at 25°C

Preparation of Medium: Dissolve the sulfanilamide in the ethanol. Add all components to distilled/deionized water and bring volume to 1.0L. Mix thoroughly. Autoclave for 15 min at 15 psi pressure–121°C. Distribute in 1.8mL volumes to sterile Petri dishes with tight-fitting lids and an absorbent filter.

Use: For the cultivation and enumeration of coliform bacteria by the delayed-incubation total coliform procedure. For use as a holding or transport medium to keep coliform bacteria viable between sampling and laboratory culture.

Stan 4 Agar

Composition per liter:

Solution B	650.0mL
Solution A	350.0mL

Solution A:

Composition per 350mL:

$CaCl_2 \cdot 2H_2O$	1.0g
KNO_3	1.0g
$MgSO_4 \cdot 7H_2O$	1.0g
Trace element solution	1.0mL

Preparation of Solution A: Add components to distilled/deionized water and bring volume to 350.0mL. Mix thoroughly. Gently heat and bring to boiling. Autoclave for 15 min at 15 psi pressure–121°C. Cool to 45–50°C.

Trace Element Solution:

Composition per liter:

EDTA	8.0g
$MnCl_2 \cdot 4H_2O$	0.1g
$CoCl_2$	0.02g
KBr	0.02g
$ZnCl_2$	0.02g
$CuSO_4$	0.01g
H_3BO_3	0.01g
$NaMoO_4 \cdot 2H_2O$	0.01g
$BaCl_2$	5.0mg
LiCl	5.0mg
$SnCl_2 \cdot 2H_2O$	5.0mg

Preparation of Trace Element Solution: Add components to distilled/deionized water and bring volume to 1.0L. Mix thoroughly.

Solution B:

Composition per 650mL:

Agar	10.0g
K_2HPO_4	1.0g

Preparation of Solution B: Add components to distilled/deionized water and bring volume to 650.0mL. Mix thoroughly. Gently heat and bring to boiling. Autoclave for 15 min at 15 psi pressure–121°C. Cool to 45–50°C.

Preparation of Medium: Aseptically combine 350.0mL of cooled, sterile solution A and 650.0mL of cooled sterile solution B. Mix thoroughly. Pour into sterile Petri dishes or distribute into sterile tubes.

Use: For the cultivation of myxobacteria.

Stan 5 Agar

Composition per liter:

Solution B	650.0mL
Solution A	350.0mL

Solution A:

Composition per 350mL:

$CaCl_2 \cdot 2H_2O$	1.0g
$(NH_4)_2SO_4$	1.0g
$MgSO_4 \cdot 7H_2O$	1.0g
Trace element solution	1.0mL

Preparation of Solution A: Add components to distilled/deionized water and bring volume to 350.0mL. Mix thoroughly. Gently heat and bring to boiling. Autoclave for 15 min at 15 psi pressure–121°C. Cool to 45–50°C.

Trace Element Solution:

Composition per liter:

EDTA	8.0g
$MnCl_2 \cdot 4H_2O$	0.1g
$CoCl_2$	0.02g
KBr	0.02g
$ZnCl_2$	0.02g
$CuSO_4$	0.01g
H_3BO_3	0.01g
$NaMoO_4 \cdot 2H_2O$	0.01g
$BaCl_2$	5.0mg
LiCl	5.0mg
$SnCl_2 \cdot 2H_2O$	5.0mg

Preparation of Trace Element Solution: Add components to distilled/deionized water and bring volume to 1.0L. Mix thoroughly.

Solution B:

Composition per 650mL:

Agar	10.0g
K_2HPO_4	1.0g

Preparation of Solution B: Add components to distilled/deionized water and bring volume to 650.0mL. Mix thoroughly. Gently heat and bring to boiling. Autoclave for 15 min at 15 psi pressure–121°C. Cool to 45–50°C.

Preparation of Medium: Aseptically combine 350.0mL of cooled, sterile solution A and 650.0mL of cooled sterile solution B. Mix thoroughly. Pour into sterile Petri dishes or distribute into sterile tubes.

Use: For the cultivation of myxobacteria.

Stan 6 Agar

Composition per liter:

Agar	10.0g
$CaCl_2 \cdot 2H_2O$	1.0g
K_2HPO_4	1.0g
$MgSO_4 \cdot 7H_2O$	1.0g
$(NH_4)_2SO_4$	1.0g
$FeCl_3$	0.2g
$MnSO_4 \cdot 7H_2O$	0.1g
Yeast extract	0.02g
Trace element solution	1.0mL

Trace Element Solution:

Composition per liter:

Disodium EDTA	8.0g
$MnCl_2 \cdot 4H_2O$	0.1g
$CoCl_2 \cdot 6H_2O$	0.02g
KBr	0.02g
KI	0.02g
$ZnCl_2$	0.02g
$CuSO_4$	0.01g
H_3BO_3	0.01g
$Na_2MoO_4 \cdot 2H_2O$	0.01g
LiCl	5.0mg
$SnCl_2 \cdot 2H_2O$	5.0mg

Preparation of Trace Element Solution: Add components to distilled/deionized water and bring volume to 1.0L. Mix thoroughly.

Preparation of Medium: Add components to distilled/deionized water and bring volume to 1.0L. Mix thoroughly. Gently heat and bring to boiling. Distribute into tubes or flasks. Autoclave for 15 min at 15 psi pressure–121°C. Pour into sterile Petri dishes or leave in tubes.

Use: For the isolation and cultivation of *Cytophaga* species, *Herpetosiphon* species, *Saprospira* species, and *Flexithrix* species.

Stan 5 Mineral Medium

Composition per liter:

K_2HPO_4	1.0g
$(NH_4)_2SO_4$	1.0g
$MgSO_4 \cdot 7H_2O$	0.2g
$CaCl_2 \cdot 2H_2O$	0.1g
$FeCl_3$	0.02g

pH 7.0–7.5 at 25°C

Preparation of Medium: Add components to distilled/deionized water and bring volume to 1.0L. Mix thoroughly. Distribute into tubes or flasks. Autoclave for 15 min at 15 psi pressure–121°C.

Use: For the isolation and cultivation of *Cytophaga* species, *Herpetosiphon* species, *Saprospira* species, and *Flexithrix* species.

Standard Agar with Methanol and Yeast Extract

Composition per liter:

Base solution	982.0mL
Methanol	10.0mL
Solution B	5.0mL
Solution A	1.0mL
Solution C	1.0mL
Solution D	1.0mL

pH 4.0–4.5 at 25°C

Base Solution:

Composition per 982mL:

Agar	20.0g
Yeast extract	0.5g

Preparation of Base Solution: Add components to distilled/deionized water and bring volume to 982.0mL. Mix thoroughly. Gently heat and bring to boiling. Autoclave for 15 min at 15 psi pressure–121°C. Cool to 45–50°C.

Methanol:

Composition per 10mL:

Methanol	10.0mL

Preparation of Methanol: Filter sterilize.

Solution A:

Composition per liter:

K_2HPO_4	87.09g
KH_2PO_4	68.05g

Preparation of Solution A: Add components to distilled/deionized water and bring volume to 1.0L. Mix thoroughly. Filter sterilize.

Solution B:

Composition per liter:

NH_4Cl	152.28g

Preparation of Solution B: Add NH_4Cl to distilled/deionized water and bring volume to 1.0L. Mix thoroughly. Filter sterilize.

Solution C:

Composition per liter:

$CaCl_2 \cdot 6H_2O$	5.47g

Preparation of Solution C: Add $CaCl_2 \cdot 6H_2O$ to distilled/deionized water and bring volume to 1.0L. Mix thoroughly. Filter sterilize.

Solution D:

Composition per liter:

$MgSO_4 \cdot 7H_2O$	71.2g
$FeSO_4 \cdot 7H_2O$	5.0g
$MnSO_4 \cdot 4H_2O$	0.81g
$CuSO_4 \cdot 5H_2O$	0.79g
$ZnSO_4 \cdot 7H_2O$	0.44g
$Na_2MoO_4 \cdot 2H_2O$	0.25g

Preparation of Solution D: Add components to distilled/deionized water and bring volume to 1.0L. Mix thoroughly. Filter sterilize.

Preparation of Medium: To 982.0mL of cooled sterile Base solution, aseptically add 10.0mL of sterile methanol, 1.0mL of sterile solution A, 5.0mL of sterile solution B, 1.0mL of sterile solution C, and 1.0mL of sterile solution D. Mix thoroughly. Adjust pH to 4.0–4.5 if necessary. Pour into sterile Petri dishes or distribute into sterile tubes.

Use: For the cultivation and maintenance of *Acetobacter methanolicus*.

Standard Methods Agar (Tryptone Glucose Yeast Agar) (Plate Count Agar)

Composition per liter:

Agar	15.0g
Pancretic digest of casein	5.0g
Yeast extract	2.5g
Glucose	1.0g

pH 7.0 ± 0.1 at 25°C

Source: Available as a premixed powder from BBL Microbiology Systems and Difco Laboratories.

Preparation of Medium: Add components to distilled/deionized water and bring volume to 1.0L. Mix thoroughly. Gently heat and bring to boiling. Distribute into tubes or flasks. Autoclave for 15 min at 15 psi pressure–121°C. Pour into sterile Petri dishes or leave in tubes.

Use: For the cultivation and enumeration by microbial plate counts of microorganisms isolated from water and other specimens.

Standard Methods Agar with Lecithin and Polysorbate 80

Composition per liter:

Agar	15.0g
Pancreatic digest of casein	5.0g
Polysorbate 80	5.0g
Yeast extract	2.5g
Glucose	1.0g
Lecithin	0.7g

pH 7.0 ± 0.2 at 25°C

Source: This medium is available as a premixed powder from BBL Microbiology Systems.

Preparation of Medium: Add components to distilled/deionized water and bring volume to 1.0L. Mix thoroughly. Gently heat and bring to boiling. Distribute into tubes or flasks. Autoclave for 15 min at 15 psi pressure–121°C. Pour into sterile Petri dishes or leave in tubes.

Use: For determination of the sterility of surfaces.

Standard Methods Broth (m–Standard Methods Broth) (Tryptone Glucose Yeast Broth) (m–Plate Count Broth)

Composition per liter:

Pancreatic digest of casein	10.0g
Yeast extract	5.0g
Glucose	2.0g

pH 7.0 ± 0.2 at 25°C

Source: Available as a premixed powder from BBL Microbiology Systems and Difco Laboratories.

Preparation of Medium: Add components to distilled/deionized water and bring volume to 1.0L. Mix thoroughly. Distribute into tubes or flasks. Autoclave for 15 min at 15 psi pressure–121°C.

Use: For the enumeration of total number of microorganisms by the membrane filter method.

Standard II Nutrient Agar

Composition per liter:

Agar	13.0g
Tryptose	7.0g
NaCl	5.0g

pH 7.5 ± 0.2 at 25°C

Source: This medium is available as a premixed powder from Difco Laboratories.

Preparation of Medium: Add components to distilled/deionized water and bring volume to 1.0L. Mix thoroughly. Gently heat and bring to boiling. Distribute into tubes or flasks. Autoclave for 15 min at 15 psi pressure–121°C. Pour into sterile Petri dishes or leave in tubes.

Use: For the cultivation of nonfastidious microorganisms. For the maintenance of cultures of a wide variety of nonfastidious bacteria. May also be used as a base for blood and other enrichments for the cultivation of fastidious microorganisms. May be used to determine indole production.

Stanier's Basal Medium with Pyridoxine and Yeast Extract

Composition per liter:

KH_2PO_4	2.78g
Na_2HPO_4	2.78g
$(NH_4)_2SO_4$	1.0g
Yeast extract	0.2g
Hutner's mineral base	20.0mL
Pyridoxine solution	10.0mL

pH 6.8 ± 0.2 at 25°C

Pyridoxine Solution:

Composition per 10mL:

Pyridoxine	2.0g

Preparation of Pyridoxine Solution: Add pyridoxine to distille/deionized water and bring volume to 10.0mL. Mix thoroughly. Filter sterilize.

Hutner's Mineral Base:

Composition per liter:

$MgSO_4 \cdot 7H_2O$	29.7g
Nitrilotriacetic acid	10.0g
$CaCl_2 \cdot 2H_2O$	3.34g
$FeSO_4 \cdot 7H_2O$	99.0mg
$(NH_4)_2MoO_4$	9.25mg
Metals "44"	50.0mL

Preparation of Hutner's Mineral Base: Add nitrilotriacetic acid to 500.0mL of distilled/deionized water. Dissolve by adjusting pH to 6.5 with KOH. Add remaining components. Readjust pH to 7.2 with H_2SO_4 or KOH. Add distilled/deionized water to 1.0L. Store at 5°C.

Metals "44":

Composition per100mL:

$ZnSO_4 \cdot 7H_2O$	1.1g
$FeSO_4 \cdot 7H_2O$	0.5g
EDTA	0.25g
$MnSO_4 \cdot 7H_2O$	0.154g
$CuSO_4 \cdot 5H_2O$	0.04g
$Co(NO_3)_2 \cdot 6H_2O$	0.025g
$Na_2B_4O_7 \cdot 10H_2O$	0.018g

Preparation of Metals "44": Add a few drops of H_2SO_4 to distilled/deionized water to inhibit precipitate formation. Add components to acidified distilled/deionized water and bring volume to 100.0mL. Mix thoroughly.

Preparation of Medium: Add components, except pyridoxine solution, to distilled/deionized water and bring volume to 990.0mL. Mix thoroughly. Adjust pH to 6.8 with 1*N* KOH. Autoclave for 15 min at 15 psi pressure–121°C. Cool to 45–50°C. Aseptically add sterile pyridoxine solution. Mix thoroughly. Aseptically distribute into sterile tubes or flasks.

Use: For the cultivation and maintenance of *Pseudomonas* species.

Stanier's Basal Medium with Trichlorophenoxyacetate

Composition per liter:

KH_2PO_4	2.78g
Na_2HPO_4	2.78g
$(NH_4)_2SO_4$	1.0g
2,4,5-trichlorophenoxyacetate	1.0g
Hutner's mineral base	20.0mL

Hutner's Mineral Base:

Composition per liter:

$MgSO_4 \cdot 7H_2O$	29.7g
Nitrilotriacetic acid	10.0g
$CaCl_2 \cdot 2H_2O$	3.34g
$FeSO_4 \cdot 7H_2O$	0.10g
$(NH_4)_2MoO_4$	9.25mg
Metals "44"	50.0mL

Preparation of Hutner's Mineral Base: Add nitrilotriacetic acid to 500.0mL of distilled/deionized water. Dissolve by adjusting pH to 6.5 with KOH. Add remaining components. Readjust pH to 7.2 with H_2SO_4 or KOH. Add distilled/deionized water to 1.0L. Store at 5°C.

Metals "44":

Composition per 100mL:

$ZnSO_4 \cdot 7H_2O$	1.1g
$FeSO_4 \cdot 7H_2O$	0.5g
EDTA	0.25g
$MnSO_4 \cdot 7H_2O$	0.154g
$CuSO_4 \cdot 5H_2O$	0.04g
$Co(NO_3)_2 \cdot 6H_2O$	0.025g
$Na_2B_4O_7 \cdot 10H_2O$	0.018g

Preparation of Metals "44": Add a few drops of H_2SO_4 to distilled/deionized water to inhibit precipitate formation. Add components to acidified distilled/deionized water and bring volume to 100.0mL. Mix thoroughly.

Preparation of Medium: Add components to distilled/deionized water and bring volume to 1.0L. Mix thoroughly. Distribute into tubes or flasks. Autoclave for 15 min at 15 psi pressure–121°C.

Use: For the cultivation and maintenance of *Pseudomonas cepacia*.

Starch Agar

Composition per liter:

Starch, soluble	20.0g
Agar	10.0g
$NaNO_3$	2.5g
K_2HPO_4	1.0g
$MgSO_4 \cdot 7H_2O$	0.6g
$CaCl_2 \cdot 2H_2O$	0.1g
NaCl	0.1g
$FeCl_3$	1mg

pH 7.2 ± 0.2 at 25°C

Preparation of Medium: Add components to distilled/deionized water and bring volume to 1.0L. Mix thoroughly. Gently heat and bring to boiling. Distribute into tubes or flasks. Autoclave for 15 min at 15 psi pressure–121°C. Pour into sterile Petri dishes or leave in tubes.

Use: For the cultivation of myxobacteria.

Starch Agar

Composition per liter:

Agar	15.0g
Potato starch	10.0g
Pancreatic digest of gelatin	5.0g
Beef extract	3.0g

pH 6.8 ± 0.2 at 25°C

Preparation of Medium: Add components, except potato starch, to distilled/deionized water and bring volume to 500.0mL. Mix thoroughly. Gently heat and bring to boiling. Add potato starch to distilled/deionized water and bring volume to 250.0mL. Gently heat and bring to boiling. Combine the two solutions and bring the volume to 1.0L with distilled/deionized water. Autoclave for 15 min at 15 psi pressure–121°C. Pour into sterile Petri dishes.

Use: For the cultivation and differentiation of aerobic *Actinomycetes* species based on amylase production. After incubation, starch hydrolysis is determined by the addition of Gram's or Lugol's iodine solution. Organisms which produce amylase appear as colonies surrounded by a clear zone.

Starch Agar

Composition per liter:

Agar	12.0g
Soluble starch	10.0g
Beef extract	3.0g

pH 7.5 ± 0.2 at 25°C

Source: This medium is available as a premixed powder from Difco Laboratories.

Preparation of Medium: Add components to distilled/deionized water and bring volume to 1.0L. Mix thoroughly. Gently heat and bring to boiling. Distribute into tubes or flasks. Autoclave for 15 min at 15 psi pressure–121°C. Pour into sterile Petri dishes.

Use: For the cultivation and differentiation of a variety of microorganisms based on amylase production. After incubation, starch hydrolysis is determined by the addition of Gram's or Lugol's iodine solution. Organisms which produce amylase appear as colonies surrounded by a clear zone.

Starch Agar Medium for *Pseudomonas*

Composition per liter:

Agar	15.0g
Peptone	5.0g
Yeast extract	5.0g
Soluble starch	3.0g

pH 7.0 ± 0.2 at 25°C

Preparation of Medium: Add components to distilled/deionized water and bring volume to 1.0L. Mix thoroughly. Gently heat and bring to boiling. Distribute into tubes or flasks. Autoclave for 15 min at 15 psi pressure–121°C. Pour into sterile Petri dishes or leave in tubes.

Use: For the cultivation and maintenance of *Pseudomonas* species and *Erwinia herbicola*.

Starch Casein Agar

Composition per liter:

Agar	15.0g
Soluble starch	10.0g
K_2HPO_4	2.0g
KNO_3	2.0g
NaCl	2.0g
Casein	0.3g
$MgSO_4 \cdot 7H_2O$	0.05g
$CaCO_3$	0.02g
$FeSO_4 \cdot 7H_2O$	0.01g

Preparation of Medium: Add components to distilled/deionized water and bring volume to 1.0L. Mix thoroughly. Gently heat and bring to boiling. Distribute into tubes or flasks. For bottom layers, distribute into tubes in 15.0mL volumes. For top layers, distribute into tubes in 17.0mL volumes. Autoclave for 15 min at 15 psi pressure–121°C. Pour into sterile Petri dishes or leave in tubes.

Use: For the cultivation and enumeration of *Actinomycetes* species from water and soil samples by the double-layer agar technique.

Starch Casein KNO_3 Agar

Composition per liter:

Agar	18.0g
Starch	10.0g
KNO_3	2.0g
K_2HPO_4	2.0g
NaCl	2.0g
Casein	0.3g
$MgSO_4 \cdot 7H_2O$	0.05g
$CaCO_3$	0.02g
$FeSO_4 \cdot 7H_2O$	0.01g

Preparation of Medium: Add components to distilled/deionized water and bring volume to 1.0L. Mix thoroughly. Gently heat and bring to boiling. Distribute into tubes or flasks. Autoclave for 15 min at 15 psi pressure–121°C. Pour into sterile Petri dishes or leave in tubes.

Use: For the selective isolation and cultivation of streptomycetes.

Starch Fermentation Broth

Composition per 225.2mL:

Heart infusion broth	5.0mL
Bromcresol purple solution	0.2mL
Starch solution	20.0mL

pH 7.8 ± 0.2 at 25°C

Heart Infusion Broth:

Composition per liter:

Beef heart, infusion from	500.0g
Tryptose	10.0g
NaCl	5.0g

pH 7.4 ± 0.2 at 25°C

Source: Heart infusion broth is available as a premixed powder from Difco Laboratories.

Preparation of Heart Infusion Broth: Add components to distilled/deionized water and bring volume to 1.0L. Mix thoroughly. Distribute into tubes or flasks. Autoclave for 15 min at 15 psi pressure–121°C.

Bromcresol Purple Solution:

Composition per 10mL:

Bromcresol purple	0.1g
Ethanol (95% solution)	10.0mL

Preparation of Bromcresol Purple Solution: Add bromcresol purple to 10.0mL ethanol. Mix thoroughly.

Starch Solution:

Composition per 20mL:

Starch	0.4g

Preparation of Starch Solution: Add starch to distilled/deionized water and bring volume to 20.0mL. Mix thoroughly. Gently heat while stirring and bring to boiling.

Preparation of Medium: Combine 5.0mL heart infusion, 0.2mL of bromcresol purple solution, 200.0mL of distilled/deionized water and 20.0mL of starch solution. Mix thoroughly. Distribute into tubes or flasks. Autoclave for 15 min at 15 psi pressure–121°C. Pour into sterile Petri dishes or leave in tubes.

Use: For the cultivation of *Corynebacterium* species.

Starch Hydrolysis Agar

Composition per liter:

Beef heart, infusion from	500.0g
Soluble starch	20.0g
Agar	15.0g
Tryptose	10.0g
NaCl	5.0g

pH 7.4 ± 0.2 at 25°C

Preparation of Medium: Add components to distilled/deionized water and bring volume to 1.0L. Mix thoroughly. Gently heat and bring to boiling. Distribute into tubes or flasks. Autoclave for 15 min at 15 psi pressure–121°C. Pour into sterile Petri dishes or leave in tubes.

Use: For the cultivation and differentiation of a variety of microorganisms based on amylase production. After incubation, starch hydrolysis is determined by the addition of Gram's or Lugol's iodine solution. Organisms that produce amylase appear as colonies surrounded by a clear zone.

Starkey's Medium C, Modified

Composition per liter:

Sodium lactate	3.5g
$MgSO_4 \cdot 7H_2O$	2.0g
Na_2SO_4	1.0g
NH_4Cl	1.0g
Yeast extract	1.0g
KH_2PO_4	0.5g
$CaCl_2 \cdot 2H_2O$	0.1g
Ferrous ammonium sulfate solution	50.0mL
Cysteine·HCl·H_2O solution	10.0mL

pH 7.5 ± 0.2 at 25°C

Ferrous Ammonium Sulfate Solution:

Composition per 100mL:

$Fe(NH_4)_2(SO_4)_2 \cdot 6H_2O$	1.0g

Preparation of Ferrous Ammonium Sulfate Solution: Add $Fe(NH_4)_2(SO_4)_2 \cdot 6H_2O$ to distilled/deionized water and bring volume to 100.0mL. Mix thoroughly. Filter sterilize.

Cysteine·HCl·H_2O Solution:

Composition per 10mL:

Cysteine·HCl·H_2O	0.75g

Preparation of Cysteine·HCl·H_2O Solution: Add cysteine·HCl·H_2O to distilled/deionized water and bring volume to 10.0mL. Mix thoroughly. Filter sterilize.

Preparation of Medium: Add components, except ferrous ammonium sulfate solution and cysteine·HCl·H_2O solution, to tap water and bring volume to 940.0mL. Mix thoroughly. Gently heat and bring to boiling. Autoclave for 15 min at 15 psi pressure–121°C. Cool to 45–50°C. Aseptically add 50.0mL of sterile ferrous ammonium sulfate solution and 10.0mL of sterile cysteine·HCl·H_2O solution. Mix thoroughly. Adjust pH to 7.5 with filter-sterilized 2*N* NaOH. Pour into sterile Petri dishes or distribute into sterile tubes.

Use: For the cultivation and maintenance of *Desulfotomaculum* species and *Desulfovibrio* species.

Starkey's Medium C, Modified with Salt

Composition per liter:

NaCl	25.0g
Sodium lactate	3.5g
$MgSO_4 \cdot 7H_2O$	2.0g
Na_2SO_4	1.0g
NH_4Cl	1.0g
Yeast extract	1.0g
KH_2PO_4	0.5g
$CaCl_2 \cdot 2H_2O$	0.1g
Ferrous ammonium sulfate solution	50.0mL
Cysteine·HCl·H_2O solution	10.0mL

pH 7.5 ± 0.2 at 25°C

Ferrous Ammonium Sulfate Solution:

Composition per 100mL:

$Fe(NH_4)_2(SO_4)_2 \cdot 6H_2O$	1.0g

Preparation of Ferrous Ammonium Sulfate Solution: Add $Fe(NH_4)_2(SO_4)_2 \cdot 6H_2O$ to distilled/deionized water and bring volume to 100.0mL. Mix thoroughly. Filter sterilize.

Cysteine·HCl·H$_2$O Solution:
Composition per 10mL:

Cysteine·HCl·H$_2$O	0.75g

Preparation of Cysteine·HCl·H$_2$O Solution: Add cysteine·HCl·H$_2$O to distilled/deionized water and bring volume to 10.0mL. Mix thoroughly. Filter sterilize.

Preparation of Medium: Add components, except ferrous ammonium sulfate solution and cysteine·HCl·H$_2$O solution, to tap water and bring volume to 940.0mL. Mix thoroughly. Gently heat and bring to boiling. Autoclave for 15 min at 15 psi pressure–121°C. Cool to 45–50°C. Aseptically add 50.0mL of sterile ferrous ammonium sulfate solution and 10.0mL of sterile cysteine·HCl·H$_2$O solution. Mix thoroughly. Adjust pH to 7.5 with filter-sterilized 2*N* NaOH. Pour into sterile Petri dishes or distribute into sterile tubes.

Use: For the cultivation and maintenance of halophilic *Desulfovibrio* species.

STL Broth

Composition per liter:

Casamino acids	1.0g
Glucose	1.0g
Sodium glutamate	1.0g
$CaCl_2 \cdot 2H_2O$	0.1g
KNO_3	0.1g
$MgSO_4 \cdot 7H_2O$	0.1g
Sodium glycerophosphate	0.1g
Thiamine	1.0mg
Vitamin B_{12}	1.0μg
Trace element solution	1.0mL

pH 7.5 ± 0.2 at 25°C

Trace Element Solution:
Composition per liter:

Disodium EDTA	8.0g
$MnCl_2 \cdot 4H_2O$	0.1g
$CoCl_2 \cdot 6H_2O$	0.02g
KBr	0.02g
KI	0.02g
$ZnCl_2$	0.02g
$CuSO_4$	0.01g
H_3BO_3	0.01g
$Na_2MoO_4 \cdot 2H_2O$	0.01g
LiCl	5.0mg
$SnCl_2 \cdot 2H_2O$	5.0mg

Preparation of Trace Element Solution: Add components to distilled/deionized water and bring volume to 1.0L. Mix thoroughly.

Preparation of Medium: Add components to distilled/deionized water and bring volume to 1.0L. Mix thoroughly. Gently heat and bring to boiling. Distribute into tubes or flasks. Autoclave for 15 min at 15 psi pressure–121°C. Pour into sterile Petri dishes or leave in tubes.

Use: For the isolation and cultivation of *Cytophaga* species, *Herpetosiphon* species, *Saprospira* species, and *Flexithrix* species.

Stokes Agar

Composition per liter:

Agar	12.5g
Glucose	1.0g
Peptone	1.0g
$MgSO_4 \cdot 7H_2O$	0.2g
$CaCl_2$	0.05g
$FeCl_3 \cdot 6H_2O$	0.01g

Preparation of Medium: Add components to tap water and bring volume to 1.0L. Mix thoroughly. Gently heat and bring to boiling. Distribute into tubes or flasks. Autoclave for 15 min at 15 psi pressure–121°C. Pour into sterile Petri dishes or leave in tubes.

Use: For the isolation and cultivation of *Sphaerotilus natans*.

Streptomycete Medium

Composition per liter:

Solution B	500.0mL
Solution A	400.0mL
Solution C	100.0mL

Solution A:
Composition per 400mL:

Glucose	20.0g
Agar	4.0g
Yeast extract	1.2g
$MgSO_4 \cdot 7H_2O$	0.25g
Bromcresol Purple	0.012g

Preparation of Solution A: Add components to distilled/deionized water and bring volume to 400.0mL. Mix thoroughly. Autoclave for 15 min at 15 psi pressure–121°C. Cool to 45–50°C.

Solution B:
Composition per 500mL:

$Na_2HPO_4 \cdot 2H_2O$	534.0mg
KH_2PO_4	272.0mg

Preparation of Solution B: Add components to distilled/deionized water and bring volume to 500.0mL. Mix thoroughly. Autoclave for 15 min at 15 psi pressure–121°C. Cool to 45–50°C.

Solution C:

Composition per 100mL:

$CaCO_3$	1.0g

Preparation of Solution C: Add $CaCO_3$ to distilled/deionized water and bring volume to 100.0mL. Mix thoroughly.

Preparation of Medium: Distribute solution C into test tubes in 0.2mL volumes. Autoclave for 15 min at 15 psi pressure–121°C. Cool to 45–50°C. Combine cooled, sterile solution A and cooled, sterile solution B. Mix thoroughly. Add 1.8mL of solution A-B to each test tube containing sterile solution C. Mix thoroughly to distribute the $CaCO_3$. Cool tubes rapidly in an ice water bath.

Use: For the cultivation and differentiation of streptomycetes based on their formation of organic acids. Bacteria that form organic acids turn the medium yellow and dissolve the $CaCO_3$.

Streptomycete Medium

Composition per liter:

Glycerol	5.0g
Agar	4.0g
NaCl	2.0g
KNO_3	1.0g
$Na_2HPO_4 \cdot 2H_2O$	0.534g
$MgSO_4 \cdot 7H_2O$	0.5g
KH_2PO_4	0.272g
Trace elements solution	1.0mL

pH 6.8 ± 0.2 at 25°C

Trace Elements Solution:

Composition per 100mL:

$FeSO_4 \cdot 7H_2O$	0.1g
$MnCl_2 \cdot 4H_2O$	0.1g
$ZnSO_4 \cdot 7H_2O$	0.1g

Preparation of Trace Elements Solution: Add components to distilled/deionized water and bring volume to 100.0mL. Mix thoroughly.

Preparation of Medium: Add components to distilled/deionized water and bring volume to 1.0L. Mix thoroughly. Gently heat and bring to boiling. Distribute into tubes in 1.0mL volumes. Autoclave for 15 min at 15 psi pressure–121°C.

Use: For the cultivation and differentiation of streptomycetes based on their reduction of nitrate to nitrite. Bacteria that reduce nitrate to nitrite for a red color after the addition of Griess-Ilosvay reagent.

Streptomycete Medium

Composition per liter:

Agar	12.0g
NaCl	5.0g
$Na_2HPO_4 \cdot 2H_2O$	1.98g
KH_2PO_4	1.51g
Glucose	1.0g
Pancreatic digest of casein	1.0g
$MgSO_4 \cdot 7H_2O$	0.5g
Phenol red	0.012g
Urea solution	100.0mL

pH 6.8 ± 0.2 at 25°C

Urea Solution:

Composition per 100mL:

Urea	20.0g

Preparation of Urea Solution: Add urea to distilled/deionized water and bring volume to 100.0mL. Mix thoroughly. Filter sterilize.

Preparation of Medium: Add components, except urea solution, to distilled/deionized water and bring volume to 900.0mL. Mix thoroughly. Gently heat and bring to boiling. Autoclave for 15 min at 15 psi pressure–121°C. Cool to 45–50°C. Aseptically add 100.0mL of sterile urea solution. Mix thoroughly. Aseptically distribute into sterile tubes. Allow tubes to cool in a slanted position.

Use: For the cultivation and differentiation of streptomycetes based on their ability to produce urease.

Streptomycete Medium

Composition per liter:

Sodium hippurate	10.0g
Na_2HPO_4	5.0g
Glucose	2.0g
Meat extract	2.0g
Peptone	2.0g
Yeast extract	2.0g

pH 7.0 ± 0.2 at 25°C

Preparation of Medium: Add components to distilled/deionized water and bring volume to 1.0L. Mix thoroughly. Distribute into tubes in 3.0mL volumes. Autoclave for 15 min at 15 psi pressure–121°C.

Use: For the cultivation and differentiation of streptomycetes based on their ability to hydrolyze hippurate.

Streptomycin Terramycin® Malt Extract Agar

Composition per liter:

Malt extract	30.0g
Agar	15.0g
Peptone	5.0g

Streptomycin solution	100.0mL
Terramycin solution	100.0mL

pH 5.4 ± 0.2 at 25°C

Streptomycin Solution:
Composition per 100mL:

Streptomycin	0.07g

Preparation of Streptomycin Solution: Add streptomycin to distilled/deionized water and bring volume to 100.0mL. Mix thoroughly. Filter sterilize.

Terramycin Solution:
Composition per 10mL:

Terramycin	0.07g

Preparation of Terramycin Solution: Add terramycin to distilled/deionized water and bring volume to 100.0mL. Mix thoroughly. Filter sterilize.

Preparation of Medium: Add components, except streptomycin solution and terramycin solution, to distilled/deionized water and bring volume to 800.0mL. Mix thoroughly. Gently heat and bring to boiling. Autoclave for 15 min at 15 psi pressure–121°C. Cool to 45–50°C. Aseptically add 100.0mL of sterile streptomycin solution and 100.0mL of sterile terramycin solution. Mix thoroughly. Pour into sterile Petri dishes in 20.0mL volumes.

Use: For the cultivation and enumeration of fungi isolated from sewage and polluted waters.

Stuart *Leptospira* Broth, Modified

Composition per liter:

NaCl	1.93g
Na_2HPO_4	0.66g
NH_4Cl	0.34g
$MgCl_2 \cdot 6H_2O$	0.19g
L-Asparagine	0.13g
KH_2PO_4	0.08g
Glycerol	5.0mL
Rabbit serum, inactivated at 56°C, 30 min	100.0mL

pH 7.4 ± 0.2 at 25°C

Preparation of Medium: Add each component, except rabbit serum, to distilled/deionized water in separate flasks and bring each volume to 100.0mL. Mix thoroughly. Combine the seven solutions, except the rabbit serum. Mix thoroughly. Gently heat and bring to boiling. Autoclave for 15 min at 15 psi pressure–121°C. Cool to 45–50°C. Aseptically add sterile rabbit serum. Mix thoroughly. Aseptically distribute into sterile tubes or flasks.

Use: For the cultivation of *Leptospira* species.

Stuart Medium Base

Composition per 1100mL:

NaCl	1.8g
Na_2HPO_4	0.67g
$MgCl_2 \cdot 6H_2O$	0.41g
NH_4Cl	0.27g
Asparagine	0.13g
KH_2PO_4	0.09g
Phenol red	0.01g
Glycerol	5.0mL
Leptospira enrichment (Difco)	100.0mL

pH 7.6 ± 0.2 at 25°C

Source: This medium is available as a premixed powder from Difco Laboratories. *Leptospira* enrichment contains rabbit serum and hemoglobin and is available from Difco Laboratories.

Preparation of Medium: Add components, except glycerol and *Leptospira* enrichment, to distilled/deionized water and bring volume to 995.0mL. Mix thoroughly. Add glycerol. Mix thoroughly. Autoclave for 15 min at 15 psi pressure–121°C. Cool to 45–50°C. Aseptically add *Leptospira* enrichment. Mix thoroughly. Aseptically distribute into sterile screw-capped tubes in 10.0mL volumes.

Use: For the cultivation of *Leptospira* species.

Sucrose Peptone Agar

Composition per liter:

Sucrose	20.0g
Agar	12.0g
Peptone	5.0g
K_2HPO_4	0.5g
$MgSO_4 \cdot 7H_2O$	0.25g

pH 7.2–7.4 at 25°C

Preparation of Medium: Add components to distilled/deionized water and bring volume to 1.0L. Mix thoroughly. Gently heat and bring to boiling. Adjust pH to 7.2–7.4. Distribute into tubes or flasks. Autoclave for 15 min at 15 psi pressure–121°C. Pour into sterile Petri dishes or leave in tubes.

Use: For the cultivation and maintenance of *Pseudomonas solanacearum* and *Xanthomonas albilineans*.

Sucrose Teepol Tellurite Agar (STT Agar)

Composition per liter:

Agar	20.0g
Beef extract	1.0g
Peptone	1.0g
Sucrose	1.0g
NaCl	0.5g

Bromthymol blue (0.2% solution)	2.5mL
Tellurite solution	2.5mL
Sodium lauryl sulfate (Teepol—0.1% solution)	0.2mL

pH 8.0 ± 0.2 at 25°C

Tellurite Solution:

Composition per 100mL:

K_2TeO_3	0.05g

Preparation of Tellurite Solution: Add the K_2TeO_3 to distilled/deionized water and bring the volume to 100.0mL. Mix thoroughly. Filter sterilize. Use freshly prepared solution.

Caution: Potassium tellurite is toxic.

Preparation of Medium: Add components to distilled/deionized water and bring volume to 1.0L. Mix thoroughly. Gently heat and bring to boiling. Do not autoclave. Pour into sterile Petri dishes.

Use: For the selective isolation, cultivation and differentiation of *Vibrio* species based on their ability to ferment sucrose. *Vibrio cholerae* appears as flat yellow colonies. *Vibrio parahaemolyticus* appears as levated green-yellow mucoid colonies.

Sulfate API Broth

Composition per liter:

NaCl	10.0g
Sodium lactate	5.2g
Yeast extract	1.0g
$MgSO_4 \cdot 7H_2O$	0.2g
Ascorbic acid	0.1g
$Fe(NH_4)_2(SO_4)_2 \cdot 6H_2O$	0.1g
K_2HPO_4	0.01g

pH 7.5± 0.2 at 25°C

Source: This medium is available as a premixed powder from Difco Laboratories.

Preparation of Medium: Add the components to distilled/deionized water and bring volume to 1.0L. Mix thoroughly until dissolved. Distribute into tubes in 9.0mL volumes. Autoclave for 10 min at 15 psi pressure–121°C.

Use: For the detection, differentiation and estimation of sulfate-reducing bacteria.

Sulfate-Reducing Bacteria Enrichment Medium

Composition per 1018mL:

Solution 1	970.0mL
Solution 4	30.0mL
Solution 6A, 6B, 6C, 6D or 6E	10.0mL
Solution 5	3.0mL
Solution 2	1.0mL
Solution 3	1.0mL
Solution 7	1.0mL
Solution 8	1.0mL
Solution 9	1.0mL

pH 7.2 ± 0.2 at 25°C

Solution 1:

Composition per 970mL:

Na_2SO_4	3.0g
NaCl	1.2g
$MgCl_2 \cdot 6H_2O$	0.4g
KCl	0.3g
NH_4Cl	0.3g
KH_2PO_4	0.2g
$CaCl_2 \cdot 2H_2O$	0.15g

Preparation of Solution 1: Add components to distilled/deionized water and bring volume to 970.0mL. Mix thoroughly. Autoclave for 30 min at 15 psi pressure–121°C. Cool to 25°C under 90% N_2 + 10% CO_2.

Solution 2:

Composition per liter:

$FeCl_2 \cdot 4H_2O$	1.5g
$CoCl_2 \cdot 6H_2O$	0.12g
$MnCl_2 \cdot 4H_2O$	0.1g
$ZnCl_2$	0.07g
H_3BO_3	0.06g
$Na_2MoO_4 \cdot 2H_2O$	0.025g
$NiCl_2 \cdot 6H_2O$	0.025g
$CuCl_2 \cdot 2H_2O$	0.015g
HCl (25% solution)	6.5mL

Preparation of Solution 2: Add components to distilled/deionized water and bring volume to 1.0L. Mix thoroughly. Autoclave for 15 min at 15 psi pressure–121°C. Cool to 25°C.

Solution 3:

Composition per liter:

NaOH	0.5g
Na_2SeO_3	3.0mg

Preparation of Solution 3: Add components to distilled/deionized water and bring volume to 1.0L. Mix thoroughly. Autoclave for 15 min at 15 psi pressure–121°C. Cool to 25°C.

Solution 4:

Composition per 100mL:

$NaHCO_3$	8.5g

Preparation of Solution 4: Add $NaHCO_3$ to distilled/deionized water and bring volume to 100.0mL. Mix thoroughly. Saturate with 100% CO_2. Filter sterilize. Aseptically add solution to sterile, gas-tight, screw-capped bottles.

Solution 5:

Composition per 100mL:

$Na_2S \cdot 9H_2O$ 12.0g

Preparation of Solution 5: Add $Na_2S \cdot 9H_2O$ to distilled/deionized water and bring volume to 100.0mL. Mix thoroughly. Add solution to gas-tight, screw-capped bottles. Gas under 100% N_2 for 20 min. Close caps tightly. Autoclave for 15 min at 15 psi pressure–121°C. Cool to 25°C.

Solution 6A:

Composition per 100mL:

Sodium acetate·$3H_2O$ 20.0g

Preparation of Solution 6A: Add sodium acetate·$3H_2O$ to distilled/deionized water and bring volume to 100.0mL. Autoclave for 15 min at 15 psi pressure–121°C. Cool to 25°C.

Solution 6B:

Composition per 100mL:

n-Butyric acid 8.0g

Preparation of Solution 6B: Add *n*-butyric acid to distilled/deionized water and bring volume to 100.0mL. Adjust pH to 9.0 with NaOH. Autoclave for 15 min at 15 psi pressure–121°C. Cool to 25°C.

Solution 6C:

Composition per 100mL:

Propionic acid 7.0g

Preparation of Solution 6C: Add propionic acid to 100.0mL of distilled/deionized water. Adjust pH to 9.0 with NaOH. Autoclave for 15 min at 15 psi pressure–121°C. Cool to 25°C.

Solution 6D:

Composition per 100mL:

Benzoic acid 5.0g

Preparation of Solution 6D: Add benzoic acid to distilled/deionized water and bring volume to 100.0mL. Adjust pH to 9.0 with NaOH. Autoclave for 15 min at 15 psi pressure–121°C. Cool to 25°C.

Solution 6E:

Composition per 100mL:

n-Palmitic acid 5.0g
NaOH 0.78g

Preparation of Solution 6E: Add *n*-palmitic acid and NaOH to distilled/deionized water and bring volume to 100.0mL. Heat in a water bath until clear. Autoclave for 15 min at 15 psi pressure–121°C. Cool to 25°C.

Solution 7:

Composition per 100mL:

Thiamine 0.01g
p-Aminobenzoic acid 5.0mg
Vitamin B_{12} 5.0mg
Biotin 1.0mg

Preparation of Solution 7: Add components to distilled/deionized water and bring volume to 100.0mL. Mix thoroughly. Filter sterilize.

Solution 8:

Composition per liter:

Succinic acid 0.6g
Isobutyric acid 0.5g
2-Methylbutyric acid 0.5g
3-Methylbutyric acid 0.5g
Valeric acid 0.5g
Caproic acid 0.2g

Preparation of Solution 8: Add components to distilled/deionized water and bring volume to 100.0mL. Mix thoroughly. Adjust pH to 9.0 with NaOH. Autoclave for 15 min at 15 psi pressure–121°C. Cool to 25°C.

Solution 9:

Composition per 100mL:

$Na_2S_2O_4$ 3.0g

Preparation of Solution 9: Add $Na_2S_2O_4$ to 100.0mL of O_2-free distilled/deionized water. Mix thoroughly. Anaerobically filter sterilize.

Preparation Medium: To 970.0mL of cooled, sterile solution 1, aseptically and anaerobically add 1.0mL of sterile solution 2, 1.0mL of sterile solution 3, 30.0mL of sterile solution 4, and 3.0mL of sterile solution 5. Mix thoroughly. Adjust pH to 7.2 with sterile HCl solution or sterile Na_2CO_3 solution. Aseptically and anaerobically distribute into sterile screw-capped bottles in 100.0mL volumes. Add 1.0mL of solution 6A, 6B, 6C, 6D, or 6E to each bottle containing 100.0mL of basal medium. Add 0.1mL of solution 7, 0.1mL of solution 8, and 0.1mL of solution 9 to each bottle containing 100.0mL of basal medium. Mix thoroughly.

Use: For the isolation, cultivation and enrichment of sulfate-reducing bacteria.

Sulfate-Reducing Bacteria Medium with Lactate

Composition per liter:

Solution 1 .... 980.0mL
Solution 2 .... 10.0mL
Solution 3 .... 10.0mL

pH 7.4 ± 0.2 at 25°C

Solution 1:

Composition per 980mL:

Sodium lactate (70% solution) 3.5g
$MgSO_4 \cdot 7H_2O$ 2.0g

NH_4Cl 1.0g
Na_2SO_4 .... 1.0g
Yeast extract 1.0g
K_2HPO_4 .... 0.5g
$CaCl_2 \cdot 2H_2O$ 0.1g

Preparation of Solution 1: Add components to distilled/deionized water and bring volume to 980.0mL. Mix thoroughly. Autoclave for 15 min at 15 psi pressure–121°C. Cool to 50°C.

Solution 2:
Composition per 10mL:
$FeSO_4 \cdot 7H_2O$ 0.5g

Preparation of Solution 2: Add $FeSO_4 \cdot 7H_2O$ to distilled/deionized water and bring volume to 10.0mL. Mix thoroughly. Autoclave for 15 min at 15 psi pressure–121°C. Cool to 50°C.

Solution 3:
Composition per 10mL:
Ascorbic acid 0.1g
Sodium thioglycollate 0.1g

Preparation of Solution 2: Add components to distilled/deionized water and bring volume to 10.0mL. Mix thoroughly. Autoclave for 15 min at 15 psi pressure–121°C. Cool to 50°C.

Preparation of Medium: Aseptically combine 980.0mL of cooled sterile solution 1, 10.0mL of cooled, sterile solution 2, and 10.0mL of cooled, sterile solution 3. Mix thoroughly. Aseptically distribute into sterile tubes or flasks.

Use: For the enrichment and isolation of sulfate-reducing bacteria.

Sulfate–Reducing Medium

Composition per liter:
Sodium lactate 3.5g
$MgSO_4 \cdot 7H_2O$ 2.0g
Peptone 2.0g
Na_2SO_4 .... 1.5g
Beef extract 1.0g
K_2HPO_4 .... 0.5g
$CaCl_2$ 0.10g
$Fe(NH_4)_2(SO_4)_2 \cdot 6H_2O$ solution 10.0mL
Sodium ascorbate solution 10.0mL
pH 7.5 ± 0.3 at 25°C

$Fe(NH_4)_2(SO_4)_2 \cdot 6H_2O$ Solution:
Composition per 100mL:
$Fe(NH_4)_2(SO_4)_2 \cdot 6H_2O$ 3.92g

Preparation of $Fe(NH_4)_2(SO_4)_2 \cdot 6H_2O$ Solution: Add $Fe(NH_4)_2(SO_4)_2 \cdot 6H_2O$ to distilled/deionized water and bring volume to 100.0mL. Mix thoroughly. Filter sterilize. Use freshly prepared medium.

Sodium Ascorbate Solution:
Composition per 100mL:
Sodium ascorbate 0.050g

Preparation of Sodium Ascorbate Solution: Add sodium ascorbate to distilled/deionized water and bring volume to 100.0mL. Mix thoroughly. Filter sterilize. Use freshly prepared medium.

Preparation of Medium: Add components, except $Fe(NH_4)_2(SO_4)_2 \cdot 6H_2O$ solution and sodium ascorbate solution, to distilled/deionized water and bring volume to 980.0mL. Mix thoroughly. Distribute into screw-capped tubes in 10.0mL volumes. Autoclave for 15 min at 15 psi pressure–121°C. Tubes must be filled to capacity after inoculation so prepare extra medium and sterilize in a screw-capped flask or bottle. Prior to inoculation, aseptically add 0.1mL of freshly prepared sterile $Fe(NH_4)_2(SO_4)_2 \cdot 6H_2O$ solution for each 10.0mL of medium in the tubes. Also aseptically add 0.1mL of freshly prepared sterile sodium ascorbate solution for each 10.0mL of medium in the tubes. Inoculate tubes. Fill tubes to capacity with extra sterile medium. Screw caps tight.

Use: For the isolation, cultivation, and enumeration of iron and sulfur bacteria.

Sulfite Agar

Composition per liter:
Agar 20.0g
Pancreatic digest of casein 10.0g
Na_2SO_3 .... 1.0g
Iron nails 66
pH 7.6 ± 0.2 at 25°C

Source: This medium is available as a premixed powder from Difco Laboratories.

Preparation of Medium: Add components to distilled/deionized water and bring volume to 1.0L. Mix thoroughly. Gently heat and bring to boiling. Distribute into screw-capped tubes in 15.0mL volumes. Add a clean iron nail to each tube. Autoclave for 15 min at 15 psi pressure–121°C. Cool to 45–50°C until ready to inoculate.

Use: For the detection and cultivation of thermophilic anaerobes that can produce H_2S from sulfite. Sulfite reduction appears as a blackening of the medium.

Sulfolobus Medium

Composition per liter:
$(NH_4)_2SO_4$ 1.3g
Yeast extract 1.0g
KH_2PO_4 .... 0.28g

$MgSO_4 \cdot 7H_2O$	0.25g
$CaCl_2 \cdot 2H_2O$	0.07g
$FeCl_3 \cdot 6H_2O$	0.02g
$Na_2B_4O_7 \cdot 10H_2O$	4.5mg
$MnCl_2 \cdot 4H_2O$	1.8mg
$ZnSO_4 \cdot 7H_2O$	0.22mg
$CuCl_2 \cdot 2H_2O$	0.05mg
$Na_2MoO_4 \cdot 2H_2O$	0.03mg
$VOSO_4 \cdot 2H_2O$	0.03mg
$CoSO_4$	0.01mg

pH 2.0 ± 0.2 at 25°C

Preparation of Medium: Add components to distilled/deionized water and bring volume to 1.0L. Mix thoroughly. Adjust pH at 25°C to 2.0 with 10*N* H_2SO_4. Filter sterilize. Aseptically distribute into tubes or flasks.

Use: For the cultivation and maintenance of *Sulfolobus acidocaldarius*.

Sulfolobus Medium, Revised

Composition per liter:

$(NH_4)_2SO_4$	1.3g
Tryptone	1.0g
KH_2PO_4	0.28g
$MgSO_4 \cdot 7H_2O$	0.25g
$CaCl_2 \cdot 2H_2O$	0.07g
Yeast Extract	0.05g
$FeCl_3 \cdot 6H_2O$	0.02g
$Na_2B_4O_7$	4.5mg
$MnCl_2 \cdot 4H_2O$	1.8mg
$ZnSO_4 \cdot 7H_2O$	0.22mg
$CuCl_2 \cdot H_2O$	0.05mg
$Na_2MoO_4 \cdot H_2O$	0.03mg
$VOSO_4 \cdot 2H_2O$	0.03mg
$CoSO_4$	0.01mg

pH 3.0 ± 0.2 at 25°C

Preparation of Medium: Add components to distilled/deionized water and bring volume to 1.0L. Mix thoroughly. Adjust pH at 25°C to 3.0 with 10*N* H_2SO_4. Filter sterilize. Aseptically distribute into tubes or flasks.

Use: For the cultivation and maintenance of *Sulfolobus* species.

Sulfolobus solfataricus Medium

Composition per liter:

KH_2PO_4	3.1g
$(NH_4)_2SO_4$	2.5g
Casamino acids	1.0g
Yeast extract	1.0g
$CaCl_2 \cdot 2H_2O$	0.25g
$MgSO_4 \cdot 7H_2O$	0.2g
$Na_2B_4O_7 \cdot 10H_2O$	4.5mg
$MnCl_2 \cdot 4H_2O$	1.8mg
$ZnSO_4 \cdot 7H_2O$	0.22mg
$CuCl_2 \cdot 2H_2O$	0.05mg
$Na_2MoO_4 \cdot 2H_2O$	0.03mg
$VOSO_4 \cdot 2H_2O$	0.03mg
$CoSO_4 \cdot 7H_2O$	0.01mg

pH 4.0–4.2 at 25°C

Preparation of Medium: Add components to distilled/deionized water and bring volume to 1.0L. Mix thoroughly. Adjust pH at 25°C to 4.0–4.2 with 10*N* H_2SO_4. Filter sterilize. Aseptically distribute into tubes or flasks.

Use: For the cultivation and maintenance of *Sulfolobus solfataricus*.

Sulfur Medium

Composition per liter:

Sulfur, elemental	10.0g
KH_2PO_4	3.0g
$MgSO_4 \cdot 7H_2O$	0.5g
$(NH_4)_2SO_4$	0.3g
$CaCl_2 \cdot 2H_2O$	0.25g
$FeCl_3 \cdot 6H_2O$	0.02g

pH 4.8± 0.2 at 25°C

Preparation of Medium: Add components, except sulfur, to distilled/deionized water and bring volume to 1.0L. Mix thoroughly. Add 1.0g sulfur to each of ten 250.0mL flasks. Add 100.0mL of medium to each flask. Autoclave for 30 min at 0 psi pressure–100°C on 3 consecutive days.

Use: For isolation, cultivation and enumeration of iron and sulfur bacteria.

SW 2 Agar

Composition per liter:

Agar	15.0g
NH_4Cl	1.0g
Sodium acetate	0.02g
Artificial seawater	1.0L

Artificial Seawater:

Composition per liter:

NaCl	24.7g
$MgSO_4 \cdot 7H_2O$	6.3g
$MgCl_2 \cdot 6H_2O$	4.6g
$CaCl_2$	1.0g
KCl	0.7g
$NaHCO_3$	0.2g

Preparation of Artificial Seawater: Add components to distilled/deionized water and bring volume to 1.0L. Mix thoroughly.

Preparation of Medium: Add solid components to 1.0L of artificial seawater. Mix thoroughly. Gently heat and bring to boiling. Distribute into tubes or flasks. Autoclave for 15 min at 15 psi pressure–121°C. Pour into sterile Petri dishes or leave in tubes.

Use: For the isolation and cultivation of *Cytophaga* species, *Herpetosiphon* species, *Saprospira* species, and *Flexithrix* species.

Swampy Medium

Composition per liter:

Agar	10.0g
$CaCO_3$	10.0g
Peptone	0.5g
Yeast extract	0.5g

Preparation of Medium: Add components to seawater and bring volume to 1.0L. Mix thoroughly. Gently heat and bring to boiling. Distribute into tubes or flasks. Autoclave for 15 min at 15 psi pressure–121°C. Pour into sterile Petri dishes or leave in tubes.

Use: For the cultivation and maintenance of *Vibrio liquefaciens.*

Sweet E Broth for Anaerobes

Composition per 100mL:

Gelatin	0.3g
Cellobiose	0.1g
Fructose	0.1g
Glucose	0.1g
L-Arabinose	0.1g
Maltose	0.1g
Starch	0.1g
Agar	0.075g
Peptone	0.05g
L-Cysteine·HCl·H_2O	0.05g
$(NH_4)_2SO_4$	0.05g
Yeast extract	0.05g
Salts solution	50.0mL
Rumen fluid	30.0mL
Resazurin solution	0.4mL
Pyruvic acid	0.01mL

pH 6.5 ± 0.2 at 25°C

Salts Solution:

Composition per liter:

$NaHCO_3$	10.0g
NaCl	2.0g
K_2HPO_4	1.0g
KH_2PO_4	1.0g
$CaCl_2$, anhydrous	0.2g
$MgSO_4 \cdot 7H_2O$	0.2g

Preparation of Salts Solution: Add $CaCl_2$ and $MgSO_4 \cdot 7H_2O$ to distilled/deionized water and bring volume to 300.0mL. Mix thoroughly. Bring volume to 800.0mL with distilled/deionized water. Add remaining components and mix. Bring volume to 1.0L with distilled/deionized water. Mix thoroughly. Store at 4°C.

Resazurin Solution:

Composition per 44mL:

Resazurin	0.011g

Preparation of Resazurin Solution: Add resazurin to distilled/deionized water and bring volume to 44.0mL. Mix thoroughly.

Preparation of Medium: Add components to distilled/deionized water and bring volume to 100.0mL. Mix thoroughly. Gently heat and bring to boiling under O_2-free 97% N_2 + 3% H_2. Adjust the pH to 6.5 if necessary. Continue boiling until the medium turns yellow. Distribute into tubes or flasks under O_2-free 97% N_2 + 3% H_2. Cap tubes with rubber stoppers. Place tubes in a press. Autoclave for 15 min at 15 psi pressure–121°C with fast exhaust.

Use: For the cultivation and maintenance of *Clostridium cocleatum* and *Cl.spiroforme.*

SWMTY Marine Medium

Composition per liter:

Marine salts mix	38.0g
Agar	15.0g
Pancreatic digest of casein	2.0g
Yeast extract	2.0g
Tris(hydroxymethyl)aminomethane buffer	1.0g
KNO_3	0.5g
Sodium glycerophosphate	0.1g
Trace element solution HO-LE	1.0mL

pH 7.0 ± 0.2 at 25°C

Trace Element Solution HO-LE:

Composition per liter:

H_3BO_3	2.85g
$MnCl_2 \cdot 4H_2O$	1.8g
Sodium tartrate	1.77g
$FeSO_4 \cdot 7H_2O$	1.36g
$CoCl_2 \cdot 6H_2O$	0.04g
$CuCl_2.2H_2O$	0.027g
$Na_2MoO_4 \cdot 2H_2O$	0.025g
$ZnCl_2$	0.020g

Preparation of Trace Element Solution HO-LE: Add components to distilled/deionized water and bring volume to 1.0L. Mix thoroughly. Filter sterilize.

Preparation of Medium: Add components to distilled/deionized water and bring volume to 1.0L. Mix thoroughly. Gently heat and bring to boiling. Distribute into tubes or flasks. Autoclave for 15 min at 15 psi pressure–121°C. Pour into sterile Petri dishes or leave in tubes.

Use: For the cultivation and maintenance of a variety heterotrophic marine bacterial species.

SY Broth

Composition per liter:

$(NH_4)_2SO_4$	2.0g
$Na_2HPO_4 \cdot 2H_2O$	1.4g
KH_2PO_4	0.7g
$MgSO_4 \cdot 7H_2O$	0.2g
$FeSO_4$	5.0mg
$MnSO_4$	5.0mg
Glucose solution	100.0mL

Glucose Solution:

Composition per 100mL:

D-Glucose	10.0g

Preparation of Glucose Solution: Add D-glucose to distilled/deionized water and bring volume to 100.0mL. Mix thoroughly. Autoclave for 15 min at 15 psi pressure–121°C. Cool to 25°C.

Preparation of Medium: Add components, except glucose solution, to distilled/deionized water and bring volume to 900.0mL. Mix thoroughly. Gently heat and bring to boiling. Autoclave for 15 min at 15 psi pressure–121°C. Cool to 45–50°C. Aseptically add sterile glucose solution. Mix thoroughly. Aseptically distribute into sterile tubes or flasks.

Use: For the isolation and cultivation of *Cytophaga* species, *Herpetosiphon* species, *Saprospira* species, and *Flexithrix* species.

Synthetic Seawater Medium

Composition per liter:

NaCl	27.0g
$MgSO_4 \cdot 7H_2O$	7.0g
Monosodium glutamate	5.0g
Tris(hydroxymethyl)aminomethane buffer	2.0g
Glucose	1.0g
KCl	0.6g
$CaCl_2$	0.3g
Sodium glycerophosphate	0.2g
Vitamin B_{12}	1.0μg

pH 7.5 ± 0.2 at 25°C

Preparation of Medium: Add components to distilled/deionized water and bring volume to 1.0L. Mix thoroughly. Adjust pH to 7.5. Distribute into tubes or flasks. Autoclave for 15 min at 15 psi pressure–121°C.

Use: For the cultivation and maintenance of *Leucothrix mucor*.

T_1N_0 Broth (Tryptone Broth)

Composition per liter:

Pancreatic digest of casein	10.0g

pH 7.1 ± 0.2 at 25°C

Preparation of Medium: Add components to distilled/deionized water and bring volume to 1.0L. Mix thoroughly. Gently heat and bring to boiling. Distribute into tubes or flasks. Autoclave for 15 min at 15 psi pressure–121°C.

Use: For the cultivation of *Vibrio cholerae* and other *Vibrio* species.

T_1N_1 Agar (Tryptone Salt Agar)

Composition per liter:

Agar	20.0g
NaCl	10.0g
Pancreatic digest of casein	10.0g

pH 7.1 ± 0.2 at 25°C

Preparation of Medium: Add components to distilled/deionized water and bring volume to 1.0L. Mix thoroughly. Gently heat and bring to boiling. Distribute into tubes or flasks. Autoclave for 15 min at 15 psi pressure–121°C. Pour into sterile Petri dishes or leave in tubes. Allow tubes to cool in a slanted position.

Use: For the cultivation of *Vibrio cholerae* and other *Vibrio* species.

T_1N_2 Agar (Tryptone Salt Agar)

Composition per liter:

Agar	20.0g
NaCl	20.0g
Pancreatic digest of casein	10.0g

pH 7.1 ± 0.2 at 25°C

Preparation of Medium: Add components to distilled/deionized water and bring volume to 1.0L. Mix thoroughly. Gently heat and bring to boiling. Distribute into tubes or flasks. Autoclave for 15 min at 15 psi pressure–121°C. Pour into sterile Petri dishes or leave in tubes. Allow tubes to cool in a slanted position.

Use: For the cultivation of *Vibrio cholerae* and other *Vibrio* species.

T_1N_1 Broth (Tryptone Salt Broth)

Composition per liter:

NaCl 10.0g
Pancreatic digest of casein 10.0g
pH 7.1 ± 0.2 at 25°C

Preparation of Medium: Add components to distilled/deionized water and bring volume to 1.0L. Mix thoroughly. Gently heat and bring to boiling. Distribute into tubes or flasks. Autoclave for 15 min at 15 psi pressure–121°C.

Use: For the cultivation of *Vibrio cholerae* and other *Vibrio* species.

T_1N_3 Broth (Tryptone Salt Broth)

Composition per liter:

NaCl 30.0g
Pancreatic digest of casein 10.0g
pH 7.1 ± 0.2 at 25°C

Preparation of Medium: Add components to distilled/deionized water and bring volume to 1.0L. Mix thoroughly. Gently heat and bring to boiling. Distribute into tubes or flasks. Autoclave for 15 min at 15 psi pressure–121°C.

Use: For the cultivation of *Vibrio cholerae* and other *Vibrio* species.

T_1N_6 Broth (Tryptone Salt Broth)

Composition per liter:

NaCl 60.0g
Pancreatic digest of casein 10.0g
pH 7.1 ± 0.2 at 25°C

Preparation of Medium: Add components to distilled/deionized water and bring volume to 1.0L. Mix thoroughly. Gently heat and bring to boiling. Distribute into tubes or flasks. Autoclave for 15 min at 15 psi pressure–121°C.

Use: For the cultivation of *Vibrio cholerae* and other *Vibrio* species.

T_1N_8 Broth (Tryptone Salt Broth)

Composition per liter:

NaCl 80.0g
Pancreatic digest of casein 10.0g
pH 7.1 ± 0.2 at 25°C

Preparation of Medium: Add components to distilled/deionized water and bring volume to 1.0L. Mix thoroughly. Gently heat and bring to boiling. Distribute into tubes or flasks. Autoclave for 15 min at 15 psi pressure–121°C.

Use: For the cultivation of *Vibrio cholerae* and other *Vibrio* species.

T_1N_{10} Broth (Tryptone Salt Broth)

Composition per liter:

NaCl 100.0g
Pancreatic digest of casein 10.0g
pH 7.1 ± 0.2 at 25°C

Preparation of Medium: Add components to distilled/deionized water and bring volume to 1.0L. Mix thoroughly. Gently heat and bring to boiling. Distribute into tubes or flasks. Autoclave for 15 min at 15 psi pressure–121°C.

Use: For the cultivation of *Vibrio cholerae* and other *Vibrio* species.

T2 Medium for *Thiobacillus*

Composition per liter:

Solution A 250.0mL
Solution B 250.0mL
Solution C 250.0mL
Solution D 250.0mL
pH 7.0 ± 0.2 at 25°C

Solution A:

Composition per 250mL:

$Na_2S_2O_3 \cdot 5H_2O$ 5.0g
KNO_3 .... 2.0g
NH_4Cl 1.0g

Preparation of Solution A: Add components to distilled/deionized water and bring volume to 250.0mL. Mix thoroughly. Filter sterilize.

Solution B:

Composition per 250mL

KH_2PO_4 .... 2.0g

Preparation of Solution B: Add KH_2PO_4 to distilled/deionized water and bring volume to 250.0mL. Mix thoroughly. Filter sterilize.

Solution C:

Composition per 250mL

$NaHCO_3$ 2.0g

Preparation of Solution C: Add $NaHCO_3$ to distilled/deionized water and bring volume to 250.0mL. Mix thoroughly. Filter sterilize.

Solution D:

Composition per 250mL

$MgSO_4 \cdot 7H_2O$	0.8g
$FeSO_4 \cdot 7H_2O$ (2%, w/v, in *N* HCl)	1.0mL
Trace metal solution	1.0mL

Preparation of Solution D: Add components to distilled/deionized water and bring volume to 250.0mL. Mix thoroughly. Filter sterilize.

$FeSO_4 \cdot 7H_2O$ Solution:

Composition per 100mL

$FeSO_4 \cdot 7H_2O$	2.0g
HCl (1*N* solution)	100.0mL

Preparation of $FeSO_4 \cdot 7H_2O$ Solution: Add the $FeSO_4 \cdot 7H_2O$ to the HCl solution. Mix thoroughly.

Trace Metal Solution:

Composition per liter:

EDTA	50.0g
$ZnSO_4$	22.0g
$CaCl_2$	5.54g
$MnCl_2$	5.06g
$FeSO_4 \cdot 7H_2O$	4.99g
$CoCl_2$	1.61g
$CuSO_4$	1.57g
$(NH_4)_2MoO_4$	1.10g

Preparation of Trace Metal Solution: Add components to distilled/deionized water and bring volume to 1.0L. Mix thoroughly. Adjust pH to 6.0 with KOH.

Preparation of Medium: Aseptically combine the four sterile solutions: solution A, solution B, solution C and solution D. Adjust the pH to 7.0. Aseptically distribute into sterile tubes or flasks.

Use: For the cultivation and maintenance of *Thiobacillus denitrificans* and other thiobacilli.

T7 Agar Base (m–T7 Agar Base)

Composition per liter:

Lactose	20.0g
Agar	15.0g
Polyoxyethylene ether W-1	5.0g
Yeast extract	3.0g
Pancreatic digest of casein	2.5g
Peptic digest of animal tissue	2.5g
Sodium heptadecyl sulfate	0.1g
Bromthymol blue	0.1g
Bromcresol purple	0.1g

pH 7.4 ± 0.2 at 25°C

Source: Available as a premixed powder from BBL Microbiology Systems and Difco Laboratories.

Preparation of Medium: Add components to distilled/deionized water and bring volume to 1.0L. Mix thoroughly. Gently heat while stirring and bring to boiling. Distribute into tubes or flasks. Autoclave for 15 min at 15 psi pressure–121°C. Cool to 45–50°C. The medium may be made more selective by adding 1.0mg penicillin G per liter. Pour into sterile Petri dishes or leave in tubes.

Use: For the selective recovery and differential identification of injured coliform microorganisms from chlorinated water by the membrane filter method. Also, for rapid estimation of the bacteriological quality of water using the membrane filter method.

Tap Water Agar

Composition per liter:

Agar	15.0g
Tap water	1.0L

Preparation of Medium: Add agar to 1.0L of tap water. Mix thoroughly. Gently heat and bring to boiling. Autoclave for 15 min at 15 psi pressure–121°C. Pour into sterile Petri dishes.

Use: For the cultivation and differentiation of fungi and aerobic actinomycetes based on filament and aerial hyphae morphology.

TCBS Agar (Thiosulfate Citrate Bile Salt Sucrose Agar)

Composition per liter:

Sucrose	20.0g
Agar	14.0g
NaCl	10.0g
Sodium citrate	10.0g
$Na_2S_2O_3$	10.0g
Yeast extract	5.0g
Pancreatic digest of casein	5.0g
Peptic digest of animal tissue	5.0g
Oxgall	5.0g
Sodium cholate	3.0g
Ferric citrate	1.0g
Thymol blue	0.04g
Bromthymol blue	0.04g

pH 8.6 ± 0.2 at 25°C

Source: This medium is available as a premixed powder from BBL Microbiology Systems and Difco Laboratories.

Preparation of Medium: Add components to distilled/deionized water and bring volume to 1.0L. Mix thoroughly. Gently heat while stirring and bring to boiling. Do not autoclave. Cool to 45–50°C. Pour into sterile Petri dishes or distribute into sterile tubes.

Use: For the selective isolation of *Vibrio cholerae* and *V. parahaemolyticus.*

TDC Medium

Composition per liter:

Agar	20.0g
$CaCO_3$	10.0g
Glucose	5.0g
K_2HPO_4	1.0g
$MgSO_4$	1.0g

Preparation of Medium: Add components to tap water and bring volume to 1.0L. Mix thoroughly. Gently heat and bring to boiling. Distribute into tubes or flasks. Autoclave for 15 min at 15 psi pressure–121°C. Pour into sterile Petri dishes or leave in tubes.

Use: For the cultivation and maintenance of *Azotobacter beijerinckii* and other *Azotobacter* species.

TEC Agar (m–TEC Agar)

Composition per liter:

Agar	15.0g
Lactose	10.0g
NaCl	7.5g
Proteose peptone	5.0g
K_2HPO_4	3.3g
Yeast extract	3.0g
KH_2PO_4	1.0g
Sodium lauryl sulfate	0.2g
Sodium deoxycholate	0.1g
Bromcresol purple	0.08g
Bromphenol red	0.08g

pH 5.0 ± 0.2 at 25°C

Source: Available as a premixed powder from Difco Laboratories.

Preparation of Medium: Add components to distilled/deionized water and bring volume to 1.0L. Mix thoroughly. Gently heat and bring to boiling. Adjust pH to 5.0. Sterilization is unnecessary. Pour into sterile Petri dishes or distribute into sterile tubes or flasks. Store at 2–8°C. Use within 1 week.

Use: For detection of *Escherichia coli* in recreational waters by the membrane filter method. This agar is used in conjunction with a urea substrate to detect urease production. After addition of the urea substrate, *E. coli* appears as yellow-yellow/brown colonies when viewed under a fluorescent lamp.

Tech Agar

Composition per liter:

Pancreatic digest of gelatin	20.0g
Agar	13.6g
$K_2SO_4 \cdot 7H_2O$	10.0g
$MgCl_2 \cdot 6H_2O$	1.4g
Glycerol	10.0mL

pH 7.2 ± 0.2 at 25°C

Source: This medium is available as a premixed powder from BBL Microbiology Systems.

Preparation of Medium: Add components, except glycerol, to distilled/deionized water and bring volume to 990.0mL. Mix thoroughly. Add glycerol. Gently heat and bring to boiling. Distribute into tubes or flasks. Autoclave for 15 min at 15 psi pressure–121°C. Pour into sterile Petri dishes or leave in tubes.

Use: For the production of pyocyanin pigment by *Pseudomonas* species.

Teepol Broth, Enriched (m–Teepol Broth, Enriched)

Composition per liter:

Peptone	40.0g
Lactose	30.0g
Yeast extract	6.0g
Phenol red	0.2g
Sodium lauryl sulfate (Teepol—0.1% solution)	4.0mL

pH 7.4 ± 0.2 at 25°C

Preparation of Medium: Add components to distilled/deionized water and bring volume to 1.0L. Mix thoroughly. Distribute into tubes or flasks. Autoclave for 15 min at 15 psi pressure–121°C.

Use: For the enumeration of coliform organisms and *Escherichia coli* in water by the membrane filter method.

Tergitol 7 Agar

Composition per liter:

Agar	15.0g
Lactose	10.0g
Yeast extract	3.0g
Pancreatic digest of casein	2.5g
Peptic digest of animal tissue	2.5g
Tergitol 7	0.1g
Bromthymol blue	25.0mg
TTC solution	3.0mL

pH 6.9 ± 0.2 at 25°C

Source: This medium is available as a premixed powder from BBL Microbiology Systems and Difco Laboratories.

TTC Solution:

Composition per 100mL:

Triphenyltetrazolium chloride	1.0g

Preparation of TTC Solution: Add triphenyltetrazolium chloride to distilled/deionized water and bring volume to 100.0mL. Mix thoroughly. Filter sterilize.

Preparation of Medium: Add components to distilled/deionized water and bring volume to 997.0mL. Mix thoroughly. Gently heat and bring to boiling. Autoclave for 15 min at 15 psi pressure–121°C. Cool to 50°C. Aseptically add 3.0mL of sterile TTC solution. Mix thoroughly. Pour into sterile Petri dishes or distribute into sterile tubes.

Use: For the selective isolation and differentiation of of coliform bacteria based on lactose fermentation. Lactose-fermenting bacteria appear as yellow colonies. Lactose-nonfermenting bacteria appear as blue colonies.

Tergitol 7 Broth

Composition per liter:

Lactose	10.0g
Yeast extract	3.0g
Pancreatic digest of casein	2.5g
Peptic digest of animal tissue	2.5g
Tergitol 7	0.1g
Bromthymol blue	25.0mg
TTC solution	3.0mL

pH 6.9 ± 0.2 at 25°C

Source: This medium is available as a premixed powder from BBL Microbiology Systems and Difco Laboratories.

TTC Solution:

Composition per 100mL:

Triphenyl tetrazolium chloride	1.0g

Preparation of TTC Solution: Add triphenyltetrazolium chloride to distilled/deionized water and bring volume to 100.0mL. Mix thoroughly. Filter sterilize.

Preparation of Medium: Add components to distilled/deionized water and bring volume to 997.0mL. Mix thoroughly. Gently heat while stirring and bring to boiling. Autoclave for 15 min at 15 psi pressure–121°C. Cool to 25°C. Aseptically add 3.0mL of sterile TTC solution. Mix thoroughly.

Use: For the isolation and cultivation of coliforms, *Salmonella* and other enteric bacteria.

Tetrathionate Broth

Composition per liter:

$Na_2S_2O_3$	40.7g
$CaCO_3$	25.0g
NaCl	4.5g
Peptone	4.5g
Yeast extract	1.8g
Beef extract	0.9g
Iodine solution	20.0mL

Source: This medium is available as a premixed powder from Oxoid Unipath.

Iodine Solution:

Composition per 20mL:

Iodine	6.0g
KI	5.0g

Preparation of Iodine Solution: Add iodine and KI to distilled/deionized water and bring volume to 20.0mL. Mix thoroughly.

Preparation of Medium: Add components, except iodine solution, to distilled/deionized water and bring volume to 980.0mL. Mix thoroughly. Gently heat and bring to boiling. Do not autoclave. Cool to 40°C. Add 20.0mL of iodine solution. Mix thoroughly. Distribute into tubes in 10.0mL volumes. Use medium the same day it is prepared.

Use: For the selective isolation and enrichment of *Salmonella typhi* and other salmonellae from fecal specimens, sewage and other specimens.

Tetrathionate Broth (m–Tetrathionate Broth) (m–TT Broth)

Composition per liter:

$Na_2S_2O_3$	30.0g
$CaCO_3$	10.0g
Pancreatic digest of casein	2.5g
Peptic digest of animal tissue	2.5g
Iodine-iodide solution	20.0mL

pH 8.0 ± 0.2 at 25°C

Iodine-Iodide Solution:

Composition per 20mL:

Iodine	6.0g
KI	5.0g

Preparation of Iodine-Iodide Solution: Add iodine and KI to distilled/deionized water and bring volume to 20.0mL. Mix thoroughly.

Preparation of Medium: Add components, except iodine-iodide solution, to distilled/deionized water and bring volume to 980.0mL. Mix thoroughly. Gently heat and bring to boiling. Do not autoclave.

Cool to 40°C. Add 20.0mL of iodine-iodide solution. Mix thoroughly. Distribute into tubes in 10.0mL volumes. Use medium the same day it is prepared.

Use: For the selective isolation in the membrane filter method of *Salmonella* species from specimens of sanitary importance.

Tetrathionate Broth (m–Tetrathionate Broth)

Composition per liter:

$Na_2S_2O_3$	30.0g
Proteose peptone	5.0g
Bile salts	1.0g
Iodine solution	20.0mL

pH 8.0 ± 0.2 at 25°C

Source: This medium is available as a premixed powder from Difco Laboratories.

Iodine Solution:

Composition per 20mL:

Iodine	6.0g
KI	5.0g

Preparation of Iodine Solution: Add iodine and KI to distilled/deionized water and bring volume to 20.0mL. Mix thoroughly.

Preparation of Medium: Add components, except iodine solution, to distilled/deionized water and bring volume to 980.0mL. Mix thoroughly. Gently heat and bring to boiling. Do not autoclave. Cool to 40°C. Add 20.0mL of iodine solution. Mix thoroughly. Use medium the same day it is prepared.

Use: For the enrichment of *Salmonella* species in the membrane filter method prior to placing the filter on selective media such as brilliant green broth.

Tetrathionate Broth, Hajna (TT Broth, Hajna)

Composition per liter:

$Na_2S_2O_3$	38.0g
$CaCO_3$	25.0g
Casein/meat peptone (50/50)	18.0g
NaCl	5.0g
D–Mannitol	2.5g
Yeast extract	2.0g
Glucose	0.5g
Sodium deoxycholate	0.5g
Brilliant green	0.01g
Iodine solution	40.0mL

pH 7.5–7.8 at 25°C

Source: This medium is available as a premixed powder from Difco Laboratories.

Iodine Solution:

Composition per 40mL:

KI	8.0g
Iodine	5.0g

Preparation of Iodine Solution: Add iodine and KI to distilled/deionized water and bring volume to 40.0mL. Mix thoroughly.

Preparation of Medium: Add components, except iodine solution, to distilled/deionized water and bring volume to 960.0mL. Mix thoroughly. Gently heat and bring to boiling. Do not autoclave. Cool to 40°C. Add 40.0mL of iodine solution. Mix thoroughly. Distribute into tubes in 10.0mL volumes. Use medium the same day it is prepared.

Use: For the isolation of *Salmonella* species, except *Salmonella typhi*, and *Arizona* species from specimens of sanitary significance.

Tetrathionate Broth, USA (TT Broth, USA)

$Na_2S_2O_3$	30.0g
$CaCO_3$	10.0g
Casein peptone	2.5g
Meat peptone	2.5g
Bile salts	1.0g

Source: This medium is available as a premixed powder from Oxoid Unipath.

Iodine-Iodide Solution:

Composition per 20mL:

Iodine	6.0g
KI	5.0g

Preparation of Iodine-Iodide Solution: Add iodine and KI to distilled/deionized water and bring volume to 20.0mL. Mix thoroughly.

Preparation of Medium: Add components, except iodine solution, to distilled/deionized water and bring volume to 980.0mL. Mix thoroughly. Gently heat and bring to boiling. Do not autoclave. Cool to 40°C. Add 20.0mL of iodine solution. Mix thoroughly. Distribute into tubes in 10.0mL volumes. Use medium the same day it is prepared.

Use: For the selective enrichment of *Salmonella* species from specimens of sanitary importance.

Tetrathionate Broth with Novobiocin

Composition per liter:

$Na_2S_2O_3$	38.0g
$CaCO_3$	25.0g

Casein/meat peptone (50/50) 18.0g
NaCl 5.0g
Yeast extract 2.0g
D-Mannitol 0.5g
Glucose 0.5g
Sodium deoxycholate 0.5g
Brilliant green 0.01g
Novobiocin 4.0mg
Iodine solution 40.0mL

pH 7.5–7.8 at 25°C

Iodine Solution:

Composition per 40mL:

KI 8.0g
Iodine 5.0g

Preparation of Iodine Solution: Add iodine and KI to distilled/deionized water and bring volume to 20.0mL. Mix thoroughly.

Preparation of Medium: Add components, except iodine solution, to distilled/deionized water and bring volume to 960.0mL. Mix thoroughly. Gently heat and bring to boiling. Do not autoclave. Cool to 40°C. Add 40.0mL of iodine solution. Mix thoroughly. Distribute into tubes in 10.0mL volumes. Use medium the same day it is prepared.

Use: For the isolation of *Salmonella* species, except *Salmonella typhi*, and *Arizona* species from specimens of sanitary importance. Novobiocin suppresses the growth of *Proteus* species.

Tetrathionate Crystal Violet Enhancement Broth

Composition per liter:

Potassium tetrathionate 20.0g
Casein/meat peptone (50/50) 8.6g
NaCl 6.4g
Crystal violet 0.005g

pH 6.5 ± 0.2 at 25°C

Preparation of Medium: Add components to distilled/deionized water and bring volume to 1.0L. Mix thoroughly. Distribute into tubes or flasks. Autoclave for 15 min at 15 psi pressure–121°C.

Use: For the isolation of *Salmonella* (except *Salmonella typhi*) and *Arizona* from specimens of sanitary significance.

Tetrathionate Reductase Medium

Composition per tube:

Solution I 10.0mL
Solution III 0.2mL
Solution II 0.1mL
Solution IV 0.1mL

Solution I:

Composition per liter:

$Na_2HPO_4 \cdot 12H_2O$ 3.6g
KH_2PO_4 .... 1.0g
NH_4Cl 0.5g
Peptone 0.25g
Yeast Extract 0.25g
$MgSO_4 \cdot 7H_2O$ 0.03g

Preparation of Solution I: Add components to distilled/deionized water and bring volume to 1.0L. Mix thoroughly. Gently heat and bring to boiling. Distribute into tubes in 10.0mL volumes. Autoclave for 15 min at 15 psi pressure–121°C. Cool to 25°C.

Solution II:

Composition per 100mL:

$CaCl_2 \cdot 2H_2O$ 0.1g
Ferric ammonium citrate 0.05g

Preparation of Solution II: Add components to distilled/deionized water and bring volume to 100.0mL. Mix thoroughly. Gently heat and bring to boiling. Autoclave for 15 min at 15 psi pressure–121°C. Cool to 25°C.

Solution III:

Composition per 100mL:

Sodium succinate 15.0g

Preparation of Solution III: Add sodium succinate to distilled/deionized water and bring volume to 100.0mL. Mix thoroughly. Gently heat until dissolved. Autoclave for 15 min at 15 psi pressure–121°C. Cool to 25°C.

Solution IV

Composition per 100mL:

$Na_2S_4O_6 \cdot 2H_2O$ 10.0g

Preparation of Solution IV: Add $Na_2S_4O_6 \cdot 2H_2O$ to distilled/deionized water and bring volume to 100.0mL. Mix thoroughly. Sterilize by filtration. Store at 4°C.

Preparation of Medium: To each tube containing 10.0mL of sterile solution I, aseptically add 0.1mL of sterile solution II, 0.2mL of sterile solution III, and 0.1mL of sterile solution IV. Mix thoroughly. Use immediately.

Use: For the cultivation and differentiation of hydrogen-oxidizing bacteria based on their production of tetrathionate reductase.

Tetrathionate Reductase Test Medium

Composition per 1025mL:

$K_2S_4O_6$ 5.0g

Peptone water 1.0L
Bromthymol blue (0.2% solution) 25.0mL
pH 7.4 ± 0.2 at 25°C

Peptone Water:
Composition per liter:
Peptone 10.0g
NaCl 5.0g

Preparation of Peptone Water: Add components to distilled/deionized water and bring volume to 1.0L. Mix thoroughly.

Preparation of Medium: Combine components. Mix thoroughly. Adjust pH to 7.4. Filter sterilize. Dispense into tubes in 1.0mL volumes or into wells of sterile microculture plates for replica inoculation.

Use: For the cultivation and identification of *Serratia* species based on their ability to reduce tetrathionate. Bacteria that reduce tetrathionate turn the medium yellow.

TGE Broth

Composition per liter:
Pancreatic digest of casein 10.0g
Beef extract 6.0g
Glucose 2.0g
pH 7.0 ± 0.2 at 25°C

Source: This medium is available as a premixed powder from Difco Laboratories.

Preparation of Medium: Add components to distilled/deionized water and bring volume to 1.0L. Mix thoroughly. Distribute into tubes or flasks. Autoclave for 15 min at 15 psi pressure–121°C.

Use: For the enumeration of bacteria by the membrane filter method.

TGY Medium (Tryptone Glucose Yeast Extract Medium)

Composition per liter of tap water:
Agar 20.0g
Pancreatic digest of casein 5.0g
Yeast extract 5.0g
Glucose 1.0g
K_2HPO_4 .... 1.0g
pH 7.0 ± 0.2 at 25°C

Preparation of Medium: Add components to distilled/deionized water and bring volume to 1.0L. Mix thoroughly. Gently heat and bring to boiling. Distribute into tubes or flasks. Autoclave for 15 min at 15 psi pressure–121°C. Pour into sterile Petri dishes or leave in tubes.

Use: For the cultivation and maintenance of a variety of bacteria including *Bacillus* species, *Corynebacterium* species, *Enterococcus* species and *Pseudomonas* species.

Thermoactinomyces Medium

Composition per liter:
Agar 20.0g
Malt extract 10.0g
Yeast extract 4.0g
Glucose 4.0g
pH 7.3 ± 0.2 at 25°C

Preparation of Medium: Add components to distilled/deionized water and bring volume to 1.0L. Mix thoroughly. Gently heat and bring to boiling. Distribute into tubes or flasks. Autoclave for 15 min at 15 psi pressure–121°C. Pour into sterile Petri dishes or leave in tubes.

Use: For the cultivation and maintenance of *Thermoactinomyces sacchari.*

Thermoactinopolyspora Medium

Composition per liter:
Maltose 20.0g
Agar 15.0g
Papaic digest of soybean meal 15.0g
Yeast extract 2.0g
pH 7.2 ± 0.2 at 25°C

Preparation of Medium: Add components to tap water and bring volume to 1.0L. Mix thoroughly. Gently heat and bring to boiling. Distribute into tubes or flasks. Autoclave for 15 min at 15 psi pressure–121°C. Pour into sterile Petri dishes or leave in tubes.

Use: For the cultivation and maintenance of *Thermoactinomyces* and *Thermoactinopolyspora* species.

Thermoanaerobacter ethanolicus Medium

Composition per liter:
Glucose 8.0g
$Na_2HPO_4 \cdot 12H_2O$ 4.2g
Yeast extract 2.0g
KH_2PO_4 .... 1.5g
NH_4Cl 0.5g
$MgCl_2 \cdot 6H_2O$ 0.18g
Reducing solution 40.0mL
Wolfe's modified mineral solution 5.0mL
Resazurin (0.1% solution) 1.0mL
Vitamin solution 0.5mL

Caution: This medium contains Na_2S, and H_2S production will occur, especially upon prolonged boiling. H_2S is hazardous and preparation of this medium should be done in a chemical fume hood.

Reducing Solution:

Composition per 200mL:

Cysteine·HCl·H_2O	2.5g
$Na_2S·9H_2O$	2.5g
NaOH (0.2*N* solution)	200.0mL

Preparation of Reducing Solution: Gently heat the NaOH solution and bring to boiling. Gas with 95% N_2 + 5% H_2. Cool to room temperature. Add the cysteine·HCl·H_2O and $Na_2S·9H_2O$. Anaerobically distibute into tubes. Cap with rubber stoppers. Autoclave for 15 min at 15 psi pressure–121°C.

Vitamin Solution:

Composition per 500 mL:

Pyridoxine HCl	0.1g
p-Aminobenzoic acid	0.05g
Calcium pantothenate	0.05g
Nicotinic acid	0.05g
Thioctic acid	0.05g
Biotin	0.02g
Folic acid	0.02g
Riboflavin	5.0mg
Thiamine·HCl	5.0mg
Vitamin B_{12}	1.0mg

Preparation of Vitamin Solution: Add components to distilled/deionized water and bring volume to 500.0mL. Mix thoroughly. Store solution in the dark at −10°C.

Wolfe's Modified Mineral Solution:

Composition per liter:

$MgSO_4·7H_2O$	3.0g
Nitrilotriacetic acid	1.5g
NaCl	1.0g
$MnSO_4·H_2O$	0.5g
$CaCl_2$ (anhydrous)	0.1g
$Co(NO_3)_2·6H_2O$	0.1g
$FeSO_4·7H_2O$	0.1g
$ZnSO_4·7H_2O$	0.1g
$AlK(SO_4)_2$ (anhydrous)	0.01g
$CuSO_4·5H_2O$	0.01g
H_3BO_3	0.01g
$Na_2MoO_4·2H_2O$	0.01g
Na_2SeO_3 (anhydrous)	1.0mg

Preparation of Wolfe's Modified Mineral Solution: Add nitrilotriacetic acid to 500.0mL of distilled/deionized water. Dissolve by adjusting pH to 6.5 with KOH. Add remaining components. Add distilled/deionized water to 1.0L.

Preparation of Medium: Add components, except reducing solution, to distilled/deionized water and bring volume to 1.0L. Gently heat and bring to boiling under 95% N_2 + 5% H_2. Continue boiling until color changes from blue to pink. Add the reducing solution. The pink color will disappear, indicating that the solution has been reduced. Distribute into tubes or flasks under 95% N_2 + 5% H_2 using anerobic techniques. Cap tubes with rubber stoppers. Autoclave for 15 min at 15 psi pressure–121°C.

Use: For the cultivation and maintenance of thermophilic anaerobes such as *Thermoanaerobacter* species and some *Clostridium* species.

Thermoanaerobium brockii Medium

Composition per liter:

Pancreatic digest of casein	10.0g
Yeast extract	3.0g
K_2HPO_4	1.5g
NH_4Cl	0.9g
NaCl	0.9g
KH_2PO_4	0.75g
$MgCl_2·6H_2O$	0.2g
Glucose solution	25.0mL
$Na_2S·9H_2O$ (10% solution)	10.0mL
Trace element solution	9.0mL
Vitamin solution	5.0mL
Resazurin (0.025% solution)	4.0mL
$FeSO_4·7H_2O$ (10% solution)	0.03mL

pH 7.3 ± 0.2 at 25°C

Glucose Solution:

Composition per 100mL:

Glucose	20.0g

Preparation of Glucose Solution: Add glucose to distilled/deionized water and bring volume to 100.0mL. Mix thoroughly. Filter sterilize.

Trace Element Solution:

Composition per liter:

Nitrilotriacetic acid	12.5g
NaCl	1.0g
$FeCl_3·4H_2O$	0.2g
$MnCl_2·4H_2O$	0.1g
$CaCl_2·2H_2O$	0.1g
$ZnCl_2$	0.1g
$CuCl_2$	0.02g
Na_2SeO_3	0.02g
$CoCl_2·6H_2O$	0.017g
H_3BO_3	0.01g
$Na_2MoO_4·2H_2O$	0.01g

Preparation of Trace Element Solution: Add nitrilotriacetic acid to 500.0mL of distilled/deionized water. Adjusting pH to 6.5 with KOH. Add remaining components. Add distilled/deionized water to 1.0L.

Wolfe's Vitamin Solution:

Composition per liter:

Pyridoxine·HCl	10.0mg
Thiamine·HCl	5.0mg
Riboflavin	5.0mg
Nicotinic acid	5.0mg
Calcium pantothenate	5.0mg
p-Aminobenzoic acid	5.0mg
Thioctic acid	5.0mg
Biotin	2.0mg
Folic acid	2.0mg
Cyanocobalamin	100.0μg

Preparation of Wolfe's Vitamin Solution: Add components to distilled/deionized water and bring volume to 1.0L. Mix thoroughly.

Preparation of Medium: Add components, except glucose solution, to distilled/deionized water and bring volume to 975.0mL. Mix thoroughly. Autoclave for 15 min at 15 psi pressure–121°C. While still hot, aseptically add 25.0mL of the sterile glucose solution under 97% N_2 + 3% H_2. Adjust pH to 7.3 if necessary. Aseptically and anaerobically distribute into tubes. Cap with rubber stoppers.

Use: For the cultivation and maintenance of *Thermoanaerobium brockii*.

Thermobacterium Medium

Composition per liter:

Agar	20.0g
$(NH_4)_2SO_4$	1.3g
Yeast extract	1.0g
Pancreatic digest of casein	1.0g
KH_2PO_4	0.28g
$MgSO_4 \cdot 7H_2O$	0.247g
$CaCl_2 \cdot 2H_2O$	0.074g
$FeCl_3.6H_2O$	0.019g
Salt solution	1.0mL

pH 8.5 + 0.2 at 25°C

Salt Solution:

Composition per liter:

$Na_2B_4O_7 \cdot 10H_2O$	4.4g
$MnCl_2 \cdot 4H_2O$	1.8g
$ZnSO_4 \cdot 7H_2O$	0.22g
$CuCl_2 \cdot H_2O$	0.05g
$Na_2MoO_4.2H_2O$	0.03g
$VOSO_4 \cdot 2H_2O$	0.03g

Preparation of Salt Solution: Add components to distilled/deionized water and bring volume to 1.0L. Mix thoroughly. Adjust pH to 2.0 with H_2SO_4.

Preparation of Medium: Add components to distilled/deionized water and bring volume to 1.0L. Mix thoroughly. Gently heat and bring to boiling. Distribute into tubes in 11.0–12.0mL volumes. Autoclave for 15 min at 15 psi pressure–121°C. Allow tubes to solidify in a slanted position.

Use: For the cultivation and maintenance of *Thermomicrobium roseum*.

Thermodesulfotobacterium Agar

Composition per liter:

Na_2SO_4	30.0g
Agar	20.0g
Sodium lactate	4.0g
Yeast extract	1.0g
Mineral solution 2	50.0mL
Na_2CO_3 solution	50.0mL
Mineral solution 1	25.0mL
Cysteine-sulfide reducing agent	20.0mL
Wolfe's Mineral solution	10.0mL
Wolfe's Vitamin solution	10.0mL
Resazurin (0.025% solution)	4.0mL

pH 7.2 ± 0.2 at 25°C

Mineral Solution 1:

Composition per liter:

K_2HPO_4	6.0g

Preparation of Mineral Solution 1: Add K_2HPO_4 to distilled/deionized water and bring volume to 1.0L. Mix thoroughly.

Mineral Solution 2:

Composition per liter:

NaCl	12.0g
KH_2PO_4	6.0g
$(NH_4)_2SO_4$	6.0g
$MgSO_4 \cdot 7H_2O$	2.4g
$CaCl_2 \cdot 2H_2O$	1.6g

Preparation of Mineral Solution 2: Add components to distilled/deionized water and bring volume to 1.0L. Mix thoroughly.

Na_2CO_3 Solution:

Composition per 100mL:

Na_2CO_3	8.0g

Preparation of Na_2CO_3 Solution: Add Na_2CO_3 to distilled/deionized water and bring volume to 100.0mL. Mix thoroughly.

Cysteine-Sulfide Reducing Agent:

Composition per 20mL:

L-Cysteine·HCl·H_2O	300.0mg
$Na_2S \cdot 9H_2O$	300.0mg

Preparation of Cysteine-Sulfide Reducing Agent: Add L-cysteine·HCl·H_2O to 10.0mL of distilled/deionized water. Mix thoroughly. In a separate tube, add $Na_2S \cdot 9H_2O$ to 10.0mL of distilled/deionized water. Mix thoroughly. Gas both solutions with 100% N_2 and cap tubes. Autoclave both solutions for 15 min at 15 psi pressure–121°C using fast exhaust. Cool to 50°C. Aseptically combine the two solutions under 100% N_2.

Wolfe's Mineral Solution:

Composition per liter

$MgSO_4 \cdot 7H_2O$	3.0g
Nitriloacetic acid	1.5g
NaCl	1.0g
$MnSO_4 \cdot H_2O$	0.5g
$FeSO_4 \cdot 7H_2O$	0.1g
$CoCl_2 \cdot 6H_2O$	0.1g
$CaCl_2$	0.1g
$ZnSO_4 \cdot 7H_2O$	0.1g
$CuSO_4 \cdot 5H_2O$	0.01g
$AlK(SO_4)_2 \cdot 12H_2O$	0.01g
H_3BO_3	0.01g
$Na_2MoO_4 \cdot 2H_2O$	0.01g

Preparation of Wolfe's Mineral Solution: Add nitrilotriacetic acid to 500.0mL of distilled/deionized water. Dissolve by adjusting pH to 6.5 with KOH. Add remaining components. Add distilled/deionized water to 1.0L.

Wolfe's Vitamin Solution:

Composition per liter:

Pyridoxine·HCl	10.0mg
Thiamine·HCl	5.0mg
Riboflavin	5.0mg
Nicotinic acid	5.0mg
Calcium pantothenate	5.0mg
p-Aminobenzoic acid	5.0mg
Thioctic acid	5.0mg
Biotin	2.0mg
Folic acid	2.0mg
Cyanocobalamin	100.0μg

Preparation of Wolfe's Vitamin Solution: Add components to distilled/deionized water and bring volume to 1.0L. Mix thoroughly. Filter sterilize.

Preparation of Medium: Add components, except vitamin solution and cysteine-sulfide reducing agent, to distilled/deionized water and bring volume to 970.0mL. Mix thoroughly. Gently heat and bring to boiling. Autoclave for 15 min at 15 psi pressure–121°C. Cool to 50–55°C under 80% N_2 + 20% CO_2. Aseptically add the sterile vitamin solution and then the sterile cysteine-sulfide reducing agent. Adjust the pH to 7.2. Distribute aseptically and anaerobically into sterile tubes.

Use: For the cultivation and maintenance of *Thermodesulfobacterium commune* and other *Thermodesulfobacterium* species.

Thermodesulfotobacterium Broth

Composition per liter:

Na_2SO_4	30.0g
Sodium lactate	4.0g
Yeast extract	1.0g
Mineral solution 2	50.0mL
Na_2CO_3 solution	50.0mL
Mineral solution 1	25.0mL
Cysteine-sulfide reducing agent	20.0mL
Wolfe's mineral solution	10.0mL
Wolfe's vitamin solution	10.0mL
Resazurin (0.025% solution)	4.0mL

pH 7.2 ± 0.2 at 25°C

Mineral Solution 1:

Composition per liter:

K_2HPO_4	6.0g

Preparation of Mineral Solution 1: Add K_2HPO_4 to distilled/deionized water and bring volume to 1.0L. Mix thoroughly.

Mineral Solution 2:

Composition per liter:

NaCl	12.0g
KH_2PO_4	6.0g
$(NH_4)_2SO_4$	6.0g
$MgSO_4 \cdot 7H_2O$	2.4g
$CaCl_2 \cdot 2H_2O$	1.6g

Preparation of Mineral Solution 2: Add components to distilled/deionized water and bring volume to 1.0L. Mix thoroughly.

Na_2CO_3 Solution:

Composition per 100mL:

Na_2CO_3	8.0g

Preparation of Na_2CO_3 Solution: Add Na_2CO_3 to distilled/deionized water and bring volume to 100.0mL Mix thoroughly.

Cysteine-Sulfide Reducing Agent:

Composition per 20mL:

L-Cysteine·HCl·H_2O	300.0mg
$Na_2S \cdot 9H_2O$	300.0mg

Preparation of Cysteine-Sulfide Reducing Agent: Add L-Cysteine·HCl·H_2O to 10.0mL of distilled/deionized water. Mix thoroughly. In a separate tube, add $Na_2S \cdot 9H_2O$ to 10.0mL of distilled/deionized water. Mix thoroughly. Gas both solutions with 100% N_2 and cap tubes. Autoclave both solutions for 15 min at 15 psi pressure–121°C using fast exhaust.

Cool to 50°C. Aseptically combine the two solutions under 100% N_2.

Wolfe's Mineral Solution:

Composition per liter

$MgSO_4 \cdot 7H_2O$	3.0g
Nitriloacetic acid	1.5g
NaCl	1.0g
$MnSO_4 \cdot H_2O$	0.5g
$FeSO_4 \cdot 7H_2O$	0.1g
$CoCl_2 \cdot 6H_2O$	0.1g
$CaCl_2$	0.1g
$ZnSO_4 \cdot 7H_2O$	0.1g
$CuSO_4 \cdot 5H_2O$	0.01g
$AlK(SO_4)_2 \cdot 12H_2O$	0.01g
H_3BO_3	0.01g
$Na_2MoO_4 \cdot 2H_2O$	0.01g

Preparation of Wolfe's Mineral Solution: Add nitrilotriacetic acid to 500.0mL of distilled/deionized water. Dissolve by adjusting pH to 6.5 with KOH. Add remaining components. Add distilled/deionized water to 1.0L.

Wolfe's Vitamin Solution:

Composition per liter:

Pyridoxine·HCl	10.0mg
Thiamine·HCl	5.0mg
Riboflavin	5.0mg
Nicotinic acid	5.0mg
Calcium pantothenate	5.0mg
p-Aminobenzoic acid	5.0mg
Thioctic acid	5.0mg
Biotin	2.0mg
Folic acid	2.0mg
Cyanocobalamin	100.0μg

Preparation of Wolfe's Vitamin Solution: Add components to distilled/deionized water and bring volume to 1.0L. Mix thoroughly. Filter sterilize.

Preparation of Medium: Add components, except vitamin solution and cysteine-sulfide reducing agent, to distilled/deionized water and bring volume to 970.0mL. Mix thoroughly. Autoclave for 15 min at 15 psi pressure–121°C. Cool under 80% N_2 + 20% CO_2. Aseptically add the sterile vitamin solution and then the sterile cysteine-sulfide reducing agent. Adjust the pH to 7.2. Distribute aseptically and anaerobically into sterile tubes.

Use: For the cultivation and maintenance of *Thermodesulfobacterium commune* and other *Thermodesulfobacterium* species.

Thermophilic *Bacillus* Medium

Composition per liter:

Peptone	8.0g
Yeast extract	4.0g
NaCl	3.0g

pH 7.5 ± 0.2 at 25°C

Preparation of Medium: Add components to distilled/deionized water and bring volume to 1.0L. Mix thoroughly. Distribute into tubes or flasks. Autoclave for 15 min at 15 psi pressure–121°C.

Use: For the cultivation and maintenance of a variety of thermophilic *Bacillus* species.

Thermophilic Maintenance Medium

Composition per liter:

$NaHCO_3$	3.0g
Yeast extract	1.0g
NH_4Cl	1.0g
KH_2PO_4	0.4g
K_2HPO_4	0.4g
$MgSO_4 \cdot 7H_2O$	0.1g
Cysteine-sulfide reducing solution	40.0mL
Fructose solution	25.0mL
Wolfe's vitamin solution	10.0mL
Wolfe's mineral solution	10.0mL
Resazurin (0.01% solution)	1.0mL

pH 5.6 ± 0.2 at 25°C

Cysteine-Sulfide Reducing Solution:

Composition per 100mL:

$Cysteine \cdot HCl \cdot H_2O$	1.25g
$Na_2S \cdot 9H_2O$	1.25g

Preparation of Cysteine-Sulfide Reducing Solution: Add $Cysteine \cdot HCl \cdot H_2O$ and $Na_2S \cdot 9H_2O$ to distilled/deionized water and bring volume to 100.0mL. Mix thoroughly.

Fructose Solution:

Composition per 100mL:

Fructose	20.0g

Preparation of Fructose Solution: Add fructose to distilled/deionized water and bring volume to 100.0mL. Mix thoroughly. Filter sterilize.

Wolfe's Vitamin Solution:

Composition per liter:

Pyridoxine·HCl	0.01g
Thiamine·HCl	5.0mg
Riboflavin	5.0mg
Nicotinic acid	5.0mg
Calcium pantothenate	5.0mg
p-Aminobenzoic acid	5.0mg
Thioctic acid	5.0mg
Biotin	2.0mg
Folic acid	2.0mg
Cyanocobalamin	100.0μg

Preparation of Wolfe's Vitamin Solution: Add components to distilled/deionized water and bring volume to 1.0L. Mix thoroughly.

Wolfe's Mineral Solution:

Composition per liter

$MgSO_4 \cdot 7H_2O$	3.0g
Nitriloacetic acid	1.5g
NaCl	1.0g
$MnSO_4 \cdot H_2O$	0.5g
$FeSO_4 \cdot 7H_2O$	0.1g
$CoCl_2 \cdot 6H_2O$	0.1g
$CaCl_2$	0.1g
$ZnSO_4 \cdot 7H_2O$	0.1g
$CuSO_4 \cdot 5H_2O$	0.01g
$AlK(SO_4)_2 \cdot 12H_2O$	0.01g
H_3BO_3	0.01g
$Na_2MoO_4 \cdot 2H_2O$	0.01g

Preparation of Wolfe's Mineral Solution: Add nitrilotriacetic acid to 500mL of distilled/deionized water. Dissolve by adjusting pH to 6.5 with KOH. Add remaining components. Add distilled/deionized water to 1.0L.

Preparation of Medium: Add components, except fructose solution, to distilled/deionized water and bring volume to 935.0mL. Mix thoroughly. Gently heat and bring to boiling. Continue boiling until resazurin turns colorless indicating reduction. Add 40.0mL of the cysteine-sulfide reducing solution. Autoclave for 15 min at 15 psi pressure–121°C. Cool to 50°C under O_2-free 90% N_2 + 10% CO_2. Add 25.0mL of the sterile fructose solution. Adjust the pH to 5.6 if necessary. Aseptically and anaerobically distribute into sterile tubes. Cap with rubber stoppers.

Use: For the cultivation and maintenance of a variety of thermophilic anaerobes including *Clostridium thermoautotrophicum.*

Thermophilic Streptomycete Medium

Composition per liter:

Agar	20.0g
Maltose	20.0g
Soybean meal	5.0g
Yeast extract	2.0g

pH 6.5 ± 0.2 at 25°C

Preparation of Medium: Add components to tap water and bring volume to 1.0L. Mix thoroughly. Gently heat and bring to boiling. Distribute into tubes or flasks. Autoclave for 15 min at 15 psi pressure–121°C. Pour into sterile Petri dishes or leave in tubes.

Use: For the isolation and cultivation of thermophilic streptomycetes.

Thermophilic Streptomycete Medium

Composition per liter:

Soybean oil meal	20.0g
Glucose	10.0g
NaCl	10.0g
Pancreatic digest of casein	10.0g
Silica solution (Ludox)	500.0mL

Preparation of Medium: Add components, except silica solution, to distilled/deionized water and bring volume to 500.0mL. Mix thoroughly. Gently heat until dissolved. Autoclave this solution and the 500.0mL of silica solution separately for 15 min at 15 psi pressure–121°C. Cool to 25°C. Adjust the pH of both solutions to 7.0. Aseptically combine the two sterile solutions. Mix thoroughly. Pour into sterile Petri dishes in 40.0mL volumes.

Use: For the isolation and cultivation of thermophilic streptomycetes.

Thermophilic Streptomycete Medium Ia

Composition per liter:

Agar	20.0g
Sucrose	5.0g
Pancreatic digest of casein	5.0g
Yeast extract	3.0g
$MgSO_4 \cdot 7H_2O$	0.5g
$FeSO_4 \cdot 7H_2O$	0.01g
Dung extract	5.0mL
Molasses	5.0mL
Trace element solution	1.0mL

pH 7.2 ± 0.2 at 25°C

Dung Extract:

Composition per 100mL:

Sheep manure, dried	25.0g

Preparation of Dung Extract: Add dried sheep manure to 100.0mL of tap water. Mix thoroughly. Autoclave for 30 min at 15 psi pressure–121°C. Filter through Whatman #1 filter paper. Store at 4°C under toluene.

Trace Element Solution:

Composition per 100mL:

$Fe(NH_4)_2SO_4$	0.1g
$ZnSO_4$	0.1g
$MnSO_4$	0.05g
$CoSO_4$	0.01g
H_3BO_3	0.01g
$CuSO_4$	8.0mg

Preparation of Trace Element Solution: Add components to distilled/deionized water and bring volume to 100.0mL. Mix thoroughly.

Preparation of Medium: Add components to distilled/deionized water and bring volume to 1.0L. Mix thoroughly. Gently heat and bring to boiling. Distribute into tubes or flasks. Autoclave for 15 min at 15 psi pressure–121°C. Pour into sterile Petri dishes or leave in tubes.

Use: For the isolation and cultivation of thermophilic streptomycetes.

Thermoplasma Agar

Composition per liter:

Basal solution	450.0mL
Solution B	450.0mL
Solution C	100.0mL

pH 2.0 ± 0.2 at 25°C

Basal Solution:

Composition per 500mL:

KH_2PO_4	3.0g
Yeast extract	1.0g
$MgSO_4 \cdot 7H_2O$	0.5g
$CaCl_2 \cdot 2H_2O$	0.25g
$(NH_4)_2SO_4$	0.2g

Preparation of Basal Solution: Add components to distilled/deionized water and bring volume to 500.0mL. Mix thoroughly. Adjust pH to 2.0 with 10*N* H_2SO_4. Autoclave for 15 min at 15 psi pressure–121°C. Cool to 55°C.

Solution B:

Composition per 450mL:

Noble agar	12.0g

Preparation of Solution B: Add agar to distilled/deionized water and bring volume to 450.0mL. Mix thoroughly. Gently heat and bring to boiling. Autoclave for 15 min at 15 psi pressure–121°C. Cool to 55°C.

Solution C:

Composition per 100mL:

Glucose	10.0g

Preparation of Solution C: Add glucose to distilled/deionized water and bring volume to 100.0mL. Mix thoroughly. Filter sterilize.

Preparation of Medium: Aseptically combine the cooled sterile basal medium with the sterile solution B and sterile solution C. Mix thoroughly. Pour into sterile Petri dishes or distribute into sterile tubes.

Use: For the cultivation and maintenance of *Thermoplasma acidophilum* and other *Thermoplasma* species.

Thermoplasma Broth

Composition per liter:

Basal solution	500.0mL
Solution C	100.0mL

pH 2.0 ± 0.2 at 25°C

Basal Solution:

Composition per 500mL:

KH_2PO_4	3.0g
Yeast extract	1.0g
$MgSO_4 \cdot 7H_2O$	0.5g
$CaCl_2 \cdot 2H_2O$	0.25g
$(NH_4)_2SO_4$	0.2g

Preparation of Basal Solution: Add components to distilled/deionized water and bring volume to 500.0mL. Mix thoroughly. Adjust pH to 2.0 with 10*N* H_2SO_4.

Solution C:

Composition per 100mL:

Glucose	10.0g

Preparation of Solution C: Add glucose to distilled/deionized water and bring volume to 100.0mL. Mix thoroughly. Filter sterilize.

Preparation of Medium: Add 500.0mL of basal solution to 400.0mL of distilled/deionized water. Autoclave for 15 min at 15 psi pressure–121°C. Cool to 55°C. Aseptically add 100.0mL of sterile glucose solution. Mix thoroughly. Aseptically distribute into sterile tubes.

Use: For the cultivation and maintenance of *Thermoplasma acidophilum* and other *Thermoplasma* species.

Thermoproteus Medium

Composition per liter:

Solution A	500.0mL
Solution B	450.0mL
Solution C	50.0mL

pH 4.8–5.6 at 25°C

Solution A:

Composition per 500mL:

Glucose	10.0g
$FeSO_4 \cdot 7H_2O$	0.556g
$MgSO_4 \cdot 7H_2O$	0.492g
$CaSO_4 \cdot 2H_2O$	0.344g
$(NH_4)_2SO_4$	0.264g
Yeast extract	0.2g
KH_2PO_4	0.014g
Resazurin	1.0mg
Trace elements	10.0mL

Preparation of Solution A: Add components to distilled/deionized water and bring volume to 500.0mL. Mix thoroughly. Immediately filter sterilize.

Trace Elements:

Composition per liter:

$Na_2B_4O_7 \cdot 10H_2O$	0.45g
$MnCl_2 \cdot 4H_2O$	0.18g
$ZnSO_4 \cdot 7H_2O$	0.022g
$CuCl_2 \cdot 2H_2O$	5.0mg
$Na_2MoO_4 \cdot 2H_2O$	3.6mg
$VOSO_4 \cdot 5H_2O$	3.6mg
$CoSO_4 \cdot 7H_2O$	1.2mg

Preparation of Trace Elements: Add components to distilled/deionized water and bring volume to 1.0L. Mix thoroughly. Adjust pH to 3.0 with H_2SO_4 to retard precipitation.

Solution B:

Composition per 450mL:

Sulfur	10.0g

Preparation of Solution B: Add sulfur to 450.0mL of distilled/deionized water. Autoclave for 30 min at 0 psi pressure–100°C on three consecutive days.

Solution C:

Composition per 50mL:

$Na_2S \cdot 9H_2O$	0.85g

Preparation of Solution C: Add $Na_2S \cdot 9H_2O$ to distilled/deionized water and bring volume to 50.0mL. Mix thoroughly. Autoclave for 15 min at 15 psi pressure–121°C.

Preparation of Medium: Aseptically combine solutions A, B and C under 97% N_2 + 3% H_2. Adjust pH to 4.8–5.6 with H_2SO_4. Aseptically and anaerobically distribute into sterile tubes or flasks under 97% N_2 + 3% H_2.

Use: For the cultivation and maintenance of *Thermoproteus tenax* and other *Thermoproteus* species.

Thermus BP Medium (*Thermus* Beef Extract Polypeptone™ Medium)

Composition per liter:

Agar	25.0g
Beef extract	4.0g
Polypeptone™	4.0g
K_2HPO_4	3.0g
KH_2PO_4	1.0g

pH 7.0 ± 0.2 at 25°C

Preparation of Medium: Add components to distilled/deionized water and bring volume to 1.0L. Mix thoroughly. Gently heat and bring to boiling. Distribute into tubes or flasks. Autoclave for 15 min at 15 psi pressure–121°C. Pour into sterile Petri dishes or leave in tubes.

Use: For the cultivation and maintenance of *Thermus aquaticus* and other *Thermus* species.

Thermus Medium

Composition per liter:

Agar	30.0g
Polypeptone™	8.0g
Yeast extract	4.0g
NaCl	2.0g

pH 7.5 ± 0.2 at 25°C

Preparation of Medium: Add components to distilled/deionized water and bring volume to 1.0L. Mix thoroughly. Gently heat and bring to boiling. Distribute into tubes or flasks. Autoclave for 15 min at 15 psi pressure–121°C. Pour into sterile Petri dishes or leave in tubes.

Use: For the cultivation and maintenance of *Thermus aquaticus* and other *Thermus* species.

Thermus PMY Agar (*Thermus* Peptone Meat Extract Yeast Extract Agar)

Composition per liter:

Agar	15.0g
Peptone	5.0g
Meat extract	3.5g
Yeast extract	1.5g
NaCl	1.5g

pH 7.0 ± 0.2 at 25°C

Preparation of Medium: Add components to distilled/deionized water and bring volume to 1.0L. Mix thoroughly. Gently heat and bring to boiling. Distribute into tubes or flasks. Autoclave for 15 min at 15 psi pressure–121°C. Pour into sterile Petri dishes or leave in tubes.

Use: For the cultivation and maintenance of *Thermus aquaticus* and other *Thermus* species.

Thermus PMY Broth (*Thermus* Peptone Meat Extract Yeast Extract Broth)

Composition per liter:

Peptone	5.0g
Meat extract	3.5g
Agar	3.0g

Yeast extract..1.5g
NaCl...1.5g

pH 7.0 ± 0.2 at 25°C

Preparation of Medium: Add components to distilled/deionized water and bring volume to 1.0L. Mix thoroughly. Distribute into tubes or flasks. Autoclave for 15 min at 15 psi pressure–121°C.

Use: For the cultivation and maintenance of *Thermus aquaticus* and other *Thermus* species.

Thiobacillus A2 Agar (T3 Agar)

Composition per 1100mL:

Solution A 100.0mL
Solution B 1.0L

pH 8.5 ± 0.2 at 25°C

Solution A:

Composition per 100mL:

$Na_2S_2O_3 \cdot 5H_2O$ 5.0g
Na_2HPO_4...4.2g
KH_2PO_4..1.5g
NH_4Cl...1.0g
Phenol red (0.2% solution) 1.0mL

Preparation of Solution A: Add components to distilled/deionized water and bring volume to 100.0mL. Mix thoroughly. Adjust pH to 9.0. Autoclave for 15 min at 15 psi pressure–121°C. Cool to 45–50°C.

Solution B:

Composition per liter:

Agar...15.0g
$MgSO_4 \cdot 7H_2O$ 0.1g
Trace metal solution......................................5.0mL

Preparation of Solution B: Add components to distilled/deionized water and bring volume to 1.0mL. Mix thoroughly. Autoclave for 15 min at 15 psi pressure–121°C. Cool to 45–50°C.

Trace Metal Solution:

Composition per liter:

EDTA 50.0g
$ZnSO_4$..22.0g
$CaCl_2$...5.54g
$MnCl_2$...5.06g
$FeSO_4 \cdot 7H_2O$..4.99g
$CoCl_2$...1.61g
$CuSO_4$ 1.57g
$(NH_4)_2MoO_4 \cdot 4H_2O$ 1.10g

Preparation of Trace Metal Solution: Add components to distilled/deionized water and bring volume to 1.0L. Mix thoroughly. Adjust pH to 6.0 with KOH.

Preparation of Medium: Aseptically add 100.0mL of sterile solution A to 1.0L of sterile solution B. Mix thoroughly. Adjust pH to 8.5 if necessary. Pour into sterile Petri dishes or distribute into sterile tubes.

Use: For the cultivation and maintenance of *Thiobacillus versutus* and other *Thiobacillus* species.

Thiobacillus A2 Broth (T3 Broth)

Composition per 1100mL:

Solution B 1.0L
Solution A 100.0mL

pH 8.5 ± 0.2 at 25°C

Solution A:

Composition per 100mL:

$Na_2S_2O_3 \cdot 5H_2O$ 5.0g
Na_2HPO_4...4.2g
KH_2PO_4..1.5g
NH_4Cl...1.0g
Phenol red (0.2% solution) 1.0mL

Preparation of Solution A: Add components to distilled/deionized water and bring volume to 100.0mL. Mix thoroughly. Adjust pH to 9.0. Autoclave for 15 min at 15 psi pressure–121°C. Cool to 45–50°C.

Solution B:

Composition per liter:

$MgSO_4 \cdot 7H_2O$ 0.1g
Trace metal solution......................................5.0mL

Preparation of Solution B: Add components to distilled/deionized water and bring volume to 1.0mL. Mix thoroughly. Autoclave for 15 min at 15 psi pressure–121°C. Cool to 45–50°C.

Trace Metal Solution:

Composition per liter:

EDTA 50.0g
$ZnSO_4$..22.0g
$CaCl_2$...5.54g
$MnCl_2$...5.06g
$FeSO_4 \cdot 7H_2O$..4.99g
$CoCl_2$...1.61g
$CuSO_4$ 1.57g
$(NH_4)_2MoO_4 \cdot 4H_2O$ 1.10g

Preparation of Trace Metal Solution: Add components to distilled/deionized water and bring volume to 1.0mL. Mix thoroughly. Adjust pH to 6.0 with KOH.

Preparation of Medium: Aseptically add 100.0mL of sterile solution A to 1.0L of sterile solution B. Mix thoroughly. Adjust pH to 8.5 if necessary. Distribute into sterile tubes or flasks.

Use: For the cultivation and maintenance of *Thiobacillus versutus* and other *Thiobacillus* species.

Thiobacillus denitrificans Medium

Composition per liter:

KNO_3	5.0g
$Na_2S_2O_3 \cdot 5H_2O$	5.0g
$NaHCO_3$	1.0g
K_2HPO_4	0.2g
$MgCl_2$	0.1g

pH 7.0 ± 0.2 at 25°C

Preparation of Medium: Add components to distilled/deionized water and bring volume to 1.0L. Mix thoroughly. Distribute into tubes or flasks. Autoclave for 15 min at 15 psi pressure–121°C.

Use: For the cultivation of *Thiobacillus denitrificans*.

Thiobacillus denitrificans Medium

Composition per liter:

$Na_2S_2O_3 \cdot 5H_2O$	5.0g
KNO_3	2.0g
KH_2PO_4	2.0g
$NaHCO_3$	2.0g
NH_4Cl	1.0g
$MgSO_4 \cdot 7H_2O$	0.8g
Trace metals solution	1.0mL

pH 6.8–7.0 at 25°C

Trace Metals Solution:

Composition per liter:

Disodium EDTA	50.0g
NaOH	11.0g
$CaCl_2 \cdot 2H_2O$	7.34g
$FeSO_4 \cdot 7H_2O$	5.0g
$MnCl_2 \cdot 2H_2O$	2.5g
$ZnSO_4 \cdot 7H_2O$	2.2g
$CoCl_2 \cdot 6H_2O$	0.5g
$(NH_4)_6Mo_7O_{24} \cdot 4H_2O$	0.5g
$CuSO_4 \cdot 5H_2O$	0.2g

Preparation of Trace Metals Solution: Add EDTA to distilled/deionized water and bring volume to 500.0mL. Mix thoroughly. Adjust pH to 6.0 with NaOH. Add remaining components, one by one. Maintain the pH at 6.0. After dissolution of all the salts, adjust the pH to 4.0 with HCl. Store at 4°C.

Preparation of Medium: Add components to distilled/deionized water and bring volume to 1.0L. Mix thoroughly. Distribute into tubes or flasks. Autoclave for 15 min at 15 psi pressure–121°C.

Use: For the isolation and cultivation of *Thiobacillus denitrificans*.

Thiobacillus ferrooxidans Medium

Composition per liter:

$Al_2(SO_4)_3 \cdot 12H_2O$	1.4g
NaCl	1.0g
KH_2PO_4	0.4g
$MgSO_4 \cdot 7H_2O$	0.1g
$(NH_4)_2SO_4$	0.1g
$CaCl_2$	0.03g
$MnSO_4 \cdot 4H_2O$	0.02g
$FeSO_4 \cdot 7H_2O$ solution	100.0mL

$FeSO_4 \cdot 7H_2O$ Solution:

Composition per 100mL:

$FeSO_4 \cdot 7H_2O$	10.0g
H_2SO_4, concentrated	0.09mL

Preparation of $FeSO_4 \cdot 7H_2O$ Solution: Add $FeSO_4 \cdot 7H_2O$ and H_2SO_4 to distilled/deionized water and bring volume to 100.0mL. Mix thoroughly. Autoclave for 15 min at 15 psi pressure–121°C.

Preparation of Medium: Add components, except $FeSO_4 \cdot 7H_2O$ solution, to distilled/deionized water and bring volume to 900.0mL. Mix thoroughly. Gently heat and bring to boiling. Distribute into flasks in 90.0mL volumes. Autoclave for 15 min at 15 psi pressure–121°C. Cool to 25°C. Aseptically add 10.0mL of sterile $FeSO_4 \cdot 7H_2O$ solution to each flask. Mix thoroughly.

Use: For the cultivation of *Thiobacillus ferrooxidans*.

Thiobacillus ferrooxidans Medium

Composition per liter:

Solution I	400.0mL
Solution III	400.0mL
Solution II	200.0mL

Solution I:

Composition per 500mL:

K_2HPO_4	0.5g
$MgSO_4 \cdot 7H_2O$	0.5g
$(NH_4)_2SO_4$	0.5g
1*N* H_2SO_4	5.0mL

Preparation of Solution I: Add components to distilled/deionized water and bring volume to 500.0mL. Mix thoroughly. Autoclave for 15 min at 15 psi pressure–121°C. Cool to 45–50°C.

Solution II:

Composition per liter:

$FeSO_4 \cdot 7H_2O$	167.0g
1*N* H_2SO_4	50.0mL

Preparation of Solution II: Add components to distilled/deionized water and bring volume to 1.0L. Mix thoroughly. Filter sterilize. Warm to 45–50°C.

Solution III:

Composition per liter:

Agar.......10.0g

Preparation of Solution III: Add agar to distilled/deionized water and bring volume to 1.0L. Mix thoroughly.

Preparation of Medium: Aseptically combine 400.0mL of sterile solution I, 200.0mL of sterile solution II and 400.0mL of sterile solution III. Mix thoroughly. Aseptically distribute into sterile tubes or flasks.

Use: For the isolation and cultivation of *Thiobacillus ferrooxidans.*

Thiobacillus Heterotrophic Medium

Composition per liter:

Glucose.......5.0g
$MgSO_4 \cdot 7H_2O$.......0.5g
$(NH_4)_2SO_4$.......0.15g
KH_2PO_4.......0.1g
KCl.......0.05g
$Ca(NO_3)_2$.......0.01g

pH 3.0 ± 0.2 at 25°C

Preparation of Medium: Add components to distilled/deionized water and bring volume to 1.0L. Mix thoroughly. Filter sterilize.

Use: For the cultivation and maintenance of *Thiobacillus organoparus* and other heterotrophic *Thiobacillus* species.

Thiobacillus intermedius Medium

Composition per 1010mL:

$Na_2S_2O_3 \cdot 5H_2O$.......10.0g
Solution I.......1.0L
Solution II.......10.0mL

Solution I:

Composition per liter:

NH_4Cl.......1.0g
K_2HPO_4.......0.6g
$MgCl_2 \cdot 6H_2O$.......0.5g
KH_2PO_4.......0.4g
$MgSO_4$.......0.3g
$CaCl_2 \cdot 2H_2O$.......0.2g
$FeCl_3 \cdot 6H_2O$.......0.02g

Preparation of Solution I: Add components to distilled/deionized water and bring volume to 1.0L. Mix thoroughly.

Solution II:

Composition per liter:

$CaCl_2 \cdot 2H_2O$.......0.1g
$ZnSO_4 \cdot 7H_2O$.......0.09g
$CuSO_4 \cdot 5H_2O$.......0.04g
$MnSO_4$.......0.02g
$Na_2B_4O_7$.......0.01g
$(NH_4)_6Mo_7O_{24} \cdot 4H_2O$.......5.0mg

Preparation of Solution II: Add components to distilled/deionized water and bring volume to 1.0L. Mix thoroughly.

Preparation of Medium: Combine solution I, 10.0mL of solution II and 10.0g of $Na_2S_2O_3 \cdot 5H_2O$. Mix thoroughly. Filter sterilize. Aseptically distribute into sterile tubes or flasks.

Use: For the isolation and autotrophic cultivation of *Thiobacillus intermedius.*

Thiobacillus intermedius Medium

Composition per 1010mL:

Glucose.......10.0g
$Na_2S_2O_3 \cdot 5H_2O$.......10.0g
Solution I.......1.0L
Solution II.......10.0mL

Solution I:

Composition per liter:

NH_4Cl.......1.0g
K_2HPO_4.......0.6g
$MgCl_2 \cdot 6H_2O$.......0.5g
KH_2PO_4.......0.4g
$MgSO_4$.......0.3g
$CaCl_2 \cdot 2H_2O$.......0.2g
$FeCl_3 \cdot 6H_2O$.......0.02g

Preparation of Solution I: Add components to distilled/deionized water and bring volume to 1.0L. Mix thoroughly.

Solution II:

Composition per liter:

$CaCl_2 \cdot 2H_2O$.......0.1g
$ZnSO_4 \cdot 7H_2O$.......0.09g
$CuSO_4 \cdot 5H_2O$.......0.04g
$MnSO_4$.......0.02g
$Na_2B_4O_7$.......0.01g
$(NH_4)_6Mo_7O_{24} \cdot 4H_2O$.......5.0mg

Preparation of Solution II: Add components to distilled/deionized water and bring volume to 1.0L. Mix thoroughly.

Preparation of Medium: Combine 1.0L of solution I, 10.0mL of solution II, 10.0g of glucose and 10.0g of $Na_2S_2O_3 \cdot 5H_2O$. Mix thoroughly. Filter sterilize. Aseptically distribute into sterile tubes or flasks.

Use: For the isolation and mixotrophic cultivation of *Thiobacillus intermedius.*

Thiobacillus intermedius Medium

Composition per 1010mL:

Glucose	10.0g
Yeast extract	0.3g
Solution I	1.0L
Solution II	10.0mL

Solution I:

Composition per liter:

NH_4Cl	1.0g
K_2HPO_4	0.6g
$MgCl_2 \cdot 6H_2O$	0.5g
KH_2PO_4	0.4g
$MgSO_4$	0.3g
$CaCl_2 \cdot 2H_2O$	0.2g
$FeCl_3 \cdot 6H_2O$	0.02g

Preparation of Solution I: Add components to distilled/deionized water and bring volume to 1.0L. Mix thoroughly.

Solution II:

Composition per liter:

$CaCl_2 \cdot 2H_2O$	0.1g
$ZnSO_4 \cdot 7H_2O$	0.09g
$CuSO_4 \cdot 5H_2O$	0.04g
$MnSO_4$	0.02g
$Na_2B_4O_7$	0.01g
$(NH_4)_6Mo_7O_{24} \cdot 4H_2O$	5.0mg

Preparation of Solution II: Add components to distilled/deionized water and bring volume to 1.0L. Mix thoroughly.

Preparation of Medium: Combine 1.0L of solution I, 10.0mL of solution II, 10.0g of glucose, and 0.3g of yeast extract. Mix thoroughly. Filter sterilize. Aseptically distribute into sterile tubes or flasks.

Use: For the isolation and heterotrophic cultivation of *Thiobacillus intermedius.*

Thiobacillus Medium

Composition per 100mL:

$Na_2S_2O_3 \cdot 5H_2O$	1.0g
KH_2PO_4	0.1g
NH_4Cl	0.1g
$MgCl_2 \cdot 7H_2O$	0.05g

pH 6.8 ± 0.2 at 25°C

Preparation of Medium: Add components to distilled/deionized water and bring volume to 1.0L. Mix thoroughly. Distribute into tubes or flasks. Autoclave for 15 min at 15 psi pressure–121°C.

Use: For the cultivation of *Thiobacillus thioparus* and *T. thiooxidans.*

Thiobacillus Medium

Composition per liter:

$Na_2S_2O_3 \cdot 5H_2O$	10.0g
K_2HPO_4	4.0g
KH_2PO_4	4.0g
$CaCl_2$	0.1g
$MgSO_4 \cdot 7H_2O$	0.1g
$(NH_4)_2SO_4$	0.1g
$FeCl_3 \cdot 6H_2O$	0.02g
$MnSO_4 \cdot 4H_2O$	0.02g

pH 6.6 ± 0.2 at 25°C

Preparation of Medium: Add components to distilled/deionized water and bring volume to 1.0L. Mix thoroughly. Distribute into flasks in 100.0mL volumes. Autoclave for 60 min at 0 psi pressure–100°C on three consecutive days.

Use: For the cultivation of nonaciduric *Thiobacillus* species.

Thiobacillus Medium (ATCC Medium 64)

Composition per 500mL:

Solution A	400.0mL
Solution B	100.0mL

pH 2.8 ± 0.2 at 25°C

Solution A:

Composition per 400mL:

$(NH_4)_2SO_4$	0.4g
KH_2PO_4	0.2g
$MgSO_4 \cdot 7H_2O$	0.08g

Preparation of Solution A: Add components to distilled/deionized water and bring volume to 400.0mL. Mix thoroughly. Autoclave for 15 min at 15 psi pressure–121°C. Cool to 45–50°C.

Solution B:

Composition per 100mL:

$FeSO_4 \cdot 7H_2O$	10.0g
H_2SO_4 (1*N* solution)	1.0mL

Preparation of Solution B: Add components to distilled/deionized water and bring volume to 100.0mL. Mix thoroughly. Autoclave for 15 min at 15 psi pressure–121°C. Cool to 45–50°C.

Preparation of Medium: Aseptically add 100.0mL of cooled sterile solution B to 400.0mL of cooled sterile solution A. Mix thoroughly. Adjust pH to 2.8. Aseptically distribute into sterile tubes or flasks.

Use: For the cultivation and maintenance of a variety of *Thiobacillus* species.

Thiobacillus Medium

Composition per liter:

$Na_2S_2O_3 \cdot 5H_2O$	10.0g
$Na_2HPO_4 \cdot 7H_2O$	7.9g
Sodium formate	6.8g
Glucose	3.6g
KNO_3	2.0g
KH_2PO_4	1.5g
NH_4Cl	0.3g
$MgSO_4 \cdot 7H_2O$	0.1g
Trace metals solution	5.0mL

pH 7.6–8.5 at 25°C

Trace Metals Solution:

Composition per liter:

Disodium EDTA	50.0g
NaOH	11.0g
$CaCl_2 \cdot 2H_2O$	7.34g
$FeSO_4 \cdot 7H_2O$	5.0g
$MnCl_2 \cdot 2H_2O$	2.5g
$ZnSO_4 \cdot 7H_2O$	2.2g
$CoCl_2 \cdot 6H_2O$	0.5g
$(NH_4)_6Mo_7O_{24} \cdot 4H_2O$	0.5g
$CuSO_4 \cdot 5H_2O$	0.2g

Preparation of Trace Metals Solution: Add EDTA to distilled/deionized water and bring volume to 500.0mL. Mix thoroughly. Adjust pH to 6.0 with NaOH. Add remaining components, one by one. Maintain the pH at 6.0. After dissolution of all the salts, adjust the pH to 4.0 with HCl. Store at 4°C.

Preparation of Medium: Add components to distilled/deionized water and bring volume to 1.0L. Mix thoroughly. Adjust pH to 7.6–8.5. Filter sterilize. Aseptically distribute into sterile tubes or flasks.

Use: For the isolation and anaerobic cultivation of *Thiobacillus* species.

Thiobacillus Medium

Composition per liter:

$Na_2S_2O_3 \cdot 5H_2O$	10.0g
$Na_2HPO_4 \cdot 7H_2O$	7.9g
Sodium formate	6.8g
Glucose	3.6g
KH_2PO_4	1.5g
NH_4Cl	0.3g
$MgSO_4 \cdot 7H_2O$	0.1g
Trace metals solution	5.0mL

pH 7.6–8.5 at 25°C

Trace Metals Solution:

Composition per liter:

Disodium EDTA	50.0g
NaOH	11.0g
$CaCl_2 \cdot 2H_2O$	7.34g
$FeSO_4 \cdot 7H_2O$	5.0g
$MnCl_2 \cdot 2H_2O$	2.5g
$ZnSO_4 \cdot 7H_2O$	2.2g
$CoCl_2 \cdot 6H_2O$	0.5g
$(NH_4)_6Mo_7O_{24} \cdot 4H_2O$	0.5g
$CuSO_4 \cdot 5H_2O$	0.2g

Preparation of Trace Metals Solution: Add EDTA to distilled/deionized water and bring volume to 500.0mL. Mix thoroughly. Adjust pH to 6.0 with NaOH. Add remaining components, one by one. Maintain the pH at 6.0. After dissolution of all the salts, adjust the pH to 4.0 with HCl. Store at 4°C.

Preparation of Medium: Add components to distilled/deionized water and bring volume to 1.0L. Mix thoroughly. Adjust pH to 7.6–8.5. Filter sterilize. Aseptically distribute into sterile tubes or flasks.

Use: For the isolation and aerobic cultivation of *Thiobacillus* species.

Thiobacillus Medium (ATCC Medium 125)

Composition per liter:

Sulfur	10.0g
KH_2PO_4	3.0g
$MgSO_4 \cdot 7H_2O$	0.5g
$CaCl_2$	0.25g
$(NH_4)_2SO_4$	0.2g
$FeSO_4 \cdot 7H_2O$	5.0mg

Preparation of Medium: Add components, except sulfur, to tap water and bring volume to 1.0L. Mix thoroughly. Add 1.0g of sulfur to each of 10 flasks. Distribute the broth in 100.0mL volumes into the flasks. Pour the broth down the side of the flask so that the sulfur is not wetted. Autoclave for 30 min at 0 psi pressure–100°C on three consecutive days. Be sure that sulfur remains on the surface of the broth during the sterilization.

Use: For the cultivation and maintenance of a variety of *Thiobacillus* species.

Thiobacillus Medium (ATCC Medium 152)

Composition per liter:

Agar	15.0g
$Na_2S_2O_3 \cdot 5H_2O$	10.0g

NH_4Cl	1.0g
Yeast extract	1.0g
K_2HPO_4	0.6g
$MgCl_2$	0.5g
KH_2PO_4	0.4g
Chlorophenol red	0.08g
$FeCl_3$	0.02g

Preparation of Medium: Add components to distilled/deionized water and bring volume to 1.0L. Mix thoroughly. Gently heat and bring to boiling. Distribute into tubes or flasks. Autoclave for 15 min at 15 psi pressure–121°C. Pour into sterile Petri dishes or leave in tubes.

Use: For the cultivation and maintenance of a variety of *Thiobacillus* species.

Thiobacillus Medium (ATCC Medium 426)

Composition per liter:

$Na_2S_2O_3 \cdot 5H_2O$	10.0g
$Na_2HPO_4 \cdot 7H_2O$	7.9g
KH_2PO_4	1.5g
NH_4Cl	0.3g
$MgSO_4 \cdot 7H_2O$	0.1g
Phenol red	2.0mg
Trace metal solution	5.0mL

pH 8.5 ± 0.2 at 25°C

Trace Metal Solution:
Composition per liter:

EDTA	50.0g
$ZnSO_4$	22.0g
$CaCl_2$	5.54g
$MnCl_2$	5.06g
$FeSO_4 \cdot 7H_2O$	4.99g
$CoCl_2$	1.61g
$CuSO_4$	1.57g
$(NH_4)_2MoO_4 \cdot 4H_2O$	1.10g

Preparation of Trace Metal Solution: Add components to distilled/deionized water and bring volume to 1.0mL. Mix thoroughly. Adjust pH to 6.0 with KOH.

Preparation of Medium: Add components to distilled/deionized water and bring volume to 1.0L. Mix thoroughly. Adjust pH to 8.5 with 10% Na_2CO_3. Distribute into tubes or flasks. Autoclave for 15 min at 15 psi pressure–121°C. Adjust pH to 8.5 with sterile 10% Na_2CO_3 if necessary. The broth should be pink.

Use: For the cultivation and maintenance of a variety of *Thiobacillus* species.

Thiobacillus Medium (ATCC Medium 528)

Composition per liter:

$Na_2S_2O_3$	10.0g
Yeast extract	5.0g
NH_4Cl	1.0g
K_2HPO_4	0.6g
$MgCl_2$	0.5g
KH_2PO_4	0.4g
$MgSO_4$	0.3g
Bromthymol blue	0.03g
$FeCl_3$	0.02g
Heavy metal solution	30.0mL

pH 6.8 ± 0.2 at 25°C

Heavy Metal Solution:
Composition per liter:

Ethylenediamine tetraacetate	1.5g
$FeSO_4 \cdot 7H_2O$	0.2g
$ZnSO_4 \cdot 7H_2O$	0.1g
$MnCl_2 \cdot 4H_2O$	0.02g
Modified Hoagland trace element solution	6.0mL

Preparation of Heavy Metal Solution: Add EDTA to approximately 900.0mL of distilled/deionized water. Dissolve by adjusting pH to 7.0 with NaOH. Bring volume to 1.0L with distilled/deionized water.

Modified Hoagland Trace Element Solution:
Composition per 3.6 liters:

H_3BO_3	11.0g
$MnCl_2 \cdot 4H_2O$	7.0g
$AlCl_3$	1.0g
$CoCl_2$	1.0g
$CuCl_2$	1.0g
KI	1.0g
$NiCl_2$	1.0g
$ZnCl_2$	1.0g
$BaCl_2$	0.5g
KBr	0.5g
LiCl	0.5g
Na_2MoO_4	0.5g
$SeCl_4$	0.5g
$SnCl_2 \cdot 2H_2O$	0.5g
$NaVO_3 \cdot H_2O$	0.1g

Preparation of Modified Hoagland Trace Element Solution: Prepare each component as a separate solution. Dissolve each salt in approximately 100.0mL of distilled /deionized water. Adjust the pH of each solution to below 7.0. Combine all the salt solutions and bring the volume to 3.6L with distilled/deionized water. Adjust the pH to 3–4. A yellow precipitate may form after mixing. After a few days, it will turn into a fine white precipitate. Mix the solution thoroughly before using.

Preparation of Medium: Add components to distilled/deionized water and bring volume to 1.0L. Mix thoroughly. Distribute into tubes or flasks. Autoclave for 15 min at 15 psi pressure–121°C.

Use: For the cultivation and maintenance of a variety of *Thiobacillus* species.

Thiobacillus Medium B

Composition per liter:

Noble agar	15.0g
$Na_2S_2O_3.5H_2O$	5.0g
KH_2PO_4	3.0g
NH_4Cl	0.1g
$MgCl_2$	0.1g
$CaCl_2$	0.1g

pH 4.2 ± 0.2 at 25°C

Preparation of Medium: Add components to distilled/deionized water and bring volume to 1.0L. Mix thoroughly. Gently heat and bring to boiling. Distribute into tubes or flasks. Autoclave for 15 min at 15 psi pressure–121°C. Pour into sterile Petri dishes or leave in tubes.

Use: For the cultivation and maintenance of *Thiobacillus thiooxidans* and *Streptomyces scabies*.

Thiobacillus neapolitanus Medium

Composition per 1002mL:

Solution I	1.0L
Solution II	2.0mL

pH 6.2–7.0 at 25°C

Solution I:

Composition per liter:

$Na_2S_2O_3 \cdot 5H_2O$	10.0g
KH_2PO_4	4.0g
K_2HPO_4	4.0g
$MgSO_4 \cdot 7H_2O$	0.8g
$KHCO_3$	0.7g
NH_4Cl	0.4g

Preparation of Solution I: Add components to distilled/deionized water and bring volume to 1.0L. Mix thoroughly.

Solution II:

Composition per liter:

Disodium EDTA	50.0g
NaOH	11.0g
$CaCl_2 \cdot 2H_2O$	7.34g
$FeSO_4 \cdot 7H_2O$	5.0g
$MnCl_2 \cdot 2H_2O$	2.5g
$ZnSO_4 \cdot 7H_2O$	2.2g
$CoCl_2 \cdot 6H_2O$	0.5g
$(NH_4)_6Mo_7O_{24} \cdot 4H_2O$	0.5g
$CuSO_4 \cdot 5H_2O$	0.2g

Preparation of Solution II: Add EDTA to distilled/deionized water and bring volume to 500.0mL. Mix thoroughly. Adjust pH to 6.0 with NaOH. Add remaining components, one by one. Maintain the pH at 6.0. After dissolution of all the salts, adjust the pH to 4.0 with HCl. Store at 4°C.

Preparation of Medium: Aseptically combine 1.0L of solution I and 2.0mL of solution II. Mix thoroughly. Adjust pH to 6.2–7.0. Distribute into tubes or flasks. Autoclave for 15 min at 15 psi pressure–121°C.

Use: For the isolation and cultivation of *Thiobacillus neapolitanus*.

Thiobacillus novellus Medium

Composition per liter:

$Na_2S_2O_3 \cdot 5H_2O$	10.0g
K_2HPO	4.0g
KH_2PO_4	1.5g
$MgSO_4 \cdot 7H_2O$	0.5g
$(NH_4)_2SO_4$	0.3g
Yeast extract	0.3g
Trace metals solution	10.0mL

pH 6.8–7.2 at 25°C

Trace Metals Solution:

Composition per liter:

Disodium EDTA	50.0g
NaOH	11.0g
$CaCl_2 \cdot 2H_2O$	7.34g
$FeSO_4 \cdot 7H_2O$	5.0g
$MnCl_2 \cdot 2H_2O$	2.5g
$ZnSO_4 \cdot 7H_2O$	2.2g
$CoCl_2 \cdot 6H_2O$	0.5g
$(NH_4)_6Mo_7O_{24} \cdot 4H_2O$	0.5g
$CuSO_4 \cdot 5H_2O$	0.2g

Preparation of Trace Metals Solution: Add EDTA to distilled/deionized water and bring volume to 500.0mL. Mix thoroughly. Adjust pH to 6.0 with NaOH. Add remaining components, one by one. Maintain the pH at 6.0. After dissolution of all the salts, adjust the pH to 4.0 with HCl. Store at 4°C.

Preparation of Medium: Add components to distilled/deionized water and bring volume to 1.0L. Mix thoroughly. Distribute into tubes or flasks. Autoclave for 15 min at 15 psi pressure–121°C.

Use: For the isolation and cultivation of *Thiobacillus novellus*.

Thiobacillus tepidarius Medium

Composition per liter:

Agar	10.0g
$Na_2S_2O_3 \cdot 5H_2O$	4.96g

$MgSO_4 \cdot 7H_2O$	0.8g
NH_4Cl	0.4g
Phosphate solution	100.0mL
Bromcresol purple, saturated solution	2.0mL
Trace metals A-5	1.0mL

Phosphate Solution:

Composition per 100mL:

KH_2PO_4	4.0g
K_2HPO_4	4.0g

Preparation of Phosphate Solution: Add components to distilled/deionized water and bring volume to 100.0mL. Mix thoroughly. Autoclave for 15 min at 15 psi pressure–121°C.

Trace Metals A-5:

Composition per liter:

H_3BO_3	2.86g
$MnCl_2 \cdot 4H_2O$	1.81g
$Na_2MoO_4 \cdot 2H_2O$	0.39g
$ZnSO_4 \cdot 7H_2O$	0.222g
$CuSO_4 \cdot 5H_2O$	0.079g
$Co(NO_3)_2 \cdot 6H_2O$	49.4mg

Preparation of Trace Metals A-5: Add components to distilled/deionized water and bring volume to 1.0L. Mix thoroughly.

Preparation of Medium: Add components, except phosphate solution, to distilled/deionized water and bring volume to 900.0mL. Autoclave for 15 min at 15 psi pressure–121°C. Aseptically add 100.0mL of the sterile phosphate solution. Mix thoroughly. Aseptically distribute into sterile tubes or flasks.

Use: For the cultivation and maintenance of *Thiobacillus tepidarius*.

Thiobacillus thiooxidans Medium

Composition per liter:

Sulfur, powdered	10.0g
KH_2PO_4	5.0g
$MgSO_4 \cdot 7H_2O$	0.5g
$CaCl_2$	0.25g
$(NH_4)_2SO_4$	0.2g
$FeSO_4$	0.01g

pH 7.0 ± 0.2 at 25°C

Preparation of Medium: Add components, except sulfur, to distilled/deionized water and bring volume to 1.0L. Mix thoroughly. Distribute into flasks in 100.0mL volumes. Add 1.0g of sulfur to each flask. Autoclave for 30 min at 0 psi pressure–100°C on three consecutive days.

Use: For the cultivation of *Thiobacillus thiooxidans*.

Thiobacillus thiooxidans Medium

Composition per liter:

Flowers of sulfur	5.0g
K_2HPO	3.5g
$MgSO_4 \cdot 7H_2O$	0.5g
$(NH_4)_2SO_4$	0.3g
$CaCl_2$	0.25g
$FeSO_4 \cdot 7H_2O$	0.02g

pH 4.5 ± 0.2 at 25°C

Preparation of Medium: Add components, except flowers of sulfur, to distilled/deionized water and bring volume to 1.0L. Mix thoroughly. Gently heat and bring to boiling. Distribute into flasks or bottles in 100.0mL volumes. Add 0.5g of flowers of sulfur to each flask or bottle. Autoclave for 15 min at 15 psi pressure–121°C.

Use: For the isolation and cultivation of *Thiobacillus thiooxidans*.

Thiobacillus thioparus Medium

Composition per liter:

$Na_2S_2O_3 \cdot 5H_2O$	5.0g
K_2HPO_4	4.0g
$MgSO_4 \cdot 7H_2O$	0.5g
$(NH_4)_2SO_4$	0.4g
$CaCl_2$	0.25g
$FeSO_4$	0.01g

pH 7.0 ± 0.2 at 25°C

Preparation of Medium: Add components to distilled/deionized water and bring volume to 1.0L. Mix thoroughly. Distribute into tubes or flasks. Autoclave for 15 min at 15 psi pressure–121°C.

Use: For the cultivation of *Thiobacillus thioparus*.

Thiocapsa Medium

Composition per 127mL:

Solution 1	76.2mL
Solution 2 + Solution 3	44.8mL
Solution 4	6.0mL

Solution 1:

Composition per 2.5L:

NaCl	39.68g
$CaCl_2$	2.0g

Preparation of Solution 1: Add components to distilled/deionized water and bring volume to 2.5L. Distribute in 80.0mL volumes into 127mL screw-capped bottles. Autoclave for 15 min at 15 psi pressure–121°C.

Solution 2:

Composition per 100mL:

Sodium ascorbate	2.4g
KCl	1.0g
KH_2PO_4	1.0g
$MgCl_2 \cdot 6H_2O$	0.8g
NH_4Cl	0.8g
Heavy metal solution	50.0mL
Vitamin solution	15.0mL
Vitamin B_{12} solution	3.0mL

Preparation of Solution 2: Add components to distilled/deionized water and bring volume to 100.0mL. Mix thoroughly.

Heavy Metal Solution:

Composition per liter:

Ethylenediamine tetraacetate (EDTA)	1.5g
$FeSO_4 \cdot 7H_2O$	0.2g
$ZnSO_4 \cdot 7H_2O$	0.1g
$MnCl_2 \cdot 4H_2O$	0.02g
Modified Hoagland trace element solution	6.0mL

Preparation of Heavy Metal Solution: Dissolve EDTA in approximately 800.0mL of distilled/deionized water. Add remaining components. Bring volume to 1.0L with distilled/deionized water. Mix thoroughly.

Modified Hoagland Trace Element Solution:

Composition per 3.6L:

H_3BO_3	11.0g
$MnCl_2 \cdot 4H_2O$	7.0g
$AlCl_3$	1.0g
$CoCl_2$	1.0g
$CuCl_2$	1.0g
KI	1.0g
$NiCl_2$	1.0g
$ZnCl_2$	1.0g
$BaCl_2$	0.5g
KBr	0.5g
LiCl	0.5g
Na_2MoO_4	0.5g
$SeCl_4$	0.5g
$SnCl_2 \cdot 2H_2O$	0.5g
$NaVO_3 \cdot H_2O$	0.1g

Preparation of Modified Hoagland Trace Element Solution: Prepare each component as a separate solution. Dissolve each salt in approximately 100.0mL of distilled/deionized water. Adjust the pH of each solution to below 7.0. Combine all the salt solutions and bring the volume to 3.6L with distilled/deionized water. Adjust the pH to 3–4. A yellow precipitate may form after mixing. After a few days, it will turn into a fine white precipitate. Mix the solution thoroughly before using.

Vitamin Solution:

Composition per 100mL:

Pyridoxamine·2HCl	5.0mg
Nicotinic acid	2.0mg
Thiamine	1.0mg
Pantothenic acid	0.5mg
Biotin	0.2mg
p-Aminobenzoic acid	0.1mg

Preparation of Vitamin Solution: Add components to distilled/deionized water and bring volume to 100.0mL. Mix thoroughly.

Vitamin B_{12} Solution:

Composition per 100mL:

Vitamin B_{12} (cyanocobalamin)	2.0mg

Preparation of Vitamin B_{12} Solution: Add vitamin B_{12} to distilled/deionized water and bring volume to 100.0mL. Mix thoroughly.

Solution 3:

Composition per 900mL:

$NaHCO_3$	4.5g

Preparation of Solution 3: Add $NaHCO_3$ to distilled/deionized water and bring volume to 900.0mL. Mix thoroughly. Bubble 100% CO_2 through the solution for 30 min. After CO_2 saturation of solution 3, add solution 2 and immediately filter the mixture through a Seitz filter (or a Millipore) using positive CO_2 pressure to push the liquid through.

Solution 4:

Composition per 200mL:

$Na_2S \cdot 9H_2O$	3.0g

Preparation of Solution 4: Add $Na_2S \cdot 9H_2O$ to distilled/deionized water and bring volume to 200.0mL. Add a magnetic stir bar to the flask. Autoclave for 15 min at 15 psi pressure–121°C. On a magnetic stirrer, slowly add 2.0mL of sterile 2*M* H_2SO_4. This partially neutralizes the solution. The solution should turn yellow. H_2S gas will be liberated—neutralization and distribution of the solution should be done as rapidly as possible under adequate ventilation.

Preparation of Medium: To the 80.0mL of sterile solution 1 in screw-capped bottles, add combined solutions 2 and 3 immediately after filtration and fill bottles to capacity. Mix thoroughly. Aseptically remove 6.0mL of the medium from the bottles and replace it with 6.0mL of neutralized solution 4. Let stand for 24 hr. The medium should form a fine white precipitate before using. To inoculate, remove 6.0mL of the completed medium from the bottles and replace it with 6.0mL of inoculum.

Use: For the cultivation and maintenance of a variety of *Thiocapsa* species.

Thiocyanate Utilization Medium

Composition per 1225mL:

Basal solution	1.0L
Solution C	200.0mL
Solution B	20.0mL
Solution A	5.0mL

Basal Solution:

Composition per liter:

Na_2HPO_4	4.8g
KH_2PO_4	4.4g
$MgSO_4·7H_2O$	0.5g

Preparation of Basal Solution: Add components to distilled/deionized water and bring volume to 1.0L. Mix thoroughly. Autoclave for 15 min at 15 psi pressure–121°C. Cool to 45–50°C.

Solution A:

Composition per100mL:

$FeCl_3·6H_2O$	1.0g
$CaCl_2$	0.1g

Preparation of Solution A: Add components to distilled/deionized water and bring volume to 100.0mL. Mix thoroughly. Filter sterilize.

Solution B:

Composition per 100mL:

D-Glucose	10.0g

Preparation of Solution B: Add glucose to distilled/deionized water and bring volume to 100.0mL. Mix thoroughly. Filter sterilize.

Solution C:

Composition per 200mL:

NaSCN	1.0g

Preparation of Solution C: Add NaSCN to distilled/deionized water and bring volume to 200.0mL. Mix thoroughly. Filter sterilize.

Preparation of Medium: To 1.0L of cooled, sterile basal solution, aseptically add 5.0mL of sterile solution A, 20.0mL of sterile solution B, and 200.0mL of sterile solution C. Mix thoroughly. Aseptically distribute into sterile tubes or flasks.

Use: For the cultivation and maintenance of a variety of microorganisms which can utilize thiocyanate as sole source of nitrogen and sulfur.

Thiogel® Medium

Composition per liter:

Gelatin	50.0g
Pancreatic digest of casein	17.0g
Glucose	6.0g
Papaic digest of soybean meal	3.0g
NaCl	2.5g
Sodium thioglycollate	0.5g
Agar	0.7g
Na_2SO_3	0.1g
L-Cystine	0.25g

pH 7.0 ± 0.2 at 25°C

Source: This medium is available as a premixed powder from BBL Microbiology Systems.

Preparation of Medium: Add components to distilled/deionized water preheated to 50°C and bring volume to 1.0L. Mix thoroughly. Let stand for 5 min. Gently heat while stirring and bring to boiling. Distribute into tubes filling them half full. Autoclave for 15 min at 13 psi pressure–118°C. Pour into sterile Petri dishes or leave in tubes.

Use: For the differentiation of microorganisms based on their ability to liquefy gelatin.

Thioglycollate Gelatin Medium

Composition per liter:

Gelatin	50.0g
Pancreatic digest of casein	15.0g
Yeast extract	5.0g
NaCl	2.5g
Glucose	2.0g
Agar	0.75g
L-Cystine	0.25g
Na_2SO_3	0.1g
Thioglycollic acid	0.3mL

pH 7.0 ± 0.2 at 25°C

Source: This medium is available as a premixed powder from Difco Laboratories.

Preparation of Medium: Add components to distilled/deionized water and bring volume to 1.0L. Mix thoroughly. Gently heat and bring to 50°C. Let stand 5 min. Gently heat and bring to boiling. Distribute into tubes or flasks. Autoclave for 15 min at 15 psi pressure–121°C.

Use: For the determination of gelatin liquefaction by aerobes, microaerophiles and anaerobes without special incubation.

Thioglycollate Medium, Brewer Modified

Composition per liter:

Pancreatic digest of casein	17.5g
Glucose	10.0g
NaCl	5.0g
Papaic digest of soybean meal	2.5g
K_2HPO_4	2.0g
Sodium thioglycollate	1.0g

Agar	0.5g
Methylene blue	0.002g

pH 7.2 ± 0.2 at 25°C

Source: This medium is available as a premixed powder from BBL Microbiology Systems.

Preparation of Medium: Add components to distilled/deionized water and bring volume to 1.0L. Mix thoroughly. Gently heat while stirring and bring to boiling. Distribute into tubes or flasks, filling them half full. Autoclave for 15 min at 15 psi pressure–121°C.

Use: For the cultivation of obligate anaerobes, microaerophiles and facultative organisms.

Thioglycollate Medium, Enriched (THIO Medium) (Thioglycollate Medium with Vitamin K_1 and Hemin)

Composition per liter:

Thioglycollate medium without indicator	1.0L
Hemin solution	0.5mL
Vitamin K_1 solution	0.1mL

pH 7.0 ± 0.2 at 25°C

Thioglycollate Medium without Indicator:
Composition per liter:

Pancreatic digest of casein	17.0g
Glucose	6.0g
Papaic digest of soybean meal	3.0g
NaCl	2.5g
Agar	0.7g
Sodium thioglycollate	0.5g
L-Cystine	0.25g
Na_2SO_3	0.1g

Source: Thioglycollate medium without indicator is available as a premixed powder from Oxoid Unipath and BBL Microbiology Systems.

Preparation of Thioglycollate Medium without Indicator: Add components to distilled/deionized water and bring volume to 1.0L. Mix thoroughly.

Vitamin K_1 Solution:
Composition per 100mL:

Vitamin K_1	1.0g

Preparation of Vitamin K_1 Solution: Add vitamin K_1 to 99.0mL of absolute ethanol. Mix thoroughly.

Hemin Solution:
Composition per 100mL:

Hemin	1.0g
NaOH (1*N* solution)	20.0mL

Preparation of Hemin Solution: Add hemin to 20.0mL of 1*N* NaOH solution. Mix thoroughly. Bring volume to 100.0mL with distilled/deionized water.

Preparation of Medium: Add 0.5mL of hemin solution and 0.1mL of vitamin K_1 solution to 1.0L of thioglycollate medium without indicator. Mix thoroughly. Distribute into screw-capped tubes or flasks. Autoclave for 15 min at 15 psi pressure–121°C. Cool tubes or flasks under 85% N_2 + 10% H_2 + 5% CO_2. Tighten caps.

Use: For the isolation, cultivation and identification of a wide variety of obligate anaerobic bacteria.

Thioglycollate Medium without Glucose

Composition per liter:

Pancreatic digest of casein	15.0g
Yeast extract	5.0g
NaCl	2.5g
Agar	0.75g
L-Cystine	0.25g
Methylene blue	2.0mg
Thioglycollic acid	0.3mL

pH 7.2 ± 0.2 at 25°C

Source: This medium is available as a premixed powder from Difco Laboratories.

Preparation of Medium: Add components to distilled/deionized water and bring volume to 1.0L. Mix thoroughly. Gently heat and bring to boiling. Distribute into tubes or flasks. Autoclave for 15 min at 15 psi pressure–121°C. If medium becomes oxidized before use (methylene blue turns blue) heat in a boiling water bath to expel absorbed O_2. Cool to 25°C.

Use: For the cultivation of anaerobic, microaerophilic and aerobic microorganisms. For use in sterility testing of a variety of specimens.

Thioglycollate Medium without Glucose

Composition per liter:

Pancreatic digest of casein	20.0g
NaCl	2.5g
K_2HPO_4	1.5g
Sodium thioglycollate	0.6g
Agar	0.5g
L-Cystine	0.4g
Na_2SO_3	0.2g
Methylene blue	2.0mg

pH 7.2 ± 0.2 at 25°C

Source: This medium is available as a premixed powder from BBL Microbiology Systems.

Preparation of Medium: Add components to distilled/deionized water and bring volume to 1.0L. Mix thoroughly. Gently heat while stirring and bring to boiling. Distribute into tubes or flasks, filling them half full. Autoclave for 15 min at 15 psi pressure–121°C.

Use: Use as a base for fermentation studies of anaerobic bacteria and for the promotion of endospore formation.

Thioglycollate Medium without Glucose and Indicator

Composition per liter:

Pancreatic digest of casein	15.0g
Yeast extract	5.0g
NaCl	2.5g
Agar	0.75g
L-Cystine	0.25g
Thioglycollic acid	0.3mL

pH 7.2 ± 0.2 at 25°C

Source: This medium is available as a premixed powder from Difco Laboratories.

Preparation of Medium: Add components to distilled/deionized water and bring volume to 1.0L. Mix thoroughly. Gently heat and bring to boiling. Distribute into tubes or flasks. Autoclave for 15 min at 15 psi pressure–121°C. If medium becomes oxidized before use, heat in a boiling water bath to expel absorbed O_2. Cool to 25°C.

Use: For the cultivation of anaerobic, microaerophilic and aerobic microorganisms. For use in sterility testing of a variety of specimens.

Thioglycollate Medium without Indicator

Composition per liter:

Pancreatic digest of casein	15.0g
Yeast extract	5.0g
Glucose	5.0g
NaCl	2.5g
Agar	0.75g
Sodium thioglycollate	0.5g
L-Cystine	0.25g

pH 7.2 ± 0.2 at 25°C

Source: This medium is available as a premixed powder from Difco Laboratories.

Preparation of Medium: Add components to distilled/deionized water and bring volume to 1.0L. Mix thoroughly. Gently heat and bring to boiling. Distribute into tubes or flasks. Autoclave for 15 min at 15 psi pressure–121°C. If medium becomes oxidized before use, heat in a boiling water bath to expel absorbed O_2. Cool to 25°C.

Use: For the cultivation of anaerobic, microaerophilic and aerobic microorganisms. For use in sterility testing of a variety of specimens.

Thiomicrospira denitrificans Agar

Composition per 1001mL:

Solution A	940.0mL
Solution B	40.0mL
Solution C	20.0mL
Solution D	1.0mL

pH 7.0 ± 0.2 at 25°C

Solution A:

Composition per 940mL:

Agar	15.0g
KH_2PO_4	2.0g
KNO_3	2.0g
NH_4Cl	1.0g
$MgSO_4 \cdot 7H_2O$	0.8g
Trace element solution SL-4	2.0mL

Preparation of Solution A: Add components to distilled/deionized water and bring volume to 940.0mL. Mix thoroughly. Gently heat and bring to boiling. Adjust pH to 7.0 with NaOH. Autoclave for 15 min at 15 psi pressure–121°C. Cool to 45–50°C.

Trace Element Solution Sl-4:

Composition per liter:

EDTA	0.5g
$FeSO_4 \cdot 7H_2O$	0.2g
Trace element solution SL-6	100.0

Preparation of Trace Elements Solution SL-4: Add components to distilled/deionized water and bring volume to 1.0L. Mix thoroughly.

Trace Elements Solution SL-6:

Composition per liter:

H_3BO_3	0.3g
$CoCl_2 \cdot 6H_2O$	0.2g
$ZnSO_4 \cdot 7H_2O$	0.10g
$MnCl_2 \cdot 4H_2O$	0.03g
$Na_2MoO_4 \cdot H_2O$	0.03g
$NiCl_2 \cdot 6H_2O$	0.02g
$CuCl_2.2H_2O$	0.01g

Preparation of Trace Elements Solution SL-6: Add components to distilled/deionized water and bring volume to 1.0L. Mix thoroughly. Adjust pH to 3.4.

Solution B:

Composition per 40mL:

$Na_2S_2O_3 \cdot 5H_2O$ 5.0g

Preparation of Solution B: Add $Na_2S_2O_3 \cdot 5H_2O$ to distilled/deionized water and bring volume to 40.0mL. Mix thoroughly. Autoclave for 15 min at 15 psi pressure–121°C. Cool to 45–50°C.

Solution C:

Composition per 20mL:

$NaHCO_3$ 1.0g

Preparation of Solution C: Add $NaHCO_3$ to distilled/deionized water and bring volume to 20.0mL. Mix thoroughly. Filter sterilize.

Solution D:

Composition per liter:

$FeSO_4 \cdot 7H_2O$ 2.0mg
H_2SO_4 (0.1*N* solution) 1.0mL

Preparation of Solution D: Add $FeSO_4 \cdot 7H_2O$ to 1.0mL of 0.1*N* H_2SO_4 solution. Mix thoroughly. Autoclave for 15 min at 15 psi pressure–121°C. Cool to 45–50°C.

Preparation of Medium: Aseptically add 40.0mL of sterile solution B, 20.0mL of sterile solution C and 1.0mL of sterile solution D to 940.0mL of sterile solution A. Mix thoroughly. Aseptically and anaerobically distribute into sterile tubes under 100% N_2.

Use: For the cultivation and maintenance of *Thiomicrospira denitrificans*.

Thiomicrospira denitrificans Broth

Composition per 1001mL:

Solution A 940.0mL
Solution B 40.0mL
Solution C 20.0mL
Solution D 1.0mL

pH 7.0 ± 0.2 at 25°C

Solution A:

Composition per 940mL:

KH_2PO_4 .... 2.0g
KNO_3 .... 2.0g
NH_4Cl 1.0g
$MgSO_4 \cdot 7H_2O$ 0.8g
Trace element solution SL-4 .... 2.0mL

Preparation of Solution A: Add components to distilled/deionized water and bring volume to 940.0mL. Mix thoroughly. Adjust pH to 7.0 with NaOH. Autoclave for 15 min at 15 psi pressure–121°C. Cool to 45–50°C.

Solution B:

Composition per 40mL:

$Na_2S_2O_3 \cdot 5H_2O$ 5.0g

Preparation of Solution B: Add $Na_2S_2O_3 \cdot 5H_2O$ to distilled/deionized water and bring volume to 40.0mL. Mix thoroughly. Autoclave for 15 min at 15 psi pressure–121°C. Cool to 45–50°C.

Solution C:

Composition per 20mL:

$NaHCO_3$ 1.0g

Preparation of Solution C: Add $NaHCO_3$ to distilled/deionized water and bring volume to 20.0mL. Mix thoroughly. Filter sterilize.

Solution D:

Composition per liter:

$FeSO_4 \cdot 7H_2O$ 2.0mg
H_2SO_4 (0.1*N* solution) 1.0mL

Preparation of Solution D: Add $FeSO_4 \cdot 7H_2O$ to 1.0mL of 0.1*N* H_2SO_4 solution. Mix thoroughly. Autoclave for 15 min at 15 psi pressure–121°C. Cool to 45–50°C.

Trace Element Solution Sl-4:

Composition per liter:

EDTA 0.5g
$FeSO_4 \cdot 7H_2O$ 0.2g
Trace element solution SL-6 .... 100.0mL

Preparation of Trace Element Solution SL-4: Add components to distilled/deionized water and bring volume to 1.0L. Mix thoroughly.

Trace Element Solution SL-6:

Composition per liter:

H_3BO_3 .... 0.3g
$CoCl_2 \cdot 6H_2O$ 0.2g
$ZnSO_4 \cdot 7H_2O$ 0.10g
$MnCl_2 \cdot 4H_2O$ 0.03g
$Na_2MoO_4 \cdot H_2O$ 0.03g
$NiCl_2 \cdot 6H_2O$ 0.02g
$CuCl_2 \cdot 2H_2O$ 0.01g

Preparation of Trace Element Solution SL-6: Add components to distilled/deionized water and bring volume to 1.0L. Mix thoroughly. Adjust pH to 3.4.

Preparation of Medium: Aseptically add 40.0mL of sterile solution B, 20.0mL of sterile solution C and 1.0mL of sterile solution D to 940.0mL of sterile solution A. Mix thoroughly. Aseptically and anaerobically distribute into sterile tubes under 100% N_2.

Use: For the cultivation and maintenance of *Thiomicrospira denitrificans*.

Thiomicrospira denitrificans Medium

Composition per liter:

Part I	500.0mL
Part II	500.0mL

pH 7.0–8.0 at 25°C

Part I:

Composition per liter:

NaCl	20.0g
KNO_3	4.0g
$(NH_4)_2SO_4$	2.0g
$MgSO_4 \cdot 7H_2O$	1.5g
K_2HPO_4	0.6g
KH_2PO_4	0.4g
$FeSO_4$ solution	2.0mL
Trace metals solution	2.0mL
HCl, concentrated	1.0mL

Preparation of Part I: Add components to distilled/deionized water and bring volume to 1.0L. Mix thoroughly. Autoclave for 15 min at 15 psi pressure–121°C. Cool to 25°C.

$FeSO_4$ Solution:

Composition per 100mL:

$FeSO_4 \cdot 7H_2O$	0.5g
HCl (1*N* solution)	100.0mL

Preparation of $FeSO_4$ Solution: Combine the $FeSO_4 \cdot 7H_2O$ and 100.0mL of HCl solution. Mix thoroughly.

Trace Metals Solution:

Composition per liter:

Disodium EDTA	50.0g
NaOH	11.0g
$CaCl_2 \cdot 2H_2O$	7.34g
$MnCl_2 \cdot 2H_2O$	2.5g
$ZnSO_4 \cdot 7H_2O$	2.2g
$CoCl_2 \cdot 6H_2O$	0.5g
$(NH_4)_6Mo_7O_{24} \cdot 4H_2O$	0.5g
$CuSO_4 \cdot 5H_2O$	0.2g

Preparation of Trace Metals Solution: Add EDTA to distilled/deionized water and bring volume to 500.0mL. Mix thoroughly. Adjust pH to 6.0 with NaOH. Add remaining components, one by one. Maintain the pH at 6.0. After dissolution of all the salts, adjust the pH to 4.0 with HCl. Store at 4°C.

Part II:

Composition per liter:

$Na_2S_2O_3 \cdot 5H_2O$	10.0g
$NaHCO_3$	3.0g
NaOH	0.05g

Preparation of Part II: Add components to distilled/deionized water and bring volume to 1.0L. Mix thoroughly. Autoclave for 15 min at 15 psi pressure–121°C. Cool to 25°C.

Preparation of Medium: Aseptically combine 500.0mL of sterile part I and 500.0mL of sterile part II. Mix thoroughly. Aseptically distribute into sterile tubes or flasks.

Use: For the isolation and cultivation of *Thiomicrospira denitrificans*.

Thiomicrospira Medium (ATCC Medium 1036)

Composition per liter:

NaCl	25.0g
$Na_2S_2O_3 \cdot 5H_2O$	8.0g
$MgSO_4 \cdot 7H_2O$	1.5g
$(NH_4)_2SO_4$	1.0g
K_2HPO_4	0.5g
$CaCl_2$	0.3g
Vitamin B_{12}	15.0μg
Vishniac and Santer trace metals	0.2mL
Bromcresol purple (0.05% solution)	0.1mL

pH 7.2 ± 0.2 at 25°C

Vishniac and Santer Trace Metals:

Composition per liter:

Ethylenediamine tetraacetic acid (EDTA)	50.0g
$ZnSO_4 \cdot 7H_2O$	22.0g
$CaCl_2$	5.54g
$MnCl_2 \cdot 4H_2O$	5.06g
$FeSO_4 \cdot 7H_2O$	4.99g
$CoCl_2 \cdot 6H_2O$	1.61g
$CuSO_4 \cdot 5H_2O$	1.57g
$(NH_4)_6Mo_7O_{24} \cdot 4H_2O$	1.10g

Preparation of Vishniac and Santer Trace Metals: Add components to distilled/deionized water and bring volume to 1.0L. Adjust pH to 6.0 with KOH. Mix thoroughly.

Preparation of Medium: Add components to distilled/deionized water and bring volume to 1.0L. Mix thoroughly. Adjust pH to 7.2. Filter sterilize. Aseptically distribute into sterile tubes or flasks.

Use: For the cultivation and maintenance of *Thiomicrospira* species.

Thiomicrospira Medium (ATCC Medium 1422)

Composition per liter:

NaCl	25.1g
Tris·HCl	3.07g
$Na_2S_2O_3 \cdot 5H_2O$	2.48g
$MgSO_4 \cdot 7H_2O$	1.5g
$(NH_4)_2SO_4$	1.0g
KH_2PO_4	0.42g
$CaCl_2 \cdot 2H_2O$	0.29g
$NaHCO_3$	0.20g
Phenol red (0.5% solution)	1.0mL
Vishniac and Santer trace metals	0.2mL

pH 7.5 ± 0.2 at 25°C

Vishniac and Santer Trace Metals:
Composition per liter:

Ethylenediamine tetraacetic acid (EDTA)	50.0g
$ZnSO_4 \cdot 7H_2O$	22.0g
$CaCl_2$	5.54g
$MnCl_2 \cdot 4H_2O$	5.06g
$FeSO_4 \cdot 7H_2O$	4.99g
$CoCl_2 \cdot 6H_2O$	1.61g
$CuSO_4 \cdot 5H_2O$	1.57g
$(NH_4)_6Mo_7O_{24} \cdot 4H_2O$	1.10g

Preparation of Vishniac and Santer Trace Metals: Add components to distilled/deionized water and bring volume to 1.0L. Adjust pH to 6.0 with KOH. Mix thoroughly.

Preparation of Medium: Add components to distilled/deionized water and bring volume to 1.0L. Mix thoroughly. Adjust pH to 7.5. Filter sterilize. Aseptically distribute into sterile tubes or flasks.

Use: For the cultivation and maintenance of *Thiomicrospira* species.

Thiomicrospira pelophila Medium

Composition per liter:

NaCl	25.0g
Agar	10.0g
$Na_2S_2O_3 \cdot 5H_2O$	5.0-8.0g
$MgSO_4 \cdot 7H_2O$	1.5g
$(NH_4)_2SO_4$	1.0g
K_2HPO_4	0.5g
$CaCl_2$	0.3g
Vitamin B_{12}	0.15mg
Trace metals solution	0.2mL

Trace Metals Solution:
Composition per liter:

Disodium EDTA	50.0g
NaOH	11.0g
$CaCl_2 \cdot 2H_2O$	7.34g
$FeSO_4 \cdot 7H_2O$	5.0g
$MnCl_2 \cdot 2H_2O$	2.5g
$ZnSO_4 \cdot 7H_2O$	2.2g
$CoCl_2 \cdot 6H_2O$	0.5g
$(NH_4)_6Mo_7O_{24} \cdot 4H_2O$	0.5g
$CuSO_4 \cdot 5H_2O$	0.2g

Preparation of Trace Metals Solution: Add EDTA to distilled/deionized water and bring volume to 500.0mL. Mix thoroughly. Adjust pH to 6.0 with NaOH. Add remaining components, one by one. Maintain the pH at 6.0. After dissolution of all the salts, adjust the pH to 4.0 with HCl. Store at 4°C.

Preparation of Medium: Add components to distilled/deionized water and bring volume to 1.0L. Mix thoroughly. Gently heat and bring to boiling. Distribute into tubes or flasks. Autoclave for 15 min at 15 psi pressure–121°C. Pour into sterile Petri dishes or leave in tubes.

Use: For the cultivation of *Thiomicrospira pelophila.*

Thiosphaera Agar

Composition per liter:

Agar	15.0g
Na_2HPO_4	4.2g
KH_2PO_4	1.5g
NH_4Cl	0.3g
$MgSO_4 \cdot 7H_2O$	0.1g
KNO_3	0.1g
Vishniac and Santer trace metals	2.0mL

pH 8.0–8.2 at 25°C

Vishniac and Santer Trace Metals:
Composition per liter:

Ethylenediamine tetraacetic acid (EDTA)	50.0g
$ZnSO_4 \cdot 7H_2O$	22.0g
$CaCl_2$	5.54g
$MnCl_2 \cdot 4H_2O$	5.06g
$FeSO_4 \cdot 7H_2O$	4.99g
$CoCl_2 \cdot 6H_2O$	1.61g
$CuSO_4 \cdot 5H_2O$	1.57g
$(NH_4)_6Mo_7O_{24} \cdot 4H_2O$	1.10g

Preparation of Vishniac and Santer Trace Metals: Add components to distilled/deionized water and bring volume to 1.0L. Adjust pH to 6.0 with KOH. Mix thoroughly.

Preparation of Medium: Add components, except agar, to distilled/deionized water and bring volume to 500.0mL. Mix thoroughly. Adjust pH to 8.0–8.2. Filter sterilize. Warm to 45–50°C. Add agar to distilled/deionized water and bring volume to 500.0mL. Mix thoroughly. Gently heat and bring to boiling. Autoclave for 15 min at 15 psi pressure–

121°C. Cool to 45–50°C. Aseptically combine the two sterile solutions. Mix thoroughly. Pour into sterile Petri dishes or distribute into sterile tubes.

Use: For the cultivation and maintenance of *Thiosphaera pantotropha.*

Thiosphaera Broth

Composition per liter:

Na_2HPO_4	4.2g
KH_2PO_4	1.5g
NH_4Cl	0.3g
$MgSO_4 \cdot 7H_2O$	0.1g
KNO_3	0.1g
Vishniac and Santer trace metals	2.0mL

pH 8.0–8.2 at 25°C

Vishniac and Santer Trace Metals:

Composition per liter:

Ethylenediamine tetraacetic acid (EDTA)	50.0g
$ZnSO_4 \cdot 7H_2O$	22.0g
$CaCl_2$	5.54g
$MnCl_2 \cdot 4H_2O$	5.06g
$FeSO_4 \cdot 7H_2O$	4.99g
$CoCl_2 \cdot 6H_2O$	1.61g
$CuSO_4 \cdot 5H_2O$	1.57g
$(NH_4)_6Mo_7O_{24} \cdot 4H_2O$	1.10g

Preparation of Vishniac and Santer Trace Metals: Add components to distilled/deionized water and bring volume to 1.0L. Adjust pH to 6.0 with KOH. Mix thoroughly.

Preparation of Medium: Add components to distilled/deionized water and bring volume to 1.0L. Mix thoroughly. Adjust pH to 8.0–8.2. Filter sterilize. Aseptically distribute into sterile tubes or flasks.

Use: For the cultivation and maintenance of *Thiosphaera pantotropha.*

Thiosulfate–Oxidizing Medium

Composition per liter:

K_2HPO_4	2.0g
$MgSO_4 \cdot 7H_2O$	0.1g
$CaCl_2 \cdot 2H_2O$	0.1g
$FeCl_3 \cdot 6H_2O$	0.02g
$(NH_4)_2SO_4$ solution	100.0mL
Thiosulfate solution	100.0mL

pH 7.8 ± 0.2 at 25°C

$(NH_4)_2SO_4$ Solution:

Composition per 100mL:

$(NH_4)_2SO_4$	0.1g

Preparation of $(NH_4)_2SO_4$ Solution: Add the $(NH_4)_2SO_4$ to distilled/deionized water and bring volume to 100.0mL. Mix thoroughly. Autoclave for 15 min at 15 psi pressure–121°C. Cool to 45–50°C

Thiosulfate Solution:

Composition per 100mL:

$Na_2S_2O_3 \cdot 5H_2O$	10.0g

Preparation of Thiosulfate Solution: Add the $Na_2S_2O_3 \cdot 5H_2O$ to distilled/deionized water and bring volume to 100.0mL. Mix thoroughly. Autoclave for 15 min at 15 psi pressure–121°C. Cool to 45–50°C.

Preparation of Medium: Add components, except $(NH_4)_2SO_4$ solution and thiosulfate solution, to distilled/deionized water and bring volume to 800.0mL. Mix thoroughly. Autoclave for 15 min at 15 psi pressure–121°C. Cool to 45–50°C. Aseptically add the sterile $(NH_4)_2SO_4$ solution and the sterile thiosulfate solution. Mix thoroughly. Adjust the pH to 7.8 if necessary. Aseptically distribute into sterile tubes or flasks.

Use: For the isolation and cultivation of iron and sulfur bacteria.

TPGY Medium (Thioglycollate Peptone Glucose Yeast Extract Medium)

Composition per liter:

Pancreatic digest of casein	50.0g
Peptone	5.0g
Yeast extract	5.0g
Glucose	1.0g
Sodium thioglycollate	1.0g

pH 7.1 ± 0.2 at 25°C

Preparation of Medium: Add components to distilled/deionized water and bring volume to 1.0L. Mix thoroughly. Distribute into tubes or flasks. Autoclave for 15 min at 15 psi pressure–121°C.

Use: For the cultivation of a variety of anaerobic bacteria.

TPL Medium (Thioglycollate Potato Liver Medium)

Composition per liter:

Potato	200.0g
Yeast extract	31.0g
Liver	25.0g
Glycerol	15.0g
Agar	15.0g
Meat extract	5.5g

Glucose 7.5g
Peptone 2.5g
NaCl 2.5g
Sodium thioglycollate 0.5g
Methylene blue 1.0mg
pH 7.0 ± 0.2 at 25°C

Preparation of Medium: Add peeled, sliced potato to approximately 500.0mL of distilled/deionized water. Gently heat and bring to boiling. Continue boiling for 30 min. Filter through cheesecloth. Cut up liver into small pieces and to approximately 150.0mL of distilled/deionized water. Gently heat and bring to boiling. Continue boiling for 30 min. Filter through cheesecloth. Add boiled potato solids, boiled liver solids and remaining components to distilled/deionized water and bring volume to 1.0L. Mix thoroughly. Gently heat and bring to boiling. Distribute into tubes or flasks. Make sure each of the tubes receive a few pieces of liver. Autoclave for 15 min at 15 psi pressure–121°C.

Use: For the cultivation and maintenance of *Pseudomonas* species.

Trace Element Solution HO–LE

Composition per liter:

H_3BO_3 .... 2.85g
$MnCl_2 \cdot 4H_2O$ 1.8g
Sodium tartrate 1.77g
$FeSO_4 \cdot 7H_2O$ 1.36g
$CoCl_2 \cdot 6H_2O$ 0.04g
$CuCl_2.2H_2O$ 0.027g
$Na_2MoO_4 \cdot 2H_2O$ 0.025g
$ZnCl_2$ 0.020g

Preparation of Trace Element Solution HO-LE: Add components to distilled/deionized water and bring volume to 1.0L. Mix thoroughly. Filter sterilize.

Use: For the enrichment of other media requiring added trace metals.

Tributyrin Agar

Composition per liter:

Agar 15.0g
Tributyrin (glyceryl tributyrate) 10.0g
Peptone 5.0g
Yeast extract 3.0g
pH 7.5 ± 0.2 at 25°C

Source: Available as a prepared medium from Oxoid Unipath.

Preparation of Medium: Add components to distilled/deionized water and bring volume to 1.0L. Mix thoroughly. Gently heat and bring to boiling. Distribute into tubes or flasks. Autoclave for 15 min at 15 psi pressure–121°C. Pour into sterile Petri dishes.

Use: For the cultivation and enumeration of lipolytic fungi and bacteria, especially *Staphylococcus* species, *Flavobacterium* species, *Clostridium* species and *Pseudomonas* species from butter. Lipolytic bacteria appear as colonies surrounded by a clear zone.

Trichlorophenol Medium

Composition per liter:

Pancreatic digest of casein 8.5g
NaCl 2.5g
Papaic digest of soybean meal 1.5g
K_2HPO_4 .... 1.25g
Glucose 1.25g
2,4,6-Trichlorophenol 1.25g
pH 7.3 ± 0.2 at 25°C

Preparation of Medium: Add components to distilled/deionized water and bring volume to 1.0L. Mix thoroughly. Gently heat until dissolved. Distribute into tubes or flasks. Autoclave for 15 min at 15 psi pressure–121°C.

Use: For the cultivation and maintenance of *Arthrobacter* species and other microorganisms which can degrade chlorinated phenols.

Triple Sugar Iron Agar (TSI Agar)

Composition per liter:

Peptone 20.0g
Agar 12.0g
Lactose 10.0g
Sucrose 10.0g
NaCl 5.0g
Beef extract 3.0g
Yeast extract 3.0g
Glucose 1.0g
Ferric citrate 0.3g
$Na_2S_2O_3$ 0.3g
Phenol red 0.025g
pH 7.4 ± 0.2 at 25°C

Source: This medium is available as a premixed powder from Difco Laboratories and Oxoid Unipath.

Preparation of Medium: Add components to distilled/deionized water and bring volume to 1.0L. Mix thoroughly. Gently heat and bring to boiling. Distribute into tubes or flasks. Autoclave for 15 min at 15 psi pressure–121°C. Allow tubes to cool in a slanted position to form a 1.0 in. butt.

Use: For the differentiation of members of the Enterobacteriaceae based on their fermentation of lactose, sucrose and glucose and the production of H_2S.

Triple Sugar Iron Agar (TSI Agar)

Composition per liter:

Agar	13.0g
Pancreatic digest of casein	10.0g
Peptic digest of animal tissue	10.0g
Lactose	10.0g
Sucrose	10.0g
NaCl	5.0g
Glucose	1.0g
$Fe(NH_4)_2(SO_4)_2 \cdot 6H_2O$	0.2g
$Na_2S_2O_3$	0.2g
Phenol red	0.025g

pH 7.3 ± 0.2 at 25°C

Source: This medium is available as a premixed powder from BBL Microbiology Systems.

Preparation of Medium: Add components to distilled/deionized water and bring volume to 1.0L. Mix thoroughly. Gently heat and bring to boiling. Distribute into tubes or flasks. Autoclave for 15 min at 15 psi pressure–121°C. Allow tubes to cool in a slanted position to form a 1.0 inch butt.

Use: For the differentiation of members of the Enterobacteriaceae based on their fermentation of lactose, sucrose and glucose and the production of H_2S.

Tris YP Agar (Tris Yeast Extract Peptone Agar)

Composition per liter:

Agar	19.0g
Yeast extract	3.0g
Glucose	1.0g
Peptone	0.6g
Tris-buffer (0.05*M*, pH 7.5)	1.0L

pH 7.5 ± 0.2 at 25°C

Preparation of Medium: Add components to distilled/deionized water and bring volume to 1.0L. For top layer agar add 6.0g of agar instead of 19.0g. Mix thoroughly. Gently heat and bring to boiling. Distribute into tubes or flasks. Autoclave for 15 min at 15 psi pressure–121°C. Pour into sterile Petri dishes.

Use: For the cultivation and maintenance of *Bdellovibrio* species.

Tris YP Broth (Tris Yeast Extract Peptone Broth)

Composition per liter:

Yeast extract	3.0g
Glucose	1.0g
Peptone	0.6g
Tris-buffer (0.05*M*, pH 7.5)	1.0L

pH 7.5 ± 0.2 at 25°C

Preparation of Medium: Add components to distilled/deionized water and bring volume to 1.0L. Mix thoroughly. Distribute into tubes or flasks. Autoclave for 15 min at 15 psi pressure–121°C.

Use: For the cultivation and maintenance of *Bdellovibrio* species.

Tryptic Digest Broth

Composition per liter:

Tryptic digest of beef heart	10.0g
NaCl	5.0g
Glucose	1.0g

pH 7.6 ± 0.2 at 25°C

Source: This medium is available as a premixed powder from Difco Laboratories.

Preparation of Medium: Add components to distilled/deionized water and bring volume to 1.0L. Mix thoroughly. Distribute into tubes or flasks. Autoclave for 15 min at 15 psi pressure–121°C.

Use: For use as a base medium to which enrichments are added for cultivation of fastidious microorganisms.

Tryptic Nitrate Medium

Composition per liter:

Tryptose	20.0g
Na_2HPO_4	2.0g
Agar	1.0g
Glucose	1.0g
KNO_3	1.0g

pH 7.6 ± 0.2 at 25°C

Source: This medium is available as a premixed powder from Difco Laboratories.

Preparation of Medium: Add components to distilled/deionized water and bring volume to 1.0L. Mix thoroughly. Gently heat and bring to boiling. Distribute into tubes in 10.0mL volumes. Autoclave for 15 min at 15 psi pressure–121°C.

Use: For the cultivation and differentiation of *Pseudomonas* and related genera. For the differentiation of bacteria based on their reduction of nitrate to nitrite. After incubation of the bacterium in tryptic nitrate medium for 18–24 hr, sulfanillic acid and α-naphthol reagents are added. Nitrate reduction is indicated by the development of a red to violet color.

Trypticase™ Agar Base

Composition per liter:

Pancreatic digest of casein	20.0g
Agar	3.5g
Phenol red	0.02g

pH 7.4 ± 0.2 at 25°C

Source: This medium is available as a premixed powder from BBL Microbiology Systems, Difco Laboratories and Oxoid Unipath.

Preparation of Medium: Add components to distilled/deionized water and bring volume to 1.0L. Mix thoroughly. Gently heat and bring to boiling. Distribute into tubes or flasks. Autoclave for 15 min at 15 psi pressure–121°C. Pour into sterile Petri dishes or leave in tubes.

Use: For the differentiation of microorganisms based on their motility.

Trypticase™ Agar Base with Carbohydrate

Composition per liter:

Pancreatic digest of casein	20.0g
Carbohydrate	5.0g
Agar	3.5g
Phenol red	0.02g

pH 7.4 ± 0.2 at 25°C

Preparation of Medium: Add components to distilled/deionized water and bring volume to 1.0L. Mix thoroughly. Gently heat and bring to boiling. Distribute into tubes. Autoclave for 15 min at 13 psi pressure–118°C. Do not overheat. Pour into sterile Petri dishes or leave in tubes.

Use: For differentiation of microorganisms based on their motility and fermentation reactions. Fermentation of carbohydrate changes the medium yellow.

Trypticase™ Broth, Supplemented

Composition per liter:

Pancreatic digest of casein	20.0g
$MgSO_4 \cdot 7H_2O$	0.015g
$FeCl_3$	7.0mg

pH 7.2 ± 0.2 at 25°C

Preparation of Medium: Add components to distilled/deionized water and bring volume to 1.0L. Mix thoroughly. Distribute into tubes or flasks. Autoclave for 15 min at 15 psi pressure–121°C.

Use: For the cultivation of *Bacillus stearothermophilus.*

Trypticase™ Glucose Extract Agar

Composition per liter:

Agar	15.0g
Pancreatic digest of casein	5.0g
Beef extract	3.0g
Glucose	1.0g

pH 7.0 ± 0.2 at 25°C

Source: This medium is available as a premixed powder from BBL Microbiology Systems.

Preparation of Medium: Add components to distilled/deionized water and bring volume to 1.0L. Mix thoroughly. Gently heat and bring to boiling. Distribute into tubes or flasks. Autoclave for 15 min at 15 psi pressure–121°C. Pour into sterile Petri dishes or leave in tubes.

Use: For enumeration of bacteria in water and other specimens.

Trypticase™ Soy Agar

Composition per liter:

Pancreatic digest of casein	17.0g
Agar	15.0g
NaCl	5.0g
Papaic digest of soybean meal	3.0g
K_2HPO_4	2.5g
Glucose	2.5g

pH 7.3 ± 0.2 at 25°C

Preparation of Medium: Add components to distilled/deionized water and bring volume to 1.0L. Mix thoroughly. Gently heat and bring to boiling. Distribute into tubes or flasks. Autoclave for 15 min at 15 psi pressure–121°C. Pour into sterile Petri dishes or leave in tubes.

Use: For the cultivation and maintenance of a wide variety of heterotrophic microorganisms.

Trypticase™ Soy Agar (ATCC Medium 18)

Composition per liter:

Pancreatic digest of casein	17.0g
Agar	15.0g
NaCl	5.0g
Papaic digest of soybean meal	3.0g

K_2HPO_4	2.5g
Glucose	2.5g

pH 7.3 ± 0.2 at 25°C

Preparation of Medium: Add components to distilled/deionized water and bring volume to 1.0L. Mix thoroughly. Distribute into tubes or flasks. Autoclave for 15 min at 15 psi pressure–121°C.

Use: For the cultivation of a wide variety of fastidious and nonfastidious microorganisms from clinical and nonclinical specimens. Also used for the rapid estimation of the bacteriological quality of water.

Trypticase™ Soy Agar (Tryptic Soy Agar) (Soybean Casein Digest Agar) (ATCC Medium 77)

Composition per liter:

Pancreatic digest of casein	15.0g
Agar	15.0g
Papaic digest of soybean meal	5.0g
NaCl	5.0g

pH 7.3 ± 0.2 at 25°C

Source: This medium is available as a premixed powder from BBL Microbiology Systems and Difco Laboratories.

Preparation of Medium: Add components to distilled/deionized water and bring volume to 1.0L. Mix thoroughly. Gently heat and bring to boiling. Distribute into tubes or flasks. Autoclave for 15 min at 15 psi pressure–121°C. Do not overheat. Pour into sterile Petri dishes or leave in tubes.

Use: For the isolation and cultivation of a wide variety of fastidious as well as nonfastidious microorganisms.

Trypticase™ Soy Agar, Modified

Composition per liter:

Pancreatic digest of casein	17.0g
Agar	15.0g
NaCl	5.0g
Yeast extract	4.0g
Papaic digest of soybean meal	3.0g
K_2HPO_4	2.5g

Preparation of Medium: Add components to distilled/deionized water and bring volume to 1.0L. Mix thoroughly. Gently heat and bring to boiling. Distribute into tubes or flasks. Autoclave for 15 min at 15 psi pressure–121°C. Pour into sterile Petri dishes or leave in tubes.

Use: For the cultivation and maintenance of the *Simonsiella* species.

Trypticase™ Soy Agar, Modified with Horse Serum

Composition per liter:

Pancreatic digest of casein	17.0g
Agar	15.0g
NaCl	5.0g
Yeast extract	4.0g
Papaic digest of soybean meal	3.0g
K_2HPO_4	2.5g
Horse serum	100.0mL

pH 7.3 ± 0.2 at 25°C

Preparation of Medium: Add components, except horse serum, to distilled/deionized water and bring volume to 900.0mL. Mix thoroughly. Gently heat and bring to boiling. Autoclave for 15 min at 15 psi pressure–121°C. Cool to 45–50°C. Aseptically add sterile horse serum. Mix thoroughly. Pour into sterile Petri dishes or distribute into sterile tubes.

Use: For the cultivation and maintenance of the *Simonsiella* species, the *Alysiella* species, and the *Moraxella* species.

Trypticase™ Soy Agar with NaCl

Composition per liter:

NaCl	30.0g
Pancreatic digest of casein	15.0g
Agar	15.0g
Papaic digest of soybean meal	5.0g
Bile salts No. 3	1.0g

pH 7.3 ± 0.2 at 25°C

Preparation of Medium: Add components to distilled/deionized water and bring volume to 1.0L. Mix thoroughly. Gently heat and bring to boiling. Distribute into tubes or flasks. Autoclave for 15 min at 15 psi pressure–121°C. Pour into sterile Petri dishes or leave in tubes.

Use: For the cultivation and maintenance of *Vibrio alginolyticus*.

Trypticase™ Soy Agar with 3% NaCl (TSA NaCl)

Composition per liter:

NaCl	30.0g
Agar	15.0g
Pancreatic digest of casein	15.0g
Papaic digest of soybean meal	5.0g

pH 7.3 ± 0.2 at 25°C

Preparation of Medium: Add components to distilled/deionized water and bring volume to 1.0L. Mix thoroughly. Gently heat and bring to boiling. Distribute into tubes or flasks. Autoclave for 15 min at 15 psi pressure–121°C. Pour into sterile Petri dishes or leave in tubes.

Use: For the cultivation of halophilic microorganisms isolated from foods.

Trypticase™ Soy Agar with NaCl, Horse Serum and Penicillin

Composition per liter:

Pancreatic digest of casein	15.0g
Agar	15.0g
Papaic digest of soybean meal	5.0g
NaCl	35.0g
Horse serum, inactivated	100.0mL
Penicillin solution	10.0mL

pH 7.3 ± 0.2 at 25°C

Penicillin Solution:

Composition per 10mL:

Penicillin	1,000,000U

Preparation of Penicillin Solution: Add penicillin to distilled/deionized water and bring volume to 10.0mL. Mix thoroughly. Filter sterilize.

Preparation of Medium: Add components, except horse serum and penicillin solution, to distilled/deionized water and bring volume to 890.0mL. Mix thoroughly. Gently heat and bring to boiling. Autoclave for 15 min at 15 psi pressure–121°C. Do not overheat. Cool to 50°C. Aseptically add 100.0mL of sterile horse serum and 10.0mL of sterile penicillin solution. Mix thoroughly. Pour into sterile Petri dishes or distribute into sterile tubes.

Use: For the isolation and cultivation of fungi.

Trypticase™ Soy Agar Yeast Extract (TSAYE)

Composition per liter:

Pancreatic digest of casein	17.0g
Agar	15.0g
Yeast extract	6.0g
NaCl	5.0g
Papaic digest of soybean meal	3.0g
K_2HPO_4	2.5g
Glucose	2.5g

pH 7.3 ± 0.2 at 25°C

Preparation of Medium: Add components to distilled/deionized water and bring volume to 1.0L. Mix thoroughly. Gently heat and bring to boiling. Distribute into tubes or flasks. Autoclave for 15 min at 15 psi pressure–121°C. Pour into sterile Petri dishes or leave in tubes.

Use: For the cultivation and maintenance of a wide variety of heterotrophic microorganisms.

Trypticase™ Soy Broth (Soybean Casein Digest Broth, USP)

Composition per liter:

Pancreatic digest of casein	17.0g
NaCl	5.0g
Papaic digest of soybean meal	3.0g
K_2HPO_4	2.5g
Glucose	2.5g

pH 7.3 ± 0.2 at 25°C

Source: This medium is available as a premixed powder from BBL Microbiology Systems and Difco Laboratories.

Preparation of Medium: Add components to distilled/deionized water and bring volume to 1.0L. Mix thoroughly. Distribute into tubes or flasks. Autoclave for 15 min at 15 psi pressure–121°C.

Use: For the cultivation of a wide variety of fastidious and nonfastidious microorganisms from clinical and nonclinical specimens. Also used for the rapid estimation of the bacteriological quality of water.

Tryptone Soya Agar

Composition per liter:

Agar	15.0g
Pancreatic digest of casein	15.0g
NaCl	5.0g
Pancreatic digest of soybean meal	5.0g

pH 7.3 ± 0.2 at 25°C

Source: This medium is available as a premixed powder from Oxoid Unipath.

Preparation of Medium: Add components to distilled/deionized water and bring volume to 1.0L. Mix thoroughly. Gently heat and bring to boiling. Distribute into tubes or flasks. Autoclave for 15 min at 15 psi pressure–121°C. Pour into sterile Petri dishes or leave in tubes.

Use: For cultivation and maintenance of a wide variety of microorganisms.

Tryptone Soya Broth

Composition per liter:

Pancreatic digest of casein	17.0g
NaCl	5.0g
Pancreatic digest of soybean meal	3.0g
K_2HPO_4	2.5g
Glucose	2.5g

pH 7.3 ± 0.2 at 25°C

Source: This medium is available as a premixed powder from Oxoid Unipath.

Preparation of Medium: Add components to distilled/deionized water and bring volume to 1.0L. Mix thoroughly. Distribute into tubes or flasks. Autoclave for 15 min at 15 psi pressure–121°C.

Use: For the cultivation of a wide variety of microorganisms.

Tryptone Water Broth (Tryptone Broth)

Composition per liter:

Pancreatic digest of casein	10.0g
NaCl	5.0g

pH 7.5 ± 0.2 at 25°C

Source: This medium is available as a premixed powder from Oxoid Unipath.

Preparation of Medium: Dissolve 15.0g in 1L of distilled water and distribute into final containers. Sterilize by autoclaving at 121°C for 15 min.

Use: For the cultivation of production of indole by microorganisms.

Tryptose Broth

Composition per liter:

Pancreatic digest of casein	10.0g
Peptic digest of animal tissue	10.0g
NaCl	5.0g
Glucose	1.0g

pH 7.2 ± 0.2 at 25°C

Source: This medium is available as a premixed powder from Difco Laboratories.

Preparation of Medium: Add components to distilled/deionized water and bring volume to 1.0L. Mix thoroughly. Distribute into tubes or flasks. Autoclave for 15 min at 15 psi pressure–121°C.

Use: For the cultivation of fastidious aerobic and facultative microorganisms including streptococci.

TS Soil Extract (Trypticase™ Soy Soil Extract)

Composition per liter:

Pancreatic digest of casein	17.0g
Agar	15.0g
NaCl	5.0g
Papaic digest of soybean meal	3.0g
K_2HPO_4	2.5g
Glucose	2.5g
Soil extract	250.0mL

Soil Extract:

Composition per 400mL:

African violet soil	154.0g
Na_2CO_3	0.4g

Preparation of Soil Extract: Add components to tap water and bring volume to 400.0mL. Autoclave for 60 min at 15 psi pressure–121°C. Filter through Whatman filter paper.

Preparation of Medium: Add components to tap water and bring volume to 1.0L. Mix thoroughly. Gently heat and bring to boiling. Distribute into tubes or flasks. Autoclave for 15 min at 15 psi pressure–121°C. Pour into sterile Petri dishes or leave in tubes.

Use: For cultivation and maintenance of *Bacillus xerothermodurans*.

Tween™ 80 Hydrolysis Broth

Composition per liter:

Na_2HPO_4	5.79g
NaH_2PO_4	3.53g
Neutral red	0.02g
Tween™ 80	5.00mL

pH 7.0 ± 0.2 at 25°C

Preparation of Medium: Add components to distilled/deionized water and bring volume to 1.0L. Mix thoroughly. Distribute into tubes or flasks. Autoclave for 15 min at 15 psi pressure–121°C.

Use: For the differentiation of *Mycobacterium* species. Strains that hydrolyze Tween™ 80 within 5 days turn the medium pink to red.

Tween™ 80 Hydrolysis Broth

Composition per 125mL:

Neutral red	0.1g
Solution 1	38.9mL
Solution 2	61.1mL
Tween™ 80	25.0mL

pH 7.0 ± 0.2 at 25°C

Solution 1:

Composition per 400mL:

KH_2PO_4 .... 22.7g

Preparation of Solution 1: Add KH_2PO_4 to distilled/deionized water and bring volume to 400.0mL. Mix thoroughly.

Solution 2:

Composition per 400mL:

Na_2HPO_4 .... 23.8g

Preparation of Solution 2: Add Na_2HPO_4 to distilled/deionized water and bring volume to 400.0mL. Mix thoroughly.

Preparation of Medium: Combine components. Mix thoroughly. Distribute into tubes or flasks. Autoclave for 15 min at 15 psi pressure–121°C.

Use: For the differentiation of *Mycobacterium* species. Strains that hydrolyze Tween™ 80 within 5 days turn the medium pink to red.

Tween™ 80 Hydrolysis Broth

Composition per 102.5mL:

$NaHPO_4$ (0.066*M* solution) 61.1mL
KH_2PO_4 (0.066*M* solution) 38.9mL
Neutral red (0.1% solution) 2.0mL
Tween™ 80 .... 0.5mL

pH 7.0 ± 0.2 at 25°C

Preparation of Medium: Combine components. Mix thoroughly. Distribute into tubes or flasks. Autoclave for 15 min at 15 psi pressure–121°C.

Use: For the differentiation of *Mycobacterium* species. Strains that hydrolyze Tween™ 80 within 5 days turn the medium pink to red.

Tween™ 80 Hydrolysis Medium

Composition per liter:

Agar 12.0g
Peptone 10.0g
NaCl 5.0g
$CaCl_2$ 0.1g
Tween™ 80 .... 10.0mL

pH 7.2–7.4 at 25°C

Preparation of Medium: Add components to distilled/deionized water and bring volume to 1.0L. Mix thoroughly. Gently heat and bring to boiling. Distribute into tubes or flasks. Autoclave for 15 min at 15 psi pressure–121°C. Pour into sterile Petri dishes.

Use: For the cultivation and differentiation of *Pseudomonas* species based on their ability to hydrolyze Tween™ 80. Bacteria that hydrolyze Tween™ 80 appear as colonies surrounded by an opaque zone.

TY Medium, 2X

Composition per liter:

Pancreatic digest of casein 16.0g
Yeast extract 10.0g
NaCl 5.0g

pH 7.0 ± 0.2 at 25°C

Preparation of Medium: Add components to distilled/deionized water and bring volume to 1.0L. Mix thoroughly. Distribute into tubes or flasks. Autoclave for 25 min at 15 psi pressure–121°C.

Use: For the cultivation of *Escherichia coli.*

TYE HES Medium

Composition per 950mL:

NaCl 49.7g
$MgSO_4 \cdot 7H_2O$ 49.3g
Noble agar 10.0g
Yeast extract 0.5g
Pancreatic digest of casein 0.5g
$CaCl_2 \cdot 2H_2O$ solution 50.0mL

pH 7.2 ± 0.2 at 25°C

$CaCl_2 \cdot 2H_2O$ Solution:

Composition per 100mL:

$CaCl_2 \cdot 2H_2O$ 0.3g

Preparation of $CaCl_2 \cdot 2H_2O$ Solution: Add the $CaCl_2 \cdot 2H_2O$ to distilled/deionized water and bring volume to 100.0mL. Mix thoroughly. Autoclave for 15 min at 15 psi pressure–121°C. Cool to 45–50°C.

Preparation of Medium: Add components, except $CaCl_2 \cdot 2H_2O$ solution, to distilled/deionized water and bring volume to 950.0mL. Mix thoroughly. Gently heat and bring to boiling. Autoclave for 15 min at 15 psi pressure–121°C. Cool to 45–50°C. Aseptically add 50.0mL of sterile $CaCl_2 \cdot 2H_2O$ solution. Mix thoroughly. Adjust pH to 7.2. Pour into sterile Petri dishes or distribute into sterile tubes.

Use: For the cultivation of *Planococcus* species.

TYEG Medium (Trypticase™ Yeast Extract Glucose Medium)

Composition per 1050mL:

NaCl 100.0g
Pancreatic digest of casein 10.0g
$Na_2HPO_4 \cdot 7H_2O$ 2.1g
NH_4Cl 1.0g
KH_2PO_4 .... 0.3g
$MgCl_2 \cdot 6H_2O$ 0.2g
Glucose solution 50.0mL
$Na_2S \cdot 7H_2O$ solution 25.0mL

Trace mineral solution II	10.0mL
Wolfe's vitamin solution	10.0mL
Yeast extract solution	5.0mL
Resazurin (0.2% solution)	1.0mL
$FeSO_4 \cdot 9H_2O$ (2.5% solution)	25.0µl

pH 7.3 ± 0.1 at 25°C

Glucose Solution:

Composition per 100mL:

D-Glucose	10.0g

Preparation of Glucose Solution: Add glucose to distilled/deionized water and bring volume to 100.0mL. Mix thoroughly. Filter sterilize. Aseptically bubble with 90% N_2 + 10% CO_2 to reduce.

$Na_2S \cdot 7H_2O$ Solution:

Composition per 100mL:

$Na_2S \cdot 7H_2O$	2.5g

Preparation of $Na_2S \cdot 7H_2O$ Solution: Add $Na_2S \cdot 7H_2O$ to distilled/deionized water and bring volume to 100.0mL. Mix thoroughly. Autoclave for 15 min at 15 psi pressure–121°C. Use freshly prepared solution.

Trace Mineral Solution II:

Composition per liter:

Nitrilotriacetic acid	12.8g
$CoCl_2 \cdot 6H_2O$	0.17g
$CaCl_2 \cdot 2H_2O$	0.1g
FeSO4·7H2O	0.1g
$MnCl_2 \cdot 4H_2O$	0.1g
NaCl	0.1g
$ZnCl_2$	0.1g
$NiSO_4 \cdot 6H_2O$	0.026g
$CuCl_2 \cdot 2H_2O$	0.02g
Na_2SeO_3	0.017g
H_3BO_3	0.01g
$Na_2MoO_4 \cdot 2H_2O$	0.01g

Preparation of Trace Mineral Solution II: Add nitrilotriacetic acid to 500.0mL of distilled/deionized water. Dissolve by adjusting pH to 6.5 with KOH. Add remaining components. Add distilled/deionized water to 1.0L. Filter through Whatman filter paper. Store under N_2.

Wolfe's Vitamin Solution:

Composition per liter:

Pyridoxine·HCl	10.0mg
Thiamine·HCl	5.0mg
Riboflavin	5.0mg
Nicotinic acid	5.0mg
Calcium pantothenate	5.0mg
p-Aminobenzoic acid	5.0mg
Thioctic acid	5.0mg
Biotin	2.0mg
Folic acid	2.0mg
Cyanocobalamin	100.0µg

Preparation of Wolfe's Vitamin Solution: Add components to distilled/deionized water and bring volume to 1.0L. Mix thoroughly. Filter sterilize. Aseptically bubble with 90% N_2 + 10% CO_2 to reduce.

Yeast Extract Solution:

Composition per 100mL:

Yeast extract	10.0g

Preparation of Yeast Extract Solution: Add yeast extract to distilled/deionized water and bring volume to 100.0mL. Mix thoroughly. Filter sterilize. Aseptically bubble with 90% N_2 + 10% CO_2 to reduce.

Preparation of Medium: Add components—except glucose solution, yeast extract solution, and Wolfe's vitamin solution—to distilled/deionized water and bring volume to 960.0mL. Mix thoroughly. Adjust pH to 7.3. Gently heat and bring to boiling under 90% N_2 + 10% CO_2. Autoclave for 15 min at 15 psi pressure–121°C. Cool to 45–50°C. Aseptically and anaerobically add 50.0mL of sterile glucose solution, 10.0mL of sterile Wolfe's vitamin solution, and 5.0mL of sterile yeast extract solution. Aseptically and anaerobically distribute into sterile tubes in 5.0mL volumes. Immediately prior to inoculation aseptically add 0.125mL of sterile $Na_2S \cdot 9H_2O$ solution per tube.

Use: For the cultivation and maintenenace of *Halobacteroides acetoethylicus*.

TYG Medium (Tryptone Yeast Extract Glucose Medium) (ATCC Medium 741)

Composition per liter:

Agar	20.0g
Pancreatic digest of casein	3.0g
Yeast extract	3.0g
Glucose	3.0g
K_2HPO_4	1.0g

pH 7.4 ± 0.2 at 25°C

Preparation of Medium: Add components to distilled/deionized water and bring volume to 1.0L. Mix thoroughly. Gently heat and bring to boiling. Distribute into tubes or flasks. Autoclave for 15 min at 15 psi pressure–121°C. Pour into sterile Petri dishes or leave in tubes.

Use: For the cultivation and maintenenace of *Thermomonospora fusca*.

TYGS Medium (Tryptone Yeast Extract Glucose Salt Medium)

Composition per liter:

Agar	15.0g
Pancreatic digest of casein	10.0g
NaCl	8.0g
Yeast extract	1.0g
$CaCl_2 \cdot 2H_2O$ solution	100.0mL
Glucose solution	100.0mL

$CaCl_2 \cdot 2H_2O$ Solution:

Composition per 100mL:

$CaCl_2 \cdot 2H_2O$	0.3g

Preparation of $CaCl_2 \cdot 2H_2O$ Solution: Add the $CaCl_2 \cdot 2H_2O$ to distilled/deionized water and bring volume to 100.0mL. Mix thoroughly. Filter sterilize.

Glucose Solution:

Composition per 100mL:

D-Glucose	1.0g

Preparation of Glucose Solution: Add glucose to distilled/deionized water and bring volume to 100.0mL. Mix thoroughly. Filter sterilize.

Preparation of Medium: Add components, except $CaCl_2 \cdot 2H_2O$ solution and glucose solution, to distilled/deionized water and bring volume to 800.0mL. Mix thoroughly. Gently heat and bring to boiling. Autoclave for 15 min at 15 psi pressure–121°C. Cool to 45–50°C. Aseptically add the sterile $CaCl_2 \cdot 2H_2O$ solution and sterile glucose solution. Mix thoroughly. Pour into sterile Petri dishes or distribute into sterile tubes.

Use: For the cultivation and maintenance of a variety of bacteria.

TYN Medium

Composition per liter:

$Na_2S_2O_3 \cdot 5H_2O$	10.0g
Pancreatic digest of casein	1.0g
Yeast extract	1.0g
Na_2SO_4	1.0g

Preparation of Medium: Add components to distilled/deionized water and bring volume to 1.0L. Mix thoroughly. Distribute into tubes or flasks. Autoclave for 15 min at 15 psi pressure–121°C.

Use: For cultivation and maintenance of *Thiobacillus* species.

Tyrosine Agar

Composition per liter:

Solution 1	900.0mL
Solution 2	100.0mL

pH 7.0 ± 0.2 at 25°C

Solution 1:

Composition per 900mL:

Agar	15.0g
Pancreatic digest of gelatin	5.0g
Beef extract	3.0g

Preparation of Solution 1: Add components to distilled/deionized water and bring volume to 900.0mL. Mix thoroughly. Gently heat and bring to boiling.

Solution 2:

Composition per 100mL:

Tyrosine	5.0g

Preparation of Solution 2: Add tyrosine to distilled/deionized water and bring volume to 100.0mL. Mix thoroughly. Gently heat and bring to boiling.

Preparation of Medium: Combine solutions 1 and 2. Mix thoroughly. Distribute into tubes or flasks. Autoclave for 15 min at 15 psi pressure–121°C. Pour into sterile Petri dishes or leave in tubes.

Use: For the differentiation of aerobic *Actinomycete* species. Clearing around a colony indicates utilization of tyrosine. *Streptomyces* and *Actinomadura* species utilize tyrosine. *Nocaridia asteroides, N. caviae,* and *Mycobacterium fortuitum* do not utilize tyrosine.

Tyrosine Casein Nitrate Medium (TCN Medium)

Composition per liter:

Sodium caseinate	25.0g
Agar	15.0g
$NaNO_3$	10.0g
L-Tyrosine	1.0g

Preparation of Medium: Add components to tap water and bring volume to 1.0L. Mix thoroughly. Gently heat and bring to boiling. Distribute into tubes or flasks. Autoclave for 15 min at 15 psi pressure–121°C. Pour into sterile Petri dishes or leave in tubes.

Use: For the isolation and cultivation of streptomycetes from infected plants.

Urea Agar (Urease Test Agar) (Urea Agar Base, Christensen)

Composition per liter:

Urea	20.0g
Agar	15.0g
NaCl	5.0g
KH_2PO_4	2.0g
Peptone	1.0g
Glucose	1.0g
Phenol red	0.012g

pH 6.8 ± 0.2 at 25°C

Source: This medium is available as a premixed powder from BBL Microbiology Systems and Difco Laboratories.

Preparation of Medium: Add components, except agar, to distilled/deionized water and bring volume to 100.0mL. Mix thoroughly. Filter sterilize. Add agar to distilled/deionized water and bring volume to 900.0mL. Mix thoroughly. Gently heat and bring to boiling. Autoclave for 15 min at 15 psi pressure–121°C. Cool to 50°C. Aseptically add the 100.0mL of sterile basal medium. Mix thoroughly. Distribute into sterile tubes. Allow tubes to solidify in a slanted position.

Use: For the differentiation of a variety of microorganisms, especially members of the Enterobacteriaceae, aerobic actinomycetes, streptococci and nonfermenting Gram-negative bacteria, on the basis of urease production.

Urea Agar Base

Composition per liter:

Agar	15.0g
NaCl	5.0g
Na_2HPO_4	1.2g
Peptone	1.0g
Glucose	1.0g
KH_2PO_4	0.8g
Phenol red	0.012g
Urea solution	50.0mL

pH 6.8 ± 0.2 at 25°C

Source: This medium is available as a premixed powder from Oxoid Unipath.

Urea Solution:

Composition per 100mL:

Urea	40.0g

Preparation of Urea Solution: Add urea to distilled/deionized water and bring volume to 100.0mL. Mix thoroughly. Filter sterilize.

Preparation of Medium: Add components, except urea solution, to distilled/deionized water and bring volume to 950.0mL. Mix thoroughly. Gently heat and bring to boiling. Autoclave for 20 min at 10 psi pressure–115°C. Cool to 50°C. Aseptically add 50.0mL of sterile urea solution. Mix thoroughly. Pour into sterile Petri dishes or distribute into sterile tubes. Allow tubes to solidify in a slanted position.

Use: For the detection of *Proteus* species based on rapid urease activity and the identification of other members of the Enterobacteriaceae based on urease activity. Urease positive bacteria turn the medium pink.

Urea Broth Base

Composition per liter:

NaCl	5.0g
Na_2HPO_4	1.2g
Peptone	1.0g
Glucose	1.0g
KH_2PO_4	0.8g
Phenol red	0.012g
Urea solution	50.0mL

pH 6.8 ± 0.2 at 25°C

Source: This medium is available as a premixed powder from Oxoid Unipath.

Urea Solution:

Composition per 100mL:

Urea	40.0g

Preparation of Urea Solution: Add urea to distilled/deionized water and bring volume to 100.0mL. Mix thoroughly. Filter sterilize.

Preparation of Medium: Add components, except urea solution, to distilled/deionized water and bring volume to 950.0mL. Mix thoroughly. Autoclave for 20 min at 10 psi pressure–115°C. Cool to 50°C. Aseptically add 50.0mL of sterile urea solution. Mix thoroughly. Aseptically distribute into sterile tubes or flasks.

Use: For the differentiation of members of the Enterobacteriaceae based on urease production. Urease positive bacteria turn the medium pink.

Urea R Broth (Urea Rapid Broth)

Composition per liter:

Urea	20.0g
Yeast extract	0.1g
Na_2HPO_4	0.095g
KH_2PO_4	0.091g
Phenol red	0.01g

pH 6.9 ± 0.2 at 25°C

Source: Available as a prepared medium from Difco Laboratories.

Preparation of Medium: Add components to distilled/deionized water and bring volume to 1.0L. Mix thoroughly. Filter sterilize. Aseptically distribute into sterile tubes or flasks.

Use: For the differentiation of members of the Enterobacteriaceae based on rapid detection of urease activity. Urease positive bacteria turn the medium cerise.

Urea Semisolid Medium

Composition per liter:

Solution A	400.0mL
Solution B	50.0mL

Solution A:

Composition per 400mL:

Pancreatic digest of casein	6.0g
Yeast extract	2.0g
NaCl	1.0g
Yeast extract	0.8g
Agar	0.3g
L-Cystine	0.1g
Thioglycollic acid	0.12mL

pH 7.2 ± 0.2 at 25°C

Preparation of Solution A: Add components to distilled/deionized water and bring volume to 400.0mL. Mix thoroughly. Gently heat and bring to boiling. Autoclave for 15 min at 15 psi pressure–121°C. Cool to 60°C.

Solution B:

Composition per 50mL:

Urea	8.0g
Na_2HPO_4	3.8g
KH_2PO_4	3.64g
Yeast extract	0.04g
Phenol red	4.0mg

Preparation of Solution B: Add components to distilled/deionized water and bring volume to 50.0mL. Mix thoroughly. Filter sterilize.

Preparation of Medium: Aseptically combine 400.0mL of sterile solution A and 50.0mL of sterile solution B. Mix thoroughly. Aseptically distribute into sterile screw-capped tubes in 7.0mL volumes. Pass the tubes into an anaerobic chamber containing 85% N_2 + 10% H_2 + 5% CO_2 for 60 min. Close screw caps tightly.

Use: For the cultivation and differentiation of anaerobic bacteria based on their production of urease. Bacteria that produce urease turn the medium bright red.

Urea Test Broth

Composition per liter:

Urea	20.0g
Na_2HPO_4	9.5g
KH_2PO_4	9.1g
Yeast extract	0.1g
Phenol red	0.01g
Urea solution	100.0mL

Urea Solution:

Composition per 100mL:

Urea	20.0g

Preparation of Urea Solution: Add urea to distilled/deionized water and bring volume to 100.0mL. Mix thoroughly. Filter sterilize.

Preparation of Medium: Add components, except urea solution, to distilled/deionized water and bring volume to 900.0mL. Mix thoroughly. Autoclave for 15 min at 15 psi pressure–121°C. Cool to 45–50°C. Aseptically add sterile urea solution. Mix thoroughly. Aseptically distribute into sterile tubes in 3.0mL volumes.

Use: For the cultivation and differentiation of members of the Enterobacteriaceae and aerobic actinomycetes based on their production of urease. Bacteria that produce urease turn the medium bright red.

Urease Test Broth (Urea Broth)

Composition per liter:

Urea	20.0g
Na_2HPO_4	9.5g
KH_2PO_4	9.1g
Yeast extract	0.1g
Phenol red	0.01g

pH 6.8 ± 0.2 at 25°C

Source: This medium is available as a premixed powder from BBL Microbiology Systems and Difco Laboratories.

Preparation of Medium: Add components to distilled/deionized water and bring volume to 1.0L. Mix thoroughly. Filter sterilize. Aseptically distribute into sterile tubes or flasks.

Use: For the differentiation of organisms, especially the Enterobacteriaceae, on the basis of urease production. Urease positive bacteria turn the medium pink.

Ustilago Medium

Composition per liter:

Yeast extract	11.0g
Glucose	10.0g

NH_4NO_3	1.5g
Salt solution	62.5mL
Vitamin solution	10.0mL

Salt Solution:

Composition per liter:

KH_2PO_4	16.0g
KCl	8.0g
Na_2SO_4	4.0g
$MgSO_4 \cdot 7H_2O$	2.0g
$CaCl_2$	1.0g
Trace elements solution	8.0mL

Preparation of Salt Solution: Add components to distilled/deionized water and bring volume to 1.0L. Mix thoroughly.

Trace Elements Solution:

Composition per 500mL:

$CuSO_4 \cdot 5H_2O$	0.2g
$ZnCl_2$	0.2g
$MnCl_2 \cdot 4H_2O$	0.07g
$FeCl_3 \cdot 6H_2O$	0.05g
H_3BO_3	0.03g
$Na_2MoO_4 \cdot 2H_2O$	0.02g

Preparation of Trace Elements Solution: Add components to distilled/deionized water and bring volume to 500.0mL. Mix thoroughly.

Vitamin Solution:

Composition per liter:

Inositol	0.4g
Calcium pantothenate	0.2g
Choline chloride	0.2g
Nicotinic acid	0.2g
Thiamine	0.1g
Pyridoxine	0.05g
Riboflavin	0.05g

Preparation of Vitamin Solution: Add components to distilled/deionized water and bring volume to 1.0L. Mix thoroughly. Filter sterilize.

Preparation of Medium: Add components, except vitamin solution, to distilled/deionized water and bring volume to 990.0mL. Mix thoroughly. Gently heat and bring to boiling. Autoclave for 15 min at 15 psi pressure–121°C. Cool to 45–50°C. Aseptically add 10.0mL of sterile vitamin solution. Mix thoroughly. Aseptically distribute into sterile tubes or flasks.

Use: For the cultivation of *Ustilago* species.

Ustilago Minimal Medium

Composition per liter:

Glucose	10.0g
KNO_3	3.0g
Salt solution	62.5mL

Salt Solution:

Composition per liter:

KH_2PO_4	16.0g
KCl	8.0g
Na_2SO_4	4.0g
$MgSO_4 \cdot 7H_2O$	2.0g
$CaCl_2$	1.0g
Trace elements solution	8.0mL

Preparation of Salt Solution: Add components to distilled/deionized water and bring volume to 1.0L. Mix thoroughly.

Trace Elements Solution:

Composition per 500mL:

$CuSO_4 \cdot 5H_2O$	0.2g
$ZnCl_2$	0.2g
$MnCl_2 \cdot 4H_2O$	0.07g
$FeCl_3 \cdot 6H_2O$	0.05g
H_3BO_3	0.03g
$Na_2MoO_4 \cdot 2H_2O$	0.02g

Preparation of Trace Elements Solution: Add components to distilled/deionized water and bring volume to 500.0mL. Mix thoroughly.

Preparation of Medium: Add components to distilled/deionized water and bring volume to 1.0L. Mix thoroughly. Distribute into tubes or flasks. Autoclave for 15 min at 15 psi pressure–121°C.

Use: For the cultivation of *Ustilago* species.

V–8™ Agar

Composition per liter:

Agar	20.0g
$CaCO_3$	4.0g
V-8™ canned vegetable juice	200.0mL

pH 7.3 ± 0.2 at 25°C

Preparation of Medium: Add components to distilled/deionized water and bring volume to 1.0L. Mix thoroughly. Gently heat and bring to boiling. Distribute into tubes or flasks. Autoclave for 15 min at 15 psi pressure–121°C. Pour into sterile Petri dishes or leave in tubes.

Use: For the isolation and cultivation of *Actinomadura* species, *Actinopolyspora* species, *Excellospora* species and *Microspora* species.

V Agar

Composition per liter:

Agar	13.5g
Pancreatic digest of casein	12.0g
Peptone	10.0g
Peptic digest of animal tissue	5.0g
NaCl	5.0g

Beef extract 3.0g
Yeast extract 3.0g
Cornstarch 1.0g
Human blood, anticoagulated 50.0mL
pH 7.4 ± 0.2 at 25°C

Source: This medium is available as a prepared medium from BBL Microbiology Systems.

Preparation of Medium: Add components, except human blood, to distilled/deionized water and bring volume to 950.0mL. Mix thoroughly. Gently heat and bring to boiling. Distribute into tubes or flasks. Autoclave for 15 min at 15 psi pressure–121°C. Cool to 50°C. Aseptically add 50.0mL of human blood. Mix thoroughly. Pour into sterile Petri dishes or leave in tubes.

Use: For the isolation and differentiation of *Gardnerella vaginalis* from clinical specimens. Plates are incubated under an atmosphere with 3-10% CO_2. *G. vaginalis* appears as small white colonies with diffuse β-hemolysis.

Van Niel's Medium, Modified

Composition per liter:

Yeast extract 10.0g
$MgSO_4$ 0.1g
EDTA 2.0mg
Trace elements solution 10.0mL
K_2HPO_4 solution 2.5mL
pH 7.1 ± 0.2 at 25°C

Trace Elements Solution:
Composition per 100mL:

$CaCl_2 \cdot 2H_2O$ 0.3g
Ferric ammonium citrate 0.2g

Preparation of Trace Elements Solution: Add components to distilled/deionized water and bring volume to 100.0mL. Mix thoroughly. Filter sterilize.

K_2HPO_4 Solution:
Composition per 100mL:

K_2HPO_4 ....4.0g

Preparation of K_2HPO_4 Solution: Add K_2HPO_4 to distilled/deionized water and bring volume to 100.0mL. Mix thoroughly. Filter sterilize.

Preparation of Medium: Add components, except K_2HPO_4 and trace elements solution, to distilled/deionized water and bring volume to 987.5mL. Mix thoroughly. Autoclave for 15 min at 15 psi pressure–121°C. Cool to 50°C. Aseptically add trace elements and K_2HPO_4 solutions. Mix thoroughly. Distribute into sterile tubes or flasks.

Use: For the cultivation and maintenance of *Rhodobacter sphaeroides*.

Van Niel's Yeast Medium with Pyruvate, Modified

Composition per liter:

Yeast extract 10.0g
$MgSO_4$ 0.1g
EDTA 2.0mg
Sodium pyruvate solution 100.0mL
Trace elements solution 10.0mL
K_2HPO_4 solution 5.0mL
Trace metal A-5 solution 1.0mL
pH 7.1± 0.1 at 25°C

Sodium Pyruvate Solution:
Composition per 100mL:

Sodium pyruvate 1.1g

Preparation of K_2HPO_4 Solution: Add sodium pyruvate to distilled/deionized water and bring volume to 100.0mL. Mix thoroughly. Filter sterilize.

Trace Elements Solution:
Composition per 100 mL:

$CaCl_2 \cdot 2H_2O$ 0.3g
Ferric ammonium citrate 0.2g

Preparation of Trace Elements Solution: Add components to distilled/deionized water and bring volume to 100.0mL. Mix thoroughly. Filter sterilize.

K_2HPO_4 Solution:
Composition per 100mL:

K_2HPO_4 ....4.0g

Preparation of K_2HPO_4 Solution: Add K_2HPO_4 to distilled/deionized water and bring volume to 100.0mL. Mix thoroughly. Filter sterilize.

Trace Metal A-5 Solution:
Composition per liter:

H_3BO_3 ....2.86g
$MnCl_2 \cdot 4H_2O$ 1.81g
$Na_2MoO_4 \cdot 2H_2O$ 0.39g
$ZnSO_4 \cdot 7H_2O$ 0.222g
$CuSO_4 \cdot 5H_2O$ 0.079g
$Co(NO_3)_2 \cdot 6H_2O$ 0.049.g

Preparation of Trace Metal A-5 Solution: Add components to distilled/deionized water and bring volume to 1.0L. Mix thoroughly. Filter sterilize.

Preparation of Medium: Add components, except sodium pyruvate, trace elements, K_2HPO_4 and trace metal A-5 solutions, to distilled/deionized water and bring volume to 884.0mL. Mix thoroughly. Autoclave for 15 min at 15 psi pressure–121°C. Cool to 25°C. Aseptically add 100.0mL of sterile sodium pyruvate solution, 10.0mL of sterile trace elements soultion, 5.0mL of sterile K_2HPO_4 solution, and 1.0mL of sterile trace metal A-5 solution. Mix thoroughly. Adjust pH to 7.1± 0.1. Aseptically distribute into sterile tubes or flasks.

Use: For the cultivation and maintenance of photosynthetic bacteria, such as *Heliobacillus mobilis* and *Rhodopseudomonas palustris*.

Veal Infusion Agar

Composition per liter:

Agar	15.0g
Veal, infusion from	10.0g
Pancreatic digest of casein	5.0g
Peptic digest of animal tissue	5.0g
NaCl	5.0g

pH 7.4 ± 0.2 at 25°C

Source: This medium is available as a premixed powder from BBL Microbiology Systems.

Preparation of Medium: Add components to distilled/deionized water and bring volume to 1.0L. Mix thoroughly. Gently heat and bring to boiling. Distribute into tubes or flasks. Autoclave for 15 min at 15 psi pressure–121°C. Pour into sterile Petri dishes or leave in tubes.

Use: For the cultivation and maintenance of a variety of microorganisms. Can be used for cultivation of fastidious microorganisms when enriched with blood or serum.

Veal Infusion Agar (ATCC Medium 521)

Composition per liter:

Veal, infusion from	500.0g
Agar	15.0g
Pancreatic digest of casein	5.0g
Peptic digest of animal tissue	5.0g
NaCl	5.0g

pH 7.4 ± 0.2 at 25°C

Source: This medium is available as a premixed powder from Difco Laboratories.

Preparation of Medium: Add components to distilled/deionized water and bring volume to 1.0L. Mix thoroughly. Gently heat and bring to boiling. Distribute into tubes or flasks. Autoclave for 15 min at 15 psi pressure–121°C. Pour into sterile Petri dishes or leave in tubes.

Use: For the cultivation and maintenance of a variety of microorganisms. Can be used for cultivation of fastidious microorganisms when enriched with blood or serum.

Veal Infusion Broth (ATCC Medium 521)

Composition per liter:

Veal, infusion from	500.0g
Pancreatic digest of casein	5.0g
Peptic digest of animal tissue	5.0g
NaCl	5.0g

pH 7.4 ± 0.2 at 25°C

Source: This medium is available as a premixed powder from Difco Laboratories.

Preparation of Medium: Add components to distilled/deionized water and bring volume to 1.0L. Mix thoroughly. Distribute into tubes or flasks. Autoclave for 15 min at 15 psi pressure–121°C. Use freshly prepared solution.

Use: For the cultivation and maintenance of *Arthrobacter* species, streptococci and other microorganisms.

Violet Red Bile Agar (VRB Agar)

Composition per liter:

Agar	15.0g
Lactose	10.0g
Pancreatic digest of gelatin	7.0g
NaCl	5.0g
Yeast extract	3.0g
Bile salts	1.5g
Neutral red	0.03g
Crystal violet	2.0mg

pH 7.4 ± 0.2 at 25°C

Source: This medium is available as a premixed powder from BBL Microbiology Systems, Difco Laboratores and Oxoid Unipath.

Preparation of Medium: Add components to distilled/deionized water and bring volume to 1.0L. Mix thoroughly. Gently heat while stirring and bring to boiling. Distribute into tubes or flasks. Autoclave for 15 min at 15 psi pressure–121°C. Pour immediately into sterile Petri dishes or leave in tubes.

Use: For the detection of coliform bacteria in water and food.

VY Agar

Composition per liter:

Agar	15.0g
Baker's yeast	10.0g
$CaCl_2 \cdot 2H_2O$	1.0g
Cyanocobalamin	5.0mg

pH 7.2 ± 0.2 at 25°C

Preparation of Medium: Add components to distilled/deionized water and bring volume to 1.0L. Mix thoroughly. Gently heat and bring to boiling. Distribute into tubes or flasks. Autoclave for 15 min at 15 psi pressure–121°C. Pour into sterile Petri dishes or leave in tubes.

Use: For the cultivation and maintenance of myxobacteria.

VY2 Agar

Composition per liter:

Agar	15.0g
Baker's yeast	5.0g
$CaCl_2 \cdot 2H_2O$	1.0g
Cyanocobalamin	5.0mg

pH 7.2 ± 0.2 at 25°C

Preparation of Medium: Add components to distilled/deionized water and bring volume to 1.0L. Mix thoroughly. Gently heat and bring to boiling. Distribute into tubes or flasks. Autoclave for 15 min at 15 psi pressure–121°C. Pour into sterile Petri dishes or leave in tubes.

Use: For the cultivation and maintenance of *Myxococcus amylovorans.*

VY5 Agar

Composition per liter:

Agar	15.0g
Baker's yeast	2.0g
$CaCl_2 \cdot 2H_2O$	1.0g
Cyanocobalamin	5.0mg

pH 7.2 ± 0.2 at 25°C

Preparation of Medium: Add components to distilled/deionized water and bring volume to 1.0L. Mix thoroughly. Gently heat and bring to boiling. Distribute into tubes or flasks. Autoclave for 15 min at 15 psi pressure–121°C. Pour into sterile Petri dishes or leave in tubes.

Use: For the cultivation and maintenance of myxobacteria.

Waksman's Glucose Agar

Composition per liter:

Agar	12.5g
Glucose	10.0g
Peptone	5.0g
Beef extract	5.0g
NaCl	5.0g

pH 7.4-7.6 at 25°C

Preparation of Medium: Add components to distilled/deionized water and bring volume to 1.0L. Mix thoroughly. Gently heat and bring to boiling. Distribute into tubes or flasks. Autoclave for 15 min at 15 psi pressure–121°C. Pour into sterile Petri dishes or leave in tubes.

Use: For the cultivation and maintenance of *Streptomyces* species.

Waksman's Sulfur Medium

Composition per liter:

KH_2PO_4	3.0g
$MgSO_4 \cdot 7H_2O$	0.5g
$(NH_4)_2SO_4$	0.2g
$CaCl_2 \cdot 2H_2O$	0.2g
$Fe_2(SO_4)_3$	0.1mg

Preparation of Medium: Add components to distilled/deionized water and bring volume to 1.0L. Mix thoroughly. It is not necessary to sterilize this medium. Distribute into sterile tubes or flasks.

Use: For the cultivation of sulfate-reducing microorganisms from soil.

Walsby Medium

Composition per liter:

$MgSO_4 \cdot 7H_2O$	0.075g
K_2HPO_4	0.039g
Na_2CO_3	0.02g
$CaCl_2 \cdot 2H_2O$	0.018g
H_3BO_3	2.8mg
$MnSO_4 \cdot 4H_2O$	2.0mg
$ZnSO_4$	0.22mg
MoO_3	0.18mg
$CuSO_4 \cdot 5H_2O$	0.08mg
$Co(NO_3)_2 \cdot 6H_2O$	0.05mg
Iron-EDTA solution	1.0mL

pH 8.5 ± 0.2 at 25°C

Iron-EDTA Solution:

Composition per liter:

EDTA	12.7g
$FeSO_4 \cdot 7H_2O$	4.98g

Preparation of Iron-EDTA Solution: Add components to distilled/deionized water and bring volume to 1.0L. Mix thoroughly.

Preparation of Medium: Add components to distilled/deionized water and bring volume to 1.0L. Mix thoroughly. Distribute into tubes or flasks. Autoclave for 15 min at 15 psi pressure–121°C.

Use: For the isolation and cultivation of planktonic gas-vacuolate cyanobacteria.

Water Agar

Composition per liter:

Agar	20.0g

Preparation of Medium: Add agar to distilled/deionized water and bring volume to 1.0L. Mix thoroughly. Gently heat and bring to boiling. Distribute into tubes or flasks. Autoclave for 15 min at 15 psi pressure–121°C. Pour into sterile Petri dishes or leave in tubes.

Use: For the cultivation and of some fungi.

Water Agar

Composition per liter:

Agar	15.0g
$CaCl_2 \cdot 2H_2O$	1.0g

pH 7.2 ± 0.2 at 25°C

Preparation of Medium: Add components to distilled/deionized water and bring volume to 1.0L. Mix thoroughly. Gently heat and bring to boiling. Distribute into tubes or flasks. Autoclave for 15 min at 15 psi pressure–121°C. Pour into sterile Petri dishes or leave in tubes.

Use: For the cultivation of myxobacteria.

Waxy Maize Starch Medium

Composition per liter:

Agar	20.0g
Waxy maize starch	5.0g
Pancreatic digest of casein	5.0g
Yeast extract	5.0g
$CoCl_2 \cdot 6H_2O$	0.1g
$CaCl_2 \cdot 2H_2O$	0.1g
Maltose solution	100.0mL

pH 6.7 ± 0.2 at 25°C

Maltose Solution:

Composition per 100mL:

Maltose	10.0g

Preparation of Maltose Solution: Add maltose to distilled/deionized water and bring volume to 100.0mL. Mix thoroughly. Filter sterilize.

Preparation of Medium: Add components, except maltose solution, to distilled/deionized water and bring volume to 900.0mL. Mix thoroughly. Gently heat and bring to boiling. Adjust pH to 6.7. Distribute into tubes or flasks. Autoclave for 15 min at 15 psi pressure–121°C. Aseptically add maltose solution. Pour into sterile Petri dishes or leave in tubes.

Use: For the cultivation and maintenance of *Bacillus* species.

WCX Agar

Composition per liter:

Agar	15.0g
$CaCl_2 \cdot 2H_2O$	1.0g
Cycloheximide solution	100.0mL

pH 7.2 ± 0.2 at 25°C

Cycloheximide Solution

Composition per 100mL:

Cycloheximide	2.5mg

Preparation of Cycloheximide Solution: Add components to distilled/deionized water and bring volume to 100.0mL. Mix thoroughly. Filter sterilize.

Preparation of Medium: Add components, except cycloheximide solution, to distilled/deionized water and bring volume to 900.0mL. Mix thoroughly. Gently heat and bring to boiling. Autoclave for 15 min at 15 psi pressure–121°C. Cool to 45–50°C. Aseptically add sterile cycloheximide solution. Mix thoroughly. Pour into sterile Petri dishes or distribute into sterile tubes.

Use: For the cultivation of myxobacteria.

Wickerham Broth

Composition per 100mL:

Carbohydrate	10.0g
Yeast nitrogen base	100.0mL

Yeast Nitrogen Base, 10X:

Composition per liter:

Glucose	10.0g
KH_2PO_4	1.0g
$MgSO_4 \cdot 7H_2O$	0.5g
NaCl	0.1g
$CaCl_2 \cdot 2H_2O$	0.1g
DL-Methionine	0.02g
DL-Tryptophan	0.02g
L-Histidine·HCl	0.01g
Inositol	2.0mg
H_3BO_3	0.5mg
$ZnSO_4 \cdot 7H_2O$	0.4mg
$MnSO_4 \cdot 4H_2O$	0.4mg
Thiamine·HCl	0.4mg
Pyridoxine	0.4mg
Niacin	0.4mg
Calcium pantothenate	0.4mg
p-Aminobenzoic acid	0.2mg
Riboflavin	0.2mg
$FeCl_3$	0.2mg
$Na_2MoO_4 \cdot 4H_2O$	0.2mg
KI	0.1mg
$CuSO_4 \cdot 5H_2O$	0.04mg

Folic acid 2.0μg
Biotin 2.0μg

pH 4.5 ± 0.2 at 25°C

Preparation of Yeast Nitrogen Base: Add components to distilled/deionized water and bring volume to 1.0L. Mix thoroughly.

Preparation of Medium: To 100.0mL of yeast nitrogen base, add 10.0g of carbohydrate. Mix thoroughly. Filter sterilize. Aseptically distribute 0.5mL into tubes containing 4.5mL of sterile, distilled/deionized water.

Use: For the cultivation and differentiation of bacteria based on carbohydrate assimilation.

Wickerham Broth

Composition per 100mL:

KNO_3 .... 0.78g
Yeast carbon base 100.0mL

pH 4.5 ± 0.2 at 25°C

Yeast Carbon Base:

Composition per liter:

Glucose 10.0g
KH_2PO_4 .... 1.0g
$MgSO_4 \cdot 7H_2O$ 0.5g
NaCl 0.1g
$CaCl_2 \cdot 2H_2O$ 0.1g
DL-Methionine 0.02g
DL-Tryptophan 0.02g
L-Histidine·HCl 0.01g
Inositol 2.0mg
H_3BO_3 .... 0.5mg
$ZnSO_4 \cdot 7H_2O$ 0.4mg
$MnSO_4 \cdot 4H_2O$ 0.4mg
Thiamine·HCl 0.4mg
Pyridoxine 0.4mg
Niacin 0.4mg
Calcium pantothenate 0.4mg
p-Aminobenzoic acid 0.2mg
Riboflavin 0.2mg
$FeCl_3$ 0.2mg
$Na_2MoO_4 \cdot 4H_2O$ 0.2mg
KI 0.1mg
$CuSO_4 \cdot 5H_2O$ 0.04mg
Folic acid 2.0μg
Biotin 2.0μg

Preparation of Yeast Carbon Base: Add components to distilled/deionized water and bring volume to 1.0L. Mix thoroughly.

Preparation of Medium: To 100.0mL of yeast carbon base, add 0.78g of KNO_3 (or peptone). Mix thoroughly. Filter sterilize. Aseptically distribute 0.5mL into tubes containing 4.5mL of sterile, distilled/deionized water.

Use: For the cultivation and differentiation of bacteria based on nitrate assimilation.

Wickerham Broth with Raffinose

Composition per 100mL:

Raffinose 20.0g
Yeast nitrogen base 100.0mL

Yeast Nitrogen Base, 10 X:

Composition per liter:

Glucose 10.0g
KH_2PO_4 .... 1.0g
$MgSO_4 \cdot 7H_2O$ 0.5g
NaCl 0.1g
$CaCl_2 \cdot 2H_2O$ 0.1g
DL-Methionine 0.02g
DL-Tryptophan 0.02g
L-Histidine·HCl 0.01g
Inositol 2.0mg
H_3BO_3 .... 0.5mg
$ZnSO_4 \cdot 7H_2O$ 0.4mg
$MnSO_4 \cdot 4H_2O$ 0.4mg
Thiamine·HCl 0.4mg
Pyridoxine 0.4mg
Niacin 0.4mg
Calcium pantothenate 0.4mg
p-Aminobenzoic acid 0.2mg
Riboflavin 0.2mg
$FeCl_3$ 0.2mg
$Na_2MoO_4 \cdot 4H_2O$ 0.2mg
KI 0.1mg
$CuSO_4 \cdot 5H_2O$ 0.04mg
Folic Acid 2.0μg
Biotin 2.0μg

pH 4.5 ± 0.2 at 25°C

Preparation of Yeast Nitrogen Base: Add components to distilled/deionized water and bring volume to 1.0L. Mix thoroughly.

Preparation of Medium: To 100.0mL of yeast nitrogen base, add 20.0g of raffinose. Mix thoroughly. Filter sterilize. Aseptically distribute 0.5mL into tubes containing 4.5mL of sterile, distilled/deionized water.

Use: For the cultivation and differentiation of bacteria based on carbohydrate assimilation.

Wilbrinck's Agar for *Xanthomonas albilineans*

Composition per liter:

Agar 20.0g
Sucrose 10.0g
Peptone 5.0g
K_2HPO_4 .... 0.5g

$MgSO_4 \cdot 7H_2O$ 0.25g
Na_2SO_3 (anhydrous) 0.05g
pH 7.2 ± 0.2 at 25°C

Preparation of Medium: Add components to distilled/deionized water and bring volume to 1.0L. Mix thoroughly. Gently heat and bring to boiling. Distribute into tubes or flasks. Autoclave for 15 min at 15 psi pressure–121°C. Pour into sterile Petri dishes or leave in tubes.

Use: For the cultivation and maintenance of *Xanthomonas albilineans* and other *Xanthomonas* species.

Winogradsky's Medium, Modified

Composition per liter:

$CaCO_3$ 5.0g
$(NH_4)_2SO_4$ 1.0g
K_2HPO_4 .... 1.0g
NaCl 1.0g
$MgSO_4 \cdot 7H_2O$ 0.5g
$FeSO_4$ 0.4g

Preparation of Medium: Add components to distilled/deionized water and bring volume to 1.0L. Mix thoroughly. Gently heat until dissolved. Do not autoclave. Distribute into tubes or flasks. Swirl flask while dispensing to suspend precipitate.

Use: For the cultivation of nitrifying bacteria.

Winogradsky's N–Free Medium

Composition per liter:

Agar 20.0g
$CaCO_3$ 5.0mg
Sugar solution 100.0mL
Concentrated salt solution 5.0mL
pH 7.2 ± 0.2 at 25°C

Sugar Solution:
Composition per 100mL:

Sucrose or glucose 10.0g

Preparation of Sugar Solution: Add sugar to 100.0mL distilled/deionized water. Mix thoroughly. Autoclave for 10 min at 10 psi pressure–115°C. Cool to 50°C.

Concentrated Salt Solution:
Composition per liter:

KH_2PO_4 .... 50.0g
$MgSO_4 \cdot 7H_2O$ 25.0g
NaCl 25.0g
$FeSO_4 \cdot 7H_2O$ 1.0g
$MnSO_4 \cdot 4H_2O$ 1.0g
$Na_2MoO_4 \cdot 4H_2O$ 1.0g

Preparation of Concentrated Salt Solution: Add components to tap water and bring volume to 1.0L. Mix thoroughly. Filter sterilize.

Preparation of Medium: Add components, except sugar solution, to distilled/deionized water and bring volume to 900.0mL. Mix thoroughly. Gently heat and bring to boiling. Distribute into tubes or flasks. Autoclave for 15 min at 15 psi pressure–121°C. Cool to 50°C. Aseptically add sugar solution. Adjust pH to 7.2. Mix thoroughly. Pour into sterile Petri dishes or leave in tubes.

Use: For the cultivation and maintenance of *Azomonas insignis*.

Winogradsky's Nitrite Medium

Composition per liter:

Agar 15.0g
$NaNO_2$ 2.0g
Na_2CO_3, anhydrous 1.0g
K_2HPO_4 .... 0.5g

Preparation of Medium: Add components to distilled/deionized water and bring volume to 1.0L. Mix thoroughly. Gently heat and bring to boiling. Distribute into tubes. Autoclave for 15 min at 15 psi pressure–121°C.

Use: For the selective isolation and cultivation of *Nocardia* species and *Rhodococcus* species.

Wort Agar

Composition per liter:

Agar 15.0g
Malt extract 15.0g
Maltose 12.75g
Dextrin 2.75g
Glycerol 2.35g
K_2HPO_4 .... 1.0g
NH_4Cl 1.0g
Pancreatic digest of gelatin 0.78g
pH 4.8 ± 0.2 at 25°C

Source: This medium is available as a premixed powder from BBL Microbiology Systems, Difco Laboratories and Oxoid Unipath.

Preparation of Medium: Add components to distilled/deionized water and bring volume to 1.0L. Mix thoroughly. Gently heat and bring to boiling. Boil for 1 min with mixing. Distribute into tubes or flasks. Autoclave for 15 min at 15 psi pressure–121°C. Do not overheat as this will result in hydrolysis of the agar. An additional 5.0g of agar can be used to make a firmer agar. Pour into sterile Petri dishes or leave in tubes.

Use: For the cultivation and enumeration of yeasts. The low pH of the agar selectively inhibits bacterial growth.

Xanthine Agar

Composition per liter:

Solution 1	900.0mL
Solution 2	100.0mL

pH 7.0 ± 0.2 at 25°C

Solution 1:

Composition per 900mL:

Agar	15.0g
Pancreatic digest of gelatin	5.0g
Beef extract	3.0g

Preparation of Solution 1: Add components to distilled/deionized water and bring volume to 900.0mL. Mix thoroughly. Gently heat and bring to boiling.

Solution 2:

Composition per 100mL:

Xanthine	4.0g

Preparation of Solution 2: Add xanthine to distilled/deionized water and bring volume to 100.0mL. Mix thoroughly. Gently heat and bring to boiling.

Preparation of Medium: Combine solutions 1 and 2. Mix thoroughly. Distribute into tubes or flasks. Autoclave for 15 min at 15 psi pressure–121°C. Pour into sterile Petri dishes or leave in tubes.

Use: For the differentiation of aerobic *Actinomycete* species. Clearing around a colony indicates utilization of xanthine. *Streptomyces* species utilize xanthine; most *Nocardia* and *Actinomadura* species do not utilize xanthine.

Xanthomonas Agar

Composition per liter:

Agar	15.0g
Pancreatic digest of gelatin	10.0g
Sucrose	10.0g
Beef extract	6.0g

pH 6.8 ± 0.2 at 25°C

Preparation of Medium: Add components to distilled/deionized water and bring volume to 1.0L. Mix thoroughly. Gently heat and bring to boiling. Distribute into tubes or flasks. Autoclave for 15 min at 15 psi pressure–121°C. Pour into sterile Petri dishes or leave in tubes.

Use: For cultivation and maintenance of *Xanthomonas* species.

Xanthomonas Medium

Composition per liter:

Pancreatic digest of gelatin	10.0g
Sucrose	10.0g
Beef extract	6.0g

pH 6.8 ± 0.2 at 25°C

Preparation of Medium: Add components to distilled/deionized water and bring volume to 1.0L. Mix thoroughly. Gently heat with mixing. Distribute into screw cap test tubes. Autoclave for 15 min at 15 psi pressure–121°C.

Use: For cultivation and maintenance of *Xanthomonas* species.

Xanthomonas TYG Agar (*Xanthomonas* Tryptone Yeast Extract Glucose Agar)

Composition per liter:

Agar	20.0g
Pancreatic digest of casein	5.0g
Glucose	5.0g
Yeast extract	3.0g
K_2HPO_4	0.7g
$MgSO_4 \cdot 7H_2O$	0.25g

Preparation of Medium: Add components to distilled/deionized water and bring volume to 1.0L. Mix thoroughly. Gently heat and bring to boiling. Distribute into tubes or flasks. Autoclave for 15 min at 15 psi pressure–121°C. Pour into sterile Petri dishes or leave in tubes.

Use: For the cultivation and maintenance of *Xanthomonas* species.

Xylan Medium

Composition per liter:

Xylan	30.0g
Agar	12.0g
Peptone	2.0g
Yeast extract	0.5g
Cysteine·HCl·H_2O	0.25g
$Na_2S \cdot 9H_2O$	0.25g
Rumen fluid	400.0mL
$NaHCO_3$ solution	40.0mL
Mineral solution I	25.0mL
Mineral solution II	25.0mL
Wolfe's vitamin solution	10.0mL
VFA solution	10.0mL
Hemin solution	10.0mL
Trace elements solution SL-6	1.0mL

pH 7.0 ± 0.2 at 25°C

$NaHCO_3$ Solution:

Composition per 100mL:

$NaHCO_3$	3.96g

Preparation of $NaHCO_3$ Solution: Add $NaHCO_3$ to distilled/deionized water and bring volume to 100.0mL. Mix thoroughly. Gas with 100% CO_2.

Mineral Solution I:

Composition per liter:

K_2HPO_4	3.0g

Preparation of Mineral Solution I: Add K_2HPO_4 to distilled/deionized water and bring volume to 1.0L. Mix thoroughly.

Mineral Solution II:

Composition per liter:

Sodium citrate	20.0g
NaCl	12.0g
KH_2PO_4	6.0g
$MgCl_2{\cdot}6H_2O$	2.0g
$CaCl_2$	1.2g

Preparation of Mineral Solution II: Add components to distilled/deionized water and bring volume to 1.0L. Mix thoroughly.

Wolfe's Vitamin Solution:

Composition per liter:

Pyridoxine·HCl	10.0mg
Thiamine·HCl	5.0mg
Riboflavin	5.0mg
Nicotinic acid	5.0mg
Calcium pantothenate	5.0mg
p-Aminobenzoic acid	5.0mg
Thioctic acid	5.0mg
Biotin	2.0mg
Folic acid	2.0mg
Cyanocobalamin	100.0μg

Preparation of Wolfe's Vitamin Solution: Add components to distilled/deionized water and bring volume to 1.0L. Mix thoroughly.

VFA Solution:

Composition per liter:

Acetic acid	178.3mL
Propionic acid	59.6mL
n-Butyric acid	38.4mL
Isobutyric acid	9.5mL
n-Valeric acid	9.4mL
Isovaleric acid	9.3mL
DL α-Methylbutyric acid	4.4mL

Preparation of VFA Solution: Add components to distilled/deionized water and bring volume to approximately 500.0mL. Adjust pH to 7.5 with NaOH. Mix thoroughly. Bring volume to 1.0L with distilled/deionized water.

Hemin Solution:

Composition per 100mL:

Hemin	0.01g

Preparation of Hemin Solution: Add hemin to 100.0mL of 0.01*N* NaOH. Mix thoroughly.

Trace Elements Solution SL-6:

Composition per liter:

H_3BO_3	0.30g
$CoCl_2{\cdot}6H_2O$	0.20g
$ZnSO_4{\cdot}7H_2O$	0.10g
$MnCl_2{\cdot}4H_2O$	0.03g
$Na_2MoO_4{\cdot}H_2O$	0.03g
$NiCl_2{\cdot}6H_2O$	0.02g
$CuCl_2.2H_2O$	0.01g

Preparation of Trace Elements Solution SL-6: Add components to distilled/deionized water and bring volume to 1.0L. Mix thoroughly. Adjust pH to 3.4.

Preparation of Medium: Add components, except cysteine·HCl·H_2O, $Na_2S{\cdot}9H_2O$, and sodium bicarbonate solution, to distilled/deionized water and bring volume to 1.0L. Mix thoroughly. Gently heat and bring to boiling. Cool under 80% N_2 + 20% CO_2. Add cysteine·HCl·H_2O and $Na_2S{\cdot}9H_2O$. Add sufficient sodium bicarbonate solution to bring pH to 7.2 under 80% N_2 + 20% CO_2. Anaerobically distribute into tubes under 80% N_2 + 20% CO_2. Autoclave for 15 min at 15 psi pressure–121°C.

Use: For cultivation and maintenance of *Clostridium xylanolyticum* and other microorganisms that can utilize xylan as a carbon source.

YA Halophile Medium

Composition per liter:

NaCl	100.0g
Agar	15.0g
Sodium acetate·$3H_2O$	10.0g
Na_2HPO_4	3.8g
KH_2PO_4	1.3g
$Mg(NO_3)_2{\cdot}6H_2O$	1.0g
$(NH_4)_2SO_4$	1.0g
Yeast extract	1.0g

pH 7.2 ± 0.2 at 25°C

Preparation of Medium: Add components except magnesium nitrate to tap water and bring volume to 1.0L. Mix thoroughly. Distribute into tubes or flasks. Autoclave for 15 min at 15 psi pressure–121°C. Aseptically add magnesium nitrate. Adjust pH 7.2 with sterile KOH. Pour into sterile Petri dishes or leave in tubes.

Use: For the cultivation and maintenance of halophilic microorganisms, including *Bacillus halodenitrificans*.

YB Medium (Yeast Extract Beef Extract Medium)

Composition per liter:

Agar	20.0g
Peptone	10.0g
Beef extract	7.0g
Yeast extract	5.0g
NaCl	3.0g
Thiourea	0.1g
Methanol	20.0mL

pH 7.2 ± 0.2 at 25°C

Preparation of Medium: Add components except methanol to distilled/deionized water and bring volume to 1.0L. Mix thoroughly. Distribute into tubes or flasks. Autoclave for 15 min at 15 psi pressure–121°C. Aseptically add filter sterilized methanol. Pour into sterile Petri dishes or leave in tubes.

Use: For the cultivation and maintenance of bacteria that can utilize methanol as a carbon source, including *Achromobacter methanolophila, Methanomonas methylovora, Methylobacterium* species and *Pseudomonas methanolica.*

YDC Medium

Composition per liter:

$CaCO_3$	20.0g
Glucose	20.0g
Agar	15.0g
Yeast extract	10.0g

pH 7.2 ± 0.2 at 25°C

Preparation of Medium: Add components to distilled/deionized water and bring volume to 1.0L. Mix thoroughly. Gently heat and bring to boiling. Distribute into tubes or flasks. Autoclave for 15 min at 15 psi pressure–121°C. Pour into sterile Petri dishes or leave in tubes.

Use: For the cultivation of *Bdellovibrio* species.

Yeast Agar, Van Niel's

Composition per liter:

Agar	20.0g
Yeast extract	10.0g
K_2HPO_4	1.0g
$MgSO_4 \cdot 7H_2O$	0.5g

pH 7.0–7.2 at 25°C

Preparation of Medium: Add components to tap water and bring volume to 1.0L. Mix thoroughly. Gently heat and bring to boiling. Distribute into tubes or flasks. Autoclave for 15 min at 15 psi pressure–121°C. Pour into sterile Petri dishes or leave in tubes.

Use: For the cultivation and maintenance of a variety of microorganisms including *Cytophaga* species, *Heliobacterium chlorum, Lysobacter enzymogenes, Rhodobacter* species, *Rhodocyclus gelatinosus, Rhodomicrobium vannielii, Rhodopseudomonas palustris* and *Rhodospirillum rubrum.*

Yeast Agar, Van Niel's with Glutamate

Composition per liter:

Agar	20.0g
Yeast extract	10.0g
K_2HPO_4	1.0g
$MgSO_4$	0.5g
Monosodium glutamate	0.85g

pH 7.0–7.2 at 25°C

Preparation of Medium: Add components to tap water and bring volume to 1.0L. Mix thoroughly. Gently heat and bring to boiling. Distribute into tubes or flasks. Autoclave for 15 min at 15 psi pressure–121°C. Pour into sterile Petri dishes or leave in tubes.

Use: For the cultivation and maintenance of a variety of bacteria, including *Bacillus firmus, Cytophaga johnsonae, Heliobacterium chlorum, Lysobacter enzymogenes, Rhodobacter capsulatus, Rhodobacter sphaeroides, Rhodocyclus gelatinosus, Rhodocyclus gelatinosus, Rhodomicrobium vannielii, Rhodopseudomonas palustris,* and *Rhodospirillum rubrum.*

Yeast Agar, Van Niel's with 2.5% NaCl (ATCC Medium 1370)

Composition per liter:

NaCl	25.0g
Agar	20.0g
Yeast extract	10.0g
K_2HPO_4	1.0g
$MgSO_4$	0.5g

pH 7.0–7.2 at 25°C

Preparation of Medium: Add components to tap water and bring volume to 1.0L. Mix thoroughly. Gently heat and bring to boiling. Distribute into tubes or flasks. Autoclave for 15 min at 15 psi pressure–121°C. Pour into sterile Petri dishes or leave in tubes.

Use: For the cultivation and maintenance of *Chromatium vinosum* and *Rhodopseudomonas* species.

Yeast Agar, Van Niel's with 25% NaCl

Composition per liter:

NaCl	250.0g
Agar	20.0g
Yeast extract	10.0g
K_2HPO_4	1.0g
$MgSO_4$	0.5g

pH 7.0–7.2 at 25°C

Preparation of Medium: Add components to tap water and bring volume to 1.0L. Mix thoroughly. Gently heat and bring to boiling. Distribute into tubes or flasks. Autoclave for 15 min at 15 psi pressure–121°C. Pour into sterile Petri dishes or leave in tubes.

Use: For the cultivation and maintenance of halophilic bacteria including *Haloarcula vallismortis, Halococcus morrhuae* and *Halobacterium* species.

Yeast Agar, Van Niel's with Succinate

Composition per liter:

Agar	20.0g
Yeast extract	10.0g
Sodium succinate	1.35
K_2HPO_4	1.0g
$MgSO_4 \cdot 7H_2O$	0.5g

pH 7.0–7.2 at 25°C

Preparation of Medium: Add components to tap water and bring volume to 1.0L. Mix thoroughly. Gently heat and bring to boiling. Distribute into tubes or flasks. Autoclave for 15 min at 15 psi pressure–121°C. Pour into sterile Petri dishes or leave in tubes.

Use: For the cultivation and maintenance of *Rhodobacter capsulatus.*

Yeast Ascospore Agar

Composition per liter:

Agar	30.0g
Potassium acetate	10.0g
Yeast extract	2.5g
Glucose	1.0g

Preparation of Medium: Add components to distilled/deionized water and bring volume to 1.0L. Mix thoroughly. Gently heat and bring to boiling. Distribute into tubes or flasks. Autoclave for 15 min at 15 psi pressure–121°C.

Use: For the cultivation and observation of ascospore formation of yeast.

Yeast Carbon Base, 10X (Wickerham Carbon Base Broth)

Composition per liter:

Glucose	10.0g
KH_2PO_4	1.0g
$MgSO_4 \cdot 7H_2O$	0.5g
NaCl	0.1g
$CaCl_2 \cdot 2H_2O$	0.1g
DL-Methionine	0.02g
DL-Tryptophan	0.02g
L-Histidine·HCl	0.01g
Inositol	2.0mg
H_3BO_3	0.5mg
$ZnSO_4 \cdot 7H_2O$	0.4mg
$MnSO_4 \cdot 4H_2O$	0.4mg
Thiamine·HCl	0.4mg
Pyridoxine	0.4mg
Niacin	0.4mg
Calcium pantothenate	0.4mg
p-Aminobenzoic acid	0.2mg
Riboflavin	0.2mg
$FeCl_3$	0.2mg
$Na_2MoO_4 \cdot 4H_2O$	0.2mg
KI	0.1mg
$CuSO_4 \cdot 5H_2O$	0.04mg
Folic acid	2.0μg
Biotin	2.0μg

pH 5.5 ± 0.2 at 25°C

Source: This medium is available as a premixed powder from Difco Laboratories.

Preparation of Medium: Add components to distilled/deionized water and bring volume to 1.0L. Mix thoroughly. Filter sterilize.

Use: Used as a base to which different nitrogen sources may be added. For the cultivation and differentiation of bacteria based on their ability to utilize diverse added nitrogen sources.

Yeast Dextrose Agar

Composition per liter:

Agar	15.0g
Glucose	10.0g
Yeast extract	10.0g

pH 7.0 ± 0.2 at 25°C

Preparation of Medium: Add components to distilled/deionized water and bring volume to 1.0L. Mix thoroughly. Gently heat and bring to boiling. Adjust pH to 7.0. Distribute into tubes or flasks. Autoclave for 15 min at 15 psi pressure–121°C.

Use: For the cultivation of a variety of heterotrophic microorganisms.

Yeast Extract Agar

Composition per liter:

Agar	15.0g
Peptone	5.0g
Yeast extract	3.0g

pH 7.2 ± 0.2 at 25°C

Source: This medium is available as a premixed powder from Oxoid Unipath.

Preparation of Medium: Add components to distilled/deionized water and bring volume to 1.0L. Mix thoroughly. Gently heat and bring to boiling. Distribute into tubes or flasks. Autoclave for 15 min at 15 psi pressure–121°C. Pour into sterile Petri dishes or leave in tubes.

Use: For the enumeration of microorganisms in potable and freshwater samples.

Yeast Extract Agar

Composition per liter:

Agar	15.0g
Peptone	9.5g
Yeast extract	7.0g
Beef extract	5.0g
NaCl	5.0g

pH 7.0 ± 0.2 at 25°C

Preparation of Medium: Add components to distilled/deionized water and bring volume to 1.0L. Mix thoroughly. Gently heat and bring to boiling. Distribute into tubes or flasks. Autoclave for 15 min at 15 psi pressure–121°C. Pour into sterile Petri dishes or leave in tubes.

Use: For the cultivation of *Aeromonas salmonicida*.

Yeast Extract Agar

Composition per liter:

Agar	20.0g
Yeast extract	1.0g
Buffer solution	2.0mL

pH 6.0 ± 0.2 at 25°C

Buffer Solution:

Composition per 400mL:

KH_2PO_4	60.0g
Na_2HPO_4	40.0g

Preparation of Buffer Solution: Add 40.0g Na_2HPO_4 to 300.0mL of distilled/deionized water. Mix thoroughly. Add 60.0g of KH_2PO_4. Mix thoroughly. Adjust pH to 6.0.

Preparation of Medium: Add components to distilled/deionized water and bring volume to 1.0L. Mix thoroughly. Autoclave for 15 min at 15 psi pressure–121°C. Pour into sterile Petri dishes.

Use: For identification of *Histoplasma capsulatum, Blastomyces dermatitidis* and *Coccidioides immitis*.

Yeast Extract Agar

Composition per liter:

Agar	15.0g
Proteose peptone	10.0g
NaCl	5.0g
Yeast extract	3.0g

Preparation of Medium: Add components to distilled/deionized water and bring volume to 1.0L. Mix thoroughly. Gently heat and bring to boiling. Distribute into tubes or flasks. Autoclave for 15 min at 15 psi pressure–121°C. Pour into sterile Petri dishes or leave in tubes.

Use: For the cultivation of a variety of heterotrophic microorganisms.

Yeast Extract Glucose Calcium Carbonate Agar

Composition per liter:

$CaCO_3$	20.0g
Glucose	20.0g
Agar	15.0g
Yeast extract	10.0g

Preparation of Medium: Add components to distilled/deionized water and bring volume to 1.0L. Mix thoroughly. Gently heat and bring to boiling. Distribute into tubes or flasks. Autoclave for 15 min at 15 psi pressure–121°C. Pour into sterile Petri dishes or leave in tubes.

Use: For the isolation and cultivation of *Erwinia* species.

Yeast Extract Glucose Medium

Composition per liter:

Agar	15.0g
Yeast extract	10.0g
Glucose	10.0g

Preparation of Medium: Add components to tap water and bring volume to 1.0L. Mix thoroughly. Gently heat and bring to boiling. Distribute into tubes or flasks. Autoclave for 15 min at 15 psi pressure–121°C. Pour into sterile Petri dishes or leave in tubes.

Use: For the cultivation of a variety of bacteria, including *Streptomyces* species, *Rhodococcus* species and others.

Yeast Extract Malt Extract Glucose Agar

Composition per liter:

Agar	20.0g
Glucose	10.0g
Neopeptone	5.0g
Malt extract	3.0g
Yeast extract	3.0g

Preparation of Medium: Add components to distilled/deionized water and bring volume to 1.0L. Mix thoroughly. Gently heat and bring to boiling. Distribute into tubes or flasks. Autoclave for 15 min at 15 psi pressure–121°C. Pour into sterile Petri dishes or leave in tubes.

Use: For the isolation and cultivation of yeasts.

Yeast Extract Mannitol Agar

Composition per liter:

Agar	15.0g
Mannitol	10.0g
$CaCO_3$	4.0g
K_2HPO_4	0.5g
Yeast extract	0.4g
$MgSO_4 \cdot 7H_2O$	0.2g
NaCl	0.1g

pH 6.8–7.0 ± 0.2 at 25°C

Preparation of Medium: Add components to distilled/deionized water and bring volume to 1.0L. Omit $CaCO_3$ if a clear solution is needed. Mix thoroughly. Gently heat and bring to boiling. Distribute into tubes or flasks. Autoclave for 15 min at 15 psi pressure–121°C. Pour into sterile Petri dishes or leave in tubes.

Use: For the cultivation of members of the Rhizobiaceae.

Yeast Extract Medium

Composition per liter:

Yeast extract	10.0g

Preparation of Medium: Add yeast extract to distilled/deionized water and bring volume to 1.0L. Mix thoroughly. Distribute into tubes or flasks. Autoclave for 15 min at 15 psi pressure–121°C.

Use: For the cultivation of *Pseudomonas cepacia.*

Yeast Extract Peptone Starch Agar

Composition per liter:

Agar	18.0g
Soluble starch	10.0g
Peptone	10.0g
$CaCO_3$	5.0g
Sodium acetate	5.0g
Yeast extract	3.0g
KH_2PO_4	0.5g
K_2HPO_4	0.5g
$MgSO_4 \cdot 7H_2O$	0.3g
Sodium citrate	0.027g
NaCl	0.01g
$MnSO_4 \cdot 5H_2O$	0.01g
$CuSO_4 \cdot 5H_2O$	1.0mg
$CoCl_2 \cdot 6H_2O$	1.0mg
$FeSO_4 \cdot 7H_2O$	1.0mg

Preparation of Medium: Add components to tap water and bring volume to 1.0L. Mix thoroughly. Gently heat and bring to boiling. Distribute into tubes or flasks. Autoclave for 15 min at 15 psi pressure–121°C. Pour into sterile Petri dishes or leave in tubes.

Use: For the cultivation and maintenance of *Bacillus* species that utilize starch as a carbon source.

Yeast Fermentation Broth

Composition per liter:

Carbohydrate	10g
Pancreatic digest of gelatin	7.5g
Yeast extract	5.5g
Bromcresol purple	16.0mg

Source: This medium is available as a premixed powder from BBL Microbiology Systems.

Preparation of Medium: Add components to distilled/deionized water and bring volume to 1.0L. Mix thoroughly. Distribute into test tubes, each containing an inverted Durham tube. Autoclave for 15 min at 15 psi pressure–121°C.

Use: For fermentation tests of specific carbohydrates used in the characterization and identification of yeasts. Gas accumulation in the Durham tube and a color change of the medium to yellow indicates carbohydrate fermentation.

Yeast Fermentation Medium

Composition per liter:

Peptone	7.5g
Yeast extract	4.5g
Bromthymol blue (1.6% solution)	1.0mL
Carbohydrate solution	1.0mL

Carbohydrate Solution:

Composition per 10mL:

Carbohydrate	0.6g

Preparation of Carbohydrate Solution: Add carbohydrate to distilled/deionized water and bring volume to 10.0mL. Glucose, maltose, lactose, galactose or trehalose may be used. If raffinose is used, prepare a 12% solution. Mix thoroughly. Filter sterilize.

Preparation of Medium: Add components, except carbohydrate solution, to distilled/deionized water and bring volume to 1.0L. Mix thoroughly. Gently heat and bring to boiling. Distribute in 2.0mL volumes into test tubes that contain an inverted Durham tube. Autoclave for 15 min at 15 psi pressure–121°C. Cool to 45–50°C. Aseptically add 1.0mL of sterile carbohydrate solution. Mix thoroughly.

Use: For the cultivation and differentiation of yeast based on carbohydrate fermentation patterns. Yeast that can ferment a specific carbohydrate turn the medium yellow.

Yeast Malate Medium

Composition per liter:

Yeast extract	5.0g
Sodium malate	1.0g

pH 7.0 ± 0.2 at 25°C

Preparation of Medium: Add components to distilled/deionized water and bring volume to 1.0L. Mix thoroughly. Distribute into tubes or flasks. Autoclave for 15 min at 15 psi pressure–121°C.

Use: For the cultivation of *Rhodopseudomonas viridis.*

Yeast Malt Extract Agar (YM Agar)

Composition per liter:

Agar	20.0g
Glucose	10.0g
Peptone	5.0g
Yeast extract	3.0g
Malt extract	3.0g

pH 6.2 ± 0.2 at 25°C

Source: This medium is available as a premixed powder from Difco Laboratories.

Preparation of Medium: Add components to distilled/deionized water and bring volume to 1.0L. Mix thoroughly. Gently heat and bring to boiling. Distribute into tubes or flasks. Autoclave for 15 min at 15 psi pressure–121°C. The medium may be rendered selective by adjusting the pH to 3.0–4.0 at 45–55°C or by the addition of antibiotics at 45–50°C or below. Pour into sterile Petri dishes or leave in tubes.

Use: For the cultivation of fungi, including yeasts, and other aciduric microorganisms such as *Actinoplanes* species, *Streptomyces* species, *Streptoverticillium* species, and *Nocardia* species.

Yeast Malt Extract Broth (YM Broth)

Composition per liter:

Glucose	10.0g
Peptone	5.0g
Yeast extract	3.0g
Malt extract	3.0g

pH 6.2 ± 0.2 at 25°C

Source: This medium is available as a premixed powder from Difco Laboratories.

Preparation of Medium: Add components to distilled/deionized water and bring volume to 1.0L. Mix thoroughly. Distribute into tubes or flasks. Autoclave for 15 min at 15 psi pressure–121°C. The medium may be rendered selective by adjusting the pH to 3.0–4.0 at 45–55°C or by the addition of antibiotics at 45–50°C or below.

Use: For the cultivation of yeasts, molds and other aciduric microorganisms such as *Actinoplanes* species, *Streptomyces* species, *Streptoverticillium* species, and *Nocardia* species.

Yeast Malt Extract Catalase Agar (YM Catalase Agar)

Composition per liter:

Agar	15.0g
K_2HPO_4	5.74g
Malt extract	5.0g
Yeast extract	5.0g
$NH_4H_2PO_4$	1.15g
$MgSO_4 \cdot 7H_2O$ solution	10.0mL
Catalase solution	10.0mL

Catalase Solution:

Composition per 10mL:

Catalase	60.0mg

Preparation of Catalase Solution: Add catalase to distilled/deionized water and bring volume to 10.0mL. Mix thoroughly. Filter sterilize.

Magnesium Sulfate Solution:

Composition per 10mL:

$MgSO_4 \cdot 7H_2O$	205.0mg

Preparation of Magnesium Sulfate Solution: Add $MgSO_4 \cdot 7H_2O$ to distilled/deionized water and bring volume to 10.0mL. Mix thoroughly. Filter sterilize.

Preparation of Medium: Add components, except catalase and magnesium sulfate, to distilled/deionized water and bring volume to 980.0mL. Mix thoroughly. Gently heat and bring to boiling. Distribute into tubes or flasks. Autoclave for 15 min at 15 psi pressure–121°C. Cool to 50°C. Aseptically add filter-sterilized catalase and $MgSO_4 \cdot 7H_2O$ solutions.

Use: For the cultivation and maintenance of *Rarobacter faecitabidus.*

Yeast Mannitol Agar

Composition per liter:

Agar	15.0g
Mannitol	10.0g
K_2HPO_4	0.5g
Yeast extract	0.4g
$MgSO_4 \cdot 7H_2O$	0.2g
NaCl	0.1g

Preparation of Medium: Add components to distilled/deionized water and bring volume to 1.0L. Mix thoroughly. Gently heat and bring to boiling. Distribute into tubes or flasks. Autoclave for 15 min at 15 psi pressure–121°C. Pour into sterile Petri dishes or leave in tubes.

Use: For the cultivation of *Rhizobium* and *Azorhizobium* species.

Yeast Nitrogen Base

Composition per liter:

$(NH_4)_2SO_4$	5.0g
KH_2PO_4	1.0g
$MgSO_4 \cdot 7H_2O$	0.5g
NaCl	0.1g
$CaCl_2 \cdot 2H_2O$	0.1g
DL-Methionine	0.02g
DL-Tryptophan	0.02g
L-Histidine·HCl	0.01g
Inositol	2.0mg
KI	0.1mg
H_3BO_3	0.5mg
$ZnSO_4 \cdot 7H_2O$	0.4mg
$MnSO_4 \cdot 4H_2O$	0.4mg
Thiamine·HCl	0.4mg
Pyroxidine·HCl	0.4mg
Niacin	0.4mg
Calcium pantothenate	0.4mg
p-Aminobenzoic acid	0.2mg
Riboflavin	0.2mg
$FeCl_3$	0.2mg
$Na_2MoO_4 \cdot 4H_2O$	0.2mg
$CuSO_4 \cdot 5H_2O$	0.04mg
Folic acid	2.0μg
Biotin	2.0μg

pH 5.5 ± 0.2 at 25°C

Source: This medium is available as a premixed powder from BBL Microbiology Systems and Difco Laboratories.

Preparation of Medium: Add components to distilled/deionized water and bring volume to 1.0L. Mix thoroughly. Distribute into tubes or flasks. Autoclave for 15 min at 15 psi pressure–121°C. Alternately for carbon assimilation tests, prepare a 10× concentrated solution by adding components to distilled/deionized water and bring volume to 100.0mL. Mix thoroughly. Distribute into tubes or flasks. Autoclave for 15 min at 15 psi pressure–121°C. Prepare a carbohydrate solution by adding 0.5g of carbohydrate to 90.0mL of distilled/deionized water. Mix thoroughly. Filter sterilize. Aseptically add 0.5mL of the 10× concentrated solution to 4.5mL of the filter sterilized carbohydrate solution. Mix thoroughly.

Use: For carbohydrate assimilation tests in the characterization and identification of yeasts.

Yeast Nitrogen Base Glucose Broth

Composition per liter:

Yeast nitrogen base	25.0mL
Glucose solution	25.0mL

pH 5.6 ± 0.2 at 25°C

Yeast Nitrogen Base:

Composition per 500mL:

$(NH_4)_2SO_4$	5.0g
KH_2PO_4	1.0g
$MgSO_4 \cdot 7H_2O$	0.5g
NaCl	0.1g
$CaCl_2 \cdot 2H_2O$	0.1g
DL-Methionine	0.02g
DL-Tryptophan	0.02g
L-Histidine·HCl	0.01g
Inositol	2.0mg
H_3BO_3	0.5mg
$ZnSO_4 \cdot 7H_2O$	0.4mg
$MnSO_4 \cdot 4H_2O$	0.4mg
Thiamine·HCl	0.4mg
Pyroxidine·HCl	0.4mg
Niacin	0.4mg
Calcium pantothenate	0.4mg
p-Aminobenzoic acid	0.2mg
Riboflavin	0.2mg
$FeCl_3$	0.2mg
$Na_2MoO_4 \cdot 4H_2O$	0.2mg
KI	0.1mg
$CuSO_4 \cdot 5H_2O$	0.04mg
Folic acid	2.0μg
Biotin	2.0μg

Source: Yeast nitrogen base is available as a premixed powder from BBL Microbiology Systems.

Preparation of Yeast Nitrogen Base: Add components to distilled/deionized water and bring volume to 500.0mL. Mix thoroughly. Filter sterilize.

Glucose Solution:

Composition per 500mL:

Glucose	10.0g

Preparation of Glucose Solution: Add glucose to distilled/deionized water and bring volume to 500.0mL. Mix thoroughly. Filter sterilize.

Preparation of Medium: Aseptically combine 25.0mL of sterile yeast nitrogen base and 25.0mL of sterile glucose solution. Mix thoroughly.

Use: For the cultivation and enrichment of yeast from sewage and polluted waters.

Yeast Nitrogen Base with Carbohydrate

Composition per liter:

Carbohydrate	5.0g
$(NH_4)_2SO_4$	5.0g
KH_2PO_4	1.0g
$MgSO_4·7H_2O$	0.5g
NaCl	0.1g
$CaCl_2·2H_2O$	0.1g
DL-Methionine	0.02g
DL-Tryptophan	0.02g
L-Histidine·HCl	0.01g
Inositol	2.0mg
KI	0.1mg
H_3BO_3	0.5mg
$ZnSO_4·7H_2O$	0.4mg
$MnSO_4·4H_2O$	0.4mg
Thiamine·HCl	0.4mg
Pyroxidine·HCl	0.4mg
Niacin	0.4mg
Calcium pantothenate	0.4mg
p-Aminobenzoic acid	0.2mg
Riboflavin	0.2mg
$FeCl_3$	0.2mg
$Na_2MoO_4·4H_2O$	0.2mg
$CuSO_4·5H_2O$	0.04mg
Folic acid	2.0μg
Biotin	2.0μg

pH 5.6 ± 0.2 at 25°C

Preparation of Medium: Add components to distilled/deionized water and bring volume to 1.0L. Mix thoroughly. Filter sterilize. Aseptically distribute into tubes or flasks.

Use: For carbohydrate assimilation tests in the characterization and identification of yeasts.

Yeast Nitrogen Base, 10X with Asparagine and Glucose

Composition per liter:

Glucose	10.0g
$(NH_4)_2SO_4$	5.0g
L-Asparagine	1.5g
KH_2PO_4	1.0g
$MgSO_4·7H_2O$	0.5g
NaCl	0.1g
$CaCl_2·2H_2O$	0.1g
DL-Methionine	0.02g
DL-Tryptophan	0.02g
L-Histidine·HCl	0.01g
Inositol	2.0mg
H_3BO_3	0.5mg
$ZnSO_4·7H_2O$	0.4mg
$MnSO_4·4H_2O$	0.4mg
Thiamine·HCl	0.4mg
Pyridoxine·HCl	0.4mg
Niacin	0.4mg
Calcium pantothenate	0.4mg
p-Aminobenzoic acid	0.2mg
Riboflavin	0.2mg
$FeCl_3$	0.2mg
$Na_2MoO_4·4H_2O$	0.2mg
KI	0.1mg
$CuSO_4·5H_2O$	0.04mg
Folic acid	2.0μg
Biotin	2.0μg

pH 5.6 ± 0.2 at 25°C

Preparation of Medium: Add components to distilled/deionized water and bring volume to 1.0L. Dilute 100.0mL of 10× medium with 900.0mL of distilled/deionized water. Mix thoroughly. Filter sterilize. Aseptically distribute into sterile tubes or flasks.

Use: For susceptibility tests with yeasts and fungi.

Yeast Peptone Broth

Composition per liter:

Yeast extract	2.5g
Peptone	2.5g

pH 7.0 ± 0.2 at 25°C

Preparation of Medium: Add components to distilled/deionized water and bring volume to 1.0L. Mix thoroughly. Distribute into tubes or flasks. Autoclave for 15 min at 15 psi pressure–121°C.

Use: For the cultivation of *Rhodopseudomonas* species.

Yeast Synthetic Minimal Medium

Composition per liter:

D-Glucose	20.0g

Agar	15.0g
$(NH_4)_2SO_4$	5.0g
KH_2PO_4	1.0g
$MgSO_4 \cdot 7H_2O$	0.5g
NaCl	0.1g
$CaCl_2 \cdot 2H_2O$	0.1g
Inositol	2.0mg
H_3BO_3	0.5mg
$ZnSO_4 \cdot 7H_2O$	0.4mg
$MnSO_4 \cdot 4H_2O$	0.4mg
Thiamine·HCl	0.4mg
Pyridoxine·HCl	0.4mg
Niacin	0.4mg
Calcium pantothenate	0.4mg
p-Aminobenzoic acid	0.2mg
Riboflavin	0.2mg
$FeCl_3$	0.2mg
$Na_2MoO_4 \cdot 4H_2O$	0.2mg
KI	0.1mg
$CuSO_4 \cdot 5H_2O$	0.04mg
Folic acid	2.0μg
Biotin	2.0μg

pH 5.6 ± 0.2 at 25°C

Preparation of Medium: Add agar to 900.0mL of distilled/deionized water. Mix thoroughly. Gently heat and bring to boiling. Distribute into tubes or flasks. Autoclave for 15 min at 15 psi pressure–121°C. Cool to 45–50°C. In a separate flask, add remaining components to 100.0mL of distilled/deionized water. Mix thoroughly. Filter sterilize. Aseptically combine the two sterile solutions. Mix thoroughly. Pour into sterile Petri dishes.

Use: For the cultivation of a wide variety of heterotrophic microorganisms.

Yeast Water Agar

Composition per liter:

Glucose	20.0g
Agar	15.0g
Casein hydrolysate	5.0g
Yeast extract	4.0g
KH_2PO_4	0.55g
KCl	0.40g
$CaCl_2$	0.13g
$MgCl_2 \cdot 7H_2O$	0.13g
$FeCl_3 \cdot 6H_2O$	2.5mg
$MnSO_4 \cdot 4H_2O$	2.5mg
Bromcresol green solution	1.0mL

Bromcresol Green Solution:

Composition per 10mL:

Bromcresol green	0.22g
Ethanol	10.0mL

Preparation of Bromcresol Green Solution: Add bromcresol green to 10.0mL of ethanol. Mix thoroughly. Filter sterilize.

Preparation of Medium: Add components to distilled/deionized water and bring volume to 1.0L. Mix thoroughly. Gently heat and bring to boiling. Distribute into tubes or flasks. Autoclave for 15 min at 15 psi pressure–121°C. Pour into sterile Petri dishes or leave in tubes.

Use: For the cultivation of *Zymomonas* species.

YEP Galactose Agar

Composition per liter:

Agar	20.0g
Galactose	20.0g
Peptone	20.0g
Yeast extract	10.0g

Preparation of Medium: Add components to distilled/deionized water and bring volume to 1.0L. Mix thoroughly. Gently heat and bring to boiling. Distribute into tubes or flasks. Autoclave for 15 min at 15 psi pressure–121°C. Pour into sterile Petri dishes or leave in tubes.

Use: For the cultivation of a variety of heterotrophic microorganisms.

YEPD Medium

Composition per liter:

Agar	20.0g
Glucose	20.0g
Peptone	20.0g
Yeast extract	10.0g

Preparation of Medium: Add components to distilled/deionized water and bring volume to 1.0L. Mix thoroughly. Gently heat and bring to boiling. Distribute into tubes or flasks. Autoclave for 15 min at 15 psi pressure–121°C. Pour into sterile Petri dishes or leave in tubes.

Use: For the cultivation of a variety of heterotrophic microorganisms.

YEPP Medium (Yeast Extract Proteose Peptone Medium)

Composition per liter:

Agar	15.0g
Proteose peptone	10.0g
NaCl	5.0g
Yeast extract	3.0g

pH 7.2-7.4 ± 0.2 at 25°C

Preparation of Medium: Add components to distilled/deionized water and bring volume to 1.0L. Mix thoroughly. Gently heat and bring to boiling. Distribute into tubes or flasks. Autoclave for 15 min at 15 psi pressure–121°C. Pour into sterile Petri dishes or leave in tubes.

Use: For the cultivation and maintenance of *Pseudomonas* species.

YGC Medium (Yeast Extract Glucose Carbonate Medium) (ATCC Medium 73)

Composition per liter:

Agar	20.0g
Glucose	20.0g
$CaCO_3$	20.0g
Yeast extract	10.0g

Preparation of Medium: Add components to distilled/deionized water and bring volume to 1.0L. Mix thoroughly. Gently heat and bring to boiling. Distribute into tubes or flasks. Autoclave for 30 min at 10 psi pressure–115°C. Cool to 48°C. Mix thoroughly. Pour into sterile Petri dishes or leave in tubes.

Use: For the cultivation of *Xanthomonas* species, *Erwinia* species, *Kluyvera* species, *Rhodococcus* species, *Streptomyces* species, *Pseudomonas psueudoalcaligenes* and *Xylophilus ampelinus*.

YGC Medium (Yeast Extract Glucose Citrate Medium) (ATCC Medium 216)

Composition per liter:

Beef extract	10.0g
Glucose	10.0g
Peptone	10.0g
Ammonium citrate	5.0g
Yeast extract	5.0g
Sodium acetate	2.0g
Tween™ 80	1.0g
$MgSO_4 \cdot 7H_2O$	0.2g
$MnSO_4 \cdot 4H_2O$	0.05g

pH 6.5 ± 0.2 at 25°C

Preparation of Medium: Add components to distilled/deionized water and bring volume to 1.0L. Mix thoroughly. Distribute into tubes or flasks. Autoclave for 15 min at 15 psi pressure–121°C.

Use: For the isolation and cultivation of *Leuconostoc* species.

YGC Medium with Glutamic Acid (Yeast Extract Glucose Carbonate Medium with Glutamic Acid)

Composition per liter:

Agar	20.0g
Glucose	20.0g
$CaCO_3$	20.0g
Agar	20.0g
Yeast extract	10.0g
Glutamic acid	0.1g

Preparation of Medium: Add components to distilled/deionized water and bring volume to 1.0L. Mix thoroughly. Gently heat and bring to boiling. Distribute into tubes or flasks. Autoclave for 30 min at 10 psi pressure–115°C. Cool to 48°C. Mix thoroughly. Pour into sterile Petri dishes or leave in tubes.

Use: For the cultivation and maintenance of *Xanthomonas campestris*.

YGCP Medium (Yeast Extract Glucose Carbonate Peptone Medium)

Composition per liter:

Glucose	20.0g
Agar	17.5g
$CaCO_3$	10.0g
Yeast extract	2.5g
Peptone	2.5g
NaCl	1.0g
K_2HPO_4	1.0g
$MgSO_4$	0.5g

Preparation of Medium: Add components, except calcium carbonate, to distilled/deionized water and bring volume to 1.0L. Mix thoroughly. Gently heat and bring to boiling. Adjust pH to 7.2. Add calcium carbonate. Mix thoroughly. Distribute into tubes or flasks. Autoclave for 15 min at 15 psi pressure–121°C. Pour into sterile Petri dishes or leave in tubes.

Use: For the cultivation and maintenance of *Xanthomonas campestris* and *X. oryzae*.

YNA Medium (Yeast Extract Nutrient Agar Medium)

Composition per liter:

Agar	15.0g
NaCl	5.0g
Peptone	5.0g
Meat extract	4.0g
Yeast extract	2.5g

pH 7.0 ± 0.2 at 25°C

Preparation of Medium: Add components to distilled/deionized water and bring volume to 1.0L. Mix thoroughly. Gently heat and bring to boiling. Distribute into tubes or flasks. Autoclave for 15 min at 15 psi pressure–121°C. Pour into sterile Petri dishes or leave in tubes.

Use: For the isolation and cultivation of *Kurthia* species according to the agar streak method.

YNG Medium (Yeast Extract Nutrient Gelatin Medium)

Composition per liter:

Gelatin	100.0g
NaCl	5.0g
Peptone	5.0g
Meat extract	4.0g
Yeast extract	2.5g

pH 7.0 ± 0.2 at 25°C

Preparation of Medium: Add components to distilled/deionized water and bring volume to 1.0L. Mix thoroughly. Gently heat until dissolved. Distribute into tubes or flasks. Autoclave for 30 min at 10 psi pressure–115°C.

Use: For the isolation and cultivation of *Kurthia* species using the gelatin streak method.

Yopp's Medium

Composition per liter:

NaCl	116.88g
$MgCl_2 \cdot 6H_2O$	10.68g
$MgSO_4 \cdot 7H_2O$	10.0g
KCl	2.0g
$CaNO_3 \cdot 4H_2O$	1.0g
Glycyl-glycine buffer	0.5g
$K_2HPO_4 \cdot 3H_2O$	0.065g
Ferric EDTA	5.0mg
Trace metal solution	1.0mL

pH 7.8 ± 0.2 at 25°C

Trace Metal Solution :
Composition per liter:

$MnCl_2 \cdot 4H_2O$	2.0g
H_3BO_3	0.5g
$ZnNO_3 \cdot 6H_2O$	0.5g
$Co(NO_3)_2 \cdot 6H_2O$	0.025g
$CuCl_2 \cdot 2H_2O$	0.025g
$Na_2MoO_4 \cdot 2H_2O$	0.025g
$VOSO_4 \cdot 6H_2O$	0.025g
HCl	3.0mL

Preparation of Trace Metal Solution: Add components to distilled/deionized water and bring volume to 1.0L. Mix thoroughly.

Preparation of Medium: Add components to distilled/deionized water and bring volume to 1.0L. Mix thoroughly. Gently heat and bring to boiling. Distribute into tubes or flasks. Autoclave for 15 min at 15 psi pressure–121°C.

Use: For the isolation and cultivation of halophilic cyanobacteria.

YPSC Agar (Yeast Extract Peptone Sulfate Cysteine Agar)

Composition per liter:

Agar	15.0g
Yeast extract	1.0g
Peptone	1.0g
Sodium acetate·$3H_2O$	0.5g
$MgSO_4 \cdot 7H_2O$	0.25g
$CaCl_2 \cdot 2H_2O$	0.25g
L-Cysteine·HCl·H_2O	0.05g

pH 7.5 ± 0.2 at 25°C

Preparation of Medium: Add components to distilled/deionized water and bring volume to 1.0L. Mix thoroughly. Gently heat and bring to boiling. Distribute into tubes or flasks. Autoclave for 15 min at 15 psi pressure–121°C. Adjust pH to 7.5 with sterile 10M NaOH. Pour into sterile Petri dishes or leave in tubes.

Use: For the cultivation and maintenance of *Bdellovibrio* species.

YPSC Agar, Cation–Supplemented

Composition per liter:

Sodium acetate·$3H_2O$	50.0g
Agar	15.0g
Peptone	10.0g
Yeast extract	10.0g
$MgSO_4 \cdot 7H_2O$	0.74g

$CaCl_2 \cdot 2H_2O$ 0.29g
L-Cysteine·HCl·H_2O 0.05g
Bacitracin solution 10.0mL

Bacitracin Solution:
Composition per 10mL:
Bacitracin 6,000U

Preparation of Bacitracin Solution: Add bacitracin to distilled/deionized water and bring volume to 10.0mL. Mix thoroughly. Filter sterilize.

Preparation of Medium: Add components, except bacitracin solution, to distilled/deionized water and bring volume to 990.0mL. Mix thoroughly. Gently heat and bring to boiling. Autoclave for 15 min at 15 psi pressure–121°C. Cool to 45–50°C. Aseptically add sterile bacitracin solution. Mix thoroughly. Pour into sterile Petri dishes or distribute into sterile tubes.

Use: For the cultivation and enumeration of *Bdellovibrio* species.

YPSC Medium (Yeast Extract Peptone Sulfate Cysteine Medium)

Composition per liter:
Yeast extract 1.0g
Peptone 1.0g
Sodium acetate·$3H_2O$ 0.5g
$MgSO_4 \cdot 7H_2O$ 0.25g
$CaCl_2 \cdot 2H_2O$ 0.25g
L-Cysteine·HCl·H_2O 0.05g
pH 7.5 ± 0.2 at 25°C

Preparation of Medium: Add components to distilled/deionized water and bring volume to 1.0L. Mix thoroughly. Distribute into tubes or flasks. Autoclave for 15 min at 15 psi pressure–121°C. Adjust pH to 7.5 with sterile 10*M* NaOH.

Use: For the cultivation and maintenance of *Bdellovibrio* species.

YPSC Soft Agar (Yeast Extract Peptone Sulfate Cysteine Soft Agar)

Composition per liter:
Agar 6.0g
Yeast extract 1.0g
Peptone 1.0g
Sodium acetate·$3H_2O$ 0.5g
$MgSO_4 \cdot 7H_2O$ 0.25g
$CaCl_2 \cdot 2H_2O$ 0.25g
L-Cysteine·HCl·H_2O 0.05g
pH 7.5 ± 0.2 at 25°C

Preparation of Medium: Add components to distilled/deionized water and bring volume to 1.0L. Mix thoroughly. Gently heat and bring to boiling. Distribute into tubes or flasks. Autoclave for 15 min at 15 psi pressure–121°C. Adjust pH to 7.5 with sterile 10*M* NaOH. Pour into sterile Petri dishes or leave in tubes.

Use: For the cultivation and maintenance of *Bdellovibrio* species.

Zoogloea Medium

Composition per liter:
Agar 15.0g
Pancreatic digest of casein 5.0g
Glycerol 5.0g
Yeast autolysate 1.0g
Sodium lactate 0.5g

Preparation of Medium: Add components to distilled/deionized water and bring volume to 1.0L. Mix thoroughly. Gently heat and bring to boiling. Distribute into screw cap test tubes. Autoclave for 15 min at 15 psi pressure–121°C. Pour into sterile Petri dishes or leave in tubes.

Use: For cultivation and maintenance of *Zoogloea ramigera* and other *Zoogloea* species.